Flora of Nepal
नेपालका वनस्पति

Volume ③

Magnoliaceae
Lauraceae
Fumariaceae
Papaveraceae
Capparaceae
Cruciferae
Crassulaceae
Saxifragaceae
Grossulariaceae
Rosaceae

2011

This book is dedicated to

Hiroshi Hara, Samar B. Malla, Tirtha B. Shrestha and William T. Stearn
whose vision we follow and on whose shoulders we stand

Contributors to families in Volume 3

Annonaceae	Puran P. Kurmi
Capparaceae	Sajan Dahal, Krishna K. Shrestha, Gordon C. Tucker
Cleomaceae	Jyoti P. Gajurel, Krishna K. Shrestha, Gordon C. Tucker
Crassulaceae	Hideaki Ohba, Keshab R. Rajbhandari
Crucifereae	Ihsan A. Al-Shehbaz, Mark F. Watson
Droseraceae	Anjana Giri, Mark F. Watson
Fumariaceae	Magnus Lidén
Grossulariaceae	J. Crinan M. Alexander, Bandana Shakya
Hamamelidaceae	Mark F. Watson
Hydrangaceae	Sirjana Shrestha, Colin A. Pendry, Krishna K. Shrestha
Lauraceae	Colin A. Pendry
Magnoliaceae	Yumiko Baba, Colin A. Pendry, Ram C. Poudel
Moringaceae	Stephen Blackmore, Mark F. Watson
Myristicaceae	Kamal Maden, Mark F. Watson
Papaveraceae	Paul A. Egan, Colin A. Pendry, Sangita Shrestha
Parnassiaceae	Bhaskar Adhikari, Shinobu Akiyama
Pittosporaceae	Martin R. Pullan, Mark F. Watson
Rosaceae	J. Crinan M. Alexander, David E. Boufford, Anthony R. Brach, Hiroshi Ikeda, Nirmala Joshi, Cathy King, Vidya K. Manandhar, Cliodhna D. Ní Bhroin, Hideaki Ohba, Colin A. Pendry, Sangeeta Rajbhandary, Mohan Siwakoti, Mark. F. Watson
Saxifragaceae	Bhaskar Adhikari, Shinobu Akiyama, Richard J. Gornall, Colin A. Pendry, Mark F. Watson
Schisandraceae	Colin A. Pendry
Tetracentraceae	Kamal Maden, Colin A. Pendry

Cover *Potentilla fruticosa* L. var. *arbuscula* (D. Don) Maxim. From Wallich, N. (1832) *Plantae Asiaticae Rariores* 3 [10]: pl. 228 [as *P. arbuscula* D.Don]. Drawing by Gorachand, engraving by Maxim Gauci.

Frontispiece *Sorbus himalaica* Gabrieljan: a, flowering branch; b, flower; c, petal; d, sepal; e, gynoecium; f, stamen. *Sorbus microphylla* Wenz.: g, fruiting branch. Drawing by Neera Joshi Pradhan.

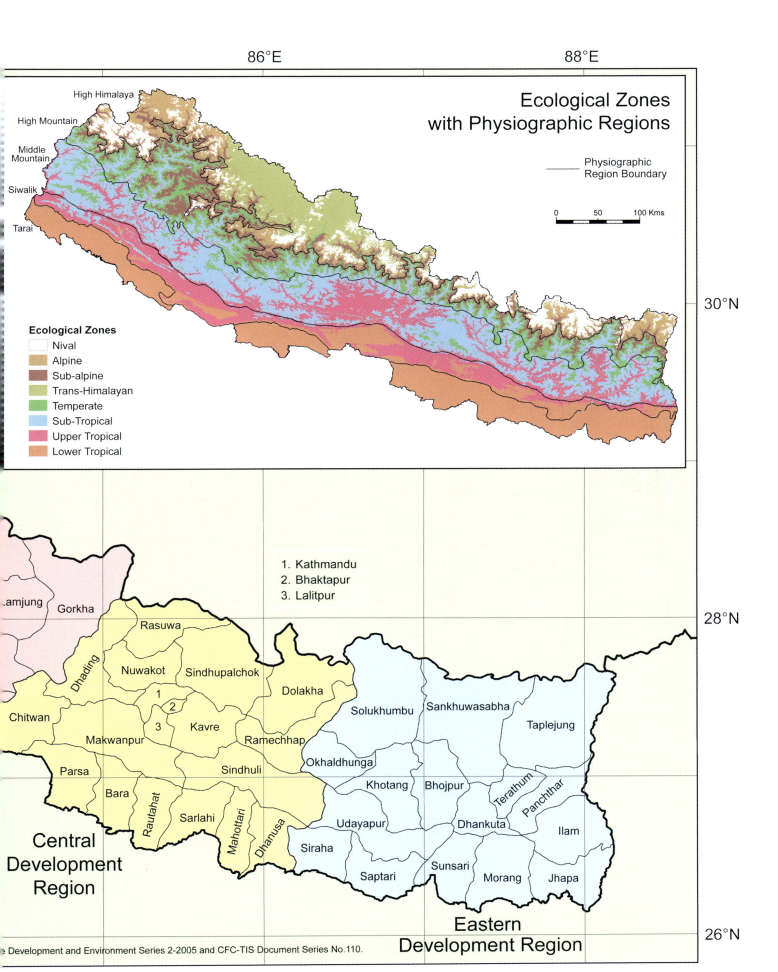

Ecological Zones
with Physiographic Regions

High Himalaya
High Mountain
Middle Mountain
Siwalik
Tarai

Physiographic
Region Boundary

0   50   100 Kms

Ecological Zones
- Nival
- Alpine
- Sub-alpine
- Trans-Himalayan
- Temperate
- Sub-Tropical
- Upper Tropical
- Lower Tropical

30°N

1. Kathmandu
2. Bhaktapur
3. Lalitpur

28°N

Lamjung | Gorkha
Rasuwa
Dhading
Nuwakot | Sindhupalchok
Dolakha
1
2
Chitwan
3 | Kavre
Makwanpur
Ramechhap
Solukhumbu | Sankhuwasabha
Taplejung
Parsa
Sindhuli
Okhaldhunga
Bara
Rautahat
Khotang | Bhojpur
Terathum
Panchthar
Sarlahi
Mahottari
Dhanusa
Udayapur
Dhankuta
Ilam
Siraha
Sunsari
Saptari
Morang
Jhapa

**Central
Development
Region**

**Eastern
Development Region**

26°N

Development and Environment Series 2-2005 and CFC-TIS Document Series No.110.

86°E   88°E

Flora of Nepal
नेपालका वनस्पति

Volume ③

Magnoliaceae
Lauraceae
Fumariaceae
Papaveraceae
Capparaceae
Cruciferae
Crassulaceae
Saxifragaceae
Grossulariaceae
Rosaceae

Edited by

Mark F. Watson
Hiroshi Ikeda
Keshab R. Rajbhandari

Shinobu Akiyama
Colin A. Pendry
Krishna K. Shrestha

Royal
Botanic Garden
Edinburgh

2011

Flora of Nepal
नेपालका वनस्पति

Volume 3

ISBN 978-1-906129-78-1 (Volume 3)
ISBN 978-1-906129-79-8 (Entire work)
www.rbge.org.uk/publications/floraofnepal/volume3

First published 2011 by the Royal Botanic Garden Edinburgh, 20A Inverleith Row, Edinburgh EH3 5LR, UK

© text and images, except where individually credited,

Nepal Academy of Science and Technology
   Khumaltar, Lalitpur, Kathmandu, Nepal

Royal Botanic Garden Edinburgh
   20a Inverleith Row, Edinburgh, EH3 5LR, UK

The Society of Himalayan Botany
   University Museum, University of Tokyo, Hongo 7-3-1, Tokyo 113-0033, Japan

2011

Nepal Academy of
Science and Technology

Ministry of Forests and
Soil Conservation

Royal
Botanic Garden
Edinburgh

Tribhuvan
University

Printed by Scotprint, Scotland.
Text printed on 120gsm Essential Offset and cover printed on 130gsm Galerie silk, PEFC certified.

This volume was printed on 1st September 2011.

PEFC

PEFC/16-33-107

Contents

Foreword

It gives us great pleasure and personal satisfaction to see the publication of this, the first volume of the *Flora of Nepal*, a book that takes floristic knowledge of this spectacular Himalayan country to new heights. Decades have passed since Nepalese systematic botanists first set their sights on writing a Flora, but time has not been wasted as the passing years have seen the building-up of reference collections and the development of the expertise needed to reach this goal. Many Nepalese and foreign botanists have dedicated their careers to researching the plants of Nepal, sometimes in collaboration and sometimes apart, but always to the common cause of a national Flora. This book is testament to their work.

The publication of the *Enumeration of the Flowering Plants of Nepal*, at the beginning of the 1980s, was a major achievement that took us over the mid hills along this metaphorical journey up to the high Himal. The current volume is dedicated to four great botanists involved in that project, which produced a meticulous, authoritative checklist that has truly stood the test of time. But a checklist has neither descriptions nor keys, and so is little use for identifying unknown plants – a primary need for those who study the natural world. A full Flora was clearly needed, but this required a great deal more research, in the field, herbaria and libraries, and the pooling of expertise and resources through international collaboration. As experience has shown, producing a checklist is one thing, writing a comprehensive Flora is quite another.

The drive to take things forward was started in 1993 at the Fifteenth International Botanical Congress in Japan, when an international team of Himalayan botanists pledged to work together on the *Flora of Nepal*. Subsequent meetings reinforced this pledge, and in 1997, at a meeting held at the Natural History Museum, London, the present *Advisory Board* was established. We have been privileged to serve as leaders of this initiative, encouraging and directing as plans were developed and agreements signed. Our institutes were appointed as lead organisations and were joined by others, forming a partnership with an impressive track record in Himalayan botanical research, extensive collections and staff expertise – their logos appear earlier in this book and on the back cover. A strong editorial team was constituted, and it is the six editors who deserve special praise for taking us to where we are today – they proudly stand on the summit as this first volume is published. The difficulties were many, and the obstacles were tremendous. The fact that they were overcome is thanks to the dogged perseverance of each member of the team.

This printed book is a great achievement in itself, not only for Nepal but also for the wider Himalaya. About a third of the plants in the pan-Himalayan region are found in Nepal, and with the publication of this book one of the last glaring gaps in the knowledge of Himalayan plants has begun to be filled. However, the *Flora of Nepal* is far more than just a book, as the project is at the forefront of floristic research as we embrace the digital age. Biodiversity informatics are used extensively and, as we are able to make all our data available electronically, we are finding new markets for our work. In particular, primary occurrence data is greatly sought after by environmental modellers such as climate change and earth system scientists.

Himalayan botanists and environmentalists have long felt the need for a *Flora of Nepal*, and in the last decade the Government of Nepal designated this as a priority for fulfilling Nepal's commitments under the Convention of Biological Diversity in its *National Biodiversity Strategy*. The publication of the current volume is the first stage in meeting these requirements, but much more work needs to be done to complete the remaining nine volumes. Although there are many projects contributing to the *Flora of Nepal*, there is no large-scale funding to support this vital work. Institutional commitment has kept us going thus far, but progress is slow and needs to be accelerated. The publication of this volume in printed and digital formats will be of great value to conservation, sustainable use and developing climate change mitigation strategies. These are very pressing issues and the information contained within the *Flora of Nepal* is needed now. We hope that large-scale funding can be secured soon so that we can accelerate this vital baseline research.

Dayananda Bajracharya, Stephen Blackmore and Hideaki Ohba
Advisory Board, Flora of Nepal
July 2011

Preface

The *Flora of Nepal* is the first comprehensive record of all the vascular plants of Nepal. It includes keys and descriptions to aid identification, distribution maps to assess rarity and geographic spread, authoritative scientific names to promote communication, synonyms to help interpret previous works and supplementary information such as ecology, phenology, ethnobotanical uses and discussion on taxonomic issues. These provide the baseline data needed for environmental studies, climate change modelling, biodiversity inventories, conservation planning and the sustainable use of natural resources.

In economic terms Nepal has a low gross domestic product (GDP), but it has a great natural wealth and a very rich flora of an estimated 6500 to 7000 species of ferns, conifers and flowering plants. The Himalayan range is recognized as one of the 'hottest' of the Global Biodiversity Hotspots and, as about a third of all species in the wider Himalayan range occur in Nepal, the plants of Nepal are of international as well as local importance.[1] Nepal also provides many ecosystem services, such as water supply and carbon management, benefiting its neighbours and the global community. The Hindu Kush-Himalaya is often referred to as the 'water tower of Asia', with 1.3 billion people dependent on the water that flows from its mountains and glaciers. The economic value of this and other ecosystem services is only now being assessed and beginning to be included in national financial profiles. Plants are of fundamental importance to the functioning of these ecosystems, and so promoting and conserving plant biodiversity within them is an essential part of maintaining their health. At a human level plants are vital to the everyday livelihoods of the vast majority of Nepalese people, as sources of food and medicine, shelter and fire and for numerous other purposes such as dyes, fibres, beverages and animal fodder.

Plant biodiversity is vulnerable to habitat destruction and over-exploitation. Habitats need to be conserved and species used sustainably if they are to survive for future generations to use and enjoy. Tourism, particularly eco-tourism, is one of Nepal's main revenue sources and many people depend on the trekking industry for their income. Tourists are attracted by Nepal's spectacular landscapes and the fascinating plants and animals of which they are a part. If these natural resources are lost then Nepal's economy and the livelihoods of the local people will also suffer. Conservation and sustainable use are important at all levels – local, national and international – and for sound economic, scientific and sociological reasons, and the Government of Nepal recognizes this in its *National Biodiversity Strategy*.[2] However, effective planning for conservation and sustainable use is dependent upon reliable data relating to the species found in Nepal, their distribution and their ecological requirements. For the most part this baseline information is very limited, even for Nepal's extensive network of protected areas. Environmental research into habitats, ecosystems and plant use is reliant on accurate species identification, and so tools are needed to aid this process. Recent growth in climate change research has also highlighted the urgent need for accurate biodiversity occurrence data to contribute to modelling current changes and predicting the future. The *Flora of Nepal* forms a major component of these requirements and the Government of Nepal has designated it a top priority project in its *National Biodiversity Strategy Implementation Plan* and essential in addressing Nepal's obligations under the international *Convention on Biological Diversity*.[3]

Physiography and species diversity

Nepal is situated at the centre of the Himalaya, between latitudes 26° 22' and 30° 27' N, and longitudes 80° 40' and 88° 22' E. The average length of the country is 885 km from east to west, and its width varies from 145 km to 241 km north to south. A little more than 80 per cent of the total land area is covered by hills and high mountains, the remainder comprising the flat lowlands of the Tarai at less than 300 m in elevation. Altitude varies from around 60 m above sea level at the lowest point in the southern border regions of eastern Nepal, to 8848 m at the summit of the world's highest mountain, Sagarmatha (Mount Everest).

Nepal is a relatively small country, with a land area of 147,181 km^2, but it has a level of species diversity that far exceeds its size. Although covering less than 0.01 per cent of the total global land surface area, it is home to 2.2 per cent of the world's flowering plants. Nepal owes its botanical richness to the combination of several environmental and geographic factors that produces a complex mosaic of ecological regions. Elevation has a fundamental effect with tropical vegetation in the lowland Tarai giving way with increasing altitude to evergreen warm temperate forests, cool temperate deciduous forest, conifers at the tree line and alpines beyond. These changes are often abrupt, with just 150 km between the lowest areas in Nepal, 57 m above sea level, and the highest point on Earth. Weather patterns have a major effect with the summer monsoon sweeping east to west across the country, depositing its heaviest rainfall in the east and gradually reducing its influence westwards. The high east–west mountain chains interrupt the path of monsoon weather systems, depositing rainfall on their southern slopes and creating desert conditions in rain shadow areas to the north. Major rivers cut deep valleys running south through the mountain chains, allowing typically southern lowland plants to penetrate northwards far into the interior of Nepal. Aspect is also important in this highly dissected landscape, adding to the huge diversity of vegetation types encountered within very small areas. Dobremez recognized 118 major ecological zones in Nepal,[4] and even reducing this diversity to just eight vegetation types produces a very complex pattern (see map inside front cover).

Environmental conditions are only part of the explanation of high species diversity, and geological history and ancient species migrations also play a part. Nepal is a geologically young country, created when the Indian tectonic plate collided with the Eurasian plate some 40 to 50 million years ago. Since then a series of mountain-building events has resulted in the main east–west mountain ranges increasing in altitude northwards from the Tarai to the snowy peaks of the high Himal. Nepal lies at the crossroads of several floristic realms and plants migrated from neighbouring regions taking advantage of the new ecological niches, sometimes resulting in the rapid evolution of new species. At the centre of Nepal lies a transition zone where species typical of the Western Himalaya and Mediterranean (the Western Asiatic element) give way to species of the Eastern Himalaya and China (the Sino-Japanese element). The Tibetan Plateau to the north contributes some of its characteristic high-altitude plants, especially in the dry areas of Nepal in deep rain shadow, and species from the tropical Gangetic Plain extend across the southern border into the Tarai. The influence of Southeast Asia is seen in the lowlands of eastern Nepal where plants typical of Myanmar and Thailand reach Nepal via Assam. This simplified overview will be expanded upon in greater detail in the Introductory Volume, and more information can be found on the *Flora of Nepal* website (www.floraofnepal.org), but this should be sufficient to enable interpretation of the distribution maps within *Flora of Nepal*.

Botanical exporation and collecting

The botanical wealth of Nepal started to become known to the scientific world in the first decades of the nineteenth century through the pioneering collections of Francis Buchanan (later Hamilton), Edward Gardner and Nathaniel Wallich. In 1802 Buchanan-Hamilton (as he is now known to botanists) was the first to collect specimens of plants during his 12-month stay in the Kathmandu Valley. David Don used these collections and those of Gardner (wrongly attributed to Wallich) when preparing *Prodromus Florae Nepalensis* (1825),[5] the first account of the plants of Nepal. Don documented over 800 Nepalese species, nearly 80 per cent of them newly described, but he was aware that his 'forerunner' was incomplete as he did not have access to Buchanan-Hamilton's top set of specimens, paintings or notes. Around the same time Wallich was also publishing his Nepalese discoveries in *Tentamen Florae Napalensis* (the first botanical book published in India to be lithographically illustrated),[6] the sumptuously illustrated *Plantae Asiaticae Rariorum*,[7] and in the '*Wallich Catalogue*' of the vast herbarium that he assembled in Calcutta (Kolkata), in which he recorded over 1800 species from Nepal.[8] In homage to this illustrious history we have used one of Wallich's magnificent coloured drawings for the cover of our work.

Political restrictions prevented further scientific exploration in Nepal, especially by foreigners, although notable collections were made by J.D. Hooker (1848), the Schlagintweit brothers (1857), J. Scully (1876), J.F. Duthie (1884–1886), I.H. Burkill (1907), Lall Dhwoj (1927–1937), K.N. Sharma (1935–1937) and M.L. Banerji (1947–1967). The Government of Nepal began to relax restrictions in the late 1940's and natural historians such as O. Polunin and D.G. Lowndes resumed botanical explorations by joining mountaineering parties in 1949 and 1950. Since then there have been numerous Nepalese and international botanical expeditions exploring many parts of the country, with the latter predominantly led by British and Japanese scientists. Prominent figures include M.S. Bista, D.P. Joshi, S.B. Malla, S.B. Rajbhandari, P.R. Shakya, T.B. Shresha, M. Subedi and A.V. Upadhya from Nepal; L. Beer, C. Grey-Wilson, D.G. Long, A. D. Schilling, J.D.A. Stainton, W.R. Sykes and L.H.J. Williams from the UK; H. Hara, H. Kanai, M. Minaki, F. Miyamoto, S. Nakao, S. Noshiro, H. Ohashi, H. Ohba, M. Suzuki and M. Wakabayashi from Japan; J.F. Dobremez and A. Maire from France; and D.H. Nicolson from the USA. The editors of this volume have also all been heavily involved in fieldwork in Nepal. See Sutton in Hara *et al.* Volume 1 and Rajbhandari for further details on plant collectors in Nepal.[9, 10]

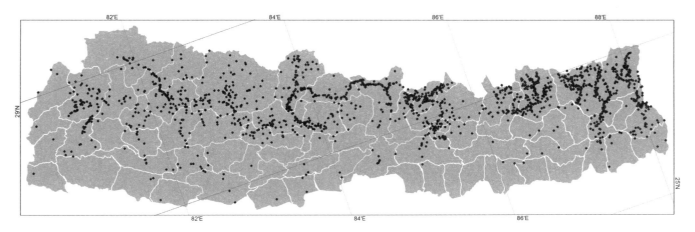

Map of Nepal showing the distribution of herbarium collections used in preparation of *Flora of Nepal* Volume 3.

Hundreds of thousands of herbarium specimens have been collected from Nepal and form the basis of floristic work, yet further fieldwork is needed. Many areas have only been visited once and so information on some species is very incomplete, especially for endemic plants, many of which are only known from one or two historic collections. The district-level distribution maps given for each species in this volume highlight the knowledge gaps and can be used to target future fieldwork. The patchiness of plant collections in Nepal is clearly illustrated by the map above, which shows the distribution of specimens used in writing Volume 3. The mountainous regions have attracted the most attention from botanists, but there are many mid elevation and highland areas of Nepal that are under-collected. Although the southern lowland areas are less explored, the situation is not as severe as indicated by this map as relatively few species included in this volume are distributed in the Tarai.

History of *Flora of Nepal*

The evolution and development of the *Flora of Nepal* project has a long and complex history that has been summarized by Watson *et al.* and will be fully recounted in the forthcoming Introductory Volume.[11] Producing a Flora for Nepal was a prime objective of the Government of Nepal's Department of Medicinal Plants (now Department of Plant Resources) soon after its establishment in 1961, but it was soon realized that extensive fieldwork would be needed to build expertise and provide the research materials required for this. Indeed, it took nearly 20 years before collections and experience were considered extensive enough for botanists in Japan, Britain and Nepal to join forces to produce the first detailed checklist of the plants of Nepal. The *Enumeration of the Flowering Plants of Nepal* (usually referred to as 'EPN' or the

'*Enumeration*') provided a comprehensive listing of accepted names and synonyms, with altitude range and crude distribution using three zones in Nepal – West, Central and East.[9] This groundbreaking work established a solid nomenclature, but was of limited use for identification as it included keys to genera and species for very few families and there were no descriptions. The *Enumeration* highlighted geographic regions and taxonomic groups that needed further study, and the following decades saw extensive collaborative projects targeting these knowledge gaps.

The next major landmark was the publication of the *Annotated Checklist of the Flowering Plants of Nepal* (ACFPN) in 2000.[12] This was produced by the *Plant Information and Technology Transfer for Nepal* project, which took the *Enumeration* into the electronic age, incorporating and updating the information using a database from which ACFPN was generated. Although presented in a different format from the *Enumeration*, ACFPN contained similar information with the addition of life form for each species. An internet version of ACFPN was later produced and made available through the eFloras website (www.efloras.org). At about the same time, Nepalese workers published *Flowering Plants of Nepal (Phanerogams)* presenting their checklist information in a more condensed and abbreviated format, and *Pteridophytes of Nepal*, the first comprehensive checklist of 534 species of ferns and fern allies.[13, 14]

Whilst progress was being made on the country-wide checklists, botanists in Nepal and Japan also undertook revisions of specific groups and worked on regional Floras. These extended the information available for a selected number of species, providing keys, descriptions, distributions and supplementary data.

Notable publications include: the *Flora of Eastern Himalaya* [15] series and other works by the University of Tokyo and the Society of Himalayan Botany; *Flora of Langtang*,[16] *Flora of the Kathmandu Valley* [17] and the *Fascicles of Flora of Nepal* [18] series by the Department of Plant Resources, based primarily on collections in the National Herbarium (KATH); and the recent *Flora of Mustang*.[19]

The *Flora of Nepal* project in its present form dates back to the first editorial meeting convened in Edinburgh in 2002. The *Advisory Board* comprises senior representatives of the three lead organizations in each country: D. Bajracharya, Nepal Academy of Science and Technology (NAST, formerly Royal Nepal Academy of Science and Technology), Nepal; S. Blackmore, Royal Botanic Garden Edinburgh (RBGE), UK; and H. Ohba, University of Tokyo/Society of Himalayan Botany, Japan. RBGE is the administrative centre, and the activities of the three Nepalese partner institutes, the *Nepal Chapter*, are coordinated by NAST and directed by the *National Coordination and Steering Committee*. The *Nepal Chapter* includes: NAST, the Department of Plant Resources, and the Central Department of Botany, Tribhuvan University. The *Editorial Board* is responsible for the day-to-day running of the project and comprises the editors and invited experts. Membership of each of these groups is listed at the start of this book.

New floristic approaches

Looking back over half a century, the *Flora of Nepal* is now very different from how it was originally envisaged. Over the years it has been shaped by changing demands from a wider and more diverse spectrum of end users, and developments in biodiversity informatics have radically altered the approach taken. Although the ten printed volumes will continue to be important as landmark publications, the electronic *Flora of Nepal Knowledge Base* is the primary focus of our work today. The printed volumes are generated from this, and it is used to produce downloadable digital versions (and revisions) of published accounts, and present data on the *Flora of Nepal* website. Completed accounts of new families and genera will be made available as pdf downloads via the website in advance of the printed volume. The *Flora of Nepal Knowledge Base* is stored and managed using RBGE's Padme database system. It was initiated using an import from the ACFPN (eFloras) database, and now comprises over 20,000 scientific names and over 25,000 occurrence records from Nepal. Both primary and synthesized floristic data are made freely available via the project website, giving access to the most up-to-date species-level information to those who need it.

Floristic research has embraced the electronic age and the internet is being used both to facilitate research and to disseminate outputs. Early examples of online Floras were simply electronic versions of the printed works, offering easier access with basic search and navigation capabilities. More sophisticated online Floras began to deliver their data from a database, enabling users to query and filter on a selection of criteria, such as listing only those species recorded from a particular province or between certain altitudes. The eFloras website (www.efloras.org) has led the way, and its *Flora of China* dataset is a good example of what can be done using existing information. However, these datasets are mainly derived retrospectively from the published works, and so are constrained in the information they can deliver. Traditional Floras have dealt with the space limitations of the printed format by taking a highly synoptic approach with a liberal use of abbreviations. This results in the loss of a great deal of detailed information which needs to be gathered to write a floristic account but cannot be included in the printed work. The *Flora of Nepal* seeks to address this and take electronic Floras onto a new level by recording all the primary data and making this available electronically alongside the synthesized information. Technical data for nomenclature and classification are good examples of this, as although a summary is given in the printed work, full data on places of publication, basionym relationships, homonyms and explicit statements of name misapplications are available online.

The *Flora of Nepal* also takes an evidence-based approach where, for example, all statements of occurrence are based either on a herbarium specimen, a reliable field observation or an authenticated literature record. Using Padme we are able to move the point of electronic data capture as close as possible to the point when it is recorded, so that data and images on new collections are entered in the database whether in the field or the herbarium. Distribution maps are generated from these records, and this rich source of geo-referenced occurrence data is now being valued by other researchers as primary biodiversity data for geo-spatial and temporal analyses such as climate change modelling, conservation prioritization and earth system science.

Flora of Nepal Volume 3

Volume 3 of the *Flora of Nepal* includes 21 families with 123 genera, 600 species, 19 subspecies, 31 varieties and 4 forma. The major families Rosaceae, Saxifragaceae and Cruciferae alone account for over half of its contents. The largest genera in the volume are *Saxifraga* (87 species), *Corydalis* (45), *Rubus*

(32), *Potentilla* (31), *Draba* (23), *Rhodiola* (22) and *Meconopsis* (22). It is noteworthy that total numbers of species in most of these genera have remained rather similar to totals in previous national checklists, with a rough balance between of the inclusion of new species hitherto unknown in Nepal and the reduction of names to synonymy following critical examination. The greatest reduction of doubtful species to synonymy is seen in *Cotoneaster* where 34 very poorly differentiated species have been reduced to a more practical 13. By contrast, the critical re-evaluation of species limits in *Meconopsis* has led to the inclusion of half as many species again compared with the previous number (22 compared to 15 species), several of which are rather narrow endemics.

Although Nepal has high species diversity, the level of endemism is not particularly high. Opinions vary on the numbers of strict endemics to Nepal, with estimates ranging from 246 to 342 species, or 4 to 5 per cent of vascular plants.[20, 21] The low level of endemism is a reflection of Nepal's political borders and the diverse phytogeographic regions they span; if neighbouring areas were included then a considerable number of 'near endemics' could be added. There are many examples of primarily Nepalese species whose distributions extend just over the eastern border into Sikkim, such as *Chrysosplenium singalilense*, *Rodgersia nepalensis*, and *Rubus griffithii*, and there are similar examples for the northern and western borders. ACFPN estimates that as many as 30 per cent of the vascular plants of Nepal are endemic to the Himalayas as a whole.[12]

Flora of Nepal Volume 3 is surprisingly rich in strict endemics as it includes 64 species and eight infraspecific endemics: over 10 per cent, or twice as many as would be expected. This may be a reflection of the included families having unusually high levels of recent species evolution, or that the critical taxonomic revision of Nepalese plants undertaken by world experts for the *Flora of Nepal* is uncovering higher levels of endemism than previously realized. See page xvii for a full list of endemic species included in Volume 3.

Conservation assessments

Nepal has no 'Red Data Book' in which all its rare and threatened plants are documented and standard IUCN threat categories assigned. However, *Rare, Endemic and Endangered Plants of Nepal* is an excellent precursor to this, listing 246 endemic and a further 60 non-endemic taxa that are threatened in Nepal.[20] Human activities such as habitat destruction from changing land use and over-collection for local use or commercial exploitation were given as the main threats to conservation. Now, 15 years later, the largely unknown effects of climate change must be added to these. Nine species of flowering plants were suspected to be extinct from Nepal, only one of which, *Persea blumei* Kosterm., is included in the current volume. *Persea blumei* was only known from the type specimen collected in Nepal by Wallich in 1820, but this is now considered a synonym of *Machilus pubescens* Blume, itself a Nepalese endemic but known from several districts.

The threat status of the vast majority of taxa included in *Rare, Endemic and Endangered Plants of Nepal* was recorded as 'unknown' because of a lack of information – many endemics are only known from one or two historic collections, some nearly 200 years old. Although data on Nepal's endemics has recently been updated,[21] information on populations of these plants in the field is still lacking and they would almost all be listed as 'Data Deficient' using current IUCN threat categories. Research towards a comprehensive Red Data Book must include targeted fieldwork to assess the current conservation status of populations of these plants, and data from the *Flora of Nepal* will make a significant contribution towards this.

M.F. Watson, S. Akiyama, H. Ikeda, C.A. Pendry, K.R. Rajbhandari & K.K. Shrestha
July 2011

References

1   **Meyers, N., Mittermeier, R.A., Mittermeier, C.G., de Fonseca, G.A.B. & Kent, J.** (2000). 'Biodiversity hotspots for conservation priorities', *Nature* 403: 853–858.

2   **GoN/MFSC** (2002). *Nepal Biodiversity Strategy*, Government of Nepal, Ministry of Forests and Soil Conservation, Singha Durbar, Kathmandu.

3   **GoN/MFSC** (2006). *Nepal Biodiversity Strategy Implementation Plan*, Government of Nepal, Ministry of Forests and Soil Conservation, Singha Durbar, Kathmandu.

4   **Dobremez, J.-F.** (1972). 'Les grandes divisions phytogéographiques du Népal et de l'Himalaya', *Bull. Soc. Bot. France* 119: 111–120.

5   **Don, D.** (1825). *Prodromus Florae Nepalensis*, London.

6   **Wallich, N.** (1824–1826). *Tentamen Florae Napalensis Illustratae*, Calcutta and Serampore.

7   **Wallich, N.** (1829–1832). *Plantae Asiaticae Rariores*, London.

8   **Wallich, N.** (1828–1849). *A Numerical List of Dried Specimens of Plants in the East India Company's Museum*, London.

9   **Hara, H., Stearn, W.T., Williams, W.T. & Chater, A.O.** (1978, 1979, 1982). *An Enumeration of the Flowering Plants of Nepal*, 3 volumes. Trustees of the British Museum (Natural History), London.

10 **Rajbhandari, K.R.** (2002). 'Flora of Nepal: 200 years' march', in: Noshiro, S. & Rajbhandari, K.R., 'Himalayan Botany in the Twentieth Century': 76–93, Society of Himalayan Botany, Tokyo.

11 **Watson, M.F., Pendry, C.A. & Ikeda, H.** (2010). 'Publication of Flora of Nepal', in: Ikeda, H. & Noshiro, S. (eds), 'Himalaya Hotspot of Biodiversity', 75–96, The University Museum, University of Tokyo.

12 **Press, R., Shrestha, K.K. & Sutton, D.A.** (2000). *Annotated Checklist of Flowering Plants of Nepal*, Natural History Museum, London & Tribhuvan University, Nepal.

13 **Singh, A.P., Bista, M.S., Adhikari, M.K. & Rajbhandari, K.R.** (2001). *Flowering Plants of Nepal (Phanergams)*, HM Government of Nepal, Ministry of Forests and Soil Conservation, Department of Plant Resources, Kathmandu.

14 **Thapa, N., Bista, M.S., Adhikari, M.K. & Rajbhandari, K.R.** (2002). *Pteridophytes of Nepal*, HM Government of Nepal, Ministry of Forests and Soil Conservation, Department of Plant Resources, Kathmandu.

15 **Hara, H.** (1966, 1971). *Flora of Eastern Himalaya, 1 & 2*; **Ohashi, H.** (1975). *Flora of Eastern Himalaya, 3*, University of Tokyo Press, Tokyo.

16 **Malla, S.B., Shrestha, A.B., Rajbhandari, S.B., Shrestha, T.B., Adhikari, P.M. & Adhikari, S.R.** (1976). *Flora of Langtang and Cross Section Vegetation Survey (Central Zone)*, HM Government of Nepal, Ministry of Forests, Department of Medicinal Plants, Kathmandu.

17 **Malla, S.B., Rajbhandari, S.B., Shrestha, T.B., Adhikari, P.M., Adhikari, S.R. & Shakya, P.R.** (1986). *Flora of Kathmandu Valley*, HM Government of Nepal, Ministry of Forests and Soil Conservation, Department of Medicinal Plants, Kathmandu.

18 **Bista, M.S. *et al.*** (1995–ongoing) *Fascicles of Flora of Nepal*, HM Government of Nepal, Ministry of Forests and Soil Conservation, Department of Plant Resources, Kathmandu.

19 **Ohba, H., Iokawa, Y. & Sharma, L.R.** (2008). *Flora of Mustang, Nepal*, Kodansha Scientific Ltd., Tokyo.

20 **Shrestha, T.B. & Joshi, R.M.** (1996). *Rare, Endemic and Endangered Plants of Nepal*, WWF Nepal Program, Kathmandu.

21 **Rajbhandari, K.R. & Adhikari, M.K.** (2009). *Endemic Flowering Plants of Nepal Part 1*, Government of Nepal, Ministry of Forests and Soil Conservation, Department of Plant Resources, Kathmandu.

Introduction

The *Flora of Nepal* will be published in ten volumes and will account for all the estimated 6000 to 6500 species of conifers and flowering plants, and 530 species of ferns and fern allies known to occur in Nepal. The taxa treated will include all native and fully naturalized vascular plants, including brief references to agricultural and horticultural plants as appropriate.

The *Flora of Nepal* takes an innovative approach to Flora writing, with an underlying database system managing the *Flora of Nepal Knowledge Base* from which the printed volumes and the 'online Flora' (www.floraofnepal.org) are generated. The Internet-accessible dataset augments the printed Flora by presenting all herbarium specimen data, detailed taxonomic information (such as full nomenclatural references and typification), distribution maps with point occurrences and images used when preparing the Flora. Much of this information is accumulated as a normal part of taxonomic working practices when undertaking a floristic revision, but it is usually lost to a wider audience as it is rarely included in the traditional printed Flora. The *Flora of Nepal* website will also give access to a set of Flora accounts in pdf format for families and genera that have been finalized, including those in volumes yet to be printed. These pdf documents will be periodically updated in numbered versions, permanently available and citable, and so will continue to provide the definitive account of the vascular plants of Nepal.

For pragmatic reasons the arrangement of families in the printed volumes of the *Flora of Nepal* follows a modified Englerian sequence, closely following that of the *Flora of China* and, to a lesser extent, the *Flora of Bhutan*.[1, 2] A guide to the placement of families in *Flora of Nepal* volumes is given at the end of this book. In recent years the world view on the arrangement of families has radically changed following overwhelming phylogenetic evidence. The emergent family-level classification, now in its third iteration as APG III, is reasonably stable and widely accepted.[3] It has not been possible to alter the family sequence in *Flora of Nepal* printed volumes midway through the project, but as the data are stored separately in a database, the families can be reorganized electronically at a later date to reflect alternative classifications. Circumscription of families and genera, however, generally does follow a contemporary understanding of their relationships (e.g.

Saxifragaceae *s.l.* is segregated into Saxifragaceae *s.s.*, Parnassiaceae, Hydrangeaceae and Grossulariaceae), except where group experts advise otherwise (e.g. Fumariaceae and Papaveraceae). Genera and species are treated in taxonomic order, or if there is disagreement then morphologically similar species are usually grouped together or occasionally listed alphabetically. Infraspecific taxa are always presented in alphabetical order. Intermediate ranks, such as subfamily, tribe, subgenus, section and series, are only employed when they are useful in the treatment of large families or genera (e.g. *Saxifraga*).

Information on nomenclature and classification is given for all accepted scientific names and synonyms pertaining to Nepal and nearby regions. Emphasis is given to those names listed in the primary checklists for Nepal: *Enumeration of the Flowering Plants of* Nepal,[4] *Annotated Checklist of the Flowering Plants of* Nepal,[5] and *Flowering Plants of Nepal (Phanerogams)*.[6] At the generic level, synonyms widely used in the Asian literature are included. Full bibliographic citation with authorship is given for all accepted names and their basionyms at the rank of genus and below. As far as possible, the bibliographic citations of all accepted names and their basionyms have been verified with the original literature. The basionym precedes all other synonyms, which are listed alphabetically. Misapplied names (misidentifications encountered in the literature) are not included in synonymy, but are discussed in the supporting information at the end of a taxon. Authors of plant names follow the standard forms given in *Authors of Plant Names* and its continuously updated online supplement (www.ipni.org).[7] Bibliographic references are given using the standard abbreviations in BPH-2 for serial publications (journals and periodicals) and in TL-2 (and its supplements) for books.[8, 9] In some cases books were published in several fascicles on different dates, sometimes over different years, but not indicated as such in the printed work. Date of publication is critical for establishing nomenclatural priority, and so it is important to be precise when citing names published in such works. The fascicle composition and publication dates of these often complex cases are clearly explained in TL-2, but the standard abbreviation does not differentiate between them. In these instances the TL-2 abbreviation has been amended with brackets to clearly indicate which fascicle is being referred to, for example Wallich, N., Pl. As. Rar. 2[8]. 1831. Books and periodicals not included in these two standard references have been abbreviated according to the recommendation in Appendix A of BPH-2.

Where a taxon has a widely recognized local name this is given in Devanagari script, followed by its transliteration into the Latin alphabet and the language of the vernacular name in parentheses '( )'. One local name is given in the printed Flora, whereas multiple alternative vernacular names in different languages may be included in the *Flora of Nepal Knowledge Base* and made available online. Separate indexes to vernacular names in Devanagari, their Latin transliterations and scientific names are included at the end of each volume.

Descriptions are given for all taxa (family, genus, species, infraspecies and occasionally intermediate ranks) and wherever possible are based on primary observations and measurements made on specimens from Nepal. If no such material was available to authors, descriptions are taken from specimens from adjacent countries or secondary sources, and annotated as such. Most descriptions are about 150 words long, but exceptionally they are shorter or longer depending on the complexity of the taxon being described. For species with more than one infraspecific taxon, a full description is given for the species and short diagnoses for the lower taxa. Descriptions aim to be consistent and parallel between taxa of the same rank within a higher taxon. Authors were asked to standardize descriptive terms using the definitions given in *Plant Identification Terminology*.[10] If a single measurement is given it refers to length, and if width is also given it is in the format length × width. Ranges are separated by an en-dash (–) and discontinuous states by the word 'or'. Exceptional measurements are given in parentheses '( )'. Taxon statistics and short statements on worldwide distribution are provided for families and genera, with summary statistics of lower taxa represented in Nepal.

Identification keys are dichotomous and presented in a bracketed format, with all elements strictly parallel between the two leads of each couplet. Keys are artificial and not intended to reflect any taxonomic classification. There is usually a single key to genera within a family, combining flowering, fruiting and vegetative characters, but where this is unwieldy separate keys are given for flowering and fruiting material (e.g. Cruciferae, Rosaceae). Keys are also given for species within a genus and taxa within a species. Figures are provided to aid identification by illustrating the diagnostic characters of each family and genus,

and for large genera variation in major morphological features is represented.

The geographic distribution within Nepal is indicated for each species and infraspecific taxon at the political district level by a shaded distribution map: a map illustrating the 75 Districts and five Development Regions in Nepal is given on the inside front cover. The distribution maps are evidence-based, produced from the *Flora of Nepal Knowledge Base* using locality information taken from authenticated herbarium specimens and records of plants made *in situ* by credible observers. Ideally all specimens identified by authors should be geo-referenced and databased when they are preparing *Flora of Nepal* accounts, but where this is not possible a minimum of one specimen per district is required. Sometimes the distribution of a species is greater than the sum of the distribution maps of its infraspecific taxa. This is a result of some herbarium specimens only being identifiable to species level. Occasionally species are known only from poorly localised collections, especially those from the early nineteenth century. For example, Wallich often only gave 'Napalia' as the locality for many of his 1820–1821 collections. These specimens are most likely to have come from the Kathmandu Valley, known as the 'Nepal Valley' or just 'Nepal' at that time, but they might also have been collected during his inward and outward journeys from India via Hetauda, or by pilgrims going north to 'Gossainthan' (Gossaikund). It is therefore impossible to be sure of the correct district and in such cases this is noted in the supporting information and the map omitted. The *Flora of Nepal* website gives access to the underlying distribution and specimen information through an interactive dot map plotting all geo-referenced occurrence records and a listing of all material recorded.

Distribution for species and infraspecific taxa occurring outside Nepal is indicated by a list of geographical regions, with the resolution becoming coarser with increasing distance from Nepal. In order to utilise information contained within other published Floras these areas are defined according to political borders, with countries or provinces grouped to form regions that have some underlying biogeographic basis. For example, although the Tibetan Plateau extends into parts of Sichuan and Yunnan, we limit it to Xizang and Qinghai. *Flora of Nepal* takes no stance on any politically disputed border areas and is following the current international mapping convention of using the 'lines of control' to delineate its regions. The names used for the regions are intended to be descriptive and non-political. The regions are illustrated by the map on the inside back cover, and listed opposite:

W Himalaya	NW India (Jammu & Kashmir [including Siachen Glacier, regions disputed with Pakistan], Himachal Pradesh, Uttarakhand), N Pakistan (Khyber Pakhtunkhwa [previously known as North West Frontier Province], Gilgit-Baltistan [also known as Northern Areas, region disputed with India], Azad Kashmir [region disputed with India]).
E Himalaya	N India (Sikkim & Darjeeling), Bhutan, NE India (Arunachal Pradesh [disputed border with S Tibet, China]).
Tibetan Plateau	China (Xizang [including Aksai Chin and Shaksgam Valley regions disputed with India], Qinghai).
Assam-Burma	NE India (Assam, Nagaland & Manipur), Myanmar.
S Asia	E Pakistan (Punjab, Sind, Islamabad), peninsular India, Sri Lanka, Bangladesh, Maldives.
E Asia	China (excluding Xizang, Xinjiang, Qinghai), N & S Korea, Japan, Taiwan.
SE Asia	Thailand, Laos, Cambodia, Vietnam, Malaysia, Indonesia, Philippines, New Guinea.
N Asia	China (Xinjiang), Russia, Mongolia.
C Asia	Kazakhstan, Uzbekistan, Turkmenistan, Tajikistan, Kyrgyzstan.
SW Asia	Afghanistan, W Pakistan (Baluchistan, Federally Administered Tribal Areas), Iran, Middle East, Arabian Peninsula, Turkey, Azerbaijan, Armenia, Georgia.
Asia	collective term for all above areas of Asia.
Europe	includes Ukraine, Belarus, Baltic republics.
Africa	includes Madagascar.
N America	includes C America south to Panama.
S America	south of Panama.
Australasia	Australia, New Zealand, Pacific Islands.
Cosmopolitan	collective term for a generally worldwide distribution.

Altitudes (elevation above sea level) are based on herbarium specimen data or records from credible observers. They are given to the nearest 100 m rounded up or down, with exceptional altitudes given in parentheses '( )'. Likewise, flowering and fruiting times are based on specimens collected from Nepal, or on material from adjacent regions if these data are lacking and a note is provided to explain this. The short statement on the ecological preference of each species and infraspecific taxon is mostly taken from herbarium specimen data. Currently these often lack detail, a reflection of the shortcomings of poor-quality data recorded by the past collectors of herbarium material, but these will improve with more field studies.

Supplementary information is given at the end of a taxon account discussing taxonomic issues, highlighting spot characters useful for identification, noting similar species that could cause confusion, and detailing the misapplication of names. Summary information is provided for ethnobotanical and other uses, but this is not intended to be exhaustive and is derived from secondary sources, such as *Plants and People of Nepal* and *A Compendium of Medicinal Plants in Nepal*.[11, 12]

Abbreviations

Standard abbreviations for the International System of Units (SI) are used for measurements. Herbaria are cited using the standard abbreviation in *Index Herbariorum*.[13] Other abbreviations used in the text include:

C	central.
ca.	*circa* – about, approximately.
comb. nov.	*combinatio nova* – new combination of name and epithet.
dbh	diameter at breast height – measured on tree trunks at 1.3 m above the ground.
E	east, eastern.
et al.	*et alia* – and others.
fig.	figure.
N	north, northern.

nom. cons.	*nomen conservandum* – name officially conserved in ICBN.[14]
nom. illegit.	*nomen illegitimum* – illegitimate name, according to ICBN.[14]
nom. inval.	*nomen invalidum* – invalid name, according to ICBN.[14]
nom. nud.	*nomen nudum* – name lacking a description, or reference to an effectively published description, and so invalid according to ICBN.[14]
nom. rej.	*nomen rejiciendum* – name officially rejected in ICBN.[14]
nom. superfl.	*nomen superfluum* – name superfluous when published, and so illegitimate according to ICBN.[14]
pl.	plate.
q.v.	*quod vide* – which see.
S	south, southern.
s.l.	*sensu lato* – for a taxon treated in a broad sense.
s.s.	*sensu stricto* – for a taxon treated in a narrow sense.
sect.	section.
subfam.	subfamily.
subgen.	subgenus.
subsp.	subspecies.
subvar.	subvariety.
syn.	synonym
var.	variety.
W	west, western.
>	greater than
<	less than

References

1 **Wu, Z.Y., Raven, P.H. & Hong, D.Y.** (1994–ongoing). *Flora of China*, Science Press (Beijing) & Missouri Botanical Garden Press, St Louis [available online at flora.huh.harvard.edu/china].

2 **Grierson, A.J.C., Long, D.G. & Noltie, H.J.** (1983–2002). *Flora of Bhutan*, Royal Botanic Garden Edinburgh, Edinburgh.

3 **Angiosperm Phylogeny Group III** (2009). 'An update of the Angiosperm Phylogeny Group classification for the orders and families of flowering plants': APG III. Bot. J. Linn. Soc. 161: 105–21.

4 **Hara, H., Stearn, W.T., Williams, W.T. & Chater, A.O.** (1978, 1979, 1982). *An Enumeration of the Flowering Plants of Nepal*, 3 volumes, Trustees of the British Museum (Natural History), London.

5 **Press, R., Shrestha, K.K. & Sutton, D.A.** (2000). *Annotated Checklist of Flowering Plants of Nepal*, Natural History Museum: London & Tribhuvan University, Kathmandu [updated version available online at efloras.org].

6 **Singh, A.P., Bista, M.S., Adhikari, M.K. & Rajbhandari, K.R.** (2001). *Flowering Plants of Nepal (Phanerogams)*, HM Government of Nepal, Ministry of Forests and Soil Conservation, Department of Plant Resources, Kathmandu.

7 **Brummit, R.K. & Powell, C.E.** (1992). *Authors of Plant Names*, Royal Botanic Gardens, Kew, London [available online with revisions at www.ipni.org].

8 **Bridson, G.D.R. & Smith, E.R.** (1991). *Botanico-Periodicum-Huntianum*, ed. 2, Hunt Institute for Botanical Documentation, Pittsburgh.

9 **Stafleu, F.A., Cowan, R.S. & Mennega, E.** (1973–1988). *Taxonomic Literature*, ed. 2 (TL-2), Bonn, Scheltma & Holkema, Utrecht/Antwerpen; dr. W. Junk b.v., The Hague/Boston [available online at tl2.idcpublishers.info].

10 **Harris, J.G. & Harris, M.W.** (2001). *Plant Identification Terminology*, ed. 2, Spring Lake Publishing, Utah.

11 **Manandhar, N.P.** (2002). *Plants and People of Nepal*, Timber Press, Oregon.

12 **Baral, S.R. & Kurmi, P.P.** (2006). *A Compendium of Medicinal Plants in Nepal*, Mass Printing Press, Kathmandu.

13 **Holmgren, P.K., Holmgren, N.H. & Barnett, L.C.** (eds) (1990). *Index Herbariorum. Part 1: The Herbaria of the World*. ed. 8. New York Botanic Garden: New York. [available online with revisions at sweetgum.nybg.org/ih].

14 **McNeill, J., Barrie, F.R., Burdet, H.M., Demoulin, V., Hawksworth, D.L., Marhold, K., Nicolson, D.H., Prado, J., Silva, P.C., Skog, J.E., Wiersema, J.H. & Turland, N.J.** (eds) (2006). *International Code of Botanical Nomenclature (Vienna Code)*, Regnum Vegetabile 146. Gantner, Ruggell.

Strict Endemics in Nepal Included in Volume 3

Aphragmus hinkuensis (Kats.Arai, H.Ohba & Al-Shehbaz) Al-Shehbaz & S.I.Warwick

Aphragmus nepalensis (H.Hara) Al-Shehbaz

Corydalis calycina Lidén

Corydalis clavibracteata Ludlow

Corydalis elegans subsp. *robusta* Lidén

Corydalis megacalyx Ludlow

Corydalis simplex Lidén

Corydalis spicata Lidén

Corydalis stipulata Lidén

Corydalis terracina Lidén

Corydalis uncinata Lidén

Corydalis uncinatella Lidén

Draba macbeathiana Al-Shehbaz

Draba poluniniana Al-Shehbaz

Draba staintonii Jafri ex H.Hara

Geum elatum forma *rubrum* Ludlow

Lepidostemon williamsii (H.Hara) Al-Shehbaz

Litsea doshia (D.Don) Kosterm.

Machilus pubescens Blume

Meconopsis autumnalis P.A.Egan

Meconopsis chankheliensis Grey-Wilson

Meconopsis dhwojii G.Taylor ex Hay

Meconopsis ganeshensis Grey-Wilson

Meconopsis gracilipes G.Taylor

Meconopsis grandis subsp. *jumlaensis* Grey-Wilson

Meconopsis manasluensis P.A.Egan

Meconopsis napaulensis DC.

Meconopsis regia G.Taylor

Meconopsis simikotensis Grey-Wilson

Meconopsis staintonii Grey-Wilson

Meconopsis taylorii L.H.J.Williams

Noccaea nepalensis Al-Shehbaz

Parnassia chinensis var. *ganeshii* S.Akiyama & H.Ohba

Potentilla cardotiana var. *nepalensis* H.Ikeda & H.Ohba

Potentilla eriocarpa var. *major* Kitam.

Potentilla makaluensis H.Ikeda & H.Ohba

Potentilla peduncularis var. *ganeshii* H.Ikeda & H.Ohba

Potentilla turfosoides H.Ikeda & H.Ohba

Prunus himalaica Kitam.

Prunus jajarkotensis H.Hara

Prunus taplejunica H.Ohba & S.Akiyama

Prunus topkegolensis H.Ohba & S.Akiyama

Rhodiola nepalica (H.Ohba) H.Ohba

Rosularia marnieri (Raym.-Hamet ex H.Ohba) H.Ohba

Rubus x seminepalensis Naruh.

Saxifraga alpigena Harry Sm.

Saxifraga amabilis H.Ohba & Wakab.

Saxifraga cinerea Harry Sm.

Saxifraga excellens Harry Sm.

Saxifraga ganeshii H.Ohba & S.Akiyama

Saxifraga harae H.Ohba & Wakab.

Saxifraga hypostoma Harry Sm.

Saxifraga jaljalensis H.Ohba & S.Akiyama

Saxifraga lowndesii Harry Sm.

Saxifraga mallae H.Ohba & Wakab.

Saxifraga micans Harry Sm.

Saxifraga mira Harry Sm.

Saxifraga namdoensis Harry Sm.

Saxifraga neopropagulifera H.Hara

Saxifraga poluninana Harry Sm.

Saxifraga rhodopetala Harry Sm.

Saxifraga rolwalingensis H.Ohba

Saxifraga roylei Harry Sm.

Saxifraga staintonii Harry Sm.

Saxifraga williamsii Harry Sm.

Saxifraga zimmermannii Baehni

Sedum obtusipetalum subsp. *dandyanum* Raym.-Hamet ex H.Ohba

Sedum pseudomulticaule H.Ohba

Sibbaldia emodi H.Ikeda & H.Ohba

Solms-laubachia haranensis (Al-Shehbaz) J.P.Yue, Al-Shehbaz & H.Sun

Solms-laubachia nepalensis (H.Hara) J.P.Yue, Al-Shehbaz & H.Sun

Sorbus sharmae M.F.Watson, V.Manandhar & Rushforth

Nomenclatural Novelties

It is not intended to publish names for new taxa in *Flora of Nepal* – these should be previously effectively and validly published elsewhere. However, new combinations and other nomenclatural acts such as lecto- or neotypification can be included and will be listed at the start of each volume before the main family accounts.

The following new combination is needed for Saxifragaceae in *Flora of Nepal* Volume 3:

Micranthes gageana (W.W.Sm.) Gornall & H.Ohba, comb. nov.
  Basionym: *Saxifraga gageana* W.W.Sm., Rec. Bot. Surv. India 4: 265 (1911).
  Types: India, Sikkim, Chola Range, 14000–15000 ft. W.W. Smith, 3809 & 3989.

Corydalis magni Pusalkar (Fumaraceae) will be formally published in *Kew Bull*. 66: in press (2011). However, as there are technical difficulties with the production of this issue it will not be published before 1st September 2011, the publication date for the current volume of *Flora of Nepal*. We are not intending to publish *Corydalis magni* in *Flora of Nepal*, and it will remain invalidly published until the appearance of the relevant issue of the *Kew Bulletin*.

Acknowledgements

The *Flora of Nepal* has had a long gestation from its formal inception in 1993 at the 15th International Botanical Congress in Yokohama, Japan. Over the years many people have made important contributions through discussions and workshops, and the team that currently carries the torch forward owes a great debt of gratitude to their foresight and enthusiasm. This first volume is dedicated to four prominent figures from these formative years: Hiroshi Hara, Samar Bahadur Malla, Tirtha Bahadur Shrestha and William T. Stearn. To these we can add many names, most of whom continue their involvement and are mentioned in the following paragraphs. The *Flora of Nepal* would not have been possible without the efforts of hundreds of botanists spending thousands of hours on expeditions, collecting hundreds of thousands of specimens, often in arduous conditions. These herbarium specimens are our raw materials, the lifeblood of floristic research, and the names of their collectors are recorded for posterity on specimen labels, information now available through our website (www.floraofnepal.org). A complete account of the history of the *Flora of Nepal*, including the contributions of these individuals and their institutes, will be given in the Introductory Volume.

Many people have helped in the preparation of the current volume, notably the 34 authors of accounts listed at the start of this book. We deeply appreciate their contribution to the *Flora of Nepal*, especially their willingness to contribute data beyond that normally required when preparing floristic accounts. This extra information is needed to make our work as evidence-based as possible, and to enable us to provide electronic access to all the underlying information. We thank Bernhard Dickoré and Gerwin Kasperek for their comments as reviewers on the *Cotoneaster* account, likewise Lars Chatrou and Tanawat Chaowasku for their review of the Annonaceae, and Jane Nyberg for her work in editing the Cruciferae. Royal Botanic Garden Edinburgh (RBGE) was the editorial centre for the current volume.

We gratefully acknowledge the significant support for the project from the following institutes which provide working space, access to collections and resources, and dedicated staff time: RBGE – the administrative centre for the project; University of Tokyo (TI) and the Society of Himalayan Botany (SHB) in Japan; Nepal Academy of Science and Technology (NAST) – the coordinating centre in Nepal; the Government of Nepal, Ministry of Forests and Soil Conservation, Department of Plant Resources (DPR); and the Central Department of Botany, Tribhuvan University (CDB-TU). The British and Nepalese ambassadors, their embassy staff in Kathmandu and London, the Minister of Forest and Soil Conservation and The British Council in Kathmandu are warmly thanked for their support and encouragement for all of our floristic research and capacity building activities in Nepal and the UK.

The editors greatly appreciate the guidance and support from members of the *Advisory Board*, and fellow members on the *Editorial Board*: their names are listed at the start of this volume and so are not repeated here. Likewise we acknowledge the sterling service and sage advice from the *National Coordination and Steering Committee* in Nepal and their strong support for the wider activities of the *Flora of Nepal* project, especially fieldwork and capacity building. The current members are listed at the start of this book, and to these should be added the following who have served on this committee in the past: Soorya Lal Amatya, Hom Nath Bhattarai, Madhusudan Bista, Govinda Prasad Sharma Ghimire, Pramod Kumar Jha, Kamal Krishna Joshi, Sanu Devi Joshi, Surya Prasad Joshi, Krishna P. Manandhar, Krishna Chandra Paudel, Saman Bahadur Rajbhandari, Hari Krishna Saiju, Lokendra Raj Sharma, Madhav Prasad Sharma, Udaya Raj Sharma, Chandi Prasad Shrestha, Puspa Raj Shrestha, Dilip Subba, Bishal Nath Upreti, Yogesh Nandan Vaidya and Kayo Devi Yami. Many other Nepalese botanists have contributed to the *Flora of Nepal* leading up to the publication of this first volume. Some of these are already listed as co-authors of family accounts and also we gratefully acknowledge the contribution of: Sunil Kumar Acharya, Mahesh Kumar Adhikari, Devendra Mananda Bajracharya, Siddhartha Bajra Bajracharya, Khem Bhattarai, Nakul Chettri, Bimala Devi Devkota, Suraj Ketan Dhungana, Lajimina Joshi, Umesh Koirala, Lalit Narayan Manadar, Nirmala Pandey, Lokesh Ratna Shakya, Indira Sharma, Keshab Shrestha, Mahendra Nath Subedi, Batu Krishna Uprety, Rajesh Kumar Uprety and Sheetal Vaidya. Several international botanists have also helped shape the Flora of Nepal, particularly in its early stages, and we deeply thank the following for their contributions at workshops and editorial meetings: J. Crinan M. Alexander, Ihsan A. Al-Shehbaz, Makoto Amano, David E. Boufford, Mary Gibby, J. Charles Jarvis, Hiroo Kanai,

David G. Long, Henry J. Noltie, Hiroyoshi Ohashi, J. Robert Press, Peter H. Raven, Mitsuo Suzuki, Dan H. Nicolson, Nicholas J. Turland and Toshio Yoshida. A special mention must go to Nima Thudu Sherpa and Dawa Thundup Sherpa who for many years have expertly organized and participated on British- and Japanese-led fieldwork expeditions to far flung places in Nepal. The success of these expeditions would not have been possible without the great help and support of their trek teams and Sherpa guides, many of whom have worked on numerous collecting trips and can justly call themselves 'parataxonomists'.

Funding for the *Flora of Nepal* project has been largely through institutional commitment in staff time and resources. We greatly appreciate the generous additional funding recently provided by the Board of Trustees of RBGE, which has made it possible for us to take this first volume though to publication. The UK Darwin Initiative is also thanked for funding the *Plant Information and Technology Transfer for Nepal* project (Ref. 162/06/052, 1997–1999) and the three-year project *Capacity Building for Plant Biodiversity, Inventory and Conservation in Nepal* (Ref. 162/12/030, 2003–2006). Plant names from the electronic dataset produced during the first of these was used to seed the *Flora of Nepal Knowledge Base* at the start of our project. Many of the trainees in the second project appear as co-authors of family accounts in this volume, and others are working on accounts for future volumes. We are also pleased to acknowledge the generous support of the Andrew W. Mellon Foundation for establishing specimen digitization in the National Herbarium of Nepal (KATH) and enabling Nepal to be the first Asian country to join the Global Plant Initiative partnership.

It would have been impossible to produce the *Flora of Nepal* in its present formats without the skills and dedication of Martin R. Pullan, the creator and developer of RBGE's Padme taxonomic database system. Padme manages all the information underlying the Flora of Nepal, the *Flora of Nepal Knowledge Base*, and makes it available in print and over the internet. We sincerely thank him for his enormous contribution to the project, especially in developing outputs to produce pdfs of accounts, the whole volume in html mark-up ready for typesetting and the website itself.

We warmly acknowledge the important contribution of Bhaskar Adhikari who coordinated the production and composition of the plates of illustrations, in addition to drawing some of the figures himself. We also thank Claire Banks, Jane Nyberg, Louise Olley, Mutsuko Nakajima, Neera Joshi Pradhan and Cliodhna Ni Bhroin for their original line drawings. Many of the plates in this volume are composed from illustrations previously published in the *Flora of China*, and we are very grateful to Peter H. Raven and Hong De-Yuan for arranging for their inclusion, and Victoria C. Holliwell, Beth Parada, Nicholas J. Turland and Libing Zhang for facilitating this process. Several illustrations were previously published in the *Flora of Bhutan* and we thank RBGE for permission to re-use them. Accreditation for all the artwork used in this volume is given at the end of this book. We also extend our appreciation to Martin R. Pullan, Mike Shand and David Defew for designing the maps used for the end papers, to Jens-Peter Barnekow Lillesø for permission to reproduce the vegetation map inset in the front end paper, to McBride Design and Caroline Muir for the cover design and to Jamie Aguilar for scanning artwork.

We wish to sincerely thank the directors and curators of the following herbaria for their support and the access they have allowed to their collections during visits to the herbaria and through loans of specimens: A, B, BM, CAL, CBM, E, G, GH, GOET, GZU, HNWP, K, KATH, KYO, KUN, L, LD, LINN, MAK, MB, MBK, MO, PE, P, PR, S, TCD, TI, TNS, TUCH, TUS, UPS, US, W and WHB. In particular we are very grateful to John Hunnex, Hidetoshi Nagamasu and Akiko Shimizu for their assistance with seeking out and imaging specimens at BM, KYO and TI, respectively. Takahide Kurosawa is thanked for distributing unmounted specimens from TI. We must recognize the huge contribution made by Anshu Bhattarai, Catherine Cooke, Neil Gordon and account authors who have spent a great deal of time geo-referencing and entering herbarium specimen data into Padme: without them we would have no species distribution maps. We have also been able to draw on data held in the SHB Flora of Nepal database at the University of Tokyo managed by Hidehisa Koba and have learned from his experience with the preparation of distribution maps for the *Newsletter of Himalayan Botany*.

Hamish Adamson (RBGE) has expertly steered this book through all stages of publication, with copy editing by Anna Stevenson, typesetting by David Defew (Fakenham Prepress Solutions) and indexing by Faith Williams. Bhaskar Adhikari undertook the checking and indexing of Nepalese local names in both Devanagari and Latin transliteration. Caroline Muir and Joanna Jolly are thanked for design input and general comments on the manuscript.

M.F. Watson, S. Akiyama, H. Ikeda, C.A. Pendry, K.R. Rajbhandari and K.K. Shrestha
July 2011.

Magnoliaceae

Ram C. Poudel, Yumiko Baba & Colin A. Pendry

Evergreen or deciduous trees, often aromatic. Stipules large, caducous, convolute, enclosing the bud, usually adnate to the petiole and leaving a longitudinal scar there and a circular scar around the twig. Leaves alternate, simple, petiolate, leathery or papery, pinnately-veined, usually glabrous and shiny above and more or less pubescent below. Inflorescences terminal or axillary, consisting of a single flower enclosed by 1–4 convolute, caducous bracts leaving circular scars around the peduncle. Flowers bisexual, often large, showy and fragrant. Perianth of fleshy tepals spirally arranged in 2–4 more or less differentiated whorls. Stamens numerous, spirally arranged, free, poorly differentiated into anthers and filament, with the connective produced into an apical appendage. Anthers 2-locular, opening by longitudinal slits. Gynoecium with or without short gynophore, composed of numerous free or united carpels spirally arranged on an elongated receptacle. Carpels superior, 1-locular, with 2 or more ovules. Styles short. Stigma ventrally decurrent. Fruit of fleshy or dry, dorsally or laterally dehiscent follicles. Seeds solitary or few, pendulous from thread-like funicles.

Worldwide two genera and 221 species in tropical to warm temperate regions. One genus and eight or possibly nine species native to Nepal.

Recent molecular evidence suggests that the Magnoliaceae should be treated as two monogeneric subfamilies, with Magnolioideae containing the genus *Magnolia* and Liriodendroideae containing *Liriodendron* (Azuma *et al.*, Proc. Int. Symp. Fam. Magnoliac.: 21. 2000; Figlar & Nooteboom, Blumea 49: 87. 2004; Nooteboom in Fl. China 7: 48. 2008; Nie *et al.*, Mol. Phylogenet. Evol. 48: 1027. 2008) and this is the approach followed in this account.

1. *Magnolia* L., Sp. Pl. 1: 535 (1753).

Manglietia Blume; *Michelia* L.; *Oyama* (Nakai) N.H.Xia & C.Y.Wu; *Sampacca* Kuntze; *Talauma* Juss.; *Yulania* Spach.

Description as for Magnoliaceae.

219 species from the Himalaya to E Asia, SE Asia, N America and tropical America. Eight or possibly nine species native to Nepal.

Magnoliaceae are widely cultivated for their attractive flowers, and Herklots (J. Roy. Hort. Soc. 89: 295. 1964) reported that *Magnolia grandiflora* L., *M. coco* (Lour.) DC., *M. liliiflora* Desr. and *M. figo* (Lour.) DC. (syn. *Michelia figo* (Lour.) Spreng.) are cultivated in the Kathmandu Valley. These species are included in the key, but are not otherwise treated. *Magnolia pterocarpa* Roxb. was reported from Nepal (Hooker & Thomson in Fl. Brit. Ind. 1: 41. 1872), based on Wallich's collection from central Nepal. This collection is the type of *Michelia macrophylla* D.Don (= *M. sphenocarpa* Roxb.), but its identity cannot be confirmed as the specimen has not been traced (Dandy, Enum. Fl. Pl. Nepal 2: 24. 1979). *Magnolia pterocarpa* is therefore also included in the key to species but not treated further.

Key to Species

1a	Flowers on short axillary shoots, enclosed by 2–4 bracts. Gynophore present. Fruit a lax spike	2
b	Flowers terminal, usually enclosed by a single bract. Gynophore absent. Fruit dense, compact	6
2a	Inner tepals longer than outer tepals. Petioles to 5 mm	***M. figo***
b	Inner tepals shorter than outer tepals. Petioles 8–45 mm	3
3a	Young twigs densely tomentose. Petioles, axillary shoots and pedicels persistently pilose	**7. *M. lanuginosa***
b	Young twigs glabrous or sericeous and soon glabrescent. Petioles, axillary shoots and pedicels glabrous or glabrescent	4
4a	Stipular scar less than half length of petiole. Flowers white to creamy white	**5. *M. doltsopa***
b	Stipular scar more than half length of petiole. Flowers yellow or orange	5
5a	Leaves ovate to narrowly ovate. Outer tepals 3.5–4.5 cm. Fruiting spikes more than 6 cm	**4. *M. champaca***
b	Leaves elliptic to obovate. Outer tepals 2.5–3 cm. Fruiting spikes up to 6 cm	**6. *M. kisopa***
6a	Carpels with 4 or more ovules	**3. *M. insignis***
b	Carpels with 2 ovules	7

7a	Carpels dehiscing transversely around the base, leaving only the lower portion of each carpel persistent on the receptacle. Leaves narrowly obovate or obovate-oblong .... **8. *M. hodgsonii***
b	Carpels dehiscing along dorsal sutures, persistent on receptacle. Leaves ovate to elliptic-oblong or obovate 8

8a	Leaves 18–35 cm with 10–20 pairs of secondary veins 9
b	Leaves up to 22 cm, rarely to 28 cm, with up to 10 pairs of secondary veins..10

9a	Fruiting receptacle cylindrical. Carpels without a beak. Leaves pubescent or silky below. Stipular scar less than half length of petiole...**1. *M. campbellii***
b	Fruiting receptacle cylindrical. Carpels with a prominent beak to 3.5 cm long. Leaves glabrous beneath. Stipular scar almost as long as petiole... ***M. pterocarpa***

10a	Tepals 6–10 cm. Petioles without a stipular scar ....***M. grandiflora***
b	Tepals 2–6 cm. Petioles with a stipular scar 11

11a	Stipular scar about half the length of petiole. Tepals purplish white...***M. liliiflora***
b	Stipular scar reaching almost to apex of petiole. Tepals greenish white or yellowish white...12

12a	Leaves and twigs rusty villous. Peduncle with a solitary bract scar ....**2. *M. globosa***
b	Leaves and twigs glabrous, glossy. Peduncle with 2–4 bract scars .... ***M. coco***

1. *Magnolia campbellii* Hook.f. & Thomson, Fl. Brit. Ind. 1[1]: 41 (1872).
Yulania campbellii (Hook.f. & Thomson) D.L.Fu.

घोगे चाँप Ghoge chanp (Nepali).

Deciduous trees to 25 m. Twigs greyish, glabrous, smooth or slightly lenticellate. Petioles swollen at the base, 1.8–4.5 cm, densely pilose, with the stipular scar 3–6 mm. Leaves ovate to elliptic-oblong, 12–30 × 5–16 cm, base rounded, oblique or cordate, apex acute, secondary veins 10–17 pairs, shiny and almost glabrous above, pubescent or silky below, denser along midrib. Inflorescences terminal. Buds ovoid. Bract 1, ovate, 4–8 × 2.5–5 cm, golden yellow sericeous. Flowers red, pink or white, erect, appearing before the leaves, opening fully, 6–15 cm across. Peduncles ca. 2 cm. Tepals 12–15, in 3–4 whorls, elliptic, subequal, 6–10 × 4–7 cm. Filaments purplish, 5–8 mm. Anthers pink with white margins, latrorse, 1.5–2.4 cm, connective with pointed appendage. Gynophore absent. Carpels purplish, free, densely crowded on receptacle, obovoid, laterally compressed. Ovules 2. Style 2 mm. Fruit cylindrical, 10–15cm, follicles dehiscing dorsally. Seeds usually solitary in each carpel, sometimes 2, bright red, shiny, obovoid, laterally compressed with a pointed base, ca. 10 × 5 mm.
Fig. 1a–c.

Distribution: Nepal, E Himalaya, Tibetan Plateau, Assam-Burma and E Asia.

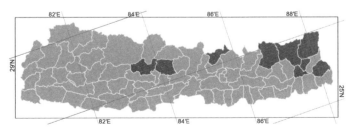

Altitudinal range: 2200–2800 m.

Ecology: Temperate forests.

Flowering: March–May. **Fruiting:** July–October.

2. *Magnolia globosa* Hook.f. & Thomson, Fl. Brit. Ind. 1[1]: 41 (1872).
Oyama globosa (Hook.f. & Thomson) N.H.Xia & C.Y.Wu.

खुकी चाँप Khuki chanp (Nepali).

Deciduous trees to 10 m. Twigs brown, villous when young, glabrescent, smooth or slightly lenticellate. Petioles swollen at the base, 1.5–5 cm, glabrous to densely villous, with stipular scar as long as petiole. Leaves, ovate, 5.5–22 × 2.7–14 cm, base rounded, apex acute or shortly mucronate, secondary veins up to 10 pairs, prominent, glabrous above, rusty villous below especially along veins. Inflorescences terminal. Buds ellipsoid. Bract 1, ca. 2 × 1.5 cm, pubescent above, glabrous below. Flowers white, pendulous, appearing with the leaves, opening fully, ca. 8 cm across. Peduncles ca. 7 cm, densely brown pubescent. Tepals 9, in 3–4 whorls, broadly obovate, subequal, 3–6 × 2–5 cm. Filaments crimson, ca. 3 mm. Anthers latrorse, ca. 1.3 cm, oblong truncate at apex. Gynophore absent. Carpels compressed, free, densely crowded on receptacle, obovoid, laterally compressed, rounded at base, beaked at apex. Ovules 2. Style 2–3 mm. Fruit ellipsoid, 6–8 cm, follicles dehiscing dorsally. Seeds 2 in each carpel, shiny, red, three angled, base pointed, tip rounded, ca. 7 × 9 mm.

Distribution: Nepal, E Himalaya, Tibetan Plateau, Assam-Burma and E Asia.

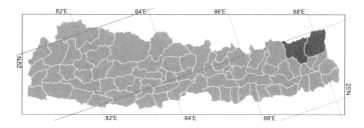

Altitudinal range: 2800–3400 m.

Ecology: Subalpine forest.

Flowering: May–June. **Fruiting:** July–September.

3. *Magnolia insignis* Wall., Tent. Fl. Napal. [Fasc. 1]: 3, pl. 1 (1824).
Manglietia insignis (Wall.) Blume.

बन कमल Ban kamal (Nepali).

Evergreen trees to 20 m. Twigs glabrous, smooth or slightly lenticellate. Petioles swollen at the base, ca. 3 cm with stipular scar ca. 1.8 cm, glabrous. Leaves elliptic, narrowly ovate to narrowly obovate, 15–22 × 6–8.5 cm, base cuneate, apex acute-acuminate, secondary veins 10–15 pairs, mid rib raised beneath, brown and sometimes white hairs on veins below, shiny above and below. Inflorescences terminal. Flower buds ovoid to ellipsoid. Bract 1, pubescent. Flowers opening fully, 10–15 cm across. Peduncles 1 or 2 cm. Perianth of 12 tepals slightly differentiated with 3 in outer whorl and 9 in inner whorls, the outer tepals slightly larger. Outer tepals pink, obovate, 4–6 × 2–4 cm apex obtuse; inner tepals pinkish white, obovate, 3.7–5.5 × 1.5–4 cm. Anthers yellow, introrse, 8–10 mm, connective produced into a acicular appendage. Gynophore absent. Gynoecium 1.5–3.5 cm. Carpels free, densely crowded on receptacle, beaked, ca. 1 cm. Ovules 4 or more. Fruit narrowly ovoid to subcylindric, 4–6 × 2.5–3.5 cm, bright purple when young and fresh. Follicles dehiscing dorsally. Seeds 1 or 2, bright red, three angled.
Fig. 1d–e.

Distribution: Nepal, W Himalaya, Assam-Burma, E Asia and SE Asia.

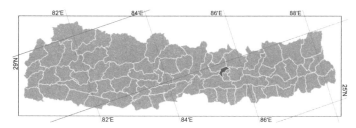

Altitudinal range: 1500–2700 m.

Ecology: Lower temperate forest.

Flowering: May–July. **Fruiting:** September–January.

Poorly known in Nepal and in need of more collections.

4. *Magnolia champaca* (L.) Baill. ex Pierre, Fl. Forest. Cochinch. 1: pl. 3 (1880).
Michelia champaca L., Sp. Pl. 1: 536 (1753); *M. aurantiaca* Wall.

सुन चाँप Sun chanp (Nepali).

Evergreen trees to 20 m. Twigs irregularly lenticellate, pubescent, young twigs sericeous. Petioles scarcely swollen at base, 2.5–4.5 cm, pubescent to subglabrous, with 2–4 cm stipular scar. Leaves ovate or elliptic-ovate, 12–27 × 6–9.5 cm, base cuneate, apex acute to long-acuminate, secondary veins 13–18 pairs, glabrous and shiny above, glabrous to puberulous beneath, denser on veins. Inflorescences terminal on ca. 5 mm, densely sericeous axillary shoots. Flower buds ovoid. Bracts 2–4, broadly ovate, ca. 2 cm, orange brown silky. Flowers yellow, opening fully, 3.5–5 cm across. Peduncles ca. 1.2 cm, tomentose. Tepals 12–15, somewhat differentiated into outer and inner series, the inner tepals shorter; outer tepals oblong or acute, 3.5–4.5 × 0.7–1 cm; inner tepals linear 2–2.5 × 0.7–0.9 cm. Stamens ca. 10 mm. Anthers latrorse, 6–8 mm, with connective appendage ca. 1 mm. Gynoecium on 4 mm gynophore, exerted beyond the androecium. Carpels free, somewhat undeveloped towards apex, ovoid-oblong, ca. 2 mm, densely sericeous, containing up to 8 ovules. Style ca. 1 mm. Fruit a lax spike of carpels, 10–15 cm, on the somewhat elongated gynophore, the follicles dehiscing dorsally. Seeds 3–8 per carpel, red, shiny, triangular with angular edge.
Fig. 1f.

Distribution: Nepal, E Himalaya, Assam-Burma, S Asia, E Asia and SE Asia.

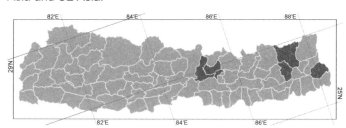

Altitudinal range: 900–1900 m.

Ecology: Evergreen broad-leaved forests and cultivated in temple gardens or village margins in subtropical areas.

Flowering: May–July. **Fruiting:** June–December.

Wallich (Tent. Fl. Napal. [Fasc. 1]: 7, t. 3. 1824) misapplied the name *Michellia doltsopa* Buch.-Ham. ex DC. to this species.
The timber is valuable. The species has a wide range of medicinal uses including the treatment of fever, coughs, bronchitis, gastritis and renal diseases.

5. *Magnolia doltsopa* (Buch.-Ham. ex DC.) Figlar, Proc. Int. Symp. Fam. Magnoliac.: 21 (2000).
Michelia doltsopa Buch.-Ham. ex DC., Syst. Nat. 1: 448 (1817); *Magnolia excelsa* Wall.; *Michelia excelsa* (Wall.) Blume; *Sampacca excelsa* (Wall.) Kuntze.

रानी चाँप Rani chanp (Nepali).

Deciduous tree to 30 m. Twigs irregularly lenticellate, glabrous. Petioles scarcely swollen at base, 1–3 cm, glabrous, with 0.3–1 cm stipular scar. Leaves elliptic to elliptic-ovate or narrowly elliptic, 13.5–25 × 5–7 cm, base cuneate-rounded, apex acute-acuminate, glabrous to shiny above, silky brown or densely pubescent below, lateral veins 7–13 pairs, main vein prominent. Inflorescences terminal on 1–5 mm, densely orange or pale brown villous axillary shoots. Flower buds ellipsoid. Bracts 2–4, ovate, ca. 5.4 × 2.4 cm, dark red brown villous. Flower white or creamy white, opening fully, 5–10 cm across. Peduncles 5–7 mm, densely orange or pale brown villous. Tepals 9–12, somewhat differentiated into outer and inner series, the inner tepals shorter; outer tepals narrowly obovate, 5–6 × 2–3 cm, inner tepals progressively narrower than outer, 3.5–5.5 × 1.5–2.5 cm. Stamens 10–15 mm, yellowish. Anthers latrorse, 8–14 mm, with connective appendage ca. 1 mm. Gynoecium on 6–8 mm gynophore, exerted beyond the androecium. Carpels free, somewhat undeveloped towards apex, 2–3 mm, compressed, densely sericeous. Ovules 2–several. Style 2–3 mm. Fruit a lax spike of carpels, 7–10 cm, on the somewhat elongated gynophore, the follicles dehiscing dorsally. Seeds 2 or more per carpel, red, compressed.
Fig. 1g–h.

Distribution: Nepal, E Himalaya, Tibetan Plateau, Assam-Burma and E Asia.

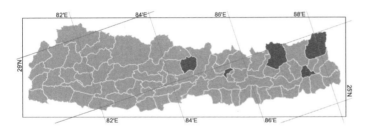

Altitudinal range: 1900–2600 m.

Ecology: Temperate forests.

Flowering: March–May. **Fruiting:** September–October.

A collection by Buchanan-Hamilton from the Kathmandu Valley (BM) was identified by Dandy (J. Bot. 65: 278. 1927) as a hybrid between *M. doltsopa* and *M. champaca* (L.) Baill. ex Pierre.

6. *Magnolia kisopa* (Buch.-Ham. ex DC.) Figlar, Proc. Int. Symp. Fam. Magnoliac.: 22 (2000).
Michelia kisopa Buch.-Ham. ex DC., Syst. Nat. 1: 448 (1817); *M. tila* Buch.-Ham. ex Wall. nom. nud.; *Sampacca kisopa* (Buch.-Ham. ex DC.) Kuntze.

बन चाँप Ban chanp (Nepali).

Deciduous trees to 25 m. Twigs irregularly lenticellate, glabrous, except for appressed grey sericeous young parts. Petioles scarcely swollen at base, 1.5–3.5 cm, minutely appressed sericeous, glabrescent with stipular scar 0.7–2.4 cm. Leaves elliptic or oblong to narrowly ovate or obovate, 9–18 × 3–7.5 cm, base cuneate to rounded, apex acute-acuminate, secondary veins 10–14 pairs, glabrous and shiny above, glabrous or minutely sericeous below. Inflorescences terminal on 1–5 mm, sericeous axillary shoots. Flower buds ovoid. Bracts 2–4, ovate, pubescent, greyish brown. Flowers yellow, opening fully, 2–3 cm across. Peduncles ca. 6 mm. Tepals 9–12, somewhat differentiated into outer and inner series, the inner tepals shorter; outer tepals obovate, 2.5–3 × 0.7–1 cm, inner tepals slightly narrower, 2–2.2 × 0.6–0.8 cm. Stamens ca. 10 mm. Anthers latrorse, ca. 6 mm, with connective appendages to 3 mm. Gynoecium on 2–3 mm gynophore, scarcely or not exerted beyond the androecium. Carpels free, somewhat undeveloped towards apex, ca. 2 mm, sessile, globose or ovoid, densely sericeous. Ovules 2–several. Style recurved, ca. 1 mm. Fruit a lax spike of carpels, 3–6 cm, on the somewhat elongated gynophore, the follicles dehiscing dorsally. Seeds 1 or 3 per carpel, red, spheroidal.
Fig. 1i–j.

Distribution: Nepal, W Himalaya, E Himalaya and Tibetan Plateau.

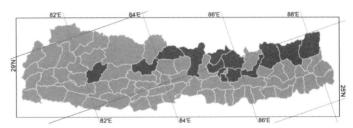

Altitudinal range: 1000–3000 m.

Ecology: Subtropical and lower temperate forest.

Flowering: June–October. **Fruiting:** August–November.

The glabrous, dark twigs with densely grey sericeous, ovoid buds are characteristic of *Magnolia kisopa*.
    The foliage is used as fodder and the powdered bark is mixed with corn flour and baked.

7. *Magnolia lanuginosa* (Wall.) Figlar & Noot., Blumea 49: 96 (2004).
Michelia lanuginosa Wall., Tent. Fl. Napal. [Fasc. 1]: 8, pl. 5 (1824); *Magnolia velutina* (DC.) Figlar later homonym, non P.Parm.; *Michelia lanceolata* E.H.Wilson; *M. velutina* DC.; *Sampacca lanuginosa* (Wall.) Kuntze.

गोगल चाँप Gogal chanp (Nepali).

Deciduous tree to 15 m. Twigs irregularly lenticellate, densely grey tomentose. Petioles scarcely swollen at base, 0.8–2.5 cm, densely pilose, with 4–11 mm stipular scar up to half of the length of petiole. Leaves elliptic or oblong to narrowly ovate, 10–25 × 3.5–6.5 cm, base cuneate-attenuate to round, apex acute to abruptly acute-acuminate, secondary veins 13–20 pairs, glabrous and shiny above, sometimes slightly hairy along veins, sparsely to densely appressed

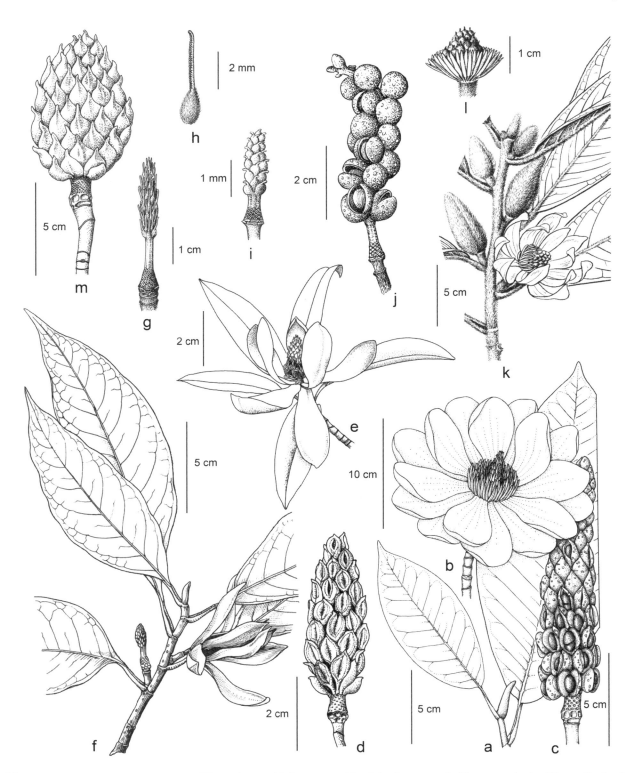

FIG. 1. Magnoliaceae. **Magnolia campbellii**: a, shoot with leaves and bud; b, flower; c, fruiting receptacle. **Magnolia insignis**: d, fruiting receptacle; e, flower. **Magnolia champaca**: f, shoot with leaves and flower. **Magnolia doltsopa**: g, flower with tepals and stamens removed; h, pistil. **Magnolia kisopa**: i, flower with tepals and stamens removed; j, fruiting receptacle. **Magnolia lanuginosa**: k, shoot with leaves, buds and flower; l, flower with tepals removed. **Magnolia hodgsonii**: m, fruiting receptacle.

pilose below, especially on veins. Inflorescences terminal on 3–6 mm, densely pilose axillary shoots. Flower buds ovoid to elongated ellipsoid. Bracts 2–4, 3.5 × 1.2 cm, densely pilose. Flowers yellowish white, opening fully, 5–7 cm across. Peduncles 3–6 mm, densely pilose. Tepals 12–18, somewhat differentiated into outer and inner series, the inner tepals shorter; outer tepals narrowly obovate, 3–4.5 × 0.6–1.1 cm; inner tepals narrowly elliptic, 2.5–3.5 × 0.5–0.7 cm. Stamens ca. 11 mm. Anthers latrorse, to 8 mm, with connective appendage 1 mm. Gynoecium on 5–8 mm gynophore, exerted beyond the androecium. Carpels free, somewhat undeveloped towards apex, 2–2.5 mm obovate. Ovules 2–several. Style 1–2 mm. Fruit a lax spike of carpels, 8–12 cm, on the somewhat elongated gynophore, the follicles dehiscing dorsally. Seeds 1–3 per carpel, orange, angular.
Fig. 1k–l.

Distribution: Nepal, E Himalaya, Tibetan Plateau, Assam-Burma and E Asia.

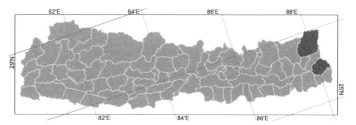

Altitudinal range: 300–2100 m.

Ecology: Subtropical forest.

Flowering: August–December. **Fruiting:** September–January.

The type of *Michelia lanuginosa* Wall. (*Wallich 6493*) was collected in 1821 and is most probably from the Kathmandu Valley, but could have been collected on his journey back to Calcutta. As there is no precise locality it has not been included on the distribution map.
    Used for timber and firewood.

8. ***Magnolia hodgsonii*** (Hook.f. & Thomson) H.Keng, Gard. Bull. Singapore 31: 129 (1976).
Talauma hodgsonii Hook.f. & Thomson, Fl. Ind. 1: 74 (1855).

भालु काठ Bhalu kath (Nepali).

Evergreen trees to 12 m. Twigs warty, slightly glaucous, glabrous, young shoots fleshy and pink. Petioles swollen at the base 2–6 cm, with stipular scar as long as petiole, glabrous. Leaves narrowly obovate or obovate-oblong, 17–53 × 8–16 cm, base acute, apex acute to obtuse, secondary veins prominent, 7–22 pairs, glabrous above and below. Inflorescences terminal. Flower buds subglobose. Bracts 1 or 2, acute, glabrous, ca. 4 × 1.5 cm. Flowers never opening fully, ca. 5–6 cm across. Peduncles 1–2 cm. Perianth of 9–15 subequal tepals in 3–4 whorls, the inner tepals the smallest; outer tepals white, oblong, ca. 3.5 × 3.5 cm; inner tepals purplish, greenish white at base, ovate. Stamens yellowish, filament short. Anthers introrse, linear, ca. 1.5 cm, connective exerted ca. 2 mm beyond anther. Gynophore absent. Carpels united, tapering to sharp beak. Ovules 2. Fruit composed of overlapping woody carpels dehiscing laterally around the base with the lower portion persistent on the receptacle, ovoid, 10–15 cm long, breaking up as the 2–3 cm carpels dehisce. Seed 1, suspended from the persistent base of carpel, red, shiny, ca. 8 × 6 mm, oblong, plano-convex.
Fig. 1m.

Distribution: Nepal, E Himalaya, Tibetan Plateau, Assam-Burma, E Asia and SE Asia.

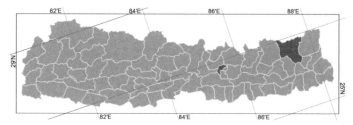

Altitudinal range: 900–1800 m.

Ecology: Subtropical forest, often on rocky slopes.

Flowering: April–June. **Fruiting:** August–September.

Poorly known in Nepal and in need of more collections.

Schisandraceae

Colin A. Pendry

Dioecious woody vines, climbing by twining and not by any specialized organs, glabrous throughout. Stipules absent. Leaves alternate, petiolate, elliptic, ovate, or obovate, base cuneate to broadly cuneate, apex acute to acuminate, margin sparsely denticulate, papery to leathery. Flowers unisexual, solitary or rarely paired or clustered, axillary to leaves or to caducous bracts at the base of shoots, fragrant. Peduncles with or without bracteoles, elongating in fruit. Calyx and corolla not differentiated, tepals 6–17, spirally arranged, obovate to suborbicular, often somewhat cucullate, rather fleshy, outermost and innermost smaller. Male flowers with 8–50 stamens free, or connate to form a fleshy synandrium. Female flowers apocarpic with 18–120 free carpels on a columnar receptacle. Carpels with a stigmatic crest extended into a subulate pseudostyle above and irregular appendages below. Ovary with 2(or 3) ventrally attached ovules. Fruit an aggregate of red or purplish, ellipsoid to obovoid fleshy carpels borne on a greatly elongated receptacle (torus). Seeds (1 or)2(or 3) per carpel, smooth to rugulose.

Worldwide two genera and about 40 species, mainly in E and SE Asia, with one species of *Schisandra* in N America. One genus and three species in Nepal.

1. *Schisandra* Michx. nom cons., Fl. Bor.-Amer. 2: 218 (1803).

Description as for Schisandraceae.

Key to Species

1a Stamens connate into a synandrium. Peduncles 0.5–1.7 cm, with 1–5 bracteoles to 1 mm ....**3. *S. propinqua***
 b Stamens free. Peduncles 1.3–5.5 cm, ebracteolate, or rarely with a solitary bracteole to 3 mm......................................2

2a Leaves obovate. Male flowers with laxly arranged, narrowly triangular anthers with narrow, rounded connectives. Female flowers with 70–120 carpels .... **1. *S. grandiflora***
 b Leaves ovate. Male flowers with densely clustered, oblong anthers with broad, truncate connectives. Female flowers with 20–45 carpels ....**2. *S. neglecta***

1. *Schisandra grandiflora* (Wall.) Hook.f. & Thomson, Fl. Brit. Ind. 1[1]: 44 (1872).
Kadsura grandiflora Wall., Tent. Fl. Napal. [Fasc. 1]: 10, pl. 14 (1824); *Sphaerostema grandiflorum* (Wall.) Blume.

सिगुण्टा लहरा Sigunta lahara (Nepali).

Climber to 5 m. Petioles 0.8–2.5 cm. Leaves obovate, 6–13 × 2–5 cm, base cuneate, attenuate, or rarely broadly cuneate, apex shortly to long abruptly acuminate, margin denticulate, papery, secondary veins 5–8 pairs. Flowers solitary, axillary to caducous bracts or rarely axillary to leaves. Peduncle 1.2–4.7 cm. Bracteoles absent or rarely at peduncle base, ca. 3 mm. Tepals 6–9, white, cream-white, or sometimes pink-tinged, obovate, up to 1.2–2.1 × 0.7–1.2 cm. Male flowers with androecium conical, ca. 10 mm. Stamens free, 30–50, narrowly triangular, ca. 3.5 mm, with narrow, rounded connective. Female flowers with gynoecium ellipsoid, 3–6 mm. Carpels 70–120, obovoid. Fruit with peduncle 2.5–8 cm, torus 2–21 cm, mature fruiting carpels red, 7–9 × 5–6 mm. Seeds smooth.
Fig. 2a–f.

Distribution: Nepal, W Himalaya, E Himalaya and Tibetan Plateau.

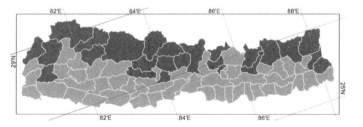

Altitudinal range: 2100–3400 m.

Ecology: In the understorey and margins of *Pinus wallichiana*, *Picea*, *Rhododendron* and *Quercus* forests.

Flowering: June–July. **Fruiting:** September–October.

The obovate leaves drying papery, with obvious reticulations below and rather frequent denticulations are characteristic of *Schisandra grandiflora*. It is by far the most commonly collected *Schisandra* species in Nepal.

Schisandraceae

2. *Schisandra neglecta* A.C.Sm., Sargentia 7: 127, fig. 17g (1947).
Schisandra elongata Hook.f. & Thomson later homonym, non Baill.

Climber to 3 m. Petioles 0.7–3 cm. Leaves broadly ovate to elliptic, 5–11 × 2–6.5 cm, base cuneate to rounded, apex acuminate to acute, margin remotely denticulate, papery to rarely subleathery, secondary veins 5–6 pairs. Flowers solitary or rarely in pairs, axillary to leaves or caducous bracts. Peduncle 1.3–5.5 cm. Bracteoles absent. Tepals 6–11, white, yellow, orange, or cream, pink at base within, suborbicular to broadly elliptic or obovate, up to 5–9 × 5–8 mm. Male flowers with androecium obovoid or subglobose, 6–7 mm. Stamens free, 12–40, oblong, ca. 3 mm, with broad, truncate connective. Female flowers with gynoecium subglobose, 4–6 mm. Carpels 20–45, falcate-ellipsoid. Fruit with peduncle 3.5–7.5 cm, torus 4–11 cm, mature fruiting carpels red, 5–8 × 4–5 mm. Seeds smooth, rugulose, or rarely somewhat tuberculate.
Fig. 2g–i.

Distribution: Nepal, E Himalaya, Assam-Burma and E Asia.

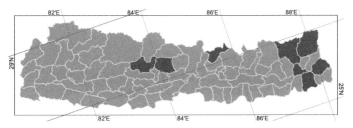

Altitudinal range: 1700–3100 m.

Ecology: Forest understorey and margins, often by rivers.

Flowering: May–July. **Fruiting:** July–November.

Hooker & Thomson (Fl. Ind. 1: 85. 1855) misapplied the name *Sphaerostema elongatum* Blume to this species.

The long, slender peduncles of *Schisandra neglecta*, often less than 1 mm thick in dried material, are the most obvious feature distinguishing it from the other Nepalese species.

3. *Schisandra propinqua* (Wall.) Baill., Hist. Pl. 1: 148, fig. 183 (1868).
Kadsura propinqua Wall., Tent. Fl. Napal. [Fasc. 1]: 11, pl. 15 (1824); *Sphaerostema propinquum* (Wall.) Blume.

पहेँलो सिंगुल्टो Pahenlo singulto (Nepali).

Climber to 4 m. Petioles 0.8–1.6 cm. Leaves elliptic to ovate or narrowly ovate, 6–14 × 2–5 cm, base rounded to cuneate, apex shortly acuminate, margin entire to subentire or remotely denticulate, papery to leathery, secondary veins 5–8 pairs. Flowers solitary or in clusters, axillary to leaves or sometimes to caducous bracts. Peduncle 0.5–1.7 cm. Bracteoles 1–5, ca. 1 mm. Tepals 9–16, cream, yellow, or white, sometimes red within, up to 7–9 × 5–7 mm in male flowers, up to 9–15 × 7–11 mm in female flowers. Male flowers with androecium subglobose, ca. 7 mm, composed of 6–16 stamens, completely fused into a fleshy synandrium. Female flowers with gynoecium subglobose, 3–6 mm. Carpels 25–45, ellipsoid or obovoid. Fruit with peduncle 0.5–3 cm, torus 2–15 cm, mature fruiting carpels red to purple, 6–10 × 5–7 mm. Seeds smooth.
Fig. 2j–l.

Distribution: Nepal, W Himalaya, E Himalaya, Tibetan Plateau, Assam-Burma, E Asia and SE Asia.

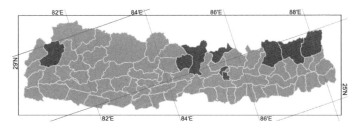

Altitudinal range: 1600–2200 m.

Ecology: Forest understorey, stream-sides and disturbed areas.

Flowering: June–July. **Fruiting:** September–October.

Nepalese material is referable to subsp. *propinqua*, with other subspecies found in China and SE Asia. The fleshy synandrium is quite different from the androecium of free stamens in the other species. It is superficially similar to the androecium of *Kadsura heteroclita* (Roxb.) Craib, but the latter has entire leaves and the fruiting carpels are borne in subglobose clusters. *Kadsura heteroclita* is known from Sikkim, but it has not yet been recorded in Nepal.
The juice of the roots is used to treat fever.

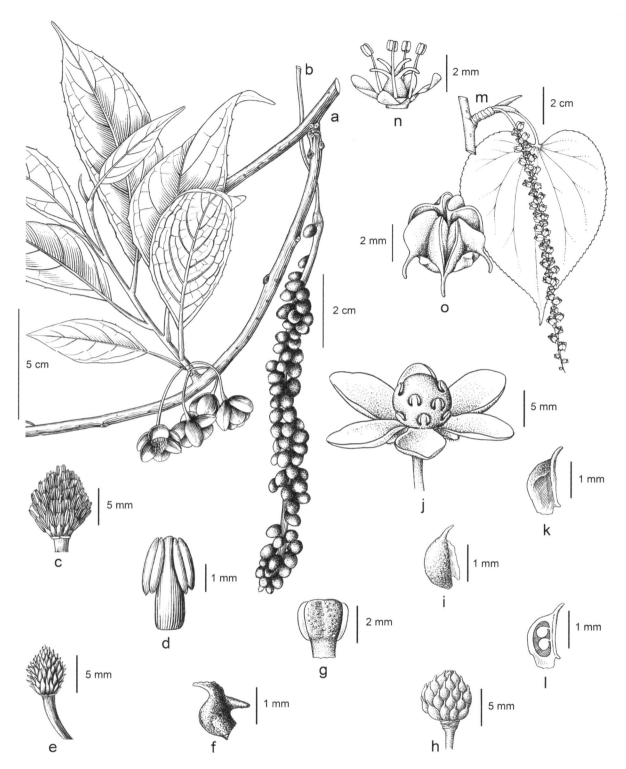

FIG. 2. SCHISANDRACEAE. **Schisandra grandiflora**: a, flowering branch; b, fruit; c, androecium; d, stamen; e, gynoecium; f, pistil. **Schisandra neglecta**: g, stamen; h, gynoecium; i, pistil. **Schisandra propinqua**: j, male flower; k, pistil; l, longitudinal section of pistil. TETRACENTRACEAE. **Tetracentron sinense**: m, branch with short-shoot, leaf and inflorescence; n, flower; o, dehisced fruit.

Tetracentraceae

Colin A. Pendry & Kamal Maden

Deciduous forest trees with leaves borne on alternately arranged woody short-shoots densely covered with the scars of fallen leaves, bud scales and inflorescences and terminated by a bud which is initially enclosed by the leaf base and later free. Leaves borne singly at the apex of the shortshoots, simple, exstipulate, serrate, palmately veined, petiolate. Inflorescence a long, slender, pendulous catkin borne close to the petiole near the apex of the short-shoot. Flowers subtended by a minute bract, small, numerous, spirally arranged or in clusters of 3 or 4. Sepals 4. Petals absent. Stamens 4, opposite the sepals; anthers dehiscing by a lateral slit. Carpels superior, fused only at the base. Styles subulate, erect when the bud opens and then recurved, with the stigma along the adaxial surface. Fruit a cluster of 4 follicles fused at the base, and with the style almost basal due to unequal growth of carpel post fertilisation. Seeds ca. 5 per follicle, compressed ellipsoidal and with a short wing at each end.

One genus with a single species in the Himalayas, E Asia and SE Asia.

1. *Tetracentron* Oliv., Hooker's Icon. Pl. 19: pl. 1892 (1889).

Generic description as for Tetracentraceae.

1. *Tetracentron sinense* Oliv., Hooker's Icon. Pl. 19: pl. 1892 (1889).
Tetracentron sinense var. *himalense* H.Hara & H.Kanai.

Trees to 30 m, glabrous. Woody portion of short-shoots 0.8–1.5(–4) cm. Buds reddish brown, oblong, scale margins entire, apex acute, 1–1.5 cm. Leaves broadly ovate or rarely somewhat oblong, 7–19 × 4–11 cm, length:width ratio 1.6–2.1, green above, pale below, base cordate or rarely truncate, margins serrulate, apex acuminate; veins 5–7. Petioles red, 1.5–4 cm. Inflorescences 6–19 cm, shortly pedunculate. Bracts flap-like, broader than long, 0.3–0.5 mm. Flowers yellowish green, sessile, ca. 2 mm across. Sepals ovate, 1–1.5 mm, entire, apex rounded. Stamens exserted, 2.5–3.5 mm. Carpels up to 1 mm. Styles to 1.5 mm. Fruits brown, follicles 3–4 mm. Seeds 2–3 mm including the wings.
Fig. 2m–o.

Distribution: Nepal, E Himalaya, Assam-Burma, E Asia and SE Asia.

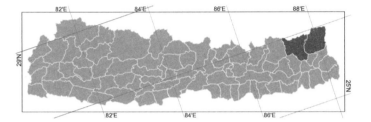

Altitudinal range: 1900–2900(–3200) m.

Ecology: Evergreen and deciduous temperate forests with Fagaceae, Lauraceae, *Tsuga* and *Alnus*.

Flowering: June–August(–October). **Fruiting:** July–November(–March).

When compared with Chinese material the Nepalese specimens tend to have larger leaves which are caudately-acuminate at the apex and more minutely and acutely serrate, leading earlier authors to distinguish them as var. *himalense*.

Annonaceae

Puran P. Kurmi

Trees or shrubs, erect or climbing. Glabrous or with an indumentum of simple or stellate hairs. Leaves simple, distichous, entire, pinnately veined, exstipulate. Inflorescences terminal, axillary or extra-axillary, leaf-opposed or cauliflorous, fascicles or sympodial cymes, or the flowers solitary. Flowers usually bisexual, large or small, often scented, actinomorphic, hypogynous. Perianth of dissimilar sepals and petals. Sepals 3, free, usually valvate. Petals 6, in two whorls, sometimes up to 9, rarely the inner whorl absent, valvate or imbricate, outer whorls sometimes sepaloid. Stamens numerous, minute, free, spirally arranged on the torus. Filaments short and thick. Anthers adnate, extrorse, connective apically prolonged, concealing the anthers. Carpels free or subconnate, spirally arranged, minute, usually numerous, sometime few, superior. Fruiting carpels usually many, free and radiating umbellately from the receptacle, dry or fleshy, indehiscent berries, globose, ellipsoid to cylindric, sometimes moniliform, rarely united into a many-celled fruit (*Annona*). Seeds 1–many, in 1 or 2 rows; endosperm ruminate.

About 120 genera and 2300 species in the tropics and subtropics with the greatest diversity in the Old World. Six genera and 11 species in Nepal.

Key to Genera

1a	Scrambling shrubs. Peduncles flattened and hooked	**4. *Artabotrys***
b	Scrambling or erect shrubs or trees. Peduncles not flattened, nor hooked	2
2a	Leaves stellate hairy beneath. Petals imbricate in bud	**1. *Uvaria***
b	Leaves glabrous or with simple hairs beneath. Petals valvate in bud	3
3a	Inner petals shorter than outer petals	4
b	Inner petals as long as or longer than outer petals	5
4a	Erect or scrambling trees or shrubs. Petals not thick and fleshy, inner ones half or more times the length of the outer. Fruiting carpels free, moniliform	**3. *Desmos***
b	Trees or shrubs. Petals thick and fleshy, inner ones much reduced or absent. Fruiting carpels fused to form a many-loculed syncarpous fruit	**6. *Annona***
5a	Sepals and outer petals equal and alike	**2. *Miliusa***
b	Sepals much shorter and different from petals	**5. *Polyalthia***

1. *Uvaria* L., Sp. Pl. 1: 536 (1753).

Scandent shrubs. Leaves and young branches with indumentum of stellate hairs. Flowers usually bisexual, terminal, axillary, extra-axillary, leaf-opposed, solitary or fascicled. Sepals 3, valvate, often connate at base. Petals 6 in 2 whorls, free, sometimes connate at base, imbricate. Torus depressed, hairy. Stamens numerous, outer stamens sterile. Carpels many, rarely few, linear-oblong; ovules many, 2-seriate. Fruiting carpels many, stalked, dry or baccate. Seeds many, 2-seriate.

About 190 species in tropical Asia and Africa. One species in Nepal.

1. *Uvaria hamiltonii* Hook.f. & Thomson, Fl. Ind. 1: 96 (1855).

रबु लहरा Rabu lahara (Nepali).

Large scandent shrubs. Indumentum densely rufous stellate tomentose to stellate pubescent throughout. Petioles 4–10 mm, rufous stellate tomentose. Leaves elliptic-oblong or obovate, 15–25 × 7–11 cm, base rounded or slightly cordate, apex acuminate, adpressed-pubescent, ultimately glabrescent above, rufous stellate-pubescent beneath, secondary veins 12–22 pairs. Flowers solitary or 2, extra-axillary, leaf opposed. Pedicels 2.5–3.7 cm. Bract solitary, suborbicular, 9–10 × 7.5–8 mm. Sepals obovate, connate at the base, ca. 1 × 1 cm. Petals red, ovate, tomentose on both surfaces. Outer petals as long as inner, 2–3.3 × 1.1–1.7 cm, apex rounded. Stamens numerous, linear, ca. 4 mm. Carpels many, ca. 4 mm, pubescent; stigmas glabrous. Fruiting carpels ca. 10, 1.5–2.5 cm, densely rufous stellate tomentose, on 2–3 cm stalks.
Fig. 3a–b.

Distribution: Nepal, Assam-Burma, S Asia and SE Asia.

Annonaceae

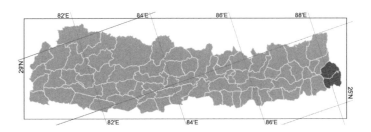

Altitudinal range: 300–600 m.

Ecology: Tropical forest.

Flowering: May–June. **Fruiting:** August–September.

2. *Miliusa* Lesch. ex A.DC., Mém. Soc. Phys. Genève 5: 213 (1832).

Trees or shrubs, dioecious or polygamous. Indumentum of simple hairs. Leaves ovate or elliptic, pubescent or glabrescent. Flowers bisexual or rarely unisexual, axillary, extra-axillary, solitary, fascicled or in cymes. Pedicels bracteate. Bracts ovate, elliptic or obovate, upper bracts often leaf-like. Sepals 3, valvate. Petals 6 in 2 whorls, free. Outer petals sepaloid and subequal to the sepals. Inner petals larger, cohering when young by their margins, becoming free, subsaccate at base, apex usually revolute. Torus elevated, usually with long pubescence. Stamens definite or indefinite. Carpels numerous, linear-oblong; ovules 1-10; stigma club-shaped. Fruiting carpels many, globose or oblong, stalked or subsessile. Seeds 1–5.

About 40 species in Asia and Australia. Four species in Nepal.

Key to Species

1a	Leaves hairy	2
b	Leaves glabrous	3
2a	Leaves pubescent beneath, nearly glabrous above. Fruiting carpels 3–5-seeded	**3. *M. tomentosa***
b	Leaves velvety-tomentose on both surfaces. Fruiting carpels 1–2-seeded	**4. *M. velutina***
3a	Pedicels 1.5–2.5 cm long, pubescent	**1. *M. globosa***
b	Pedicels 4–8 cm long, glabrous	**2. *M. macrocarpa***

1. *Miliusa globosa* (A.DC.) Panigrahi & S.Misra, Taxon 33: 713 (1984).
Guatteria globosa A.DC., Mém. Soc. Phys. Genève 5: 217 (1832); *Hyalostemma roxburghiana* Wall.; *Miliusa roxburghiana* (Wall.) Hook.f. & Thomson; *Uvaria dioica* Roxb.

Small trees or large shrubs, dioecious or polygamous. Young branches pubescent. Petioles 1–2 mm. Leaves usually oblong-lanceolate or ovate-lanceolate, 8–18 × 2.5–6 cm, base rounded or acute, apex abruptly acuminate, secondary veins 8–12 pairs, upper surface glabrous, lower at first pubescent, glabrous with age. Flowers red, bisexual or unisexual, solitary or 2–3 together, axillary, extra-axillary or leaf-opposed. Pedicels 1.5–2.5 cm, pubescent. Bracts 1–3, at the base of the pedicel or slightly above, linear-lanceolate, acuminate. Sepals 3, linear-lanceolate, acuminate, 3–6 × 1 mm, tomentose outside. Petals in 2 whorls. Outer petals similar to sepals. Inner petals ovate, 1–1.5 × 0.4–0.8 cm, saccate at base, sparsely pubescent. Stamens ca. 1 mm; connective dome-shaped at top. Carpels ca. 2 mm long, intermixed with hairs, ovate or oblong. Fruiting carpels many, ovoid, 7–10 mm; stalks slender, ca. 2 cm. Seed usually 1, ovoid.

Distribution: Nepal, E Himalaya, Assam-Burma and S Asia.

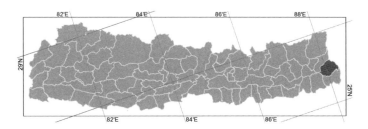

Altitudinal range: 100–300 m.

Ecology: Tropical forest.

Flowering: December–May. **Fruiting:** August–September.

2. *Miliusa macrocarpa* Hook.f. & Thomson, Fl. Ind. 1: 150 (1855).

कालि काठ Kali kath (Nepali).

Small trees, ca. 10 m. Branches glabrous. Petioles 3–4 mm, glabrous. Leaves lanceolate or narrowly oblong, 6–15 × 3–5 cm, base acute, apex acuminate, secondary veins ca. 10 pairs glabrous. Flowers bisexual, solitary or in few-flowered, extra-axillary or leaf-opposed cymes. Pedicels 4–8 cm, glabrous. Bracts absent or small. Sepals and outer petals alike, ovate, acute, fleshy, glabrous, warty outside, dull rusty-pubescent inside. Inner petals obovate, erect, reddish-brown with red veins, glabrous, except a few fugacious hairs near margin and top of inner surface. Stamens numerous; anthers linear; connective apiculate. Carpels silky. Fruiting carpels ca. 2. 5 cm on ca. 2.5 cm stalks. Seed usually 1.

Distribution: Nepal, E Himalaya and Assam-Burma.

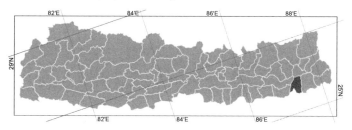

Altitudinal range: 1200–1700 m.

Ecology: Tropical forest.

Flowering: March–May. **Fruiting:** August–November.

3. *Miliusa tomentosa* (Roxb.) J.Sinclair, Gard. Bull. Straits Settlem. Ser. 3 14: 378 (1955).
Uvaria tomentosa Roxb., Pl. Coromandel 1[2]: pl. 35 (1795); *Saccopetalum tomentosum* (Roxb.) Hook.f. & Thomson.

Large deciduous trees. Bark longitudinally fissured. Young parts tomentose, glabrate. Petioles 2–4 mm, tomentose. Leaves ovate or oblong-ovate, 5–12 × 3–6 cm, base rounded or subacute, apex obtuse or acute, tomentose above when young, nearly glabrous with age, the midrib always pubescent, tomentose beneath. Flowers bisexual, in leaf-opposed or subterminal cymes, greenish. Pedicels 3–6 cm, tomentose. Sepals linear-lanceolate, ca. 1 mm, tomentose. Outer petals sepaloid, linear-lanceolate, ca. 3 mm. Inner petals oblong-ovate, 8–10 × 5 mm, tomentose. Stamens, ca. 1mm, connectives apiculate. Carpels many, ovate; stigma sessile, globose. Fruiting carpels 10–15, subglobose, stalks 1–1.5 cm, stout, tomentose when young. Seeds 4–5.

Distribution: Nepal and S Asia.

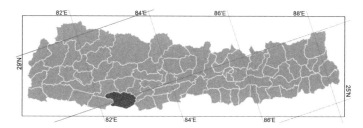

Altitudinal range: 200–400 m.

Ecology: Tropical forest.

Flowering: March–May. **Fruiting:** May–July.

4. *Miliusa velutina* (Dunal) Hook.f. & Thomson, Fl. Ind. 1: 151 (1855).
Uvaria velutina Dunal, Monogr. Anonac.: 91 (1817).

Deciduous small trees. Young parts densely grey-tomentose. Petioles 2–5 mm, tomentose. Leaves ovate or oblong or oblong-elliptic, 10–24 × 7–12 cm, base rounded or cordate, apex acute or apiculate, velvety–tomentose on both surfaces. Flowers bisexual, pale yellow or greenish, solitary or in pairs on simple or branched tomentose pedicels. Sepal ovate, acute, 3–4 mm, tomentose. Outer petals sepaloid 4–6 × ca. 2 mm. Inner petals broadly ovate, slightly acute, 10–17 × 7 mm, densely tomentose outside, glabrous inside. Stamens ca. 1 mm; connective bluntly apiculate. Carpels many, oblong, velutinous. Fruiting carpels many, ovoid or ellipsoid, tomentose, stalked. Seeds 1–2.

Distribution: Nepal, Assam-Burma, S Asia and SE Asia.

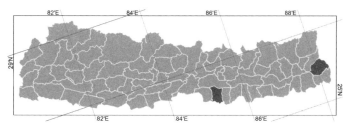

Altitudinal range: 100–500 m.

Ecology: Subtropical forest.

Flowering: February–May. **Fruiting:** July–October.

3. *Desmos* Lour., Fl. Cochinch. 1: 352 (1790).

Erect or scandent shrubs or small trees. Leaves often glaucous beneath, shiny above. Flowers bisexual, solitary, extra-axillary or leaf opposed, usually pendulous. Sepals 3, valvate, free. Petals 6, in 2 whorls, valvate, outer petals usually larger than inner, basally clawed. Torus flat or apex slightly concave. Stamens numerous, apex obtuse, truncate or acuminate. Carpels numerous, pubescent; style oblong or ovoid; stigma oblong, ovoid or clavate, opening U-shaped, grooved; ovules 1–8 per locule. Fruiting carpels many, moniliform, with 2–6 segments. Seeds solitary in each segment.

About 30 species in tropical Asia and Australia. One species in Nepal.

1. *Desmos chinensis* Lour., Fl. Cochinch. 1: 352 (1790).
Unona discolor Vahl; *U. undulata* Wall.

माले लहरा Male lahara (Nepali).

Spreading or sarmentose shrubs to 3 m. Young branches densely lenticellate, more or less pubescent. Petioles 3–5 mm. Leaves oblong-elliptic or oblong-lanceolate, 7–15 × 3–6 cm, base subcordate, rounded or subcuneate, apex bluntly acute to acuminate, secondary veins 7–12 pairs, prominent beneath, glabrous and shiny above, more or less pubescent beneath. Flowers yellowish-green or cream-coloured. Pedicels 2.5–3 cm, pubescent. Bracts on lower half of pedicel. Sepals elliptic-ovate to ovate-lanceolate, 9–11 × 5–7 mm, obtuse or acute, almost glabrous. Petals elliptic-lanceolate, narrowed and clawed at base, obtuse at apex. Outer petals 5.5–7.5 × 1.5–2.9 cm. Inner petals 4.5–6.5 × 1.1–1.7 cm. Stamens sessile, ca. 1 mm long. Carpels ca. 1.5 mm. Fruiting carpels moniliform, ca. 3 cm, stalks to 5 mm.
Fig. 3c–d.

Distribution: Nepal, E Himalaya, Assam-Burma, S Asia, E Asia and SE Asia.

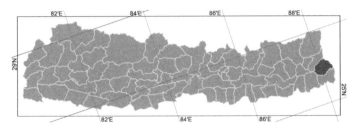

Altitudinal range: 300–600 m.

Ecology: Tropical forest.

Flowering: April–June. **Fruiting:** July–October.

4. *Artabotrys* R.Br., Bot. Reg. 5: pl. 423 (1820).

Shrubs, climbing by means of persistent hook-like peduncles, glabrous. Leaves shiny, coriaceous. Flower bisexual, solitary or several fascicled on woody, hooked-recurved leaf-opposed peduncles. Sepals 3, small, valvate, base connate. Petals 6–9, in 2 whorls, pale yellowish, subequal, valvate, clawed, base concave; inner petals connivent at base, covering stamens and carpels. Torus flat or concave. Stamens oblong, cuneate; connectives flat apically. Carpels few; style oblong or columnar; ovules 2 per carpel, basal. Fruiting carpels free, sessile or shortly stipitate, elliptic-obovate to globose, fleshy and berry-like. Seeds 1 or 2.

About 110 species in S Asia, SE Asia, Australia and Africa. One species cultivated and naturalized in Nepal.

1. *Artabotrys hexapetalus* (L.f.) Bhandari, Baileya 12: 149 (1965).
Annona hexapetala L.f., Suppl. Pl.: 270 (1782); *A. uncinata* Lam.; *Artabotrys odoratissimus* R.Br.; *A. uncinatus* (Lam.) Merr.; *Uvaria odoratissima* (R.Br.) Roxb.

कनक चम्पा Kanak champa (Nepali).

Bushy shrubs, often scandent. Petioles 0.3–0.6 cm. Leaves oblong-lanceolate, 7.5–16.5 × 2.5–5 cm, base acute, apex acuminate, secondary veins 6–14 pairs, coriaceous, glabrous. Flowers sweet-scented, on hooked peduncles, solitary or in 2-flowered cymes, pendulous. Pedicels 0.9–1.4 cm. Sepals ovate, 5–7 × 5–6 mm, connate at base, apex acute and reflexed, puberulent. Petals greenish yellow to yellow at maturity, limb lanceolate, saccate at base, ovate-lanceolate, spreading from saccate coherent base. Outer petals 2.0–3.2 × 0.7–0.9 cm. Inner petals 2.3–2.5 × 0.5–0.6 cm. Stamens ca. 2 mm long; prolonged connective triangular, peltate. Carpels sickle-shaped, ca. 2 mm long; style curved; stigma blunt. Fruiting carpels 8–22, yellowish at maturity, ovoid or obovoid, apiculate, 2.2–3 cm, glabrous, subsessile. Seeds oblong, brown.
Fig. 3e–g.

Distribution: Nepal, Assam-Burma, S Asia and SE Asia.

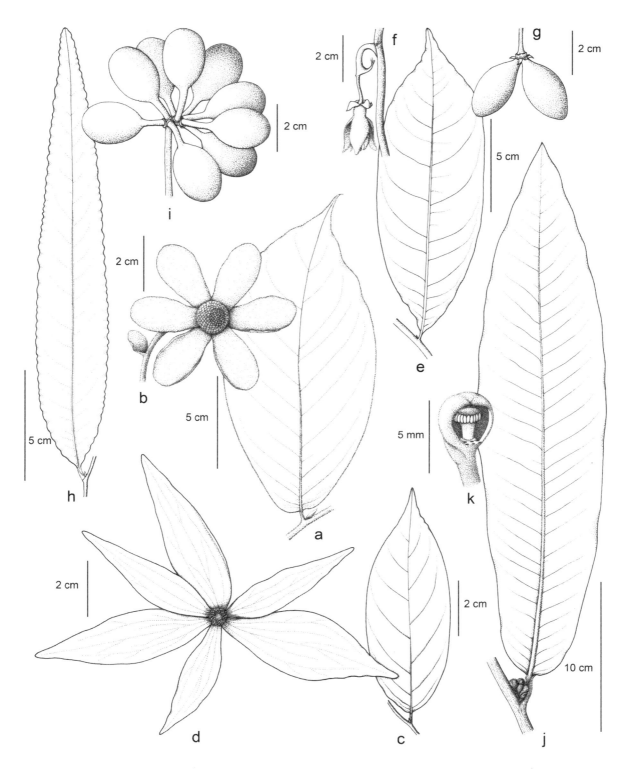

FIG. 3. ANNONACEAE. **Uvaria hamiltonii**: a, leaf; b, flower. **Desmos chinensis**: c, leaf; d, flower. **Artabotrys hexapetalus**: e, leaf; f, inflorescence; g, infructescence. **Polyalthia longifolia**: h, leaf; i, infructescence. MYRISTICACEAE. **Knema tenuinervia**: j, leaf; k, flower.

Annonaceae

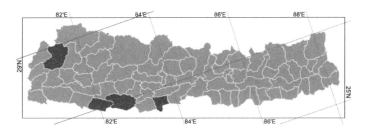

Altitudinal range: 100–1300 m.

Ecology: Cultivated and naturalized.

Flowering: April–July. **Fruiting:** September–February.

Said to be native of South India.

5. *Polyalthia* Blume, Fl. Javae [28-29]: 68 (1830).

Trees or shrubs. Leaves glabrous or pubescent with simple hairs. Inflorescences axillary, internodal, or leaf-opposed, few- to many-flowered cymes or subumbellate clusters. Flowers bisexual, sessile or short pedunculate. Sepals 3, usually small, distinct, valvate. Petals 6, in 2 whorls, valvate, subequal, similar, flat and spreading. Torus convex. Stamens numerous; anther locules cuneate. Carpels indefinite, oblong, cylindrical or angled; ovules 1 or 2; styles absent; stigmas mostly dilated. Fruiting carpels few to many, stipitate, ellipsoid to ovoid, fleshy, indehiscent. Seed solitary.

About 100 species in the Old World tropics. Two species cultivated in Nepal.

Polyalthia is highly polyphyletic (Mols *et al.*, Amer. J. Bot. 91: 590–600. 2004) and in the future it is likely to undergo much reorganization and change in nomenclature.

Key to Species

1a   Leaves neither pendulous nor undulate along margin, lateral veins not prominent .... **1. *P. fragrans***
b   Leaves pendulous and strongly undulate along margin, lateral veins prominent ....**2. *P. longifolia***

1. *Polyalthia fragrans* (Dalzell) Hook.f. & Thomson, Fl. Brit. Ind. 1[1]: 63 (1872).
Guatteria fragrans Dalzell, Hooker's J. Bot. Kew Gard. Misc. 3: 206 (1851).

Trees to 20 m. Young branches densely minutely tomentose. Petioles 0.8–1.1 cm, glabrous. Leaves elliptic to oblong-lanceolate, 7.5–19 × 3.5–8.5 cm, base acute or rounded, apex acute to shortly acuminate, glabrous, pubescent on nerves when young, secondary veins 12–19 pairs, prominent. Flowers fragrant, greenish, in few to many-flowered branched cymes from axils of fallen leaves or on tubercles. Pedicels 1.2–4 cm, tomentose; bracts tomentose. Sepals ca. 3 mm, orbicular. Petals linear-lanceolate, 1.5–2 × 0.2–0.3 cm, coriaceous, inner petals slightly longer. Stamens numerous, cuneate, ca. 1 mm; connectives dome-shaped at apex. Carpels many, oblong-ovoid, ca. 1.5 mm. Fruiting carpels many, obliquely ellipsoid to broadly ovoid, 3–5 × 1.5–2 cm; stalks 2.5–5 cm.

Distribution: Nepal, S Asia and E Asia.

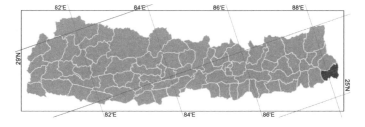

Altitudinal range: ca. 300 m.

Ecology: Cultivated.

Flowering: May–June. **Fruiting:** July–August.

Native to the Western Ghats of India, cultivated elsewhere in tropical areas.
     Wood is used for timber.

2. *Polyalthia longifolia* (Sonn.) Thwaites, Enum. Pl. Zeyl. [5]: 398 (1864).
Uvaria longifolia Sonn., Voy. Indes Orient.: 260, pl. 131 (1782).

नक्कली अशोक Nakkali ashok (Nepali).

Tree to 15–20 m. Young branches glabrous or minutely puberulous. Petioles 0.3–1 cm, glabrous. Leaves oblong-lanceolate to narrow-lanceolate, 9–20 × 1.2–5 cm, base obtuse or cuneate, apex gradually long acuminate, strongly undulate along margins, glabrous, lateral veins up to–24 pairs, faint. Flowers greenish yellow, numerous, in subumbellate clusters at axils of fallen leaves. Pedicels slender, 1.5–3.5 cm, usually bracteolate, slightly pubescent. Sepals connate at the base, broadly ovate-triangular, acute, ca. 2 mm wide, pubescent. Petals greenish yellow, subequal, narrowly lanceolate, 8–20 × 2–4 mm, base broad, tapering gradually to acute apex. Stamens numerous, ca. 1 mm. Carpels many, linear, ca. 1 mm long. Fruiting carpels many, broadly ellipsoid, 2–3 × 1.5–2 cm; stalks 1–1.5 cm. Fig. 3h–i.

Distribution: Nepal, S Asia and E Asia.

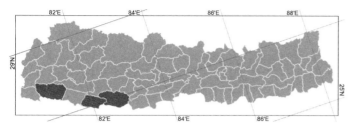

Altitudinal range: 100–700 m.

Ecology: Cultivated.

Flowering: March–May. **Fruiting:** July–September.

Native to India and Sri Lanka, grown in subtropical areas especially as a street tree.
Wood is useful for timber.

6. *Annona* L., Sp. Pl. 1: 536 (1753).

Trees or shrubs, with simple hairs. Inflorescences terminal, leaf-opposed, extra-axillary, or cauliflorous, solitary or fascicled. Flowers bisexual. Sepals 3, small, valvate. Petals 6, in 2 whorls, free or connate at the base. Outer petals lanceolate or ovate, valvate, fleshy but leathery when dry, connivent or spreading, inside basally concave, margin thick. Inner petals rudimentary or absent. Torus convex. Anther connective ovoid apically. Carpels numerous, subconnate; ovule solitary, basal. Styles oblong, stigmas entire clavate, muriculate. Fruiting carpels fleshy, confluent into a syncarpous, many-seeded, ovoid or globose fruit, surface covered with reticulate ridges or bulges and deep grooves.

About 175 species, mostly in tropical America with a few species in tropical Africa. Two species cultivated and naturalized in Nepal.

Key to Species

1a   Leaves acuminate. Outer petals puberulent. Fruiting carpels separated by reticulation of raised ridges
....**1. *A. reticulata***
b   Leaves obtuse. Outer petals glabrous with age. Fruiting carpels separated by deep grooves ....**2. *A. squamosa***

1. *Annona reticulata* L., Sp. Pl. 1: 537 (1753).

रामफल Ramphal (Nepali).

Small trees to 4–6 m. Branches pubescent when young, glabrous with age. Leaves oblong-lanceolate, 8–18 × 3–5 cm, base rounded, apex acute to acuminate, secondary veins 8–18 pairs, minutely pubescent beneath when young, glabrous with age. Flowers solitary, terminal or in leaf-opposed or internodal 1–4-flowered cymes. Pedicels 0.8–1.5 cm, bracteolate, glabrous; bracteoles minute, basal or halfway along the pedicel. Sepals broadly ovate, shortly acuminate, ca. 3 × 3–3.5 mm, pubescent outside, glabrous inside. Petals yellowish green. Outer petals 3 narrowly oblong, triquetrous, 1.7–2.1 cm, fleshy, pubescent outside, glabrous inside, base concave. Inner petals absent, sometimes minute remnants present. Stamens ca. 1 mm long, connective truncate. Carpels ovoid to linear, ca. 1 mm long. Fruit yellow to reddish, ovoid to subglobose, 5–10 cm in diameter, carpels separated by reticulation of raised ridges. Seeds black-brown.

Distribution: Asia, Africa, N America, S America and Australasia.

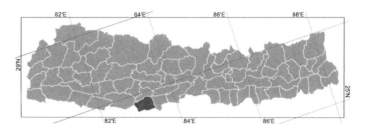

Altitudinal range: 200–1100 m.

Ecology: Cultivated, sometimes naturalized in disturbed habitats.

Flowering: May–July. **Fruiting:** September–January.

Cultivated throughout the tropics, native to the West Indies.
Ripe fruits are edible (custard apple: bullock's heart). Bark and fruits are used to treat diarrhoea and dysentery and oil extracted from the seeds has insecticidal properties.

Annonaceae

2. *Annona squamosa* L., Sp. Pl. 1: 537 (1753).

सरीफा Saripha (Nepali).

Small tree or shrubs to 3–6 m. Branches pubescent when young, glabrous with age. Leaves elliptic-lanceolate, narrowly elliptic, or oblong, 5–15 × 2.5–5 cm, base obtuse to rounded, apex obtuse to slightly acute, lateral veins 8–12 pairs, glabrous, pubescent on nerves when young. Flowers solitary, terminal, or in leaf-opposed or internodal 1–4-flowered cymes. Pedicels 1–1.4 cm, bracteolate, glabrous; bracteoles minute, basal or halfway along the pedicel. Sepals broadly ovate, shortly acuminate, 2–3 × 3–3.5 mm, pubescent outside, glabrous inside. Petals yellowish green. Outer petals 3 narrowly oblong, triquetrous, 1–1.4 cm, fleshy, pubescent when young, glabrous with age, base concave. Inner petals absent. Stamens ca. 1 mm long, connective broad. Carpels ovoid, ca. 1 mm long. Fruit greenish yellow, ovoid to globose, 7–10 cm in diameter, carpels separated by deep grooves. Seeds black-brown.

Distribution: Asia, Africa, N America, S America and Australasia.

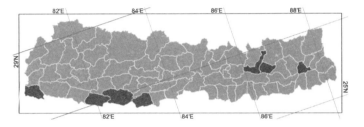

Altitudinal range: 100–1000 m.

Ecology: Cultivated, sometimes naturalized in disturbed habitats.

Flowering: May–July. **Fruiting:** September–January.

Cultivated throughout the tropics, native to the West Indies. Misspelt '*squamata*' by Whitmore (Enum. Fl. Pl. Nepal 2: 27. 1979) and copied in some Nepalese publications (e.g. Annot. Checkl. Fl. Pl. Nepal: 10. 2000; Fl. Pl. Nepal: 24. 2001).

Fruits are edible (custard apple: sweet sop). The leaves are used as a fish poison and the plant also has insecticidal properties. A paste of the root is used to relieve headaches. Unripe fruit are used to treat diarrhoea and dysentery.

Myristicaceae

Mark F. Watson & Kamal Maden

Dioecious (Nepalese species) evergreen trees, often stellate-pubescent, inner bark producing a reddish juice when cut. Stipules absent. Leaves simple, alternate, pinnately veined, entire. Flowers small, in axillary clusters or panicles, unisexual, actinomorphic. Bracts caducous, bracteoles on pedicels at base of perianth, or absent. Perianth fused and shortly cup-shaped at base, 3(–4)-lobed, lobes triangular, thick, valvate, sepals and petals indistinguishable. Male flowers with filaments fused into a column bearing anthers. Female flowers with a superior, sessile, 1-celled ovary, ovules 1, basal, style short or absent, stigma 2-lobed. Fruit a dehiscent drupe, pericarp fleshy, skin leathery, 2–4-valved, single-seeded. Seed very large, partially or completely enveloped by a bright coloured, fleshy laciniate or subentire aril, endosperm ruminate, often with volatile oil.

Worldwide 19 or 20 genera, ca. 500 species, an ancient family of lowland rainforests in tropical regions of Asia, Africa and America. Two genera and two species in Nepal.

Key to Genera

1a   Leaves not whitish below. Flowers in long-stalked, much-branched panicles. Pedicels without bracteole. Staminal column globose .... **1. *Horsfieldia***

  b   Leaves whitish below. Flowers in short-stalked axillary clusters. Pedicel with bracteole. Staminal column forming a peltate disc .... **2. *Knema***

1. *Horsfieldia* Willd., Sp. Pl., ed. 5[as 4], 4(2): 872 (1806).

Leaves thinly leathery, essentially glabrous, lower surface not white glaucous. Flowers numerous in subglobose axillary panicles. Bracteoles absent. Male flowers globose, glabrous, staminal column globose bearing 18–20 elongated anthers over its surface. Female flowers obovoid, ovary, pubescent, stigma sessile, small. Fruit ovoid or ellipsoid, glabrous, perianth persistent, aril subentire.

About 100 species from Indomalesia to Australia. One species in Nepal.

1. *Horsfieldia kingii* (Hook.f.) Warb., Nova Acta Phys.-Med. Acad. Caes. Leop.-Carol. Nat. Cur. 68: 308 (1897).
Myristica kingii Hook.f., Fl. Brit. Ind. 5[13]: 106 (1886).

रुन्चे पात Runche paat (Nepali).

Tree, 10–20 m, bark smooth, grey. Branches dark brown, often crowded towards the top of trunk, hollow, smooth or with small lenticels, puberulent to glabrous. Petioles 1.5–3 cm, stout, glabrous. Leaves oblong–obovate, 12–35(–45) × 5–9(–12) cm, base rounded or cuneate, apex acute or shortly acuminate, both surfaces of mature leaves glabrous except mid vein sometimes pubescent below, lateral veins 13–18-pairs, not prominent above. Male panicles lax, 9-15 cm, pedicels slender 1–2 mm, flowers 3–4 mm diameter, clustered Female panicles more compact, branched, 3–7 cm, flowers ca. 5 mm diameter, not clustered. Fruit 4–5 × 2.3–2.8 cm, edible.

Distribution: Nepal, E Himalaya, Assam-Burma, E Asia and SE Asia.

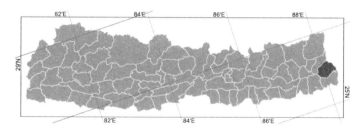

Altitudinal range: 300–600 m.

Ecology: Lowland subtropical forests.

Flowering: June–?August. **Fruiting:** ?October–March.

Very poorly collected in Nepal, and only known from two specimens at the BM (not represented at KATH). Phenological data are extrapolated from Li & Wilson (Fl. China 7: 100. 2008).

2. *Knema* Lour., Fl. Cochinch. 2: 604 (1790).

Leaves somewhat leathery, essentially glabrous, glaucous whitish beneath Flowers numerous in dense axillary clusters or on short thick branchlets. Bracteoles present, inserted near base of perianth. Male flowers obovoid, rusty-brown tomentose, staminal column disc-like bearing 9–18 anthers around the rim. Female flowers globose with ovate rusty-brown pubescent ovary, stigma on a short, thick style. Fruit ovoid or ellipsoid, densely tomentose with rusty-brown dendritic hairs, perianth caducous, aril red, lacinate at apex.

About 100 species in Indomalesia. One species in Nepal.

1. *Knema tenuinervia* W.J.de Wilde, Blumea 25(2): 405 (1979).

रातो भलायो Rato bhalayo (Nepali).

Large tree, 10–20m, bark light grey, vertically fissured. Young branchlets brownish, stellate-tomentose, becoming greyish, glabrescent and drooping. Petioles 1.5–2.5 cm, stout, rusty brown pubescent when young, glabrescent. Leaves oblong to narrowly elliptic, 20–65 × 8–16 cm, base rounded or shallowly cordate, apex acute, lateral veins 20–35 pairs, flat to slightly prominent above. Male flowers usually arising in axils of fallen leaves, pedicels stout, 10–15 mm, flower buds 7–10 × 5–6 mm, perianth segments red within, 7–10 mm. Female flowers sessile or on very short pedicels, buds globose, ca. 6 mm, style ca. 1.5 mm, stigma lobulate. Infructescence short, usually with 1 or 2 fruit, pedicels 3–5 mm. Fruit ovoid or ellipsoid, 3.5–4.5 × 2.2–3 cm.
Fig. 3j–k.

Distribution: Nepal, E Himalaya, Assam-Burma and SE Asia.

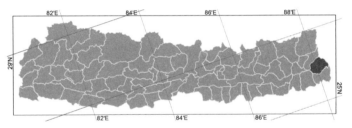

Altitudinal range: 100–1000 m.

Ecology: Moist subtropical forests.

Flowering: April–?July. **Fruiting:** ?October–?February.

Very poorly collected in Nepal, and only known from one specimen at the BM (not represented at KATH). Phenological data are extrapolated from Li & Wilson (Fl. China 7: 97. 2008).

Nepalese material includes the type and so represents the typical subspecies when an infraspecific classification is used. *Knema errarica* (Hook.f. & Thomson) Sinclair is recorded below 200 m in Darjeeling District and may occur in the lowland forests of border areas in East Nepal. It has smaller leaves (usually less than 20 × 5 cm) with cuneate bases and smaller fruit (less than 2.5 × 1.5 cm).

Lauraceae

Colin A. Pendry

Deciduous or evergreen shrubs or trees, often aromatic. Terminal buds often very large and surrounded by numerous scales which leave dense clusters of scars in rings around the twigs (perulate buds). Leaves usually alternate, rarely opposite or whorled, pinnately veined or more or less strongly 3-veined. Stipules absent. Flowers usually small, in panicles, cymes or umbels surrounded by decussate bracts, with the umbels themselves solitary, in clusters on lateral short-shoots or rarely in racemes. Flowers bisexual or unisexual, trimerous or rarely dimerous, actinomorphic, with a more or less pronounced hypanthium. Fertile stamens usually 9, rarely 6 or 12, in whorls of 3, with the innermost whorl usually with a pair of basal glands. Inner whorl of staminodes sometimes present. Female flowers with 9 or rarely 6 or 12 staminodes with the innermost glandular. Anthers dehiscing by 2 or 4 valves. Ovary superior, 1-locular with a single apical ovule. Style simple. Fruit a drupe borne on the more or less enlarged cup-like remains of the perianth, rarely the perianth completely enclosing the fruit. Tepals persistent or caducous.

Worldwide 52 genera and 2550 species throughout the tropics and subtropics. Eleven genera and 46 species in Nepal.

Generic limits within the family are the subject of much debate, and many species have complicated synonymy due to transfer between genera (see Long, Notes Roy. Bot. Gard. Edinburgh 41: 505-525. 1984). When comparing with past works it should be noted that several specimens cited in Kostermans & Chater (Enum. Fl. Pl. Nepal 3: 182-187. 1982) were not reliably identified and are now assigned to different species: *Cinnadenia paniculata* (Hook.f.) Kosterm. is excluded for this reason (see comment under *Litsea doshia* (D.Don) Kosterm.). In pinnately veined leaves the lowest pair of veins is always weaker than those above, while in 3-veined leaves the lowest pair is the only or the strongest pair. In some species the distinction is not so clear and the lowest pair is only slightly stronger than those above. These species are referred to as 'weakly 3-veined'.

Key to Genera

1a   Flowers solitary or several in umbels surrounded by persistent bracts, the umbels solitary, clustered on leafless short axillary shoots or racemosely arranged on short shoots. Flowers unisexual (bisexual in *Dodecadenia*)..........................2
   b   Flowers in panicles or cymes not surrounded by bracts. Flowers bisexual (unisexual in one species of *Actinodaphne*) 6

2a   Flowers bisexual, flowers usually solitary, rarely in an inflorescence of up to 3 flowers. Fruits 1(–3) on thickened pedicels .... **10. *Dodecadenia***
   b   Flowers unisexual, several to many per umbel. Fruits usually more than 3 per inflorescence, pedicels not thickened or only slightly thickened 3

3a   Flowers dimerous ....**9. *Neolitsea***
   b   Flowers trimerous..4

4a   Leaves whorled at tips of twigs ....**11. *Actinodaphne***
   b   Leaves alternate, rarely subopposite or opposite, sometimes clustered towards tips of twigs, but not whorled 5

5a   Anthers 2-valved ....**7. *Lindera***
   b   Anthers 4-valved ....**8. *Litsea***

6a   Inflorescence irregularly paniculate, the ultimate branches subopposite. Anthers 2-locular 7
   b   Inflorescence cymose, the ultimate branches opposite. Anthers 4-locular 8

7a   Flowers with a deep, tubular hypanthium, this enclosing the ovary and later the fruit ....**1. *Cryptocarya***
   b   Flowers with a cup-shaped hypanthium, the fruit unprotected on the pedicel or infrequently with small, persistent tepals at the base...**2. *Beilschmiedia***

8a   Leaves opposite, 3-veined ....**3. *Cinnamomum***
   b   Leaves alternate, pinnately veined or 3-veined 9

9a   Leaves 3-veined. Inflorescences somewhat condensed and not strictly cymose.............................**4. *Neocinnamomum***
   b   Leaves pinnately veined. Inflorescence regularly cymose...10

10a   Perianth segments persistent in fruit. Leaves often clustered at the tips of the twigs......................................11
    b   Perianth segments not persistent in fruit. Leaves usually not clustered at the tips of the twigs (except *Cinnamomum tenuipile*). 12

11a   Perianth segments of flowers rigid, erect; in fruit stiff, clasping the base of the fruit.........................**5. Phoebe**

   b   Perianth segments of flowers soft, often spreading; spreading to reflexed in fruit .... **6. Machilus**

12a   Fruit up to 1.5 cm long, with thin mesocarp. Wild. ....**3. Cinnamomum**

   b   Fruit 8–18 cm long, with fleshy, edible mesocarp. Cultivated (avocado). .... **Persea**

1. *Cryptocarya* R.Br., Prodr. Fl. Nov. Holland.: 402 (1810).

Large evergreen trees. Perulate buds absent. Leaves subopposite or alternate, evenly spaced along the twig, secondary venation pinnate. Inflorescences paniculate, axillary or terminal. Flowers small, bisexual, trimerous. Hypanthium elongate, completely enclosing the ovary. Tepals 6. Stamens 9, in 3 whorls, 2-valved. Outer whorls introrse, inner whorl extrorse. Stamens eglandular, glands present at edge of hypanthium. Staminodes 3, fleshy. Fruit completely enclosed by the enlarged hypanthium which terminates in a small circular scar, tepals not persistent.

Worldwide 300 species in Asia and Australia. One species in Nepal.

1. Cryptocarya amygdalina Nees, in Wall., Pl. Asiat. Rar. 2[8]: 69 (1831).
Cryptocarya floribunda Nees; *Laurus amygdalina* Buch.-Ham. ex Wall. nom. nud.; *L. floribunda* Wall. nom. nud.

Trees to 25 m. Twigs mid brown, minutely tomentose, glabrescent, smooth. Perulate buds absent. Leaves alternate or subopposite, evenly distributed along twig, elliptic or oblong to ovate or obovate, 4.5–22 × 3–9 cm, length:width ratio 1.5–3.2, base rounded to cuneate or attenuate, apex acuminate to obtuse or emarginate, underside glaucous or not, glabrous, secondary veins 4–12 pairs, tertiary venation reticulate-scalariform. Petioles 1–1.8 cm. Inflorescences to 14 cm, tomentose. Pedicels 1 mm. Flowers yellow, 3–5 mm, tomentose outside. Tepals narrowly ovate, 1.5–2 mm tomentose within. Stamens 1.5–2 mm, the inner whorl longest. Staminodes triangular, 1 mm. Ovary 1 mm, glabrous or sparsely hairy at apex. Style 1 mm, glabrous. Fruit on 3–5 mm pedicels, globose to ellipsoid or ovoid, 10–16 mm, conspicuously ridged.
Fig. 4a–d.

Distribution: Nepal, E Himalaya, Tibetan Plateau and Assam-Burma.

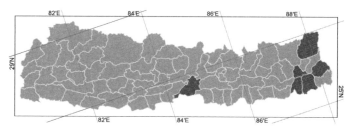

Altitudinal range: 100–1600 m.

Ecology: Evergreen and mixed subtropical forests.

Flowering: February–March. **Fruiting:** April–June.

The elongate hypanthium in flower, which goes on to completely enclose the fruit, is quite unlike any other species of Lauraceae in Nepal.

2. *Beilschmiedia* Nees, in Wall., Pl. Asiat. Rar. 2[8]: 69 (1831).

Large evergreen trees. Perulate buds present or absent. Leaves opposite or subopposite, slightly clustered or evenly spaced along the twig, secondary venation pinnate. Inflorescences paniculate, solitary in leaf axils or clustered at base of new shoots. Flowers small, bisexual, trimerous. Hypanthium cup-shaped, not enclosing the ovary. Tepals 6. Stamens 9, in 3 whorls, 2-valved. Outer whorls eglandular, introrse, inner whorl glandular, lateral. Staminodes 3, fleshy. Fruit usually lacking persistent perianth remains, rarely the unenlarged tepals persistent.

Worldwide 200 species throughout the tropics and as far south as Chile and New Zealand. Two species in Nepal.

Key to Species

1a   Perulate buds present, leaving conspicuous scars at the base of the shoots. Flowers and inflorescences
      glabrous.. **1. B. gammieana**

   b   Perulate buds absent, shoots without conspicuous scars at the base. Flowers and inflorescences
      sericeous .... **2. B. roxburghiana**

1. *Beilschmiedia gammieana* King ex Hook.f., Fl. Brit. Ind. 5[13]: 124 (1886).

Trees to 20 m. Twigs glabrous, initially dark reddish brown or blackish, smooth becoming paler and lenticellate. Perulate buds present. Leaves slightly clustered, elliptic or ovate, 6.5–15.5 × 2–5 cm, length:width ratio 2.6–3.3, base cuneate, apex acuminate, underside non glaucous, glabrous, secondary veins 9–11 pairs, tertiary venation reticulate. Petioles 1–1.7 cm. Inflorescences 1–2 cm, glabrous. Pedicels 2–4 mm. Flowers yellow, 2.5 mm, glabrous outside. Tepals ovate, 2.5 mm, glabrous within. Stamens 1.5–2 mm, the inner whorl the longest. Staminodes 1 mm. Ovary 1 mm, glabrous. Style 1 mm, glabrous. Fruits solitary, globose to obovoid, to 3 cm on a 1.5–2.5 cm peduncle.

Distribution: Nepal and E Himalaya.

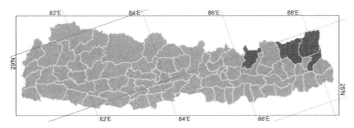

Altitudinal range: 1000–2000 m.

Ecology: Lower temperate broad-leaved forest.

Flowering: April. **Fruiting:** July–March.

2. *Beilschmiedia roxburghiana* Nees, in Wall., Pl. Asiat. Rar. 2[8]: 69 (1831).
Laurus bilocularis Roxb.

Trees to 20 m. Twigs glabrous, initially mid brown and smooth, often becoming whitish and slightly lenticellate. Perulate buds absent. Leaves evenly spaced or slightly clustered, elliptic to slightly obovate or slightly ovate, 10–20 × 4–7.5 cm, length:width ratio 2.3–2.9, base cuneate, apex acute or slightly acuminate, underside not glaucous, glabrous, secondary veins 9–15 pairs, tertiary venation reticulate. Petioles 1.5–2.3 cm. Inflorescences 1–2 cm, sericeous. Pedicels 2–3 mm. Flowers pale yellow or yellowish, 3–4 mm, sericeous outside. Tepals oblong or ovate, 3 mm, sericeous within. Outer whorls of stamens 2 mm. Inner whorl 3 mm. Staminodes triangular, 1 mm. Ovary 1 mm, glabrous. Style 2 mm, glabrous. Fruits solitary, oblong, to 3 cm on a 1.5–2 cm peduncle.

Distribution: Nepal, E Himalaya, Tibetan Plateau, Assam-Burma, S Asia and SE Asia.

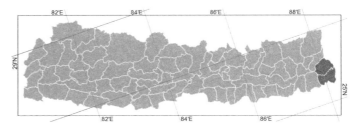

Altitudinal range: 200–400 m.

Ecology: Mixed moist subtropical forest.

Flowering: April. **Fruiting:** ?July.

3. *Cinnamomum* Schaeff. nom cons., Bot. Exped.: 74 (1760).

Evergreen or deciduous shrubs to large trees. Perulate buds present or absent. Leaves alternate, subopposite or opposite or very rarely whorled, evenly spaced along the twig or more or less clustered, 3-veined or with pinnate secondary venation. Inflorescences cymose, axillary or rarely terminal. Flowers small, bisexual, trimerous. Hypanthium cup-shaped, not enclosing the ovary. Tepals 6. Stamens 9, in 3 whorls, 4-valved. Outer whorls eglandular, introrse, inner whorl glandular, extrorse. Staminodes 3, fleshy. Fruit on an enlarged, fleshy perianth cup, tepals caducous or enlarged and persistent.

Worldwide 250 species in Asia, Australia and tropical America. Five species in Nepal.

Cinnamon of international commerce comes from the bark of *Cinnamomum verum* J.Presl, native to Sri Lanka and SW India, but several other species are used locally. *Cinnamomum glaucecens* (Nees) Hand.-Mazz. and *C. parthenoxylon* (Jack.) Meisn. are excluded from this account as they are seen to be based on misidentifications of *Cinnamomum glanduliferum* (see note under this species). *C. camphora* (L.) J.Presl (*Laurus camphora* L.) from Japan is a widespread street tree in Nepal. It has not yet been reported as naturalized so is included in the key but not described further.

Key to Species

1a   Leaves pinnately veined, the veins clearly different from the midvein. Perulate buds present 2

b   Leaves 3-veined(or 5-veined) at the base, the secondary veins reaching to the apex of the leaf and similar in appearance to the midvein. Perulate buds absent...4

2a   Leaves villous below, lacking glands in axils of veins. Inflorescence and flowers tomentose..................... **5. *C. tenuipile***

b   Leaves glabrous below, with glands present in axils of veins. Inflorescence and flowers glabrous.................................3

3a Leaves elliptic to obovate. Secondary veins all similar .... **4. *C. glanduliferum***
b Leaves ovate or rarely elliptic. Lowest pair of secondary veins distinctly stronger than the others .... ***C. camphora***

4a Flowers 3–4 mm long. Leaves broadly elliptic, to 30 cm long. Petioles 1.3–2.0 cm. Length:width ratio
2.0–2.5(–2.9) ....**2. *C. bejolghota***
b Flowers 5–7 mm long. Leaves more or less ovate, to 20 cm long. Petioles 0.7–1.3 cm. Length:width ratio 2.4–3.6.......**5**

5a Upper surface of leaves flat; veins not sunken. Perianth cup of fruits with persistent tepals.........................**1. *C. tamala***
b Upper surface of leaves raised between the sunken veins. Perianth cup of fruits with entire rim
.... **3. *C. impressinervium***

1. *Cinnamomum tamala* (Buch.-Ham.) Nees & Eberm.,
Handb. Med.-Pharm. B. 2: 426 (1831).
Laurus tamala Buch.-Ham., Trans. Linn. Soc. London, Bot.
13: 555 (1822); *Cinnamomum albiflorum* Nees; *C. soncaurium*
Buch.-Ham. ex Nees & Eberm.; *Laurus albiflora* Wall. nom.
nud.; *L. soncaurium* Buch.-Ham.; *L. tazia* Buch.-Ham.

तेजपात Tejpat (Nepali).

Evergreen shrub or small tree, occasionally to 20 m. Twigs
red-brown, smooth, sericeous when young, soon glabrescent,
rarely glaucous. Perulate buds absent. Leaves 3-veined,
opposite or subopposite, evenly spaced along twig, usually
at least slightly ovate, rarely elliptic, 8.5–20 × 3.0–6.5 cm,
base cuneate to rounded, apex acute to acuminate, margin
slightly inrolled, underside more or less glaucous, minutely
sparsely sericeous below, tertiary venation scalariform. Glands
not present in vein axils. Petioles 0.7–1 cm. Inflorescences
to 9 cm, more or less sericeous. Flowers white, 5–7 mm
long, sericeous. Pedicels 4–5 mm. Tepals oblong or narrowly
ovate, 4–6 mm, sericeous within. Fertile stamens 3–4 mm,
the innermost whorl usually slightly longer. Staminodes 2 mm.
Ovary 1–1.5 mm, hairy or sparsely hairy. Style 2–3 mm, hairy
or sparsely hairy. Fruit obovoid or ellipsoid, 10–14 mm long.
Tepals persistent on rim of cupule.
Fig. 4e–g.

Distribution: Nepal, W Himalaya, E Himalaya, Assam-Burma
and E Asia.

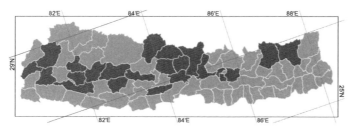

Altitudinal range: 400–2300 m.

Ecology: Subtropical and lower temperate broad-leaved
forests, often in gullies and north-facing slopes.

Flowering: (February–)April–June. **Fruiting:** May–October.

Don (Prodr. Fl. Nepal.: 67. 1825) misapplied the name
Cinnamomum cassia Blume to this species.

Cinnamomum tamala is distinguished from *C. bejolghota*
(Buch.-Ham.) Sweet by its smaller leaves which are always
ovate and not elliptic, its longer flowers and fruits with
partially, instead of completely persistent tepals. Its leaves
are completely flat above, in contrast with those of *C.
impressinervium* Meisn. which always have sunken veins and
furthermore has fruits in which the tepals are completely lost,
leaving an entire rim.

The bark and leaves are used as spices and are exported
from Nepal. They are also used to treat colic and diarrhoea.

2. *Cinnamomum bejolghota* (Buch.-Ham.) Sweet, Hort. Brit.,
ed. 1, 2: 344 (1826).
Laurus bejolghota Buch.-Ham., Trans. Linn. Soc. London, Bot.
13: 559 (1822); *Cinnamomum bazania* (Buch.-Ham.) Nees; *C.
obtusifolium* Roxb. ex Nees; *Laurus bazania* Buch.-Ham.; *L.
obtusifolia* Roxb. nom. nud.

सिन्काउली Sinkauli (Nepali).

Evergreen shrub or small trees to 12 m. Twigs pale brown, green
or reddish, smooth, glabrous. Perulate buds absent. Leaves
3-veined, opposite, subopposite or whorled, evenly spaced
along twig, elliptic or rarely slightly ovate, 10–27 × 4.0–10 cm,
base cuneate to rounded, apex obtuse to acute or slightly
acuminate, margin flat or inrolled, underside not or slightly
glaucous, glabrous, tertiary venation scalariform or slightly
reticulate. Glands not present in vein axils. Petioles 1.3–2.0 cm.
Inflorescences to 24 cm, glabrous. Flowers yellow, 3–4 mm
long, sericeous inside and out. Pedicels 3–6 mm. Tepals broadly
ovate, 3 mm, sericeous within. Fertile stamens 2–2.5 mm, the
innermost whorl usually slightly longer. Staminodes 1.5 mm.
Ovary 1 mm, glabrous. Style 1.5 mm, glabrous. Fruit obovoid or
ellipsoid, 15 mm long. Tepals persistent on rim of cupule.

Distribution: Nepal, E Himalaya, Assam-Burma, S Asia, E
Asia and SE Asia.

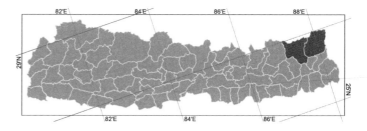

Altitudinal range: 1200–1900 m.

Ecology: Subtropical and temperate broad-leaved forests.

Flowering: May–July. **Fruiting:** August.

The smaller flowers and larger, more coriaceous and usually elliptic leaves distinguish *Cinnamomum bejolghota* from *C. tamala* (Buch.-Ham.) Nees & Eberm.

3. *Cinnamomum impressinervium* Meisn., in DC., Prodr. 15(1): 21 (1864).

Evergreen shrub or small trees to 12 m. Twigs dark reddish brown, sericeous when young, soon glabrescent, smooth. Perulate buds absent. Leaves 3-veined, opposite or subopposite, evenly spaced, elliptic or ovate, length 8.5–13.5 × 2.0–5.5 cm, base rounded to cuneate, apex acuminate, margin flat; underside slightly glaucous, villous on midrib below and minutely sparsely sericeous below, tertiary venation scalariform. Glands not present in vein axils. Petioles 0.8–1.3 cm. Flowers not seen. Infructescences to 12 cm. Fruits globose. Tepals not persistent, rim of cupule entire.

Distribution: Nepal, E Himalaya and Assam-Burma.

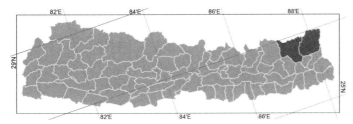

Altitudinal range: 1700–2100 m.

Ecology: *Castanopsis* forest.

Flowering: May. **Fruiting:** August.

The leaves of *Cinnamomum impressinervium* always have sunken veins on the upper surface in contrast with the leaves of *C. tamala* (Buch.-Ham.) Nees which are completely flat and usually larger. The entire rim of the cupule of *C. impressinervium* also distinguishes it from *C. tamala* whose cupules have persistent tepals. Kosterman & Chater's (Enum. Fl. Pl. Nepal 3: 183. 1982) record of *C. impressinervium* from C Nepal is erroneous and is based on Banerji's (Rec. Bot. Surv. India 19(2): 80 1965) citation of a specimen from Taplejung.

4. *Cinnamomum glanduliferum* (Wall.) Meisn., in DC., Prodr. 15(1): 25 (1864).
Laurus glandulifera Wall., Trans. Med. Soc. Calcutta 1: 45 & 51, pl. 1 (1825); *Camphora glandulifera* (Wall.) Nees.

माला गिरी Mala giri (Nepali).

Evergreen shrub or small trees to 14 m. Twigs reddish brown,

glabrous, smooth. Perulate buds present. Leaves pinnate-veined, alternate or rarely subopposite, evenly spaced along the twig, elliptic to obovate, 6–13 × 2.5–5.0 cm, base cuneate, apex acute to acuminate or apiculate, margin flat or slightly inrolled, underside glaucous or not, glabrous, secondary veins 4–6 pairs, tertiary venation scalariform. Glands present in the axils of the veins. Petioles 1–2.8 cm. Inflorescences to 12 cm, glabrous or very sparsely villous especially in upper parts. Pedicels 2–4 mm. Flowers yellow, 2–3 mm long, glabrous or sparsely villous outside. Tepals ovate, 2 mm, tomentose within. Fertile stamens 1–1.5 mm, the innermost whorl sometimes slightly longer. Staminodes 0.5–1 mm. Ovary 1 mm, glabrous. Style 1–1.5 mm, glabrous. Fruit globose, 12 mm long. Tepals not persistent, rim of cupule entire.

Distribution: Nepal, E Himalaya, Tibetan Plateau, Assam-Burma, E Asia and SE Asia.

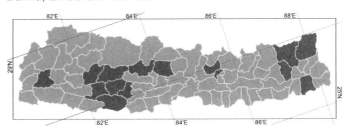

Altitudinal range: 700–2600 m.

Ecology: Temperate forests and scrub with *Quercus*, *Castanopsis* and *Magnolia*.

Flowering: April–May. **Fruiting:** June–July.

Kostermans & Chater (Enum. Fl. Pl. Nepal 3: 183. 1982) recorded *Cinnamomum glaucescens* (Nees) Hand.-Mazz. based on *Stainton 6197 & 6532*, but both specimens (BM) are *C. glanduliferum* and so this name is misapplied and this species is not yet confirmed in Nepal. Likewise *C. parthenoxylon* (Jack) Meisn. is based on a misdentification of *Stainton, Sykes & Williams 4265* (BM, E, KATH, TI) which is also *C. glanduliferum*.

Cinnamomum glaucescens differs from *C. glanduliferum* in having clustered inflorescences with a dense brown tomentose instead of widely spaced, almost glabrous inflorescences in the axils of the leaves. *Cinnamomum parthenoxylon* is clearly distinguished from *C. glanduliferum* by having no glands in the axils of its veins.

5. *Cinnamomum tenuipile* Kosterm., Reinwardtia 8(1): 74 (1970).
Alseodaphne mollis W.W.Sm., Notes Roy. Bot. Gard. Edinburgh 13: 153 (1921).

Deciduous trees to 25 m. Twigs brown, sometimes reddish, smooth, tomentose when young, glabrescent. Perulate buds present. Leaves pinnate-veined, alternate, more or less clustered towards the tips of the twigs, ovate, 6.5–17 × 2.5–8.0 cm, base rounded to cuneate, apex acuminate, margin flat; underside not glaucous or slightly glaucous, sparsely or densely villous below, especially on midrib and veins, secondary veins 4–7 pairs, tertiary venation scalariform or somewhat reticulate. Glands not present in vein axils. Petioles 1.5–3.5 cm. Inflorescences to 12 cm, densely tomentose. Flowers greenish yellow, 3–4 mm long, sericeous or tomentose. Pedicels 2–10 mm. Tepals ovate or oblong, 2–3 mm, sericeous or tomentose within. Fertile stamens 1.5–2 mm, the innermost whorl longer. Staminodes 1–1.5 mm. Ovary 1–1.5 mm, glabrous. Style 1–1.5 mm, glabrous. Fruit globose, 15 mm long. Tepals not persistent, rim of cupule entire.

Distribution: Nepal, E Himalaya and E Asia.

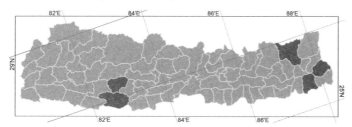

Altitudinal range: 400–1300 m.

Ecology: Temperate forests and evergreen gullies in *Shorea* forests.

Flowering: March–April. **Fruiting:** August.

Closest to *Cinnamomum glanduliferum* (Wall.) Meisn., but easily distinguished from it by its more clustered leaves with villous undersides and the absence of glands in the axils of the veins.

4. *Neocinnamomum* H.Liu, Laurac. Chine & Indochine: 82 (1934).

Evergreen small trees. Perulate buds absent. Leaves opposite or alternate, evenly spaced along the twig, 3-veined. Inflorescences cymose, axillary. Flowers small, bisexual, trimerous. Hypanthium cup-shaped, not enclosing the ovary. Tepals 6. Stamens 9, in 3 whorls, 4-valved. Outer whorls eglandular, introrse, inner whorl glandular, extrorse. Staminodes 3, fleshy. Fruit on an enlarged, fleshy cup with enlarged and persistent tepals.

Worldwide six species in China and SE Asia. One species in Nepal.

1. *Neocinnamomum caudatum* (Nees) Merr., Contr. Arnold Arbor. 8: 64 (1934).
Cinnamomum caudatum Nees, in Wall., Pl. Asiat. Rar. 2[8]: 76 (1831); *Laurus caudata* Wall. nom. nud.

Small trees to 10 m. Twigs dark reddish brown, young twig sericeous, soon glabrescent, smooth. Perulate buds absent. Leaves suborbicular to broadly elliptic or elliptic, 6.5–10 × 3–6.5 cm, length:width ratio 1.6–2.0, base rounded, apex caudate or acuminate, 3-veined above the base, tertiary venation scalariform, underside non glaucous, glabrous. Petioles 0.7–1.2 cm. Inflorescences 1–6 cm, sericeous. Pedicels 4–7 mm, sericeous. Flowers greenish, 2 mm. Tepals broadly ovate, 2 mm, sericeous inside and out. Stamens almost equal, ca. 1 mm. Staminodes up to 0.5 mm. Ovary 1 mm, glabrous. Style subsessile. Pedicel elongating to 2–3.5 cm in fruit. Fruit ellipsoid, 2–2.5 cm.

Distribution: Nepal, E Himalaya, Assam-Burma, E Asia and SE Asia.

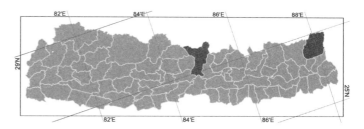

Altitudinal range: 900–2600 m.

Ecology: Subtropical and lower temperate forests. Often on riversides.

Flowering: July. **Fruiting:** August.

Neocinnamomum is close to *Cinnamomum*, but differs in its somewhat contracted inflorescence, which sometimes has small leaves present amongst the flowers, and subsequently develops into a leafy twig with very large, solitary, axillary fruits. Uniquely among the Nepalese species, *N. caudatum* has anthers with two valves opening introrsely and two opening extrorsely.

5. *Phoebe* Nees, Syst. Laur.: 98 (1836).

Evergreen small to large trees. Perulate buds present or absent. Leaves alternate, evenly spaced along the twig or weakly to strongly clustered at the tips, secondary venation pinnate. Inflorescences cymose, axillary. Flowers small, bisexual, trimerous. Hypanthium cup-shaped, not enclosing the ovary. Tepals 6. Stamens 9, in 3 whorls, 4-valved. Outer whorls eglandular, introrse, inner whorl glandular, extrorse. Staminodes 3, fleshy. Fruit enclosed at base by the stiff, upright, somewhat enlarged tepals.

Worldwide 100 species in Indomalesia. Five species in Nepal.

Key to Species

1a	Bud scale scars in diffuse clusters along twig	2
b	Bud scale scars in dense rings around twig	3
2a	Inflorescence, young twigs and undersides of leaves tomentose	**2. *P. cathia***
b	Inflorescence, young twigs and undersides of leaves glabrous or minutely sparsely sericeous	**3. *P. pallida***
3a	Leaves narrowly elliptic to slightly ovate or obovate, length:width ratio 3.4–6.0. Secondary veins 7–12 pairs	**1. *P. lanceolata***
b	Leaves broadly obovate or elliptic, length:width ratio 2.5–3.6. Secondary veins 12–24 pairs	4
4a	Young twigs tomentose. Flowers to 4 mm long on 2 mm pedicels which are not articulated in their lower half	**4. *P. attenuata***
b	Young twigs glabrous or minutely sparsely sericeous. Flowers to 6–7 mm long on 5–8 mm pedicels articulated in their lower half	**5. *P. hainesiana***

1. *Phoebe lanceolata* (Nees) Nees, Syst. Laur.: 109 (1836).
Ocotea lanceolata Nees, in Wall., Pl. Asiat. Rar. 2[8]: 71 (1831); *Laurus lanceolaria* Roxb.; *L. lanceolata* Roxb. ex Wall. nom. nud.; *L. salicifolia* Buch.-Ham. nom. nud.; *L. salicifolia* Buch.-Ham. ex Nees.

झाँक्री काठ Jhankri kath (Nepali).

Trees to 20 m. Twigs whitish, glabrous or sparsely villous and soon glabrescent, smooth or slightly lenticellate. Perulate buds present. Leaves more or less clustered at the ends of the twigs, elliptic to slightly ovate or slightly obovate, 7–26 × 2–8 cm, length:width ratio 3.4–6.0, base cuneate to attenuate or somewhat rounded, rarely oblique, apex usually long acuminate, margin flat; underside not or slightly glaucous, minutely sparsely sericeous below, secondary veins 7–12 pairs, tertiary venation reticulate-scalariform. Petioles 0.8–2.2 cm. Inflorescences to 20 cm, glabrous. Pedicels 2–7 mm. Flowers yellow, 3–4 mm long, glabrous outside. Tepals ovate, 3–3.5 mm, more or less sericeous within. Stamens 2–2.5 mm, almost equal. Staminodes 1.5 mm. Ovary 1.5 mm, glabrous. Fruit globose, 10–12 mm long, on a 6–10 mm pedicel.

Distribution: Nepal, E Himalaya, Assam-Burma, S Asia, E Asia and SE Asia.

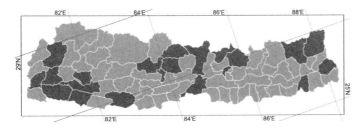

Altitudinal range: 100–2900 m.

Ecology: Evergreen broadleaved forests including *Schima-Castanopsis* forest and evergreen gullies in *Shorea* forest.

Flowering: April–June(–August). **Fruiting:** (February–) May–November.

The most widespread species of *Phoebe* in Nepal. *Phoebe lanceolata* is characterized by its long, narrow leaves and glabrous inflorescences.

2. *Phoebe cathia* (D.Don) Kosterm., Nat. Hist. Bull. Siam Soc. 25(3-4): 44 (1975).
Cinnamomum cathia D.Don, Prodr. Fl. Nepal.: 66 (1825); *Laurus cathia* Buch.-Ham. nom. nud.; *L. paniculata* Wall. nom. nud.; *Ocotea paniculata* Nees; *O. pubescens* Nees; *Phoebe paniculata* (Nees) Nees; *P. pubescens* (Nees) Nees.

Trees to 20 m. Twigs mid to dark brown, tomentose, smooth. Perulate buds absent. Leaves evenly spaced along twig or slightly clustered towards the tips, elliptic or slightly obovate, 8–17 × 2.0–5.5 cm, length:width ratio 2.5–3.9, base cuneate, apex acuminate, margin flat or slightly inrolled, underside not or slightly glaucous, densely tomentose on veins below and sparsely tomentose on lamina, secondary veins 5–10 pairs, tertiary venation scalariform. Petioles 0.4–2 cm. Inflorescences to 10 cm, tomentose. Pedicels 2–3 mm. Flowers pale yellow, 3.5–4 mm long, tomentose or sericeous outside. Tepals ovate or broadly ovate, 3–3.5 mm, tomentose or sericeous within. Stamens 2–2.5 mm, almost equal. Staminodes 1–1.5 mm.

Ovary 1 mm, glabrous. Fruit ovoid, 10 mm long on a 3–4 mm pedicel.
Fig. 4h.

Distribution: Nepal, E Himalaya, Assam-Burma and S Asia.

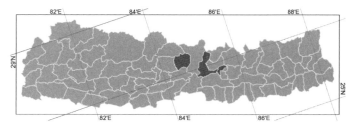

Altitudinal range: 1300–1900 m.

Ecology: Evergreen broad-leaved forests and evergreen gullies in *Shorea* forest.

Flowering: May–September.

The absence of perulate buds and the tomentose indumentum on the young twigs, inflorescences and undersides of the leaves are found only in *Phoebe cathia*. The secondary venation of the leaves is often clearly impressed on the upper surface, and the scalariform tertiary venation rather prominent below. Most of the specimens cited in Enum. Fl. Pl. Nepal 3: 187 (1982) are misidentified *Machilus* or *Persea* species.

3. *Phoebe pallida* (Nees) Nees, Syst. Laur.: 112 (1836).
Ocotea pallida Nees, in Wall., Pl. Asiat. Rar. 2[8]: 71 (1831); *Laurus pubescens* Wall. nom. nud.

Trees to 10 m. Twigs blackish or dark reddish brown, glabrous or minutely sparsely sericeous and then soon glabrescent. Perulate buds absent. Leaves evenly spaced along the twig, elliptic or oblong, 8–17 × 1.5–3.5 cm, length:width ratio 4.4–4.6, base rounded or cuneate, apex acuminate, margin flat, underside slightly glaucous, minutely sparsely sericeous below, secondary veins 10–15 pairs, tertiary venation reticulate-scalariform. Petioles 0.5–1.1 cm. Inflorescences to 7 cm, sparsely sericeous, glabrescent in lower parts. Pedicels 2 mm. Flowers cream, ca. 2 mm long, glabrous or very sparsely pubescent outside. Tepals broadly ovate, 2.5 mm, tomentose within. Stamens ca. 1.5 mm. Staminodes 1 mm. Ovary 1 mm, glabrous. Fruit not seen.

Distribution: Nepal, W Himalaya and Assam-Burma.

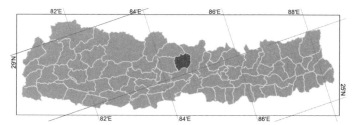

Altitudinal range: 1800–2000 m.

Ecology: Temperate forests.

Flowering: May. **Fruiting:** June.

A rather poorly known species characterized by its small flowers and twigs which are almost glabrous, smooth and dark when young, apparently becoming paler when older.

4. *Phoebe attenuata* (Nees) Nees, Syst. Laur.: 104 (1836).
Ocotea attenuata Nees, in Wall., Pl. Asiat. Rar. 2[8]: 71 (1831); *Laurus attenuata* Wall. nom. nud.

Trees to 30 m. Twigs very pale or dark brown, tomentose, glabrescent, smooth, rugose or lenticellate. Perulate buds present. Leaves alternate, densely clustered at the tips of the twigs, obovate, 10–22 × 3.5–8.5 cm, length:width ratio 2.8–3.4, base cuneate to attenuate, apex acuminate or apiculate, margin flat; underside non glaucous, villous below, especially on veins, secondary veins 14–24 pairs, tertiary venation scalariform. Petioles 0.8–2 cm. Inflorescences to 15 cm, tomentose. Pedicels 1–2 mm. Flowers yellow, ca. 4 mm long, sericeous outside. Tepals broadly ovate, 3–3.5 mm, sericeous within. Stamens 2–3.5 mm, those of the second series the longest. Staminodes 2 mm. Ovary 1.5 mm, sparsely hairy at apex. Fruit ellipsoid, to 16 mm long on a 1–2 mm pedicel.

Distribution: Nepal, E Himalaya and Assam-Burma.

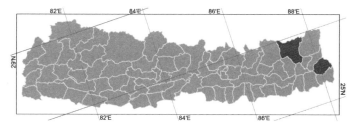

Altitudinal range: 1000–1400 m.

Ecology: Mixed moist subtropical forest.

Flowering: May.

Similar in overall appearance to *Phoebe hainesiana* Brandis, but differing from it in its more obviously hairy leaves and the tomentose inflorescence with smaller flowers. *Stainton 6255 & 6854* (BM), recorded in Enum. Fl. Pl. Nepal 3: 187 (1982) as *P. attenuata*, are *P. hainesiana*.

5. *Phoebe hainesiana* Brandis, Hooker's Icon. Pl.: pl. 2803 (1906).

Trees to 30 m. Twigs pale brown or whitish, glabrous or minutely sparsely sericeous and then soon glabrescent, lenticellate or rugose. Perulate buds present. Leaves clustered at the ends of the twigs, usually obovate, sometimes elliptic or oblong, 8–19 × 3–8 cm, length:width ratio 2.5–3.6, base cuneate, sometimes oblique, apex acute to acuminate or apiculate, margin flat; underside not or slightly glaucous, glabrous or minutely sparsely sericeous below, secondary veins 12–18 pairs, tertiary venation scalariform or reticulate-scalariform. Petioles 1.2–4.5 cm. Inflorescences to 15 cm,

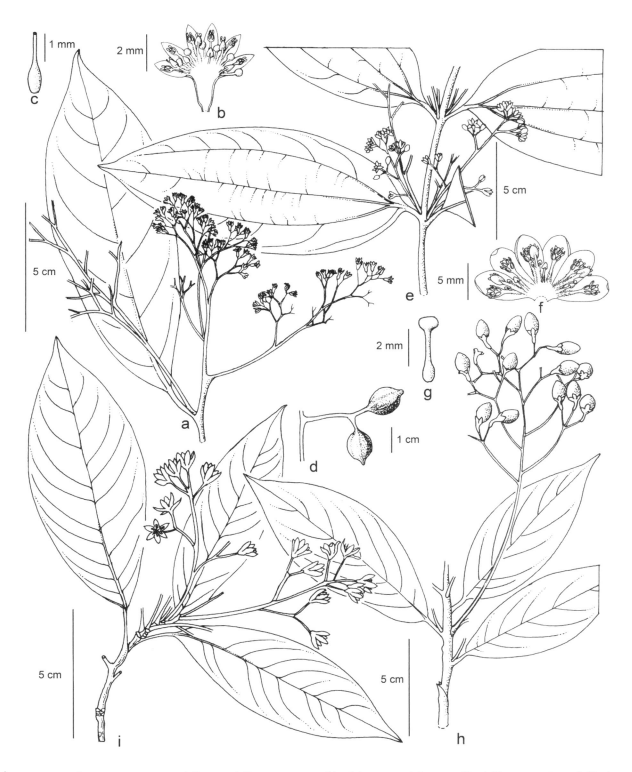

FIG. 4. Lauraceae. **Cryptocarya amygdalina**: a, inflorescence and leaf; b, opened flower with pistil removed; c, pistil; d, infructescence. **Cinnamomum tamala**: e, inflorescence and leaves; f, opened flower with pistil removed; g, pistil. **Phoebe cathia**: h, infructescence and leaves. **Machilus odoratissima**: i, inflorescence and leaves.

sparsely sericeous, glabrescent in lower parts. Pedicels 5–8 mm. Flowers yellow or pale green, 6–7 mm long, sericeous outside. Tepals broadly ovate, 4–6 mm, tomentose within. Stamens 4.5–5 mm. Staminodes 3 mm. Ovary 2.5 mm, glabrous. Fruit ellipsoid, to 25 mm long on a 15 mm pedicel.

Distribution: Nepal and E Himalaya.

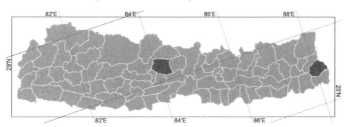

Altitudinal range: 1300–1600 m.

Ecology: Wet subtropical and temperate forests.

Flowering: April–May.

Phoebe hainesiana is similar to *P. attenuata* (Nees) Nees, but has larger flowers and a generally glabrous appearance. The pedicel is articulated in the lower half, with the stump persisting on the inflorescence when flowers are shed.

6. *Machilus* Nees, in Wall., Pl. Asiat. Rar. 2[6]: 70 (1831).

Persea Mill.

Evergreen small to large trees. Perulate buds present or absent. Leaves alternate or subopposite, evenly spaced along the twig or somewhat clustered at the tips, secondary venation pinnate. Inflorescence cymose, axillary. Flowers small, bisexual, trimerous. Hypanthium cup-shaped, not enclosing the ovary. Tepals 6. Stamens 9, in 3 whorls, 4-valved. Outer whorls eglandular, introrse, inner whorl glandular, extrorse. Staminodes 3, fleshy. Fruit borne on small cupule with soft, spreading or reflexed tepals.

Worldwide 100 species in tropical and subtropical Asia. Seven species in Nepal.

Molecular evidence (Rohwer *et al.*, Taxon 58: 1153. 2009) supports the separation of Asian *Machilus* from *Persea* which is distributed in tropical America and Macaronesia. Thus all the Nepalese species are now included within *Machilus*, and the only species of *Persea* known from Nepal is the cultivated *P. americana* Mill., the avocado.

Key to Species

1a	Perulate buds absent. Leaves clustered and leaving scars at base of current year's growth, but scars not forming dense ring around twig	2
b	Perulate buds present. Twigs ringed with dense clusters of scars from bud scales	3
2a	Indumentum of inflorescence rachis appressed, grey, sericeous. Leaves glabrous or appressed sericeous on midrib below. Secondary venation almost equally prominent above and below	**2. M. sericea**
b	Indumentum of inflorescence rachis spreading, rusty-brown tomentose. Leaves with tomentose indumentum on midrib below. Secondary venation clearly more prominent below than above	**3. M. glaucescens**
3a	Inflorescence rachis completely glabrous, sometimes glaucous	4
b	Inflorescence rachis sericeous or tomentose, at least around the axils of rachis	5
4a	Leaves broadly elliptic to slightly obovate or slightly ovate, length:width ratio 2.0–4.8 Secondary veins 6–14 pairs	**1. M. odoratissima**
b	Leaves narrowly elliptic or oblong to slightly obovate, length:width ratio 4.2–6.0. Secondary veins 13–20 pairs	**5. M. clarkeana**
5a	Leaves reddish-brown villous below. Inflorescence peduncle spreading rusty-brown tomentose	**6. M. pubescens**
b	Leaves glabrous or greyish sericeous below. Inflorescence peduncle grey rather appressed sericeous	6
6a	Leaves 10–23 cm long, length:width ratio 3.5–5.3. Secondary veins 11–18 pairs	**4. M. duthiei**
b	Leaves 5–16 cm long, length:width ratio 2.7–4.6. Secondary veins 6–12(–15) pairs	**7. M. gamblei**

1. *Machilus odoratissima* Nees, in Wall., Pl. Asiat. Rar. 2[8]: 70 (1831).
Laurus odoratissima Wall. nom. nud.; *Persea odoratissima* (Nees) Kosterm.

Trees to 25 m. Twigs dark reddish brown or dark brown, smooth or slightly ridged, glabrous, sometimes glaucous. Perulate buds present. Leaves alternate, slightly clustered or evenly spaced, elliptic to slightly obovate or slightly ovate, 6–19 × 2–5.5 cm, length:width ratio 4.2–5.9, base cuneate to rounded, apex acute or slightly acuminate, margin flat or slightly inrolled, underside slightly glaucous, glabrous, secondary veins 6–14, tertiary venation reticulate-scalariform, not particularly prominent below. Petioles 1–2.2 cm. Inflorescences 5–9 cm, glabrous often glaucous. Pedicels 4–9 mm. Flowers pale yellow to pale green, 4–6 mm long, more or less glabrous outside. Tepals dimorphic, ovate to elliptic or oblong, 3–5.5 mm, glabrous to tomentose within. Fertile stamens 4–4.5 mm, the innermost whorl the longest. Staminodes 2 mm. Ovary and style glabrous. Fruits obovoid or ellipsoid, 15–16 mm, glabrous, somewhat glaucous. Fig. 4i.

Distribution: Nepal, E Himalaya, Assam-Burma and SE Asia.

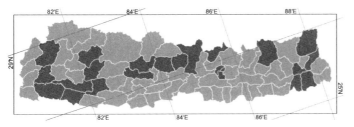

Altitudinal range: 300–2200 m.

Ecology: Gullies in *Shorea* forest and mixed moist subtropical and lower temperate forests with *Quercus*, *Castanopsis*, *Magnolia* and *Pinus*.

Flowering: March–June. **Fruiting:** April–September.

According to Kostermans (Bibliogr. Lauracearum: 640. 1964) Loureiro (Fl. Cochinch. 1: 253. 1790) misapplied *Laurus indica* L. to this species.

The glabrous and often glaucous, reddish twigs of *Machilus odoratissima* are distinctive, particularly in combination with the glabrous flowers and often glaucous fruits.

2. *Machilus sericea* Blume, Mus. Bot. 1: 330 (1851).
Laurus sericea Wall. nom. nud.; *Ocotea sericea* Nees later homonym, non Kunth; *Persea wallichii* D.G.Long; *Phoebe sericea* Nees nom. illegit.

Trees to 15 m. Twigs dark brown or dark reddish brown, smooth to rugose or slightly ridged, sometimes lenticellate, sericeous, glabrescent. Perulate buds absent. Leaves alternate, slightly or strongly clustered, oblong or elliptic to slightly ovate or slightly obovate, 6.5–13 × 2–5 cm, length:width ratio 2.0–3.3, base cuneate to rounded, apex acute or obtuse, margin flat, underside slightly glaucous, glabrous or minutely sparsely sericeous below, secondary veins 7–12, tertiary venation reticulate-scalariform, not particularly prominent below. Petioles 1.4–2.8 cm. Inflorescences 5–16 cm, sericeous or glabrescent in lower parts of peduncle, not glaucous. Pedicels 3–4 mm. Flowers green or yellow, 4–6 mm long, sericeous outside. Tepals elliptic or oblong, 3–5 mm sericeous within. Fertile stamens 2–3.5 mm, the innermost whorl the longest. Staminodes 1–1.5 mm. Ovary and style glabrous. Fruits globose, ca. 10 mm, glabrous, not glaucous.

Distribution: Nepal and W Himalaya.

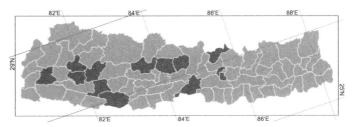

Altitudinal range: 200–2500 m.

Ecology: Subtropical and mixed lower temperate forests with *Quercus* and *Magnolia*.

Flowering: March–May. **Fruiting:** June–August.

With its lack of perulate buds *Machilus sericea* could only be confused with *M. glaucescens* (Nees) Meisn, but the latter has leaves which are clearly tomentose on the midrib and veins below and a much more spreading, tomentose indumentum on its inflorescences.

3. *Machilus glaucescens* (Nees) Wight, Icon. Pl. Ind. Orient. 5[2]: 12 (1852).
Ocotea glaucescens Nees, in Wall., Pl. Asiat. Rar. 2[8]: 71 (1831); *Laurus glaucescens* Roxb. ex Wall. nom. nud.; *L. villosa* Roxb.; *L. villosa* Roxb. nom. nud.; *Machilus villosa* (Roxb.) Hook.f.; *Persea glaucescens* (Nees) D.G.Long; *P. villosa* (Roxb.) Kosterm.; *Phoebe glaucescens* (Nees) Nees; *P. villosa* (Roxb.) Wight.

Trees to 15(–25) m. Twigs dark brown or grey brown, smooth or slightly ridged, tomentose, glabrescent. Perulate buds absent. Leaves alternate, slightly or strongly clustered, elliptic or oblong to slightly ovate or obovate, 8.5–19 × 2.5–6.5 cm, length:width ratio 2.8–3.5, base cuneate, apex acuminate or apiculate, margin slightly inrolled, underside more or less glaucous, tomentose on midrib and veins below and sparsely pubescent on lamina, secondary veins 5–9, tertiary venation scalariform, prominent below. Petioles 0.9–2.5 cm. Inflorescences 15–25 cm, tomentose. Pedicels 3–9 mm. Flowers yellow or greenish-yellow, 4–5 mm long, tomentose outside. Tepals dimorphic, broadly ovate, 3–5 mm, tomentose within. Fertile stamens 2–2.5 mm, almost equal. Staminodes 1 mm. Ovary and style glabrous. Fruits globose, ca. 7 mm, glabrous or sparsely puberulous, not glaucous.

Distribution: Nepal, E Himalaya and Assam-Burma.

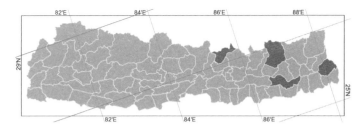

Altitudinal range: 300–2400 m.

Ecology: Temperate forest with *Quercus*, *Lithocarpus*, evergreen gullies in *Shorea* forest.

Flowering: February–March. **Fruiting:** March–April.

Machilus glaucescens shares the absence of perulate buds with *M. sericea* Blume, but the latter has leaves which are almost glabrous below and has an appressed sericeous indumentum on its inflorescences.

4. *Machilus duthiei* King ex Hook.f., Fl. Brit. Ind. 5[16]: 861 (1890).
Persea duthiei (King ex Hook.f.) Kosterm.

Large trees to 25 m, 1 m dbh. Twigs dark reddish brown, glabrous and sometimes glaucous, lenticellate or rugose or smooth, becoming grey brown or whitish and rugose or lenticellate. Perulate buds present. Leaves alternate or occasionally subopposite, more or less clustered, oblong or elliptic to slightly obovate, 10–23 × 2–6 cm, length:width ratio 3.5–5.3, base cuneate, apex more or less acuminate, margin slightly inrolled; underside more or less glaucous, glabrous or minutely sparsely sericeous below, secondary veins 11–18, tertiary venation reticulate-scalariform, not particularly prominent below. Petioles 0.9–2.3 cm. Inflorescences 4–16 cm, sericeous. Pedicels 4–6 mm. Flowers pale greenish-yellow or rarely white, 5–6 mm long, sericeous outside. Tepals dimorphic, oblong or ovate, 4.5–5.5 mm, glabrous or sericeous inside. Fertile stamens 3.5–5 mm, the innermost whorl the longest. Staminodes 1.5–2 mm. Ovary and style glabrous. Fruits globose, 10–15 mm, glabrous, not glaucous.

Distribution: Nepal, W Himalaya, E Himalaya, Tibetan Plateau and Assam-Burma.

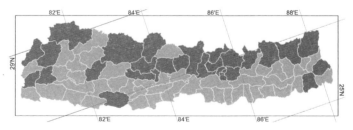

Altitudinal range: 1500–3200 m.

Ecology: In primary and secondary temperate forests with *Quercus*, *Magnolia*, *Rhododendron*, *Aesculus* and *Pinus*. Sometimes becoming dominant.

Flowering: March–May. **Fruiting:** May–November.

The large, narrow leaves of *Machilus duthei*, often with glaucous undersides and numerous pairs of secondary veins, are distinctive. The leaves are larger than those of *M. clarkeana* King ex Hook.f., which has glabrous inflorescences. *M. duthei* is most likely to be confused with *M. gamblei* King ex Hook.f., but the latter has shorter, relatively broader leaves.

5. *Machilus clarkeana* King ex Hook.f., Fl. Brit. Ind. 5[13]: 137 (1886).
Machilus gammieana King ex Hook.f.; *Persea clarkeana* (King ex Hook.f.) Kosterm.; *P. gammieana* (King ex Hook.f.) Kosterm.

Trees to 15 m, 80 cm dbh. Twigs pale brown or whitish, rugose, glabrous. Perulate buds present. Leaves alternate, markedly clustered, elliptic or oblong to slightly obovate, 6–14 × 1.2–3 cm, length:width ratio 4.2–5.9, base cuneate, apex acute to acuminate, margin inrolled; underside more or less glaucous, glabrous; secondary veins 13–20, tertiary venation reticulate or reticulate-scalariform, not particularly prominent below. Petioles 0.6–1.8 cm. Inflorescences 5–9 cm, glabrous. Pedicels 6–10 mm. Flowers greenish, 8 mm long, glabrous or sparsely pubescent outside. Tepals narrowly ovate, 6–7 mm, glabrous to tomentose inside. Fertile stamens ca. 6 mm almost equal. Staminodes 2 mm. Ovary and style, glabrous. Fruits globose or broadly ellipsoid, 25 mm, glabrous, not glaucous.

Distribution: Nepal, E Himalaya and Assam-Burma.

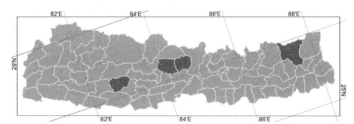

Altitudinal range: 1600–2900 m.

Ecology: Temperate *Quercus*-Lauraceae forests.

Flowering: March–April. **Fruiting:** June–September.

The combination of strongly clustered, very narrow leaves with numerous pairs of rather indistinct secondary veins and a glabrous inflorescence is found only in *Machilus clarkeana*.

6. *Machilus pubescens* Blume, Mus. Bot. 1(21): 330 (1851).
Cinnamomum tomentosum D.Don; *Persea blumei* Kosterm.; *P. tomentosa* (D.Don) Spreng.

Trees to 20 m. Twigs pale brown to dark brown or blackish, sometimes slightly reddish, tomentose, glabrescent, smooth or slightly ridged or rugose, more or less lenticellate. Perulate buds present. Leaves alternate, strongly clustered, elliptic or oblong to slightly ovate or slightly obovate, 6–15 × 1.8–5 cm, length:width ratio 2.4–4.3, base cuneate, apex acute or acuminate, margin flat or inrolled, underside glaucous or not, tomentose or villous on midrib and veins and villous on lamina below, secondary veins 8–12, tertiary venation reticulate-scalariform or reticulate, not particularly prominent below. Petioles 0.7–1.5 cm. Inflorescence 4–9 cm, tomentose, with spreading red-brown hairs. Pedicels 3–5 mm. Flowers yellow or greenish, 5–6 mm long, sericeous or tomentose outside. Tepals dimorphic, oblong, ovate or narrowly ovate, 3–5 mm, glabrous or slightly sericeous within. Fertile stamens 2.5–4 mm, the innermost whorl longest. Staminodes 1 mm. Ovary and style glabrous. Fruits globose, ca. 10 mm, glabrous, not glaucous.

Distribution: Endemic to Nepal.

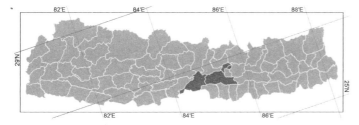

Altitudinal range: 100–2500 m.

Ecology: Lower temperate forests with *Quercus*, *Magnolia* and *Rhododendron* and evergreen gullies in *Shorea* forest.

Flowering: March–April. **Fruiting:** March–June.

It is possible to confuse some forms of *Machilus gamblei* King ex Hook.f. with *M. pubescens*, but the inflorescence of *M. pubescens* always has a denser, more spreading indumentum, and furthermore some collections also have narrow 1 cm-long caducous bracts. Although *M. pubescens* (as *Persea tomentosa*) was synonymised under *Phoebe cathia* (D.Don) Kosterm. in Enum. Fl. Pl. Nepal: 3. 187. 1982) the two species are clearly distinct as *P. cathia* has smaller flowers and lacks perulate buds.

7. *Machilus gamblei* King ex Hook.f., Fl. Brit. Ind. 5[13]: 138 (1886).
Machilus bombycina King ex Hook.f.; *Persea bombycina* (King ex Hook.f.) Kosterm.; *P. gamblei* (King ex Hook.f.) Kosterm.

Trees to 20 m, 80 cm dbh. Twigs dark reddish brown, glabrous or young growth tomentose and soon glabrescent, smooth or slightly ridged, sometimes lenticellate. Perulate bud present. Leaves alternate, more or less clustered, oblong, elliptic or obovate, 5–16 × 1.5–5 cm, length:width ratio 2.7–4.6, base cuneate, apex acute or acuminate, margin flat or slightly inrolled, underside slightly glaucous or non glaucous, glabrous or densely sericeous below, secondary veins 6–12(–15), tertiary venation reticulate or scalariform, not particularly prominent below. Petioles 0.6–1.8 cm. Inflorescences 4–11 cm, sericeous or tomentose or tomentose in axils of peduncle and otherwise glabrous. Pedicels 4–8 mm. Flowers greenish-yellow, 5–8 mm, sericeous outside. Tepals oblong, acute, 5–6 mm, sericeous within. Fertile stamens ca. 4 mm. Staminodes 1 mm. Ovary and style glabrous. Fruit globose or obovoid, 6 mm, glabrous, not glaucous.

Distribution: Nepal, E Himalaya, Tibetan Plateau, Assam-Burma, E Asia and SE Asia.

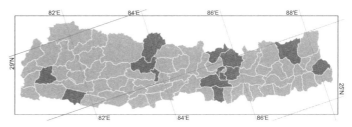

Altitudinal range: 700–2400 m.

Ecology: Temperate forests with Fagaceae and evergreen gullies in *Shorea* forest.

Flowering: March–May. **Fruiting:** March–July.

Glabrous forms of *Machilus gamblei* are most similar to *M. duthiei* King ex Hook.f but have relatively broader leaves with fewer pairs of secondary veins. The more pubescent forms of *M. gamblei* may be confused with *M. pubescens* Blume but the inflorescence of the latter has a denser indumentum of longer, more spreading hairs.

7. *Lindera* Thunb. nom cons., Nov. Gen. Pl. 1: 64 (1781).

Evergreen or deciduous shrubs or small trees, rarely large trees. Perulate buds present or absent. Leaves alternate, evenly spaced along the twig, secondary venation pinnate or weakly 3-veined or strongly 3-veined. Inflorescences umbellate, enclosed by decussate bracts, solitary or in clusters on lateral short-shoots. Flowers small, unisexual, trimerous. Hypanthium cup-shaped, not enclosing the ovary. Tepals 6. Male flowers with 9 stamens in 3 whorls. Stamens 2-valved, introrse, the outer whorls eglandular, inner whorl glandular. Rudimentary ovary present. Female flowers with 9 filiform staminodes in 3 whorls, the outer 6 eglandular and the inner 3 glandular. Fruit on a little-enlarged perianth cup, tepals persistent or not.

Worldwide 100 species primarily in tropical and subtropical Asia. Six species in Nepal.

In species in which the umbels grow on axillary short-shoots, these may or may not develop into twigs. If they do develop further the fruiting umbels appear to be solitary at the base of the twig, though their true origin can be seen by the presence of clusters

Lauraceae

of bud scale scars rather than leaf scars at the base of the twigs. *Lindera melastomacea* (Nees) Fern.-Vill. is excluded from this account as all Nepalese material (relating to *Laurus cuspidata* D.Don) is now treated as *Lindera pulcherrima* (Nees.) Hook.f. *Lindera melastomacea* is currently known eastwards from Bhutan and Assam.

Key to Species

1a Leaves pinnate-veined, the basal pair of lateral veins weaker than those above 2
 b Leaves 3-veined, with either a single pair of lateral veins, or the basal pair of lateral veins stronger than those above 4

2a Petioles at least 3 cm. Leaves 17–26 cm ....**6. *L. bootanica***
 b Petioles to 1 cm. Leaves up to 18 cm...3

3a Perulate buds absent. Umbels on slender pedicels...**2. *L. assamica***
 b Perulate buds present. Umbels sessile..**5. *L. nacusua***

4a Evergreen. Leaves coriaceous. Secondary venation consisting of 1 pair of lateral veins, occasionally with another pair of very weak veins towards the apex ....**3. *L. pulcherrima***
 b Deciduous. Leaves membranous. Secondary venation consisting of 3–6 pairs of lateral veins 5

5a Leaves suborbicular or broadly obovate, hairy below, strongly 3-veined, with the basal pair clearly stronger than those above ....**1. *L. heterophylla***
 b Leaves ovate, glabrous below, weakly 3-veined, with the basal pair only slightly stronger than those above ....**4. *L. neesiana***

---

1. *Lindera heterophylla* Meisn., in A.DC., Prodr. 15(1): 246 (1864).

Deciduous shrub or small trees to 5(–10) m. Twigs prominently rusty villous when young, soon glabrescent and then tomentose to glabrous, grey brown, sometimes reddish, smooth or lenticellate. Perulate buds present. Leaves suborbicular or broadly oblong to slightly ovate or slightly obovate, 4–11 × 3–8 cm, length:width ratio 1.2–1.5, base rounded to slightly cordate, apex acute or obtuse and apiculate, strongly 3-veined, secondary veins 3–6 pairs, tertiary venation scalariform, densely villous below especially on veins and midrib, underside not or slightly glaucous. Petioles 0.6–2 cm. Inflorescences to 2 cm. Umbels sessile and solitary or 2-3 grouped on shortshoots to 8 mm. Male umbels with ca. 10 flowers. Pedicels 6–9 mm, sericeous. Flowers ca. 7 mm. Tepals oblong or narrowly obovate, 6 mm, sericeous outside, glabrous within. Stamens 2 mm. Female umbels with 6–12 flowers. Pedicels 8–13 mm, sericeous. Flowers ca. 5 mm. Tepals linear, ca. 3 mm, long sericeous outside and glabrous inside. Ovary 2 mm with almost sessile style. Infructescences to 3.5 cm, with 7–16 fruit. Fruits ellipsoid, 8 mm, on 7–25 mm pedicels.

Distribution: Nepal and E Himalaya.

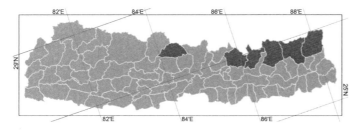

Altitudinal range: 1900–3200 m.

Ecology: Mixed broad-leaved forests and scrub with *Rhododendron*, *Abies* and *Larix*.

Flowering: April. **Fruiting:** May–September.

Kostermans & Chater (Enum. Fl. Pl. Nepal 3: 185. 1982) report *Litsea confertiflora* (Meisn.) Kosterm. from Nepal, but this is based on a misidentication of *Beer, Lancaster & Morris 12329* which is *Lindera heterophylla*.
   Litsea heterophylla is characterized by its broad, villous, 3-veined leaves which are unlike those of any other species. Like *L. neesiana* (Nees) Kurz it flowers when leafless, but the latter has much finer twigs, and flowers up until December.

2. *Lindera assamica* (Meisn.) Kurz, Prelim. Rep. Forest Pegu App. A: 103 (1875).
Aperula assamica Meisn., in A.DC., Prodr. 15(1): 240 (1864).

Evergreen trees to 8(–15) m. Twigs tomentose, dark brown, sometimes reddish, smooth or slightly lenticellate. Perulate buds absent. Leaves elliptic, length 9–18 × 2–5 cm, length:width ratio 3.7–4.2, base cuneate to rounded, apex acute to acuminate, pinnate veined, secondary veins 6–9 pairs, tertiary venation reticulate-scalariform, villous on veins below and sparsely villous on lamina, underside not or slightly glaucous. Petioles 0.5–1 cm. Inflorescences to 2.5 cm, with 1–3 umbels on 3–10 mm short-shoots. Male umbels with 8–11 flowers. Peduncles 9–16 mm, subglabrous. Pedicels 3–4 mm, sericeous. Flowers 4 mm. Tepals linear, sparsely sericeous, 3 mm. Stamens to 3 mm. Female umbels with 8–15 flowers. Peduncles 5–9 mm, subglabrous. Pedicels 2–3 mm, sericeous. Female flowers 3 mm. Tepals linear, glabrous, 2 mm. Ovary shorter than style, glabrous. Infructescences to 4 cm, with 3–6 fruit. Peduncles 7–9 mm. Fruits ellipsoid, 7–9 mm, on 15–20 mm pedicels.

Distribution: Nepal, E Himalaya and Assam-Burma.

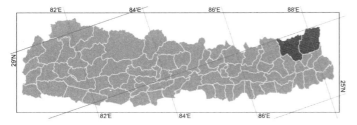

Altitudinal range: 2000–2800 m.

Ecology: Mixed temperate forests.

Flowering: November.

Lindera assamica is most similar to *L. nacusua* (D.Don) Merr., but is easily distinguished from it by having pedunculate rather than sessile umbels and larger leaves.

3. *Lindera pulcherrima* (Nees) Hook.f., Fl. Brit. Ind. 5[13]: 185 (1886).
Daphnidium pulcherrimum Nees, in Wall., Pl. Asiat. Rar. 2[8]: 63 (1831); *Laurus cuspidata* D.Don; *Lindera pulcherrima* var. *attenuata* C.K.Allen; *Tetranthera pulcherrima* Wall. nom. nud.; *Tomex bolo* Buch.-Ham. nom. nud.

खराने Kharane (Nepali).

Evergreen shrub or trees to 12 m. Twigs densely white sericeous when young, glabrescent and then dark reddish brown or dark brown, smooth or lenticellate. Perulate buds present. Leaves elliptic, oblong or slightly ovate, 5–16 × 2–5 cm, length:width ratio 2.5–4.8, base rounded to cuneate, apex acuminate to caudate, strongly 3-veined, secondary veins 1–2 pairs, tertiary venation scalariform, glabrous or sparsely villous, usually glaucous below. Petioles 0.6–1.7 cm. Inflorescences to 1.2 cm, with 1–3 almost sessile umbels on 2–7 mm short-shoots. Male umbels with 3–6 flowers. Pedicels 3–9 mm, sericeous. Flowers yellow or green, 3–4 mm. Tepals oblong or elliptic, 3–4 mm, sericeous outside and glabrous within. Stamens 2.5–4 mm. Female umbels with ca. 5 flowers. Pedicels 3.5–5 mm, sericeous. Flowers yellow or green, ca. 3 mm. Tepals elliptic or oblong, 3–3.5 mm, sparsely sericeous outside and glabrous within. Ovary and style equal, hairy. Infructescences to 2 cm, with 2–7 fruit. Fruit ellipsoid, 8–9 mm, on 8–12 mm pedicels.

Distribution: Nepal, W Himalaya, E Himalaya, Tibetan Plateau, Assam-Burma and E Asia.

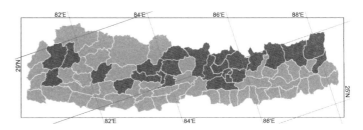

Altitudinal range: 700–3600 m.

Ecology: Locally frequent in temperate broad-leaved forest and scrub with *Quercus*, *Lithocarpus*, *Rhododendron*, *Acer* and *Magnolia*.

Flowering: March–April(–November). **Fruiting:** (March–) May–August(–December).

The *Cinnamomum*-like venation of the leaves is only found in *Lindera pulcherrima*, so this species is quite distinct from the other members of the genus in Nepal.
  Leafy branches are used as fodder and the wood is used for fuel.

4. *Lindera neesiana* (Nees) Kurz, Prelim. Rep. Forest Pegu App. A: 103 (1875).
Benzoin neesianum Nees, in Wall., Pl. Asiat. Rar. 2[8]: 63 (1831); *Aperula neesiana* (Nees) Blume; *Tetranthera neesiana* Wall. nom. nud.

ठेकी फूल Theki phul (Nepali).

Deciduous shrub or trees to 10(–12) m. Twigs dark or mid brown, sometimes reddish, smooth, glabrous. Perulate buds present, but bud scales leaving diffuse scars. Leaves ovate, 3.5–21 × 1.5–13 cm, length:width ratio 1.6–2.4, base rounded, apex acute or slightly acuminate, weakly 3-veined, secondary veins 3–9 pairs, tertiary venation somewhat reticulate, glabrous, underside glaucous or not. Petioles 0.5–2.2 cm. Inflorescences to 1.5 cm with 1–6 umbels on 3–9 mm short-shoots. Male umbels with 5–8 flowers. Peduncles 5–8 mm, glabrous. Pedicels 2–3 mm, sericeous. Flowers cream, 2.5 mm. Tepals dimorphic, 2 mm, glabrous. Stamens 2 mm. Female umbels with 6–8 flowers. Peduncles 4–5 mm, glabrous. Pedicels 2 mm, glabrous. Flowers 2 mm. Tepals triangular, 1 mm, glabrous. Ovary and style equal, glabrous. Infructescences to 1.5 cm, with 1–8 fruit. Peduncles 4–9 mm. Fruit globose or ellipsoid, 5–7 mm on 4–7 mm pedicels.

Distribution: Nepal, E Himalaya, Assam-Burma and E Asia.

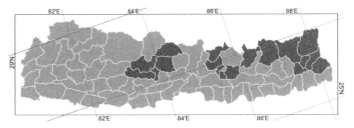

Altitudinal range: 800–2900 m.

Ecology: Woodland and scrub with *Quercus* and *Pinus*. Often associated with disturbance.

Flowering: August–December. **Fruiting:** April–October.

The glabrous, ovate leaves and perulate buds leaving diffuse clusters of scales at the base of the twigs are features unique to *Lindera neesiana* among the Nepali species of *Lindera*. Flowering specimens without leaves may appear superficially similar to *Litsea sericea* (Wall. ex Nees) Hook.f., but the latter is a spring- and early summer-flowering species.

The plant is aromatic and carminative. Powder of root and bark are used to relieve pain. Leaves and fruits are used in the treatment of skin disease. Seeds are efficacious antidote.

5. *Lindera nacusua* (D.Don) Merr., Lingnan Sci. J. 15: 419 (1936).
Laurus nacusua D.Don, Prodr. Fl. Nepal.: 64 (1825); *Daphnidium bifarium* Nees; *Laurus naucushia* Buch.-Ham. nom. nud.; *L. umbellata* Buch.-Ham. ex D.Don later homonym, non Thunb.; *Lindera bifaria* (Nees) Hook.f.; *Tetranthera bifaria* Wall. nom. nud.; *T. vestita* Wall. nom. nud.

पहेंलो खपटे Pahenlo khapate (Nepali).

Evergreen shrub or small trees to 7 m. Young twigs densely rusty villous, later glabrous, grey brown or blackish, smooth or slightly lenticellate. Perulate buds present. Leaves elliptic to ovate, 4–18 × 2–6.5 cm, length:width ratio 1.9–2.8, base cuneate to rounded, apex acute to acuminate, pinnate veined, secondary veins 6–10 pairs, tertiary venation rather scalariform, villous below, especially on veins, underside more or less glaucous. Petioles 0.3–0.9 cm. Inflorescences to 1.1 cm, umbels solitary or up to 3 on 2–5 mm short-shoots. Male umbels with 6–10 flowers. Peduncles 1–3 mm, sericeous. Pedicels 3–5 mm, sericeous. Flowers yellow or green, 3 mm. Tepals obovate or oblong, 3 mm, glabrous inside and sparsely sericeous outside. Stamens 3 mm. Female umbels with 4–6 flowers. Peduncles 2–4 mm, sericeous. Pedicels 2–4 mm, sericeous. Flowers 2 mm. Ovary 2 mm glabrous, with style almost sessile. Infructescences to 1.5 cm, with 2–18 fruit. Fruit globose, 6–7 mm, on 2–6 mm pedicels.

Distribution: Nepal, W Himalaya, E Himalaya, Tibetan Plateau, Assam-Burma and E Asia.

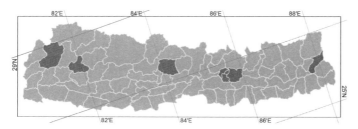

Altitudinal range: 1200–2500 m.

Ecology: Broad-leaved forest.

Flowering: January–April. **Fruiting:** February–October.

Lindera nacusua is most similar to *L. assamica* (Meisn.) Kurz, but differs from it in having sessile rather than pedunculate umbels and generally smaller leaves.

6. *Lindera bootanica* Meisn., in A.DC., Prodr. 15(1): 245 (1864).
Daphnidium venosum Meisn.; *Lindera venosa* (Meisn.) Hook.f.

Deciduous trees to 25 m. Twigs grey brown, smooth, glabrous. Perulate buds present. Leaves evenly spaced, elliptic or slightly obovate, 17–26 × 6–10 cm, length:width ratio 2.6–2.9, base rounded to cuneate, apex slightly acuminate, pinnate veined, secondary veins 8–12 pairs, tertiary venation reticulate, villous on veins below and sparsely villous on lamina; underside slightly glaucous. Petioles 3–4 cm. Inflorescences to 2 cm, 2–3 umbels on 3–5 mm short-shoots. Peduncles 10–13 mm, sericeous. Mature flowers not seen. Infructescences to 5 cm. Fruits globose, ca. 15 mm, on 12–16 mm pedicels.

Distribution: Nepal and E Himalaya.

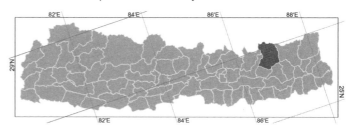

Altitudinal range: 1800–2000 m.

Ecology: Broad-leaved forest.

Flowering: October.

With its large pinnately-veined leaves and large inflorescences and fruits, *Lindera bootanica* is unlikely to be confused with any other species. When treating *L. venosa* as conspecific with this species Long (Notes Roy. Bot. Gard. Edinburgh 41: 507. 1984) selected *L. bootanica* as the correct name.

8. *Litsea* Lam., Encycl. 3(2): 574 (1792).

Evergreen or deciduous shrubs, small trees or large trees. Perulate buds present or absent. Leaves alternate, evenly spaced along the twig or more or less clustered at the tips, secondary venation pinnate. Inflorescences umbellate, enclosed by decussate bracts, solitary or on lateral short-shoots. Flowers small, unisexual, trimerous. Hypanthium cup-shaped, not enclosing the ovary. Tepals usually 6, absent in one species. Male flowers usually with 9 stamens in 3 whorls. Stamens 4-valved, introrse, the outer whorls eglandular, inner whorl glandular. Rudimentary ovary present. Female flowers usually with 9 filiform staminodes in 3 whorls, the outer 6 eglandular and the inner 3 glandular. Fruit on a scarcely or distinctly enlarged perianth cup, tepals persistent or not.

Worldwide 300 species in tropical and subtropical Asia, Australia and America. Ten species in Nepal.

While umbels are in axillary clusters or in apparently racemose inflorescences, the rachis of these inflorescences are actually shoots with vegetative terminal buds. In some cases the buds develop into shoots, leaving the individual umbellate infructescences apparently solitary on the previous year's wood.

The following three species reported in Kostermans & Chater (Enum. Fl. Pl. Nepal 3: 185. 1982) are excluded from this account: *Litsea confertiflora* (Meisn.) Kosterm., based on *Beer, Lancaster & Morris 12329* (BM), determined in this revision as *Lindera heterophylla* Meisn.; *Litsea glabrata* (Wall. ex Nees) Hook.f., based on *Stainton 6518* (BM), determined as *L. panamanja* (Buch.-Ham. ex Nees) Hook.f.; and *Litsea lancifolia* (Roxb. ex Nees) Hook.f. included on the basis of the report by Shrestha (Bull. Dep. Med. Pl. 1: 42. 1967), itself apparently based on *Pradhan 4459* (KATH), determined here as *Lindera nacusua* (D.Don) Merr.

Key to Species

1a Perulate buds present, twigs with clusters of scars from bud scales (rather diffuse in *L. kingii*)..2
  b Perulate buds absent, twigs lacking clusters of scars from bud scales..4

2a Umbels with reflexed peduncles, solitary or several on short-shoots. Leaves glabrous, conspicuously shiny above..**2. *L. kingii***
  b Umbels with straight peduncles, solitary. Leaves more or less villous below, not conspicuously shiny above................3

3a Deciduous, leaves absent at time of flowering. Pedicels 2–9 mm in flower, 10–22 mm in fruit. Perulate bud scales glabrous...**1. *L. sericea***
  b Evergreen, leaves present at time of flowering. Pedicels to 2 mm in flower, 4–6 mm in fruit. Perulate bud scales sericeous ....**3. *L. elongata***

4a Umbels more or less racemosely arranged in open inflorescences (2–)3–9 cm long..5
  b Umbels in dense axillary clusters to 1.5 cm long or in dense inflorescences to 2 cm long on short, thick, persistent shoots 7

5a Umbels arranged in a subumbellate cluster at the end of 1–3 cm stalk...................................**11. *L. glutinosa***
  b Umbels racemosely arranged along length of stalk 6

6a Leaves not or scarcely glaucous below. Inflorescences 2–5 cm. Outer stamens 7–8 mm. Style 2.5 mm. Fruit ellipsoid or ovoid, not enclosed by enlarged cupule ....**4. *L. doshia***
  b Leaves glaucous below. Inflorescences 3–9 cm. Outer stamens 2–3.5 mm. Style 1 mm. Fruit globose, half enclosed by enlarged cupule ....**5. *L. panamanja***

7a Umbels densely arranged on persistent 2–3 mm-thick shoots to 15 mm long..8
  b Umbels in dense axillary clusters or on 1 mm-thick shoots to 8 mm long...9

8a Leaves broadly elliptic or obovate. Length:width ratio 1.3–2.5. Stamens 9..**6. *L. monopetala***
  b Leaves oblong or elliptic or slightly obovate. Length:width ratio 2–3. Stamens 12 ....**7. *L. hookeri***

9a Inflorescences with 6–15 umbels in axillary clusters or on shoots to 5 mm long. Secondary veins 10–13 pairs. Tertiary venation scalariform, faint or rather prominent..**8. *L. salicifolia***
  b Inflorescences with up to 6 umbels on shoots to 8 mm long. Secondary veins 6–10 pairs. Tertiary venation reticulate, rather faint..10

10a Leaves membranous, narrowly ovate or rarely elliptic. Fruit globose, to 6 mm on a scarcely enlarged perianth cup to 3 mm across...**9. *L. cubeba***
  b Leaves coriaceous, elliptic. Fruit ellipsoid, to 17 mm on an enlarged perianth cup 7–15 mm across......**10. *L. chartacea***

Lauraceae

1. Litsea sericea (Wall. ex Nees) Hook.f., Fl. Brit. Ind. 5[13]: 156 (1886).
Tetranthera sericea Wall. ex Nees, in Wall., Pl. Asiat. Rar. 2[8]: 67 (1831); *T. sericea* Wall. nom. nud.

Deciduous shrub or small trees to 8(–12) m. Twigs dark brown or blackish, smooth or lenticellate, tomentose or glabrous. Perulate buds to 18 mm. Leaves more or less clustered, elliptic, 4–13 × 1–4 cm, base cuneate, apex acute or slightly acuminate, secondary veins 4–12 pairs, tertiary venation scalariform or reticulate, more or less villous below, sometimes villous on veins above, often glabrescent; occasionally somewhat glaucous below. Petiole 0.7–1.8 cm. Umbels solitary, to 1.2 cm, produced before leaves emerge. Umbel buds 5–6 mm in diameter. Male umbels with 9–14 flowers. Peduncles length 2–4 mm. Male flowers pale yellow to mid yellow, 3–5 mm, sericeous. Pedicels 3–9 mm. Tepals oblong, 2.5–3 mm. Stamens 9, 1.5–2 mm. Female umbels with 6–8 flowers. Peduncles 2–4 mm. Female flowers yellow, 3 mm, sericeous. Pedicels 2–6 mm. Tepals oblong, 2 mm. Staminodes 9. Style 0.5 mm, glabrous. Infructescences with 2–4 fruit. Peduncles 5–8 mm. Pedicels 10–22 mm, 0.6–1.5 mm in diameter, evenly thickened or slightly thicker beneath fruit. Cupules 2–3 mm across. Fruits globose or ellipsoid, 5–7 mm.

Distribution: Nepal, E Himalaya, Tibetan Plateau, S Asia and E Asia.

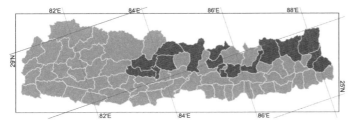

Altitudinal range: 1300–3900 m.

Ecology: Temperate forest with *Quercus*, *Rhododendron*, *Abies* and *Tsuga*.

Flowering: March–May(–July). **Fruiting:** June–August.

The combination of glabrous, perulate buds with blackish twigs and solitary umbels with long pedicels is characteristic of *Litsea sericea*. Some collections with reddish twigs and slender petioles are superficially similar to *L. cubeba* (Lour.) Pers., but the latter has glabrous leaves and the umbels are always clustered. Flowering specimens appear similar to *L. kingii* Hook.f., but the latter has shorter, glabrous pedicels and its young leaves are also glabrous. Fruiting material also has similarities with *Lindera nacasua* (D.Don) Meisn. and *Lindera assamica* (Meisn.) Kurz, but the former have shorter pedicels (2–6 mm) and the leaves of the latter have a shorter indumentum and more pronounced and scalariform tertiary venation. Umbel buds apparently develop over the summer and remain dormant till the spring flowering. Specimens with rufous indumentum on the lower surface of the leaves may prove to be distinct, and should be investigated further.

The plant is used as fodder.

2. Litsea kingii Hook.f., Fl. Brit. Ind. 5[13]: 156 (1886).

Deciduous trees to 15 m. Twigs dark reddish brown or blackish, smooth or slightly ridged, glabrous. Perulate buds to 17 mm. Leaves evenly spaced, elliptic or slightly obovate, 6–13 × 1.8–5 cm, base cuneate to attenuate, apex acute to acuminate or apiculate, secondary veins 9–12 pairs, tertiary venation reticulate, glabrous; underside glaucous or not. Petiole 0.5–1 cm. Inflorescences to 1.3 cm, produced before leaves emerge. Umbels solitary or 2–4 clustered on slender, 3–10 mm shoots. Umbel buds 4–5 mm in diameter. Male umbels with 5–8 flowers. Peduncles 5–11 mm. Male flowers yellow, 3 mm, glabrous. Pedicels 1–2 mm. Tepals ovate or obovate, sometimes clawed, 1.5–3 mm. Stamens 9, 2–3 mm. Female umbels with 4–5 flowers. Peduncles 6–11 mm. Female flowers 2–2.5 mm, glabrous. Pedicels 2–3 mm. Tepals ovate or obovate sometimes clawed, 2–3 mm. Staminodes 5–10. Style 1–1.5 mm, glabrous. Infructescences with 1–4 fruit. Pedicels 9–14 mm, 1–1.5 mm diameter, evenly thickened or slightly thicker beneath fruit. Cupules ca. 2 mm across. Fruit ovoid, obovoid or globose, 10 mm.
Fig. 5a.

Distribution: Nepal, E Himalaya, Tibetan Plateau, Assam-Burma and E Asia.

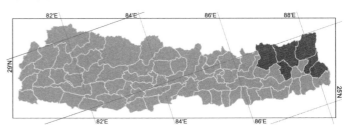

Altitudinal range: 1900–2900 m.

Ecology: Temperate forests with *Quercus* and *Alnus*.

Flowering: March–May. **Fruiting:** May–October.

Although treated as as a synonym of *Litsea cubeba* (Lour.) Pers. by Kostermans & Chater (Enum. Fl. Pl. Nepal 3: 185. 1982) *L. kingii* is a distinctive species. With its reflexed peduncles and leaves with shiny upper surfaces, *L. kingii* is unlikely to be confused with any other species of *Litsea* except for flowering specimens which appear similar to *L. sericea* (Wall. ex Nees) Hook.f. The latter has longer, silkier pedicels and its young leaves are silky. Umbel buds apparently develop over the summer and remain dormant till the spring flowering. The rings of scars left by the perulate bud scales are more diffuse than those of *L. sericea* and *L. elongata* (Nees) Hook.f.

3. Litsea elongata (Nees) Hook.f., Fl. Brit. Ind. 5[13]: 165 (1886).
Daphnidium elongatum Nees, in Wall., Pl. Asiat. Rar. 2[8]: 63 (1831); *Tetranthera elongata* Wall. nom. nud.

Evergreen trees to 30 m, 60 cm dbh. Twigs dark or mid brown, slightly ridged or rugose, tomentose. Perulate buds to 40 mm. Leaves evenly spaced, elliptic, oblong or slightly obovate,

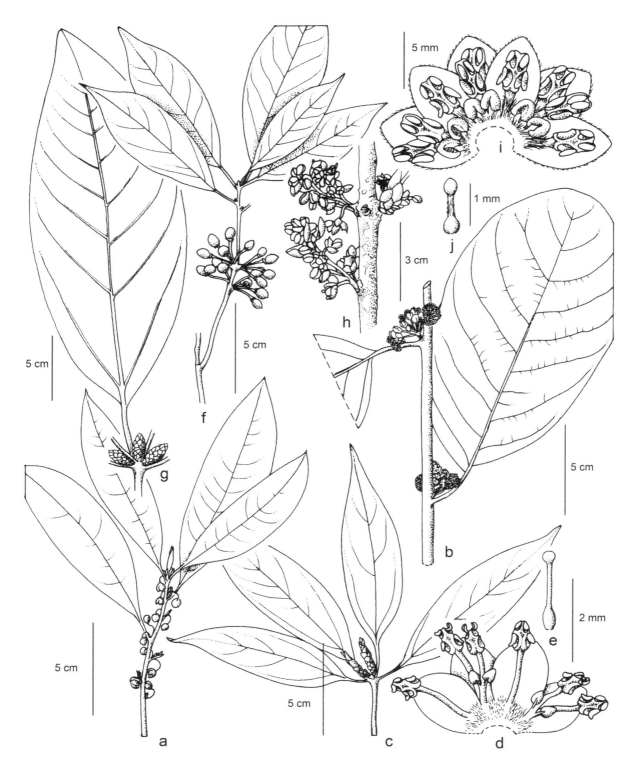

FIG. 5. LAURACEAE. **Litsea kingii**: a, shoot with flower buds. **Litsea monopetala**: b, twig with flowers. **Neolitsea foliosa**: c, shoot with perulate buds; d, opened flower with pistil removed; e, pistil. **Neolitsea pallens**: f, shoot with fruits. **Actinodaphne obovata**: g, shoot with perulate buds; h, flowering branch; i, opened flower with pistil removed; j, pistil.

6.5–15 × 1.5–4.5 cm, base cuneate, apex acute or slightly acuminate, secondary veins 5–12 pairs, tertiary venation scalariform, villous below, not glaucous. Petiole 0.8–1.5 cm. Umbels solitary, produced after leaves have emerged, to 1.5 cm long, buds 3–6 mm in diameter. Male umbels with ca. 6 flowers. Peduncles 4–5 mm. Male flowers 7 mm, sericeous. Pedicels to 2 mm. Tepals ovate or oblong, 3–4 mm. Stamens 8–11, outer stamens 4–6 mm, inner stamens 4–5 mm. Female umbels with ca. 3 flowers. Peduncles ca. 6 mm. Female flowers 3 mm, sericeous. Pedicels 1 mm. Tepals narrowly ovate, 1.5 mm. Style 2 mm, glabrous. Infructescences with 2–4 fruit. Peduncles 6–12 mm. Pedicels 4–6 mm, 1–2 mm diameter, markedly thicker beneath fruit. Cupule 6–7 mm across. Fruits 10 mm, barrel-shaped.

Distribution: Nepal, W Himalaya, E Himalaya, Tibetan Plateau, Assam-Burma and E Asia.

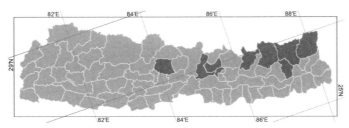

Altitudinal range: 1500–3700 m.

Ecology: Temperate forest with Fagaceae, *Rhododendron*.

Flowering: September–November. **Fruiting:** May–August.

Litsea elongata is usually easily recognisable by its leaves with very pronounced scalariform tertiary venation (secondary and even tertiary venation may be impressed above) and often densely villous undersides. It can be confused with *L sericea* (Wall. ex Nees) Hook.f., but *L. sericea* has glabrous bud scales and blackish twigs which are much less tomentose. Umbel buds apparently develop over the summer and remain dormant till the spring flowering.

4. *Litsea doshia* (D.Don) Kosterm., J. Sci. Res. [Jakarta] 1: 90 (1952).
Tetranthera doshia D.Don, Prodr. Fl. Nepal.: 65 (1825); *Litsea oblonga* (Wall. ex Nees) Hook.f.; *Tetranthera oblonga* Wall. ex Nees; *T. oblonga* Wall. nom. nud.; *Tomex doshia* Buch.-Ham. nom. nud.

काठे काउलो Kathe kaulo (Nepali).

Trees to 16 m, 70 cm dbh. Twigs pale brown or whitish, more or less lenticellate or scarcely lenticellate, glabrous. Perulate buds absent. Leaves evenly spaced, oblong or elliptic to slightly ovate or slightly obovate, 7–16 × 2–5 cm, base cuneate, apex acute, secondary veins 8–11 pairs, tertiary venation reticulate or reticulate-scalariform, glabrous, underside not or slightly glaucous. Petiole 1.4–1.9 cm. Inflorescences to 2–5 cm long, umbels racemosely arranged on twig-like 2–25 mm shoots, produced after leaves have emerged. Umbel buds 5–6 mm in diameter. Male umbels

with 4–6 flowers. Peduncles 8–18 mm. Male flowers yellow, 6–8 mm, sericeous. Pedicels to 1 mm. Tepals oblong, 2.5–4 mm. Stamens 9–13, outer stamens 7–8 mm, inner stamens c. 5 mm. Female umbels with 2–5 flowers. Peduncles 5–6 mm. Female flowers pale yellow or white, 4 mm, sericeous. Pedicels 2 mm. Tepals ovate or oblong, 1.5–2.5 mm. Staminodes 9–10. Style 2.5 mm, glabrous. Infructescences with 1–9 fruit. Peduncles 8–10(–15) mm. Pedicels 8–15(–20) mm, 2–3 mm diameter, often markedly thicker beneath fruit. Cupule ca. 9 mm across. Fruits ellipsoid or ovoid, 12–20 mm.

Distribution: Endemic to Nepal.

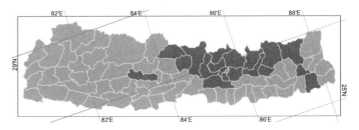

Altitudinal range: 300–2900 m.

Ecology: Temperate forests with Fagaceae, *Rhododendron*, *Betula* and *Abies*.

Flowering: (May–)August–December. **Fruiting:** May–August(–October).

Litsea doshia is apparently endemic to Nepal. Records of its presence in E Himalaya, Assam and Burma relate to *L. albescens* (Hook.f) D.G.Long (Fl. Bhutan 1: 277. 1984).

    Litsea doshia is characterized by its pale twigs and leaves with yellow midribs and secondary venation which contrast with the green of the glossy upper surfaces. With its open, racemose inflorescence, *L. doshia* is most similar to *L. panamanja*, but the latter has much larger inflorescences which are up to 20 cm in some collections from Burma. Kostermans & Chater (Enum. Fl. Pl. Nepal, 3: 183. 1982) recorded *Cinnadenia paniculata* (Hook.f.) Kosterm. from Nepal based on *Kanai & Shrestha 672656* (TI & KATH), a collection now identified as *L. doshia*.

    The plant is used as fodder.

5. *Litsea panamanja* (Buch.-Ham. ex Nees) Hook.f., Fl. Brit. Ind. 5[13]: 175 (1886).
Tetranthera panamanja Buch.-Ham. ex Nees, in Wall., Pl. Asiat. Rar. 2[8]: 67 (1831); *T. panamanja* Buch.-Ham. ex Wall. nom. nud.

Evergreen trees to 20 m. Twigs whitish, pale brown to dark brown, smooth or slightly ridged, glabrous. Perulate buds absent. Leaves evenly spaced, oblong, elliptic or slightly obovate, 9.5–28 × 3.8–10 cm, base cuneate, sometimes oblique, apex acute or obtuse, slightly acuminate, secondary veins 7–12 pairs, tertiary venation reticulate or reticulate-scalariform, glabrous; underside more or less glaucous. Petiole 1–2.5 cm. Inflorescences 3–9 cm, with 4–15 umbels racemosely arranged on 3.5–40 mm shoots, produced after leaves have emerged. Umbel buds 3–5 mm in diameter. Male umbels with 6–8 flowers. Peduncles 7–16 mm. Male flowers yellow, 7 mm, sparsely sericeous. Pedicels 1–2 mm. Tepals oblong, 2–3 mm. Stamens 11–12, outer stamens 2–3.5 mm, inner stamens 1.5–2.5 mm. Female umbels with 4–6 flowers. Peduncles length 3–10 mm. Female flowers 2–2.5 mm, sericeous. Pedicels 1 mm. Tepals obovate, 1 mm. Staminodes 12. Style 1 mm, glabrous. Infructescences with 1–3 fruit. Peduncles 5–10 mm. Pedicels 4–5 mm, 3–5 mm diameter, markedly thicker beneath fruit. Cupule 15–20 mm across. Fruit globose, 15 mm.

Distribution: Nepal, E Himalaya, Assam-Burma, E Asia and SE Asia.

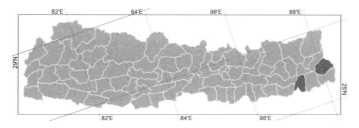

Altitudinal range: 700–800 m.

Ecology: Subtropical forest.

Flowering: March–May.

Litsea panamanja is most similar to *L. doshia* (D.Don) Kosterm., but has larger inflorescences and leaves with glaucous undersides. In some collections the inflorescence shoots develop small leaves along with the umbels.

6. *Litsea monopetala* (Roxb.) Pers., Syn. Pl. 2(1): 4 (1806).
Tetranthera monopetala Roxb., Pl. Coromandel 2[6]: 26, pl. 148 (1800); *Litsea polyantha* Juss.; *Tetranthera macrophylla* Roxb.; *T. quadriflora* Roxb.

कुटमेरो Kutmero (Nepali).

Evergreen trees to 15 m. Twigs mid brown or reddish, densely tomentose, glabrescent. Perulate buds absent. Leaves evenly spaced or slightly clustered, broadly elliptic or obovate, 11–14 × 6–11 cm, base rounded or slightly cordate, apex obtuse, secondary veins 7–12 pairs, tertiary venation scalariform, villous below; underside not glaucous. Petiole 1.8–2 cm. Inflorescences to 2 cm, with 2–8 umbels densely arranged on stout, 3–15 mm long shoots, produced after leaves have emerged. Umbel buds 3–5 mm in diameter. Male umbels with 6–8 flowers. Peduncles to 6 mm. Male flowers green or yellow, 4 mm, sericeous. Pedicels 1–1.5 mm. Tepals strap shaped, 2–2.5 mm. Stamens 9, outer stamens 3 mm, inner stamens 2 mm. Female umbels with 4–9 flowers. Peduncles 4–6 mm. Female flowers pale yellow-green, 3–3.5 mm, sericeous. Pedicels 2–2.5 mm. Tepals strap shaped, 2 mm. Staminodes 9. Style 1.5 mm, glabrous. Infructescences with 2–5 fruit. Peduncles 5–7 mm. Pedicels 6–12 mm, diameter 1–1.5 mm, slightly thickened beneath fruit. Cupules 4–5 mm across. Fruit ovoid, 10 mm.
Fig. 5b.

Distribution: Nepal, W Himalaya, E Himalaya, Assam-Burma, S Asia, E Asia and SE Asia.

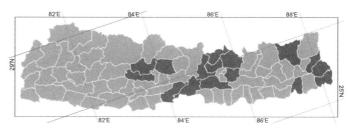

Altitudinal range: 80–2100 m.

Ecology: *Schima-Castanopsis* forest and evergreen and deciduous subtropical riverine forests.

Flowering: January–June(–October). **Fruiting:** April–July.

Litsea monopetala and *L. hookeri* (Meisn.) D.G.Long have the same type of inflorescence characterised by persistent, stout, tomentose inflorescence shoots with evident peduncle scars. *L. monopetala* has relatively broader leaves with a denser, reddish tomentum.

Leafy branches are used as fodder. The bark is astringent and used to treat diarrhoea. Powdered bark and roots are used in external applications to relieve pain.

7. *Litsea hookeri* (Meisn.) D.G.Long, Notes Roy. Bot. Gard. Edinburgh 41(3): 510 (1984).
Cylicodaphne hookeri Meisn., in A.DC., Prodr. 15(1): 209 (1864); *Litsea khasyana* Meisn.; *L. meissnerii* Hook.f. nom. superfl.; *Tetranthera khasyana* Meisn.

Evergreen trees to 20 m. Twigs pale brown, slightly ridged, tomentose. Perulate buds absent. Leaves evenly spaced, oblong or elliptic or slightly obovate, 17–22 × 5.5–9 cm, base cuneate, apex acute, secondary veins 10–12 pairs, tertiary venation reticulate-scalariform, tomentose below on midrib and veins and sparsely pubescent on lamina; underside slightly glaucous. Petiole 1.2–1.8 cm. Inflorescences to 1.5 cm, with 2–8 umbels densely arranged on stout, 4–8 mm shoots, produced after leaves have emerged. Umbel buds 3–5 mm in diameter. Male umbels with 5–8 flowers. Peduncles 10 mm. Male flowers green, 5 mm, sericeous. Pedicels 2–2.5 mm. Tepals oblong or obovate, 3–3.5 mm. Stamens 12, outer stamens 3 mm, inner stamens 1.5 mm. Female umbels with 5–8 flowers. Peduncles 6–8 mm. Female flowers yellow, 3.5 mm, tomentose outside and glabrous inside, pedicels 2.5–3 mm. Tepals ovate, 2 mm. Staminodes 10–11. Style 2 mm, glabrous. Fruit not seen.

Distribution: Nepal, E Himalaya and Assam-Burma.

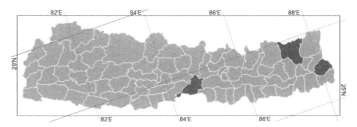

Lauraceae

Altitudinal range: 200–800 m.

Ecology: Subtropical forests.

Flowering: March–May.

Litsea hookeri is clearly close to *L. monopetala* (Roxb.) Pers. as both of them are characterized by persistent, stout, tomentose inflorescence shoots with evident peduncle scars. *L. hookeri* is distinguished by its more elongate leaves which are pale below, cuneate at the base and with a sparse, pale brown, not reddish tomentum.

8. *Litsea salicifolia* (Roxb. ex Nees) Hook.f., Fl. Brit. Ind. 5[13]: 167 (1886).
Tetranthera salicifolia Roxb. ex Nees, in Wall., Pl. Asiat. Rar. 2[8]: 66 (1831); *T. glauca* Wall. nom. nud.; *T. glauca* Wall. ex Nees.

पहेँली Pahenli (Nepali).

Evergreen trees to 10 m. Twigs pale to dark brown, smooth or slightly ridged, tomentose, glabrescent. Perulate buds absent. Leaves evenly spaced, elliptic to oblong or slightly obovate or slightly ovate, 6.2–18 × 1.6–5 cm, base cuneate or slightly rounded, apex acute to obtuse or acuminate, secondary veins 7–13 pairs, tertiary venation scalariform, sometimes sericeous below, underside glaucous or not. Petiole 1–4 cm. Inflorescences to 1.2 cm long, with 6–15 umbels in sessile clusters or densely arranged on slender, 2–5 mm shoots, produced after leaves have emerged. Umbel buds 2–4 mm in diameter. Male umbels with 4–5 flowers. Peduncles 4–7 mm. Male flowers white or yellowish green, glabrous, 4 mm. Pedicels 1–1.5 mm. Tepals ovate or oblong, 2–2.5 mm. Stamens 4–7, outer stamens 4 mm, inner stamens 2.5–4 mm. Female umbels with ca. 4 flowers. Peduncles 3–4 mm. Female flowers yellow or green, 3 mm, glabrous. Pedicels 1 mm. Tepals elliptic, sometimes slightly clawed, 2 mm. Staminodes 10. Style 2 mm, glabrous. Infructescences with 1–4 fruit. Peduncles 5–6 mm. Pedicels 3–5 mm, 1–2 mm diameter, slightly thickened beneath fruit. Cupule 3–4 mm across. Fruit ovoid, 10 mm.

Distribution: Nepal, E Himalaya, Assam-Burma, S Asia, E Asia and SE Asia.

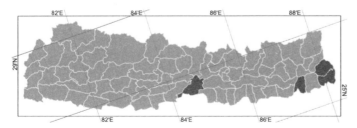

Altitudinal range: 100–800 m.

Ecology: Subtropical deciduous or evergreen forest. Often associated with rivers.

Flowering: March. **Fruiting:** May.

Momiyama (Fl. E. Himalaya: 102. 1966) misapplied the name *Litsea oblonga* (Wall. ex Nees) Hook.f. to this species.

A vegetatively variable species, *L. salicifolia* can have large leaves with glaucous undersides and prominent secondary venation or small leaves with much less distinct venation. The axillary clusters of many umbels are characteristic, as are the almost glabrous flowers. Whilst it is usually recorded as a small tree, one collection, *Troth 708* (KATH), refers to it as a large tree. The plants with large leaves and prominent scalariform venation are perhaps distinct from those with the glaucous undersides.

9. *Litsea cubeba* (Lour.) Pers., Syn. Pl. 2(1): 4 (1806).
Laurus cubeba Lour., Fl. Cochinch. 1: 252 (1790); *Actinodaphne citrata* (Blume) Hayata; *Benzoin cubeba* (Lour.) Hatus.; *Daphnidium cubeba* (Lour.) Nees; *Lindera aromatica* Brandis; *L. dielsii* Lév.; *Litsea citrata* Blume; *L. dielsii* (Lév.) Lév.; *Persea cubeba* (Lour.) Spreng.; *Tetranthera citrata* (Blume) Nees; *T. cubeba* (Lour.) Kostel.; *T. polyantha* Wall. ex Nees; *T. polyantha* Wall. nom. nud.

सिल्टीमूर Siltimur (Nepali).

Evergreen trees to 18 m, 34 cm dbh. Twigs dark, often reddish brown, smooth or slightly ridged, glabrous. Perulate buds absent. Leaves evenly spaced, narrowly ovate or elliptic, 7–11.5 × 1.8–3.5 cm, base rounded to cuneate or slightly attenuate, apex acute to acuminate, secondary veins 6–10 pairs, tertiary venation reticulate, glabrous, underside not or slightly glaucous. Petioles 1–1.6 cm. Inflorescences to 1 cm long, 1–6(–10) umbels densely arranged on slender, 2–8 mm shoots, produced after leaves have emerged. Umbel buds 2–4 mm in diameter. Male umbels with ca. 5 flowers. Peduncles 4–7 mm. Male flowers 4 mm, glabrous. Pedicels to 2 mm. Tepals oblong, 2.5 mm. Stamens 9, to 3 mm. Female umbels with 3–6 flowers. Peduncles 4–6 mm. Female flowers 1.5–3 mm, sericeous. Pedicels 1–1.5 mm. Tepals obovate, oblong or ovate, 1–1.5 mm. Staminodes 9. Style 1 mm, glabrous. Infructescences with up to 7 fruit. Peduncles 6–7 mm. Pedicels 2–6 mm, 1 mm diameter, evenly-thickened. Cupules 1.5–3 mm across. Fruits globose, 5–6 mm.

Distribution: Nepal, E Himalaya, Assam-Burma and E Asia.

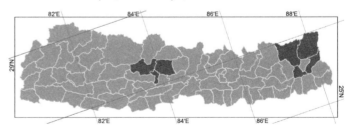

Altitudinal range: 1000–2600 m.

Ecology: Temperate forests or scrub with *Quercus*.

Flowering: November–March. **Fruiting:** July–November.

Litsea cubeba is easily recognized by its slender, smooth, often reddish twigs and slender petioles. It is most similar to *L. chartacea* (Wall. ex Nees) Hook.f., but is distinguished by its papery leaves and smaller fruits. It is distinguished from *L. salicifolia* (Wall. ex Nees) Hook.f. by its smaller, narrowly ovate leaves with fine secondary venation. Some individuals of *L. sericea* are vegetatively similar, but its leaves are always at least slightly hairy below.

The species has many medicinal properties and is used in the treatment of coughs, bronchitis and lung diseases. The edible fruits are aromatic and carminative and are used for dizziness, headache, hysteria, paralysis and loss of memory. An essential oil extracted from the seeds is used for genitourinary conditions such as cystitis and gonorrhoea.

10. *Litsea chartacea* (Wall. ex Nees) Hook.f., Fl. Brit. Ind. 5[13]: 170 (1886).

Tetranthera chartacea Wall. ex Nees, in Wall., Pl. Asiat. Rar. 2[8]: 67 (1831); *T. chartacea* Wall. nom. nud.

Evergreen trees to 8 m. Twigs dark, often reddish brown, smooth or slightly ridged, glabrous. Perulate buds absent. Leaves evenly spaced, elliptic, 8–16 × 2.5–6.5 cm, base cuneate, apex acute to acuminate, secondary veins 6–9 pairs, tertiary venation reticulate, glabrous or minutely pubescent below, underside more or less glaucous. Petioles 1–1.5 cm. Inflorescences to 1.5 cm long, 1–5 umbels densely arranged on slender, 2–5 mm shoots, produced after leaves have emerged. Umbel buds 4–6 mm in diameter. Male umbels with 5–7 flowers. Peduncles 5–12 mm. Male flowers 5 mm, sericeous. Pedicels ca. 1 mm. Tepals oblong, to 2.5 mm. Stamens 9, to 4 mm. Female flowers not seen. Infructescences with up to 4 fruit. Peduncles to 12 mm. Pedicels to 15 mm, to 2 mm diameter, thicker towards apex. Cupules 7–15 mm across. Fruits ellipsoid, to 17 mm.

Distribution: Nepal and E Himalaya.

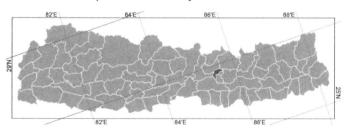

Altitudinal range: 1500–1800 m.

Ecology: Warm broad-leaved forests.

Flowering: May–June. **Fruiting:** October.

Litsea chartacea is an infrequently collected species most similar to *L. cubeba* (Lour.) Pers., but distinguished by its larger, more leathery elliptic leaves, and by its very much larger fruits.

11. *Litsea glutinosa* (Lour.) C.B.Rob., Philipp. J. Sci. 6: 321.

Sebifera glutinosa Lour., Fl. Cochinch. 2: 638 (1790); *Laurus involucrata* Roxb. later homonym, non Lam.; *Litsea sebifera* (Willd.) Pers.; *L. sebifera* var. *glabraria* Hook.f.; *L. tetranthera* (Willd.) Pers.; *Tetranthera apetala* Roxb.; *T. involucrata* (Roxb.) Nees ex Sweet; *T. laurifolia* Jacq.; *T. polycephala* Wall. ex Meisn.; *T. roxburghii* Nees; *Tomex sebifera* Willd.; *T. tetranthera* Willd.

मैदा Maida (Nepali).

Deciduous trees to 15 m. Twigs pale brown to blackish, smooth or slightly ridged, tomentose, glabrescent. Perulate buds absent. Leaves evenly spaced or slightly clustered, elliptic or oblong to ovate or obovate, 6.5–19 × 2.3–7 cm, base cuneate or rounded, rarely oblique, apex acute or obtuse, sometimes slightly acuminate, secondary veins 6–12 pairs, tertiary venation scalariform or reticulate, glabrous or pubescent below and on veins and midrib above; underside not or slightly glaucous. Petiole 1.3–3 cm. Inflorescences 2.5–6 cm, with 3–7 umbels in a subumbellate cluster at the end of 1.5–4 cm shoots, produced after leaves have emerged. Umbel buds 3–7 mm in diameter. Male umbels with ca. 12 flowers. Peduncles 12–18 mm. Male flowers green or yellow, 6 mm, sericeous. Pedicels 4 mm. Tepals absent. Stamens 18, outer stamens 4–5 mm, inner stamens 3–4 mm. Female umbels with 8–9 flowers. Peduncles 4–11 mm. Female flowers pale green, 2–2.5 mm, sericeous. Pedicels 1.5–2 mm. Tepals absent or strap-shaped, ca. 1 mm. Staminodes 12–14. Style 1.5–2 mm, glabrous. Infructescences with 2–3 fruit. Peduncles 8–11 mm. Pedicels 4–5 mm, 1–2 mm diameter, slightly thicker beneath fruit. Fruit globose, 9 mm.

Distribution: Nepal, W Himalaya, E Himalaya, Assam-Burma, S Asia, E Asia and SE Asia.

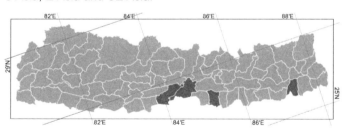

Altitudinal range: 100–1000 m.

Ecology: Subtropical mixed forests and woodlands.

Flowering: March–July. **Fruiting:** August.

Litsea glutinosa is a vegetatively variable species and Hooker (Fl. Brit. Ind. 5: 157. 1886) recognized three varieties (within *Litsea sebifera*), two of which are present in Nepal. Glabrous, oblong-leaved specimens are referable to *L. sebifera* var. *sebifera*, while the ovate-leaved more tomentose specimens are *L. sebifera* var. *glabraria*. These varieties may be transferable to *L. glutinosa*, but more research is needed. Although tepals may be irregularly formed in other species (especially *L. doshia* (D.Don) Kosterm.), *L. glutinosa* is the only species which regularly has no tepals whatsoever.

9. *Neolitsea* (Benth.) Merr. nom cons., Philipp. J. Sci. Suppl. 1: 56 (1906).

Evergreen or deciduous large or small trees. Perulate buds present. Leaves alternate or subopposite, more or less clustered at the tips, secondary venation strongly or weakly 3-veined. Inflorescences umbellate, enclosed by decussate bracts, solitary or in clusters on lateral short-shoots. Flowers small, unisexual, dimerous. Hypanthium cup-shaped, not enclosing the ovary. Tepals 4. Male flowers usually with 6 stamens in 3 whorls. Stamens 4-valved, introrse, the outer whorls eglandular, inner whorl glandular. Rudimentary ovary present. Female flowers usually with 6 filiform staminodes in 3 whorls, the outer whorls eglandular and the inner whorl glandular. Fruit on a scarcely enlarged perianth cup, tepals not persistent.

Worldwide 80 species in Indomalesia, E Asia and Australia. Three species in Nepal.

Key to Species

1a Leaves to 28 cm long, hairy below, at least around the veins, with inrolled margins, very strongly clustered at tips of twigs and below previous year's bud scars. Fruits clearly ovoid ....**1. *N. cuipala***

b Leaves to 15 cm long, completely glabrous, with flat margins, weakly clustered towards tips of twigs or rather evenly spaced along twig. Fruits globose or slightly ellipsoid...2

2a Largest leaves to 15 cm, strongly 3-veined at base .... **2. *N. foliosa***

b Largest leaves to 10 cm, weakly 3-veined at base so that the basal pair of veins is not clearly distinguished from the other veins ....**3. *N. pallens***

1. *Neolitsea cuipala* (D.Don) Kosterm., Bull. Bot. Surv. India 10: 287 (1968).
Tetranthera cuipala D.Don, Prodr. Fl. Nepal.: 65 (1825); *Laurus cuipala* Buch.-Ham. nom. nud.; *Litsea cuipala* (D.Don) Nees; *L. lanuginosa* (Nees) Nees; *Neolitsea lanuginosa* (Nees) Gamble; *Tetradenia lanuginosa* Nees; *Tetranthera lanuginosa* Wall. nom. nud.

Trees to 20 m. Twigs glabrous or minutely tomentose. Resting buds to 2 mm, glabrous or sericeous. Perulate buds 10–25 mm, sericeous. Leaves strongly clustered, elliptic to slightly obovate or slightly ovate, 8.5–28 × 2.3–6.5 cm, base cuneate, apex acute or slightly acuminate, margin flat or inrolled, more or less tomentose or strigose below, underside glaucous; clearly 3-veined secondary veins 3–6 pairs. Petioles 0.6–2.5 cm. Male inflorescences ca. 6-flowered, clusters to 1.6 cm across. Flowers 8 mm, sericeous. Pedicels 3 mm. Tepals broadly or narrowly ovate with acute apex, 4–5 mm. Outer whorl of anthers 7–9 mm, inner whorl 6–8 mm. Rudimentary ovary and style glabrous. Female inflorescence 5–6-flowered, clusters to 1.5 cm across. Flowers 3 mm long, sericeous. Pedicels 3–8 mm. Tepals narrowly triangular, 2.5–3 mm. Staminodes 1–2 mm. Ovary and style glabrous. Infructescences with up to 12 fruit. Pedicels 5–7 mm, evenly thickened. Perianth remains 4 mm in diameter. Fruits ovoid, to 13 mm.

Distribution: Nepal, W Himalaya, E Himalaya and Assam-Burma.

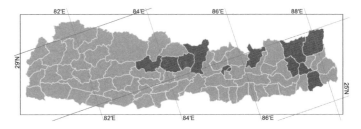

Altitudinal range: 900–2100 m.

Ecology: Warm temperate and subtropical forests. Frequently with *Castanopsis*.

Flowering: March. **Fruiting:** May–October.

Neolitsea cuipala is easily identified by its very large and strongly clustered, 3-veined leaves. Bud size was used to differentiate between *N. cuipala* and *N. foliosa* (Nees) Gamble by Long (Fl. Bhutan 1: 278. 1984) but this character is misleading as the size of the buds varies enormously according to the season. Perulate buds are most prominent between January and April, though early stages may also be seen between August and December. They are not present between May and July. Trees flower as perulate buds are developing, so flowers are on the current twig. By the time fruit has been set the bud has expanded and the new twig grown, so that fruits are on the previous year's wood.

Oil from the seed is applied to boils and scabies.

2. *Neolitsea foliosa* (Nees) Gamble, Fl. Madras 2[7]: 1240 (1925).
Tetradenia foliosa Nees, in Wall., Pl. Asiat. Rar. 2[8]: 64 (1831); *Litsea foliosa* (Nees) Nees; *Tetranthera foliosa* Wall. nom. nud.

Small trees to 10 m, or rarely large trees to 20 m. Twigs glabrous. Resting buds to 2 mm, glabrous. Perulate buds to 33 mm, sericeous. Leaves evenly spaced or strongly clustered, elliptic or oblong to slightly ovate or slightly obovate, 7–15 × 1.4–4.8 cm, base cuneate, apex acuminate, margin flat or inrolled, glabrous; underside glaucous; clearly 3-veined secondary veins 3–5 pairs. Petioles 0.8–2 cm. Male inflorescences ca. 5-flowered, clusters to 1.2 cm across. Flowers yellow, 4–6 mm, sparsely sericeous.

Pedicels 2–4 mm. Tepals ovate, 2.5–4 mm, acute or obtuse. Outer whorl of anthers ca. 5 mm, inner whorl 3–4 mm. Rudimentary ovary and style glabrous. Female inflorescences 4–10-flowered, clusters to 0.7–1 cm. Flowers pale yellow or green, ca. 3 mm long, sericeous. Pedicels 2–5 mm. Tepals broadly or narrowly ovate, 2 mm. Staminodes 1–2 mm. Ovary glabrous or hairy at apex. Style glabrous or hairy. Infructescences with 2–7 fruits. Pedicels 6–12 mm, rather thickened beneath the fruit. Perianth remains 4–5 mm in diameter. Fruits globose or ellipsoid, 10–13 mm. Fig. 5c–e.

Distribution: Nepal, E Himalaya and Assam-Burma.

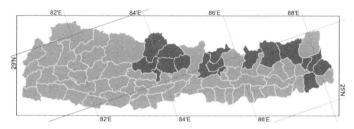

Altitudinal range: 1500–3500 m.

Ecology: Moist evergreen broad-leaved forest, often with *Quercus lamellosa*, sometimes rather abundant.

Flowering: March–April(–October). **Fruiting:** (May–) July–October.

Neolitsea foliosa and *N. pallens* (D.Don) Moimy. & H.Hara ex H.Hara are very close, and are distinguished by the larger more coriaceous leaves of the former.

3. *Neolitsea pallens* (D.Don) Momiy. & H.Hara ex H.Hara, J. Jap. Bot. 47: 269 (1972).
Tetranthera pallens D.Don, Prodr. Fl. Nepal.: 66 (1825); *Litsea consimilis* (Nees) Nees; *L. umbrosa* var. *consimilis* (Nees) Hook.f.; *Neolitsea umbrosa* (Nees) Gamble; *Tetradenia consimilis* Nees; *Tetranthera umbrosa* Wall. nom. nud.

Small trees to 10 m, or rarely large trees to 25 m. Twigs glabrous or minutely tomentose. Resting buds to 2 mm, glabrous or sericeous. Perulate buds 12–25 mm, sericeous. Leaves more or less clustered, elliptic or oblong to slightly ovate or slightly obovate, 4–9.5 × 1.2–3 cm, base cuneate or rarely rounded, apex acute to acuminate, margin flat, glabrous; underside usually glaucous; weakly 3-veined, secondary veins 3–6 pairs. Petioles 0.7–1.7 cm. Male inflorescences 4–6-flowered, clusters to 1–1.3 cm. Flowers yellow or orange-yellow, 4–7 mm, sericeous or sparsely sericeous. Pedicels 2–4 mm. Tepals ovate, 2.5–4 mm, acute or obtuse. Outer whorl of anthers 4.5–5 mm, inner whorl 3.5–6 mm. Rudimentary ovary hairy at apex. Style more or less hairy. Female inflorescence ca. 6-flowered, clusters to 0.8 cm. Flowers yellow, 2.5 mm long, sericeous. Pedicels ca. 3 mm. Tepals narrowly ovate, 2.5 mm. Staminodes 1 mm. Ovary glabrous or hairy at apex. Style slightly hairy. Infructescences with 1–7 fruits. Pedicels 6–13 mm, evenly thickened or slightly thicker beneath fruit. Perianth remains 3 mm in diameter. Fruits ellipsoid, 9–10 mm. Fig. 5f.

Distribution: Nepal, W Himalaya and Tibetan Plateau.

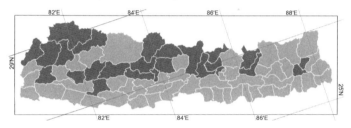

Altitudinal range: 1700–3500 m.

Ecology: *Quercus-Rhododendron* forest, with *Quercus lamellosa* or *Q. semecarpifolia*. In gullies and exposed sites, sometimes very abundant.

Flowering: (December–)February–June. **Fruiting:** May–October(–January).

Although it is close to *Neolitsea foliosa* (Nees) Gamble, *N. pallens* is distinguished by its smaller, more chartaceous leaves.
A paste of the fruit is used to treat skin diseases.

10. *Dodecadenia* Nees, in Wall., Pl. Asiat. Rar. 2[8]: 61 (1831).

Small, evergreen trees. Perulate buds present. Leaves alternate, evenly spaced along the twig, secondary venation pinnate. Flowers large, solitary or rarely in clusters of up to 3, each flower enclosed by numerous persistent bracts, bisexual or rarely unisexual, trimerous. Hypanthium cup-shaped, not enclosing the ovary. Tepals usually 6. Stamens usually 12 in 4 whorls, 4-valved, all introrse. Outer whorls eglandular, inner whorls glandular. Female flowers with 12 staminodes, the inner whorls glandular. Fruit on an enlarged perianth cup, tepals not persistent.

A monotypic genus from the Himalayas.

Lauraceae

1. *Dodecadenia grandiflora* Nees, in Wall., Pl. Asiat. Rar.
2[8]: 63 (1831).
Dodecadenia grandiflora var. *griffithii* (Hook.f.) D.G.Long;
D. griffithii Hook.f.; *Laurus macrophylla* D.Don; *Tetranthera grandiflora* Wall. nom. nud.

काउले Kaule (Nepali).

Trees to 15 m. Young twigs glabrous or greyish tomentose, often reddish, becoming grey brown, rugose and lenticellate. Perulate buds present. Leaves elliptic or oblong to slightly obovate, 7–14 × 1.5–4.5 cm, length:width ratio 3.3–4.7, apex slightly acuminate, base cuneate, secondary veins 7–11 pairs, tertiary venation reticulate, glabrous, underside glaucous or not. Petioles 0.8–1.5 cm. Flowers pale green or yellow, bisexual or rarely unisexual. Bisexual and male inflorescences to 1.5 cm, female inflorescences to 1 cm. Peduncles 2–3 mm. Pedicels 2 mm. Bisexual and male flowers 8–10 mm. Tepals 6–8, 6–10 mm, or elliptic or oblong, sericeous inside and out. Stamens 7–10 mm, those of the outer whorls longest. Ovary 2 mm, glabrous or hairy. Style 5 mm, sparsely hairy. Female flowers to 7 mm, with 4–5 mm tepals. Staminodes 3–4 mm. Infructescences to 3 cm, with 1(–3) fruit. Peduncles 2–4 mm. Pedicels stout, to 7 mm. Perianth remains 5–9 mm across. Fruit ellipsoid, 13–15 mm.

Distribution: Nepal, E Himalaya, Tibetan Plateau, Assam-Burma and E Asia.

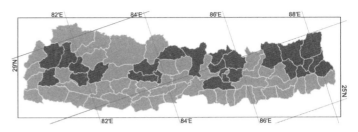

Altitudinal range: 600–3400 m.

Ecology: Temperate *Quercus*-Lauraceae forests with *Rhododendron* and *Tsuga*.

Flowering: February–April(–September). **Fruiting:** May–October.

Each flower is solitary within its bracts, but sometimes 2 or very rarely 3 are grouped together in a single cluster. Occasionally unisexual flowers are found, with male and female flowers on different trees. Female inflorescences are very much smaller (less than 1 cm long) and the flowers have 4–5 mm long tepals and 12 3–4 mm long staminodes. The male flowers have normally sized ovaries and therefore appear to be bisexual, but as ovules were not found in their ovaries they are functionally male.

11. *Actinodaphne* Nees, in Wall., Pl. Asiat. Rar. 2[8]: 68 (1831).

Large or small evergreen trees. Perulate buds present. Leaves whorled or alternate and densely clustered, secondary venation pinnate or weakly 3-veined. Flowers small, unisexual, trimerous, in panicles or umbels enclosed by decussate bracts, the umbels solitary or in cluster on lateral short-shoots. Hypanthium cup-shaped, not enclosing the ovary. Tepals 6. Male flowers usually with 9 stamens in 3 whorls. Stamens 4-valved, introrse, the outer whorls eglandular, inner whorl glandular. Rudimentary ovary present. Female flowers usually with 9 staminodes in 3 whorls. Fruit on an enlarged perianth cup, tepals not persistent.

Worldwide 100 species in Indomalesia and E Asia. Four species in Nepal.

The leaves are more strongly whorled than in *Litsea*, otherwise these genera can be rather difficult to separate. *Actinodaphne reticulata* Meisn. was reported from Sankhuwasaba by Hara (Fl. E. Himalaya: 99. 1966), but it has not been possible to verify this record and it is likely that it should be referred to *A. longipes* Kosterm. and so is excluded from this account.

Key to Species

1a Inflorescences paniculate. Leaves weakly 3-veined, more than 8 cm wide. Twigs 5–10 mm thick .... **2. *A. obovata***
b Inflorescences umbellate. Leaves pinnate veined, up to 6 cm wide. Twigs up to 5 mm thick, usually much finer 2

2a Leaves to 2.5–6 cm broad, with 5–8 pairs of secondary veins. Umbels on 3–5 mm peduncles............**1. *A. angustifolia***
b Leaves to 3 cm broad, with 9–23 pairs of secondary veins. Umbels sessile, solitary or several on short-shoots 3

3a Leaves with 11–23 pairs of secondary veins and length:width ratio > 4...**3. *A. longipes***
b Leaves with 9–11 pairs of secondary veins and length:width ratio < 4...**4. *A. sikkimensis***

1. *Actinodaphne angustifolia* (Blume) Nees, in Wall., Pl. Asiat. Rar. 3[11]: 31 (1832).
Litsea angustifolia Blume, Bijdr. Fl. Ned. Ind. [11]: 566 (1826); *Tetranthera angustifolia* (Blume) Nees.

लामपाते Lampate (Nepali).

Trees to 20 m. Twigs slender, to 4 mm in diameter, mid brown or pale brown, smooth, densely orange tomentose when young tomentose, glabrescent. Perulate buds present. Leaves elliptic, oblong or slightly obovate, 10–22 × 2.5–6 cm, base cuneate, apex more or less acuminate, margin sometimes inrolled at base, otherwise flat, pinnate veined, secondary veins 5–8 pairs, tertiary venation scalariform, rather villous on midrib and veins below, sometimes also on lamina, underside glaucous. Petioles 1–2.2 cm. Umbels on 2–7 mm short-shoots. Male inflorescence not seen. Female inflorescences with 2–4 umbels. Female flowers not seen. Infructescences with 3–12 globose, 10 mm-long fruit on perianth cups 4–5 mm across. Pedicels 3–7 mm.

Distribution: Nepal, E Himalaya and Assam-Burma.

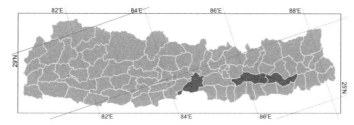

Altitudinal range: 400–700 m.

Ecology: Subtropical forests.

Flowering: October. **Fruiting:** November–February.

Actinodaphne angustifolia is distinguished by large, pinnate-veined leaves and velvety young twigs. The misspelling '*angusifolia*' by Kostermans & Chater (Enum. Fl. Pl. Nepal 3: 182. 1982) has been followed by later authors (e.g. Annot. Checkl. Fl. Pl. Nepal: 160. 2000; Enum. Fl. Pl. Nepal: 7: 2001).
The timber is used for construction.

2. *Actinodaphne obovata* (Nees) Blume, Mus. Bot. 1: 342 (1851).
Tetradenia obovata Nees, in Wall., Pl. Asiat. Rar. 2[8]: 64 (1831); *Laurus obovata* Buch.-Ham. ex Nees nom. inval.; *Tetranthera obovata* Buch.-Ham. ex Wall. nom. nud.

Trees to 15 m. Twigs stout, to 5–10 mm in diameter, blackish, smooth or lenticellate, tomentose. Perulate buds present. Leaves obovate to slightly ovate, 24–32 × 10–13 cm, base cuneate, apex acute or slightly acuminate, margin flat, weakly 3-veined, secondary veins 5–8 pairs, tertiary venation scalariform, tomentose on midrib and veins and sparsely pubescent on lamina, underside glaucous or not. Petioles 3.5–9 cm. Flowers in paniculate inflorescences. Male inflorescences 4–5 cm. Male flowers yellowish green, 8 mm, sericeous. Pedicels 1–5 mm. Tepals broadly ovate, 6 mm.

Stamens 6–7 mm. Female inflorescences to 3 cm. Tepals ca. 2.5 mm. Fruit ellipsoid, 15–20 mm long on a perianth cup ca. 10 mm across.
Fig. 5g–j.

Distribution: Nepal, E Himalaya, Tibetan Plateau, Assam-Burma and E Asia.

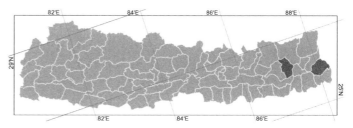

Altitudinal range: 300–1600 m.

Ecology: Subtropical forests.

Flowering: March.

Large, weakly 3-veined leaves are found only in *Actinodaphne obovata*, and with its stout twigs and paniculate inflorescences it is quite distinct from the other Nepalese *Actinodaphne* species.

3. *Actinodaphne longipes* Kosterm., Reinwardtia 9: 98 (1974).
Actinodaphne reticulata var. *glabra* Meisn.; *Laurus gushia* Buch.-Ham. nom. inval.

Trees to 15 m. Twigs slender, to 2 mm in diameter, pale brown or grey brown, glabrous, smooth or lenticellate. Perulate buds present. Leaves elliptic to slightly ovate or slightly obovate, 8–16 × 1.4–3 cm, base cuneate, apex slightly acuminate, margin slightly inrolled or inrolled at base and otherwise flat, pinnate veined, secondary veins 11–23 pairs, tertiary venation reticulate, glabrous or minutely sparsely sericeous below, underside glaucous. Petioles 0.5–1.3 cm. Umbels on 2–4 mm short-shoots, buds 4–7 mm in diameter. Male inflorescences to 1.5 cm, consisting of 1–3 sessile umbels, with ca. 6 flowers per umbel. Male flowers white, 10 mm long, densely sericeous. Pedicels 3–4 mm. Tepals oblong or narrowly ovate, 4–5 mm. Stamens 6–8 mm. Female inflorescences to 1 cm, consisting of 2 or 3 sessile umbels, with 5–6 flowers per umbel. Female flowers 4 mm long, densely sericeous. Pedicels 3–4 mm. Tepals strap-shaped, 2 mm. Infructescences with 1–5 broadly ellipsoid fruit, ca. 15 mm long, on perianth cups 9–10 mm across. Pedicels 2–3 cm.

Distribution: Nepal, E Himalaya and Assam-Burma.

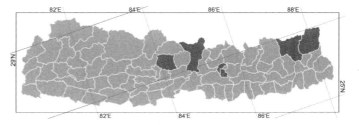

Lauraceae

Altitudinal range: 1600–2600 m.

Ecology: Temperate forest.

Flowering: March, October. **Fruiting:** April.

The very long, narrow leaves clustered at the tips of the twigs are diagnostic for *Actinodaphne longipes*.

4. *Actinodaphne sikkimensis* Meisn., in A.DC., Prodr. 15(1): 213 (1864).

खपटे Khapate (Nepali).

Trees to 6 m. Twigs slender, to 2 mm diameter, mid brown, sometimes reddish, smooth, glabrous. Leaves oblong or elliptic to slightly ovate, 7–13 × 1.5–3 cm, base cuneate, apex slightly acuminate, margin flat, pinnate veined, secondary veins 9–11 pairs, tertiary venation scalariform, glabrous, underside glaucous. Petioles 0.7–1 cm. Umbels solitary, on 2 mm short-shoots. Male inflorescences not seen. Female inflorescences not seen. Infructescences with 1–4 ellipsoid fruit, 11–14 mm long, on perianth cups 2 mm across. Pedicels 4–7 mm.

Distribution: Nepal, E Himalaya and Assam-Burma.

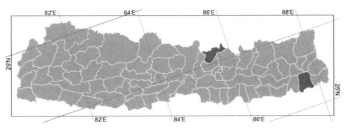

Altitudinal range: 1600–2700 m.

Ecology: Subtropical and warm temperate forests.

Flowering: October. **Fruiting:** May.

Fumariaceae

Magnus Lidén

Perennial, annual or biennial juicy herbs, usually glabrous and glaucous. Leaves alternate (very rarely opposite), cauline or in basal rosette, compound. Stipules usually absent (but see *Corydalis stipulata*). Inflorescences bracteate, terminal (often appearing leaf opposed), cymose, umbellate or racemose. Flowers dimerous, with one or two planes of symmetry (zygomorphic or bisymmetric). Sepals 2, not completely enclosing petals in bud, petaloid or sepaloid, entire or dentate. Petals 2 + 2 in dissimilar whorls, one or both of the outer petals with a basal sac or spur (except in *Hypecoum*); inner petals fused at tip and with a median joint (except in *Hypecoum*). Androecium of 8 thecae supplied by 6 vascular bundles, organised in 2 or 4 (in *Hypecoum*) 'stamens' with broad, ± hyaline filaments. Anthers opening by slits. Nectaries present at base of stamens. Ovary syncarpous, bicarpellate, 1-locular, one- to many-seeded. Fruits single-seeded nuts or variously dehiscent capsules (lomentum in *Hypecoum*).

Twenty genera and ca. 570 species in Europe, N America, Asia and Africa outside of the tropics. Four genera and 53 species in Nepal.

Key to Genera

1a	Petals free. Stamens 4. Annuals with all leaves in a basal rosette	1. *Hypecoum*
b	Inner petals united at tip. Stamens 2, each with three anthers. Annuals or perennials with leafy stems	2
2a	Flowers with two planes of symmetry. Plant climbing. Leaves with tendrils	2. *Dactylicapnos*
b	Flowers with a single plane of symmetry. Plant not climbing. Leaves without tendrils	3
3a	Flowers more than 9 mm long. Fruit a dehiscent capsule with many black seeds	3. *Corydalis*
b	Flowers less than 8 mm long. Fruit a globular nut with a single brown seed	4. *Fumaria*

1. *Hypecoum* L., Sp. Pl. 1: 124 (1753).

Low, glabrous, glaucous, tap-rooted, annual, scapose herbs. Leaves numerous in basal rosette, shortly petiolate, lanceolate, imparipinnate, with deeply divided leaflets. Flowers in few to several long-pedunculate dichasia from rosette, with two planes of symmetry. Sepals green, (sub)entire. Petals 4. Outer petals flat, entire to shallowly 3-lobed. Inner petals deeply 3-lobed; median lobe spatulate, enfolding stamens and style; lateral lobes divergent, entire. Stamens 4, with small rounded nectaries at base laterally. Style with two narrow stigmatic branches. Fruit linear, breaking up into one-seeded units.

18 species in the Mediterranean and Asia from Portugal to Mongolia and China. One species in Nepal.

1. *Hypecoum leptocarpum* Hook.f. & Thomson, Fl. Ind. 1: 276 (1855).

पर पात Par pat (Nepali).

Plants 5–40 cm. Leaves narrowly oblanceolate, pinnate, 5–20 × 1–3 cm; primary leaflets 4–9 pairs, broadly ovate, 0.4–2 cm, almost sessile, pinnatifid; lobules lanceolate, narrowly ovate to obovate, 2–4 mm, acute. Flowering stems few to many, 5–40 cm, dichotomously branched. Bracts paired, 0.5–3 cm, lower ones biternate to subbipinnate, gradually smaller and less divided upward, uppermost linear. Sepals green, ovate or ovate-lanceolate, 2–3(–4) × 1–1.5(–2) mm, margin membranous, entire, rarely toothed. Petals white or pale lavender, apically green. Outer petals obovate, 6–9 × 4–6 mm, entire. Inner petals 3-parted; lateral lobes broadly oblong, apex obtuse; middle lobe spatulate-orbicular, ca. 2 mm, shortly stalked or sessile, margin incurved, slightly denticulate. Capsule erect, linear, 30–40 × 1–1.5 mm, breaking up into 15–18 segments when mature, each segment with 1 seed. Seeds flattened, D-shaped. Fig. 6a–c.

Distribution: Nepal, E Himalaya, Tibetan Plateau, E Asia, N Asia, C Asia and SW Asia.

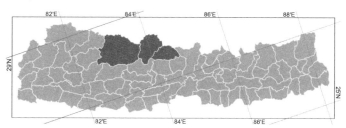

Altitudinal range: (1700–)2700–4200 m.

Ecology: River sands, gravel slopes, as a weed in fields.

Flowering: June–September. **Fruiting:** June–September.

Hypecoum parviflorum L. was reported from Nepal by Kitamura (Fauna Fl. Nepal Himalaya: 137. 1955) based on a collection by Nakao from 'Timana' (Timang near Chame, Manang District), 2700 m, 27 May 1953 (presumed in KYO). An exact match for this specimen has not been found in KYO, but two

Fumariaceae

Nepalese collections previously identified as *H. parviflorum* are present, including one collected by Nakao on the same day and at the same altitude and identified by Kitamura, but from Zoagani. Both these collections are *H. leptocarpum* and so the name *H. parviflorum* is now considered misapplied by Kitamura. Hooker.f. & Thomson (Fl. Ind.: 275. 1855) misapplied the name *H. procumbens* L. to Indian material of *H. parviflorum* further complicating historical records.

Hypecoum parviflorum is a W Himalayan and Mediterranean species which differs from *H. leptocarpum* by its pale yellow flowers, more dissected leaves with linear lobes, and pendent capsules which are 3mm broad. No specimens have been seen which confirm its occurance in Nepal, but it may yet be found in arid areas in western districts.

2. *Dactylicapnos* Wall., Tent. Fl. Napal. [Fasc. 2]: 51 (1826).

Perennial or annual glabrous glaucous herbaceous vines. Stems forming sympodium of several successive shoot generations, much-branched and leafy throughout, usually somewhat zigzag, with slightly reflexed leaves. Leaves all cauline, ternately (perennial species) or pinnately (annual species) divided; each lateral leaflet once to almost three times ternately divided with entire ultimate leaflets; apical leaflet transformed into branched tendril, except in basal-most leaves. Inflorescences terminal (appearing leaf-opposed due to displacement of the primary main shoot by the uppermost axillary shoot), racemose, subumbellate, pendent with 2–14 flowers. Flowers with two planes of symmetry. Sepals petaloid. Corolla heart-shaped to oblong, pale yellow to orange; base of outer petals saccate. Inner petals with long narrow claw fused with the outer petals and with short free limb. Stamens 2, broad, whitish-translucent, each with a central dithecal and 2 lateral monothecal anthers. Nectaries conspicuous, protruding from base of anthers into sacs of outer petals. Fruit with persistent style. Stigma flat, quadrangular to trapezoidal, with a papilla at each corner. Seeds rounded, often with a beak, black, with elaiosome.

15 species in the Himalaya and W China. Six species in Nepal.

Key to Species

1a   Annual. Sepals fringed. Fruit linear. Leaves with 3–5 primary lateral alternate leaflets and a terminal tendril. Stems angular or winged .... **1. *D. roylei***

b   Perennial. Sepals entire. Fruit ovate to lanceolate. Leaves with 2 primary lateral alternate or opposite leaflets and a terminal tendril. Stems ± terete..2

2a   Primary leaflets opposite...3

b   Primary leaflets alternate..4

3a   Wings of inner petals overtopping apex by ca. 1.5 mm. Each primary leaflet with 4–7 ultimate leaflets. Capsule fleshy. Seeds with large deeply dissected elaiosomes .... **2. *D. grandifoliolata***

b   Wings of inner petals overtopping apex by ca. 3 mm. Each primary leaflet with 10–15 ultimate leaflets. Capsule membranous. Seeds with small elaiosomes .... **3. *D. cordata***

4a   Fruit with broad base, very fleshy when fresh, becoming black in the press. Nectary sigmoidally curved, shortly acuminate.. **4. *D. scandens***

b   Fruit lanceolate, flattened, membranous, green in the press. Nectary curved upwards with thickening or small peg at middle, tip long and narrow..5

5a   Margin of fruit entire or almost so. Seeds 1.6–1.8 mm, sharply colliculate to echinulate.................... **5. *D. macrocapnos***

b   Margin of fruit grossly dentate. Seeds ca. 1.2 mm, faintly colliculate .... **6. *D. odontocarpa***

1. *Dactylicapnos roylei* (Hook.f. & Thomson) Hutch., Bull. Misc. Inform. Kew 1921: 104 (1921).
Dicentra roylei Hook.f. & Thomson, Fl. Ind. 1: 272 (1855).

Climbing, annual herb to 2–5 m. Stems, branched and leafy throughout, weak, hollow, ridged to winged. Petioles 1–3(–6) cm, with thin wing. Leaf pinnate with 3–6 leaflets alternate on zigzag rachis; primary leaflets once to twice ternate; ultimate leaflets 4–8 per primary leaflet, ovate, 5–20 × 4–10 mm, entire, acute or obtuse, mucronate. Raceme short, pendent, 2–7-flowered. Peduncle (2–)5–11 cm. Bracts lanceolate, (5–)9–15 × 2–3 mm, lacerate. Pedicels 12–25 mm.

Sepals 5–10 × 1–3 mm, with broad deeply lacerate base and narrow, sparsely dentate, apex. Corolla yellow, flat, rounded in profile, 16–19 × 8–14 mm. Outer petals broadly saccate, sharply keeled or winged in basal three quarters; sinus where pedicel is inserted wide, 1–2 mm deep. Inner petals 14–17 mm; dorsal crest broad undulate, overtopping petal apex ca. 1 mm. Nectary tapering to slender apex, 3–4 mm. Style 3–5 mm. Stigma with large basal papillae. Capsule green or slightly reddish, linear-oblong, 4–5 cm, 3–4 mm in diameter, with up to 20 seeds. Seeds faintly colliculate, ca. 2 mm, with large elaiosome.
Fig. 6d–e.

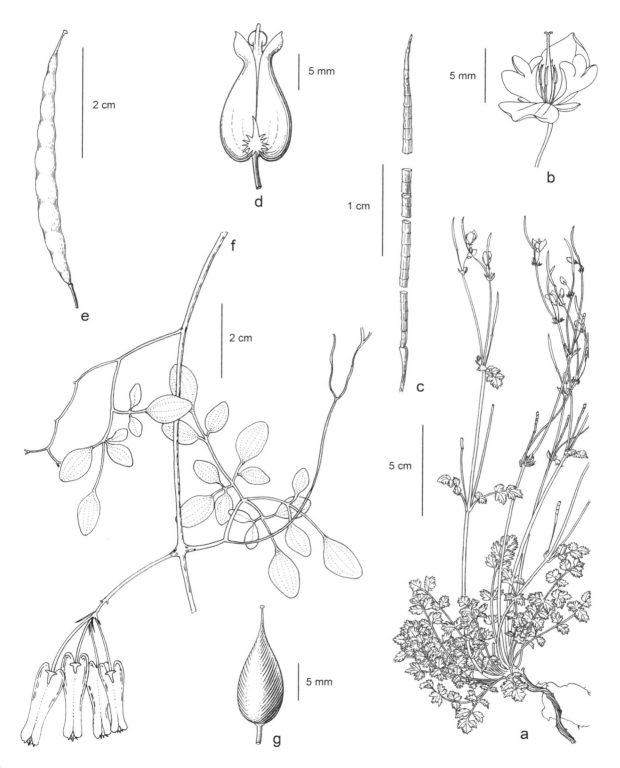

FIG. 6. FUMARIACEAE. **Hypecoum leptocarpum**: a, plant with flowers and fruits; b, flower; c, fruit. **Dactylicapnos roylei**: d, flower; e, fruit. **Dactylicapnos scandens**: f, shoot with flowers; g, fruit.

Fumariaceae

Distribution: Nepal, W Himalaya, E Himalaya, Tibetan Plateau, Assam-Burma and E Asia.

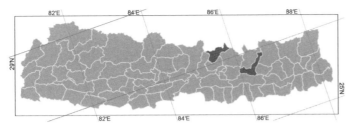

Altitudinal range: 2000–3000 m.

Ecology: Forest understorey, roadsides.

Flowering: July–November. **Fruiting:** August–December.

Self-compatible. Rare and scattered.

2. Dactylicapnos grandifoliolata Merr., Brittonia 4: 64 (1941).
Dactylicapnos ventii (Khánh) Lidén; *Dicentra ventii* Khánh.

Climbing perennial herbs. Stems to ca. 3 m, thin, solid, branched and leafy throughout. Petioles 1–2 cm. Leaf with 2 opposite leaflets and terminal branched cirrhose tendril; each primary leaflet unequally compound into 4–7 ultimate leaflets; these ovate 10–30 × 8–20 mm, subacute, entire, of a firm texture, dark green above, beneath very glaucous and with raised veins. Raceme drooping, 1–4 cm, 2–10-flowered. Peduncle 6–12 cm. Bracts linear, entire, 4–10 mm. Pedicels thin, 10–20 mm in flower, elongating to 20–35 mm in fruit. Sepals triangular-lanceolate, 2 × 1 mm, subentire. Corolla pale yellow, oblong-cordate in profile, 20–25 × 9–11 mm. Outer petals saccate, not keeled; sinus at pedicel insertion narrow, 2–3 mm deep. Dorsal crest of inner petals overtopping petal apex 1.5 mm. Nectary 2 mm, down-curved, clavate, obtuse. Style 6–7 mm. Stigma with large basal papillae. Capsule pale pink to purplish red, lanceolate in outline, 20–30 × 5–7 mm, opening with fleshy valves, with 20 to 30 seeds. Seeds 1.5 mm, evenly colliculate, slightly flattened, with small beak; elaiosome broader than seed, white, consisting of long free tubular cells. After the valves have opened, the seeds are held by the placentas and the entangled elaiosomes.

Distribution: Nepal, E Himalaya, Tibetan Plateau and Assam-Burma.

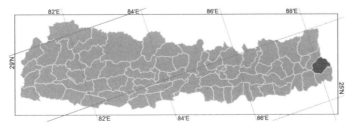

Altitudinal range: 1500–2800 m.

Ecology: Semi-shaded slopes in forests.

Flowering: October (probably June–December). **Fruiting:** October (probably June–December).

3. Dactylicapnos cordata Lidén, Nord. J. Bot. 28: 658 (2010).

Climbing perennial herbs to ca. 3 m. Stems thin, leafy and branched. Petioles 0.5–1 cm. Leaf with 2 opposite primary leaflets; each with a 2–5 cm petiolule, twice ternately compound into 6 to 11 ultimate leaflets; these broadly ovate to obovate, 10–40 × 7–30 mm, very unequally sized, thin with small recurved mucro. Raceme short, corymbose, nutant, 4 to 7-flowered. Peduncle 2–5 cm. Bracts linear, entire, 3–4 mm, acuminate. Pedicels 10–25 mm, 15–30 mm in fruit. Sepals triangular, 2–3 × 1–1.5 mm. Corolla yellow to orange yellow, cordate, 16–22 × 9–16 mm. Outer petals broadly saccate, often with basal dorsal wing; sinus at petiole insertion 3–5 mm deep. Dorsal crest of inner petals conspicuous, overtopping petal apex 3 mm. Nectary out-curved, unequally bifid. Stigma squarish. Capsule lanceolate, thin-walled, 25–30-seeded, 25–33 × 5–6 mm, excluding 3–5 mm style. Seeds ca. 1 mm, aculeate, with smooth beak. Elaiosome small.

Distribution: Nepal and E Himalaya.

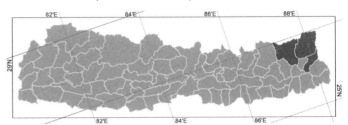

Altitudinal range: 1300–1900 m.

Ecology: Shaded places in forests.

Flowering: (May–)June–September. **Fruiting:** (May–) June–September.

4. Dactylicapnos scandens (D.Don) Hutch., Bull. Misc. Inform. Kew 1921: 105 (1921).
Diclytra scandens D.Don, Prodr. Fl. Nepal.: 198 (1825); *Dactylicapnos thalictrifolia* Wall.; *Dicentra scandens* (D.Don) Walp.; *D. thalictrifolia* (Wall.) Hook.f. & Thomson; *Dielytra scandens* G.Don orth. var.

Climbing perennial herbs, with a swollen subterranean rootstock. Stems 2–5 m, thin. Petioles 0.5–3 cm. Leaves with 1 pair of alternate primary pinnae; primary leaflets bi- to tri-ternate; ultimate leaflets 6–13 per primary leaflet, ovate, 5–30 × 4–18 mm, glaucous below, green above, subobtuse with small hooked mucro. Raceme 1.5–4 cm, 6–14-flowered, pendent. Peduncle 2–8 cm. Bracts narrowly obovate, 3–6 × 1 mm, subentire, acute. Pedicels 10–20 mm, to 30 mm in fruit. Sepals ovate-lanceolate, 3–4 × 2 mm, entire. Corolla yellow, oblong-cordate to obtusely triangular, 18–21 × 8–10 mm. Outer petals keeled or with dorsal wing at base; sinus at pedicel insertion narrow, 3–4 mm deep. Inner petals claw

11 mm; blade 5 mm; dorsal crest overtopping apex 1–2 mm.
Nectary sigmoid, acute, 3 mm. Capsule very fleshy, purple,
red, whitish or yellow (valves becoming black and papery
when dry), ovoid, 15–20 × 6–8 mm, not dehiscent by regular
valves; style 5–7 mm. Seeds colliculate, 1.7–2 mm, with
prominent beak; elaiosome small.
Fig. 6f–g.

Distribution: Nepal, E Himalaya, Tibetan Plateau, Assam-
Burma, S Asia, E Asia and SE Asia.

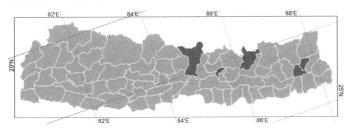

Altitudinal range: 1500–2800 m.

Ecology: Forest understorey, slopes, stony places.

Flowering: July–November. **Fruiting:** August–December.

Juice of the roots is applied to sores between toes and
leaf juice is applied to cuts and wounds. The plant is fed to
livestock as an anthelmintic.

5. *Dactylicapnos macrocapnos* (Prain) Hutch., Bull. Misc.
Inform. Kew 1921: 105 (1921).
Dicentra macrocapnos Prain, J. Asiat. Soc. Bengal, Pt. 2, Nat.
Hist. 65: 12 (1896).

Climbing perennial herbs, with stout subterranean rootstock.
Stems 2–8 m, thin. Petioles 0.5–3 cm. Leaves with 1 pair
of alternate primary leaflets; primary leaflets bi- (rarely tri-)
ternate; ultimate leaflets 5–13 per primary leaflet, ovate,
subobtuse, 10–25 × 7–20 mm, soft, ± glaucous below, with
small recurved mucro. Raceme 1–3 cm, 5–14-flowered,
obliquely pendant. Peduncle 2–4 cm. Bracts linear-lanceolate
to narrowly oblanceolate, 3–10 × 1–2 mm, entire, acute.
Pedicels 10–20 mm, in fruit to 30 mm, reflexed at base.
Sepals 2–3 × 2–2.5 mm, abruptly narrowed from broad base,
slightly dentate. Corolla yellow, oblong-cordate to obtusely
triangular, 18–20 × 8–12 mm. Outer petals keeled or winged
at base; sinus at petiole insertion narrow, 3 mm deep. Inner
petals claw 11 mm, blade 5.5 mm; dorsal crest overtopping
apex 1(–2) mm. Nectary 4–5 mm, strongly up-curved, tapering
to slender apex, with small peg at geniculation. Capsule
lanceolate, flattened, 25–35 × 6–7 mm with membranous
walls; style 5–6 mm. Seeds colliculate to aculeate,
1.6–1.9 mm, with small elaiosome.

Distribution: Nepal, W Himalaya and Tibetan Plateau.

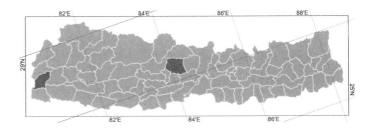

Hooker & Thomson (Fl. Ind.: 273. 1855) partially misapplied
the name *Dicentra scandens* (D.Don) Walp. to this species.

1a Margin of fruit even or almost so, rim distinct. Seeds
1.6–1.9 mm, aculeate .... subsp. ***echinosperma***
b   Margin of fruit even, rim not distinct. Seeds
1.6–1.7 mm, colliculate .... subsp. ***macrocapnos***

Dactylicapnos macrocapnos subsp. ***echinosperma*** Lidén,
Nord. J. Bot. 28: 657 (2010).

Margin of fruit even or almost so, rim distinct. Seeds
1.6–1.9 mm, aculeate.

Distribution: Nepal and Tibetan Plateau.

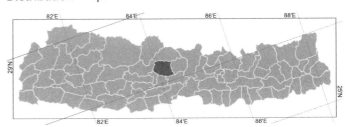

Altitudinal range: 900–2700 m.

Ecology: Forests and scrub.

Flowering: April–November. **Fruiting:** July–November.

Dactylicapnos (Prain) Hutch. ***macrocapnos*** subsp.
macrocapnos

Margin of fruit even, rim not distinct. Seeds 1.6–1.7 mm,
colliculate.

Distribution: Nepal and W Himalaya.

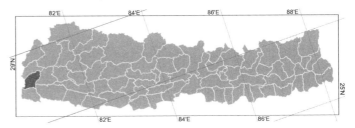

Altitudinal range: 900–2600 m.

Ecology: Forest understorey.

Flowering: April–November. **Fruiting:** July–November.

6. _Dactylicapnos odontocarpa_ Lidén, Nord. J. Bot. 28: 658 (2010).

Climbing perennial herbs, with stout subterranean rootstock. Stems 2–8 m, thin. Petioles 0.5–3 cm. Leaves with 1 pair of alternate primary leaflets; primary leaflets biternate; ultimate leaflets 5–9 per primary leaflet, ovate, subobtuse, 10–20 × 7–15 mm, soft, ± glaucous below, with small recurved mucro. Raceme 1–3 cm, 5–8-flowered, obliquely pendant. Peduncle 2–4 cm. Bracts linear to narrowly oblanceolate, ca. 5 × 1 mm, entire, subacute. Pedicels 10–20 mm, reflexed at base. Sepals 2–3 × 2–2.5 mm, abruptly narrowed from broad base, slightly dentate. Corolla yellow, oblong-cordate to obtusely triangular, 18–20 × 8–12 mm. Outer petals keeled or winged at base; sinus at petiole insertion narrow, 3 mm deep. Inner petals claw 11 mm, blade 5.5 mm; dorsal crest overtopping apex 1(–2) mm. Nectary 4–5 mm, curved, tapering to slender apex, with swelling at middle. Capsule lanceolate, flattened, 30–35 × 6–8 mm, with thin walls; style 5–6 mm. Seeds faintly colliculate, 1.2 mm, with small elaiosome.

Distribution: Nepal and E Himalaya.

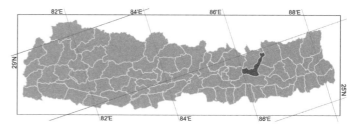

Altitudinal range: 1600–2600 m.

Ecology: Shaded places in forests.

Flowering: June–September. **Fruiting:** June–September.

3. _Corydalis_ DC. nom cons., in DC. & Lam., Fl. Franc., ed. 3, 4: 601 (1805).

Annual to perennial, soft and juicy, usually glabrous herbs. Radical leaves (i.e. not associated with flowering stems) singly from rootstock or in rosettes, pinnately or ternately divided; cauline leaves 1–several or rarely absent, alternate or rarely opposite, like radical leaves, very rarely entire. Flowers with bracts, in racemes. Sepals petaloid, entire to deeply dentate. Corolla with one plane of symmetry; upper petal spurred. Stamens 2, broad, whitish-translucent, each with a central dithecal and 2 lateral monothecal anthers. Nectary often conspicuous, protruding from base of upper anther into spur of upper petal. Fruit a many-seeded capsule with persistent style, dehiscent with valves, often explosively. Seeds black, with elaiosomes.

Worldwide about 470 species, mainly in north temperate areas (one species each in subarctic Russia/N America and mountains of tropical E Africa and three species in subtropical Indochina). 45 species in Nepal.

Key to Species

1a   Single large tuber present. Flowering stems with two opposite leaves .... **3. _C. diphylla_**
   b   Tuber absent (but see _C. lathyroides_), sometimes with a cluster of fleshy roots. Leaves usually alternate on flowering stems 2

2a   Rhizome small with a cluster of fleshy roots. Stems and petioles of radical leaves attenuate into a filiform flexuous subterranean base 3
   b   Tap-rooted or rhizomatous. Stems and petioles of radical leaves not attenuate into a filiform flexuous subterranean base...10

3a   Flowers yellow...4
   b   Flowers blue 7

4a   Cauline leaves pinnate, usually several. Stem simple or branched..5
   b   Cauline leaf entire, usually solitary. Stem simple..6

5a   Flowers 11–14 mm. Stems branched..**39. _C. lowndesii_**
   b   Flowers 14–19 mm. Stems usually simple...**40. _C. polygalina_**

6a   Bracts 5–20(–40) mm. Racemes (5–)10–30-flowered. Inner petals dark-tipped...........................**37. _C. juncea_**
   b   Bracts 30–50 mm. Racemes 2–5-flowered. Inner petals pale ....**38. _C. pseudojuncea_**

7a   All or most bracts entire. Cauline leaf trifoliate ....**41. _C. trifoliata_**
   b   All or most bracts divided. Cauline leaves more divided or absent 8

8a   Stem 4–7 cm, usually leafless. Radical leaves prominent. Upper petal 9–12 mm..........................**44. _C. jigmei_**
   b   Stem 6–30 cm, usually with 1 leaf. Radical leaves inconspicuous. Upper petal 12–21 mm............................9

| 9a | Lower petal rhombic, subacute. Leaf lobes 1–2 mm broad | **42. *C. cashmeriana*** |
| b | Lower petal very broad, apically truncate. Leaf lobes 0.5–1 mm broad | **43. *C. ecristata*** |

| 10a | Flower blue, pink or purple | 11 |
| b | Flowers yellow | 14 |

| 11a | Stems 50–130 cm, erect. Leaves tripinnate | **4. *C. flaccida*** |
| b | Stems shorter. Leaves less divided | 12 |

| 12a | Spur equalling or longer than limb of upper petal. Fruit linear. Non-alpine plant with slender, decumbent stems | **36. *C. leptocarpa*** |
| b | Spur much shorter than limb of upper petal. Fruit obovoid. Alpine plant with short crowded stems | 13 |

| 13a | Ultimate leaf lobes obovate-obtuse. Crest of upper petal long decurrent on spur. | **25. *C. latiflora*** |
| b | Ultimate leaf lobes narrow, acute. Crest short. | **26. *C. magni*** |

| 14a | Spur of upper petal much shorter than limb | 15 |
| b | Spur of upper petal equalling limb | 21 |

| 15a | Spur strongly curved (to a semi-circle). Raceme very dense, capitate to cylindrical | **45. *C. conspersa*** |
| b | Spur not or moderately curved or raceme lax | 16 |

| 16a | Bracts divided | 17 |
| b | Bracts entire | 19 |

| 17a | Stems few, ascending, branched. Racemes 2–10-flowered. Corolla 12–14 mm | **27. *C. stracheyi*** |
| b | Stems several, erect, not or sparsely branched. Racemes 5–35-flowered. Corolla 14–17 mm | 18 |

| 18a | Petiole-remnants membranous. Racemes 5–10-flowered | **14. *C. sikkimensis*** |
| b | Petiole-remnants with persistent fibres. Racemes 12–35-flowered | **28. *C. meifolia*** |

| 19a | Raceme very dense. Corolla 15–17 mm long. Leaves 2–3-pinnate | **2. *C. stricta*** |
| b | Raceme lax. Corolla 8–12 mm long. Leaves simply pinnate | 20 |

| 20a | Robust, fleshy plant 15–40 cm. Pedicels 2–4 mm | **1. *C. flabellata*** |
| b | Slender plant, 10–15 cm. Pedicels 5–20 mm | **11. *C. lathyroides*** |

| 21a | Flowers 8–12 mm. Stigma lacking basal papillae. Diffuse slender plants | 22 |
| b | Flowers 13 mm or more. Stigma usually with basal papillae. Habit variable | 27 |

| 22a | Leaves pinnate. Bracts entire. Rootstock thick | **11. *C. lathyroides*** |
| b | Leaves biternate. At least lower bracts usually divided. Rhizome thin | 23 |

| 23a | Nectary 1 mm long, ca. one fifth as long as spur. Lower petal saccate. | 24 |
| b | Nectary 1.5 mm long or longer, at least one third as long as spur. Lower petal with or without pouch | 25 |

| 24a | Outer petals long-acuminate. Inner petals pale. Stigma with 6 apical papillae | **6. *C. longipes*** |
| b | Outer petals acute. Inner petals dark-tipped. Stigma with 4 apical papillae | **8. *C. pseudolongipes*** |

| 25a | Lower petal with basal pouch | **7. *C. filiformis*** |
| b | Lower petal not saccate | 26 |

| 26a | Sepals 2–3 mm. Fruit broadly obovoid | **9. *C. calycina*** |
| b | Sepals 0.5–1 mm. Fruit linear to narrowly elliptic | **10. *C. casimiriana*** |

| 27a | Pedicels less than 10 mm long. Stems few, 15–100 cm tall, leafy and branched above | 28 |
| b | Pedicels more than 10 mm long. Stems few or several, short or tall, simple or branched | 36 |

| 28a | Bracts obovate, entire, crispate-puberulent. Upper leaves with stipule-like lower leaflets | **34. *C. stipulata*** |
| b | Bracts divided or (if entire) linear to lanceolate, glabrous. Lower leaflets of upper leaves not stipule-like | 29 |

29a	Stems zigzag. Leaves 3 or 4 times ternate with 1 mm broad acute leaflets	**19. *C. shakyae***
b	Stems not obviously zigzag. Leaves less divided or pinnate. Leaflets more than 1 mm broad	30
30a	Stems leafy only in upper half. Robust rhizomatous perennials	31
b	Stems leafy throughout. Annuals to perennials	33
31a	Bracts large, divided. Racemes lax	**32. *C. geraniifolia***
b	Bracts small, entire, linear. Racemes dense	32
32a	Stem simple, 25–50 cm, with 1–3 leaves. Lower petal with distinct pouch	**33. *C. terracina***
b	Stems branched above, 60–130 cm, with several leaves. Lower petal usually without pouch	**35. *C. chaerophylla***
33a	Lowest pair of leaflets of middle cauline leaves larger than second pair	34
b	Lowest pair of leaflets of middle cauline leaves smaller than second pair	35
34a	Crest of outer petals dentate. Pedicels 4–10 mm. Seeds smooth	**12. *C. vaginans***
b	Crest of outer petals entire. Pedicels 2–5 mm. Seeds rugose	**13. *C. cornuta***
35a	Pedicels not filiform. Outer petals crested	**20. *C. hookeri***
b	Pedicels filiform. Outer petals without dorsal crest	**21. *C. spicata***
36a	Small cushion plants. Stems less than 8 cm	37
b	Plants not forming cushions. Stems more than 8 cm.	40
37a	Leaves pinnately divided. Bracts little divided or entire	38
b	Leaves ternately divided. Bracts much divided, leaf-like	39
38a	Rootstock crowned by numerous straw-like petiole remnants. Plant of rock crevices	**18. *C. staintonii***
b	Petiole remnants absent. Scree plant	**23. *C. clavibracteata***
39a	Flowers 14–17 mm. Sepals much dentate, 2–3 mm broad. Corolla horizontal	**24. *C. megacalyx***
b	Flowers 19–25 mm. Sepals linear, less than 2 mm broad. Corolla held almost vertically	**31. *C. hendersonii***
40a	Fruiting pedicels erect, hooked only at apex. Racemes very dense	41
b	Fruiting pedicels arcuate-recurved. Racemes lax	42
41a	Bracts large, entire. Flowers 15–17 mm long	**29. *C. uncinata***
b	Bracts divided. Flowers 13–14 mm long	**30. *C. uncinatella***
42a	Most leaves cauline. Stems slender, often branched. Petiole remnants lacking	43
b	Most leaves in a basal rosette. Stems usually simple. Rootstock with petiole remnants	44
43a	Outer petals with broad blunt crest. Spur shorter than limb	**5. *C. cavei***
b	Outer petals acuminate with crest narrow, tapering to tip. Spur longer than limb	**22. *C. filicina***
44a	Plant of rock crevices. Stems several	**18. *C. staintonii***
b	Not found in rock crevices. Stems few	45
45a	Crest broad, overtopping apex. Cauline leaves not opposite	**17. *C. elegans***
b	Crest narrow or absent, not overtopping apex. Cauline leaves small, opposite	46
46a	Bracts pectinate-dentate. Stems 15–35 cm	**15. *C. govaniana***
b	Bracts entire. Stems 8–15 cm	**16. *C. simplex***

1. *Corydalis flabellata* Edgew., Trans. Linn. Soc. London 20: 30 (1851).

Perennial herbs, 15–40 cm, glabrous, very glaucous. Caudex eventually thick and dryish, crowned by fragile remnants of older leaves. Stems glaucous, terete, turgid and brittle, erect, simple to much-branched, leafy, with leaves slightly crowded towards the base. Petioles of lower leaves 4–8 cm, upper leaves short-stalked. Leaves oblong, 8–15 × 2–4 cm, pinnate with 3–5 pairs of rather distant, broadly flabellate, entire to dentate or shallowly dissected leaflets. Inflorescence often branched. Racemes spicate, becoming lax, 10–30-flowered. Bracts small, linear, long-acuminate, not or slightly longer than the pedicels. Pedicels 1–4 mm, reflexed in fruit. Sepals ovate, acute, 2–4 × 1–1.5 mm, dentate. Corolla yellow, 11–14 mm, narrow. Outer petals hardly dilated at apex, mucronate, without crest; spur of upper petal short and rounded, 3 mm, with obtuse nectary two thirds as long. Inner petals 9–12 mm. Stigma with a small transversely set ellipsoid body with 6, usually distinct marginal papillae plus 2 pairs of submarginal papillae close to style (10 papillae altogether). Capsule linear, pendent, 5–8-seeded, 14–18 × 2 mm; style 3–4 mm. Seeds in 1 row, 1.3–1.4 mm, with hilum and elaiosome set on a short protruding beak.
Fig. 7a–b.

Distribution: Nepal, W Himalaya and Tibetan Plateau.

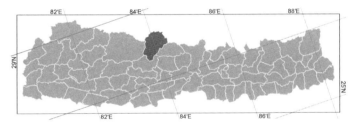

Altitudinal range: 2800–3500 m.

Ecology: Dry cliffs or dry gravel beds.

Flowering: May–August. **Fruiting:** June–September.

Self-fertile.

2. *Corydalis stricta* Steph. ex Fisch., in DC., Syst. Nat. 2: 123 (1821).
Corydalis astragalina Hook.f. & Thomson; *C. schlagentweitii* Fedde.

Perennial herbs 15–50 cm, glabrous, very glaucous. Caudex thick, dryish, only some central strands heavily lignified, crowned by remnants of dead petioles. Stems stout, rather thick, fleshy, leafy and branched. Petioles of lower leaves 5–10 cm, upper leaves subsessile. Lamina 7–15 × 2–5 cm, pinnate to bipinnate with ternately divided leaflets; ultimate lobes oblong to oblanceolate to cuneate-flabellate. Upper leaves progressively smaller. Racemes very dense, oblong, 30–60-flowered, 3–5 cm in flower, slightly elongating in fruit. Bracts 5–7(–10) mm, scarious, often long-acuminate, with the occasional tooth. Pedicels 4–6 mm, arcuately recurved

in fruit. Sepals 4–5 × 1–3 mm, sometimes long-acuminate, deeply dentate. Corolla yellow, 15–17 mm. Outer petals dilated towards apex which is sharply pointed and usually with more or less dentate margins and with short dentate crest; spur short, rounded, 4 mm, with obtuse nectary half to two thirds as long. Inner petals 12–13 mm. Stigma with a small transversely set ellipsoid body with 6 usually distinct marginal papillae plus 2 pairs of submarginal papillae close to style (10 papillae altogether). Capsule pendent, 15–18 × 3 mm, oblong, 6–9-seeded, style 4–5 mm.

Distribution: Nepal, W Himalaya, Tibetan Plateau, E Asia, N Asia and SW Asia.

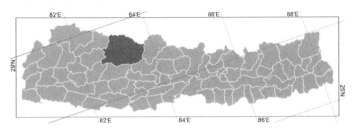

Altitudinal range: 3500–5500 m.

Ecology: Dry hillsides.

Flowering: June–August. **Fruiting:** July–September.

3. *Corydalis diphylla* Wall., Tent. Fl. Napal. [Fasc. 2]: 54 (1826).

Perennial herbs. Tuber 2–4 cm, rounded to irregular, deeply set. Stems and leaves ephemeral. Stem erect, 6–15 cm, simple or often branched. Petioles 10–40 mm. Radical leaves few, similar to cauline leaves. Cauline leaves 2, opposite, slightly glaucous, shortly stalked, triangular, 4–8 × 7–13 cm, tri-ternate; leaf lobes unequal (median one larger), obovate to broadly lanceolate. Racemes lax, 3–6-flowered, lax. Bracts 5–10 mm, enlarging in fruit, ovate-lanceolate, entire. Pedicels 10 mm in flower, 15–30 mm and reflexed in fruit. Sepals 1–2.5 mm, entire to laciniate. Corolla white or cream with purple tip. Outer petals divergent with broad obtuse apex, without crest; upper petal 18–19 mm, spur 8–13 mm, sigmoidally curved; nectary three quarters as long. Inner petals 10 mm. Capsule elliptic-oblong, on slightly arcuate pedicels, 15–20 mm incl. 4 mm style, 5–8-seeded, not explosively dehiscent. Seeds smooth, 2 mm with large elaiosomes.

Distribution: Nepal and W Himalaya.

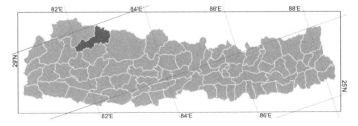

Altitudinal range: ca. 2500 m.

Fumariaceae

Ecology: Forest.

Flowering: April. **Fruiting:** April–May.

Whitmore (Enum. Fl. Pl. Nepal 2: 35. 1979) misapplied the name *Corydalis rutiflora* (Sm.) DC. (syn. *Fumaria rutifolia* Sm.) to this species in Nepal. Whitmore listed in synonymy *C. pauciflora* Edgew., *C. hamiltoniana* G.Don and the misapplied name *C. longipes* auct. non DC., but these names apply only to material outside Nepal.

Only the typical subspecies is currently known from Nepal. The other two, subsp. *occidentalis* Lidén and subsp. *murreana* (Jafri) Lidén, grow in NW Himalaya and differ in always having simple stems and minute sepals.

4. *Corydalis flaccida* Hook.f. & Thomson, Fl. Ind. 1: 269 (1855).

Perennial herbs, 50–100(–130) cm, glabrous. Rootstock long, narrow. Stems 1 to few, erect, terete, stiff, leafy and branched above. Radical leaves in lax rosette, with petioles 6–25 cm, green above, glaucous below, thin, triangular-ovate, 8–25 × 7–20 cm, tripinnate; leaflets to 1 cm, ± deeply cut into rounded mucronate lobes. Cauline leaves smaller, subsessile. Racemes 5–10 cm, elongating in fruit, 5–20-flowered. Middle and upper bracts lanceolate, acute, often with distinct stalk, entire to toothed; lower bract often like uppermost leaf. Pedicels 5–10 mm, lowermost sometimes to 30 mm. Flowers purple or variegated pink and blue. Sepals rounded, 1.5–2 mm, dentate. Outer petals without or with small crest, with narrow base and broad obtuse apex; upper petal 15–18 mm; spur down-curved or nearly straight, 5–8 mm, thin; nectary two thirds as long. Inner petals 8–10 mm. Stigma orbicular with 8 marginal papillae. Capsule linear, 20–35 × 2–3 mm, 8–15-seeded, on slightly arcuate pedicels, not explosively dehiscent. Seeds in 1 row, 2 mm.

Distribution: Nepal, E Himalaya, Tibetan Plateau, Assam-Burma and E Asia.

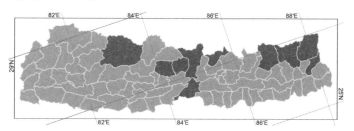

Altitudinal range: 3000–4300 m.

Ecology: Forests.

Flowering: June–July. **Fruiting:** July–August.

5. *Corydalis cavei* D.G.Long, Notes Roy. Bot. Gard. Edinburgh 42(1): 103 (1984).

Perennial herbs, 15–50 cm, glabrous or sparingly papillose-hairy. Rootstock subterranean, with slender stems branched

from soil surface. Aerial stems weak. Petioles 5–9 cm, shorter above. Leaves green above, glaucous below, ovate to triangular, 4–10 × 2–6 cm, 2 to 3 times ternate to subbipinnate with deeply divided leaflets; lobes obovate, obtuse. Racemes 5–25-flowered, rather dense at first, lax in fruit; lower bracts leaf-like, once to twice ternate to ternatifid, upper entire or dentate, usually shorter than pedicels. Pedicels 5–15 mm, down-curved in fruit. Sepals 1 mm, rounded, finely dentate. Corolla yellow to orange, tips of inner petals often purplish black. Upper petal 16–17 mm with broad entire crest reaching apex and there abruptly terminating, narrowly decurrent on spur; spur 7–8 mm, broadly triangular-obtuse; nectary half to two thirds as long; lower petal broad, acute-acuminate, shortly crested. Inner petals 9 mm. Stigma with 6 simple papillae. Capsule linear-oblong, pendent, 12–15 × 2 mm, including 2.5 mm style, 5–7-seeded, explosively dehiscent. Seeds in 1 row.

Distribution: Nepal, E Himalaya and Tibetan Plateau.

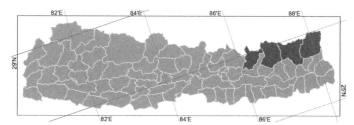

Altitudinal range: 2700–4500 m.

Ecology: Roadside gravel, scree, grassy meadows.

Flowering: June–October. **Fruiting:** June–October.

6. *Corydalis longipes* DC., Prodr. 1: 128 (1824).

Short-lived perennial, (rarely annual) herbs, 20–75 cm, glabrous or papillose-hairy. Stems leafy, slender branched, trailing, ascending to subscandent. Petiole of lower leaves 3–11 cm, upper 1–4 cm. Lamina 2–5 × 2–6 cm, biternate, leaflets deeply lobed; lobes narrowly obovate, obtuse; upper leaves smaller and less divided. Racemes long, lax, 5–15-flowered; bracts 2–9 mm, dentate to lobed; lower larger, often leaf-like; upper smaller, sometimes entire. Pedicels slender, 10–20 mm, recurved in fruit. Sepals whitish, reniform, 1 × 1–1.5 mm, deeply circumdentate. Corolla yellow, rarely white, jasmine scented. Outer petals long acuminate; upper petal 9–13 mm, crest not reaching tip, long decurrent on spur, rarely absent; spur broadly based, 5–7 mm; nectary very short, prominent, one fifth as long as spur. Inner petals 6–8 mm, never dark at apex; lower petal saccate at base. Stigma with 6 apical simple papillae. Capsule pendent, obovoid, 6–10 × 2–3 mm, 8–15-seeded, abruptly narrowed into 2–3 mm style, explosively dehiscent. Seeds in 2 rows, 1–1.2 mm, smooth.

Distribution: Nepal and Tibetan Plateau.

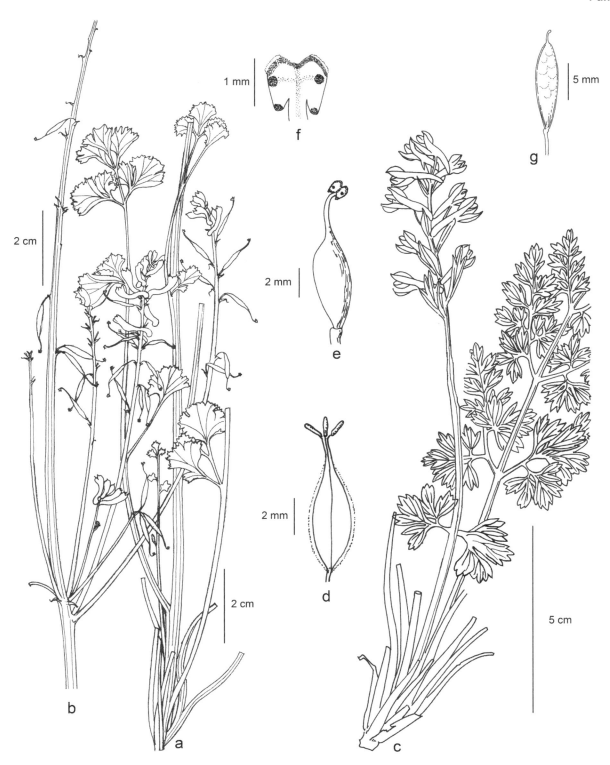

FIG. 7. FUMARIACEAE. **Corydalis flabellata**: a, stem and basal leaves; b, infructescence with flowers. **Corydalis govaniana**: c, flowering plant; d, stamens; e, pistil; f, stigma; g, capsule.

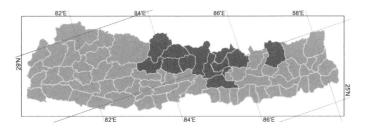

Altitudinal range: 2000–5300 m.

Ecology: Along shaded paths in forests, wet stony places, field margins, roadsides.

Flowering: May–September. **Fruiting:** May–September.

Hooker. f. & Thomson (Fl. Brit. Ind. 1: 125. 1872) misapplied the name *Corydalis sibirica* Pers. to this species.
  Self-incompatible.

7. Corydalis filiformis Royle, Ill. Bot. Himal. Mts. [1]: 68 (1833).

Ascending to subscandent annual or perennial herbs, 30–70 cm, with taproot; sometimes with fleshy adventitious roots. Petiole of lower leaves 3–11 cm, upper 1–4 cm. Leaves 2–5 × 2–6 cm, biternate; leaflets deeply lobed; ultimate lobes obovate to narrowly obovate, obtuse; upper leaves smaller and less divided. Racemes 3–12-flowered. Lower bracts much divided; upper often narrowly clavate and entire, shorter than pedicels. Pedicels slender, somewhat arcuate 10–20 mm at fruiting. Sepals large, 2–3 × 1.5–2 mm, deeply fimbriate whitish, yellow, or often reddish. Corolla yellow, 12–14 mm long. Outer petals cristate, acuminate; upper petal 9–12 mm with spur 5–7 mm and nectary a third to half as long; lower petal 7–9 mm, with prominent basal pouch and short low apical dorsal crest. Inner petals 6–8 mm, pale (in Nepal) or tipped blackish purple, but with pale dorsal wings. Stigma with 6 apical papillae. Capsule pendent, obovoid, 8–10 × 3 mm, 8–18-seeded, with 3 mm long style, explosively dehiscent. Seeds in 2 rows, 1.1–1.3 mm.

Distribution: Nepal and W Himalaya.

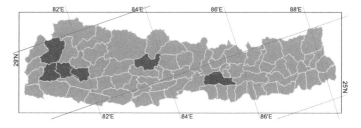

Altitudinal range: 2500–4100 m.

Ecology: Disturbed patches in alpine meadows, among bushes.

Flowering: June–September. **Fruiting:** July–September.

Similar to *Corydalis longipes* DC., but with inner petals usually tipped with blackish purple, nectary up to half as long as spur, and much larger sepals.

8. Corydalis pseudolongipes Lidén, Bull. Brit. Mus. (Nat. Hist.), Bot. 18(6): 532 (1989).

Short-lived perennial (rarely annual) herbs, 15–40 cm, glabrous. Stem weak, slender, leafy and branched throughout. Petioles 1–4(–7) cm. Leaves biternate, 1–4 cm; ultimate leaflets entire to 2–4-divided into obovate, obtuse lobes. Racemes 3–12-flowered, lax; bracts much divided or rarely uppermost entire, 3–6 mm. Pedicels 5–15 mm, slender, recurved in fruit. Sepals 1 mm, often deeply divided, rarely minute, entire. Corolla yellow; inner petals with blackish purple apex. Upper petal 10–13 mm, acute or very shortly acuminate, narrowly cristate with crest long decurrent on spur; spur up-curved, 5–7 mm; nectary thick, to 1 mm; lower petal with a basal pouch. Inner petals 5–7 mm. Stigma with 4 stalked apical papillae. Capsule pendent, obovoid, 5–7 × ca. 2 mm, smooth or minutely papillose, 4–10-seeded, explosively dehiscent; style 1.5 mm. Seeds in 2 rows, 1.2 mm, smooth.

Distribution: Nepal, E Himalaya, Tibetan Plateau and Africa.

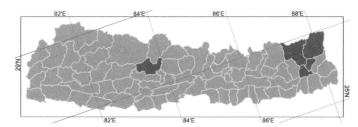

Altitudinal range: 2800–3500 m.

Ecology: Margins of *Abies* forests, slopes among shrubs.

Flowering: July–September. **Fruiting:** July–October.

Predominantly found in the Himalayan region but also disjunctly distributed on the temperate mountains of Kenya and Tanzania.

9. Corydalis calycina Lidén, Bull. Brit. Mus. (Nat. Hist.), Bot. 18(6): 532 (1989).

Annual to perennial herbs to 15–40 cm, glabrous. Stems few to several, weak, suberect, usually diffuse, branched throughout. Petioles 1–4(–7) cm. Leaves biternate, 1–4 cm, with deeply lobed leaflets; ultimate lobes obovate to narrowly obovate, obtuse. Racemes 5–14-flowered. Upper bracts small, entire; lower sometimes more or less divided, shorter than pedicels. Pedicels 10–20 mm. Sepals 2–3 × 1.5–2 mm, dentate to lacerate, whitish. Corolla 9–11 mm, yellow. Upper petal broad, truncate, mucronate; crest broad, truncate; spur 7–8 mm, slightly up-turned; nectary 3–4 mm; lower petal without crest or pouch. Inner petal 6–7 mm. Stigma with 4 stalked apical papillae. Capsule pendent, narrowly obovoid, 5–9 × 2 mm, 8–10(–14)-seeded, explosively dehiscent, with style 2–2.5 mm long. Seeds in 2 rows, 1.2 mm, shiny.

Distribution: Endemic to Nepal.

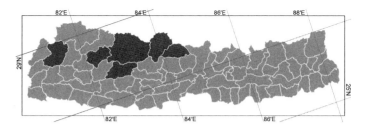

Altitudinal range: 3000–5500 m.

Ecology: Moist open grassy slopes, pastures, paths.

Flowering: June–September. **Fruiting:** July–October.

10. *Corydalis casimiriana* Duthie & Prain, J. Asiat. Soc. Bengal, Pt. 2, Nat. Hist. 65(1): 27 (1896).

Annual to short-lived perennial herbs, 20–70 cm. Stems slender, weak, much-branched from base, with many leaves. Petioles of lower leaves 2–10 cm, upper 0.5–3 cm. Leaves glaucous below, 1–4 × 1–4 cm, thin, bi-(tri-)ternate; ultimate leaflets deeply 2–5-divided; lobules obovate, obtuse. Racemes 4–12-flowered. Lower bracts divided; middle and upper bracts often entire, 2–8 mm. Pedicels 5–15 mm, slender, arcuately recurved in fruit. Sepals reniform, 0.5–1 mm, dentate. Corolla yellow; inner petals tipped with blackish purple; outer petals narrow, acute. Upper petal 9–11 mm, with narrow crest; spur often up-curved, 5–6 mm, narrow; nectary half as long; lower petal not crested, not saccate, but rarely with minute peg at base. Inner petals 4.5–5.5 mm. Stigma with 4 stalked, apical papillae. Capsule pendent, 6–10-seeded, explosively dehiscent; style 1.5–2 mm. Seeds uniseriate to subbiseriate, 1–1.1 mm, smooth.

Distribution: Nepal, W Himalaya, E Himalaya and Tibetan Plateau.

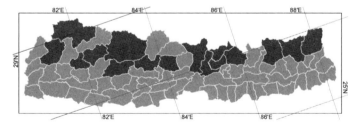

Self-compatible.

  1a Capsule 8–10 × 2 mm, narrowly obovoid, uni- to subbiseriate............................ subsp. ***brachycarpa***
   b Capsule 12–15 × 1–1.5 mm, linear, with seeds in one row .... subsp. ***casimiriana***

Corydalis casimiriana subsp. ***brachycarpa*** Lidén, Rheedea 5: 27 (1995).

Capsule 8–10 × 2 mm, narrowly obovoid, seeds uniseriate to subbiseriate.

Distribution: Nepal, E Himalaya and Tibetan Plateau.

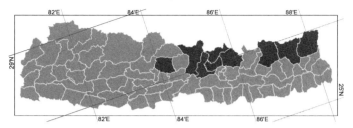

Altitudinal range: 2800–4200 m.

Ecology: Alpine slopes, open areas, among shrubs

Flowering: July–October. **Fruiting:** July–October.

Corydalis casimiriana Duthie & Prain subsp. ***casimiriana***

Capsule 12–15 × 1–1.5 mm, linear, seeds uniseriate.

Distribution: Nepal and W Himalaya.

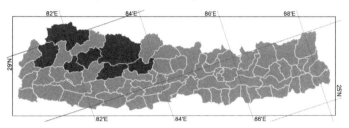

Altitudinal range: 2300–3800 m.

Ecology: Open grassy slopes, pastures, frequent in burned sites, understorey of open subalpine conifer forests, among shrubs.

Flowering: July–October. **Fruiting:** July–October.

11. *Corydalis lathyroides* Prain, J. Asiat. Soc. Bengal, Pt. 2, Nat. Hist. 65(2): 23 (1896).
Corydalis brevicalcarata Ludlow.

Delicate perennial herbs, 10–20 cm. Stems one to few, erect from cylindric or rounded tuberous rootstock, with 2–3 leaves, naked at base. Radical leaves long-petiolate, cauline subsessile. Leaves oblong, pinnate, 4–9 × 1–2 cm; leaflets sessile, 3–5(–6) pairs, entire or divided into 2–5 obovate-obtuse lobes. Racemes terminal and axillary from upper leaf, 4–12-flowered, lax, 1–4 cm. Bracts 3–4 mm, entire. Pedicels 5–15 mm, 15–20 mm in fruiting stage, erect to patent. Sepals 0.75–1.5 mm, shallowly dentate. Corolla pale yellow, 8–10(–12) mm, not or very narrowly cristate. Upper petal acute with short spur 3.5–6 mm, slightly up-curved; nectary very slender, two thirds as long; lower petal 7 mm long, acute, without pouch. Inner petals 6–7 mm. Stigma with 4 stalked apical papillae. Capsule pendent, narrowly obovoid, 6–8 × 1.5–2 mm, to 15-seeded, explosively dehiscent. Style 2 mm. Seeds in 2 rows, small, 0.7 mm long, with very long elaiosomes, glossy.

Fumariaceae

Distribution: Nepal, W Himalaya and E Himalaya.

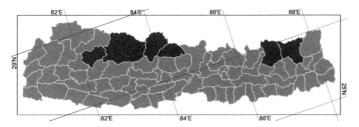

Altitudinal range: 3000–5500 m.

Ecology: Wet cliff ledges and shady rocks in moss.

Flowering: June–September. **Fruiting:** July–October.

12. *Corydalis vaginans* Royle, Ill. Bot. Himal. Mts. [1]: 69 (1833).
Corydalis ramosa Wall. ex Hook.f. & Thomson; *C. ramosa* Wall. nom. nud.; *C. ramosa* var. *nana* (Royle) Hook.f. & Thomson; *C. ramosa* var. *vaginans* (Royle) Hook.f. & Thomson.

Annual or biennial tap-rooted herbs to 15–50 cm. Stems erect or ascending, leafy and branched throughout, angular. Petioles vaginate, lower 6–15 cm, upper 1–4 cm. Leaves glaucous below, 3–8 × 3–6 cm, triangular-oblong, bipinnate to tripinnate (or first leaves ternate); ultimate lobes lanceolate. Racemes 8–15-flowered, elongating in fruit. Bracts 5–10 mm, lower deeply divided, upper entire. Pedicels 4–10 mm, recurved in fruit. Sepals 0.5 mm. Petals yellow to brownish yellow, often with brownish veins, inner petals pale at tip. Outer petals acute, with dentate dorsal crest; upper petal 14–17 mm; spur straight or slightly down-curved, 6–8 mm, slightly tapering toward apex; nectary three fifths as long; lower petal broadly saccate at base. Inner petals 6–7 mm. Stigma square with 4 apical papillae, a pair of lateral geminate papillae and basal geminate papillae on distinct basal lobes. Capsule pendent, obovoid, 9–12 × 3–4 mm, 7–11-seeded, explosively dehiscent; style 2.5 mm. Seeds in 2 rows, smooth, ca. 1.3 mm in diameter.

Distribution: Nepal and W Himalaya.

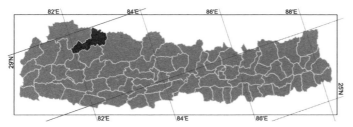

Altitudinal range: ca. 3000 m.

Ecology: Open stony slopes.

Flowering: August. **Fruiting:** August.

13. *Corydalis cornuta* Royle, Ill. Bot. Himal. Mts. [1]: 68 (1833).

Annual or biennial herbs to 15–50 cm. Stems erect or ascending, leafy and branched throughout, angular. Petioles vaginate, lower 6–15 cm, upper 1–4 cm. Leaves glaucous below, 3–8 × 3–6 cm, triangular-oblong, bipinnate to tripinnate; ultimate leaflets cuneate, obovate to oblong, mucronulate. Racemes 10–15-flowered. Bracts 4–8 mm, divided into acute lobes, or upper entire. Pedicels 2–5 mm, recurved in fruit. Sepals 0.5 mm, dentate. Petals yellow, inner petals tipped with blackish purple. Outer petals acute, with entire narrow dorsal crest; upper petal 14–16 mm; spur straight or slightly down-curved, 6–8 mm, slightly tapering toward apex; nectary three fifths as long; lower petal broadly saccate at base. Inner petals 6–7 mm. Stigma square with 4 apical papillae, a pair of lateral geminate papillae and basal geminate papillae on distinct basal lobes. Capsule pendent, obovoid, 8–9 × 2–3 mm, 8–16-seeded, explosively dehiscent; style 2.5 mm. Seeds in 2 rows, muricate, ca. 1 mm in diameter.

Distribution: Nepal, W Himalaya, E Himalaya and Africa.

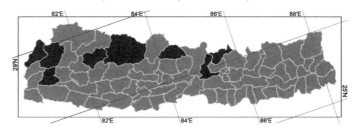

Altitudinal range: 2200–3500 m.

Ecology: Hillsides, forest margins, often in disturbed places.

Flowering: April–September. **Fruiting:** May–October.

14. *Corydalis sikkimensis* (Prain) Fedde, Repert. Spec. Nov. Regni Veg. 17: 201 (1921).
Corydalis duthiei var. *sikkimensis* Prain, J. Asiat. Soc. Bengal, Pt. 2, Nat. Hist. 65(2): 33 (1896).

Perennial herbs to 10–15(–20) cm, glabrous. Rootstock long, often multi-headed, with membranous petiole remnants. Stems several, erect, with few leaves. Radical leaves several, long-stalked with petioles 1–5 cm, 1–4 × 0.7–2 cm, ovate to oblong, bi-pinnate; pinnae 2 to 4 pairs; pinnules deeply divided; lobes narrowly lanceolate, pointed, 3–4 × 1 mm. Uppermost cauline leaves, small, sessile. Raceme 5–10-flowered, very dense, 1–3 cm, elongating in fruit. Lower bracts like upper leaf, upper smaller and less divided. Pedicels 5–12 mm, at fruiting recurved. Sepals 1 × 1 mm, dentate. Corolla yellow. Upper petal 15–16 mm, with broad crest tapering to both ends, not reaching apex; spur 5–6 mm; nectary half as long; lower petal spatulate, obtuse. Inner petals 10 mm. Stigma with small notch and two broad simple papillae at apex, geminate papillae lacking. Capsule pendent, obovoid, 8–9 × 2–2.5 mm, 3–8-seeded, explosively dehiscent; style 3–4 mm. Seeds in 2 rows, 1.0–1.3 mm, smooth.

Distribution: Nepal and E Himalaya.

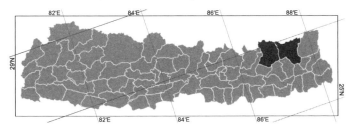

Altitudinal range: 3900–4300 m.

Ecology: Open wet rock scree.

Flowering: (May–)August. **Fruiting:** August.

15. *Corydalis govaniana* Wall., Tent. Fl. Napal. [Fasc. 2]: 55 (1826).

भूतकेश Bhutkesh (Nepali).

Perennial herbs to 15–35(–50) cm, glabrous or papillose-puberulent at base. Rootstock ± vertical, multi-stranded, with remnants of leaf sheaths and cataphylls. Stems 1 to few, erect, with (0–)2 opposite leaves below middle. Radical leaves many, with 5–10 cm, vaginate petioles; lamina ovate to triangular-ovate, bi-(tri-)pinnate, 6–14 × 3–7 cm; pinnae 4–5 pairs; pinnules deeply divided; lobes oblong to lanceolate, obtuse to acute. Cauline leaves small, vaginate. Raceme 5–15 cm, 10–25(–35)-flowered. Bracts (1–)1.5–3(–4) × 0.5–1 cm, pectinate-dentate, upper smaller and less divided. Pedicels 1–2(–3) cm, at fruiting recurved. Sepals 0.5–2 × 0.5–1.5 mm, dentate. Corolla yellow. Outer petals with rhombic-acute limb, crest 0.5–1.5 mm wide; upper petal 17–25 mm; spur tapering, 8–12 mm; nectary half as long. Inner petals 9–11 mm. Stigma square, flat, with broad diffuse simple papillae at apex, geminate papillae laterally and on basal lobes. Capsule pendent, obovoid, 10–15 × 3–4 mm, 4–8-seeded, explosively dehiscent; style 3.5–4 mm. Seeds in 2 rows, 1.5–2 mm, smooth.
Fig. 7c–g.

Distribution: Nepal, W Himalaya and Tibetan Plateau.

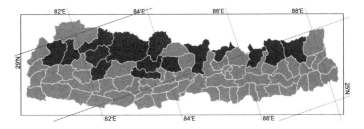

Altitudinal range: 1900–5000 m.

Ecology: Forest margins, moist slopes.

Flowering: April–July. **Fruiting:** May–August.

The roots are used against disorders from poisoning, swelling of the limbs and stomach pain from worm infection.

16. *Corydalis simplex* Lidén, Bull. Brit. Mus. (Nat. Hist.), Bot. 18(6): 514 (1989).

रोग-पो-जोम्स-स्काइज Rog-po-joms-skyes (Tibetan).

Perennial herbs 6–15 cm, glabrous. Rootstock narrow, with a few remnants of leaf sheaths and cataphylls. Stems 1 to few, erect, with 2 opposite leaves near base. Radical leaves with 2–4 cm, vaginate petioles; lamina ovate to oblong, pinnate, 3–5 × ca. 2 cm; pinnae ca. 3 pairs, deeply divided into broadly lanceolate to broadly ovate mucronate lobes. Cauline leaves small, vaginate. Raceme 2–4 cm, 4–11-flowered. Bracts lanceolate, entire, 5–10 mm. Pedicels 8–15 mm, arcuate-recurved and conspicuously elongating in fruit to 20–25 mm. Sepals 0.5 × 1 mm, dentate. Corolla yellow. Outer petals with rhombic-acute limb, crest 1 mm wide; upper petal 16–18 mm; spur tapering, 8–9 mm; nectary half as long. Inner petals 8 mm. Stigma square, flat, with broad simple papillae at apex, geminate papillae lateral (submarginal) and on basal lobes. Capsule pendent, short and broad, explosively dehiscent.

Distribution: Endemic to Nepal.

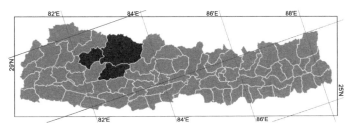

Altitudinal range: 4000–5500 m.

Ecology: Grassy slopes.

Flowering: June–July. **Fruiting:** July–August.

17. *Corydalis elegans* Wall. ex Hook.f. & Thomson, Fl. Ind. 1: 265 (1855).

Perennial herbs to 10–25 cm, very glaucous. Rootstock robust, crowned by membranous petiolar residues. Stems 2–4, simple or sparingly branched, with 0–2 leaves. Radical leaves with 5–10 cm, vaginate petioles; lamina oblong, 6–10 × 2–4 cm, pinnate to subbipinnate with 3–4 pairs of shortly stalked pinnae deeply cut into obovate lobes. Cauline leaves like radical but smaller and less divided. Racemes 6–12 cm, lax, 10–17-flowered, considerably elongating in fruit. Bracts obovate, lowermost often pinnatisect, 10–40 mm, middle and upper entire. Pedicels 12–30 mm, arcuately recurved in fruit and to 50 mm. Sepals 1–2 mm, dentate. Corolla bright yellow, inner petals darker at tip. Outer petals broadly crested (ca. 2 mm); upper petal 20–26 mm, acute; spur slightly down-curved, cylindric, 9–12 mm; nectary ca. half as long; lower petal base saccate, distal half reflexed. Inner petals 9–11 mm. Stigma square, flat, with four simple papillae at apex, inner ones small, distinct, outer broad, diffuse; geminate papillae laterally and on basal lobes. Capsule obovoid to oblanceolate, 6–12 × ca. 3 mm, explosively dehiscent; style 3–6 mm. Seeds in 2 rows.

Fumariaceae

Distribution: Nepal, W Himalaya and Tibetan Plateau.

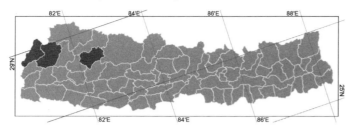

> 1a Flowers 20–23 mm. Style 5–6 mm
> .... subsp. *elegans*
> b Flowers 24–26 mm. Style 3–4 mm
> .... subsp. *robusta*

Corydalis elegans Wall. ex Hook.f. & Thomson subsp. ***elegans***

Leaflets obtuse to acute. Lower bracts to 2.5 cm. Fruiting pedicels to 3 cm. Sepals 1 × 1 mm, deeply dentate. Corolla 20–23 mm. Capsule 6–10 mm with 5–6 mm style.

Distribution: Nepal, W Himalaya and Tibetan Plateau.

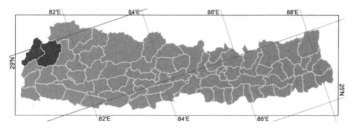

Altitudinal range: 3900–4200 m.

Ecology: Open slopes.

Flowering: July–September. **Fruiting:** August–October.

Corydalis elegans subsp. ***robusta*** Lidén, Bull. Brit. Mus. (Nat. Hist.), Bot. 18(6): 518 (1989).

Very robust. Leaflets acute. Lower bracts to 5 cm. Fruiting pedicels to 5 cm. Sepals 2 × 2–3 mm, fimbriate-dentate, Corolla 24–26 mm. Capsule 10–12 mm with 3–4 mm style.

Distribution: Endemic to Nepal.

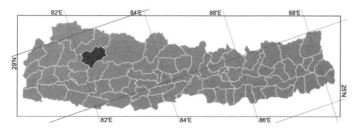

Altitudinal range: 3900–4700 m.

Ecology: Open slopes.

Flowering: July–August. **Fruiting:** August.

18. *Corydalis staintonii* Ludlow, in Ludlow & Stearn, Bull. Brit. Mus. (Nat. Hist.), Bot. 5: 65 (1975).
Corydalis chasmophila Ludlow; *C. sykesii* Ludlow.

Perennial herb, tufted, 8–15 cm. Rootstock multi-headed with dense straw-like petiole remnants. Stems numerous, 4–13(–18) cm with 0–3 leaves, simple. Radical leaves with 4–10 cm petioles, thin, 2–5 × 1–2 cm, pinnate to bipinnate; leaflets deeply cut into small obovate segments; cauline leaves small or absent. Racemes 2–17-flowered, occupying most of stem, lax. Lower bracts pinnatisect to entire, upper bracts entire, 10–30 mm. Pedicels 30–80 mm, arcuate-recurved in fruit. Sepals 0.5–2 mm, deeply laciniate-dentate. Corolla yellow with brown tips. Outer petals acute, crested; upper petal (14–)20–25 mm, subacute, crested; spur 6–13 mm, straight; nectary half as long. Stigma square, flat, with four simple papillae at apex, inner ones small, distinct, outer broad, diffuse; geminate papillae laterally and on basal lobes. Capsule linear to narrowly lanceolate, 14–20 mm × 2 mm, up to 15-seeded, apparently not explosively dehiscent; style 3–4 mm. Seeds 0.8 mm.

Distribution: Nepal and E Himalaya.

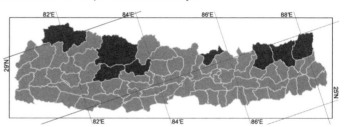

Altitudinal range: 3600–5500 m.

Ecology: Cliff crevices.

Flowering: May–July. **Fruiting:** July.

19. *Corydalis shakyae* Lidén, Bull. Brit. Mus. (Nat. Hist.), Bot. 18(6): 519 (1989).

?Perennial herb (basal parts unknown). Stems leafy, conspicuously zigzag, branched, ca. 40 cm. Lower cauline leaves with petioles to 10 cm; lamina triangular, to 10 × 10 cm (upper leaves much smaller), 3–4 times ternate to subpinnate with deeply divided leaflets; ultimate lobes narrowly oblanceolate, mucronate. Raceme 2–6 cm, 5–14-flowered. Bracts 7–15 mm, deeply ternatisect with irregularly divided segments, becoming smaller upwards. Pedicels 7–10 mm, arcuate-recurved in fruit. Sepals cordate, 1.5 mm, finely dentate. Corolla yellow, inner petals dark-tipped. Outer petals acute with dentate dorsal crests; upper petal 18–22 mm; spur 10–11 mm, tapering, slightly down-curved; nectary ⅔ as long. Inner petals 10–11 mm. Stigma square, flat, with four simple papillae at apex, inner ones small, distinct, outer broad, diffuse; geminate papillae laterally and on basal lobes. Capsule pendent, 8–10 mm, obovoid, 2–6-seeded, explosively dehiscent; style 3.5–4 mm. Seeds smooth, 1.5–2 mm.

Distribution: Nepal and W Himalaya.

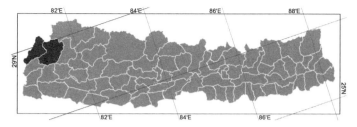

Altitudinal range: 3600–3900 m.

Ecology: Open subalpine steep slopes and rock crevices.

Flowering: July. **Fruiting:** July.

20. *Corydalis hookeri* Prain, J. Asiat. Soc. Bengal, Pt. 2, Nat. Hist. 65: 34 (1896).
Corydalis denticulatobracteata Fedde; *C. paniculata* C.Y.Wu & H.Chuang.

Perennial herbs 10–50 cm, glaucous. Rootstock slender. Stems few to several, ascending to suberect, leafy and branched throughout. Petioles shorter than to as long as blade. Lamina ovate-oblong, 6–17 × 4–10 cm, bipinnate; pinnae 3–6 pairs, petiolulate; pinnules deeply cut into 2–5 obovate obtuse to acute, often overlapping lobes. Racemes 4–12 cm, 10–30-flowered, dense, often branched at base. Bracts narrowly lanceolate to linear, 5–9 mm, lowermost often larger and divided. Pedicels 4–7 mm, recurved in fruit. Sepals rounded, 1–2 mm, finely dentate. Flowers dirty yellow to orange. Outer petals acuminate, broadly to very narrowly crested, crest reaching or overtopping apex, ± dentate; upper petal 14–17 mm; spur slightly down-curved, 6–8 mm, slightly tapering; nectary half as long; lower petal with rather broad claw, strongly reflexed, margin undulate. Inner petals 7–8 mm. Stigma square, flat, with four simple papillae at apex, inner ones small, distinct, outer broad, diffuse; geminate papillae laterally and on basal lobes. Capsule pendent, ovate or oblong, 6–8 × 2–3(–4) mm, 2–4-seeded, explosively dehiscent. Style 3–4 mm. Seeds 1.8–2 mm, smooth; elaiosome small. Fig. 8a–c.

Distribution: Nepal, E Himalaya and Tibetan Plateau.

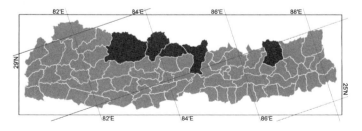

Altitudinal range: 2700–5500 m.

Ecology: Alpine grasslands, stony scree, field margins, disturbed sites.

Flowering: May–September. **Fruiting:** June–September.

21. *Corydalis spicata* Lidén, Bull. Brit. Mus. (Nat. Hist.), Bot. 18(6): 528 (1989).

?Perennial herb (basal parts unknown) to 50 cm, glaucous. Stem slender, leafy and branched. Petioles 5–11 cm. Leaves ovate-oblong, 10–15 × 4–7 cm, bipinnate; pinnae ca. 4 pairs, stalked; pinnules deeply cut into 2–5 obovate obtuse lobes. Racemes 5–10 cm, narrow, spicate, 15–40-flowered. Bracts lanceolate, entire, 3–5 mm, or lowermost larger and divided. Pedicels filiform, ca. 5 mm. Sepals minute. Corolla yellow, inner petals probably dark-tipped. Outer petals subacute, without crest; upper petal 14 mm; spur 8–9 mm, slightly tapering; nectary half as long; lower petal navicular, not reflexed. Inner petals 6 mm. Ovules 2; style 2 mm, conspicuously thickened below stigma. Stigma square, with 4 simple apical papillae, and lateral and basal geminate papillae. Fruit unknown.

Distribution: Endemic to Nepal.

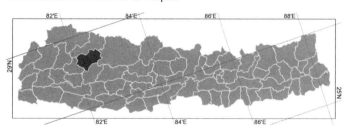

Altitudinal range: ca. 2600 m.

Ecology: Caves.

Flowering: August.

Known only from a few specimens in Jumla District.

22. *Corydalis filicina* Prain, J. Asiat. Soc. Bengal, Pt. 2, Nat. Hist. 65(2): 30 (1896).

Perennial herb 15–40 cm. Rootstock cylindric, subterranean with radical leaves. Stems few to several, slender, diffuse, usually branched with, 2–5 leaves. Petioles 3–7 cm. Leaves 4–7 × 2–3 cm, triangular to oblong, bi- to tri-ternate (rarely bipinnate). Leaflets well spaced, divided into small, linear to broadly obovate lobes. Racemes lax, 2–10 cm, 3–10(–15)-flowered. Bracts much shorter than pedicels, entire to divided. Pedicels 10–20 mm. Sepals cordate, obtuse, dentate, ca. 1 × 1.5 mm. Corolla yellow with darker veins, rarely white; inner petals dark-tipped. Outer petals acute to subacuminate, auriculate, crested; upper petal 16–20 mm; crest not reaching apex, decurrent on spur; spur 8–10 mm, tapering; nectary one third as long; lower petal rhombic in outline, shortly crested. Inner petals 10–11 mm. Stigma square, flat, with four simple papillae at apex, inner ones small, distinct, outer broad, diffuse; geminate papillae laterally and on basal lobes. Fruit (immature) linear, ca. 10-seeded. Style 3 mm.

Distribution: Nepal and E Himalaya.

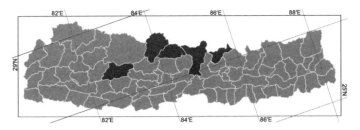

Altitudinal range: 3000–4500 m.

Ecology: Pasture, often in burnt sites, on wet organic soils.

Flowering: May–August. **Fruiting:** September.

23. *Corydalis clavibracteata* Ludlow, in Ludlow & Stearn, Bull. Brit. Mus. (Nat. Hist.), Bot. 5: 53 (1975).

Perennial herbs to 3–6(–10) cm above scree surface, very glaucous. Stems several from long, slender, deeply-buried rootstock, ± branched, with a few remnants of previous year's petioles and stems. Petioles 2–4 cm, sheathing at base. Leaves 2–4 × 1–1.5 cm, oblong, tri-pinnate; ultimate lobes linear. Racemes dense, corymbose, 10–20-flowered. Bracts 12–20 mm, clavate, entire or with 1–2 teeth at apex. Pedicels 10–20 mm, erect, in fruit down-curved at apex only. Sepals obovate, 1.5–2.5 × 1–1.5 mm, dentate, long persistent. Corolla yellow. Upper petal 14–15 mm, rhombic-acute, crested, crest attenuate to both ends, not reaching apex; spur tapering, 5–7 mm; nectary half as long. Stigma with 6 distinct marginal papillae and basal geminate papillae. Capsule pendent from erect pedicel, obovoid to oblong, ca. 5 × 1.5–2 mm, 3–6-seeded, explosively dehiscent. Style 3–3.5 mm. Seeds in one row, smooth.

Distribution: Endemic to Nepal.

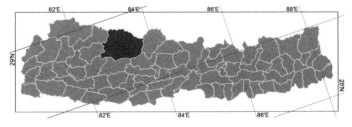

Altitudinal range: 3600–4700 m.

Ecology: Scree, gravel at stream-sides.

Flowering: July–September. **Fruiting:** August–September.

Known from several collections in Dolpa District.

24. *Corydalis megacalyx* Ludlow, in Ludlow & Stearn, Bull. Brit. Mus. (Nat. Hist.), Bot. 5: 58 (1975).

भूतकेश Bhutkesh (Nepali).

Perennial herbs to 3–6(–10) cm above scree surface, very glaucous. Stems several from long, slender, deeply buried rootstock, with a few remnants of previous season's petioles and stem-bases, ± branched. Petioles 2–3 cm, broad, often ciliate, sheathing at base. Lamina 1–3 cm, broadly triangular, bi- to tri-ternate with small deeply divided leaflets; ultimate lobes linear to obovate-lanceolate, mucronulate. Racemes dense, corymbose, 4–10-flowered. Bracts leaf-like, 10–20 mm. Pedicels 10–20 mm, erect, in fruit down-curved at apex only. Sepals 3–5 × 2–3 mm, ovate, deeply dentate, long persistent. Corolla sweetly scented, yellow to cream, rarely white; inner petals dark-tipped, upper petal with two dark spots at apex. Upper petal 14–17 mm, acute, broadly auriculate, crested, crest attenuate to both ends, not reaching apex; spur tapering, 7–8 mm; nectary half as long. Inner petals 7–8 mm. Stigma with 6 distinct marginal papillae and basal geminate papillae. Capsule pendent from erect pedicel, obovoid to oblong, 4–5 × 1.5–2 mm, 3–10-seeded, explosively dehiscent. Style 3 mm. Seeds smooth.

Distribution: Endemic to Nepal.

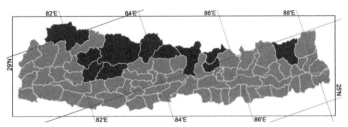

Altitudinal range: 3600–5500 m.

Ecology: Scree, often fine-grained.

Flowering: July–September. **Fruiting:** August–October.

25. *Corydalis latiflora* Hook.f. & Thomson, Fl. Ind. 1: 270 (1855).
Corydalis alburyi Ludlow & Stearn; *C. gerdae* Fedde; *C. mitae* Kitam.

Perennial herbs to 5–15 cm above scree surface, greyish purple. Rootstock long (often very long and deeply buried), narrow, crowned by a lax rosette of leaves and usually several stems. Stems above with 2 opposite leaves; basal parts of stems and petioles often thin, slender and flexuous, buried in scree. Leaves 3–7 × 2–7 cm, biternate (to bipinnate); leaflets entire to 2–5-divided; segments broadly obovate and obtuse to lanceolate and acute. Racemes corymbose, 2–8-flowered. Bracts oblanceolate-entire to flabellate-divided, 10–30 mm. Pedicels suberect, straight, 10–30 mm, at fruiting hooked at apex. Sepals ca. 0.5 × 1 mm, dentate. Corolla greyish blue to almost white, conspicuously spongy-undulate, with strong pleasant scent, apex of inner petals (excluding dorsal crests) and keels of outer petals dark-tipped. Outer petals with broadly triangular subobtuse apex, usually broadly crested; upper petal 15–20 mm, crest attenuate to spur end; spur slightly down-curved, 3(–4)mm, broad; nectary half as long; lower petal with broad subsaccate claw, limb sharply reflexed. Inner petals 10–14 mm. Stigma square, flat, with broad diffuse

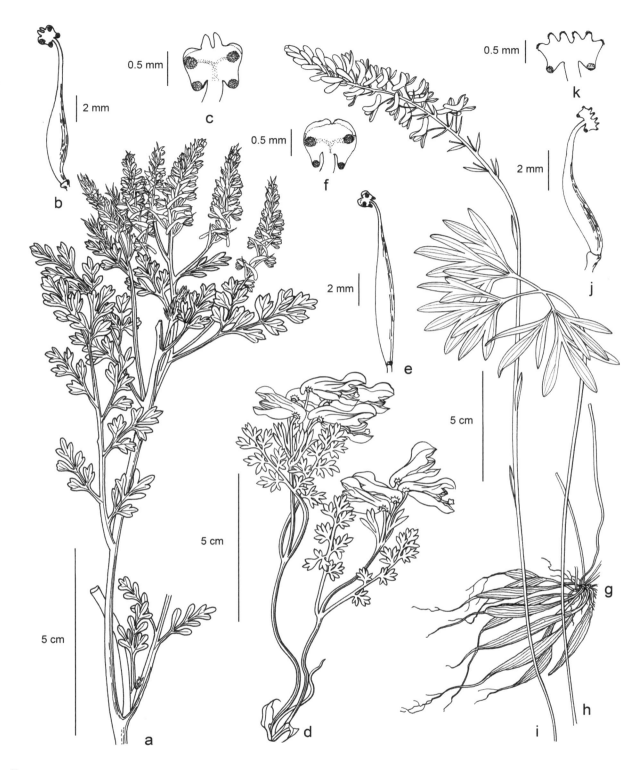

FIG. 8. FUMARIACEAE. **Corydalis hookeri**: a, upper part of plant with inflorescence; b, pistil; c, stigma. **Corydalis latiflora**: d, flowering plant; e, pistil; f, stigma. **Corydalis juncea**: g, base of plant with fleshy roots; h, leaf; i, inflorescence; j, pistil; k, stigma.

Fumariaceae

simple papillae at apex, geminate papillae laterally and on basal lobes. Capsule pendent from erect pedicels, obovoid, ca. 10 × 3–4 mm, 3–5-seeded, explosively dehiscent. Style 3 mm. Seeds smooth (rugose in type specimen). Fig. 8d–f.

Distribution: Nepal, E Himalaya and Tibetan Plateau.

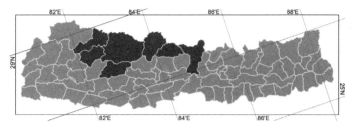

Altitudinal range: 3000–5700 m.

Ecology: Rough scree.

Flowering: July–August. **Fruiting:** August.

26. *Corydalis magni* Pusalkar, Kew Bull. 66: in press (2011).

Perennial herbs to 4–10 cm, blue-glaucous, fleshy, condensed. Rootstock long, slender, crowned by lax rosette of leaves and 1–several stems. Stems simple or branched with 3 or 4 leaves; lowermost leaves at middle of stem, often subopposite; basal parts of stems and petioles of radical leaves pale and flexuous, partly buried. Leaves triangular, 2–4 × 1–3 cm, bi- to tri-pinnate; leaflets deeply divided into oblanceolate fleshy crowded lobes. Cauline leaves narrower, shortly stalked. Racemes corymbose, 6–15-flowered. Bracts with broad cuneate base, flabellate-dissected into 3–10 linear segments, 10–25 mm. Pedicels suberect, straight, 15–25 mm, at fruiting hooked at apex. Sepals rounded, 0.5–1 × 0.5–1 mm, dentate. Corolla pale blue or white with darker apex, inner petals dark-tipped. Outer petals subacute, usually crested; upper petal 14–16 mm, crest not overtopping apex; spur slightly down-curved, 3–5(–6) mm; nectary half to two thirds as long; lower petal with rather long navicular claw and reflexed limb. Inner petals 9 mm. Stigma square, flat, with 4 simple papillae at apex (inner ones small and distinct), geminate papillae laterally and on basal lobes. Capsule pendent from erect pedicel, obovoid, ca. 6 × 2–3 mm, 5–10-seeded, probably explosively dehiscent. Style 3–3.5 mm. Seeds smooth.

Distribution: Nepal, W Himalaya and Tibetan Plateau.

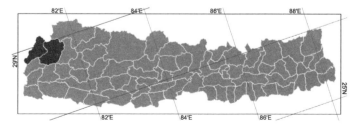

Altitudinal range: 4500–6100 m.

Ecology: Boulder scree.

Flowering: July. **Fruiting:** July–August.

Previously known in Nepal (e.g. Enum. Fl. Pl. Nepal 2: 34. 1979) by the misapplied name *Corydalis nana* Royle (syn. *C. ramosa* var. *nana* (Royle) Hook.f. & Thomson), which is a yellow-flowered, but otherwise similar, species endemic to Kumaon.

27. *Corydalis stracheyi* Duthie ex Prain, J. Asiat. Soc. Bengal, Pt. 2, Nat. Hist. 65: 37 (1896).

Perennial herb, very glaucous. Rootstock with a few remnants of previous years' petioles and stems, vertical, ± cylindric, often long, with age multi-stranded, with lax rosette of leaves and few to several, ascending, angular, leafy and branched 20–40(–60) cm stems. Petioles of basal leaves 5–12 cm, vaginate. Basal leaves narrowly ovate, 5–12 × 3–6 cm, bipinnate to almost tripinnate; ultimate segments lanceolate to linear, 3–5 mm long, acuminate. Cauline leaves short-stalked, smaller and less divided. Racemes 2–10-flowered, rather dense at first, soon elongating. Bracts 5–15 mm, pinnatifid to entire, lower more divided. Lower pedicels 10–15(–20) mm, upper shorter, arcuately recurved at fruiting. Sepals 1–2 × 1–2 mm, dentate. Corolla yellow, spur and keel of outer petals with purplish brown suffusion, or keels green. Outer petals subacute, usually crested; upper petal 11–14 mm; spur slightly down-curved, ca. 5 mm; nectary half as long. Inner petals 6–7 mm. Stigma square, flat, with broad indistinct simple papillae at apex, geminate papillae laterally and on basal lobes. Capsule pendent from recurved pedicels, broadly obovoid, 6–7 × 3–4 mm, 5–9-seeded, with 10 prominent papillose veins, explosively dehiscent. Style 2–2.5 mm. Seeds muricate, 1–1.3 mm.

Distribution: Nepal, W Himalaya, E Himalaya, Tibetan Plateau, Assam-Burma and E Asia.

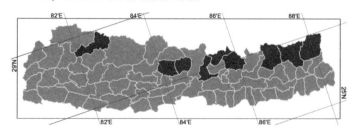

Altitudinal range: 3600–5200 m.

Ecology: Mountain slopes, often close to running water.

Flowering: June–October. **Fruiting:** July–October.

28. *Corydalis meifolia* Wall., Tent. Fl. Napal. [Fasc. 2]: 52, pl. 41 (1826).
Corydalis meifolia var. *sikkimensis* Prain.

घो-ठूल-सर्वो Gho-thul-serbo (Tibetan).

Perennial herbs 10–45 cm, very glaucous. Rootstock stout, thick at apex, branched or not, apex with very dense residual petiolar bases with persistent fibrous veins. Stems often several, simple or branched or with axillary inflorescences above, lowest cauline leaves at middle of stem, subopposite. Radical leaves with broadly vaginate petioles to 10 cm, ovate, 10–18 × 6–12 cm, finely tripinnate with 5–9 pairs of pinnae; ultimate lobes linear to lanceolate, 3–10 × 0.3–1.5 mm. Cauline leaves shortly petiolate, ca. 5 × 2 cm, bipinnate. Racemes corymbose, 3–5 cm, very broad and dense, 15–20(–35)-flowered, elongated at fruiting. Bracts pinnately to flabellately deeply divided into linear segments, about as long as pedicels. Pedicels erect, 10–30 mm, at fruiting 20–40 mm and hooked at apex. Sepals triangular-cordate, 1–3 × 1–2.5 mm, finely dentate, rarely deeply dentate. Corolla broad in profile, spongy, yellow to orange, base of corolla often brownish. Outer petals subacute; upper petal 13–18 mm, crest broad, often surpassing apex; spur 3(–4) mm, broad; nectary half as long; lower petal shallowly saccate. Inner petals 9–13 mm. Stigma square, flat, with broad indistinct simple papillae at apex, geminate papillae laterally and on basal lobes, or rarely lateral geminate papillae lacking. Capsule pendent, elliptic, 7–10(–12) × 3–4 mm, 4–6-seeded, explosively dehiscent. Style 3–4 mm. Seeds smooth.

Distribution: Nepal, W Himalaya, E Himalaya and Tibetan Plateau.

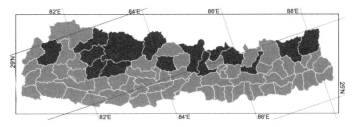

Altitudinal range: 3900–5200 m.

Ecology: Wet boulder slopes.

Flowering: June–September. **Fruiting:** July–October.

The plant is used to treat chronic fever, liver diseases, jaundice, bile fever, wounds, colds, ulcers and blood disorders. It can be used as a substitute for *Corydalis cashmeriana* Royle.

29. *Corydalis uncinata* Lidén, Bull. Brit. Mus. (Nat. Hist.), Bot. 18(6): 528 (1989).

Perennial herb 7–20 cm. Rootstock slender with few remains of previous years' growth at apex. Stems leafy, branched. Leaves mostly cauline. Petioles 4–6 cm. Leaves ovate, 3–6 × 2–4 cm, pinnate to bipinnate; leaflets deeply lobed; lobes obovate, obtuse, 3–10 mm broad, often overlapping. Racemes 5–10-flowered, very dense, elongating in fruit. Bracts entire, elliptic, 15–25 × 5–11 mm, concealing flowers in bud. Pedicels much lengthening in fruit to 20–30 mm, erect, apically hooked, sometimes slightly papillose. Flowers dull yellow, inner petals dark-tipped. Sepals rounded, 0.5–1 mm, dentate. Outer petals

acute, crested; upper petal 15–17 mm; spur down-curved, 7–8 mm; nectary half as long. Inner petals 8–9 mm. Stigma square with 4 apical papillae, a pair of lateral geminate papillae and basal geminate papillae on distinct basal lobes. Capsule pendent from erect pedicels, elliptic to obovoid, 10 × 3–4 mm, 6–10-seeded, explosively dehiscent. Style 2.5 mm. Seeds 1.6 mm, smooth.

Distribution: Endemic to Nepal.

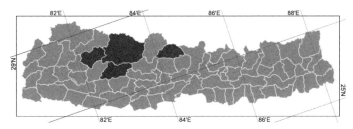

Altitudinal range: 3600–5500 m.

Ecology: Wet shale slopes, scree, field margins.

Flowering: June–August. **Fruiting:** July–August.

30. *Corydalis uncinatella* Lidén, Bull. Brit. Mus. (Nat. Hist.), Bot. 18(6): 528 (1989).

Perennial herbs to 10 cm. Rootstock slender. Stems lower part subterranean, aerial part to 10 cm, branched, leafy. Petioles 2–6 cm. Leaves ovate, 2–4 × 2–3 cm, pinnate to bipinnate; leaflets deeply lobed; lobes obovate, 2–5 mm broad, not imbricate, obtuse. Racemes 2–9-flowered, dense. Bracts 10–15 mm, flabellate, deeply lobed or the uppermost entire. Pedicels 10–15 mm, erect, apically hooked in fruit. Sepals 2–3 × 1.5 mm, acute, deeply dentate. Corolla yellow. Outer petals acute to subacuminate, broadly crested; upper petal 13–14 mm; spur down-curved, 6–7 mm; nectary two thirds as long. Inner petals 6–7 mm. Stigma square with 4 apical papillae, a pair of lateral geminate papillae and basal geminate papillae, basal lobes not developed. Capsule pendent, obovoid, 8 × 3 mm, 6–10-seeded, explosively dehiscent. Seeds 1.5 mm, smooth.

Distribution: Endemic to Nepal.

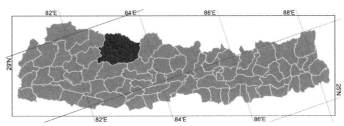

Altitudinal range: 3600–3800 m.

Ecology: Sandy scree.

Flowering: July. **Fruiting:** July.

31. *Corydalis hendersonii* Hemsl., J. Linn. Soc., Bot. 30: 109 (1894).
Corydalis nepalensis Kitam.

रेकोन Rekon (Tibetan).

Tufted perennial herbs 3–5 cm above scree surface, fleshy, brittle. Rootstock horizontal, long, crowned by dense leaf rosette, usually preceded by successive persistent leaf crowns from 1 or 2 previous years. Stem sparingly branched. Petioles flat, 1.5–3 cm, 4–5 mm broad, usually ciliate. Leaves triangular, ca. 1 × 1 cm, triternate, fleshy, very glaucous; ultimate lobes linear-oblong, 2–3 × ca. 1 mm. Racemes corymbose, very dense, 3–8-flowered. Lower bracts flabellate, much divided, 2–3 cm. Pedicels 12–18 mm, erect, apically hooked in fruit. Flowers yellow, held vertically with the spur hidden by leaves and bracts. Sepals 2–4 mm, narrowly linear. Outer petals rhombic, acute, without or with low (rarely high) crest; upper petal 19–25 mm; spur straight, conical-tapering, 10–11 mm; nectary two thirds as long; lower petal navicular. Inner petals 11–13 mm. Stigma square, flat, with four simple papillae at apex, inner ones small, distinct, outer broad, diffuse; geminate papillae laterally (submarginally), basal lobes and basal geminate papillae lacking. Capsule pendent from erect pedicel, hidden by bracts, oblong, 5–11 × 2.5–3 mm, 1–9-seeded. Style 5–6 mm. Seeds 1.8 mm, smooth.

Distribution: Nepal, W Himalaya, Tibetan Plateau and E Asia.

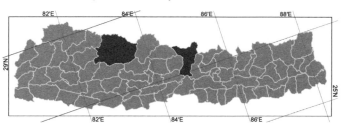

Altitudinal range: 4000–6100 m.

Ecology: Silty scree.

Flowering: June(–September). **Fruiting:** June(–September).

32. *Corydalis geraniifolia* Hook.f. & Thomson, Fl. Ind. 1: 269 (1855).
Corydalis chaerophylla var. *geraniifolia* (Hook.f. & Thomson) H.Hara.

Perennial herb 0.6–1.4 m, with stout rootstock with some scaly remains from previous years' growth at apex. Stem 0.6–1.4 m, leafy and branched in upper two thirds. Basal leaves few, very large, long-stalked, broadly triangular. Petioles of cauline leaves 2–5 cm. Cauline leaves green above, glaucous beneath, triangular, bipinnate, 8–14 × 10–15 cm, upper much smaller. Leaflets deeply dentate to pinnatifid into acute lobes. Racemes 4–10 cm, rather lax, 10–15-flowered, frequently branched at base. Bracts broadly flabellate, 7–10(–15) mm, ± deeply cut into acute lobes or the upper ovate, entire. Pedicels 4–6 mm, recurved in fruit. Sepals 1 × 1–1.5 mm, rounded, finely dentate. Corolla yellow. Outer petals without

crest, subacute; upper petal 22–23 mm, spur cylindric, slightly sigmoidally curved, 13–15 mm, nectary two thirds to three quarters as long; lower petal navicular, without or with a small pouch; distal half often reflexed. Inner petals 8–9 mm. Stigma square, flat, with four simple papillae at apex, inner ones small, distinct, outer broad, diffuse; geminate papillae laterally and on basal lobes. Capsule pendent, obovoid, 8–10 mm, 5–10-seeded, explosively dehiscent. Style 2–2.5 mm.

Distribution: Nepal and E Himalaya.

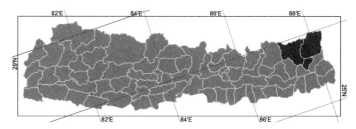

Altitudinal range: 3000–4800 m.

Ecology: On the margins and in clearings of humid forests.

Flowering: July–November. **Fruiting:** August–December.

33. *Corydalis terracina* Lidén, Bull. Brit. Mus. (Nat. Hist.), Bot. 18(6): 528 (1989).

Perennial herbs to 25–50 cm. Rootstock stout, apically with small fibrous petiole remnants. Stem erect, few, usually unbranched, with 1–3 leaves in upper half. Radical leaves several, in distinct rosette, with 5–10 cm petioles, triangular, 8–12 × 10–15 cm, ternate with pinnate to bipinnate long-stalked primary leaflets; ultimate leaflets ± deeply cut into oblong to ovate subacute lobes. Cauline leaves subsessile, 4–10 × 4–8 cm. Racemes dense, secund, 9–30-flowered, 4–15 cm. Bracts linear to narrowly obovate, 4–5 mm. Pedicels 4–5 mm, recurved in fruit. Sepals 3 × 3 mm, rounded, dentate. Corolla dirty yellow with brownish veins. Upper petal 18–24 mm, crested; spur cylindric, straight or slightly curved, 13–15 mm, nectary half to two thirds as long; lower petal saccate at base. Inner petals 10–12 mm. Stigma square, flat, with four simple papillae at apex, inner ones small, distinct, outer broad, diffuse; geminate papillae laterally and on basal lobes. Capsule pendent, obovoid, 8–10 mm, 5–10-seeded, explosively dehiscent. Style 2.5 mm. Seeds 1.5–1.9 mm, smooth.

Distribution: Endemic to Nepal.

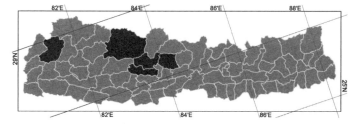

Altitudinal range: 1300–2700 m.

Ecology: Terrace banks and track-sides.

Flowering: April–October. **Fruiting:** May–November.

Easily distinguished from *Corydalis chaerophylla* DC. in growth habit. In the herbarium the unbranched stems, the longer and broader flowers, large rounded sepals and usually distinctly saccate lower petal are distinctive.

34. *Corydalis stipulata* Lidén, Bull. Brit. Mus. (Nat. Hist.), Bot. 18(6): 528 (1989).

Perennial herb to 1.2 m, crispate-puberulent throughout. Stem erect to ascending, leafy and branched. Basal parts unknown. Cauline leaves apparently with 'stipules' (leaves sessile with small lower pair of leaflets). Petiole (or rather first leaflet internode) 2–3 cm; blade 7–15 × 5–12 cm; leaflets deeply biternatifid with flabellate-dentate lobes. Upper leaves smaller, triangular, pinnate. Racemes ca. 5 cm, elongating to 10–20 cm in fruiting stage, subspicate, 12–25-flowered, simple or branched at base. Bracts obovate, 6–10 mm, entire, ciliate. Pedicels 4–5 mm, recurved in fruit. Sepals 1.5–2 × 1.5–2 mm, dentate, sometimes acuminate. Corolla yellow tinged brown; inner petals darker at apex. Outer petals acute, distinctly crested; upper petal 15–16 mm; spur slightly tapering, down-curved, 6–7 mm, nectary half to two thirds as long. Inner petals 8–9 mm. Stigma square, flat, with four simple papillae at apex, inner ones small, distinct, outer broad, diffuse; geminate papillae laterally and on basal lobes. Capsule pendent, obovoid, 8–10 mm, 5–10-seeded, explosively dehiscent. Style 2–2.5 mm. Seeds 1.3 mm, smooth.

Distribution: Endemic to Nepal.

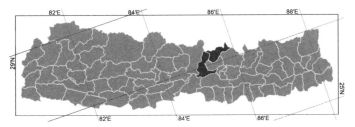

Altitudinal range: 2800–4000 m.

Ecology: Stream-sides in forest.

Flowering: August–October. **Fruiting:** September–October.

35. *Corydalis chaerophylla* DC., Syst. Nat. 2: 128 (1821).

ओखरे झार Okhare jhar (Nepali).

Perennial herbs 0.6–1.3 m. Rootstock stout with some scaly remains from previous years' growth at apex. Stem erect, leafy and branched in upper half. Radical leaves few, fern-like, with 10–30 cm petioles, triangular, 13–20 × 10–20 cm. Petioles of cauline leaves 2–6 cm. Cauline leaves green above, glaucous beneath, 7–15 × 5–12 cm; leaflets deeply dentate to pinnatifid into acute lobes. Upper leaves

subsessile, smaller, triangular, bipinnate. Racemes 5–15 cm, spicate, dense, secund, 10–40-flowered, sometimes branched. Bracts linear to narrowly obovate, 4–5 mm. Pedicels 4–5 mm, recurved in fruit. Sepals 1–1.5 × 1 mm, dentate, sometimes acuminate. Corolla pale yellow, often with brownish or greenish suffusion. Outer petals without or usually with narrow crest, subacute; upper petal 15–20 mm, spur cylindric, straight or slightly curved, 8–12 mm, nectary half to three quarters as long; lower petal navicular, not saccate, distal part often not reflexed. Inner petals 7–9 mm. Stigma square, flat, with four simple papillae at apex, inner ones small, distinct, outer broad, diffuse; geminate papillae laterally and on basal lobes. Capsule pendent, obovoid, 8–10 mm, 5–10-seeded, explosively dehiscent. Style 2–2.5 mm. Seeds 0.9–1.4 mm.

Distribution: Nepal, W Himalaya, E Himalaya and Assam-Burma.

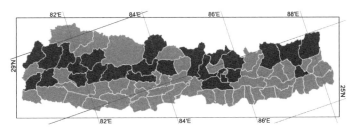

Altitudinal range: 1800–5500 m.

Ecology: Forests and clearings, often close to running water.

Flowering: April–September. **Fruiting:** May–November.

Juice of the roots is used to treat fever, indigestion and peptic ulcers.

36. *Corydalis leptocarpa* Hook.f. & Thomson, Fl. Ind. 1: 260 (1855).

Diffuse perennial herbs 15–50 cm, glabrous, flowering in its second (rarely first) year. Taproot persistent, crowned by very short, firm, knotty rhizome 5–10 × 3–5 mm, with traces of previous year's rosette leaves, apically with lax leaf rosette and a few to several stems. Stems decumbent, branched. Rosette leaves few, mostly withered at flowering, with 6–12 cm, vaginate petioles; lamina triangular, 2–8 × 2–6 cm, biternate; leaflets narrowly flabellate, entire to divided, coarsely crenate-dentate to lobed. Lamina of cauline leaves 3–7 cm. Racemes 2–4-flowered (to 7-flowered in E. Himalaya). Bracts broadly obovate to oblong-cuneate, 6–11 mm, deeply 3-divided or coarsely dentate with acute teeth to almost entire; upper entire. Pedicels 5–8 mm, in fruit elongating to 5–10 mm, only slightly arcuate. Sepals 0.5–1 mm, dentate. Corolla purple or white. Outer petals subacuminate, shortly crested; upper petal 22–25 mm; spur straight, slightly tapering, 11–12 mm; nectary one third to half as long; lower petal not reflexed, with long claw that often has a slight pouch halfway to apex. Inner petals 11–13 mm. Stigma square, flat, with four simple papillae at apex, inner ones small, distinct, outer broad, diffuse; geminate papillae laterally and on basal lobes. Capsule linear,

20–35 × ca. 2 mm, 10–20-seeded, possibly not explosively dehiscent. Style 2 mm. Seeds 1.5 mm, smooth.

Distribution: Nepal, E Himalaya, Assam-Burma, E Asia and SE Asia.

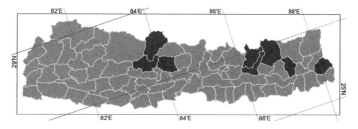

Altitudinal range: 1800–3000 m.

Ecology: Walls, roadsides, clearings in broad-leaved forest.

Flowering: May–July. **Fruiting:** May–July.

37. Corydalis juncea Wall., Tent. Fl. Napal. [Fasc. 2]: 54, pl. 42 (left side) (1826).

Perennial herb 10–45 cm, glabrous. Stems and petioles of radical leaves attenuate to filiform underground base. Storage roots several, spindle-shaped, 1–5 cm, 2–6 mm in diameter, fleshy, sessile or shortly stalked. Stems 1–5, simple, with 0–2 leaves. Radical leaves few, with 10–25 cm petioles, green above, glaucous below, deltoid, 4–7 × 4–10 cm, bi- to triternate; leaflets entire to deeply 3- or 4-divided, lobules narrowly lanceolate to broadly oblanceolate, acute. Cauline leaves entire, sessile, lanceolate, 1–6 × 0.2–1 cm. Raceme 5–20 cm, (5–)10–30(–40)-flowered. Bracts linear to lanceolate, 5–20(–40) mm, entire. Pedicels 5–15(–20) mm, recurved in fruit. Sepals 0.1–0.8 mm, dentate. Petals yellow, inner petals with sharply contrasting black purple apex. Outer petals navicular-acute, crested, upper petal 10–15 mm; spur cylindric, 5–7 mm; nectary half as long. Inner petals 6–8 mm. Stigma broad, bifid, with 8 papillae (basal ones geminate). Capsule pendent, oblong, 10–13 × 1.5–2 mm, 8–10-seeded, explosively dehiscent. Style 3 mm. Seeds ca. 1 mm, smooth. Fig. 8g–k.

Distribution: Nepal, E Himalaya, Tibetan Plateau and Assam-Burma.

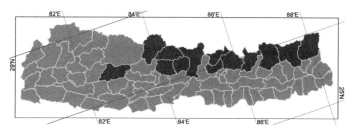

Altitudinal range: 2500–5100 m.

Ecology: Alpine pastures, often among low shrubs.

Flowering: (April–)June–October. **Fruiting:** June–October.

38. Corydalis pseudojuncea Ludlow, in Ludlow & Stearn, Bull. Brit. Mus. (Nat. Hist.), Bot. 5(2): 62 (1975).

Perennial herbs 10–30 cm, glabrous. Stems and petioles of radical leaves attenuate to filiform underground base. Storage roots several, fascicled, spindle-shaped, sessile. Stems simple, with 1 leaf. Radical leaves few, with very long thin petioles. Leaves small, 1–3 cm across, biternate, with narrow leaflets. Cauline leaf sessile, lanceolate, 2–6 cm × 4–5 mm, entire. Raceme 2–3(–6) cm, 2–6-flowered. Bracts narrowly lanceolate, 30–50 mm, entire. Pedicels 3–5 mm, recurved in fruit. Sepals minute. Petals pale yellow. Upper petal 13–18 mm, navicular-ovate, acute, crest very narrow or absent; spur slightly attenuate, 7–9 mm; nectary two thirds as long; lower petal navicular-oblanceolate. Inner petals not dark-tipped, 6–7 mm. Stigma broad, bifid, with 8 papillae (basal ones geminate). Capsule pendent, linear, 19–22 × ca. 2 mm, 8–10-seeded, explosively dehiscent. Style 2.5 mm.

Distribution: Nepal, W Himalaya and Tibetan Plateau.

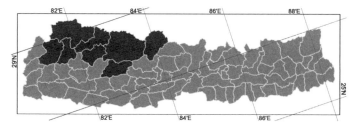

Altitudinal range: (2600–)3800–5500 m.

Ecology: Grassland on slopes.

Flowering: (?May–)June–July. **Fruiting:** (June–)July.

39. Corydalis lowndesii Lidén, Bull. Brit. Mus. (Nat. Hist.), Bot. 18(6): 491 (1989).

Perennial herb 10–20 cm, glaucous, glabrous. Stems and petioles of radical leaves attenuate to filiform underground base. Storage roots several, 1.5–5 cm × 3–4 mm, narrowed into often branched stalk. Stems 2–6, with 2–4 leaves, nearly always with late branches, ascending. Radical leaves few, with very long, thin petioles, ternate to biternate, 2–4 cm; leaf lobes linear to lanceolate. Cauline leaves sessile, 2–4 × 2–4 cm, similar to radical leaves. Racemes 1–2 cm, 2–7-flowered, dense, elongating in fruit. Lower bracts deeply divided into linear acute lobes, 8–12 mm, upper bracts often entire. Pedicels 4–10 mm, recurved in fruit. Sepals ca. 0.5 × 1 mm, dentate. Corolla dull yellow, masked with greyish ochre. Upper petal 10–14 mm, acute; crest narrow, attenuate to both ends, not reaching apex; spur slightly down-curved, 5–7 mm, attenuate; nectary ca. half as long; lower petal navicular-oblanceolate, slightly dentate, crest usually lacking. Inner petals 6–7 mm. Stigma square, flat, with four simple papillae at apex, inner ones small, distinct, outer broad, diffuse; geminate papillae laterally and on basal lobes. Capsule pendent, narrowly obovoid, ca. 1 cm, 6-seeded, explosively dehiscent. Style 2.5 mm. Seeds ca. 1.6 mm, smooth.

Distribution: Nepal and Tibetan Plateau.

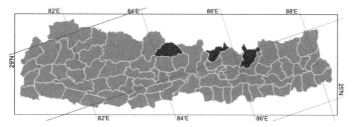

Altitudinal range: 2800–5000 m.

Ecology: Stony turf among boulders on open hillsides.

Flowering: June–September. **Fruiting:** July–September.

40. *Corydalis polygalina* Hook.f. & Thomson, Fl. Ind. 1: 263 (1855).

Perennial herbs 10–40 cm. Stems and petioles of radical leaves attenuate to filiform underground base. Storage roots long, attenuate into distinct, sometimes branched stalk. Stems few, simple or often with late branches, with 1–4 leaves in upper half. Radical leaves with long, thin petioles, ternate to biternate or pinnate; pinnae narrowly lanceolate, with prominent parallel veins below. Cauline leaves subsessile. Leaves 2–5 × 2–6 cm, pinnate; pinnae 2 or 3 pairs, linear, 1–5 cm, sometimes lowermost pinna again cleft to base. Racemes 3–4 cm, elongating at fruiting, 5–15-flowered. Bracts 5–15 mm, lanceolate, deeply pectinate to entire. Pedicels 10–20 mm, recurved in fruit. Sepals 0.5–1 × 0.5–2 mm, dentate. Outer petals yellow, often with distinct veins, often orange/brownish at tip, broadly rhombic-acute to shortly acuminate; upper petal 14–19 mm, with narrow to broad crest tapering to both ends, barely reaching apex, often long decurrent on spur; spur straight, cylindric to slightly tapering, 6–9 mm; nectary ca. half as long. Inner petals 6–7 mm. Stigma square, flat, with four simple papillae at apex, inner ones small, distinct, outer broad, diffuse; geminate papillae laterally and on basal lobes. Capsule pendent, oblong to narrowly obovoid, 5–10 mm, 2–9-seeded, explosively dehiscent. Style 2.5–3 mm. Seeds ca. 1.2 mm, smooth.

Distribution: Nepal, E Himalaya and Tibetan Plateau.

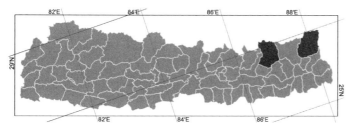

Altitudinal range: 4000–4800 m.

Ecology: Grassland, slopes.

Flowering: June–July. **Fruiting:** July–August.

41. *Corydalis trifoliata* Franch., Bull. Soc. Bot. France 33: 392 (1886).

Perennial herbs 8–25 cm, glabrous. Rhizome small, deeply buried, apically with insignificant pale bulb-like resting bud 1–2 mm. Stems and petioles of radical leaves attenuate to filiform underground base. Storage roots several, densely fascicled, spindle-shaped, 10–15 × 3–5 mm, sessile. Stems 1 or 2, erect, slender, simple, with one leaf in upper half. Radical leaves few, with 3–12 cm, slender petioles, 0.5–2 × 1–3 cm, ternate; leaflets shortly stalked to subsessile, obovate, entire or often slightly 2- or 3-divided at apex. Petioles of cauline leaves 0.5–2 cm. Leaves trifoliate; leaflets subsessile, elliptic, 1–3 × 0.5–1.5 cm. Raceme very dense, 2–5-flowered, not elongating in fruit. Bracts 10–15 mm, lowermost broadly elliptic, entire or deeply 3-fid; others ovate to lanceolate, entire. Pedicels 3–9 mm, in fruit erect and to 7–15 mm. Sepals 0.5–1 mm, fimbriate-dentate. Corolla blue with white markings. Outer petals navicular subacute, with crest narrow or absent; upper petal 12–18 mm, spur straight, cylindric, 5–9 mm; nectary half as long. Inner petals 7–9 mm. Stigma square, flat, with four simple papillae at apex; geminate papillae laterally and on basal lobes. Capsule pendent from erect pedicel, elliptic, 11–15 × 3–4 mm, 18–25-seeded, explosively dehiscent. Style 2–2.5 mm. Seeds 0.8–1 mm, smooth. Fig. 9a–c.

Distribution: Nepal, E Himalaya, Tibetan Plateau, Assam-Burma and E Asia.

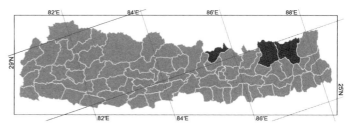

Altitudinal range: 3000–4300 m.

Ecology: Mountain slopes, in bushes, often in wet habitats.

Flowering: July–October. **Fruiting:** July–October.

Sometimes this species is misspelt '*trifoliolata*' in past publications.

42. *Corydalis cashmeriana* Royle, Ill. Bot. Himal. Mts. [2]: 69 (1834).

टोडग्रे-सिल्बा Tongre-silba (Tibetan).

Perennial herbs to 5–20 cm, glabrous. Stems and petioles of radical leaves attenuate to filiform underground base. Storage roots fascicled, 10–35 × 2–4 mm, sessile, tapering distally, set on small rhizome with apical prominent bulb-like resting bud 3–4 × 2–4 mm, composed of several pale fleshy scales. Stems erect, slender, simple, with 1 or 2 leaves in upper half. Radical leaves few, subordinate, with 4–8 cm, slender petioles, ternate, petiolules short to rather long; leaflets deeply ternatisect (rarely biternatisect); lobes narrowly elliptic, 4–11 × 1–2 mm. Cauline leaves sessile, rarely with petiole to 1 cm. Leaves 1–3 cm, bi- to tri-ternate with narrowly oblanceolate segments 1–2 mm wide. Raceme 2–5 cm, 2–8-flowered, elongating in fruit. Bracts 8–12 mm, deeply divided into 2–7 narrowly oblanceolate to linear lobes. Pedicels 10–25 mm in flower, erect and elongating to 15–30(–40) mm in fruit. Sepals minute, slightly dentate. Corolla blue, often tip of inner petals darker. Upper petal 16–21 mm, crest present or absent; spur straight to slightly down-curved, cylindric or slightly tapering, 9–12 mm; nectary half to two thirds as long; lower petal with rhombic-acute limb, surpassing upper by 1–4 mm. Inner petals 8–10 mm. Stigma square, flat, with four simple papillae at apex; geminate papillae laterally and on basal lobes. Capsule pendent, linear-oblong, 10–15 × 1.5–2 mm, 10–20-seeded, explosively dehiscent. Style 2.5 mm. Seeds smooth.

Distribution: Nepal, W Himalaya, E Himalaya and Tibetan Plateau.

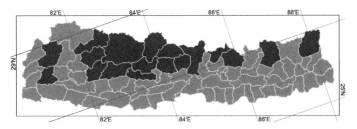

Juice of the plant is taken internally to treat chronic fever, jaundice and to relieve thirst.

1a Racemes 3–8-flowered .... subsp. **cashmeriana**
 b Racemes 2-flowered .... subsp. **longicalcarata**

Corydalis cashmeriana Royle subsp. **cashmeriana**

Racemes 3–8-flowered.

Distribution: Nepal, W Himalaya and Tibetan Plateau.

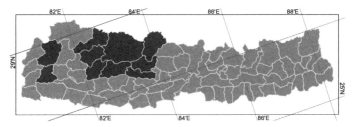

Altitudinal range: 2400–5500 m.

Ecology: Alpine meadows, mountain slopes.

Flowering: April–August. **Fruiting:** June–September.

Corydalis cashmeriana subsp. **longicalcarata** (D.G.Long) Lidén, Fl. China 7: 392 (2008).
Corydalis ecristata var. *longicalcarata* D.G.Long, Notes Roy. Bot. Gard. Edinburgh 42(1): 93 (1984); *C. cashmeriana* var. *longicalcarata* (D.G.Long) R.C.Srivastava; *C. ecristata* subsp. *longicalcarata* (D.G.Long) C.Y.Wu.

Racemes 2-flowered.

Distribution: Nepal, E Himalaya and Tibetan Plateau.

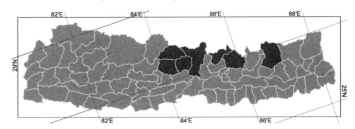

Altitudinal range: 3200–4600 m.

Ecology: Alpine meadows, mountain slopes.

Flowering: May–July. **Fruiting:** June–July.

43. Corydalis ecristata (Prain) D.G.Long, Notes Roy. Bot. Gard. Edinburgh 42(1): 91 (1984).
Corydalis cashmeriana var. *ecristata* Prain, J. Asiat. Soc. Bengal, Pt. 2, Nat. Hist. 65: 22 (1896).

Perennial herbs to 5–10 cm. Stems and petioles of radical leaves attenuate to filiform underground base. Storage roots few, fascicled, 12–30 mm, 2–4 mm in diameter, tapering distally, set on small rhizome with apical prominent bulb-like resting bud 5–8 × 5–10 mm, composed of several pale fleshy scales. Stems erect, thin, simple, with (0)1 leaf in upper third. Radical leaves few, subordinate, with 2–6 cm, thin petioles; lamina 6–15 × 8–15 mm, bi-(tri-)ternate; leaflets deeply divided into obovate to elliptic lobes. Cauline leaf sessile, 0.8–2.5 cm, deeply bi- to tri-ternatisect into linear lobes 0.5–1(–1.5) mm wide. Raceme umbellate, 2–6-flowered. Bracts 8–13 mm, deeply cut into 3–13 linear lobes. Pedicels 8–25 mm, thin, erect in fruit. Sepals minute. Petals clear blue, inner petals pale but darker at tip. Upper petal 13–17 mm, narrow, subacute, crest narrow or absent; spur straight or down-curved apically, cylindric, 6–11 mm; lower petal 10–13 mm, limb orbicular, much surpassing upper petal, 7–10 mm wide, apex truncate. Inner petals 6–9 mm. Stigma square, flat, with four simple papillae at apex; geminate papillae laterally and on basal lobes. Capsule reflexed from straight pedicel, lanceolate, 5–8 × 1.5–2 mm, ca. 10-seeded, explosively dehiscent. Style 2.5 mm.

Distribution: Nepal, E Himalaya and Tibetan Plateau.

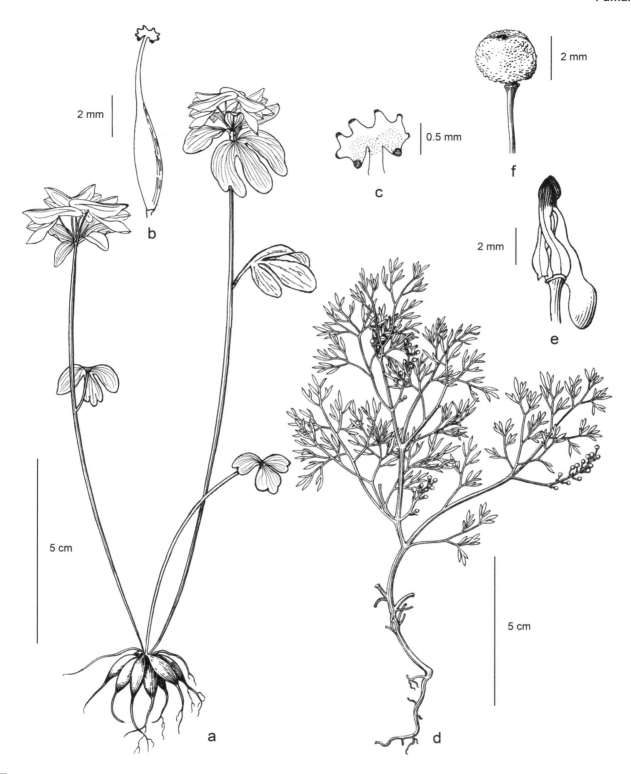

FIG. 9. FUMARIACEAE. **Corydalis trifoliata**: a, flowering plant; b, pistil; c, stigma. **Fumaria indica**: d, fruiting plant; e, flower; f, fruit.

Fumariaceae

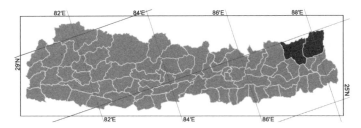

Altitudinal range: 3900–5000 m.

Ecology: Mossy rocks, fine scree.

Flowering: June–September. **Fruiting:** July–September.

44. *Corydalis jigmei* C.E.C.Fisch. & Kaul, Kew Bull. 1940: 266 (1940).

Perennial herbs 4–7(–12) cm, glabrous. Stems and petioles of radical leaves attenuate to filiform base. Storage roots fascicled, 10–15 × 1–3 mm, tapering distally. Bulb (resting bud) large, 8–12 × 10–15 mm, of several pale fleshy ovate scales. Stems 2–5, erect, thin, simple. Radical leaves conspicuous, with 1.5–4 cm petioles, grey-glaucous, sometimes with small spots, orbicular-triangular, 6–15 × 12–20 mm, fleshy, ternate; leaflets shortly stalked, once to twice deeply ternatisect into obovate lobes. Cauline leaf absent or small, shortly stalked. Raceme subumbellate, 1–2 cm, 3–5-flowered. Bracts 4–8 mm, deeply 3–8-divided into oblanceolate lobes, upward smaller and less divided. Pedicels 5–15 mm. Corolla clear blue, sometimes masked with purple, inner petals darker at tip. Upper petal navicular, acute, 9–12 mm, often narrowly crested; spur cylindric, 3–6 mm; nectary half as long; lower petal with distinct claw and rhombic acute to obtuse-mucronate limb. Inner petals 5–7 mm. Stigma with deep median sinus and 4(or 6) broad marginal papillae; geminate papillae usually only on basal lobes. Capsule reflexed from straight pedicel, narrowly elliptic, 5–6 × 1.5 mm, ca. 10-seeded, explosively dehiscent. Style 2.5 mm.

Distribution: Nepal, E Himalaya and Tibetan Plateau.

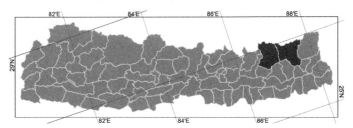

Altitudinal range: 4100–5300 m.

Ecology: Mossy cliffs, alpine meadows.

Flowering: June–September. **Fruiting:** July–October.

45. *Corydalis conspersa* Maxim., Fl. Tangut.: 49, pl. 25 (1889).
Corydalis zambuii C.E.C.Fisch. & Kaul.

Perennial herbs 10–30 cm, glabrous. Rhizome with long thick soft roots and radical leaves in 2 rows. Flowering stems 2–6, basally decumbent, ascending, simple or often branched above, with 3–7 leaves. Basal leaves with petiole as long as blade, long vaginate. Lamina 10–12 × 3-4 cm, ovate to oblong, bipinnate; pinnae 3 or 4 pairs; pinnules deeply 3-lobed; lobes often imbricate, elliptic to ovate. Cauline leaves becoming progressively smaller upward; upper leaves only 2–3 × 1.5–2 cm. Raceme capitate to subspicate, 2–4 × 2–2.5 cm, 15–30-flowered. Bracts spatulate, 6–8 mm; limb purple, membranous, finely erose-dentate. Pedicels 5–8 mm, erect in fruit. Sepals early withered, brown, 3–4 × 3–4 mm, deeply laciniate-dentate. Flowers creamy white to yellow, often with brown flecks, spur sometimes purplish, inner petals with blue dorsal crests. Outer petals obtuse, slightly dentate, with 1.5–2 mm broad short crest much overtopping apex. Upper petal 15–16 mm; spur strongly curved in a tight semicircle, 4–6 mm, apex thickened; nectary ca. half as long as spur; lower petal with long straight claw and rounded limb. Stigma slightly emarginate with 4 distinct marginal simple papillae in upper half; geminate papillae lateral (sometimes lacking) and in basal corners. Capsule pendent, oblong to obovoid, ca. 10 × 4 mm, explosively dehiscent. Style 4 mm.

Distribution: Nepal, Tibetan Plateau and E Asia.

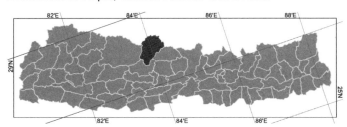

Altitudinal range: ca. 1500 m.

Ecology: Stony riversides, wet scree.

Flowering: (July–)August(–September). **Fruiting:** (July–September).

Corydalis hamata Franch. was recorded from Nepal by Whitmore (Enum. Fl. Pl. Nepal 2: 34. 1979) based only on *Stainton, Sykes & Williams 2316* (BM), but this name is considered misapplied as this collection has been reidentified as *C. conspersa*.

4. *Fumaria* L., Sp. Pl. 2: 699 (1753).

Tap-rooted glabrous summer annuals. Stems diffuse, leafy and branched throughout, without tendrils. Leaves 3–4 times ternately to pinnately divided. Inflorescence racemose, terminal (appearing leaf-opposed due to displacement of the primary main shoot by the uppermost axillary shoot) spike-like; bracts small. Sepals petaloid, dentate. Corolla with one plane of symmetry. Upper petal with short rounded spur. Stamens 2, broad, whitish-translucent, each with a central dithecal and 2 lateral monothecal anthers. Nectary small, obtuse, protruding from base of upper anther into spur of upper petal. Fruit a one-seeded nut, ± globular with 2 apical germination pores. Style whitish-translucent, early caducous.

Worldwide 51 species from the Mediterranean to the Himalaya and C Asia; a few species widely distributed as weeds. One species in Nepal.

1. *Fumaria indica* (Hausskn.) Pugsley, J. Linn. Soc., Bot. 44: 313 (1919).
Fumaria vaillantii var. *indica* Hausskn., Flora 56: 443 (1873).

ढुकुरे झार Dhukure jhar (Nepali).

Herb, 10–30 cm, summer annual, tap-rooted, squashy and glaucous. Stems angular, leafy and diffusely branched from base. Leaves alternate, finely divided into narrowly lanceolate to linear lobes to 1 mm broad. Raceme 10–20(–25)-flowered. Bracts three quarters to one and a quarter times as long as the erect, thickened 3–5 mm pedicels. Sepals small, denticulate. Corolla 5–6 mm, pink, inner petals tipped with dark purple. Outer petals truncate to emarginate. Fruit subglobose, truncate, 2–2.25 × 2.25(–2.5) mm, densely rugose.
Fig. 9d–f.

Distribution: Nepal, W Himalaya, E Himalaya, S Asia and C Asia.

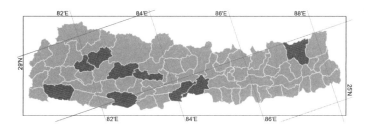

Altitudinal range: 100–2500 m.

Ecology: Open ground, fields, roadsides.

Flowering: May–September. **Fruiting:** May–September.

Wight & Arnott (Prodr. Fl. Ind. Orient. 1: 18. 1834), Banerji (J. Bombay Nat. Hist. Soc. 51: 547. 1953) and Whitmore (Enum. Fl. Pl. Nepal 2: 36. 1979) have all misapplied the name *Fumaria parviflora* Lam. to this species. Whitmore also records *F. vaillantii* Loisel. (syn. *F. parviflora* subsp. *vaillantii* (Loisel.) Hook.f. occuring in Nepal based on *TI 6302841* (TI), but this is also *F. indica*.

The plant is used as fodder. Juice of the plant is taken as an anthelmintic, and also applied to cuts and wounds.

Papaveraceae

Paul A. Egan, Colin A. Pendry & Sangita Shrestha

Herbs, annual or perennial, monocarpic or polycarpic. Latex usually copious, white, yellow or clear. Glabrous or with a more or less dense indumentum, rarely prickly. Leaves alternate, entire or lobed, often in more or less dense basal rosettes. Petioles shorter in upper leaves. Flowers solitary and scapose, terminal or axillary, or in racemes, panicles or few-flowered cymes, usually large and showy, bisexual, actinomorphic. Sepals 2 or 3, free, with or without an apical horn, soon caducous. Petals 4–6(–10), in 2 whorls on short receptacle, free, crumpled in bud. Stamens numerous, free, filaments filiform, anthers basifixed, longitudinally dehiscent. Ovary superior, 1-locular, 4–12(–18)-carpellate. Style absent to conspicuous, stigmas capitate or variously lobed. Fruit a capsule, dehiscing by subapical pores or valves leaving persistent placentae attached to stigma. Seeds small, numerous.

A cosmopolitan family of about 40 genera and 800 species, tropical to alpine. In Nepal five genera and 29 species.

Key to Genera

1a	Plants with straw-coloured prickles to 1 cm throughout. Leaves glaucous green with pale markings on venation above......	**2. Argemone**
b	Plants glabrous to setose, but not prickly. Leaves evenly coloured above......	2
2a	Leaves palmately lobed or trifoliate......	**4. Cathcartia**
b	Leaves pinnately lobed or pinnately veined	3
3a	Stigmas united into a sessile, lobed disc on the apex of the ovary. Style absent. Latex white......	**1. Papaver**
b	Stigmas united on a distinct style. Latex yellow......	4
4a	Ovary at least 4-carpellate. Sepals acute or rounded, but without apical horn	**3. Meconopsis**
b	Ovary 2-carpellate. Sepals with short apical horn......	**5. Dicranostigma**

1. *Papaver* L., Sp. Pl. 1: 506 (1753).

Colin A. Pendry

Annual herbs, glabrous or setose, glaucous or not, latex whitish, milky. Taproot slender. Stems erect, simple or branched throughout. Leaves cauline and sometimes in a lax basal rosette. Lower leaves petiolate, upper leaves sessile. Leaves evenly coloured above, entire or lobed with irregularly undulate-serrate margin or variously pinnatifid. Flowers solitary, terminal or axillary, on long leafless peduncles. Flowers nodding in bud, erect at anthesis. Sepals 2, very early caducous, apical horn absent. Petals 4(–6), mostly red, rarely white, or purplish, usually obovate, those of outer whorl larger. Ovary 6–12(–18)-carpellate, obovoid or globose, glabrous. Style absent, stigmas sessile, 6–18, united into a radiating lobed disc covering the apex of the ovary. Capsule obovoid, or globose, obviously ribbed or not, dehiscing by pores just below the lobes of the persistent stigmatic disc. Seeds reniform, pitted.

Worldwide 80–100 species, mainly in temperate regions of Europe and Asia, some in the Americas and one in S Africa. Three species in Nepal.

Papaver orientale L. is cultivated in Nepal, but is not known to naturalize. It is therefore included in the key but not treated further.

Key to Species

1a	Leaves amplexicaul, margin irregularly undulate-serrate......	**1. P. somniferum**
b	Leaves sessile or petiolate, but never amplexicaul, pinnatifid with entire or irregularly toothed margin	2
2a	Flowers 2–3 cm across. Petals 1–2 cm long	**2. P. dubium**
b	Flowers 5–10 cm across. Petals more than 2.5 cm long......	3
3a	Solitary-stemmed annual......	**3. P. rhoeas**
b	Densely tufted perennial. Cultivated	*P. orientale*

1. *Papaver somniferum* L., Sp. Pl. 1: 508 (1753).
Papaver amoenum Lindl.

अफीम Aphim (Nepali).

Annual herb, 30–60(–100) cm tall (to 1.5 m in cultivation). Stems and leaves glabrous or rarely slightly setose on stem below. Lower leaves shortly petiolate, upper leaves sessile and amplexicaul. Leaves entire or obscurely lobed, ovate or oblong, 7–25 × 8–15 cm, base cordate, apex acuminate to obtuse, margin irregularly undulate-serrate, glaucous and rather waxy on both surfaces, glabrous, veins distinct, slightly raised. Peduncle to 25 cm, glabrous or rarely sparsely setose. Flowers deeply cup-shaped, 5–12 cm across. Sepals green, broadly ovate, 1.5–3.5 cm, margin membranous, glabrous. Petals white, pink, red, purple, or various, often with a dark basal blotch, suborbicular or almost fan-shaped, 3–7(–9) × 2–10 cm, apex undulate or variously lobed. Filaments white, 1–1.5 cm, anthers yellowish or cream, oblong, 3–6 mm. Ovary, globose, 1–2 cm, glabrous. Stigmas 5–12(–18), margin of stigmatic disc deeply divided, lobes crenulate. Capsule globose, ca. 5 cm, inconspicuously ribbed, glabrous. Seeds reniform, ca. 1 mm.
Fig. 10a–b.

Distribution: Cosmopolitan.

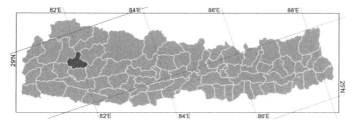

Altitudinal range: ca. 2000 m.

Ecology: Cultivated and naturalizing in disturbed areas.

Flowering: March–August. **Fruiting:** July–September.

Native to SE Europe and possibly SW Asia, but unknown in the wild. Under recorded in Nepal.

Widely cultivated throughout the world as an ornamental and as the source of legal and illegal pain relieving opiate drugs including heroin, morphine, opium and codeine. The plant is also used to cure coughs, bronchitis, diarrhoea and dysentery. The seeds are widely used as a spice, condiment, thickener or main ingredient in curries and baked products.

2. *Papaver dubium* L., Sp. Pl. 2: 1196 (1753).

Papaver dubium* subsp. *glabrum (Royle) Kadereit, Notes Roy. Bot. Gard. Edinburgh 45(2): 247 (1989).

Papaver glabrum Royle; *P. dubium* var. *glabrum* Koch; *P. dubium* var. *laevigatum* (M.Bieb.) Elkan; *P. laevigatum* M.Bieb.

Annual herb, 20–40 cm. Stems and leaves glabrous or sparsely pale, appressed setose. Leaves elliptic to narrowly obovate, pinnatifid, 4–8 × 1–2 cm, lobes acute, margin entire or with a few large teeth, venation slightly sunken above, prominent below. Peduncle 10–25 cm, glabrous or sparsely appressed setose. Sepals green, elliptic, 8–10 mm, glabrous or sparsely setose. Flowers bowl-shaped, 2–3 cm across. Petals red or orange, sometimes with a dark or blackish blotch at base, obovate, 1.0–2.0 × 1.0–1.2 cm, apex erose. Filaments black, 5–6 mm, anthers oblong, ca. 1 mm. Ovary obovoid, 6–8 mm. Stigmas 6–8, margin of stigmatic disc deeply crenate. Capsule obovoid, 14–18 mm, rather prominently ribbed, glabrous. Seeds reniform, ca. 1 mm.
Fig. 10c.

Distribution: Nepal, W Himalaya, S Asia, SW Asia, Europe and N America.

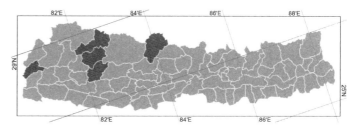

Altitudinal range: 2500–2900 m.

Ecology: Weed of cultivation (common in cornfields in Jumla).

Flowering: May–June. **Fruiting:** June–July.

Papaver dubium subsp. *glabrum* is distinguished from *P. rhoeas* L. by its narrower capsules with more prominent ribs and the peduncle which is either glabrous or with appressed rather than spreading hairs. *Papaver dubium* is a complex of five subspecies, found predominantly in Europe and into Asia as far E as Nepal. *Papaver dubium* subsp. *glabrum* is native from Afghanistan to Nepal and has been introduced as far as N. America.

Petals said to be useful medicinally to increase sweating.

3. *Papaver rhoeas* L., Sp. Pl. 1: 507 (1753).

सेती विराउली Seti birauli (Nepali).

Annual herb, 25-50(–90) cm. Stem and leaves yellowish setose. Leaves elliptic or narrowly ovate, pinnatifid, 3–12(–15) × 1–5(–9) cm, both surfaces yellowish setose, veins slightly sunken above and prominent below. Lobes narrow, pinnatilobate or more or less toothed, terminal lobes usually larger, lobules apically acuminate. Peduncle 10–15 cm, yellowish setose. Sepals green, broadly elliptic, 1–1.8 cm, setose outside. Flowers bowl-shaped, 5–9 cm across. Petals rich scarlet, with or without a basal dark blotch or flecking, orbicular, transversely broadly elliptic or broadly obovate, 2.5–4.5 × 3–3.5 cm, entire, rarely apex crenate or incised. Filaments pale green, yellow or blackish, ca. 8 mm, anthers yellow or bluish, oblong, ca. 1 mm. Ovary obovoid, 7–10 mm, glabrous. Stigmas 8–12, margin of stigmatic disc crenate. Capsule broadly obovoid, 10–18 mm, inconspicuously ribbed, glabrous. Seeds reniform-oblong, ca. 1 mm.
Fig. 10d.

Papaveraceae

Distribution: Cosmopolitan.

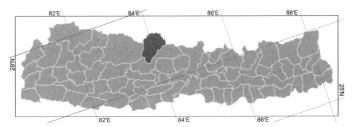

Altitudinal range: 2500–3100 m.

Ecology: Weed of cultivation.

Flowering: May–June. **Fruiting:** June–July.

Papaver rhoeas is thought to be of eastern Mediterranean origin (Kadereit, Bot. J. Linn. Soc. 103: 221. 1990), but it has been a weed of cultivation since ancient times and can now be found on all the continents except Antarctica.

2. *Argemone* L., Sp. Pl. 1: 508 (1753).

Colin A. Pendry

Annual, biennial, or perennial polycarpic herbs, prominently prickly-spiny throughout, otherwise glabrous, and somewhat glaucous, latex yellow, bitter. Taproot slender. Stems erect, branched. Leaves in lax basal rosettes and cauline. Lower leaves petiolate, upper leaves sessile. Leaves with prominent pale markings over venation above, pinnatilobed, lobes repand-dentate, teeth apex spiny. Flowers solitary, terminal and axillary, subtended by 1 or 2 leaf-like bracts. Peduncles very short. Flower buds erect, globose, ovoid or oblong. Sepals usually 3, caducous, concave, with a prominent spinous horn below apex. Petals 6 in 2 whorls, yellow, obovate. Stamens yellowish green, anthers coiling after dehiscence. Ovary 4–6 carpellate, 1-locular, ovoid, elliptical or oblong, spiny. Style obsolete or very short, stigmatic lobes 3–5, capitate or spreading. Capsule ellipsoid, 4–6-valvate (ribbed), prickly, septicidal, dehiscing from below the stigma for ¼–½ of length, rarely separating nearly to base. Seeds spherical, pitted.

About 23–29 species native to N and S America, with some widely introduced as pantropical weeds. Two introduced species in Nepal.

Key to Species

1a Stigmatic lobes broad, appressed to style. Flowers yellow ....**1. *A. mexicana***
 b Stigmatic lobes narrow, spreading. Flowers pale yellow..**2. *A. ochroleuca***

1. *Argemone mexicana* L., Sp. Pl. 1: 508 (1753).

सुँगुरे काँडा Sungure kanda (Nepali).

Herbs to 30–100(–120) cm tall. Stems rather stout, usually branching, glabrous apart from sparse, spreading, straw-coloured spines to 1 cm long. Basal leaves shortly petiolate, cauline leaves sessile and upper leaves subamplexicaul. Leaves obovate to elliptic, pinnatilobed, 5–22 × 2.5–7 cm, becoming progressively smaller up the stem, glabrous, somewhat glaucous, sparsely spiny on veins, especially below. Flower buds subglobose to slightly oblong, to 1.5 cm, sparsely spiny. Petals broadly obovate, 1.7–3 × 1.5–2 cm, apex entire. Stamen filaments ca. 7 mm, anthers 1.5–2 mm. Ovary elliptic or oblong, 7–11 mm, spiny. Style 0–1 mm, stigmas dark red, 4–6-lobed, the lobes broad, appressed to the style and completely obscuring the non-receptive surfaces between them. Capsule ovate to broadly elliptic, 2.5–4 × 1.5–2 cm, style scarcely visible. Seeds 1.5–2 mm. Fig. 10e–f.

Distribution: Nepal, W Himalaya, E Himalaya, S Asia, E Asia, Africa, N America and S America.

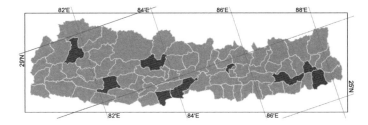

Altitudinal range: 100–1400 m.

Ecology: Disturbed open areas.

Flowering: Flowering and fruiting March–October, but can probably be found flowering somewhere in Nepal at any time.

Native to Mexico, a pantropical weed, introduced and naturalised in the lowlands of the Himalaya. Under-recorded in Nepal and of much wider distribution than the map indicates.

All parts used medicinally to treat a wide variety of conditions. The roots are used for chronic skin disease, intestinal complaints and as a poison antidote, the leaves for coughs and skin diseases, and the seed oil is a laxative and also useful for skin disorders. The seeds are poisonous.

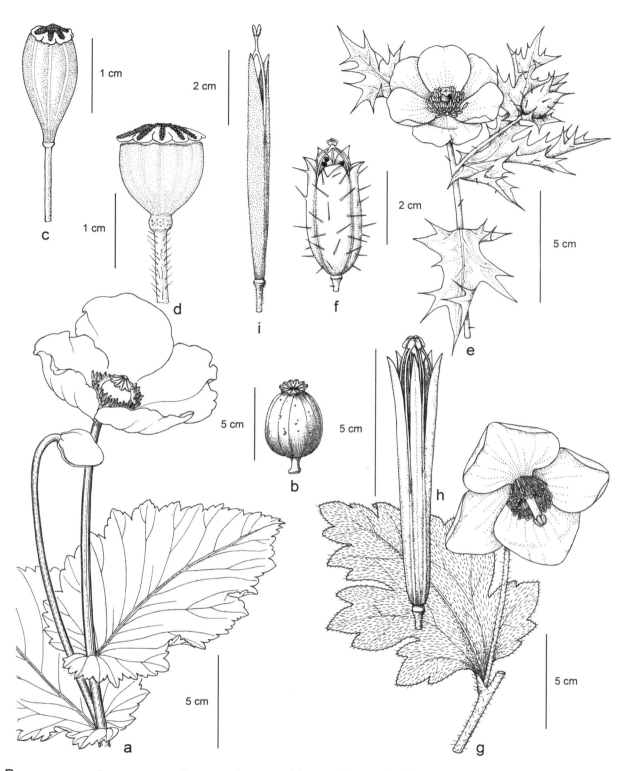

FIG. 10. Papaveraceae. **Papaver somniferum**: a, leaves and flowers; b, capsule. **Papaver dubium** subsp. **glabrum**: c, capsule. **Papaver rhoeas**: d, capsule. **Argemone mexicana**: e, leaves and flower; f, capsule. **Cathcartia villosa**: g, leaf and flower; h, capsule. **Dicranostigma lactucoides**: i, capsule.

2. *Argemone ochroleuca* Sweet, Brit. Fl. Gard. Ser.1 3(2): pl. 242 (1828).

Herbs 30–100 cm tall. Stems branching in the upper part, glabrous apart from sparse, spreading, straw-coloured spines to 9 mm long. Basal leaves shortly petiolate, cauline leaves sessile and upper leaves subamplexicaul. Leaves obovate to elliptic or ovate, pinnatilobed, 10–30 × 4–10 cm, becoming progressively smaller up the stem, glabrous, sparsely spiny on veins below. Flower buds oblong, to 1.8 cm, sparsely spiny. Petals broadly obovate, 2.8–3 × 1.5–1.8 cm, apex entire. Stamen filaments ca. 7 mm, anthers 1.5–2 mm. Ovary ovoid, 8–10 mm, spiny. Style ca. 1 mm, stigmas dark red, 4–6-lobed, deeply dissected and spreading so that the bluish non-receptive surfaces between them are visible. Capsule oblong to broadly elliptic, 2.5–4 × 1.5–2 cm, style clearly visible. Seeds 1.5–2 mm.

Distribution: Nepal, W Himalaya, S Asia, N America, S America and Australasia.

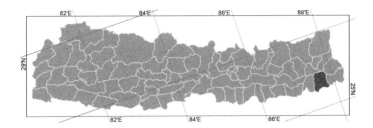

Altitudinal range: 100–200 m.

Ecology: Disturbed open areas along roads and waste places.

Flowering: April–June. **Fruiting:** April–June.

Argemone ochroleuca has been treated as a synonym of *A. mexicana* L. and is almost certainly much more widely distributed than is indicated here. The stigma is the most reliable way to distinguish the two species, with narrower, spreading lobes in *A. ochroleuca* compared with the broader, appressed lobes of *A. mexicana*.

3. *Meconopsis* Vig., Hist. Nat. Pavots Argémones: 48 (1814).

Paul A. Egan & Sangita Shrestha

Erect to ascending, monocarpic or polycarpic perennial herbs. Taproot fleshy. Latex yellow-orange. Indumentum barbellate-bristly, tomentose, or glabrous. Not glaucous. Stems solitary, simple or with divergent branches, or scapose, with withered leaf remains sometimes persistent at base. Basal leaves in rosettes, distinctly petiolate. Cauline leaves present or absent, uppermost cauline leaves sessile. Leaves evenly coloured above, elliptic, lanceolate, or oblong, rarely ovate in outline, entire to 2-pinatifid-pinnatisect. Inflorescence determinate, flowers in racemes or lax panicles of leafy axillary cymules, or solitary on basal scapes. Pedicels bracteate or ebracteate in uppermost flowers, glabrous to uniformly pubescent, sometimes with dense tufts immediately below flowers. Flower buds erect or nodding. Flowers nodding to lateral-facing. Sepals 2(–4), lacking apical horn. Corolla saucer-shaped to deeply cupulate. Petals 4–8(–10), variously coloured, rarely white, obovate to orbicular, apex rounded or rarely acute, margins entire, to denticulate or dentate. Ovary 4–8(–12)-carpellate, ellipsoid-oblong to subglobose, glabrous to densely bristly. Style usually conspicuous, rarely expanded basally into a disc surmounting apex of the ovary. Stigmatic lobes free, contiguous or connate, clavate to capitate. Capsule 4–8(–12)-valvate, dehiscing by subapical pores or slits in upper third. Seeds subreniform or ellipsoid-oblong, smooth, papillose or rugose.

Worldwide about 55 species from the Himalaya to W China. 22 species in Nepal, 11 of which are endemic and mostly very local in distribution.

Key to Species

1a	Style abruptly expanded at the base into a broad glabrous disc surmounting the ovary 2
b	Style when present of uniform thickness throughout or swollen at base, but never expanded into a disc 5
2a	Stems multiple, narrow, partially fusing above the rootstock crown ....**14.** *M. manasluensis*
b	Stem single and prominent, fleshy 3
3a	Leaves pinnately-lobed, the lobes in 3–6 subopposite pairs. Flowers deep purple to lavender .... **13.** *M. pinnatifolia*
b	Leaves entire or coarsely toothed. Flowers blue, yellow or cream...4
4a	Leaves lobed in upper third only, in 1 or 2 lateral pairs. Petals 35–50 mm long....................................**15.** *M. discigera*
b	Leaf margin sub-entire or coarsely toothed along whole margin. Petals 23–39 mm long.................**16.** *M. simikotensis*
5a	Basal rosettes lax, usually not more than 30 cm wide, dying back to a resting bud over winter 6
b	Basal rosettes dense, usually from 50 cm wide, persistent during winter 11

6a   Inflorescence scapose, with flowers borne on basal scapes, these sometimes partly aggregated, resembling a central stem...7

b   Inflorescence racemose, with flowers borne in the axils of the upper stem leaves...9

7a   Flowers borne on 1 or rarely 2 scapes. Scapes 60–110 cm. Capsules 30–50 mm ....**18. _M. simplicifolia_**

b   Flowers borne on 2–18 scapes. Scapes 5–40 cm. Capsules 5–20 mm...8

8a   Indumentum glabrous or sparse with softly barbellate bristles. Petals 4 ....**19. _M. bella_**

b   Indumentum sparse to dense with spiny bristles. Petals 6(–10)..**20. _M. horridula_**

9a   Flowers 4–8. Leaf margins coarsely or finely sinuate-lobed or subentire. Capsule sparsely to moderately hairy, bristles ascending..**21. _M. sinuata_**

b   Flowers 1–5. Leaf margins pinnatilobate, lyrate or subentire. Capsule glabrous to glabrescent 10

10a Polycarpic herb 40–100 cm. Ovary densely covered in reflexed or spreading bristles ....**17. _M. grandis_**

b   Monocarpic herb 12–35 cm. Ovary glabrous or rarely sparsely bristly.................................**22. _M. lyrata_**

11a Basal leaf margin entire or minutely serrate. 12

b   Basal leaf margins deeply to shallowly pinnately lobed...13

12a Petals yellow. Capsule oblong-ellipsoidal ....**11. _M. regia_**

b   Petals light to deep pink. Capsule narrowly clavate or ellipsoid ....**12. _M. taylorii_**

13a Petals red, purple or blue...14

b   Petals yellow or rarely white..17

14a Ovary densely appressed bristly. Capsule 22–34 mm long, with 7–10 valves.....................**9. _M. staintonii_**

b   Ovary moderately to densely bristly, the bristles suberect to spreading. Capsule 11–25 mm long, with 5–7 valves......15

15a Flowers borne in a large spreading panicle, the basal cymules usually 3–11-flowered ....**10. _M. wallichii_**

b   Flowers in racemose or subpaniculate inflorescence, the basal cymules 2–3-flowered 16

16a Basal leaves pinnatisect to pinnatifid, narrow-oblong in outline, with ca. 5 cm petioles......................**2. _M. ganeshensis_**

b   Basal leaves bipinnatisect or pinnatisect, oval in outline, with 7–16 cm petioles ....**3. _M. chankheliensis_**

17a Inflorescence strictly racemose ....**6. _M. robusta_**

b   Inflorescence paniculate, lower flowers in 2–10-flowered cymules, flowers solitary above...........................18

18a Stem usually 120–260 cm tall. Style stout. Stigma purple, 8–12-lobed...19

b   Stem generally 35–130 cm tall. Style slender. Stigma yellow-green or cream, 4–7-lobed............................20

19a Lamina noticeably parted towards the base. Stigmas yellow. Capsules mostly ellipsoidal, with spreading orange pubescence. Flowering July–September ....**7. _M. autumnalis_**

b   Lamina mostly pinnatifid or pinnatisect towards the base. Stigmas purple. Capsules mostly globose, with appressed fawn pubescence. Flowering June–July ....**8. _M. paniculata_**

20a Plants densely pubescent, with bristles conspicuously purple black at the base, reflexed to reflexed-spreading on capsule Petioles usually 2–4 cm...**4. _M. dhwojii_**

b   Plants moderately pubescent to glabrescent, the bristles uniformly coloured, spreading or glabrescent on capsule. Petioles usually 3–16 cm 21

21a Style 9–14 mm. Indumentum moderate, usually with greyish bristles. Pedicels 3–14 cm ....**1. _M. napaulensis_**

b   Style 3–7 mm. Indumentum sparse, with pale brown bristles. Pedicels 4–5.8 cm..................**5. _M. gracilipes_**

Papaveraceae

1. *Meconopsis napaulensis* DC., Prodr. 1: 121 (1824).

क्यासर Kyasar (Sherpa).

Monocarpic herb, 30–100(–160) cm, very variable in height. Stem 0.8–2.0 cm in diameter at base. Indumentum moderately dense, of stiff, greyish or fawn-coloured obliquely spreading barbellate bristles, to 12 mm, with shorter underlying hairs. Leaves cauline and in basal rosettes. Petioles 2–19 cm, cauline leaves mostly sessile or with winged petioles. Laminas distinctly pinnatisect at base to pinnatifid towards leaf apex, oblong to lanceolate, to 31 × 7.6 cm. Leaf segments in 6–9 subopposite pairs, oval to elliptic or obovate, 14–23 × 8–13 mm, subentire to acutely lobed. Inflorescence racemose or paniculate with 2- or 3-flowered lateral cymules, flowers solitary above. Pedicels 3–14 cm, conspicuously elongated by maturity. Petals 4, pale yellow, obovate, 2.4–4.7 × 2.0–3.1 cm. Filaments pale yellow, anthers yellow to orange-yellow. Ovary densely covered in ascending bristles. Style 9–14 mm, slender. Stigma yellow-green, capitate with 6 lobes. Capsule oval-oblong to narrow-ellipsoidal, 1.4–2.7 × 0.5–1.4 cm, dehiscing by 5 valves, indumentum spreading bristly to glabrescent.

Distribution: Endemic to Nepal.

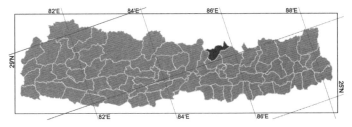

Altitudinal range: 3200–4500 m.

Ecology: Meadows and stream-sides.

Flowering: May–August. **Fruiting:** August–September.

The circumscription of *Meconopsis napaulensis* has long been problematic, and here a narrow species concept is accepted which includes only the yellow-flowered specimens from the Langtang region of Rasuwa district. Sometimes misspelt '*nipalensis*' or '*nepalensis*' in the literature. See also notes under *M. paniculata* (D.Don) Prain.

2. *Meconopsis ganeshensis* Grey-Wilson, Curtis's Bot. Mag. 23(2): 188 (2006).

Slender monocarpic herb 50–120 cm. Stem 8–16 mm in diameter at base. Indumentum moderately dense, of pale orange or fawn-coloured, 4–9 mm, spreading, barbellate bristles and accompanied by shorter underlying hairs. Leaves cauline and in basal rosettes. Petioles ca. 5 cm, cauline leaves mostly sessile or with winged petioles. Laminas pinnatisect to pinnatifid, narrow-oblong, to 14 × 4.2 cm. Leaf segments in 4–8 distinct subopposite pairs, oval to elliptic or elliptic ovate, 12–22 × 5–13 mm, subentire to shallowly pinnately-lobed. Flowers in racemose or subpaniculate inflorescences

with basal cymules 2- or 3-flowered, flowers solitary above. Pedicels 2.5–9 cm, conspicuously pubescent distally around receptacle base. Petals 4, light purple to dark red, oval to oval-obovate, 2.2–4.0 × 2.0–3.5 cm. Filaments reddish with orange-yellow anthers. Ovary with dense suberect barbellate bristles. Style 6–8 mm, slender. Stigma yellow, capitate, 6-lobate. Capsule obovoid to ellipsoid-oblong, 1.5–2.0 × 0.6–1.0 cm, dehiscing by ca. 6 valves, indumentum reflexed to spreading bristly, to glabrescent. Fig. 11a.

Distribution: Endemic to Nepal.

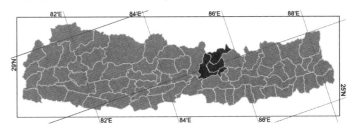

Altitudinal range: 3600–4600 m.

Ecology: Meadows.

Flowering: July–August.

Meconopsis ganeshensis is akin to a red-flowered form of the closely related *M. napaulensis* DC., though is a slightly smaller and generally more slender species.

3. *Meconopsis chankheliensis* Grey-Wilson, Curtis's Bot. Mag. 23(2): 203 (2006).

Monocarpic herb, 40–150 cm. Stem 11–16 mm in diameter at base. Indumentum sparse throughout the whole plant, of 3–6 mm, golden brown or yellowish barbellate hairs. Leaves cauline and in basal rosettes. Petiole 7–16 cm, cauline leaves shortly petiolate or sessile. Laminas pinnatisect or bipinnatisect, oval in outline, to 24 × 9.5 cm. Leaf segments in 3–7 subopposite pairs, oval to elliptic, 2.5–5.0 × 1.7–2.6 cm. Flowers in racemose or subpaniculate inflorescences, basal cymules with 2 or 3 flowers, solitary above. Pedicels 4.5–18 cm, more densely pubescent around receptacle base. Petals 4, purple to dark red, subrounded to obovate, 2.1–4.0 × 2.1–4.2 cm. Filaments purple, anthers yellow or orange. Ovary with densely appressed to ascending orange-fawn bristles. Style 4–8 mm, slender. Stigma yellowish, subcapitate, 6-lobed. Capsule unknown.

Distribution: Endemic to Nepal.

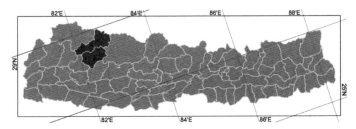

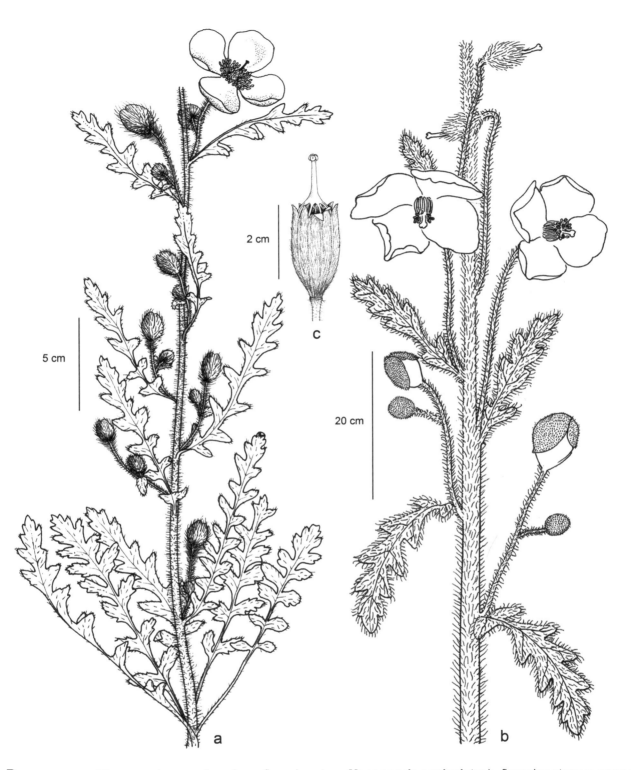

2 cm

5 cm

c

20 cm

a

b

FIG. 11. PAPAVERACEAE. **Meconopsis ganeshensis**: a, flowering stem. **Meconopsis paniculata**: b, flowering stem; c, capsule.

Altitudinal range: 3100–4600 m.

Ecology: *Quercus-Abies* forest, rocky outcrops, stream-sides.

Flowering: May–June.

4. *Meconopsis dhwojii* G.Taylor ex Hay, New Fl. & Silva 4: 225, pl. 82 (1932).

Monocarpic herb, 50–60 cm. Stem 8–13 mm in diameter at base. Indumentum dense, with 4–9 mm, golden brown barbellate bristles, widely spreading and conspicuously purple-black at base. Leaves cauline and in basal rosettes. Petioles 1.5–4 cm, cauline leaves shortly petiolate or sessile. Laminas mostly pinnatisect, pinnatifid towards leaf apex, narrowly oblong, 5–19 × 3–9.5 cm. Leaf segments in 6–9 lateral pairs, proximal pairs distinctly divided, lobes round, obtuse, or subacute. Flowers borne in leafy paniculate inflorescences, lower cymules 2–4-flowered, flowers solitary above. Pedicels 0.5–13 cm, bristly. Petals 4, yellow, obovate-orbicular, 1.5–4.0 × 1.2–3.0 cm. Filaments pale yellow, anthers golden yellow. Ovary with dense appressed bristles. Style 4–8 mm, stout. Stigma capitate, green or yellow-green, ca. 6-lobate. Capsules ellipsoid or oblong, 0.8–3.5 × 0.5–1.0 cm, dehiscing by 5–6 valves, indumentum of reflexed to reflexed-spreading bristles.

Distribution: Endemic to Nepal.

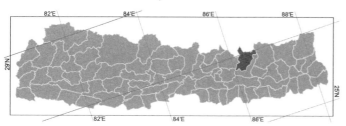

Altitudinal range: 3600–4800 m.

Ecology: Meadows, rock crevices, stream-sides.

Flowering: June–August. **Fruiting:** August–October.

5. *Meconopsis gracilipes* G.Taylor, Acc. Meconopsis: 38 (1934).

Monocarpic herb to 1 m. Stems sparsely covered with pale brown 1–9 mm, spreading bristles, later glabrescent. Leaves cauline and in basal rosettes, the latter mostly absent by flowering. Petioles 3.5–16 cm, cauline leaves shortly petiolate or sessile, auriculate at base. Laminas pinnatisect towards base, more or less deeply pinnatifid towards apex, oblong to lanceolate, 5–25.5 × 2–7.5 cm. Leaf segments in 3–5 subopposite pairs, 2.6–5 × 2–3.8 cm, lobes 3–5, acute to rounded at apex. Flowers in subpaniculate inflorescence borne on axillary branches, lower branches 3–5-flowered, upper flowers solitary. Pedicels 4–5.8 cm, bristles tufted below flowers. Petals 4, yellow, obovate-sub-orbicular, 2.0–2.5 × 1.5–2.5 cm. Filaments pale yellow, anthers orange-yellow. Ovary densely covered with appressed to obliquely spreading

hairs. Style 3–7 mm, slender. Stigma capitate. Capsule ellipsoid to oblong-ellipsoid, 1.0–2.5 × 0.5–1.0 cm, dehiscing by 4–7 valves, with sparse indumentum of spreading bristles or glabrescent.

Distribution: Endemic to Nepal.

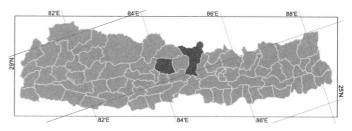

Altitudinal range: 2400–4900 m.

Ecology: Meadows, forest margins and stream-sides.

Flowering: June–July. **Fruiting:** July–September.

6. *Meconopsis robusta* Hook.f. & Thomson, Fl. Ind. 1: 253 (1855).

Monocarpic herb ca. 60–120 cm. Stem 1.0–2.0(–3.2) cm in diameter at base. Indumentum of sparse, yellowish-brown 5–6 mm, barbellate bristles, semi-glabrescent by maturity. Leaves cauline and in wide basal rosette. Petioles to 15 cm, cauline leaves mostly sessile, rarely with semi-amplexicaul base. Lamina imperfectly and deeply pinnatisect, commonly divided completely to midrib in large divisions, more pinnatifid towards apex, elliptic to elliptic-oblong, to 35 × 11 cm. Leaf segments in 3–6 lateral pairs, ovate or ovate-oblong, entire to shallowly lobed. Inflorescence strictly racemose. Pedicels 5–20 cm. Petals 4, pale to sulphur yellow, obovate, 2.8–3.3 × 1.7–2.0 cm. Filaments same colour as petals, anthers yellow-orange. Ovary moderately to densely covered in appressed or spreading bristles. Style 5–10 mm, slender. Stigma yellow, capitate, 6–9 lobed. Capsule obovoid-oblong or ellipsoidal, 1.8–2.8 × 0.9–1.3 cm, dehiscing by 6–9 valves, indumentum of sparse, spreading to ascendant bristles.

Distribution: Nepal and W Himalaya.

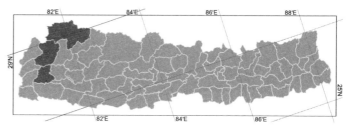

Altitudinal range: 2400–4100 m.

Ecology: Grassland and exposed alpine slopes.

Flowering: July–August. **Fruiting:** September–October.

7. *Meconopsis autumnalis* P.A.Egan, Phytotaxa 20: 48 (2011).

Monocarpic herb, 110–160 cm. Stem 2.2–3.1 cm in diameter at base. Indumentum densely puberulent with scattered, pale fawn to orange barbellate bristles 5–9 mm. Leaves cauline and in large basal rosettes. Petioles 11–25(–31) cm, in lower cauline leaves only. Upper cauline leaves sessile, deflexed, semi-amplexicaul, with large auricles prominent at the base. Lamina pinnatisect at base, pinnatifid only towards apex, broadly oblong, 33–55 × 8–16 cm. Leaf segments distinct, in 3–5 opposite to subopposite pairs, oblong or ovate, pinnatilobate at the margin, the lobes acute to obtuse. Inflorescence densely paniculate and markedly columnar in outline, cymules 3–8(–12)-flowered, the upper flowers solitary. Pedicels 6–17 cm. Petals 4(–6), pale yellow, obovate to suborbicular, 4.3–5.6 × 4.1–5.5 cm, occasionally widely spaced, only overlapping a short distance from the base. Filaments pale yellow, anthers yellow-orange. Ovary obviously stipitate and with large receptacle, densely covered with orange ascending bristles. Style 8–25 mm, stout. Stigma yellow, capitate, (3–)5–8 mm, 6–8-lobed. Capsule ovoid to ellipsoidal, 1.5–2.7 × 0.9–1.4 cm, dehiscing by 6–8 valves, indumentum moderately dense, spreading to ascending.

Distribution: Endemic to Nepal.

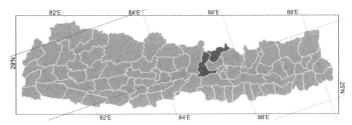

Altitudinal range: 3300–4200 m.

Ecology: Subalpine pastures, forest openings, stream margins and grassy alpine slopes.

Flowering: July–September.

Meconopsis autumnalis is somewhat intermediate between *M. napaulensis* DC. and *M. paniculata* (D.Don) Prain, but is most clearly differentiated from both by its distinctly larger stigma and late flowering period.

8. *Meconopsis paniculata* (D.Don) Prain, J. Asiat. Soc. Bengal, Pt. 2, Nat. Hist. 64: 316 (1896).
Papaver paniculatum D.Don, Prodr. Fl. Nepal.: 197 (1825); *Meconopsis longipetiolata* G.Taylor ex Hay; *M. paniculata* var. *elata* Prain.

उपल सेर्बो Upal sherbo (Sherpa).

Monocarpic herb, 80–270 cm. Stem 1.4–3.9 cm in diameter at base. Indumentum sparsely to densely clothed with grey, pale yellow or brownish barbellate bristles, to 6 mm, with dense underlay of short substellate hairs. Leaves cauline and in large basal rosettes. Petioles 4–34 cm, cauline leaves shortly petiolate, or sessile with semi-amplexicaul base. Lamina very variably divided, either shallowly pinnatilobate, or pinnatisect towards base and pinnatifid towards leaf apex, usually lanceolate or ellipsoidal-oblong, 20–60 × 5–19 cm. Leaf segments in 6–10 lateral pairs, ovate-oblong, proximalmost pairs rarely deeply divided, lobes subentire to acute. Inflorescence paniculate with upper flowers solitary, cymules 2–10-flowered. Pedicels 1.8–14 cm. Petals 4, yellow, obovate or suborbicular, 2.9–6.4 × 3–4.8 cm. Filaments pale yellow, anthers yellow-orange. Ovary densely covered with appressed pale or golden yellow bristles. Style 6–14 mm, stout. Stigma purple, subcapitate, 8–12-lobed. Capsule globose to oblong-ellipsoid, 1.1–3.9 × 0.8–2.1 cm, dehiscing by 8–12 valves, indumentum of dense to moderately dense appressed to ascendant bristles.
Fig. 11b–c.

Distribution: Nepal, E Himalaya and Tibetan Plateau.

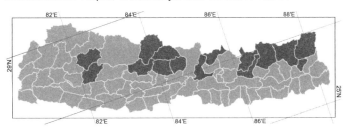

Altitudinal range: 2700–5000 m.

Ecology: Open grassy slopes, forest margins, scrubland.

Flowering: June–July. **Fruiting:** August–October.

Meconopsis napaulensis DC. has been widely misapplied in the Himalayan region as the name for *M. paniculata* because Don (Prodr. Fl. Nepal.: 197. 1825) misapplied it when listing it in synonymy under his *Papaver paniculata*. This confusion was perpetuated by Walpers (Repert. Bot. Syst. 1: 110. 1842) and Hooker (Himal. J. 2: 53. 1854) and subsequent authors until Prain made the combination *M. paniculata* and clarified its limits. Hooker & Thomson (Fl. Brit. Ind. 1: 118. 1872) also misapplied their own name *M. robusta* Hook.f. & Thomson to this species.

The plant is used in the treatment of bile disease, swelling of limbs and sores.

9. *Meconopsis staintonii* Grey-Wilson, Curtis's Bot. Mag. 23(2): 190, pl. 5-6 (2006).

Monocarpic herb to 1.8 m. Stem to 3.0 cm in diameter at base. Indumentum usually sparse, with 3–6 mm long spreading yellow barbellate and dense, short substellate hairs throughout. Leaves cauline and in wide basal rosettes, to 1 m across. Petioles 7–16 cm, cauline leaves shortly petiolate or sessile, with rounded or semi-auriculate leaf bases. Laminas pinnatisect towards leaf base, increasingly pinnatifid towards apex, oblong-oblanceolate to lanceolate, 24–52 × 6.5–18 cm. Leaf segments in 4–8 lateral pairs, oval to oblong or ovate, serrate to subentire at margins. Inflorescence paniculate, lateral cymules 3–11-flowered, pedicels 7.8–18 cm. Petals 4, red or variously pink, rarely white, obovate to suborbicular, 3.6–6.0 × 4.0–7.5 cm. Filaments the same colour as petals, anthers orange or orange-yellow. Ovary densely covered with appressed orange-yellow bristles. Style 5–8 mm, stout. Stigma dark green to yellow, capitate. Capsule ellipsoid to obovoid, 2.2–3.4 × 0.8–1.2 cm, dehiscing by 7–10 valves, indumentum dense, with semi-appressed to ascending bristles. Fig. 12a–b.

Distribution: Endemic to Nepal.

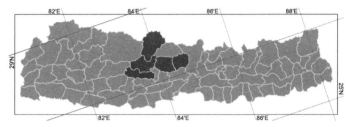

Altitudinal range: 2500–4300 m.

Ecology: Open forests, forest edges, scrub and rocky slopes.

Flowering: May–July.

10. Meconopsis wallichii Hook., Bot. Mag. 78: pl. 4668 (1852).
Meconopsis wallichii var. *fuscopurpurea* Hook.f.

Monocarpic herb to 1.4 m. Stem to ca. 2.5 cm in diameter at base. Indumentum of spreading to ascending orange-yellow to brown barbellate bristles, 3–5 mm, with underlay of shorter hairs. Leaves cauline and in rather lax basal rosettes. Petioles 2–12 cm, cauline leaves shortly petiolate or sessile to subsessile. Laminas pinnatisect towards leaf base, distally more pinnatifid, oval to oval-oblong, 11–29 × 2.5–8.5 cm long. Leaf segments in 3–7 opposite to subopposite pairs, oval to elliptic, subentire to shallowly lobed. Flowers borne in large spreading paniculate inflorescence, cymules usually 3–11-flowered, flowers solitary on upper stem. Pedicels 6–14.5 cm. Petals 4, light blue to purple or dark red, obovate to suborbicular, 2.0–3.5 × 2.3–3.8 cm. Filaments darker than petals, anthers yellow or orange. Ovary covered with appressed to ascending bristles. Style 5–9 mm, slender. Stigma greenish, capitate. Capsule ellipsoidal to broadly ovoid, 1.1–2.5 × 0.7–1.2 cm, dehiscing by 5–7 valves, with moderately dense indumentum of appressed to ascending yellow to orange bristles.

Distribution: Nepal and E Himalaya.

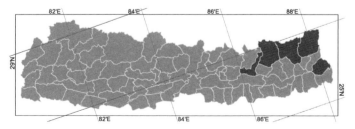

Altitudinal range: 3100–4300 m.

Ecology: Open forests, forest edges, scrub, rocky slopes and stream-sides.

Flowering: July–September.

Varieties of *Meconopsis wallichi* can be differentiated on the basis of flower colour, with var. *wallichii* having sky blue or lavender petals and var. *fuscopurpurea* Hook.f. having maroon, maroon-purple or deep red petals. The sky blue forms appear to be restricted to Sikkim and Bhutan. A rare white-flowered form is also thought to exist in Nepal.

11. Meconopsis regia G.Taylor, J. Bot. 67: 259 (1929).

Monocarpic herb usually 1.5–1.8 m. Stem to ca. 3.5 cm in diameter at base. Indumentum tomentose, whole plant densely clothed with appressed, golden or silvery 4–8 mm bristles. Leaves cauline and in densely packed, persistent basal rosettes. Petioles 5–22 cm, cauline leaves shortly petiolate, or sessile. Lamina narrowly elliptic, 31–47.5 × 6.5–11.4 cm, base attenuate, apex acuminate, margin acutely serrate or subentire. Inflorescence paniculate, lateral cymules with 2–4 flowers below, 1 or 2 flowers above. Pedicels 2–20 cm. Petals 4(–6), yellow, suborbicular, 3.5–6.6 × 3.5–7 cm. Filaments pale yellow, anthers orange. Ovary densely covered with soft, appressed yellow to yellow-orange bristles. Style 4–11 mm, stout. Stigma red-purple, capitate, 7–12 lobed. Capsule oblong-ellipsoidal, 1.0–4.5 × 0.8–1.8 cm, dehiscing by 7–12 valves, densely appressed bristly.

Distribution: Endemic to Nepal.

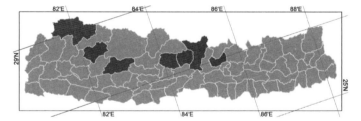

Altitudinal range: 3000–4300 m.

Ecology: Alpine meadows and low scrub.

Flowering: June–August. **Fruiting:** August–September.

A red-flowered form of *Meconopsis regia* supposedly exists in the wild. Given this occurrence, and the otherwise very close morphological relation with *M. taylori* L.H.J.Williams, it is conceivable that these species are conspecific.

12. Meconopsis taylorii L.H.J.Williams, Trans. & Proc. Bot. Soc. Edinburgh 41: 347 (1972).

Monocarpic herb to 1.5(–1.8) m. Stem to ca. 2.8 cm in diameter at base. Indumentum dense with straw-coloured, 10–15 mm, barbellate bristles and short substellate hairs.

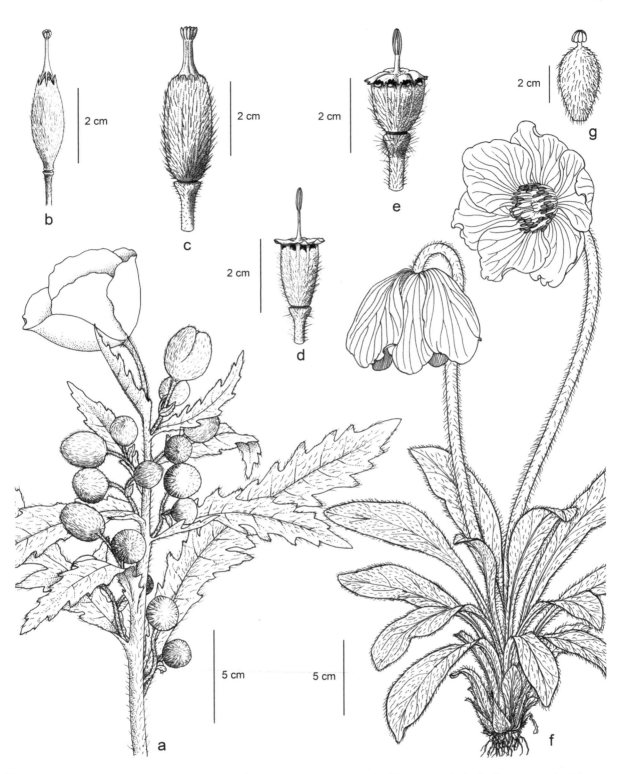

FIG. 12. PAPAVERACEAE. **Meconopsis staintonii**: a, flowering stem; b, capsule. **Meconopsis taylorii**: c, capsule. **Meconopsis pinnatifolia**: d, capsule. **Meconopsis discigera**: e, capsule. **Meconopsis simplicifolia**: f, flowering plant; g, capsule.

Leaves cauline and in large, basal rosettes to 1 m wide. Petioles 12–25 cm, cauline leaves sessile or with winged petioles. Laminas narrowly elliptic to oblong-oblanceolate, 29–60 × 7.5–16.5 cm, apex acute, base attenuate, margin subentire or crenate to shallowly lobed. Inflorescence paniculate, cymules 2–5-flowered, upper flowers solitary. Pedicels 6–18 cm, densely pubescent below ovary. Petals 4, light to deep pink, obovate to subrounded, 3.0–5.1 × 2.0–4.5 cm. Filaments whitish, anthers orange. Ovary densely pubescent, bristles orange-brown, barbellate, ascending to appressed. Style short, 4–7 mm, stout. Stigma dark brown or purple-brown, capitate. Capsule narrowly clavate or ellipsoid, 2.5–6.0 × 1.0–2.2 cm, dehiscing by 7–12 valves, densely clothed with fine, ascending to appressed bristles, eventually glabrescent.
Fig. 12c.

Distribution: Endemic to Nepal.

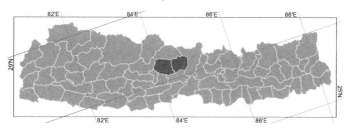

Altitudinal range: 3600–4600 m.

Ecology: Open meadows.

Flowering: July–August.

The recognition of this taxon at the species level is debatable. It is plausible that it merely constitutes a variant of *Meconopsis regia*, or a wild hybrid between *M. regia* G.Taylor and *M. staintonii* Grey-Wilson. See further notes under *M. regia*.

13. *Meconopsis pinnatifolia* C.Y.Wu & H.Chuang ex L.H.Zhou, Acta Phytotax. Sin. 17(4): 114, pl. 3 (1979).

Monocarpic herb, 40–100 cm. Stem to 2.5 cm in diameter at base. Indumentum sparsely to densely bristly throughout whole plant, bristles yellow to orange, barbellate, to ca. 4 mm. Leaves cauline and in spreading basal rosettes. Petioles 2–14 cm, conspicuously broad at base, occasionally winged, reduced or absent on upper cauline leaves. Lamina elliptic-oblong, distinctly pinnatifid, to ca. 23 × 7.5 cm. Leaf segments in 3–6 subopposite pairs, entire, to 50 × 16 mm; midribs prominent. Flowers 12–22(–30) in dense racemes. Pedicels 0.8–3.0 cm, bristles notably tufted around receptacle base. Petals 4, dark to lavender purple, rarely reddish, obovate, 4.2–7.2 × 3.8–6.8 cm. Filaments bluish, anthers yellow-orange. Ovary densely covered in spreading bristles. Style 5–11 mm; stylar disc surmounting ovary 6–9 mm in diameter. Stigma yellow, clavate, 4–5 mm. Capsule globose to subglobose, 1.5–1.6 × 0.8–0.9 cm, probably 5–7-valved, with indumentum of dense, spreading bristles.
Fig. 12d.

Distribution: Nepal and Tibetan Plateau.

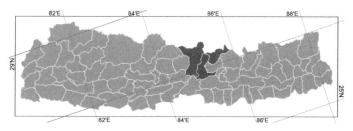

Altitudinal range: 3600–4900 m.

Ecology: Rocky grassland, stream-sides.

Flowering: June–August.

14. *Meconopsis manasluensis* P.A.Egan, Phytotaxa 20: 50 (2011).

Monocarpic herbs, 30–60 cm. Stems multiple, to 1.5 cm in diameter at base, partially fusing above the rootstock crown 2.5–3.0 cm wide. Indumentum a sparse to moderately dense cover of 3–6 mm fawn or orange barbellate bristles, present on both adaxial and abaxial leaf surfaces and petioles. Leaves cauline and in basal rosettes. Petioles 3–5 cm, winged, uppermost cauline leaves and bracts reduced and sessile. Laminas obelliptic to narrow oblanceolate, 6.5–15.5 × 1.5–3.0 cm, subacute or rounded at the apex, tapered at the leaf base, margins entire. Inflorescence racemose, ca. 30-flowered, the flowers borne singly on the central axis and on multiple, leafy lateral stems. Pedicels conspicuously elongate, 1.5–7.5 cm, decurrent on stem. Petals 4 or 5(–8), scarlet with darker, purple patches towards the base, broadly ellipsoidal to obovate, 2.0–3.0 × 1.0–2.0 cm. Filaments red-purple, anthers yellow-orange. Ovary densely covered with fawn to orange ascending bristles. Style long, 7–12 mm, slender, with stylar-disc to 8 mm in diameter projecting beyond edge of ovary. Stigma yellow, capitate, 3–4 mm, generally 6-lobed. Capsule narrowly ovoid to cylindric, 0.8–1.6 × 0.4–0.7 cm, dehiscing by 8 valves; indumentum of sparsely spreading bristles.

Distribution: Endemic to Nepal.

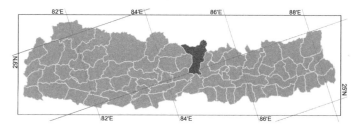

Altitudinal range: ca. 4000 m.

Ecology: Herb-rich alpine grasslands with scattered shrubs.

Flowering: July–August.

15. *Meconopsis discigera* Prain, Ann. Bot. 20: 356 (1906).

Monocarpic herb 30–60 cm, to 1 m in fruit. Stems ca. 1.5 cm in diameter at base, clothed with golden brown, barbellate, 3–5 mm spreading or reflexed bristles. Leaves mostly in dense basal rosettes. Petioles 5–14 cm, cauline leaves sessile to subsessile. Laminas oblanceolate or elliptic oblanceolate, 7.5–17 × 0.8–3 cm, base usually attenuate, apex subacute or rounded, entire to coarsely toothed or 3–5-lobed nearer apex. Flowers borne on clustered, racemose inflorescences, 10–20 flowered. Pedicels 5–10 cm, decurrent on stem. Petals 4, yellow (blue or purple), obovate, 3.5–5.0 × 2.5–4.0 cm. Filaments the same colour as petals, anthers yellow. Ovary oblong, densely yellow bristly. Style 3-7 mm, slender; stylar disc lobed, surmounting ovary, to 14 mm in diameter. Stigma yellowish, capitate, generally 6-lobed. Capsules subglobose to oblong, 0.5–1.7 × 0.5–1.3 cm, dehiscing by 6–8 valves, with indumentum of appressed bristles.
Fig. 12e.

Distribution: Nepal and E Himalaya.

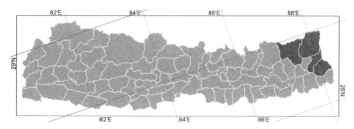

Altitudinal range: 3100–4600 m.

Ecology: Open rocky slopes, scree, cliff ledges.

Flowering: June–August. **Fruiting:** August–September.

Only the yellow-flowered form of *Meconopsis discigera* is recorded from Nepal, blue and purple forms are known in other E Himalaya countries.

16. *Meconopsis simikotensis* Grey-Wilson, Alpine Gardener 74(2): 220 (2006).

Monocarpic herb reaching to 100 cm, or longer in fruit. Stem to 2.5 cm in diameter at base, moderately to densely covered in long, spreading, pale yellow to orange, 5–13 mm, barbellate bristles, with persistent leaf remains at base. Leaves cauline and in basal rosettes. Petioles 7.5–13.5 cm, cauline leaves mostly sessile. Lamina elliptic-oblong, 15–19 × 1.4–2 cm, base attenuate, apex subacute, margins entire to coarsely toothed. Inflorescence racemose, 8–22-flowered. Pedicels 8–10 mm, densely pubescent. Petals 4, blue or purplish blue, obovate or orbicular, 2.3–3.9 × 1.7–3.5 cm. Filaments dark purple, whitish towards base; anthers yellow. Ovary covered in semi-appressed to ascending pale yellow or orange bristles. Style 2–4 mm; stylar disc surmounting ovary 9–13 mm in diameter. Stigma cream to yellow or brownish, capitate, 8 or 9 lobed. Capsule obovoid-oblong, 0.9–1.7 × 0.7–1.1 cm, dehiscing by 9 valves, sparsely bristly.

Distribution: Endemic to Nepal.

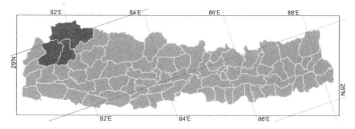

Altitudinal range: 3300–4000 m.

Ecology: Alpine slopes, stream-sides.

Flowering: June–July. **Fruiting:** August–September.

17. *Meconopsis grandis* Prain, J. Asiat. Soc. Bengal, Pt. 2, Nat. Hist. 64: 320 (1896).

चिल्दर Childar (Nepali).

Polycarpic herb, 12–100 cm. Stem to ca. 1.5 cm in diameter at base, mostly densely covered in spreading or reflexed pale yellow to orange, 3–7 mm, barbellate bristles,. Leaves cauline and in basal rosettes. Petioles 9–20 cm, commonly winged, upper cauline leaves aggregated in false whorl, sessile. Laminas oblanceolate to lanceolate or elliptic-oblong, 6.5–28 × 1.3–5.5 cm, base rounded to attenuate, apex acute or subacute, margin sub-entire to coarsely serrate. Flowers usually 1–3, solitary in the axils of uppermost leaves. Pedicels 5–27 cm. Petals usually 4–6, blue or purple, suborbicular or broadly obovate, 4.5–7 × 3.4–6 cm. Filaments whitish, anthers yellow or orange-yellow. Ovary glabrous to densely bristly, bristles appressed to spreading. Style 2–5 mm, slender. Stigma whitish or green, clavate, 4–6-lobed. Capsule narrowly ellipsoid-oblong, 1.9–4.4 × 1.2–1.5 cm, dehiscing by 4–6 valves, usually glabrous.

Distribution: Nepal, E Himalaya and Tibetan Plateau.

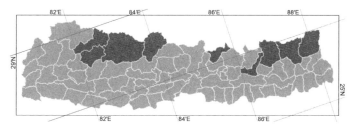

A third subspecies (subsp. *orientalis* C.Grey-Wilson) is found mostly in NE Bhutan and Arunachal Pradesh. *M. grandis* is scarcely found in C Nepal, and populations there may be the result of naturalization following anthropogenic spread of the species.

The plant is used to treat fevers associated with lung and liver diseases. Seeds are roasted and pickled by Sherpas and Tamangs.

Papaveraceae

1a Plants more than 40 cm tall. Basal leaves broad,
2–4.8 cm wide .... subsp. *grandis*

b Plants usually less than 35 cm tall. Basal leaves
narrow, up to 2.7 cm wide .... subsp. *jumlaensis*

Meconopsis grandis Prain subsp. *grandis*

Plant more than 40 cm tall. Basal leaves broad, 2–4.8 cm wide.

Distribution: Nepal and E Himalaya.

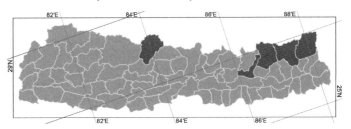

Altitudinal range: 3100–4500 m.

Ecology: Moist grasslands, shady slopes and forest margins.

Flowering: June–July. **Fruiting:** August–September.

Meconopsis grandis subsp. *jumlaensis* C.Grey-Wilson, Sibbaldia 8: 82 (2011).

Plant usually less than 35 cm tall. Basal leaves narrow, up to 2.7 cm wide.

Distribution: Endemic to Nepal.

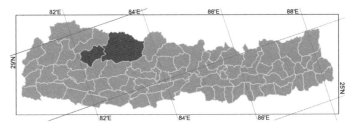

Altitudinal range: 3300–4400 m.

Ecology: Open, grassy slopes and at forest margins under *Rhododendron*, *Abies* and *Betula*.

Flowering: May–July.

18. *Meconopsis simplicifolia* (D.Don) Walp., Repert. Bot. Syst. 1(1): 110 (1842).
Papaver simplicifolium D.Don, Prodr. Fl. Nepal.: 197 (1825); *Polychaetia scapigera* Wall. nom. nud.; *Stylophorum simplicifolium* (D.Don) Spreng.

Polycarpic or monocarpic herb, 60–110 cm. Stem slender,

3–10 mm in diameter. Indumentum of sparse reflexed or spreading, soft whitish, ca. 3 mm, barbellate hairs. Leaves in basal rosettes only. Petioles variable, 5–18 cm, or leaves subsessile. Laminas very variable in size and shape, oblanceolate, spatulate, elliptic-lanceolate to oblong, 3.5–16 × 1.5–4.6 cm, base attenuate, apex subacute to obtuse, margin entire or slightly toothed to sinuate lobed. Inflorescence scapose, flowers solitary on 1 or 2, slender, leafless, suberect to erect, basal scapes. Petals usually 6–10, sky blue, purple-blue or lavender, obovate, 2.0–5.0 × 1.0–3.0 cm. Filaments dark purple, anthers golden yellow or orange. Ovary glabrescent to sparsely pubescent, bristles pale yellow sometimes with reddish base, spreading. Style 2–12 mm, slender or stout. Stigma yellow-green, capitate or subclavate, 4–9-lobed, lobes frequently distinctly decurrent. Capsule oblong-ellipsoid or oblong-obovoid, 3.1–4.9 × 0.6–1.4 cm, dehiscing by 4–9 valves, glabrescent to sparsely bristly, bristles reflexed to spreading.
Fig. 12f–g.

Distribution: Nepal and E Himalaya.

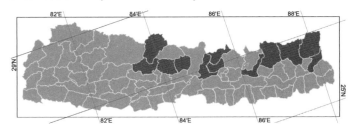

Altitudinal range: 3300–4900 m.

Ecology: Amongst low scrub and rocks on open slopes.

Flowering: May–June. **Fruiting:** September–October.

19. *Meconopsis bella* Prain, J. Asiat. Soc. Bengal, Pt. 2, Nat. Hist. 63(2): 82 (1894).

Dwarf polycarpic herb 8–20 cm. Stem to 2.5 cm in diameter. Indumentum sparse, ca. 3 mm barbellate-bristly or glabrous. Leaves all basal, numerous, crowded. Petioles 2.5–10 cm. Laminas very variable, usually elliptic, lanceolate or oblong, 1–7 × 1–4 cm, base attenuate, apex obtuse, entire to deeply and irregularly pinnatisect or bipinnatisect, ultimate segment usually 3-fid, obovate or obovate-oblong in outline. Flowers solitary on 2–8, leafless, recurved, 3–17 cm scapes. Petals 4, pale blue, pink or purple, obovate to suborbicular, 2.0–3.0 × 2.0–2.5 cm. Filaments pale to dark blue, anthers golden yellow to orange. Ovary glabrous or sparsely bristly. Style 2–3 mm, stout at base. Stigma green, capitate, 4–7 lobed. Capsules narrowly obovoid or pear shaped, 0.5–1.2 × 0.4–0.8 mm, dehiscing by 4–7 valves, glabrous or with a sparse indumentum of spreading bristles.

Distribution: Nepal and E Himalaya.

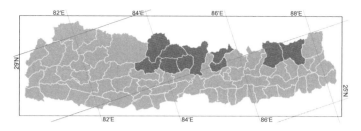

Altitudinal range: 3600–5300 m.

Ecology: Rock crevices, steep grassland, low scrub.

Flowering: June–August. **Fruiting:** September–October.

20. *Meconopsis horridula* Hook.f. & Thomson, Fl. Ind. 1: 252 (1855).
Meconopsis horridula var. *rudis* Prain; *M. prattii* (Prain) Prain; *M. rudis* (Prain) Prain; *M. sinuata* var. *prattii* Prain.

छार-गुन Chhar-gun (Tibetan).

Monocarpic herb 5–39 cm. Stem 6–15 mm in diameter at base. Indumentum sparse to dense with rigid, 4–8 mm, spreading, straw-coloured to yellow-orange bristles, commonly with purple base on buds and capsules. Leaves in basal rosettes only. Petioles 0.7–7.4 cm. Laminas lanceolate, oblanceolate or elliptic-oblong, 3.2–15.9 × 1.2–6.2 cm, base usually attenuate, apex acute or subacute, rarely to subrounded, margin entire to irregularly toothed. Inflorescence of erect to recurved scapes, 6–18-flowered, or flowers borne in a raceme composed of partly fused scapes usually also with accompanying basal scapes. Petals 6(–10), light purple to sky blue, rarely white, frequently pinkish around petal base, obovate, 2.1–4.6 × 1.5–3.0 cm. Filaments dark purple, commonly pinkish towards base, anthers yellow. Ovary densely covered with appressed to ascending bristles. Style green or reddish, 2–9 mm, stout. Stigma whitish or yellow, linear or capitate, 5–8-lobed. Capsule broadly ellipsoid-oblong to ovoid, 8–18 × 5–12 mm, dehiscing by 5–8 valves, with a moderate to dense indumentum of thick spreading bristles.

Distribution: Nepal, E Himalaya, Tibetan Plateau, Assam-Burma and E Asia.

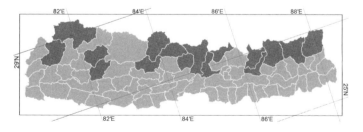

Altitudinal range: 3700–5300 m.

Ecology: Rocky slopes, among boulders and crevices.

Flowering: July–August. **Fruiting:** August–September.

Widespread in the Pan Himalayan region, but morphologically very variable and in need of taxonomic revision across its whole distribution.

The plant is used to treat fevers, colds, sinusitis, itching, wounds and lung, skin and bile diseases.

21. *Meconopsis sinuata* Prain, J. Asiat. Soc. Bengal, Pt. 2, Nat. Hist. 64: 314 (1896).

Monocarpic herb, 20–80 cm. Stem to 15 mm in diameter at base. Indumentum moderately dense, with slightly reflexed to spreading yellowish or orange-brown, 2–7 mm, spiny bristles. Leaves cauline and in sparse basal rosettes. Petioles 2–9 cm, upper cauline leaves sessile, semi-amplexicaul. Laminas narrowly oblanceolate to oblong, 4–9 × 2–3 cm, base rounded or attenuate, apex obtuse, margin coarsely or finely sinuate-lobed to subentire, lobes rounded or acute. Inflorescence racemose, flowers 4–8, on 15 cm pedicels. Petals 4, blue, purple to pale purple, obovate, 2.0–3.2 × 1.5–2.0 cm, apex rounded to subacute. Filaments the same colour as petals, anthers yellow-orange. Ovary moderately appressed bristly. Style 4–8 mm, slender. Stigma yellow, capitate or subclavate. Capsule narrowly obovoid or ellipsoidal, 1.0–4.0 × 0.5–1.0 cm, dehiscing by 3 or 4 valves, sparsely to moderately hairy, bristles ascendant.

Distribution: Nepal and E Himalaya.

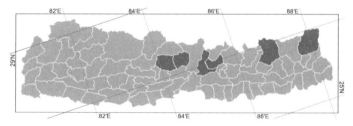

Altitudinal range: 3600–4300 m.

Ecology: Forest margins, scrub and rocky crevices.

Flowering: July–August. **Fruiting:** September–October.

22. *Meconopsis lyrata* (H.A.Cummins & Prain ex Prain) Fedde ex Prain, Kew Bull. 1915: 142 (1915).
Cathcartia lyrata H.A.Cummins & Prain ex Prain, J. Asiat. Soc. Bengal, Pt. 2, Nat. Hist. 64(3): 325 (1896); *C. polygonoides* Prain; *Meconopsis compta* Prain; *M. polygonoides* (Prain) Prain.

Monocarpic herb, erect to ascending, 10–35 cm. Stems to 5 mm in diameter at base. Indumentum sparsely yellow-brown bristly, hairs 2 or 3 mm, or glabrous. Leaves cauline and basal, few. Petioles 1.5–4 cm, bulbils sometimes present in axils, upper cauline leaves sessile. Laminas very variable, ovate, oblong, spathulate or oblanceolate, 1–4.5 × 0.5–2.5 cm, base rounded or subcordate, attenuate into petiole, apex acute or rounded, margin entire to pinnatilobate or lyrate. Inflorescence racemose with up to 5 flowers, or flowers solitary on basal scapes. Pedicels 2–15 cm. Petals 4(–6), blue, pale purple-pink or white, obovate, 0.9–1.9 × 0.4–2.0 cm, acute to subrounded at apex. Filaments the same colour as petals, anthers golden-yellow. Ovary glabrous. Style short, 2–4 mm. Stigma subclavate, 2–4 lobed. Capsules narrowly oblong or subcyclindrical, ca. 2.5 cm long, dehiscing by 3–4 valves, glabrous.

Distribution: Nepal, E Himalaya, Tibetan Plateau and E Asia.

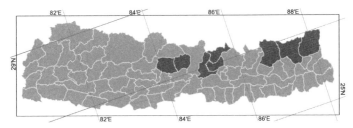

Altitudinal range: 3200–4600 m.

Ecology: Among scrub and on steep grassy slopes.

Flowering: July–August. **Fruiting:** September–October.

4. *Cathcartia* Hook.f. ex Hook., Bot. Mag. 77: pl. 4596 (1851).

Paul A. Egan

Erect to ascending, polycarpic perennial herbs. Root short, stout, branching into fibrous rootlets, crowned with persistent remains of withered leaf bases. Latex watery. Indumentum sparsely to densely barbellate-villous or glabrescent throughout. Not glaucous. Stems multiple. Basal leaves in rosettes, distinctly petiolate. Cauline leaves petiolate to sessile. Leaves evenly coloured above, palmate or trifoliate, ovate to orbicular, bi-pinnatilobate. Inflorescence determinate, simple, or rarely sparingly branched, flowers in axils of leaf-like bracts. Flower buds erect or nodding. Flowers, semi-nutant to lateral-facing. Sepals 2, lacking apical horn. Corolla saucer-shaped. Petals 4, bright yellow, obovate to suborbicular, apex rounded, margins entire to somewhat emarginate. Ovary 4–7-carpellate, narrowly oblong, glabrous or densely appressed bristly. Style very short or absent. Stigmatic lobes stellately radiating. Capsule 4–7 valvate, septicidal, dehiscing from a third to almost to base. Seeds ovoid or subreniform with rounded ends, alveolate.

A genus of two species with a disjunct distribution between E Himalaya and NE Myanmar/NW Yunnan. One species in Nepal.

Cathcartia has often been united with *Meconopsis*, but the current consensus is that the genera are distinct.

1. *Cathcartia villosa* Hook.f. ex Hook., Bot. Mag. 77: pl. 4596 (1851).
Meconopsis villosa (Hook.f. ex Hook.) G.Taylor.

Perennial, polycarpic herb to 70 cm. Stems 5–12 mm in diameter at base. Indumentum of spreading, greyish or fawn-coloured, 3–5 mm villous bristles throughout. Petioles 3–25 cm, cauline leaves more shortly petiolate or subsessile. Leaves orbicular or less commonly broadly ovate, shallowly to deeply 3- or 5-palmately segmented, 3–12 × 3–11.4 cm, base cordate or broadly cuneate. Leaf segments coarsely lobed to shallowly bi-pinnatilobate, lobes subacute to rounded at apex. Flowers 1–5 per stem, to 3.5–5 cm across. Pedicels 3–14 cm. Petals 4, bright yellow, obovate to suborbicular, 2.5–3.5 × 1.5–3.8 cm. Filaments golden-yellow with yellow anthers, these turning to dark-brown with age. Ovary glabrous, narrowly oblong. Style absent or very short. Stigma green. Capsule narrowly oblong or subcylindric, ribbed, 4.0–9.0 × 0.5–0.7 cm, glabrous.
Fig. 10g–h.

Distribution: Nepal and E Himalaya.

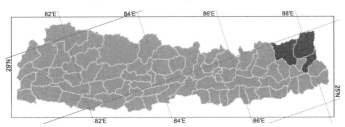

Altitudinal range: 2600–3700 m.

Ecology: Open rocky meadows, forests, damp clefts between boulders.

Flowering: June–August. **Fruiting:** August–September.

Although *Cathcartia villosa* is widely described as lacking a style, short styles have been observed in some specimens.

5. *Dicranostigma* Hook.f. & Thomson, Fl. Ind. 1: 255 (1855).

Colin A. Pendry

Biennial or short-lived perennial herbs, shortly pubescent or glabrescent, latex milky yellow or orange-yellow (staining). Taproot stout, narrowly fusiform. Stems several, ascending to spreading, usually little-branched. Basal leaves many in a dense rosette, petiolate, pinnatilobate, pinnatipartite, or bipinnatifid. Cauline leaves alternate, sessile, similar to basal leaves. Flowers terminal, solitary, 2-merous, bracteate. Pedicel slender, usually glabrous, bracteolate. Sepals 2, ovate or broadly ovate, glabrous or shortly pubescent, apiculate. Petals 4, yellow or orange, obovate or suborbicular. Stamens many, filaments filiform, anthers oblong, 2-celled, dehiscing by longitudinal slits, basifixed. Ovary 1-locular, 2-carpellate, terete, pubescent. Style short, stout, stigmas 2, ascending, capitate. Capsule terete, 2-valvate, septicidal, splitting from below style to almost the base, remaining attached to the style, shortly pubescent or glabrous. Seeds ovoid, pitted.

Worldwide three species in temperate or alpine regions in the Himalaya and China. One species in Nepal.

**1. *Dicranostigma lactucoides* Hook.f. & Thomson, Fl. Ind. 1: 255 (1855).
Stylophorum lactucoides (Hook.f. & Thomson) Benth. & Hook.f. nom. inval.

दूधे साग Dudhe saag (Nepali).

Herb 15–60 cm tall, shortly pubescent. Roots to 1.5 cm thick, apex, densely covered with persistent withered leaf bases. Stems sparsely villous or almost glabrous. Petioles of basal leaves ciliate-winged, 2–10 cm, sparsely shortly pubescent. Basal leaves glaucous above, paler below, sometimes with green or purplish blotches, elliptic to narrowly obovate, 5–15 × 1.5–6 cm (larger later in season to 25 cm), deeply pinnatilobate to pinnatisect, the basal lobes smaller, margins sinuate or with thick, mucronate teeth, both surfaces sparsely shortly pubescent. Cauline leaves sessile, 2–6 × 2.5–4 cm, similar to basal leaves but smaller, lobes 1–3 pairs. Pedicels 3.5–10(–15) cm, curved. Flower buds ovoid, 1.5–2 cm, flowers 3–5 cm diameter. Sepals ovate or broadly ovate, 1–2 cm, apex with short blunt horn, margin membranous, glabrous or shortly pubescent Petals bright yellow or orange-yellow, broadly obovate, 1.5–2.5 × 1–2 cm. Stamens 4–7 mm, anthers yellow, linear-oblong, ca. 2 mm. Ovary narrowly ovoid, 8–10 mm, shortly pubescent, style 1–2 mm, stigmas 1.5 mm. Capsule very narrowly ellipsoid, slightly broader toward base, often slightly curved, 4–8(–11) cm × 4–5 mm, beaked, shortly puberulent or glabrous. Seeds ca. 1 mm.
Fig. 10i.

Distribution: Nepal, W Himalaya, Tibetan Plateau and E Asia.

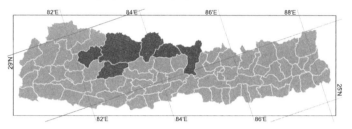

Altitudinal range: 2400–4400 m.

Ecology: Dry exposed open areas, bare earthen or rocky slopes, semi-stable scree, gravel banks by riversides, grassy areas or open shrubland.

Flowering: (May–)June–August(–September). **Fruiting:** (May–)June–September(–October).

Plant used medicinally for liver complaints, the juice used to disinfect open wounds, and powdered roots used by women to ease delivery of the after-birth.

Capparaceae

Gordon C. Tucker, Krishna K. Shrestha & Sajan Dahal

Evergreen or deciduous shrubs, trees, or woody vines. Stipules small or spine-like, sometimes absent. Branchlet base sometimes with subulate scales (cataphylls). Leaves alternate, simple or trifoliolate, petiolate. Inflorescences axillary or supra-axillary, racemose, corymbose, subumbellate, or paniculate, 2–10-flowered, or flowers solitary in leaf axil. Flowers bisexual, rarely unisexual, actinomorphic or rarely zygomorphic, often with caducous bracts. Sepals 4–6, in 1 or 2 whorls, free or basally connate. Petals absent or 4(–6), alternating with sepals, free, with or without a claw. Receptacle often extended into an androgynophore, with nectaries. Stamens 8–80; filaments on receptacle or androgynophore, free, inflexed or spiralled in bud; anthers 2-celled, introrse, basifixed or dorsifixed. Pistil 2(–8)-carpellate; gynophore about as long as stamens. Ovary 1-locular, with 2 to several parietal placentae or (3–)4-locular with axile placentation. Ovules few to many. Style obsolete or highly reduced, sometimes elongated and slender. Stigma capitate or not obvious, rarely 3-branched. Fruit a berry or drupe. Seeds 1 to many, reniform to polygonal, smooth or variously sculptured.

Worldwide about 28 genera and 650 species, mostly in tropics and subtropics with a few in temperate regions. Three genera and eight species in Nepal.

Key to Genera

1a   Leaves compound, with 3 leaflets ....**1. Crateva**

b   Leaves simple 2

2a   Petals 4. Anthers basifixed. Placentation parietal. Fruit a berry. Gynophore equalling or exceeding length of mature fruit...**2. Capparis**

b   Petals absent. Anthers dorsifixed. Placentation axile. Fruit a drupe. Gynophore less than ⅕ as long as mature fruit..**3. Stixis**

1. *Crateva* L., Sp. Pl. 1: 444 (1753).

Deciduous trees, glabrous throughout. Twigs terete or angular, with lenticels. Stipules triangular, small, caducous. Cataphylls absent. Petioles long, often with glands in upper part. Leaves trifoliolate. Petiolules short, thin when young, becoming thick in maturity. Lateral leaflet blades with asymmetric bases. Inflorescences corymbose racemes at tips of new branches. Bracts at bases of pedicels, caducous. Flowers actinomorphic, unisexual (due to failure of one sex to develop). Pedicels long. Receptacle disk-like, inner surface concave, with nectary. Sepals 4, greenish, equal, smaller than petals, deciduous. Petals 4, white, cream-coloured, or yellow, equal, clawed, blade ovate to rhomboid. Stamens 16–30; filaments basally connate to form an androgynophore. Gynophore 2–8 cm but degenerate in staminate flower. Ovary 1-locular, placentae 2, ovules many. Style short or absent, stigma inconspicuous, knob-shaped. Fruit a berry, globose or ellipsoid, drooping; pericarp drying to grey, red, purple, or brown, leathery, apically smooth or papillate. Seeds ovoid, 25–50 per berry, embedded in creamy foetid or pungent mesocarp.

About ten species worldwide in the tropics and subtropics north to Japan and south to Argentina. Two species in Nepal.

Key to Species

1a   Leaflets ovate-orbicular, apex acuminate. Fruit 1.8–2.6 cm ....**1. C. religiosa**

b   Leaflets oblong-lanceolate, apex acute. Fruit 3–4 cm ....**2. C. unilocularis**

1. *Crateva religiosa* G.Forst., Diss. Pl. Esc.: 45 (1786). *Crateva membranifolia* Miq.; *C. nurvula* Buch.-Ham.; *C. religiosa* var. *nurvula* (Buch.-Ham.) Hook.f. & Thomson.

सिप्लीकान Siplikan (Nepali).

Trees 3–15 m. Twigs with grey elongated lenticels. Petioles 5–7(–10) cm, with minute triangular glands above near apex. Petiolules 3–5(–7) mm. Leaflets dull green above with reddish midvein, grey below, ovate-orbicular, (4–)5.5–7(–10)

× (2–)3–4 cm, apex acuminate, nearly coriaceous, secondary veins 5–10 pairs. Inflorescences racemes or corymbs, 10–25-flowered. Bracts 0.8–1.5 cm, leaf-like or slender, caducous. Flowers opening as leaves emerge. Pedicels 2–5(–9) cm. Sepals ovate, 2–4.5 × 1.5–3 mm, apex acuminate. Petals white to yellow, claw 3.5–5 mm, blade 1.5–2.2 cm. Stamens 16–22(–30), filaments 3–6 cm, anthers 2–3 mm. Gynophore 3.5–6.5 cm; ovary ovoid to subcylindric, 3–4 × 1–2 mm. Fruit ovoid to obovoid, 1.8–2.6 cm; pericarp 5–10 mm thick, apically scabrous and grey to dust-coloured with nearly circular ash yellow flecks; stipe 2.5–3 mm in diameter, thickened, woody. Seeds 25–30 per fruit, dark brown, 12–18 mm, tuberculate. Fig. 13a–b.

Distribution: Nepal, E Himalaya, Assam-Burma, S Asia, E Asia, SE Asia and Australasia.

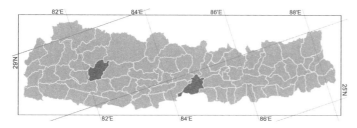

Altitudinal range: 200–1200 m.

Ecology: Roadsides, fields.

Flowering: March–May. **Fruiting:** July–August(–October).

The flowers are fragrant. Often planted as an ornamental in S China and other S and SE Asian countries.
    Dried fruit are used medicinally in China.

2. *Crateva unilocularis* Buch.-Ham., Trans. Linn. Soc. London 15: 121 (1827).

सिप्लीकान Siplikan (Nepali).

Trees 5–10(–20) m. Twigs with sparse whitish lenticels. Petioles (3–)5.5–9 cm, glands above toward apex. Petiolules (2.5–)4–7 mm. Leaflets oblong-lanceolate glossy, brown below with reddish midvein, elliptic, (6.5–)8–10 × 3–4(–5) cm, apex acuminate to abruptly acuminate, secondary veins 5–8(–10)

pairs, subcoriaceous. Inflorescences racemes or corymbs, 13–25(–35)-flowered. Pedicels 2–4 cm. Sepals linear to narrowly lanceolate, (3–)4–6 × 2–3 mm. Petals white to creamy but drying pinkish, claw 3–7 mm, blade 1.4–2.4 cm. Stamens 16–20; filaments (2–)3–4.5 cm; anthers 2–3 mm. Gynophore 4–6 cm; ovary oblong-ellipsoid, 3–4 × 1–2 mm. Fruit globose, 3–4 cm; pericarp 2–3 mm thick, apically scabrous, with nearly circular small ash yellow flecks; stipe 3–7 mm in diameter, thickened, woody. Seeds 30–50 per fruit, dull brown, lens-shaped, 8–10(–12) mm, smooth.

Distribution: Nepal, E Himalaya, Assam-Burma, S Asia, E Asia and SE Asia.

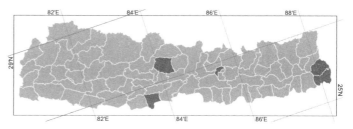

Altitudinal range: 100–1900 m.

Ecology: Wet areas, commonly cultivated.

Flowering: (December–)March–July. **Fruiting:** July–August.

Nepalese authors (Bull. Dept. Med. Pl. Nepal 7: 44. 1976) have misapplied the name *Crateva religiosa* Forst. to this species.
    Trees of *C. unilocularis* are reported to reach 30 m tall in India. Juice of the bark is taken to promote an appetite and for stomach pain. Juice of young leaves has anthelmintic properties. A paste of the fruit was used to treat smallpox. The plant is also used as fodder.

2. *Capparis* L., Sp. Pl. 1: 503 (1753).

Evergreen shrubs or small trees, erect or climbing, sometimes prostrate or hanging. Twigs with branched or simple trichomes often present (in new twigs), glabrescent or sometimes with persistent trichomes. Cataphylls sometimes present. Stipules absent or spine-like, straight or curved. Leaves simple, papery to leathery, margin entire. Inflorescences supra-axillary, axillary, or terminal, a raceme, corymb, umbel, or panicle, or row of flowers, or flower solitary. Flowers actinomorphic or zygomorphic. Bracts usually present but often caducous. Pedicels long, often twisted, resulting in apparent exchange of position of floral parts. Sepals 4, in 2 whorls. Sepals of outer whorl often thick, dissimilar to almost equal, often inwardly concave or becoming navicular, covering other flower parts, sometimes basal one becoming saccate. Sepals of inner whorl often thin, almost equal. Petals 4, imbricate; lower pair with or without a claw; upper petals with asymmetric base, revolute margin. Stamens 6–80(–200). Gynophore about as long as filaments. Ovary 1-locular, placentas 2–6(–8); ovules few to many. Fruit a berry, spheroid or elongated, often with different colour when mature or dry, usually not dehiscent, 1–many-seeded. Seeds reniform to nearly polygonal.

Worldwide about 250–400 species, mostly in tropical and subtropical regions but some in temperate regions. Five species in Nepal.

The following three species may occur in Nepal and are included in the key: *Capparis assamica* Hook.f. & Thomson, known from Yunnan, Bhutan, NE India, Laos, Myanmar, Thailand and possibly present in E Nepal below 1000 m; *C. sepiaria* L. (syn. *C. flexicaulis* Hance), a widespread species of SE and S Asia, to be expected in Nepal below 250 m; and *C. sikkimensis* Kurz (syn. *C. cathcartii* Hemsl. ex Gamble), an uncommon species of Xizang, Sikkim and W Burma which is possibly present in Nepal, from 1200–1750 m.

Capparaceae

Key to Species

1a Flowers in terminal, or axillary subumbellate or racemose clusters 2

b Flowers solitary and axillary, or in supra-axillary rows opening in a basipetal direction 4

2a Petioles 1–2 cm. Fruit 2.5–4 cm .... ***C. sikkimensis***

b Petioles 3–8 mm. Fruit 6–10 mm 3

3a Inflorescence terminal or nearly so, racemes, 1- or 2-fascicled, 10–25 cm. Leaves 12–26 cm.................. ***C. assamica***

b Inflorescence terminal on lateral shoots, subumbellate or shortly racemose, sessile, (6–)10–22(–25)-flowered. Leaves 2–7 cm .... ***C. sepiaria***

4a Flowers solitary and axillary. Seeds numerous...5

b Flowers in supra-axillary rows. Seeds 1–3 6

5a Lower sepal not galeate, 3–4.5 mm deep near apex, shorter than petals. Leaf apex acute, obtuse, or retuse but not acuminate... **3. *C. spinosa***

b Lower sepal galeate, 9–15 mm deep near apex, longer than petals. Leaf apex acuminate .... **4. *C. himalayensis***

6a Supra-axillary rows of (4–)7–10 flowers. Stamens 8–12. Leaf base cuneate or abruptly contracted ....**1. *C. multiflora***

b Supra-axillary rows of (1–)2(–3)-flowers. Stamens 34–38. Leaf base rounded................................7

7a Leaf blade ovate. Anthers ca. 2 mm. Seeds 1–3 .... **2. *C. olacifolia***

b Leaf blade elliptic-lanceolate to obovate-lanceolate, sometimes linear, or hastate. Anthers 1.2–1.5 mm. Seeds 10–20 .... **5. *C. zeylanica***

---

1. *Capparis multiflora* Hook.f. & Thomson, Fl. Brit. Ind. 1[1]: 178 (1872).

Shrubs, sometimes twining, or small trees, 3–6 m. Twigs terete, slender, spineless or sometimes with small stipular spines 1–2 mm, glabrous or with scattered, white, branched trichomes, soon glabrescent. Cataphylls subulate, 2–3 mm wide at base. Petioles 8–11 mm. Leaves broadly lanceolate to oblong, widest apically from middle, 5–10 × 2.5–3.5 cm but those subtending inflorescences 15–25 × ca. 6 cm, base cuneate to abruptly contracted, apex acuminate to abruptly acuminate, papery, midvein barely raised above, thickened below and prominent, secondary veins 7–10(–12) pairs, with slender, reticulate veins obvious and forming intra-marginal loops. Inflorescences supra-axillary rows of (4–)7–10 flowers, spaced over 5–10 mm, with several rows on a branch section, 10–20 cm between old and new leaves. Pedicels 0.6–1.5 cm. Sepals 3–4 × ca. 2 mm, slightly unequal; sepals of outer whorl round, slightly larger, navicular, glabrous or margin and outside sometimes distantly pubescent; sepals of inner whorl round, ovate, or obovate, slightly short and narrow, glabrous or with short pubescence, margin membranous. Petals white, equal, oblong, ca. 5 × 1.5–2 mm, glabrous. Stamens (8–)10–12, filaments 6–9 mm, anthers 0.7–0.8 mm. Gynophore 6–12 mm. Ovary ovoid, ca. 1.2 × 1 mm, glabrous or minutely pubescent; placentae 2; ovules several; style glabrous, stigma capitate. Fruit globose, 8–10 mm in diameter; stipe ca. 1 mm thick. Seeds 1 or 2(or 3), ca. 8 × 6 mm.

Distribution: Nepal, E Himalaya, Assam-Burma, S Asia, E Asia and SE Asia.

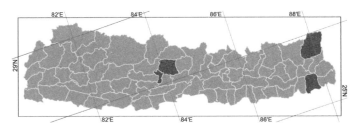

Altitudinal range: 900–1600 m.

Ecology: Forested ravines.

Flowering: June. **Fruiting:** December.

2. *Capparis olacifolia* Hook.f. & Thomson, in Hook.f., Fl. Brit. Ind. 1[1]: 178 (1872).

Shrubs or small trees, 1–5 m, with spreading branches. Twigs densely fulvous or greyish tomentose with small, ca. 5-armed, stellate trichomes. Cataphylls few, 2–3 mm. Stipular spines slender, straight, 3–5(–8) mm, ascending or rarely spreading. Petioles 5–6 mm, with stellate trichomes. Leaves ovate, 7–13 × (3.5–)4–5(–6) cm, base rounded, apex tapering and gradually acuminate, tip blunt to somewhat acute and mucronulate, ± firmly papery, rather glossy above, surfaces soon glabrescent, midvein flat, secondary veins (5 or)6 or 7(or 8) pairs, reticulate veins not distinct. Inflorescences supra-axillary rows, (1 or)2(or 3)-flowered. Pedicels 0.7–1.5 cm, with trichomes. Sepals unequal, 8–10 × (3–)4–5(–6) mm, margin membranous and tomentose; sepals of outer whorl navicular,

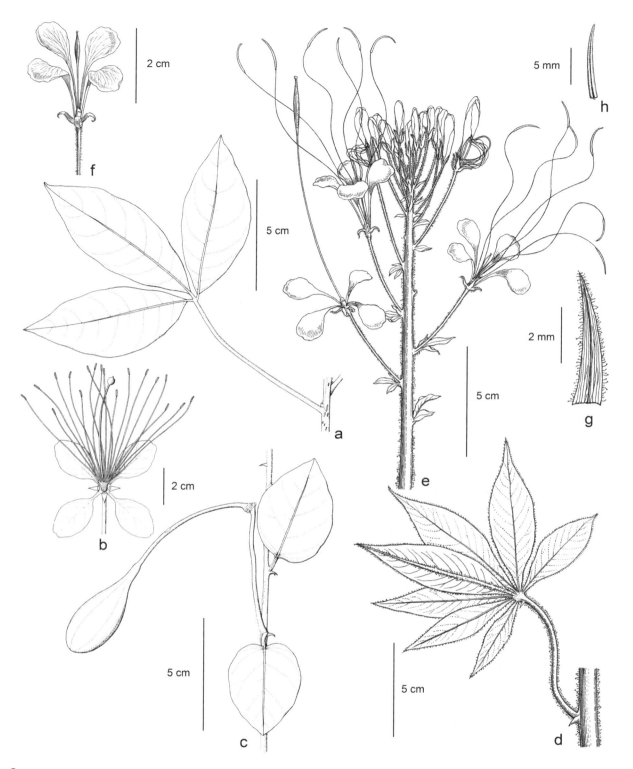

FIG. 13. CAPPARACEAE. **Crateva religiosa**: a, leaf; b, flower. **Capparis spinosa**: c, twig with fruit. CLEOMACEAE. **Cleoserrata speciosa**: d, leaf; e, inflorescence; f, flower with stamens removed; g, sepal; h, anther.

ovate, surfaces glabrous but margins tomentose, apex acute; sepals of inner whorl elliptic. Petals white, (15–)17–22 × ca. 5 mm; anterior petals obovate, ca. 2 mm longer than upper petals, outside tomentose toward apex and along margin, apex rounded; upper petals with a pale purple or yellow blotch. Stamens 34–38. Filaments 2.8–3.5 cm. Anthers ca. 2 mm. Gynophore 2.7–3.5 cm, often slightly swollen toward apex, glabrous. Ovary ellipsoid 4–6 ×1–1.5 mm, densely tomentose; placentae 2; ovules several. Style ca. 2 mm, slender, glabrous. Stigma knob-shaped. Fruit globose, 7.5–10 mm wide, apex beaked with persistent ca. 2 mm style; pericarp red, fairly thick. Seeds 1(–3), 7–8 × ca. 6 mm.

Distribution: Nepal, E Himalaya, Tibetan Plateau, Assam-Burma and S Asia.

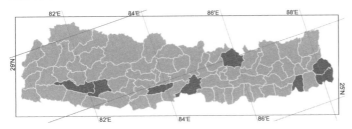

Altitudinal range: 200–1600 m.

Ecology: Moist forests.

Flowering: April–June. **Fruiting:** June–October.

Wallich (Num. List.: 234 n. 6990B. 1932) misapplied the name *Capparis acuminata* Willd. to this species.

3. *Capparis spinosa* L., Sp. Pl. 1: 503 (1753).
Capparis napaulensis DC.

बाघनङ्ग्रे Bagh nangre (Nepali).

Prostrate or hanging shrubs 50–80(–100) cm tall. Twigs glabrous to densely long or shortly white pubescent with simple trichomes, soon glabrescent. Cataphylls absent. Stipular spines 4–5 mm, ± flat, apex recurved. Petioles 1–4 mm. Leaves ovate, obovate, broadly elliptic, or suborbicular, 1.3–3 × 1.2–2 cm, base rounded, apex acute, obtuse, or retuse but not acuminate, spine-tipped, midvein prominent below but gradually becoming obscure from base to apex, fleshy when fresh but later leathery, secondary veins 4(or 5) pairs. Flowers solitary in upper axils. Pedicels 2–6(–9) cm. Calyx zygomorphic, outside ± with trichomes, inside glabrous; sepals of outer whorl 1.5–2 × 0.6–1.1 cm, navicular-lanceolate, outside with several glands, basally shallowly saccate; sepals of inner whorl 1–2 cm, not saccate, not broadest near base, 3–4.5 mm broad in upper half. Petals dimorphic, about as long or slightly longer than upper sepal; upper petals white, distinct, claw 4–7 mm, blade oblong-obovate, 1–2 cm, outside with trichomes, apex subemarginate; upper petals yellowish green to green, enclosed by sepals, thickened, margin connate from base almost to middle. Stamens ca. 80. Filaments 2–4 cm, unequal. Anthers 2–3 mm. Gynophore ca. 1 cm, sometimes basally sparsely villous.

Ovary ellipsoid, 3–4 mm, glabrous, apically with vertical thin furrow and ridge; placentae 6–8; ovules numerous. Style and stigma obscure, mound-like. Fruit dark green when dry, ellipsoid to oblong-obovoid, 1.5–4 × 0.8–1.8 cm, with 6–8 lengthwise thin ridges, dehiscent; fruiting pedicel and gynophore 3–7 cm, 1.5–2 mm in diameter, forming a right angle with each other. Seeds 40–60, 3–4 mm, smooth. Fig. 13c.

Distribution: Asia, Europe, Africa and Australasia.

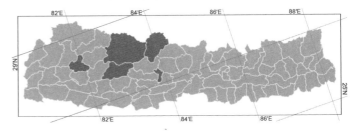

Altitudinal range: 200–2500 m.

Ecology: Plains, desert flats, open and sunny areas.

Flowering: May–August. **Fruiting:** August–October.

Whitmore (Enum. Fl. Pl. Nepal 2: 46. 1979) recorded *Capparis spinosa* from C and E Nepal based on *Wallich s.n.* (BM), (the type of *C. napaulensis*) and *TI 6301945* (TI) respectively, but it has not been possible to obtain more precise locality information for these specimens.

Buds and unripe fruits are cooked as vegetables or pickled. The ripe fruits are eaten fresh. A paste of root is applied for rheumatism, and the juice is given as an anthelmintic.

4. *Capparis himalayensis* Jafri, Pakistan J. Forest. 6: 197 (1956).
Capparis spinosa var. *himalayensis* (Jafri) Jacobs.

Prostrate or hanging shrubs 50–80(–100) cm. Twigs white pubescent with dense long to short, simple trichomes, soon glabrescent. Cataphylls absent. Stipular spines pale yellow, 4–5 mm, often flat, apex recurved. Petioles 2–4 mm. Leaves ovate to suborbicular, 1.3–3 × 1.2–2 cm, base rounded, apex shortly mucronate, secondary veins 4(or 5) pairs, reticulate veins invisible on both surfaces. Flowers solitary in upper axils. Pedicels 4–9 cm, ± pubescent. Calyx zygomorphic, with some trichomes outside, glabrous inside: sepals of outer whorl 1.7–3.2 cm, the lower one deeply saccate or galeate, broadest near apex, 0.9–1.5 cm deep in upper half; sepals of inner whorl 1.5–2 × 0.6–1.1 cm. Petals dimorphic, about as long as or slightly longer than upper sepals; upper petals white, distinct, claw 3–5 mm, blade oblong-obovate and outside with trichomes; lower petals yellowish green to green, thickened, margin connate from base almost to middle, enclosed by sepals. Stamens ca. 80; filaments 1.8–2.5 cm, unequal; anthers 2–2.5 mm. Gynophore ca. 1 cm at anthesis, sometimes basally sparsely villous. Ovary ellipsoid, 3–4 mm, glabrous, apically with vertical thin furrow and ridge; placentae 6–8; ovules numerous. Style and stigma obscure, mound-like. Fruit ellipsoid, 2.5–3 × 1.5–1.8 cm, apically with 6–8 dark red vertical thin ridges at carpel sutures, dehiscent; fruiting pedicel and gynophore 3–4 mm, 1.5–2 mm in diameter, forming a right angle with each other. Seeds 40–60, reddish brown, reniform, 3–4 mm wide, smooth.

Distribution: Nepal, W Himalaya, Tibetan Plateau, E Asia, C Asia and SW Asia.

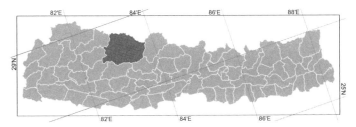

Altitudinal range: 900–2400 m.

Ecology: Plains, open and sunny areas.

Flowering: April–July. **Fruiting:** August–September.

5. *Capparis zeylanica* L., Sp. Pl., ed. 2, 1: 720 (1762).
Capparis hastigera Hance; *C. hastigera* var. *obcordata* Merr. & Metcalf; *C. horrida* L.f.

वन भेंडो Ban bhendo (Nepali).

Scandent or trailing shrubs, 2–5 m. Shoots densely pubescent with red brown to grey stellate trichomes, eventually glabrescent. Cataphylls absent. Stipular spines strong, sharp, recurved, 1–5 mm. Petioles 5–12 mm. Leaves elliptic-lanceolate to obovate-lanceolate, sometimes elliptic, linear, or hastate, 3–8(–13) × 1.5–4(–7.5) cm, base cuneate, rounded, or rarely nearly cordate, apex acute, rounded, or rarely slightly acuminate and often with an outwardly bent or recurved 2–3 mm leathery mucro, midvein flat or impressed above, prominent below, secondary veins 3–7 pairs and slender, reticulate veins obvious on both surfaces, sub-leathery, both surfaces with dense, thin, grey stellate trichomes when young but soon glabrescent. Inflorescences supra-axillary rows, (1 or)2–3(or 4)-flowered, near apex of young branches, with flowers often opened before leaf emergence and appearing racemose. Pedicels 0.5–1.8 cm, slightly stout, densely shortly red brown stellate tomentose. Sepals 8–11 × 6–8 mm, slightly unequal, outside ± reddish brown tomentose; sepals of outer whorl nearly orbicular, 1 larger, inside concave, apex acute to obtuse; sepals of inner whorl elliptic. Petals white to rarely yellowish white, oblong, 9–15 × 5–7 mm, glabrous; apical petals with red flecks on central base. Stamens 30–45. Filaments 3.3–4 cm; anthers 1.2–1.5 mm. Gynophore base grey tomentose. Ovary ellipsoid, 1.5–2 mm; placentae 4; ovules many. Stigma obvious. Fruit red to purplish red when mature, globose to ellipsoid, 2.5–4 cm in diameter, verrucose; pericarp firm when dry; fruiting gynophore 3–4.5 cm × 3–6 mm, glabrous; fruiting pedicel 3–5 mm in diameter, thickened, woody. Seeds 10–20, reddish brown, 5–8 × 4–6 mm.

Distribution: Nepal, Assam-Burma, S Asia, E Asia and SE Asia.

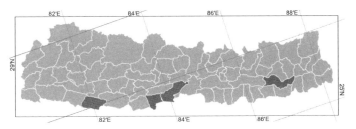

Altitudinal range: 100–300 m.

Ecology: Forest margins, thickets, limestone slopes or sandy soil, scattered grasslands.

Flowering: February–April. **Fruiting:** July–September.

Hooker & Thomson (Fl. Brit. Ind. 1: 174. 1872) misapplied *Capparis brevispina* DC. to this species.
    Internal application of the root bark is used for colic and in the treatment of cholera. A paste of leaves is applied externally to glandular swellings, piles and boils.

3. *Stixis* Lour., Fl. Cochinch. 1: 290, 295 (1790).

Roydsia Roxb.

Evergreen, woody vines or climbing shrubs. Stipular spines absent. Petioles sometimes geniculate, apically often thickened. Leaves simple, leathery, glabrous or sometimes with trichomes, midvein with small pustules above, margin entire. Inflorescences axillary racemes or terminal panicles, many-flowered. Bracts subulate, often caducous. Flowers actinomorphic, small. Pedicel short. Sepals (5 or)6, basally connate into a short tube, lobes erect, spreading, or reflexed. Petals absent. Androgynophore terete, short. Stamens (27–)40 to ca. 80. Filaments distinct, unequal with outermost shortest; anthers dorsifixed. Gynophore about as long as filaments. Ovary nearly globose or ovoid, glabrous or with trichomes, apically often with vertical grooves, 3(or 4)-loculed, placentation axile; placentae each with 4–10 ovules. Style solitary, linear, entire or divided into 3(or 4) subulate stigmas, sometimes unlobed. Fruit a drupe, ellipsoid, small, surface with lenticels, apex often with persistent style; fruiting pedicel and gynophore about equal, forming a woody stipe much shorter than fruit. Seeds 1(–3) per fruit, ellipsoid, erect.

Worldwide about seven species in E, S, and SE Asia. One species in Nepal.

Capparaceae

1. *Stixis suaveolens* (Roxb.) Pierre, Bull. Mens. Soc. Linn. Paris 1: 654 (1887).

Roydsia suaveolens Roxb., Pl. Coromandel 3[12]: 87, pl. 289 (1820).

Woody vines 1–15 m, up to 10 mm thick. Twigs pale red to pale tan-coloured when dry, stout, terete, shortly pubescent, soon glabrescent; internode length unequal, to 5 cm or longer. Petioles (1–)2–3(–5) cm, stout, with bubble-like raised structures, apically with slightly inflated pulvinus. Leaves elliptic, oblong, or oblong-lanceolate, broadest at middle but sometimes slightly basally or apically, (10–)15–28(–40) × (3.5–)4–10 cm, base cuneate to nearly rounded, apex nearly rounded to acuminate and with a 5–12 mm tip, leathery, both surfaces glabrous, secondary veins 7–9 pairs, reticulate veins obvious. Inflorescences 15–25 cm, at first erect then drooping; axis shortly pubescent to shortly tomentose. Bracts linear to ovate, 3–6 mm, trichomes like those on axis. Pedicels 2–4 mm, stout. Receptacle ca. 3.5 mm in diameter, dish-shaped. Sepals pale yellow, elliptic-oblong, (4–)5–6(–9) × 2–3 mm, erect or spreading, never reflexed, both surfaces densely tomentose, apex acute to obtuse. Androgynophore ca. 2 mm, glabrous. Filaments 4–6(–11) mm, pubescent. Anthers 0.5–0.7 mm. Gynophore 7–10 mm, with dense tan pubescence. Ovary ellipsoid, 1.7–2.5 mm, glabrous or basally sometimes with trichomes. Styles 3(or 4), apex recurved. Stigma indistinct. Fruit orange when mature, ellipsoid, 3–5 × 2.5–4 cm, surface with thin yellow verrucose flecks; fruiting pedicel plus gynophore 0.7–1.3 cm, ca. 5 mm in diameter Seed ellipsoid, 1.8–2 cm.

Distribution: Nepal, E Himalaya, Assam-Burma, S Asia, E Asia and SE Asia.

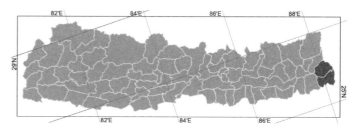

Altitudinal range: 100–1500 m.

Ecology: Thickets, open forests.

Flowering: April–May. **Fruiting:** August–October.

102

Cleomaceae

Gordon C. Tucker, Krishna K. Shrestha & Jyoti P. Gajurel

Annual or perennial herbs. Stems erect, sparsely or profusely branched, glabrous or glandular pubescent. Stipules absent, scale-like or spiny. Petioles often pulvinate at apex, petiolar spines sometimes present. Leaves spiral, palmately compound. Leaflets 3–9, venation pinnate, bases of petioles fused into a pulvinar disc. Inflorescences racemose, pedunculate. Bracts present or absent. Flowers bisexual or unisexual (due to incomplete development in upper flowers), actinomorphic or slightly zygomorphic, pedicellate, without bracteoles, rotate to campanulate. Sepals 4, equal, free or connate basally, persistent. Petals 4, free, imbricate; intrastaminal nectary-disk or glands present. Stamens 6 or 14–25. Filaments free, or fused to the gynophore, glabrous or pubescent. Anthers dehiscing by longitudinal slits. Pistil 1, bicarpellate. Ovary superior, sessile or on a short gynophore. Ovules 1–many per locule. Style 1, straight, short, thick. Stigma 1, capitate, unlobed. Fruit an elongate capsule, usually stipitate. Seeds 4–40(–100), tan to yellowish brown or brown, arillate or not, cochleate-reniform, papillose or tuberculate.

Worldwide 17 genera and about 150 species, in tropical and temperate regions. Five genera and five species in Nepal.

There is only one true *Cleome* in Nepal. Other native and non-native species formerly included in *Cleome* are placed in *Arivela* Raf., *Cleoserrata* H. H. Iltis, *Gynandropsis* DC. and *Tarenaya* Raf. following Iltis & Cochrane (Novon 17: 447. 2007).

Key to Genera

1a	Plants with stipular spines and sometimes with prickles on petioles and leaf veins.	**3. *Tarenaya***
b	Unarmed plants without spines and prickles..	2
2a	Filaments fused to lower half of gynophore (androgynophore present)	**5. *Gynandropsis***
b	Filaments free from gynophore (androgynophore absent)..	3
3a	Bracts subtending pedicels small, subulate	**2. *Cleoserrata***
b	Bracts subtending pedicels with expanded blades, trifoliate..	4
4a	Gynophore present. Stamens 6. Sepals partially connate	**1. *Cleome***
b	Gynophore absent. Stamens 14–25. Sepals free	**4. *Arivela***

1. *Cleome* L., Sp. Pl. 2: 671 (1753).

Annual herbs. Stem sparsely branched, glabrous or glandular pubescent, lacking spines. Stipules absent or scale-like, caducous. Leaflets 3. Inflorescences flat-topped or elongate, terminal or axillary. Pedicels bracteate. Flowers campanulate, slightly zygomorphic. Sepals connate initially, later free, each often subtending a basal nectary. Petals equal. Stamens 6, free. Filaments inserted on a discoid or conical receptacle (androgynophore). Gynophore slender, elongating and recurving in fruit. Ovules many. Style short. Stigma capitate. Fruit a dehiscent oblong or linear, 2-valved capsule. Seeds 4–25, reniform, arillate, striate-verrucose.

Worldwide about 20 species in warm temperate and tropical areas with centre of diversity in SW Asia. One introduced species in Nepal.

1. *Cleome rutidosperma* DC., Prodr. 1: 241 (1824).
Cleome ciliata Schumach. & Thonn.

Herbs to 30–100 cm. Stems glabrous to slightly scabrous or sometimes glandular pubescent. Stipules ca. 0.5 mm, scale-like or absent. Petioles 0.5–3.5 cm, winged proximally. Leaflets oblanceolate to rhomboid-elliptic, 1.0–3.5 × 0.5–1.7 cm, base rounded to subcuneate, apex acute to obtuse, sometimes acuminate, margins entire to serrulate-ciliate, glabrous above, with curved hairs on veins below, especially when young. Inflorescences 2–4 cm, enlarging to 8–15 cm in fruit. Bracts leaf-like, trifoliate, 1–3.5 cm. Pedicels 11–21 mm, enlarging to 18–30 mm in fruit. Sepals yellow, 2.5–4 × 0.2–0.3 mm, lanceolate, ciliate-margined, glabrous, ± persistent. Petals white or speckled with purple, the two central ones with a yellow transverse band on the back, 7–10 × 1.5–2.3 mm, oblong to narrowly ovate. Filaments yellow, 5–7 mm. Anthers purplish-brown, 1–2 mm. Pistil 2–3 mm, glabrous. Gynophore 4–12 mm. Style 0.5–1.4 mm. Fruit a capsule, 40–70 × 3–4 mm, striate. Seeds reddish brown to black, with white funicular aril, slender, 1–1.5 mm, striate-verrucose.

Distribution: Asia, Africa, N America, S America and Australasia.

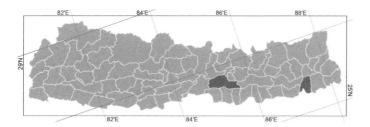

Altitudinal range: 100–500 m.

Ecology: Weed of paddy fields, stream-sides, wetlands. Grows in the open or under shade.

Flowering: June–September. **Fruiting:** June–September.

Native of Africa, introduced in several tropical and warm temperate countries.

2. *Cleoserrata* H.H.Iltis, Novon 17(4): 447 (2007).

Annual herbs. Stem sparsely or profusely branched, glabrous, sometimes with petiolar spines. Stipules absent. Petioles long or short, with pulvinus at distal or basal end. Leaflets 5–9. Inflorescences flat-topped or elongate, terminal or axillary. Pedicels with small, subulate bracts. Flowers unisexual (due to incomplete development of stamens in upper flowers), rotate. Sepals free, each often subtending a basal nectary. Petals equal. Stamens 6. Filaments inserted on a discoid or conical receptacle (androgynophore). Gynophore slender, elongating and recurving in fruit. Fruit a dehiscent oblong capsule. Seeds 10–30, reniform, not arillate.

Worldwide five species in N and S America. One species in Nepal.

1. *Cleoserrata speciosa* (Raf.) H.H.Iltis, Novon 17(4): 448 (2007).
Cleome speciosa Raf., Fl. Ludov.: 86 (1817); *C. speciosissima* Deppe ex Lindl.; *Gynandropsis speciosa* (Kunth) DC. nom. illegit.

Herbs to 0.5–1.5 m. Stems simple or sparsely branched, fluted, glabrous or sparsely glandular pubescent. Stipules absent. Petioles 2–12 cm, glandular pubescent. Leaflets 5–9, narrowly lanceolate-elliptic, 6–15 × 1–5 cm, base attenuate, entire to serrulate, glabrate to glandular-pubescent above and below. Inflorescences 15–50 cm, enlarging to 20–60 cm in fruit, glandular pubescent. Bracts ovate-cordate, 3–18 mm. Pedicels 10–50 mm, glabrous. Sepals green, lanceolate, 4–7 × 0.8–1.2 mm, cuneate, entire, glandular-pubescent, persistent. Petals brilliant pink to purple, fading to pink or white by the second day, rarely initially white, ovate, clawed, 4, 15–42 × 8–11 mm. Stamens green, filaments fused to gynophore for ⅓–½ its length, 40–85 mm. Anthers green, 6–10 mm. Pistil 6–10 mm. Gynophore with the scars from the filaments visible for about ¼ its length. Style 1–1.2 mm. Capsule cylindric, but irregularly contracted between seeds, 60–150 × 3–5 mm. Seeds pale green to brown, subspherical, 2.5–3.5 × 1.0–1.2 mm, tuberculate.
Fig. 13d–h.

Distribution: Nepal, Assam-Burma, S Asia, E Asia, N America and S America.

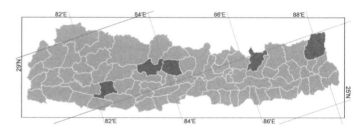

Altitudinal range: 700–1800 m.

Ecology: Cultivated and sometimes semi-naturalized on river banks or open places.

Flowering: June–December. **Fruiting:** August–November.

Native to Mexico and C America, introduced and widespread in gardens, with white-flowered plants not uncommon. Strongly resembling *Tarenaya hassleriana* (Chodat) H.H.Iltis, but considering its unique floral morphology and cytology, it is a species that is difficult to place. Although it has been placed in *Gynandropsis* it is not closely related to that genus, which has a distinctive elongated androgynophore.

3. *Tarenaya* Raf., Sylva Tellur.: 111 (1838).

Annual herbs. Stem sparsely branched, prickly, glandular pubescent. Stipules spinose. Petioles long or short, spiny, with pulvinus at distal or basal end; leaflets 5–7. Inflorescences flat-topped or elongate, terminal or axillary. Pedicels bracteate. Flowers often appearing unisexual due to incomplete development, slightly zygomorphic. Sepals free, each often subtending a basal nectary. Petals equal. Stamens 6, free. Filaments inserted on a discoid or conical receptacle (androgynophore). Gynophore slender, elongating and recurving in fruit. Style short. Fruit a dehiscent oblong capsule. Seeds 10–20, reniform, not arillate.

Worldwide 33 species native to S America and W Africa; introduced elsewhere to tropical and warm-temperate regions. One species in Nepal.

Traditionally included in a broad circumscription of *Cleome* L., *Tarenaya* is distinguished by its stipular thorns, petiolar spines, lack of arils, and seeds with a large cleft cavity. All species are native to tropical America, with the exception of *Tarenaya afrospina* (H.H. Iltis) H.H. Iltis, of West Africa. *Tarenaya hassleriana*, long known as *Cleome spinosa* and more recently as *C. hassleriana*, is a popular garden plant, and probably the most widely distributed member of the family.

1. *Tarenaya hassleriana* (Chodat) H.H.Iltis, Novon 17(4): 450 (2007).
Cleome hassleriana Chodat, Bull. Herb. Boissier 6, App. 1: 12 (1898); *C. spinosa* Jacq.

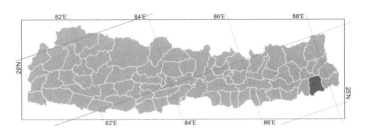

Annual herbs to (0.5–)1–2 m. Stems branched, glandular-pubescent. Stipular spines 1–3 mm. Petioles 2.5–7.5 cm, glandular pubescent, with scattered spines 1–3 mm. Leaflets 5–7, elliptic to oblanceolate, 2–6(–12) × 1–3 cm, acute, serrulate-denticulate, glandular above, glandular-pubescent below. Inflorescences 5–30 cm, enlarging in fruit to 10–80 cm. Bracts ovate, 10–25 mm. Pedicels 20–45 mm, glandular-pubescent, enlarging in fruit. Sepals green, linear-lanceolate, 5–7 × 0.8–1.3 mm, entire, acuminate, glandular-pubescent, persistent. Petals pink, purple, infrequently white, or fading to white by the second day, oblong to ovate, clawed, 20–30(–45) × 8–12 mm. Filaments purple 30–50 mm. Anthers green, 9–10 mm. Pistil 6–10 mm. Gynophore 45–80 mm. Style 0.1 mm. Capsule (25–)40–80 × 2.5–4 mm, the pedicel, glabrous. Seeds dark brown to black, triangular to subspherical, 1.9–2.1 × 1.9–2.1 mm, tuberculate.

Distribution: Nepal, E Himalaya, S Asia, E Asia, SE Asia and S America.

Altitudinal range: 100–1000 m.

Ecology: Weed of waste land and field margins. Grows in the open.

Flowering: April–December. **Fruiting:** June–December.

In horticulture and in various regional Floras, *Tarenaya hassleriana* has usually been treated under the name *Cleome spinosa* Jacq. However, this name and type apply to *T. spinosa* (Jacq.) Raf. Reports of *C. spinosa* from several S and SE Asian countries therefore probably refer to *T. hassleriana*.
   Native to southern S America. Cultivated species occasionally escaped and established.

4. *Arivela* Raf., Sylva Tellur.: 110 (1838).

Annual herbs. Stem sparsely branched, glandular pubescent or glabrous, lacking spines. Stipules absent. Petioles long or short, with pulvinus at distal or basal end. Leaflets 3–5. Inflorescences flat-topped or elongate, terminal or axillary from the upper leaves. Pedicels bracteate. Flowers zygomorphic, campanulate. Sepals free, lacking basal nectaries. Petals equal. Stamens 12–25. Filaments inserted on a discoid or conical receptacle. Gynophore absent. Fruit a partly dehiscent oblong capsule with persistent valves. Seeds 25–40(–ca. 100), spheroidal, not arillate.

Worldwide about ten species in Africa and Asia. One species in Nepal.

1. *Arivela viscosa* (L.) Raf., Sylva Tellur.: 110 (1838).
Cleome viscosa L., Sp. Pl. 2: 672 (1753); *C. icosandra* L.

वन तोरी Ban tori (Nepali).

Annual herbs to (10–) 30–100 (–160) cm. Stems simple or branched, glandular-pubescent, viscous. Stipules absent. Petioles bract-like, 1.5–4.5(–8) cm, glandular-hirsute. Leaflets 3 or 5, ovate to oblanceolate-elliptic, (0.6–) 2.0–6.0 × 0.5–3.5 cm, acute to obtuse, entire to glandular ciliate margined, glandular-hirsute above and below. Inflorescences 5–10 cm. Bracts 10–25 mm, trifoliolate, glandular-hirsute (often deciduous). Pedicels 6–30 mm, glandular hirsute. Sepals green, lanceolate, 5–10 mm × 0.8–1.2 mm, entire, glandular-hirsute, persistent. Petals bright yellow, sometimes purple basally, oblong to ovate, clawed, 7–14 × 3–4 mm. Stamens 12–25, green, 5–9 mm, dimorphic, the 4–10 adaxial stamens short with a swelling below anthers. Anthers green, 1.4–3 mm. Pistil 6–10 mm, densely glandular. Style 1–1.2 mm. Capsule sessile, dehiscing only partway from apex to base, 30–100 × 2–4 mm, glandular-pubescent. Seeds 25–40(–ca. 100), light brown, compressed spherical, 1.2–1.8 × 1.0–1.2 mm, finely ridged transversely.
Fig. 14a–f.

Distribution: Nepal, E Himalaya, S Asia, E Asia and SE Asia.

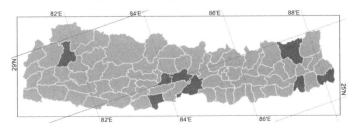

Altitudinal range: 100–1400 m.

Ecology: Weed of waste land and field margins. Grows in the open and under shade.

Flowering: April–December. **Fruiting:** April–December.

All Nepalese specimens belowng to the typical variety, which is pubescent and has an unpleasant smell. *Arivela viscosa* var. *deglabrata* (Baker) M.L.Zhang & G.C.Tucker is glabrous and lacks the unpleasant smell. This variety occurs in China, India and several other SE Asian countries and may also yet be found in Nepal.

The leaves are cooked as a vegetable and the oil-rich seeds are roasted and used in curries and pickles. A paste of the root is used to treat earache and the seeds have carminative, antihelmitic and stimulant properties.

5. *Gynandropsis* DC., Prodr. 1: 237 (1824).

Annual herbs. Stem unbranched or sparsely branched, glabrate or glandular pubescent. Stipules absent. Petioles long or short, with pulvinus at distal or basal end. Leaflets 3–5. Inflorescences elongate, terminal. Pedicels bracteate. Flowers campanulate, zygomorphic. Sepals free, each often subtending a basal nectary. Petals equal. Stamens 6. Filaments fused forming an androgynophore about as long as the petals. Gynophore slender, elongating and recurving in fruit. Style short. Fruits a dehiscent oblong capsule. Seeds 10–20, subspherical, not arillate.

Worldwide two species. One species in Nepal.

Gynandropsis is allied to *Cleome*, but distinguished by the long, conspicuous androgynophore. It has been included in *Cleome*, but most regional accounts of Capparaceae or Cleomaceae (e.g. Fl. Pakistan, Fl. China) in the Old World have given this generic status, an approach followed here.

1. *Gynandropsis gynandra* (L.) Briq., Annuaire Conserv. Jard. Bot. Geneve 17: 382 (1914).
Cleome gynandra L., Sp. Pl. 2: 671 (1753); *C. heterotricha* Burch.; *C. pentaphylla* L.; *Gynandropsis heterotricha* (Burch.) DC.; *G. pentaphylla* (L.) DC.

जुँगे फूल Junge phul (Nepali).

Annual herbs to 30–90 cm. Stems simple or sparsely branched, glabrate to glandular-pubescent. Stipules absent. Petioles 3.5–4.5(–8) cm, glandular pubescent. Leaflets 3 or 5, oblanceolate to rhombate, 2.5–4.5 × 1.2–2.5 cm, acute, serrulate-denticulate, glabrate to glandular-pubescent above and below. Inflorescences 5–20 cm, enlarging to 10–40 cm in fruit. Bracts purple, trifoliolate, 10–25 mm, pubescent beneath. Pedicels 8–15 mm, glabrous. Sepals green, lanceolate, 3.5–5 × 0.8–1.2 mm, cuneate, entire, glandular-pubescent, persistent. Petals purple or white, oblong to ovate, clawed, 7–14 × 3–4 mm. Stamens diverging at anthesis, purple, 8–30 mm. Filaments adnate to gynophore for ⅓-½ their length. Anthers green, 1–2 mm. Pistil 6–10 mm. Gynophore 10–14 mm. Style 1–1.2 mm. Fruits: capsule purple, 45–95 × 3–4 mm. Seeds reddish brown to black, subspherical, 1.4–1.6 × 1.0–1.2 mm, rugose to tuberculate. Fig. 14g.

Distribution: Nepal, E Himalaya, Assam-Burma, S Asia, E Asia, SE Asia and Africa.

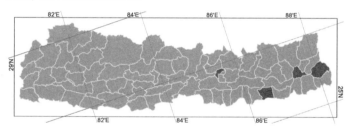

Altitudinal range: 100–1400 m.

Ecology: In open fields or cultivated land.

Flowering: February–November. **Fruiting:** July–November.

The fresh plant has a peculiar odour that is sometimes described as suggesting burning *Cannabis*. It is used to treat fever, rheumatism and scorpion stings.

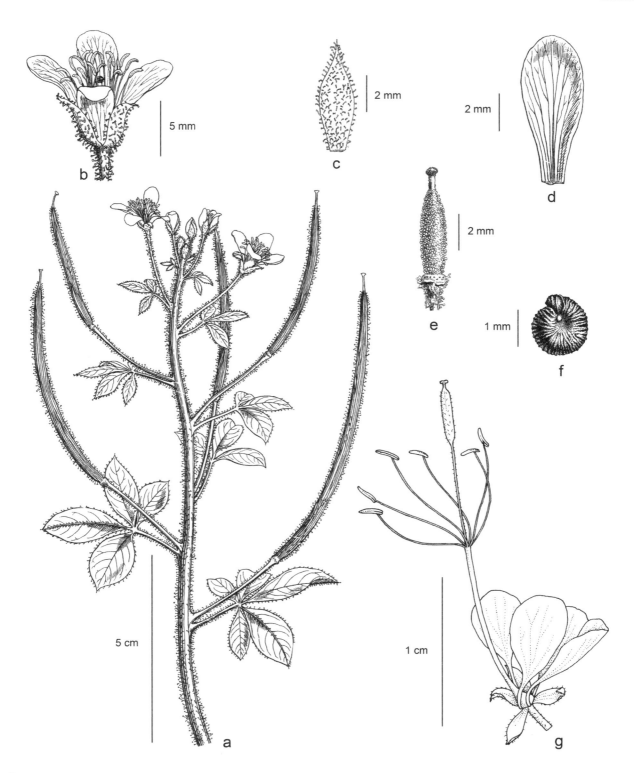

2 mm

5 mm

2 mm

2 mm

1 mm

5 cm

1 cm

a

b

c

d

e

f

g

FIG. 14. CLEOMACEAE. **Arivela viscosa**: a, inflorescence with fruits and leaves; b, flower; c, sepal; d, petal; e, pistil; f, seed. **Gynandropsis gynandra**: g, flower.

Cruciferae (Brassicaceae)

Ihsan A. Al-Shehbaz & Mark F. Watson

Annual herbs, biennial, or perennial, with a pungent watery juice. Eglandular trichomes simple, 2-many forked, stellate, dendritic, or malpighiaceous (medifixed, bifid, appressed); glandular trichomes multicellular, with uniseriate or multiseriate stalk. Leaves exstipulate, simple, entire or variously pinnately dissected, rarely trifoliolate or pinnately, palmately, or bipinnately compound, almost always alternate, rarely opposite or whorled, petiolate or sessile, sometimes only basal. Racemes bracteate or ebracteate, sometimes corymbs or panicles, or flowers solitary on long pedicels originating from axils of rosette leaves. Flowers mostly actinomorphic. Sepals 4, in 2 decussate pairs, free or sometimes united, not saccate or lateral (inner) pair saccate. Petals 4, alternate with sepals, arranged in the form of a cross, sometimes rudimentary or absent, persistent or caducous. Stamens 6, in 2 whorls, tetradynamous (outer, lateral pair shorter than inner, median 2 pairs), rarely equal or in 3 pairs of unequal length, sometimes stamens 2 or 4, very rarely to 24; filaments slender, winged, or appendaged, inner pairs free or rarely united; anthers dehiscing by longitudinal slits. Nectar glands receptacular, opposite bases of lateral filaments, median glands present or absent. Ovary superior, mostly 2-locular and with a false septum (replum) connecting 2 placentas; placentation parietal, rarely apical. Fruit typically a 2-valved capsule, generally termed silique (siliqua) when length more than 3× width, or silicle (silicula) when length less than 3× width, dehiscent or indehiscent, sometimes schizocarpic, nutlet-like, lomentaceous, or samaroid, segmented or not, terete, angled, or flattened parallel to septum (latiseptate) or at a right angle to septum (angustiseptate); replum (persistent placenta) rounded, rarely flattened or winged; septum complete, perforate, or lacking. Style 1 or absent. Stigma entire or 2-lobed, sometimes lobes decurrent and free or connate. Seeds uniseriately or biseriately arranged in each locule, winged or wingless. Cotyledons incumbent (radicle lying along back of 1 cotyledon), accumbent (radicle applied to margins of both cotyledons) or conduplicate (cotyledons folded longitudinally around radicle).

Worldwide about 330 genera and 3780 species in 44 tribes. Found in all continents except Antarctica, mainly in temperate areas, with highest diversity in Irano-Turanian, Mediterranean, and western N American regions. 33 genera and 101 species (nine endemic) in Nepal.

The sequence of genera follows their assignment to the alphabetical system of tribes as currently delimited in the new phylogenetic system proposed by Al-Shehbaz *et al.* (Pl. Syst. Evol. 259: 89–120. 2006) and its subsequent updating following Warwick *et al.* (Plant Syst. Evol. 285: 209–232. 2010).

Although fruit characters are essential in Cruciferae taxonomy and needed for the identification of some genera, many flowering specimens may be accurately identified. To facilitate identification a key emphasizing fruiting material and another emphasizing flowering material are given. The most reliable determination of genera can be achieved when the material has fruits and flowers and when both keys are used to reach the same genus. In the key to flowering material (*) indicates that fruits are necessary to proceed further in the key. The Cruciferae are rather poorly collected in Nepal, and perhaps more than most families in the Flora, the distribution maps do not necessarily accurately represent the true range of the species.

Key to Genera

1a	Stamens (8–)12–16(–24)	**28. *Megacarpaea***
b	Stamens 2–6	2
2a	Stamens 2 or 4	**27. *Lepidium***
b	Stamens 6	3
3a	Plants with multicellular glands	**17. *Dontostemon***
b	Plants eglandular	4
4a	At least some of the trichomes branched	5
b	Trichomes absent or exclusively simple	26
5a	At least some cauline leaves 1–3-pinnatisect	6
b	Cauline leaves entire, dentate, lobed, or absent	7
6a	Petals yellow. Raceme ebracteate	**16. *Descurainia***
b	Petals white. Raceme bracteate	**31. *Smelowskia***
7a	Cauline leaves absent	8
b	Cauline leaves present	13

| 8a | Flowers on solitary pedicels from basal rosette | **23. Pycnoplinthopsis** |
| b | Flowers in racemes | 9 |

| 9a | Sepals, petals, and stamens persistent | **22. Lepidostemon** |
| b | Sepals, petals, and stamens deciduous shortly after anthesis | 10 |

| 10a | Filaments flattened, subapically toothed. Ovary retrorsely pilose | **24. Solms-laubachia** |
| b | Filaments slender, toothless. Ovary glabrous or not retrorsely hairy | 11 |

| 11a | Filaments and petal bases purple. Leaves 1–2.5 cm wide | **24. Solms-laubachia** |
| b | Filaments and petal bases white or yellow. Leaves much narrower (*) | 12 |

| 12a | Fruit flattened. Seeds biseriate. Cotyledons accumbent | **4. Draba** |
| b | Fruit terete. Seeds uniseriate. Cotyledons incumbent | **19. Braya** |

| 13a | Flowers orange or yellow | 14 |
| b | Flowers white, pink, or purple | 16 |

| 14a | Trichomes exclusively sessile, medifixed, 2–4-rayed | **18. Erysimum** |
| b | Trichomes stalked, forked or stellate mixed with simple | 15 |

| 15a | Plants pubescent throughout. Ovary ovate or oblong | **4. Draba** |
| b | Plants glabrous and glaucous above middle. Ovary linear | **11. Turritis** |

| 16a | Cauline leaves auriculate or sagittate at base | 17 |
| b | Cauline leaves obtuse, cuneate, or attenuate at base | 19 |

| 17a | Ovary obtriangular or obcordate. Cauline leaves with sessile, stellate trichomes | **9. Capsella** |
| b | Ovary linear. Cauline leaves with stalked branched trichomes (*) | 18 |

| 18a | Fruit flattened. Seeds winged at least apically. Cotyledons accumbent | **3. Arabis** |
| b | Fruit terete. Seeds wingless. Cotyledons incumbent | **10. Crucihimalaya** |

| 19a | Plants perennial | 20 |
| b | Plants annual or biennial | 23 |

| 20a | Plants minutely puberulent distally | **1. Aphragmus** |
| b | Plants with distinct trichomes distally | 21 |

| 21a | Basal rosette absent. Anthers reniform | **22. Lepidostemon** |
| b | Basal rosette well developed. Anthers oblong or ovate (*) | 22 |

| 22a | Fruit flattened. Seeds biseriate. Cotyledons accumbent | **4. Draba** |
| b | Fruit terete. Seeds uniseriate. Cotyledons incumbent | **19. Braya** |

| 23a | Filaments of median stamens winged. Anthers reniform | **22. Lepidostemon** |
| b | Filaments of median stamens wingless. Anthers ovate or oblong | 24 |

| 24a | Ovaries and young fruits oblong or suborbicular | **4. Draba** |
| b | Ovaries and young fruits linear | 25 |

| 25a | Raceme ebracteate. Trichomes simple and forked | **8. Arabidopsis** |
| b | Raceme bracteate at least basally, if not then cauline leaves lobed. At least some trichomes stellate or dendritic | **10. Crucihimalaya** |

| 26a | Cauline leaves absent. Flowers solitary from basal rosette | 27 |
| b | Cauline leaves present. Flowers in distinct racemes or corymbs | 28 |

| 27a | Sepals united. Petals purplish | **24. Solms-laubachia** |
| b | Sepals free. Petals white | **26. Pegaeophyton** |

Cruciferae (Brassicaceae)

28a	Raceme bracteate throughout, rarely only basally	29
b	Raceme ebracteate	33
29a	Flowers in corymbs. Anthers apiculate. Ovaries and fruit cristate	**21. Dilophia**
b	Flowers in racemes. Anthers not apiculate. Ovaries and fruit not cristate	30
30a	Cauline leaves palmately veined and lobed	**1. Aphragmus**
b	Cauline leaves pinnately veined, not lobed	31
31a	Ovary and fruit oblong or ovate. Roots fleshy	**25. Eutrema**
b	Ovary and fruit linear. Roots not fleshy	32
32a	Plants annual. Petals white	**15. Rorippa**
b	Plants perennial. Petals purple or lavender	**24. Solms-laubachia**
33a	Petals yellow (to creamy-yellow or white in *Brassica*), rarely absent	34
b	Petals white, pink, or purple, always present	39
34a	Plants perennial, with well-developed caudex	35
b	Plants annual or biennial	36
35a	Basal rosette present. Ovaries and fruits stipitate	**6. Diplotaxis**
b	Basal rosette absent. Ovaries and fruits sessile	**20. Christolea**
36a	Uppermost cauline leaves pinnatisect or pinnatifid. Stems angular	**12. Barbarea**
b	Uppermost cauline leaves usually undivided. Stems often terete	37
37a	Stigma 2-lobed. Fruit valves 3-veined	**30. Sisymbrium**
b	Stigma entire. Fruit valves 1-veined	38
38a	Petals (6–)7–30 mm. Seeds globose. Cotyledons conduplicate	**5. Brassica**
b	Petals absent or 1.5–4.5 mm. Seeds oblong or ovate. Cotyledons accumbent	**15. Rorippa**
39a	Plants perennial, with rhizomes, tubers, or distinct caudex	40
b	Plants annual or biennial	44
40a	Plants aquatic, rooting from lower nodes	**14. Nasturtium**
b	Plants neither aquatic nor rooting from lower nodes	41
41a	Plants with tubers or rhizomes. Leaves compound, if simple then petals purple and 1–1.7 cm	**13. Cardamine**
b	Plants with a caudex. Leaves simple	42
42a	Cauline leaves auriculate. Fruit strongly flattened. Cotyledons accumbent	**29. Noccaea**
b	Cauline leaves not auriculate. Fruit terete. Cotyledons incumbent	43
43a	Ovules 30–50 per ovary. Raceme elongated in fruits. Seeds biseriate	**2. Arcyosperma**
b	Ovules 4–10 per ovary. Raceme not elongated in fruit. Seeds uniseriate	**25. Eutrema**
44a	Cauline leaves auriculate	**33. Thlaspi**
b	Cauline leaves petiolate	45
45a	Cauline leaves toothed, simple	46
b	Cauline leaves pinnatisect or compound	47
46a	Petals white, pink or purple with darker veins, 1.2–2.2 cm. Sepals 6–10 mm	**7. Raphanus**
b	Petals white, without dark veins, (2.5–)4–8(–9) mm. Sepals (2–)2.5–3.5(–4.5) mm	**32. Alliaria**
47a	Ovary linear. Ovules 10–40	**13. Cardamine**
b	Ovary oblong-ovate or elliptic. Ovules 2	**27. Lepidium**

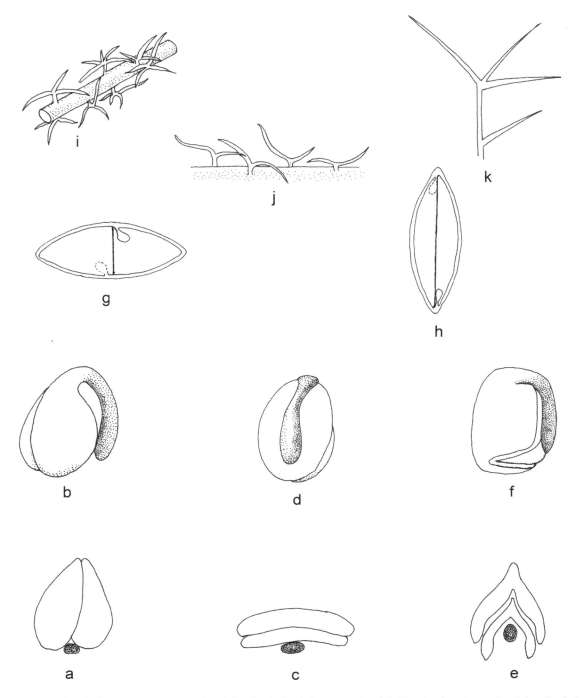

FIG. 15. CRUCIFERAE. Cotyledon types: a, accumbent (vertical view); b, accumbent (side view); c, incumbent (vertical view); d, incumbent (side view); e, conduplicate (vertical view); f, conduplicate (side view). silicula types: g, angustiseptate; h, latiseptate. Hair types: i, stellate; j, forked; k, branched.

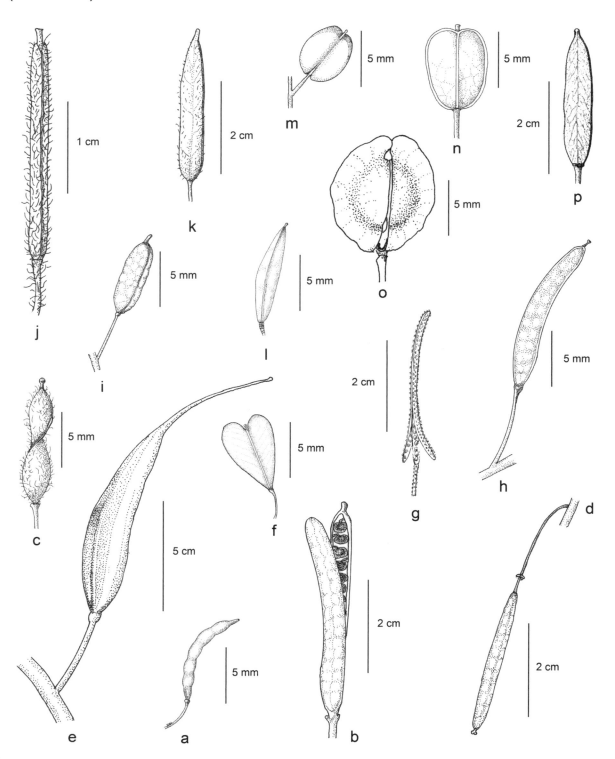

FIG. 16. CRUCIFERAE. Fruit diversity. a, **Aphragmus ohbana**; b, **Arabis pterosperma**; c, **Draba sikkimensis**; d, **Diplotaxis harra**; e, **Raphanus sativus**; f, **Capsella bursa-pastoris**; g, **Crucihimalaya lasiocarpa**; h, **Nasturtium officinale**; i, **Rorippa palustris**; j, **Braya humilis**; k, **Solms-laubachia himalayensis**; l, **Eutrema heterophyllum**; m, **Lepidium sativum**; n, **Smelowskia tibetica**; o, **Thlaspi arvense**; p, **Christolea crassifolia**.

Key to Genera (fruiting material)

1a   Fruit on solitary pedicels originating from basal rosette...2
  b   Fruit in racemes...4

2a   Plants with at least some branched trichomes. Fruit geocarpic, maturing underground ....**23. Pycnoplinthopsis**
  b   Plants glabrous or with simple trichomes only. Fruit not geocarpic 3

3a   Stigma 2-lobed. Replum rounded. Sepals united or rarely free (*S. harenensis*), often persistent at fruit
      ....**24. Solms-laubachia**
  b   Stigma entire. Replum flattened. Sepals free, caducous in fruit...**26. Pegaeophyton**

4a   Fruit strongly flattened at a right angle to septum (angustiseptate)..5
  b   Fruit flattened parallel to septum (latiseptate), terete, or angled 11

5a   At least some of the trichomes branched...6
  b   Trichomes absent or exclusively simple...7

6a   Cauline leaves sessile, auriculate or sagittate at base, entire or toothed. Raceme ebracteate. Fruit obdeltoid to
      obdeltoid-obcordate..**9. Capsella**
  b   Cauline leaves petiolate, not auriculate or sagittate, 1- or 2-pinnatisect. Raceme bracteate. Fruit oblong, elliptic,
      or suborbicular..**31. Smelowskia**

7a   Seeds (or ovules) 1 or 2 per fruit 8
  b   Seeds (or ovules) 4–24 per fruit...9

8a   Annuals. Fruit not didymous, 1.8–6(–7) mm ....**27. Lepidium**
  b   Perennials. Fruit didymous, 30–40 cm ....**28. Megacarpaea**

9a   Cauline leaves not auriculate. Roots fleshy, fusiform. Fruit cristate apically ....**21. Dilophia**
  b   Cauline leaves auriculate. Roots not fleshy, cylindric. Fruit not cristate 10

10a   Perennials. Seeds smooth ....**29. Noccaea**
  b   Annuals. Seeds concentrically striate ....**33. Thlaspi**

11a   Plants glabrous or exclusively with simple trichomes 12
  b   Plants with branched trichomes at least basally 29

12a   Plants with multicellular glands ....**17. Dontostemon**
  b   Plants eglandular...13

13a   Raceme bracteate 14
  b   Raceme ebracteate..17

14a   Cauline leaves absent, if present palmately veined and lobed..**1. Aphragmus**
  b   Cauline leaves pinnately veined, entire or dentate 15

15a   Fruit a silicle (rarely to 3× longer than broad). Replum expanded..**25. Eutrema**
  b   Fruit a silique (more than 3× longer than broad). Replum not expanded 16

16a   Fruit terete, valves readily separated from replum..**15. Rorippa**
  b   Fruit flattened, valves united with replum apically...**24. Solms-laubachia**

17a   Fruit a silicle (rarely to 3× longer than broad) 18
  b   Fruit a silique (more than 3× longer than broad) 19

18a   Cauline leaves auriculate. Seeds more than 20 per fruit..**15. Rorippa**
  b   Cauline leaves not auriculate. Seeds 4–10 per fruit...**25. Eutrema**

19a	Fruit indehiscent, corky	**7. Raphanus**
b	Fruit dehiscent, not corky	20
20a	Fruit valves without midvein, coiled at dehiscence. Replum flattened	**13. Cardamine**
b	Fruit valves with midvein, not coiled at dehiscence. Replum terete	21
21a	Leaves compound. Stems rooting from lower nodes	**14. Nasturtium**
b	Leaves simple. Stems not rooting from lower nodes	22
22a	Plants perennial. Fruits flattened	23
b	Plants annual or biennial. Fruits terete or angled	25
23a	Fruit on a gynophore 1–4 mm. Cotyledons conduplicate	**6. Diplotaxis**
b	Fruit sessile. Cotyledons incumbent	24
24a	Seeds biseriate, 30–50 per fruit. Basal rosette present	**2. Arcyosperma**
b	Seeds uniseriate, 10–20 per fruit. Basal rosette absent	**20. Christolea**
25a	Leaves palmately veined. Seeds longitudinally striate	**32. Alliaria**
b	Leaves pinnately veined. Seeds reticulate or foveolate	26
26a	Seeds globose. Cotyledons conduplicate	**5. Brassica**
b	Seeds oblong. Cotyledons accumbent or incumbent	27
27a	Cauline leaves auriculate. Stems angled throughout	**12. Barbarea**
b	Cauline leaves not auriculate. Stems not angled	28
28a	Stigma entire. Valves 1-veined. Cotyledons accumbent	**15. Rorippa**
b	Stigma 2-lobed. Valves 3-veined. Cotyledons incumbent	**30. Sisymbrium**
29a	Basal and lower cauline leaves 2- or 3-pinnatisect	**16. Descurainia**
b	Basal and lower cauline leaves entire, dentate, or rarely pinnatifid	30
30a	Fruit a silicle (less than 3× longer than broad)	31
b	Fruit a silique (more than 3× longer than broad)	33
31a	Leaves glabrous. Pedicels minutely papillate	**1. Aphragmus**
b	Leaves pubescent. Pedicels glabrous or pubescent	32
32a	Cotyledons accumbent. Seeds often biseriate. Fruits flattened or rarely terete	**4. Draba**
b	Cotyledons incumbent. Seeds uniseriate. Fruits terete	**19. Braya**
33a	Trichomes exclusively sessile, medifixed	**18. Erysimum**
b	At least some branched trichomes stalked	34
34a	Cotyledons accumbent	35
b	Cotyledons incumbent	38
35a	Fruit often much longer, rarely 15× longer than broad	36
b	Fruit rarely to 10× longer than broad	37
36a	Fruit flattened. Seeds winged at least distally. Plants not glaucous or glabrous above base	**3. Arabis**
b	Fruit subterete-quadrangular. Seeds wingless. Plants glaucous and glabrous often just above base	**11. Turritis**
37a	Seeds biseriate. Valves readily separated	**4. Draba**
b	Seeds uniseriate. Valves united with replum at fruit apex	**24. Solms-laubachia**
38a	Stems leafless	39
b	Stems leafy	40

39a	Fruit terete, 0.4–1.1 cm. Trichomes primarily dendritic	**22. Lepidostemon**
b	Fruit flattened, (0.9–)1.5–2.5(–3.5) cm. Trichomes simple and forked	**24. Solms-laubachia**
40a	Raceme ebracteate. Plants annual or biennial	41
b	Raceme bracteate at least basally. Plants perennial, sometimes annual or biennial	43
41a	Fruit (2.5–)3.5–7.5(–9.5) cm. Plants tomentose with finely branched stellate trichomes	**10. Crucihimalaya**
b	Fruit 0.4–1.5(–1.5) cm. Plants pubescent with simple and forked trichomes	42
42a	Fruit linear, glabrous. Submalpighiaceous trichomes absent	**8. Arabidopsis**
b	Fruit oblong, pubescent. Submalpighiaceous trichomes present	**22. Lepidostemon**
43a	Fruit flattened. Plants minutely puberulent along rachis and pedicels	**1. Aphragmus**
b	Fruit terete. Plants glabrous or pubescent with long trichomes on rachis and pedicels	44
44a	Plants annual or biennial. Seeds and ovules 50–160 per fruit	**10. Crucihimalaya**
b	Plants perennial. Seeds and ovules rarely to 40 per fruit	45
45a	Basal leaves rosulate, persistent. Fruit torulose	**19. Braya**
b	Basal leaves not rosulate, soon withered. Fruit not torulose	**22. Lepidostemon**

1. *Aphragmus* Andrz. ex DC., Prodr. 1: 209 (1824).

Lignariella Baehni; *Staintoniella* H.Hara.

Annual herbs, biennial or perennial. Caudex absent or thick, covered with petiolar remains of previous years, sometimes with slender rhizomes. Trichomes simple or forked, sometimes papillate, less than 0.1 mm. Stems erect, ascending, prostrate or decumbent, branched basally, often minutely puberulent. Basal leaves petiolate, rosulate, simple, entire. Cauline leaves, sometimes absent, petiolate or sessile, base cuneate or attenuate, not auriculate, margin entire, repand, or palmately 3(–5)-lobed or -toothed. Racemes few- to several-flowered, bracteate throughout, elongated or not in fruit. Sepals oblong, erect or ascending, caducous or persistent, bases saccate or not. Petals white, pink, lavender, blue, purple, orbicular to obovate or spatulate, apex obtuse or rounded; claw subequalling or shorter than sepals, sometimes absent; stamens 6, equal or slightly tetradynamous; filaments dilated or not at base; anthers obtuse or apiculate at apex; nectar glands confluent and subtending bases of stamens. Ovules 2–16 per ovary. Fruiting pedicels slender, erect, ascending, divaricate, sometimes recurved or reflexed, puberulent above. Fruit dehiscent silicles or siliques, ovate, ovoid, elliptic, oblong, lanceolate, or linear, latiseptate or terete; valves with an obscure or distinct midvein, smooth; replum flattened basally; septum absent or complete and membranous. Style 0.5–4 mm; stigma capitate, entire. Seeds uniseriate, biseriate or 1 per fruit, wingless, oblong or ovoid, plump, on filiform funicles often longer than seeds; seed coat minutely reticulate or colliculate, not mucilaginous when wetted; cotyledons incumbent.

Worldwide eleven species in the Himalayas, north-western N America, Siberia. Six species in Nepal.

Key to Species

1a	Caudex many-branched, rhizome-like. Cauline leaves absent	**3. A. nepalensis**
b	Caudex absent or present and simple or few-branched, not rhizome-like. Cauline leaves present	2
2a	All leaves entire. Plants with forked trichomes	**5. A. oxycarpus**
b	At least some leaves 3(–5)-lobed or -toothed. Plants with simple, papillate trichomes	3
3a	Fruit 1-seeded. Gynophore 0.5–1 mm in flower, 20–100 mm in fruit. Seed 3.7–4.7 mm	**1. A. hinkuensis**
b	Fruit 2–8(–12)-seeded. Gynophore obsolete in flower and fruit, rarely to 2 mm in fruit. Seed to 2.6 mm	4
4a	Petals orbicular or orbicular-obovate, (5.5–)6–8(–9) mm. Anthers 0.6–1.2 mm	**2. A. hobsonii**
b	Petals narrowly to broadly obovate, 1.5–4(–4.5) mm. Anthers 0.25–0.4 mm	5
5a	Fruit narrowly oblong to linear, 6–12 seeded. Fruiting pedicels usually straight. Petals pale lavender, 1.5–2 × 0.5–1 mm. Lower cauline leaves simple	**4. A. ohbana**
b	Fruit oblong-ovoid to broadly oblong, 2–4-seeded. Fruiting pedicels often recurved. Petals purple, (2.7–)3–4(–4.5) × 1.7–2.5 mm. All cauline leaves 3- or 5(or 7)-lobed	**6. A. serpens**

Cruciferae (Brassicaceae)

1. _Aphragmus hinkuensis_ (Kats.Arai, H.Ohba & Al-Shehbaz) Al-Shehbaz & S.I.Warwick, Canad. J. Bot. 84: 279 (2006). _Lignariella hinkuensis_ Kats.Arai, H.Ohba & Al-Shehbaz, Harvard Pap. Bot. 5(1): 117 (2000).

Annual herbs, glabrous throughout. Stems few-branched from base, ascending to decumbent, 8.5–25 cm. Cauline leaves petiolate 1.5–3.7 cm, broadly ovate to suborbicular in outline, 3–5-lobed; lobes of larger leaves ovate to oblong, 6–15 × 4–11 mm, apex acute to obtuse, entire or 1- or 2-toothed. Pedicel 1–2 cm in flower, 2–7.5 cm in fruit, strongly recurved, papillate above, glabrous below. Sepals caducous, oblong, 2.3–3 × 1.2–1.7 mm, ascending, glabrous, somewhat saccate at base, scarious at margin. Petals bluish purple, broadly obovate, 5–7 × 2.7–5.8 mm, apex rounded to somewhat retuse, base narrowed into a claw ca. 1 mm. Filaments purplish above, pale to yellowish below, 2.5–3 mm; anthers purple, oblong, cordate at base, 0.4–0.6 mm. Ovary ellipsoid, glabrous, 1–1.5 mm; ovules (2–)4 per ovary. Fruit geocarpic, curved, ovoid to ellipsoid, 2.5–7 × 2–5.5 mm; gynophore 2–10 cm, very slender; valves obscurely 1–3-veined; style 2–3 mm. Seeds 1 per fruit, ovoid-oblong, 3.7–4.7 × 1–1.2 mm.

Distribution: Endemic to Nepal.

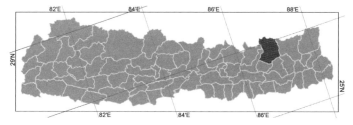

Altitudinal range: ca. 3600 m.

Ecology: Wet rocks.

Flowering: August. **Fruiting:** August.

The species is known only from the type collection, _Miyamoto et al. 9592484_ (MO, TI), from Thasing Dingma, Solu Khumbu District.

2. _Aphragmus hobsonii_ (H.Pearson) Al-Shehbaz & S.I.Warwick, Canad. J. Bot. 84: 279 (2006). _Cochlearia hobsonii_ H.Pearson, Hooker's Icon. Pl. 27: pl. 2643 (1900); _Lignariella hobsonii_ (H.Pearson) Baehni.

Short-lived perennial herbs, puberulent on stem and/or pedicels. Stems coarse, few to many from base, decumbent or rarely ascending, (6–)11–20(–30) cm, glabrous or puberulent along 1 line or throughout. Leaves with somewhat flattened petiole 4–10(–18) mm, simple or rarely deeply 3(or 5)-lobed apically, blade broadly obovate to suborbicular and often wider than long, rarely oblong, (3–)6–10(–15) × (1.5–)4–10(–15) mm, glabrous, obscurely to prominently 3- or 5-toothed, teeth sometimes restricted to subtruncate apex, when lobed central lobe broadly to narrowly oblong, slightly broader than lateral lobes. Flowers often protogynous.

Sepals oblong, glabrous, 2.5–3.5(–4) × 1–1.5 mm, margin membranous and to 0.5 mm wide, spreading,, slightly saccate at base. Petals blue to deep purple, orbicular to orbicular-obovate, (5.5–)6–8(–9) × 4–6(–7) mm, apex rounded, base abruptly narrowed into a claw 0.5–1.5 mm. Filaments purple above, whitish below, 4–5.5 mm; anthers narrowly oblong, 0.6–1.2 mm, often strongly curved after dehiscence. Ovules 4–6(–11) per ovary. Fruiting pedicels slender, strongly recurved to sigmoid, (0.7–)1.5–3(–4) cm, puberulent above with papillae to 0.1 mm. Fruit not geocarpic, linear to narrowly oblong, rarely ovoid, straight or curved, (5–)10–20 × 1.5–2 mm; gynophore ca. 0.5 mm; valves torulose; style 3–4 mm, stout. Seeds uniseriate, oblong, 2.2–2.8 × 1.2–1.4 mm.

Distribution: Nepal, E Himalaya and Tibetan Plateau.

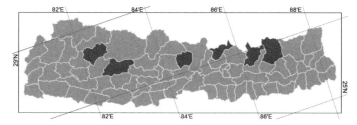

Altitudinal range: 2800–4100 m.

Ecology: River or stream banks on small stones and coarse sand, wet grass, stream-sides, mossy areas in running water, wet cliffs, scree, damp banks, wet stony slopes.

Flowering: May–July. **Fruiting:** June–September.

3. _Aphragmus nepalensis_ (H.Hara) Al-Shehbaz, Harvard Pap. Bot. 5(1): 112 (2000). _Staintoniella nepalensis_ H.Hara, J. Jap. Bot. 49: 196 (1974).

Perennial herbs, (2–)4–10 cm. Caudex many-branched, rhizome-like, with distinct internodes separating whorls of petiolar remains of successive growing seasons. Stems erect, simple, minutely papillate puberulent. Basal leaves rosulate, somewhat fleshy; petioles persistent, (0.8–)1.5–4(–6) cm, base broadly expanded to 4 mm wide; leaf blade ovate to broadly so, 4–12(–17) × 2.5–9 mm, base obtuse, apex obtuse, margin entire, glabrous. Cauline leaves absent. Fruiting racemes, elongated; bracts similar to basal leaves but very short petiolate to subsessile. Sepals often purplish, 2–3.5 × 1.2–1.5 mm, glabrous. Petals purple with darker veins, obovate, (6–)7–9 × 3–5 mm, apex rounded; claw to 2.5 mm. Filaments 3–4 mm; anthers 0.7–0.9 mm. Ovules 8–10 per ovary. Fruiting pedicels divaricate, 5–12 mm, puberulent above, glabrous below. Immature fruit ovate to elliptic, compressed; valves obscurely veined, glabrous; septum absent; style 1.5–2 mm. Mature seeds not seen.

Distribution: Endemic to Nepal.

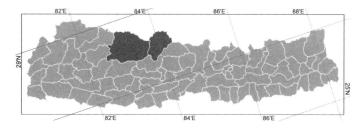

Altitudinal range: 4800–5800 m.

Ecology: Loose scree.

Flowering: June–July.

4. *Aphragmus ohbana* (Al-Shehbaz & Kats.Arai) Al-Shehbaz & S.I.Warwick, Canad. J. Bot. 84: 279 (2006).
Lignariella ohbana Al-Shehbaz & Kats.Arai, Harvard Pap. Bot. 5(1): 120 (2000).

Biennial or short-lived perennial herbs, glabrous throughout except for pedicels. Stems slender, few or rarely 1 from base, decumbent, 2–12(–15) cm. Lowermost cauline leaves simple, becoming 3-lobed in the middle and terminal portions of plant; petiole glabrous, (2–)4–8(–12) mm; blade of lowermost leaves ovate, subacute, 3–5 × 2–3 mm wide; middle lobe of bracts narrowly oblong to broadly ovate, entire, (2–)3–6(–8) × 1.3–4(–7) mm, obtuse; lateral lobes smaller and narrower, entire or rarely obscurely 1-toothed. Sepals oblong, 0.7–1.5 × 0.5–0.7 mm, not saccate at base, scarious at margin. Petals lavender, narrowly obovate, 1.5–2 × 0.5–1 mm, apex rounded, base cuneate and not clawed. Filaments white or rarely lavender, 0.8–1.2 mm; anthers purple, ovate, 0.2–0.3 mm. Ovules 6–12 per ovary. Fruiting pedicels 5–8(–13) mm, puberulent above with papillae to 0.1 mm. Fruit not geocarpic, linear to narrowly oblong, straight or slightly curved at middle, 5–10 × 0.7–1 mm; gynophore obsolete; valves torulose; style 0.5–0.8 mm, stout. Seeds uniseriate, oblong, 1–1.4 × 0.5–1 mm.
Figs 16a, 17a.

Distribution: Nepal and E Asia.

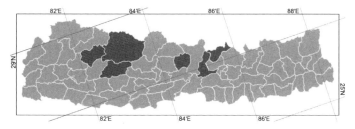

Altitudinal range: 3000–4600 m.

Ecology: Among rocks, sandy flats by stream-sides.

Flowering: July–August. **Fruiting:** July–October.

Hara (Enum. Fl. Pl. Nepal 2: 44. 1979) cited *Stainton, Sykes & Williams 6075* (BM) and *Polunin, Sykes & Williams 2369* (BM) as *Lignariella hobsonii* subsp. *serpens* (W.W.Sm.) H.Hara.

However, these collections are now identified as *Aphragmus ohbana*.

5. *Aphragmus oxycarpus* (Hook.f. & Thomson) Jafri, Notes Roy. Bot. Gard. Edinburgh 22: 96 (1956).
Braya oxycarpa Hook.f. & Thomson, J. Proc. Linn. Soc., Bot. 5: 169 (1861); *Aphragmus oxycarpus* var. *glaber* (Vassilcz.) Z.X.An; *A. oxycarpus* var. *microcarpus* Z.X.An; *A. oxycarpus* var. *stenocarpus* (O.E.Schulz) G.C.Das; *A. przewalskii* (Maxim.) A.L.Ebel; *A. stewartii* O.E.Schulz; *A. tibeticus* O.E.Schulz; *Braya foliosa* Pamp.; *B. oxycarpa* forma *glaber* Vassilcz.; *B. oxycarpa* var. *stenocarpa* O.E.Schulz; *B. rubicunda* Franch.; *Eutrema przewalskia* Maxim.; *Lignariella duthiei* Naqshi.

Perennial herbs, (1–)2–11(–18) cm. Caudex covered with petiolar bases of previous years. Stems erect to ascending, branched from base, minutely puberulent with simple or short-stalked trichomes less than 0.1 mm, rarely glabrous. Basal leaves rosulate, somewhat fleshy; petioles persistent, 0.2–2(–5) cm, base broadly expanded and to 3 mm wide; leaf blade spatulate, oblanceolate, linear, oblong, or elliptic, rarely ovate, (0.2–)0.5–2(–3) cm × 0.5–3(–5) mm, glabrous, base cuneate to attenuate, apex obtuse to acute, margin entire. Cauline leaves and bracts similar to basal leaves but narrower and sessile to short petiolate, reduced in size upward. Sepals often purplish, 1.5–2.5 × 1.2–1.5 mm, glabrous. Petals deep purple to white, broadly obovate to spatulate, 3.5–5(–6) × 1.5–3(–4) mm, apex rounded; claw 1–2.5 mm. Filaments 1.5–2.5 mm; anthers 0.3–0.6 mm. Ovules 10–16 per ovary. Fruiting raceme somewhat elongated. Fruiting pedicels erect to ascending, (1.5–)2–10(–20) mm, puberulent above, glabrous below. Fruit lanceolate to elliptic, 5–10 × 1.5–2 mm, compressed; valves obscurely veined, glabrous or rarely sparsely puberulent; gynophore to 0.7 mm; septum complete, hyaline; style 0.5–1(–2) mm. Seeds light brown, oblong, biseriate, 0.9–1.3 × 0.6–1 mm.

Distribution: Nepal, W Himalaya, E Himalaya, Tibetan Plateau, E Asia, C Asia and SW Asia.

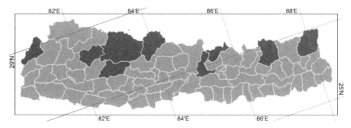

Altitudinal range: 3300–5800 m.

Ecology: Glacial moraine, among gravel, limestone rubble and cliffs, open stony slopes, alpine pastures, scree, dolomite cliffs, stream banks, peaty ground and turf.

Flowering: May–August. **Fruiting:** July–September.

Hooker & Anderson (Fl. Brit. Ind. 1: 155. 1872) misapplied the name *Braya alpina* Sternb. & Hoppe to this species.

6. Aphragmus serpens (W.W.Sm.) Al-Shehbaz &
S.I.Warwick, Canad. J. Bot. 84: 279 (2006).
Cochlearia serpens W.W.Sm., Rec. Bot. Surv. India 4: 175
(1911); *Lignariella hobsonii* subsp. *serpens* (W.W.Sm.) H.Hara;
L. obscura Jafri; *L. serpens* (W.W.Sm.) Al-Shehbaz, Kats.Arai
& H.Ohba.

Short-lived perennial herbs, glabrous throughout except for
pedicels and sometimes fruit. Stems slender, few to many from
base, decumbent, (1.5–)5–15(–30) cm. Leaves deeply 3(or
5)-lobed throughout, simple leaves absent; petiole glabrous,
(2–)5–15(–40) mm; blade of central lobe oblong to ovate,
rarely linear, (1.5–)3–8(–15) × (0.5–)1–3(–7) mm, apex obtuse
to subacute, entire or rarely 1-toothed on each side; lateral
lobes smaller, entire, 1- or 2-lobed or toothed, rarely leaves
subternate. Sepals oblong, 1.5–5 × 0.7–1 mm, not saccate
at base, scarious margin ca. 0.2 mm wide. Petals purple,
broadly obovate, (2–)3–4(–4.5) × (1–)1.5–2.5(–3) mm, apex
rounded, cuneate into a claw to 0.1 mm. Filaments purplish,
1.5–2.5 mm; anthers purple, ovate, 0.3–0.4 mm. Ovules 3–6
per ovary. Fruiting pedicels (10–)15–30(–40) mm, strongly
recurved to sigmoid, often filiform, puberulent above with
papillae to 0.1 mm. Fruit not geocarpic, ovoid to oblong,
rarely oblong-linear, often curved at middle, 2.5–8(–12) ×
1–2(–2.5) mm; gynophore 0.5–3 mm; valves slightly torulose

or not, glabrous or puberulent; style 1–1.5(–2) mm, stout.
Seeds uniseriate, oblong, (1–)1.5–2.5 × (0.5–)0.8–1.2(–
1.5) mm.

Distribution: Nepal, E Himalaya and Tibetan Plateau.

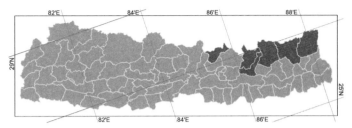

Altitudinal range: 1700–4300 m.

Ecology: Alpine peaty soil, scree, gravelly stream edges, turf
among rocks and small shrubs.

Flowering: June–August. **Fruiting:** July–October.

Hara (Fl. E. Himalaya: 109. 1966) and Photo. Fl. E. Himalaya:
25, f. 224 (1968) partially misapplied the name *Lignariella
hobsonii* (Pers.) Baehni to this species.

2. *Arcyosperma* O.E.Schulz, in Engl., Pflanzenr. IV. 105(Heft 86): 182 (1924).

Perennial herbs with caudex. Trichomes eglandular, simple. Stems erect to ascending, leafy. Basal leaves petiolate, rosulate,
simple, entire or dentate. Cauline leaves sessile, cuneate at base, not auriculate, entire or rarely denticulate. Racemes several- to
many-flowered, ebracteate, corymbose, slightly elongated in fruit; rachis straight. Sepals oblong, free, tardily deciduous to persistent,
suberect to ascending, equal, base of inner pair not saccate, margin membranous. Petals white to pinkish, erect at base with flaring
blade, longer than sepals; blade obovate-oblong, apex subemarginate; claw not differentiated from blade, glabrous. Stamens 6,
4 included, erect, slightly tetradynamous; filaments wingless, unappendaged, glabrous, free, not dilated at base; anthers oblong-
ovate, not apiculate at apex. Nectar glands confluent and subtending bases of all stamens; median nectaries present; lateral
nectaries horseshoe-shaped. Ovules 30–50 per ovary; placentation parietal. Fruiting pedicels slender, divaricate. Fruit dehiscent,
capsular siliques, broadly linear to linear-oblong, subterete, not inflated, sessile, unsegmented; valves papery, with a distinct
midvein, glabrous, not keeled, torulose, wingless, unappendaged; gynophore absent; replum rounded, visible; septum complete,
membranous, veinless; style obsolete or to 1 mm, stout; stigma capitate, entire, unappendaged. Seeds biseriate, wingless, ovate,
slightly flattened; seed coat coarsely reticulate, not mucilaginous when wetted; cotyledons obliquely incumbent.

Worldwide one species in Nepal, India and Pakistan.

1. *Arcyosperma primulifolium* (Thomson) O.E.Schulz, in
Engl., Pflanzenr. IV. 105(Heft 86): 182 (1924).
Sisymbrium primulifolium Thomson, Hooker's J. Bot. Kew
Gard. Misc. 5: 18 (1853); *Eutrema primulifolium* (Thomson)
Hook.f. & Thomson.

Herbs 4–20 cm. Caudex simple, to 1.3 cm in diameter, with
petiolar remains of previous years. Stems glabrous or pilose
with simple trichomes to 1 mm. Basal leaves rosulate; petiole
1.5–10 cm, ciliate; leaf blade obovate to oblong, 2.5–10
× 1–4.5 cm, base attenuate to cuneate, apex obtuse to
subacute, margin subentire to dentate, ciliate, glabrous on
both surfaces or pilose below. Cauline leaves 3–6, sessile,
oblong to oblong-ovate or oblanceolate, 1–4 cm × 3–12 mm,
entire or rarely denticulate, ciliate. Sepals oblong, 2.5–3 ×

1–1.5 mm, glabrous. Petals white or pinkish, obovate-oblong,
4–6 × 1.5–2 mm. Median filaments 3–4 mm; lateral filaments
2.5–3 mm; anthers oblong-ovate, 0.4–0.5 mm. Fruiting
pedicels divaricate, slender, 3–8 mm, glabrous or pilose. Fruits
broadly linear to linear-oblong, 15–24 × 1.5–2.5 mm, slightly
curved upward or straight; valves glabrous. Seeds dark brown,
ovate, 0.9–1.2 × 0.6–0.8 mm.

Distribution: Nepal, W Himalaya and E Himalaya.

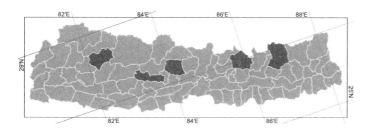

Altitudinal range: 2700–4600 m.

Ecology: Shade of rocks, damp ledges, stream-sides.

Flowering: May–June. **Fruiting:** June–July.

3. *Arabis* L., Sp. Pl. 2: 664 (1753).

Biennial, or perennial herbs. Trichomes stellate, dendritic, or stalked forked, sometimes mixed with simple ones. Stems simple or branched. Basal leaves petiolate or not, rosulate, simple, often entire, sometimes dentate, rarely lyrate-pinnatifid. Cauline leaves sessile and auriculate, sagittate, or amplexicaul, very rarely petiolate, entire or dentate. Racemes ebracteate, elongated in fruit. Sepals ovate, oblong or sublanceolate, base of lateral pair saccate or not, margin membranous. Petals white, pink, or purple, spatulate, oblong, or oblanceolate, rarely obovate, apex obtuse or emarginate; claw shorter than sepals. Stamens 6, tetradynamous; anthers oblong, or linear, obtuse at apex. Nectar glands confluent, subtending bases of all stamens; median glands sometimes tooth-like and free, rarely absent. Ovules 36–80 per ovary. Fruiting pedicels erect, ascending, divaricate, or reflexed. Fruit dehiscent siliques, linear, latiseptate, sessile or rarely shortly stipitate; valves papery, with an obscure or prominent midvein, smooth or torulose; replum rounded; septum complete; style obsolete or distinct; stigma capitate, entire or slightly 2-lobed. Seeds uniseriate, winged or margined, oblong or orbicular, flattened; seed coat smooth or minutely reticulate, not mucilaginous when wetted; cotyledons accumbent.

Worldwide about 70 species in temperate Asia, Europe, N America. Four species in Nepal.

The limits of *Arabis* are highly artificial, and the genus is defined primarily on the basis of having flattened, linear fruits, accumbent cotyledons, and branched trichomes. However, this combination of characters has evolved independently several times in the family.

Key to Species

1a	Fruits and often fruiting pedicels erect, appressed to rachis	**4. *A. pterosperma***
b	Fruits and fruiting pedicels divaricate to reflexed or ascending, fruits sometimes erect on divaricate or ascending pedicels never appressed to rachis	2
2a	Petals 3–6(–6.5) mm. Lateral sepals not saccate. Style in fruit 0.2–0.8 mm	**3. *A. paniculata***
b	Petals 7–15 mm. Lateral sepals saccate. Style in fruit 1–2.5 mm	3
3a	Plants hirsute or hispid with primarily simple, subsetose trichomes	**1. *A. amplexicaulis***
b	Plants tomentose or pilose with almost exclusively branched trichomes	**2. *A. bijuga***

1. *Arabis amplexicaulis* Edgew., Trans. Linn. Soc. London 20: 31 (1851).

Biennial or short-lived perennial herbs, (20–)30–60(–70) cm, densely to sparsely hispid or hirsute with primarily simple subsetose trichomes to 1.7 mm, these mixed with much fewer, smaller, 2–4-forked ones. Stems erect, often single from base. Basal leaves rosulate; petiole 0.5–5 cm; leaf blade lanceolate-obovate, oblong, or oblanceolate, 3–9(–11) × 1–2(–3) cm, base attenuate, apex obtuse to rounded, margin dentate or entire, ciliate. Middle cauline leaves sessile, oblong to ovate, (1–)1.5–5(–8) × 0.6–2.5(–4) cm, base cordate or amplexicaul, margin dentate to entire. Sepals oblong, 4–6 × 1.2–1.5 mm, glabrous, lateral pair saccate. Petals white, oblong to narrowly oblanceolate, (0.7–)0.9–1.2 cm × 2–3.5 mm, apex obtuse. Filaments 3.5–6 mm; anthers oblong, 0.8–1 mm. Ovules 50–70 per ovary. Fruiting pedicels, divaricate, 0.6–1.5 cm, slender, straight, glabrous. Fruit (2.5–)3.5–6.5(–7) cm × 1–1.5 mm, erect to divaricate or rarely reflexed, not appressed to rachis; valves with distinct midvein, torulose, glabrous; style 1–1.5 mm. Seeds brown, oblong, 1.2–1.5 × 0.8–1 mm, narrowly winged apically.

Distribution: Nepal, W Himalaya, E Himalaya, Tibetan Plateau and SW Asia.

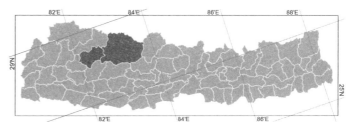

Cruciferae (Brassicaceae)

Altitudinal range: 1800–3200 m.

Ecology: Forest margins, shady places.

Flowering: April–June. **Fruiting:** May–July.

2. Arabis bijuga Watt, J. Linn. Soc., Bot. 18: 378 (1881).
Arabis macrantha C.C.Yuan & T.Y.Cheo; *A. pangiensis* Watt.

Perennial herbs, 15–40 cm, often densely tomentose or pilose, with primarily short-stalked, stellate trichomes, these sometimes mixed much fewer, simple or forked ones, rarely plants glabrescent and trichomes restricted primarily to leaf margin. Stems erect, often branched at base. Basal leaves rosulate; petiole 0.5–5 cm; leaf blade obovate-spatulate, oblong, elliptic, or oblanceolate, 1.5–5 × 0.5–1.5 cm, base attenuate, apex obtuse or rounded, margin dentate or entire. Middle cauline leaves sessile, oblong-linear, narrowly elliptic, or narrowly lanceolate, 1–4(–5.5) cm × 2–8 mm, base obtuse or auriculate, margin dentate or entire. Sepals oblong, 3.5–5 × 1.2–1.8 mm, glabrous, lateral pair saccate. Petals white or pinkish, oblong or narrowly oblanceolate, 0.9–1.4 cm × 2–4 mm, apex obtuse. Filaments 5–8 mm; anthers oblong, 1–1.5 mm. Ovules 40–70 per ovary. Fruiting pedicels divaricate or ascending, (0.8–)1.2–2.4(–3) cm, slender, straight, glabrous. Fruit 3–6 cm × 1–1.2 mm, not appressed to rachis; valves with a prominent midvein extending full length, slightly torulose, glabrous; style 1–1.5 mm. Seeds brown, oblong, ca. 1.2 × 0.5 mm, narrowly winged apically.

Distribution: Nepal, W Himalaya and E Asia.

Altitudinal range: 2400–3000 m.

Ecology: Grassy slopes, rock crevices, dry cliffs, stony pastures.

Flowering: April–June. **Fruiting:** May–July.

Reported for Nepal by Hara (Enum. Fl. Pl. Nepal: 39. 1979) under the name *Arabis pangiensis* on the basis of a collection now identified as *A. paniculata* Franch. (see note under this species).

Although no collections are now known from Nepal this species is expected to occur and so is included as a full entry without a distribution map. Except for being glabrescent, instead of moderately to densely tomentose, the type collection of *A. bijuga* is basically indistinguishable from those of *A. macrantha* and *A. pangiensis*.

3. Arabis paniculata Franch., Pl. Delavay. 1: 57 (1889).
Arabidopsis mollissima var. *yunnanensis* O.E.Schulz; *Arabis alpina* var. *parviflora* Franch.; *A. alpina* var. *rigida* Franch.; *A. alpina* var. *rubrocalyx* Franch.; *A. paniculata* var. *parviflora* (Franch.) W.T.Wang.

Biennial or short-lived perennial herbs, (10–)20–75(–110) cm, densely to sparsely hirsute with stalked, forked or 3- or 4-rayed stellate trichomes mixed at least basally with simple ones. Stems erect, simple or few from base, often branched

at middle. Basal leaves rosulate, present at anthesis; petiole absent, rarely to 2.5 cm; leaf blade oblanceolate, narrowly obovate, oblong, or lanceolate, (1–)2–6(–8) × (0.5–)1–2(–2.5) cm, base attenuate, apex obtuse to acute, margin dentate, serrate, or rarely entire. Middle cauline leaves sessile, oblong, ovate, or elliptic, rarely suboblanceolate, (0.7–)1.5–4(–6) cm × (2–)4–14(–25) mm, stellate on both surfaces or above with simple trichomes, base cordate or auriculate, rarely amplexicaul, apex obtuse to acute, margin dentate to entire. Sepals oblong to sublanceolate, 2.5–3.5(–4) × 1–1.5 mm, glabrous or sparsely pubescent, not saccate. Petals white or rarely pale pink, oblong to narrowly oblanceolate, 4–6(–6.5) × 1–2 mm, apex obtuse. Filaments 3.5–5 mm; anthers oblong, 0.8–1 mm. Ovules 36–80 per ovary. Fruiting pedicels divaricate, rarely ascending or slightly reflexed, (0.4–)0.6–1.8(–2.5) cm, slender, straight or rarely slightly curved. Fruit (1.5–)2.5–5.5(–6.5) cm × 1–1.5 mm, erect, divaricate, or rarely reflexed, not appressed to rachis; valves with distinct midvein, torulose, glabrous; style 0.4–0.8 mm. Seeds brown, oblong to ovate, 1–1.3 × 0.7–0.9 mm, narrowly winged terminally.

Distribution: Nepal, W Himalaya, Tibetan Plateau and E Asia.

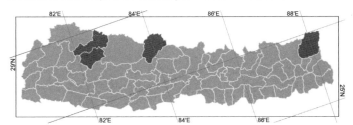

Altitudinal range: 1300–4300 m.

Ecology: Waste areas, roadsides, grassy slopes, along ditches.

Flowering: April–August. **Fruiting:** April–August.

Hara (Enum. Fl. Pl. Nepal 2: 39. 1979) listed *Stainton, Sykes & Williams 1933* (BM) as *Arabis pterosperma* Edgew., but that collection is *A. paniculata*, a species first reported for Nepal in Fl. China 8: 115. 2001. Likewise Hara listed *Stainton, Sykes & Williams 762* (BM) as *Arabis pangiensis* Watt, but his collection is also *A. paniculata*.

4. Arabis pterosperma Edgew., Trans. Linn. Soc. London 20: 33 (1851).
Arabidopsis yadungensis K.C.Kuan & Z.X.An; *Arabis alpina* var. *purpurea* W.W.Sm.; *A. latialata* Y.Z.Lan & T.Y.Cheo.

Biennial or short-lived perennial herbs, (10–)20–60(–80) cm, densely or rarely sparsely hirsute primarily with simple trichomes, these often mixed at least distally with long-stalked forked or rarely stellate ones. Stems erect, simple from base, simple or branched above. Basal leaves rosulate; petiole absent, rarely to 4 cm; leaf blade spatulate, oblanceolate, narrowly obovate, or oblong, (1–)2–7(–9) × (0.5–)1–2(–3) cm, base attenuate, apex obtuse to acute, margin dentate, serrate, or rarely entire, often ciliate. Middle cauline leaves sessile, oblong, elliptic, or ovate-lanceolate, (1–)2–4(–5) cm

× (2–)4–10(–15) mm, hirsute with predominantly simple trichomes, base auriculate, cordate, or rarely amplexicaul, apex obtuse to acute, margin dentate to entire. Sepals oblong to sublanceolate, 2.5–3.5 × 1–1.5 mm, often glabrous, not saccate. Petals purple, pink, or rarely white, oblong to narrowly oblanceolate, 6–9 × 2–3 mm, apex obtuse. Filaments 3.5–5 mm; anthers oblong, 0.6–1 mm. Ovules 40–80 per ovary. Fruiting pedicels, erect or rarely ascending, often subappressed to rachis, 4–10(–16) mm, slender, straight. Fruit (2.5–)3–5(–6) cm × 1.5–2 mm, erect, appressed to rachis; valves with a prominent midvein, torulose, glabrous; style 0.5–1 mm. Seeds brown, orbicular to ovate-orbicular, 1–1.6 mm in diameter, winged all around; wing (0.1–)0.2–0.4 mm wide.
Figs 16b, 17b–e.

Distribution: Nepal, W Himalaya, E Himalaya, Tibetan Plateau and E Asia.

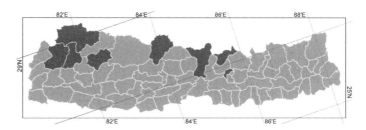

Altitudinal range: 2200–4400 m.

Ecology: Roadsides, woodlands, grassy slopes, alpine meadows.

Flowering: (March–)May–July. **Fruiting:** June–October.

Hooker & Thomson (J. Proc. Linn. Soc. Bot. 5: 141. 1861) and Hooker f. & Anderson (Fl. Brit. Ind. 1: 135. 1872) misapplied the name *Arabis alpina* L. to this species, to which Hara (Fl. E. Himalaya, Sec. Rep.: 42. 1971) misapplied the name *Arabidopsis mollissima* (C.A.Mey.) O.E.Schulz.

4. *Draba* L., Sp. Pl. 2: 642 (1753).

Perennial, sometimes annual herbs. Trichomes simple, forked, stellate, malpighiaceous, or dendritic, stalked or sessile, often mixed. Stems erect or ascending, sometimes prostrate, leafy or leafless and plants scapose. Basal leaves petiolate, often rosulate, simple, entire or toothed, rarely lobed. Cauline leaves petiolate or sessile, cuneate or auriculate at base, entire or dentate, sometimes absent. Racemes bracteate or ebracteate, elongated or not in fruit. Sepals ovate, oblong, or elliptic, base of lateral pair not saccate or subsaccate. Petals yellow, white, sometimes pink to purple, obovate, spatulate, oblong, oblanceolate, orbicular, or linear, apex obtuse, rounded, or rarely emarginate; claw obscurely to strongly differentiated from blade, or absent. Stamens 6, tetradynamous; filaments dilated or not at base; anthers ovate or oblong, obtuse at apex. Nectar glands 1, 2, or 4, distinct or confluent and subtending bases of all stamens; median glands present or absent. Ovules 4 to numerous per ovary. Fruiting pedicels slender, erect, ascending, or divaricate. Fruit dehiscent, silicles or rarely siliques, ovate, elliptic, oblong, orbicular, ovoid, globose, lanceolate, or linear, latiseptate or terete; valves distinctly or obscurely veined, glabrous or pubescent; replum rounded; septum complete; style distinct or obsolete; stigma capitate, entire or slightly 2-lobed. Seeds biseriate, wingless, oblong, ovate, or orbicular, flattened; seed coat minutely reticulate, not mucilaginous when wetted; cotyledons accumbent.

Worldwide about 350 species, primarily in the N hemisphere, especially arctic, subarctic, alpine, and subalpine regions, with about 70 species in S America. 23 species in Nepal.

Draba is the largest and taxonomically most difficult genus in the family. More than 950 binomials and nearly a fourth as many infraspecific taxa have been proposed. Numerous taxa are based on trivial characters, especially the presence or absence of trichomes on the fruit valves. Schulz (Pflanzenr. 89(IV. 105): 1–396. 1927), who was the last to monograph *Draba* on a worldwide basis, accorded varietal names to forms with glabrous and pubescent fruits, but this variation often occurs within the same population (see discussion under *D. oreades* Schrenk). By contrast, petal colour is taxonomically important and should be recorded in the field rather than from dried specimens. Another taxonomically important character is the number of ovules (seeds) per ovary (fruit). This is easily obtained by counting the total number of seeds and aborted ovules within a fruit.

Key to Species

1a Annuals..2
  b Perennials, with well-developed caudex often covered with leaf or petiolar remains of previous years 4

2a Fruits linear. Seeds and aborted ovules (30–)36–60(–72) per fruit ....**21. *D. stenocarpa***
  b Fruits ovate, oblong, or rarely suborbicular. Seeds and aborted ovules 10–24(–28) per fruit..........................3

3a Fruit ovate, glabrous, acute at apex. Cauline leaves (5–)10–24(–30) ....**22. *D. eriopoda***
  b Fruit oblong, oblong-elliptic, or rarely suborbicular, puberulent, obtuse at apex. Cauline leaves 3–12
    .... **23. *D. ellipsoidea***

Cruciferae (Brassicaceae)

4a	Cauline leaves absent	5
b	Cauline leaves 1–numerous	10
5a	Petals white	6
b	Petals yellow	8
6a	Petals 1.5–2.5(–3) × 0.6–1.2 mm. Fruit not twisted, 2.5–5 mm. Style 0.1–0.4 mm	**7. D. glomerata**
b	Petals 4–7 × 1.5–4.5 mm. Fruit often twisted, (3–)5–12 mm. Style (0.5–)0.7–1.5 mm	7
7a	Rachis of fruiting raceme often flexuous. Fruit apex acute. Petals 1.5–2.5 mm wide. Stellate trichomes sessile or subsessile, with some branched rays	**9. D. winterbottomii**
b	Rachis of fruiting raceme straight. Fruit apex obtuse. Petals 2.5–4.5 mm wide. Stellate trichomes long-stalked, with unbranched rays	**10. D. sikkimensis**
8a	Upper leaf surface primarily with simple trichomes, rarely glabrous. Fruit flattened, slightly inflated basally	**16. D. oreades**
b	Upper leaf surface with short-stalked forked and stellate trichomes. Fruit distinctly inflated at least basally	9
9a	Fruit ovoid, apex acute. Seeds and ovules 8–10 per fruit	**17. D. affghanica**
b	Fruit oblong, apex obtuse. Seeds and ovules 16–24 per fruit	**18. D. humillima**
10a	Petals yellow	11
b	Petals white, purple, or pink	17
11a	Racemes bracteate throughout or only basally	12
b	Racemes ebracteate	13
12a	Racemes 20–60(–140)-flowered. Fruiting pedicels glabrous above, pubescent below	**11. D. polyphylla**
b	Racemes 3–10(–12)-flowered. Fruiting pedicels glabrous or pubescent throughout, if glabrous above then style 1.5–3 mm	**20. D. gracillima**
13a	Fruit ovate, suborbicular, or ovate-lanceolate, slightly inflated at base	**16. D. oreades**
b	Fruit linear, linear-lanceolate, linear-elliptic, or oblong, not inflated	14
14a	Petals 6–9 × 3–4.5 mm. Style 1.5–4 mm	15
b	Petals 2.5–3.5 × 0.8–1.5 mm. Style 0.2–0.6 mm	16
15a	Plants surculose, with rhizome-like suckers to 35 cm. Seeds winged. Raceme 12–25-flowered. Fruit 3–4 mm wide. Style 2–4 mm	**12. D. radicans**
b	Plants not surculose. Seeds wingless. Raceme 2–5(–12)-flowered. Fruit 1.5–2.5 mm wide. Style 1.5–2 mm	**14. D. cholaensis**
16a	Fruit oblong, twisted, 3–4 mm wide. Fruiting pedicels with axillary surface gland	**19. D. macbeathiana**
b	Fruit not twisted, narrowly oblong, linear or linear-lanceolate, 1–2 mm wide. Fruiting pedicels without axillary surface gland	**20. D. gracillima**
17a	Seeds and aborted ovules (24–)26–60 per fruit	18
b	Seeds and aborted ovules 6–20(–24) per fruit	20
18a	Petals 2.5–3.5 mm. Fruiting pedicels (1–)2–4.5(–6) mm	**1. D. lanceolata**
b	Petals 7–12 mm. Fruiting pedicels (6–)8–24 mm	19
19a	Petals white. Fruiting pedicels divaricate, recurved. Fruit 4–5 mm wide. Raceme 6–12-flowered	**13. D. staintonii**
b	Petals pink or purple. Fruiting pedicels ascending, straight. Fruit 2–3.5 mm wide. Raceme 20–40-flowered	**15. D. amoena**

20a Petals broadly obovate, 2.5–4.5 mm wide. Sepals 2.5–3 mm. Fruit apex obtuse. Basal leaves subfloccose with slender trichomes ....**10. *D. sikkimensis***

b Petals spatulate to narrowly obovate, 1–2(–3) mm wide. Sepals 1–2(–2.5) mm. Fruit apex acute to acuminate. Basal leaves pubescent, pilose, or tomentose with short trichomes 21

21a Lower surface of basal leaves with predominantly simple trichomes. Stellate trichomes with unbranched rays ....**5. *D. altaica***

b Lower surface of basal leaves with predominantly stellate trichomes, 1 to all 4 rays of which laterally branched 22

22a Fruit 2–4(–5) mm, ovate, elliptic, or rarely oblong-ovate. Seeds and aborted ovules 6–12(–14) per fruit 23

b Fruit (5–)6–11 mm, narrowly elliptic to lanceolate-linear. Seeds and aborted ovules 12–22(–24) per fruit....................25

23a Sepals, petals, and stamens persistent in fruit ....**8. *D. stenobotrys***

b Sepals, petals, and stamens caducous before fruit maturity 24

24a Fruiting pedicels divaricate. Racemes bracteate at least basally, slightly or not elongated in fruit. ..... **6. *D. lichiangensis***

b Fruiting pedicels erect to ascending. Racemes ebracteate, distal half not elongated in fruit....................**7. *D. glomerata***

25a Petals 4–5.5 × 2–3 mm. Styles 1–1.5 mm. ....**4. *D. poluniniana***

b Petals 1.2–3.5 × 0.5–1.1 mm. Styles 0.1–0.5 mm. 26

26a Fruiting pedicels erect to ascending, forming a straight line with fruit. Petals 2–3.5 mm. Fruits lanceolate to lanceolate-linear, twisted, valves with branched trichomes. .... **2. *D. lasiophylla***

b Fruiting pedicels horizontal, often recurved, forming a distinct angle with fruit. Petals 1.2–2 mm. Fruits narrowly elliptic, untwisted, valves with simple trichomes. ....**3. *D. bagmatiensis***

1. *Draba lanceolata* Royle, Ill. Bot. Himal. Mts. [2]: 72 (1834).

Draba lanceolata var. *brachycarpa* Schult.; *D. lanceolata* var. *leiocarpa* O.E.Schulz; *D. lanceolata* var. *sonamargensis* O.E.Schulz; *D. nichanaica* O.E.Schulz; *D. stenobotrys* var. *leiocarpa* L.L.Lou & T.Y.Cheo; *D. stylaris* var. *leiocarpa* L.L.Lou & T.Y.Cheo.

Perennial herbs, 10–30(–40) cm, caespitose. Caudex slender, few-branched, ultimate branches terminated in rosettes, covered with petiolar remains of previous years. Stems erect, simple, tomentose with subsessile stellate trichomes. Basal leaves rosulate, persistent; petiole to 1.5 cm, often undifferentiated; leaf blade elliptic, oblong, or lanceolate, (0.4–)1–2(–3) cm × (1–)2–6 mm, base attenuate to cuneate, apex acute, margin entire or denticulate, tomentose with sessile or short-stalked, 4-rayed, stellate trichomes 1 or 2 rays of which often branched. Cauline leaves 6–12(–16), sessile; leaf blade ovate, oblong, elliptic, or lanceolate, 0.7–2.6(–3.5) cm × 2–7(–10) mm, base cuneate to obtuse, apex acute, margin 2–5(–9)-toothed on each side, rarely entire, pubescent as basal leaves, often pilose above with predominantly simple trichomes. Racemes (7–)14–32(–47)-flowered, ebracteate or lowermost flowers bracteate, elongated in fruit. Sepals oblong or ovate, 1–2 × 0.5–0.8 mm, erect, pilose below, base of lateral pair not saccate, margin membranous. Petals white, spatulate, 2.5–3.5 × 1–2 mm, apex subemarginate or rounded; claw absent or to 1 mm. Filaments 1.5–2 mm; anthers ovate, 0.2–0.3 mm. Ovules (26–)30–48(–56) per ovary. Fruiting pedicels (1–)2–4.5(–6) mm, erect or ascending, often appressed to rachis, straight, tomentose all around with subsessile, stellate trichomes sometimes mixed with simple ones. Fruit lanceolate to lanceolate-linear, (6–)7–11(–12) ×

1.5–2 mm, erect and often appressed to rachis, latiseptate, rarely slightly twisted; valves tomentose or very rarely glabrous, obscurely veined, apex acute; style 0.1–0.6 mm. Seeds brown, ovate, 0.8–1.1 × 0.5–0.6 mm.

Distribution: Nepal, W Himalaya, E Himalaya, Tibetan Plateau, E Asia, N Asia, C Asia and SW Asia.

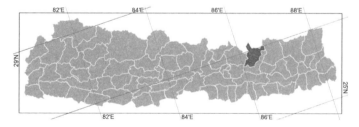

Altitudinal range: ca. 4200 m.

Ecology: Mountain slopes, meadows, scrub, gravelly streamsides, roadsides.

Flowering: June–August. **Fruiting:** June–August.

This is the first documented record of the species from Nepal, and is based on *Stainton, Sykes & Williams 1256* (BM, E, UPS). Hara (Enum. Fl. Pl. Nepal 2: 42. 1979) listed this collection as *Draba lasiophylla* Royle var. *lasiophylla*.

Cruciferae (Brassicaceae)

2. *Draba lasiophylla* Royle, Ill. Bot. Himal. Mts. [2]: 71 (1834). *Draba glomerata* var. *leiocarpa* Pamp.; *D. ladyginii* Pohle; *D. ladyginii* var. *trichocarpa* O.E.Schulz; *D. lasiophylla* var. *leiocarpa* (Pamp.) O.E.Schulz; *D. lasiophylla* var. *royleana* Pohle; *D. torticarpa* L.L.Lou & T.Y.Cheo.

Perennial herbs, (5–)10–20(–28) cm, caespitose. Caudex slender, few-branched, ultimate branches terminated in rosettes, covered with petiolar remains of previous years. Stems erect, simple, tomentose with subsessile stellate trichomes. Basal leaves rosulate; petiole to 6 mm, often undifferentiated; leaf blade lanceolate or elliptic-oblong, (3–)4–15(–18) × 1–4(–5) mm, base attenuate to cuneate, apex acute, margin entire or 1–3-toothed on each side, often ciliate at least near base, tomentose with sessile, 4-rayed stellate trichomes the rays of which having 1 or 2 lateral branches. Cauline leaves 1–4(or 5), sessile; leaf blade ovate, oblong, or elliptic, (4–)6–10(–20) × 1.5–4(–7) mm, base cuneate to obtuse, apex acute, margin entire or 1–5-toothed on each side, often subsetose ciliate at base, pubescent as basal leaves. Racemes (5–)7–20(–25)-flowered, ebracteate or lowermost flowers bracteate, at least lowermost portion elongated in fruit. Sepals oblong, 1.2–1.8 × 0.5–0.9 mm, erect, caducous, pilose below, base of lateral pair not saccate, margin membranous. Petals white, spatulate, 2–3.5 × 0.8–1.1 mm, caducous, apex subemarginate or rounded; claw absent. Filaments 1–1.8 mm, caducous; anthers ovate, ca. 0.2 mm. Ovules (12–)14–20(–22) per ovary. Fruiting pedicels (1–)2–6(–10) mm, erect or ascending, straight, often subappressed to rachis, forming a straight line with fruit, tomentose all around with subsessile, stellate trichomes rarely mixed with fewer simple ones. Fruit narrowly lanceolate to lanceolate-linear, (5–)7–10(–11) × 1.5–2 mm, erect to ascending, often appressed to rachis, latiseptate, twisted 2 or sometimes 3 turns, rarely twisted 1 or ½ turn, apex acute to acuminate; valves tomentose with branched trichomes, or glabrous, obscurely veined, base obtuse, apex acute or acuminate; style 0.1–0.5 mm. Seeds brown, ovate, 0.8–1.1 × 0.4–0.5 mm.

Distribution: Nepal, W Himalaya, E Himalaya, Tibetan Plateau, E Asia and C Asia.

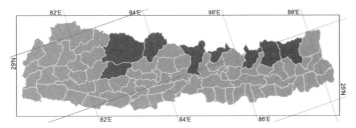

Altitudinal range: 3000–5600 m.

Ecology: Mountain slopes, crevices, *Kobresia* turf.

Flowering: June–August. **Fruiting:** July–October.

Hooker & Anderson (Fl. Brit. Ind. 1: 143. 1872) and Kitamura (Peoples Nepal Himalaya: 421. 1957) misapplied the name *Draba incana* L. to this species.

3. *Draba bagmatiensis* Al-Shehbaz, Novon 12: 317 (2002).

Perennial herbs, 1–6 cm, caespitose. Caudex slender, few- to many-branched, ultimate branches terminated in rosettes, covered with petiolar remains of previous years. Stems several to many, simple, ascending, tomentose with subsessile stellate trichomes. Basal leaves rosulate; petiole to 3 mm; leaf blade oblanceolate to oblong, 2–10 × 0.5–3 mm, base cuneate, apex obtuse, margin entire or 1-toothed on each side, often ciliate at least near base, tomentose with sessile, 4-rayed stellate trichomes the rays of which with 2 lateral branches. Cauline leaves 2–5, sessile; leaf blade ovate to oblong, 2–7 × 1–3 mm, obtuse, apex subacute, margin entire, not ciliate at base, pubescent as basal leaves. Racemes 5–13-flowered, lowermost flowers bracteate. Sepals oblong, 0.8–1 × 0.4–0.5 mm, erect, caducous, pilose below, base of lateral pair not saccate, margin membranous. Petals white, spatulate, 1.2–2 × 0.5–1 mm, caducous, apex subemarginate or rounded; claw absent. Filaments 0.8–1 mm, caducous; anthers ovate, 0.1–0.2 mm. Ovules 16–20 per ovary. Fruiting pedicels 1–5 mm, horizontal, often recurved, forming a distinct angle with fruit, tomentose all around with subsessile, stellate trichomes. Fruit narrowly elliptic, 3–5 × 1.2–1.7 mm, not appressed to rachis, latiseptate, not twisted, acute; valves puberulent with simple trichomes; style 0.1–0.4 mm long. Seeds ovate, 0.5–0.7 × 0.3–0.5 mm.

Distribution: Nepal and E Himalaya.

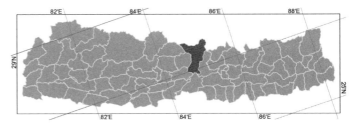

Altitudinal range: 4100–4400 m.

Ecology: Sandy soil, fine scree.

Flowering: July–August. **Fruiting:** July–August.

The type collection of the species, *F. Miyamoto et al. 9400056* (MO, TI), was collected from Ganesh Himal.

4. *Draba poluniniana* Al-Shehbaz, Harvard Pap. Bot. 8(2): 171 (2004).

Perennial herbs, 7–15 cm, caespitose. Caudex slender, many-branched, ultimate branches terminated in rosettes, covered with petiolar remains of previous years. Stems erect, simple, tomentose with subsessile stellate trichomes. Basal leaves rosulate, subsessile; leaf blade oblanceolate, 5–15 × 2–4 mm, base attenuate to cuneate, apex obtuse, margin entire or 1–3-toothed on each side, not ciliate, tomentose with sessile, 4-rayed stellate trichomes all rays of which with 1 or 2 lateral branches. Cauline leaves 2–6, widely spaced, sessile; leaf blade ovate or oblong, 5–10 × 1.5–4 mm, base cuneate to obtuse, apex acute, margin entire or 1–3-toothed on each side, not ciliate at base, pubescent as basal leaves. Racemes 5–15-flowered, ebracteate or lowermost flower bracteate, elongated in fruit. Sepals oblong, 2–3 × 1–1.3 mm, erect, caducous, tomentose below with exclusively stellate trichomes, base of lateral pair not saccate, margin membranous. Petals white, obovate, 4–5.5 × 2–3 mm, caducous, apex subemarginate or rounded; claw absent. Filaments 2.5–3.5 mm, caducous; anthers ovate, ca. 0.6 mm. Ovules 20–24 per ovary. Fruiting pedicels 4–7 mm, divaricate, straight, forming a wide angle with fruit, tomentose all around with subsessile, stellate trichomes. Fruit lanceolate-linear, 6–8 × ca. 1 mm, not appressed to rachis, latiseptate, twisted 1 turn, apex acute to acuminate; valves glabrous, obscurely veined, base obtuse, apex acute or acuminate; style 1–1.5 mm. Seeds unknown.

Distribution: Endemic to Nepal.

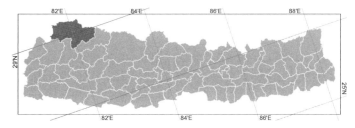

Altitudinal range: ca. 3800 m.

Ecology: Stony slopes and rock ledges.

Flowering: June.

The species is currently only known from the type collection, *Polunin, Sykes & Williams 4247* (holotype, TI; isotype, BM), a collection previously cited by Hara (Enum. Fl. Pl. Nepal 2: 43. 1979) as the only Nepalese record for *Draba tibetica* Hook.f. & Thomson var *duthiei* O.E.Schulz.

5. *Draba altaica* (C.A.Mey.) Bunge, Del. Sem. Hort. Dorpater: 8 (1841).
Draba rupestris var. *altaica* C.A.Mey., in Ledeb., Fl. Altaic. 3: 71 (1831); *D. altaica* subsp. *modesta* (W.W.Sm.) O.E.Schulz; *D. altaica* var. *foliosa* O.E.Schulz; *D. altaica* var. *glabrescens* Lipsky; *D. altaica* var. *microcarpa* O.E.Schulz; *D. altaica* var. *modesta* (W.W.Sm.) W.T.Wang; *D. altaica* var. *racemosa* O.E.Schulz; *D. modesta* W.W.Sm.; *D. rupestris* var. *pusilla* Karelin & Kirilov.

Perennial herbs, (1–)2–10(–20) cm, caespitose. Caudex few-branched, ultimate branches terminated in rosettes and covered with petiolar remains of previous years. Stems erect, simple from caudex, rarely branched above, sparsely to densely hirsute with simple, straight or crisped trichomes to 0.7 mm, these sometimes mixed with forked and stellate ones, rarely glabrescent. Basal leaves rosulate, persistent; petiole (2–)5–12 mm, becoming straw-like; leaf blade linear-lanceolate or lanceolate, rarely oblanceolate or subspatulate, (0.3–)0.6–2(–3) cm × 1–4(–7) mm, base attenuate to cuneate, apex acute, margin entire or 1–3(–6)-toothed on each side, often ciliate, pubescent exclusively or predominantly with simple

trichomes, stellate or forked trichomes mainly on midvein. Cauline leaves (1 or)2–6(–8), sessile; leaf blade ovate, oblong, elliptic, or lanceolate, (3–)4–13(–20) × 1–4.5(–6) mm, base cuneate or obtuse, apex acute, margin entire or 1–5(–7)-toothed on each side, pubescent as basal leaves. Racemes 5–15(–24)-flowered, bracteate basally, rarely ebracteate, not or only slightly elongated and subumbellate in fruit. Sepals oblong, 1–1.5 × 0.4–0.6 mm, erect, sparsely pubescent below with simple trichomes, base of lateral pair not saccate, margin not membranous. Petals white, spatulate, (1.2–)1.5–2.5 × (0.5–)0.7–1 mm, apex emarginate; claw to 0.7 mm. Filaments (0.6–)1–1.5 mm; anthers ovate, 0.1–0.2 mm. Nectar glands lateral, minute, 1 on each side of lateral stamen; median glands absent. Ovules 10–20 per ovary. Fruiting pedicels 1–4 mm, ascending to divaricate, straight, glabrous or sparsely to densely pubescent. Fruit ovate to oblong or lanceolate, (3–)4–7(–8) × 1.5–2 mm, latiseptate, not twisted apex acute to acuminate; valves glabrous, not veined, base obtuse, apex acute; style 0.1–0.2 mm. Seeds brown, ovate, 0.7–0.9 × 0.5–0.6 mm.

Distribution: Nepal, W Himalaya, E Himalaya, Tibetan Plateau, E Asia, N Asia, C Asia and SW Asia.

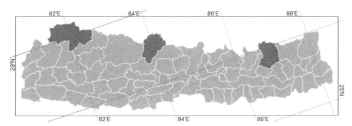

Altitudinal range: 2000–5600 m.

Ecology: Rocky slopes, gravelly areas, moraine, stream-sides, *Kobresia* turf, *Juniperus* forests, grassy slopes.

Flowering: June–August. **Fruiting:** July–September.

Hooker & Anderson (Fl. Brit. Ind. 1: 143. 1872) partially misapplied the name *Draba fladnitzensis* Wulfen to this species.

6. *Draba lichiangensis* W.W.Sm., Notes Roy. Bot. Gard. Edinburgh 11: 208 (1919).
Draba daochengensis W.T.Wang; *D. hicksii* Grierson; *D. lichiangensis* var. *microcarpa* O.E.Schulz; *D. lichiangensis* var. *trichocarpa* O.E.Schulz.

Perennial herbs, (1–)2–5(–8) cm, often densely caespitose, scapose. Caudex many-branched, ultimate branches terminated in rosettes and covered with petiolar remains of previous years. Stems erect, simple, sparsely to densely tomentose with stellate trichomes. Basal leaves rosulate, persistent; petiole to 4 mm, persistent, ciliate with simple trichomes or not ciliate; leaf blade spatulate, oblanceolate, or rarely oblong, (2–)4–8(–12) cm × (0.5–)1–2.5(–3) mm, base attenuate, apex obtuse to acute, margin entire or minutely 1–4-toothed, subglabrous above or with simple or stellate trichomes, tomentose below with stellate trichomes the rays of which with 1 or 2 lateral branches. Cauline leaves (1 or)2–8(–11), sessile; leaf blade ovate, oblong, or elliptic, similar in indumentum to basal leaves. Racemes (3–)5–10(–13)-flowered, bracteate basally or rarely throughout, slightly or not elongated in fruit. Sepals oblong, 1–1.8 × 0.6–0.8 mm, erect, sparsely pubescent below, base of lateral pair not saccate, margin not membranous, caducous before fruit maturity. Petals white, narrowly obovate to oblanceolate, 2–3 × 0.7–1.3 mm, apex subemarginate, caducous. Filaments 1–1.8 mm; anthers ovate, 0.2–0.3 mm. Ovules 8–12 per ovary. Fruiting pedicels (1–)2–8(–15) mm, divaricate, straight, sparsely to densely tomentose all around with stellate and forked trichomes. Fruit ovate to elliptic, 2–4(–5) × 1.5–2.5 mm, not inflated, latiseptate, not twisted apex acute to acuminate; valves glabrous or rarely pubescent, obscurely veined, base obtuse to acute, apex acute; style 0.1–0.4 mm. Seeds brown, ovate, (0.6–)0.9–1.3 × (0.4–)0.6–0.9 mm.

Distribution: Nepal, E Himalaya, Tibetan Plateau and E Asia.

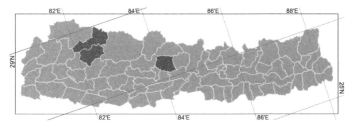

Altitudinal range: 3500–5000 m.

Ecology: Mountain slopes, gravely areas, crevices of limestone cliffs, scree, grassy hillsides, stony moist meadows.

Flowering: May–July. **Fruiting:** June–August.

This species is recorded from Nepal based on *Polunin, Sykes & Williams 4565* (BM), a collection cited by Hara (Enum. Fl. Pl. Nepal 2: 42. 1979) as *Draba altaica* Bunge.

Draba altaica is easily distinguished from *D. lichiangensis* by the dominance of simple trichomes on leaf surfaces, whereas the latter has distinctly stellate trichomes with branched rays.

7. Draba glomerata Royle, Ill. Bot. Himal. Mts. [2]: 71 (1834).
Draba glomerata var. *dasycarpa* O.E.Schulz.

Perennial herbs, (1–)2–8(–10) cm, densely caespitose, often scapose. Caudex slender, often many-branched, ultimate branches terminated in rosettes, covered with leaf remains of previous years. Stems erect, simple, tomentose with subsessile stellate trichomes. Basal leaves rosulate; petiole absent, rarely to 4 mm, rarely ciliate; leaf blade obovate, oblong, or lanceolate, 2–8 × (0.7–)1–2 mm, base attenuate to cuneate, apex obtuse, margin entire, not ciliate, densely tomentose with sessile or subsessile, very fine, 4-rayed stellate trichomes the rays of which with 1 or 2 lateral branches on each side. Cauline leaves 0–3(–5), sessile; leaf blade ovate, oblong, or elliptic, 2–5 × 0.5–2 mm, base cuneate to obtuse, apex subobtuse, margin entire or rarely with 1 or 2 minute teeth on each side, not ciliate, pubescent as basal leaves. Racemes (3–)5–10-flowered, ebracteate or

rarely lowermost flower bracteate, basally slightly elongated, remainder often subumbellate in fruit; rachis straight. Sepals oblong, 1–2 × 0.6–0.8 mm, erect, sparsely tomentose below, base of lateral pair not saccate, margin narrowly membranous, caducous before fruit maturity. Petals white, spatulate, (1.8–)2–2.7(–3) × 0.8–1.2 mm, apex subemarginate, caducous. Filaments 1–1.5 mm; anthers ovate, ca. 0.2 mm. Ovules (6–)8–12 per ovary. Fruiting pedicels 1–3(–4) mm, ascending, straight, tomentose all around with subsessile, stellate trichomes. Fruit ovate, rarely oblong-ovate, (2.5–)3–4(–5) × 1.5–2.5 mm, not inflated, latiseptate, not twisted apex acute to acuminate; valves glabrous or rarely puberulent, obscurely veined, base obtuse, apex acute; style obsolete or 0.1–0.2(–0.4) mm. Seeds brown, ovate, 0.7–1 × 0.5–0.7 mm. Fig. 17f–i.

Distribution: Nepal, W Himalaya, Tibetan Plateau and E Asia.

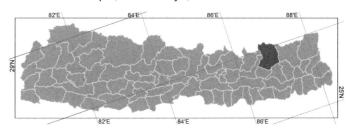

Altitudinal range: ca. 4400 m.

Ecology: Grassy areas, sandy river banks, gravelly slopes.

Flowering: June–August. **Fruiting:** July–October.

The record of *Draba glomerata* from Nepal by Hara (Enum. Fl. Pl. Nepal 2: 42. 1979) was based on misidentified specimens of *D. winterbottomii* (Hook.f. & Thomson) Pohle.

This is the first confirmed record of *D. glomerata* from Nepal, and it is based on *McCosh 317* (BM).

8. Draba stenobotrys Gilg & O.E.Schulz, in Engl., Pflanzenr. IV. 105(Heft 89): 291 (1927).
Draba ludlowiana Jafri; *D. oariocarpa* O.E.Schulz.

Perennial herbs, (4–)10–25 cm, caespitose. Caudex slender, few- to many-branched, ultimate branches terminated in rosettes, covered with petiolar remains of previous years. Stems often several from base, erect to ascending, simple, tomentose with subsessile stellate trichomes sometimes mixed with simple ones. Basal leaves rosulate; petiole to 7 mm; leaf blade oblanceolate to oblong, 6–15 × 1.5–3 mm, base attenuate to cuneate, apex obtuse, margin entire, often ciliate at least near base, tomentose on both surfaces with short-stalked to subsessile, 4-rayed stellate trichomes the rays of which having 1 or 2 lateral branches. Cauline leaves 2–10, sessile; leaf blade ovate to oblong, 3–10 × 1–4 mm, base cuneate to obtuse, apex subacute, margin entire, not ciliate at base, pubescent as basal leaves. Racemes 10–25-flowered, lowermost portion bracteate, elongated in fruit. Sepals oblong, 1.2–1.5 × 0.6–0.8 mm, erect, persistent, pilose below, base of lateral pair not saccate, margin membranous. Petals white, spatulate, 2–3 × 0.8–1.5 mm, persistent, apex

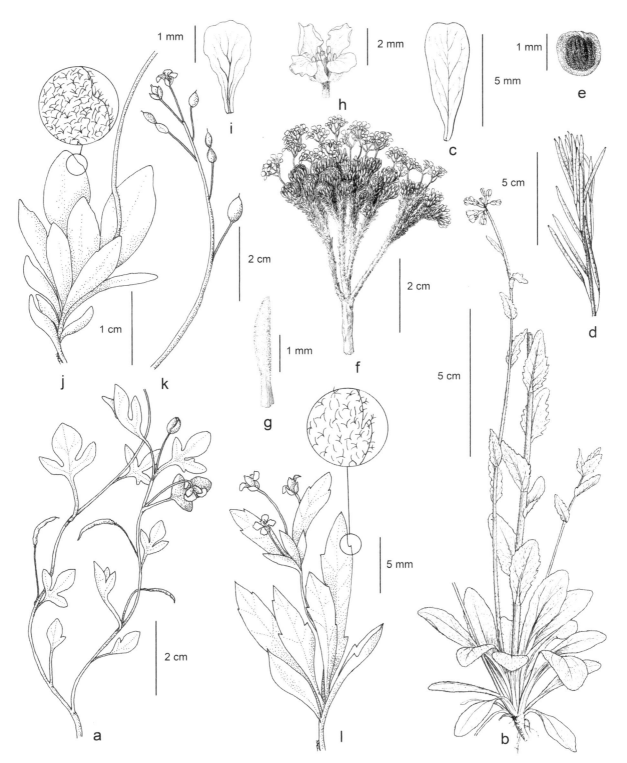

FIG. 17. CRUCIFERAE. **Aphragmus ohbana**: a, shoot with flowers and fruits. **Arabis pterosperma**: b, flowering plant; c, petal; d, fruits; e, seed. **Draba glomerata**: f, flowering plant; g, leaf; h, flower; i, petal. **Draba sikkimensis**: j, lower part of plant; k, inflorescence with fruits. **Draba macbeathiana**: l, inflorescence and leaves.

subemarginate or rounded; claw absent. Stamens persistent; filaments 1–1.5 mm, caducous; anthers ovate, ca. 0.2 mm. Ovules 8–14 per ovary. Fruiting pedicels 1–6 mm, suberect to ascending, straight, forming a straight line with fruit, tomentose all around. Fruit ovate or narrowly so, 3–5 × 1.5–2 mm, erect to ascending, sometimes appressed to rachis, latiseptate, not twisted, or rarely twisted ½ turn apex acute to acuminate; valves puberulent with simple and forked trichomes, or glabrous, not veined, base obtuse, apex acute; style obsolete to 0.5 mm. Seeds brown, ovate, 0.7–0.8 × 0.4–0.5 mm.

Distribution: Nepal and Tibetan Plateau.

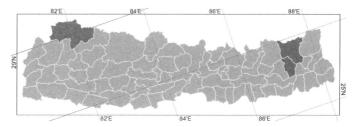

Altitudinal range: 4200–5100 m.

Ecology: Stony hillsides, open moraine scree, open turf and dwarf scrub, rock ledges.

Flowering: July. **Fruiting:** August–September.

9. Draba winterbottomii (Hook.f. & Thomson) Pohle, Repert. Spec. Nov. Regni Veg. Beih. 32: 138 (1925).
Draba tibetica var. *winterbottomii* Hook.f. & Thomson, J. Proc. Linn. Soc., Bot. 5: 152 (1861); *D. dasyastra* Gilg & O.E.Schulz; *Ptilotrichum wageri* Jafri.

Perennial herbs, (1–)2–7(–12) cm, densely or laxly caespitose, scapose. Caudex slender, often many-branched, ultimate branches terminated in rosettes, covered with straw-like petiolar remains of previous years. Stems erect, simple, tomentose with subsessile stellate trichomes. Basal leaves rosulate; petiole absent, rarely to 6 mm, rarely ciliate; leaf blade obovate, oblong, spatulate, or lanceolate, 2–10 × 0.5–2 mm, base attenuate to cuneate, apex obtuse, margin entire, not ciliate, densely tomentose with sessile or subsessile, very fine, 4-rayed stellate trichomes the rays of which having 1 or 2 lateral branches on each side. Cauline leaves absent. Racemes 5–12-flowered, ebracteate, lax and elongated in fruit; rachis often flexuous, slender. Sepals oblong, 2–3 × 1–1.5 mm, erect, sparsely tomentose below, base of lateral pair not saccate, margin narrowly membranous. Petals white, spatulate, 4–5 × 1.5–2.5 mm, apex subemarginate or rounded. Filaments 2–3 mm; anthers ovate, 0.4–0.5 mm. Ovules (10–)12–20 per ovary. Fruiting pedicels (3–)5–12(–15) mm, divaricate or rarely ascending, straight, tomentose all around with subsessile, stellate trichomes. Fruit narrowly elliptic, lanceolate, or oblong, (3–)5–9 × 1.5–2.3(–2.8) mm, not inflated, latiseptate, often twisted; valves glabrous or rarely puberulent, obscurely veined, base obtuse, apex acute; style (0.5–)0.7–1.3(–1.5) mm. Seeds brown, ovate, 0.9–1 × 0.7–0.8 mm.

Distribution: Nepal, W Himalaya and Tibetan Plateau.

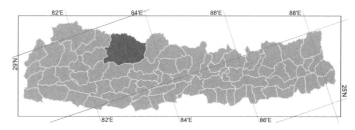

Altitudinal range: 5600–5900 m.

Ecology: Grassy slopes, gravelly areas, glacial terraces.

Flowering: June–August. **Fruiting:** July–September.

The collection *Polunin, Sykes & Williams 1186* (BM) represents the first record for Nepal. It was listed by Hara (1979) as *Draba glomerata* Royle.

10. Draba sikkimensis (Hook.f. & Thomson) Pohle, Repert. Spec. Nov. Regni Veg. Beih.: 144.
Draba sikkimensis forma *thoroldii* O.E.Schulz; *D. tibetica* var. *sikkimensis* Hook.f. & Thomson.

Perennial herbs, (4–)7–20(–26) cm, densely caespitose, scapose. Caudex several branched, ultimate branches terminated in rosettes and covered with leaves or petiolar remains of previous years. Stems erect, simple, densely tomentose towards base, sparsely so towards apex. Basal leaves rosulate, persistent; petiole 2–10 mm, persistent; leaf blade spatulate, oblanceolate, oblong, or rarely obovate, (0.3–)0.5–1.5(–2) cm × (1–)2–4(–6) mm, base attenuate to cuneate, apex obtuse to subacute, margin entire, densely subfloccose with fine, stalked, stellate trichomes with simple or branched slender rays, sometimes a few, slender, simple trichomes also present. Cauline leaves absent or 1–4, sessile; leaf blade ovate to oblong, 4–15 × 2–4 mm, apex obtuse, margin entire, indumentum as basal leaves. Racemes (2–)5–16(–20)-flowered, ebracteate, often elongated considerably in fruit; rachis straight. Sepals oblong, 2.5–3 × 1–1.5 mm, erect, sparsely pubescent below, base of lateral pair not saccate, margin narrowly membranous. Petals white, broadly obovate, 5–7 × 2.5–4.5 mm, apex rounded; claw ca. 1 mm. Filaments 2.5–3; anthers ovate, 0.5–0.6 mm. Ovules 8–14 per ovary. Fruiting pedicels (0.2–)0.7–1.7(–2) cm, erect and subappressed to rachis, or ascending, straight, tomentose all around. Fruit oblong to elliptic, 6–12 × 2.5–3 mm, not inflated, latiseptate, twisted to 2 turns, very rarely not twisted, apex obtuse; valves tomentose, not veined, base and apex obtuse; style (0.6–)1–1.5 mm. Seeds brown, ovate, 1–1.5 × 0.7–1 mm. Figs 16c, 17j–k.

Distribution: Nepal, E Himalaya and Tibetan Plateau.

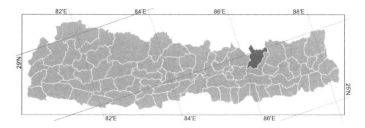

Altitudinal range: 4800–5500 m.

Ecology: Shady grassy slopes, stony slopes.

Flowering: June–July. **Fruiting:** August–September.

This is the first record of the species from Nepal, based on *Miyamoto et al. 9592300* (MO, TI), from Panch Pokhari, Dolakha District.

11. *Draba polyphylla* O.E.Schulz, in Engl., Pflanzenr. IV. 105(Heft 89): 180 (1927).

Perennial herbs, (20–)30–65(–90) cm. Caudex simple or branched. Stems erect, simple, sparsely or rarely densely pubescent with a mixture of simple and sessile stellate trichomes, apically tomentose with appressed stellate ones. Basal leaves rosulate, persistent, sessile or rarely on petiole to 4 cm; leaf blade oblanceolate to spatulate, 1.5–5 cm × 5–11 mm, base cuneate, apex obtuse, margin dentate or entire, sparsely pubescent below with stalked, 3- or 4-rayed stellate trichomes, sometimes glabrescent except for margin and midvein, with a mixture of primarily simple trichomes and fewer stellate ones above, rarely exclusively simple. Cauline leaves 5–20(–40), sessile; leaf blade oblong, lanceolate, or ovate, 1.2–3.5(–5) cm × 4–11 mm, base often auriculate or amplexicaul, apex obtuse or acute, margin dentate or subentire, pubescent as basal leaves. Racemes 20–60(–140)-flowered, bracteate basally to throughout, elongated considerably in fruit. Sepals oblong or ovate, 2–3(–3.5) × 1–2 mm, ascending, sparsely pilose below with primarily simple trichomes, base of lateral pair subsaccate, margin membranous. Petals yellow, obovate, 5–7(–8) × 2–3.5 mm, apex emarginate; claw 1–2 mm. Filaments 2.5–3.5(–4) mm; anthers oblong, 0.6–0.9 mm. Ovules (12–)16–22(–24) per ovary. Fruiting pedicels (0.3–)0.7–2(–2.5) cm, divaricate, straight, glabrous above, pubescent below with appressed stellate trichomes. Fruit oblong, elliptic, or ovate, (0.7–)1–1.5 cm × 3–5(–6) mm, latiseptate, rarely slightly twisted; valves glabrous, with a distinct midvein extending to middle or full length; style 0.5–1.5 mm. Seeds brown, ovate, 1.5–1.9 × 1–1.3 mm.

Distribution: Nepal, E Himalaya, Tibetan Plateau and E Asia.

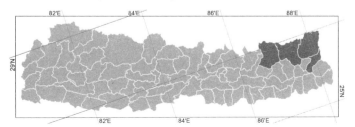

Altitudinal range: 2900–5000 m.

Ecology: Grassy slopes, scree, stream-sides, among shrubs, forests, peaty meadows.

Flowering: June–August. **Fruiting:** June–August.

Hara (Enum. Fl. Pl. Nepal 2: 42. 1979) listed *Draba polyphylla* in the synonymy of and cited *Stainton 658* (BM) as *D. elata* Hook.f. & Thomson. However, the latter species does not occur in Nepal and is known thus far only from Sikkim and adjacent Tibet. *Draba polyphylla* is easily distinguished by having at least some auriculate or amplexicaul cauline leaves, divaricate fruiting pedicels glabrous above, glabrous and slightly twisted fruits, petals 5–7(–8) mm long, and at least basally bracteate racemes 20–60(–140)-flowered. By contrast, *D. elata* has basally cuneate cauline leaves, ascending fruiting pedicels pubescent all around, pilose untwisted fruits, petals 3–4.5 mm long, and ebracteate racemes 10–25-flowered.

12. *Draba radicans* Royle, Ill. Bot. Himal. Mts. [2]: 71 (1834). *Draba radicans* var. *leiocarpa* O.E.Schulz.

Perennial herbs, 5–30 cm, surculose. Rhizome-like suckers to 35 cm, simple or branched, without leaf remains. Stems erect to ascending, simple, sparsely to densely hirsute with stout, forked and/or subsetose, simple trichomes. Basal leaves rosulate; petiole 1–10 mm; leaf blade obovate, oblanceolate, or lanceolate, 0.6–2 cm × 2–9 mm, base cuneate, apex acute, margin serrate or entire, ciliate with simple trichomes to 1 mm, sparsely to densely pubescent below with sessile or short-stalked, forked or 3- or 4-rayed stellate trichomes sometimes mixed with simple ones, hirsute above with primarily subsetose simple trichomes occasionally mixed with forked and stellate ones. Cauline leaves 4–12, sessile; leaf blade ovate to obovate, 0.7–3 × 0.3–1.4 cm, pubescent as basal leaves, margin serrate to serrulate, rarely entire. Racemes 10–25-flowered, ebracteate, elongated considerably in fruit. Sepals oblong to ovate, (2.5–)3–4 × 1.5–2 mm, erect, sparsely pilose below with simple trichomes, base of lateral pair subsaccate, margin narrowly membranous. Petals bright yellow, broadly obovate to oblong-obovate, 7–9 × 3–4.5 mm, apex subemarginate; claw absent. Filaments 4–5 mm; anthers oblong, 0.9–1.1 mm. Ovules 18–24 per ovary. Fruiting pedicels (0.6–)1–2.5 cm, slender, divaricate to horizontal, recurved or rarely straight, often subreflexed, pubescent all around, rarely subglabrous. Fruit narrowly oblong, 0.9–1.8 cm × 3–4 mm, latiseptate, not twisted; valves densely or sparsely pilose with simple or forked trichomes, base and apex subobtuse; style slender, 2–4 mm. Seeds brown, oblong, ca. 1.2 × 1 mm, winged all around; wing 0.1–0.2 mm wide.

Distribution: Nepal, W Himalaya and E Himalaya.

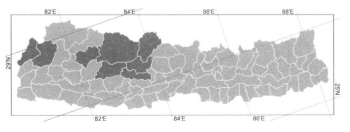

Cruciferae (Brassicaceae)

Altitudinal range: 3000–3900 m.

Ecology: Alpine meadows, forests, rocky ledges, open places, moist fields along streams.

Flowering: April–June. **Fruiting:** May–July.

Hooker & Anderson (Fl. Brit. Ind. 1: 142. 1872) partially misapplied the name *Draba alpina* L. to this species.

13. *Draba staintonii* Jafri ex H.Hara, in H.Hara & L.H.J.Williams, Enum. Fl. Pl. Nepal 2: 42 (1979).

Perennial herbs, 5–12 cm, laxly caespitose. Caudex with several long branches, ultimate ones terminated in rosettes and covered with remains of previous years. Stems erect to ascending, simple, densely pilose with subsessile stellate trichomes. Basal leaves rosulate, subsessile or with a petiole to 5 mm; leaf blade obovate to spatulate, 7–3 × 4–11 mm, base cuneate, apex obtuse to acute, margin entire, toothed, often ciliate with simple trichomes, thin, subglabrous above or distally pilose with primarily simple trichomes, pilose below with short-stalked to subsessile, 4-rayed stellate trichomes the rays of which simple or 1 or 2 of them branched at or near base and trichome appears up to 7-rayed. Cauline leaves 3–10, distal ones sessile, obovate to ovate or uppermost lanceolate, 7–17 × 3–10 mm, base subobtuse to cuneate, apex obtuse to acute, margin 2–5-toothed on each side, pubescent as basal leaves. Racemes 6–12-flowered, bracteate at least basally, soon elongated. Sepals oblong to ovate, 2.5–4 × 1.5–2.5 mm, ascending, sparsely pilose below with simple and fewer forked trichomes, base of lateral pair subsaccate, margin narrowly membranous. Petals white, broadly obovate to spatulate, 7–12 × 3–5 mm, apex rounded; claw absent. Filaments white, 3–4 mm; anthers oblong, 0.7–0.8 mm. Ovules 28–32 per ovary. Fruiting pedicels divaricate, 8–20 mm, initially straight but becoming recurved at maturity, densely stellate or pilose all around. Fruits oblong, 8–10 × 4–5 mm latiseptate, not twisted; valves pilose with simple and fewer forked trichomes, with a distinct midvein at least along lower half, base and apex obtuse to rounded; style 0.7–1.5 mm; stigma distinctly wider than style. Seeds brown, ovate, compressed, biseriate, 0.9–1 × 0.6–0.8 mm.

Distribution: Endemic to Nepal.

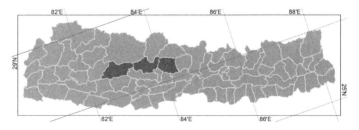

Altitudinal range: 3900–4600 m.

Ecology: Crevices in steep shale cliffs, rock ledges.

Flowering: June–July. **Fruiting:** August.

14. *Draba cholaensis* W.W.Sm., Rec. Bot. Surv. India 4: 352 (1913).
Draba cholaensis var. *leiocarpa* H.Hara.

Perennial herbs, 5–20 cm. Caudex branches slender, few to numerous, without petiolar remains of previous years. Stems erect to ascending, simple, slender, sparsely to densely pilose towards base with simple trichomes to 0.7 mm, these rarely mixed with fewer forked ones. Basal leaves subrosulate; petiole rarely to 1 cm, ciliate with simple trichomes; leaf blade narrowly obovate or oblanceolate, sometime oblong-obovate, 0.4–2 cm × 2–8 mm, base cuneate to attenuate, apex obtuse to acute, margin entire or minutely 1- or 2-toothed on each side, ciliate with simple trichomes to 1 mm, often sparsely pilose with simple trichomes to 0.6 mm, often also with stalked forked trichomes below rarely also some stellate ones. Cauline leaves 2–6, sessile, often well spaced; leaf blade oblong to subelliptic, 4–12 × 2–4 mm, with apex, leaf margin and indumentum similar to basal leaves. Racemes 2–5(–12)-flowered, ebracteate, elongated considerably and lax, subflexuous in fruit. Sepals oblong, 2.5–3 × 1.5–1.8 mm, erect, sparsely pilose below with simple trichomes, base of lateral pair subsaccate, margin narrowly membranous. Petals yellow, broadly obovate, 6–8 × 3–4 mm, apex subemarginate; claw to 1 mm. Filaments 3–4.5 mm; anthers oblong, 0.7–0.8 mm. Nectar glands confluent, subtending bases of all stamens though more developed laterally than medianly. Ovules 18–22 per ovary. Fruiting pedicels 0.6–2 cm, filiform, ascending to divaricate, slightly reflexed, straight or curved, glabrous. Fruit linear, 1.5–2.2 cm × 1.5–2.5 mm, latiseptate, not twisted; valves glabrous or pilose with simple trichomes, midvein distinct at least towards base, base and apex acute to subobtuse; style 1.5–2 mm. Seeds brown, oblong, 1–1.3 × 0.6–0.8 mm, wingless.

Distribution: Nepal, E Himalaya and Tibetan Plateau.

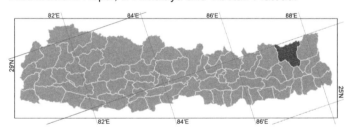

Altitudinal range: ca. 4000 m.

Ecology: Rocky ground.

Flowering: June–July. **Fruiting:** June–July.

In Nepal currently known only from the type specimen, *Stainton 639* (BM).

15. *Draba amoena* O.E.Schulz, in Engl., Pflanzenr. IV. 105(Heft 89): 188 (1927).

Perennial herbs, (20–)40–80 cm. Caudex stout, unbranched, covered with petiolar remains of previous years. Stems usually single, erect, branched above, stout, pubescent with sessile, stellate or submalpighiaceous trichomes occasionally mixed with fewer simple ones, not flexuous, leafy throughout. Basal leaves rosulate, sessile or narrowed into a petiole to 1 cm; leaf blade obovate to spatulate, 3–9 × 0.7–1.7 cm, base often attenuate, apex acute, margin dentate and often with callosites, pubescent with stalked to subsessile, cruciform trichomes often at least 1 ray of which branched and trichome appearing 5–8-rayed, rarely individual rays with 1 or 2 minute lateral branches, simple trichomes absent or restricted to ciliate base. Cauline leaves numerous, middle ones oblong to oblanceolate or lanceolate, rarely ovate, (1–)3–7 × 0.6–1.7 cm, gradually reduced upward to linear or narrowly oblong, entire bracts as small as 6 × 2 mm, base cuneate, not auriculate, apex acute to obtuse, margin toothed or entire on upper ones, pubescent as basal leaves, sometimes also ciliate at base or above with a preponderance of simple trichomes. Racemes 20–50-flowered, bracteate basally or rarely throughout, elongated considerably and lax in fruit; rachis not flexuous. Sepals oblong in median pair and ovate in lateral, 3–4.5 × 1.5–2 mm, sparsely pilose, ascending, lateral pair saccate. Petals purple to pink, obovate, 7–9 × 3–4.4 mm, apex obtuse to submarginate. Filaments not dilated at base, lateral pair 2.5–3.5 mm, median pairs 3.5–5 mm; anthers oblong, 0.7–1 mm. Nectar glands 2, lateral, well-developed, semicircular; median glands absent. Ovules 26–40 per ovary. Fruiting pedicels (0.6–)1–2.5 cm, filiform, ascending, straight, pubescent all around with sessile stellate trichomes sometimes mixed with forked ones. Fruit oblong or rarely linear, 1–2.5(–3.7) cm × 2–3.5 mm, latiseptate, twisted or not; valves puberulent or rarely glabrous, with a distinct midvein, base and apex acute to obtuse; style 1–2.5 mm, glabrous; stigma often 2-lobed. Seeds brown, ovate, compressed, 1.1–1.5 × 0.8–1 mm, wingless.

Distribution: Nepal and E Himalaya.

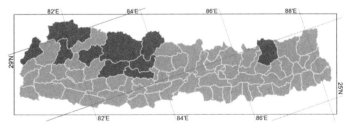

Altitudinal range: 3000–4800 m.

Ecology: Stream-sides, open hillsides, damp ground in partial shade, dry ravines.

Flowering: May–July. **Fruiting:** June–August.

16. *Draba oreades* Schrenk, Enum. Pl. Nov. 2: 56 (1842).

Draba algida var. *brachycarpa* Bunge; *D. alpicola* Klotzsch; *D. alpina* var. *rigida* Franch.; *D. kizylarti* (Korsh.) N.Busch; *D. oreades* prol. *alpicola* (Klotzsch) O.E.Schulz; *D. oreades* prol. *chinensis* O.E.Schulz; *D. oreades* prol. *exigua* O.E.Schulz; *D. oreades* prol. *pikei* O.E.Schulz; *D. oreades* var. *chinensis* O.E.Schulz; *D. oreades* var. *ciliolata* O.E.Schulz;

D. oreades var. *commutata* (Regel) O.E.Schulz; *D. oreades* var. *dasycarpa* O.E.Schulz; *D. oreades* var. *depauperata* O.E.Schulz; *D. oreades* var. *estylosa* O.E.Schulz; *D. oreades* var. *glabrescens* O.E.Schulz; *D. oreades* var. *occulata* O.E.Schulz; *D. oreades* var. *pulvinata* O.E.Schulz; *D. oreades* var. *racemosa* O.E.Schulz; *D. oreades* var. *tafelii* O.E.Schulz; *D. pilosa* var. *commutata* Regel; *D. pilosa* var. *oreades* Regel; *D. qinghaiensis* L.L.Lou; *D. rockii* O.E.Schulz; *D. tianschanica* Pohle; *Pseudobraya kizylarti* Korsh.

Perennial herbs, (0.5–)1.5–14(–20) cm, caespitose, scapose. Caudex simple to many-branched, ultimate branches terminated in rosettes and covered with petiolar remains of previous years. Stems erect, simple, often densely pubescent with a mixture of simple and subsessile forked trichomes, sometime subhirsute with almost exclusively simple trichomes, rarely glabrous. Basal leaves rosulate, persistent; petiole absent or short, rarely to 2 cm, persistent, often ciliate with simple and/or long-stalked forked trichomes; leaf blade suborbicular, obovate, spatulate, oblanceolate, or lanceolate, (0.3–)0.5–2(–3) cm × (1–)2–6(–8) mm, base cuneate to attenuate, apex obtuse to acute, margin entire or rarely 1- or 2-toothed on each side, sparsely or densely pubescent with simple trichomes, these often mixed below with stalked forked and subsessile, 3- or 4-rayed stellate ones with unbranched rays, above with predominantly simple trichomes, rarely both surfaces glabrous except for ciliate margin. Cauline leaves absent, sometimes 1, very rarely 2, sessile, similar to basal. Racemes (2–)4–15(–25)-flowered, ebracteate, subumbellate and not elongated or rarely subracemose and slightly elongated in fruit. Sepals oblong or ovate, 1.5–2.5(–3) × 0.8–1.5(–1.8) mm, erect, caducous or rarely persistent, sparsely pilose or glabrous below, base of lateral pair not saccate, margin narrowly membranous. Petals yellow, obovate to narrowly spatulate, 2.5–5(–6) × (0.9–)1.5–2.5(–3) mm, apex emarginate or rounded; claw absent, rarely 1(–2) mm. Filaments 1.5–2.5(–3) mm; anthers ovate, 0.2–0.4(–0.6) mm. Nectar glands lateral, 1 on each side of lateral stamen, sometimes these connected and glands appearing 2; median glands absent. Ovules (4–)6–12 per ovary. Fruiting pedicels 1–7(–10) mm, divaricate or divaricate-ascending, straight or rarely slightly curved, tomentose or pilose below, glabrous above or rarely throughout. Fruit ovate to suborbicular, rarely ovate-lanceolate, (3–)4–9(–12) × 1.5–4.5(–6) mm, latiseptate and basally inflated, not twisted; valves glabrous or rarely puberulent with simple or forked trichomes, not veined, base obtuse, apex acute to subacuminate; style (0.1–)0.3–0.8(–1) mm. Seeds black to dark brown, ovate, (0.7–)1–1.5 × 0.5–0.9(–1) mm.

Distribution: Nepal, W Himalaya, E Himalaya, Tibetan Plateau, E Asia, N Asia and C Asia.

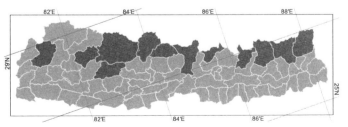

Cruciferae (Brassicaceae)

Altitudinal range: 2300–6100 m.

Ecology: Rock crevices, moraine, scree, alpine meadows and tundra, glacier margins, hillsides, grassy slopes, swampy meadows, muddy gravel, rocky outcrops, cliffs.

Flowering: June–August. **Fruiting:** June–August.

Hooker & Thomson (J. Proc. Linn. Soc. Bot. 5: 150. 1861), Hooker & Anderson (Fl. Brit. Ind. 1: 142. 1872) and Kitamura (Fauna Fl. E. Himalaya: 139. 1955) misapplied the name *Draba alpina* L. to this species.

17. *Draba affghanica* Boiss., Fl. Orient. Suppl.: 55 (1888). *Draba affghanica* var. *rostrata* O.E.Schulz; *D. affghanica* var. *subtomentosa* O.E.Schulz.

Perennial herbs, 1–5 cm, laxly caespitose, scapose. Caudex covered with leaf remains of previous years, with several prostrate branches terminated in rosettes. Stems erect, simple, pubescent 1–5 mm long, persistent, thickened, ciliate with simple trichomes. Basal leaf blade lanceolate, oblanceolate, to obovate, 5–10(–15) × 1–3(–4.5) mm, base attenuate, apex obtuse or acute, margin entire, ciliate, without setose or subsetose trichomes, pubescent with sessile or short-stalked, forked or stellate trichomes with unbranched rays, sometimes with primarily simple trichomes above. Cauline leaves absent. Racemes (3–)5–10(–13)-flowered, ebracteate, elongated in fruit. Sepals oblong-ovate, 2–3 × 1–1.5 mm, erect, sparsely pubescent below with simple and forked trichomes, base of lateral pair not saccate, margin narrowly membranous. Petals yellow, broadly obovate, 4–5.5 × 2–3.5 mm, apex rounded; claw to 1 mm long. Filaments 2–3; anthers ovate, ca. 0.4 mm long. Ovules 8–10 per ovary. Fruiting pedicels 5–10 mm, divaricate, straight, pubescent all around with stellate trichomes. Fruit ovoid, 3–6(–7) × 2–4 mm, distinctly inflated at least basally, latiseptate distally, not twisted; valves puberulent with forked trichomes mixed with much fewer simple and stellate ones, not veined, base rounded to obtuse, apex acute; style (0.5–)1–2 mm long. Seeds brown, ovoid, 1–1.3 × 0.6–0.8 mm.

Distribution: Nepal, W Himalaya and SW Asia.

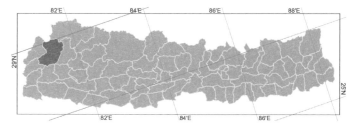

Altitudinal range: ca. 5800 m.

Ecology: Scree.

Flowering: May–July. **Fruiting:** May–July.

The single collection cited by Hara (Enum. Fl. Pl. Nepal 2: 42. 1979) from Nepal as '*Bowes Lyon 92*' is actually *Tyson 92* (BM).

18. *Draba humillima* O.E.Schulz, in Engl., Pflanzenr. IV. 105(Heft 89): 114 (1927).

Perennial herbs, 1–2 cm, densely caespitose, scapose. Caudex many-branched, ultimate branches terminated in rosettes and covered with leaf remains of previous years. Stems erect, simple, tomentose with stalked, forked and stellate trichomes. Basal leaves rosulate, persistent; petiole 2–7 mm, ciliate with simple and/or long-stalked forked trichomes; leaf blade oblong-elliptic, 3–8 × 1.5–3(–5) mm, base cuneate to attenuate, apex obtuse, margin entire, subhirsute with stalked, 3- or 4-rayed stellate trichomes with simple, rigid rays, sometimes glabrous above or sparsely pubescent distally, ciliate with simple and forked trichomes. Cauline leaves absent. Racemes 2–6-flowered, ebracteate, only slightly elongated in fruit. Sepals oblong, 2–2.5 × 0.7–1 mm, erect, sparsely pubescent below with simple trichomes, base of lateral pair not saccate, margin not membranous. Petals yellow, obovate, 3.5–4.5 × 1–1.5 mm, apex subemarginate; claw ca. 1 mm. Filaments 1.5–2 mm; anthers ovate-oblong, ca. 0.4 mm. Ovules 16–24 per ovary. Fruiting pedicels 2–5 mm, divaricate-ascending, straight, stout, tomentose all around. Fruit oblong, 5–7 × 3–4 mm, inflated, latiseptate, not twisted; valves glabrous, not veined, base and apex obtuse; style 0.5–0.6 mm. Seeds dark brown, ovate, 0.9–1 × 0.6–0.7 mm.

Distribution: Nepal, E Himalaya and Tibetan Plateau.

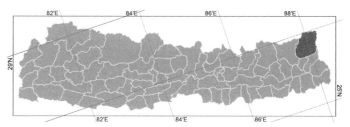

Altitudinal range: ca. 3000 m.

Ecology: Scree, sheltered shady areas below rocks, rocky slopes.

Flowering: July–September. **Fruiting:** July–September.

This is the first report of *Draba humillima* from Nepal, based on *Kanai et al. 1758* (TI), collected from Phujeng Danda, Taplejung District.

19. *Draba macbeathiana* Al-Shehbaz, Novon 12: 315 (2002).

Perennial herbs, ca. 6 cm, laxly branched. Caudex many-branched, ultimate branches terminated in few rosettes and covered with leaf remains of previous years. Stems erect, simple, tomentose with stalked, forked and stellate trichomes. Basal leaves rosulate, persistent; petiole to 10 mm, ciliate with simple and stalked forked trichomes; leaf blade oblanceolate, 5–15 × 2–5 mm, base cuneate to attenuate, apex acute, margin 1- or 2-toothed on each side, pilose with stalked, 3–4-rayed stellate trichomes with simple, soft rays. Cauline leaves few, similar to basal leaves. Racemes 3–6-flowered, ebracteate, elongated in fruit. Sepals oblong, 1.5–2 × ca. 1 mm, erect, sparsely pubescent below with stalked stellate trichomes, base of lateral pair not saccate, margin not membranous. Petals yellow, spatulate, ca. 3 × 1 mm, apex submarginate. Filaments 1.5–2 mm; anthers ovate, ca. 0.4 mm. Ovules 12–14 per ovary. Fruiting pedicels 1–2 cm, horizontal to slightly reflexed, straight, slender, tomentose all around, with a well-developed axillary gland above. Fruit oblong, 8–10 × 3–4 mm, not inflated, latiseptate, twisted; valves pubescent, not veined, base and apex obtuse; style 0.5–0.6 mm. Seeds brown, oblong, ca. 1 × 0.5 mm. Fig. 17l.

Distribution: Endemic to Nepal.

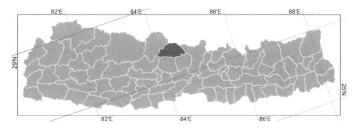

Altitudinal range: ca. 5200 m.

Ecology: Stable scree.

Flowering: July. **Fruiting:** July.

Known only from the holotype, *McBeath 1484* (E), which was collected from the Manang side of the Thorong La.

20. *Draba gracillima* Hook.f. & Thomson, J. Proc. Linn. Soc., Bot. 5: 153 (1861).
Draba granitica Hand.-Mazz.; *D. lanceolata* Klotzsch & Garcke later homonym, non Royle; *D. napalensis* Wall. nom. nud.; *D. wardii* W.W.Sm.

Perennial herbs, 5–55 cm. Caudex with several slender branches terminated in rosettes. Stems decumbent, simple, very slender, somewhat flexuous, sparsely pubescent at base with short-stalked, 3- or 4-rayed stellate trichomes, rarely densely pilose with primarily simple trichomes, glabrous towards apex. Basal leaves rosulate, persistent; leaf blade with petiole-like base to 7 mm, broadly obovate, spatulate, to oblanceolate, 0.5–2 cm × 1.5–7(–10) mm, base cuneate to attenuate, apex subacute or obtuse, margin denticulate or subentire, often ciliate, sparsely pubescent above with short-stalked, 4-rayed stellate trichomes, rarely mixed with simple and forked ones, pilose below with simple trichomes replaced with forked or stellate ones at leaf apex, rarely blade glabrous except for margin. Cauline leaves 2–6, sessile, ovate, 2–7 × 1–3 mm, base obtuse, not auriculate, apex acute, margin entire or denticulate, pubescent as basal leaves. Racemes 3–10(–12)-flowered, ebracteate or lowermost flowers bracteate, lax, elongated considerably in fruit. Sepals oblong, 1.5–2 × 0.6–1.5 mm, erect, glabrous or rarely with a few trichomes, not saccate, margin narrowly membranous. Petals pale yellow or yellowish white, narrowly obovate to oblanceolate, 2.5–3.5 × 0.8–1.5 mm, apex obtuse

or subemarginate; claw absent. Filaments 1.2–2 mm; anthers ovate, 0.2–0.4 mm. Ovules 10–20 per ovary. Fruiting pedicels 0.4–4 cm, filiform, glabrous, lowermost longest and strongly recurved, uppermost straight and ascending. Fruit narrowly oblong, linear, or linear-lanceolate, (0.5–)0.7–1.3(–1.8) cm × 1–2 mm, pendulous, latiseptate, not twisted; valves glabrous, not veined; style 0.2–0.5 mm. Seeds brown, ovate, 0.9–1.1 × 0.5–0.7 mm, subcompressed, wingless. Fl. and fr. May–August.

Distribution: Nepal, W Himalaya, E Himalaya, Tibetan Plateau and E Asia.

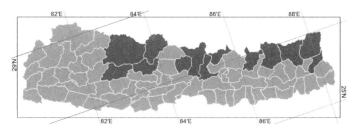

Altitudinal range: 3200–5000 m.

Ecology: Mountain slopes, grassy areas, ravines, stony slopes, scree, alpine grasslands.

Flowering: May–August. **Fruiting:** May–August.

21. *Draba stenocarpa* Hook.f. & Thomson, J. Proc. Linn. Soc., Bot. 5: 153 (1861).
Draba media Litv.; *D. media* var. *lasiocarpa* Lipsky; *D. media* var. *leiocarpa* Lipsky; *D. stenocarpa* var. *leiocarpa* (Lipsky) L.L.Lou; *D. stenocarpa* var. *media* (Litv.) O.E.Schulz; *D. stenocarpa* var. *media* subvar. *leiocarpa* O.E.Schulz.

Annual herbs, (5–)12–45(–50) cm. Stems solitary or few from base, erect, simple or few-branched near base, densely hirsute basally with subsetose simple trichomes to 1.5 mm, these rarely mixed with fewer, smaller, stalked forked trichomes, glabrous along distal half. Basal leaves rosulate, subsessile; leaf blade oblong-obovate to narrowly oblong, (0.5–)0.8–3.5(–4.2) cm × 2–8(–12) mm, base cuneate, apex subobtuse, margin entire or 1–3-toothed, ciliate with simple trichomes, with predominantly simple trichomes mixed with fewer, stalked forked ones above, densely subhirsute below with long-stalked, 3- or 4-rayed stellate trichomes with simple rays, sometimes with a few simple trichomes. Cauline leaves 2–4(–6), sessile, oblong-ovate to oblong-lanceolate, apex obtuse to subacute, margin entire or minutely 1–3-toothed, with indumentum as basal leaves. Racemes (4–)10–50(–60)-flowered, ebracteate, elongated considerably in fruit; rachis straight. Sepals ovate-oblong, 1.5–2 × ca. 1 mm, erect, hirsute below with simple trichomes, base of lateral pair not saccate, margin narrowly membranous. Petals yellow, spatulate, 3–4 × 1–1.5 mm, apex subemarginate. Filaments 1.5–2 mm; anthers ovate, 0.2–0.3 mm. Seeds and aborted ovules (30–)36–60(–72) per fruit. Fruiting pedicels (0.1–)0.5–1.8(–2.5) cm, divaricate, slightly curved upward or straight, slender, glabrous. Fruit linear, (0.6–)0.9–2(–2.5) cm × 1.5–2.5 mm, not inflated, latiseptate, not twisted, straight or slightly curved;

valves antrorsely puberulent with simple trichomes rarely mixed with forked ones, very rarely glabrous, with obscure midvein, base and apex subobtuse; style 0.1–0.2 mm. Seeds brown, ovate, 0.7–1 × 0.4–0.5 mm.

Distribution: Nepal, W Himalaya, Tibetan Plateau, E Asia, C Asia and SW Asia.

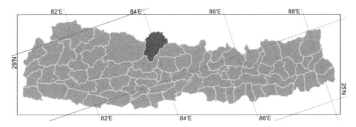

Altitudinal range: 2500–5000 m.

Ecology: Sandy areas on river sides, shady rocky areas, stony slopes, moist flats, forest margins.

Flowering: June–August. **Fruiting:** June–August.

This first record from Nepal is based on *Kriechbaum & Holtner s.n.* (WHB), from Pandag Valley, Lower Mustang.

22. ***Draba eriopoda*** Turcz., Bull. Soc. Imp. Naturalistes Moscou 15: 260 (1842).
Draba eriopoda Turcz. nom. inval.; *D. eriopoda* var. *kamensis* Pohle; *D. eriopoda* var. *sinensis* Maxim.; *D. pingwuensis* Z.M.Tan & S.C.Zhou.

Annual herbs, 4–45(–60) cm. Stems erect, simple or rarely branched above middle, sparsely to densely pubescent with a mixture of simple and subsessile stellate and forked trichomes, sometimes glabrous distally. Basal leaves subrosulate, often withered by time of flowering. Cauline leaves (5–)10–24(–30), sessile; leaf blade ovate, oblong, elliptic, or lanceolate, 0.5–2.7(–4.2) cm × 1.5–7(–15) mm, base cuneate to obtuse, apex acute, margin 1–6(–10)-toothed on each side, pubescent below with primarily 4-forked, short-stalked stellate trichomes, strigose above with simple trichomes mixed with fewer forked and smaller stellate ones. Racemes 10–45(–65)-flowered, ebracteate, elongated in fruit. Sepals oblong or ovate, 1–1.8(–2) × 0.6–0.8 mm, erect, pilose below with simple trichomes, base of lateral pair not saccate, margin membranous. Petals yellow, spatulate or narrowly obovate, 2–3(–3.5) × (0.5–)0.8–1 mm, apex emarginate; claw to 1 mm. Filaments 1–2 mm; anthers ovate, 0.1–0.2(–0.3) mm. Seeds and aborted ovules 10–24(–28) per fruit. Fruiting pedicels (0.2–)5–13(–22) mm, divaricate, straight, pubescent all around, glabrous above, or glabrous. Fruit ovate, (3–)4–9(–10) × 2–3(–4) mm, often erect, latiseptate, not twisted; valves glabrous or rarely sparsely puberulent, obscurely veined, base obtuse, apex acute; style obsolete, rarely to 0.2 mm. Seeds brown, ovate, 0.8–1.3 × 0.5–0.9 mm.

Distribution: Nepal, E Himalaya, Tibetan Plateau, E Asia and N Asia.

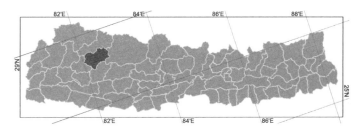

Altitudinal range: 3500–4000 m.

Ecology: Rocky slopes, grasslands, scrub, moist stream-sides, limestone cliffs, forests, river valleys.

Flowering: June–September. **Fruiting:** June–September.

23. ***Draba ellipsoidea*** Hook.f. & Thomson, J. Proc. Linn. Soc., Bot. 5: 153 (1861).

Annual herbs, (0.5–)2–12(–17) cm. Stems erect to ascending, simple, slender, flexuous, densely pubescent with short-stalked to subsessile stellate trichomes, usually glabrous distally. Basal leaves not rosulate, soon withered. Cauline leaves 3–12, obovate, elliptic-oblong, or lanceolate, (0.2–)0.5–2(–3) cm × (1.5–)3–8(–10) mm, cuneate to attenuate into a petiole-like base to 5 mm, apex obtuse or acute, margin entire or denticulate, pubescent on both surfaces with short-stalked, 4-rayed, stellate trichomes, sometimes with primarily simple trichomes mixed with fewer, forked or stellate ones above. Racemes (2–)4–10(–15)-flowered, ebracteate, lax and elongated in fruit. Sepals oblong, 0.9–1.4 × 0.4–5 mm, erect, pubescent below with simple trichomes sometimes mixed with fewer forked ones, base of lateral pair not saccate, margin not membranous. Petals white, narrowly spatulate, 0.6–1 × 0.1–0.2 mm, apex obtuse or subretuse; claw absent. Filaments 0.7–1 mm; anthers ovate, to 0.1 mm. Ovules 10–18(–22) per ovary. Fruiting pedicels (1–)3–10(–18) mm, ascending, straight, glabrous or pubescent all around with subsessile, 3- or 4-rayed stellate trichomes, slender. Fruit oblong, oblong-elliptic, or rarely suborbicular, (2–)4–8 × (1–)2–3(–4) mm, latiseptate, not twisted; valves puberulent with exclusively subsessile stellate trichomes, or with a mixture of stellate, forked, and simple trichomes, rarely only sparsely ciliate at margin, base and apex obtuse to rounded; style obsolete. Seeds blackish, ovate, 0.7–0.9 × 0.5–0.7 mm, slightly compressed, wingless.

Distribution: Nepal, W Himalaya, E Himalaya, Tibetan Plateau and E Asia.

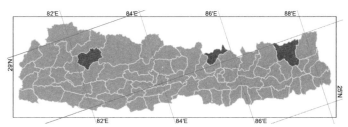

Altitudinal range: 3100–5400 m.

Ecology: Scree slopes, stream-sides, woods, pastures, alpine ledges, meadows, scrub.

Flowering: June–October. **Fruiting:** June–October.

5. *Brassica* L., Sp. Pl. 2: 666 (1753).

Annual, biennial, or perennial herbs, often glaucous. Trichomes absent or simple. Stems erect or ascending, simple or branched. Basal leaves petiolate, rosulate or not, simple, entire, dentate, lyrate-pinnatifid, or pinnatisect. Cauline leaves petiolate or sessile, base cuneate, attenuate, auriculate, sagittate, or amplexicaul, margin entire, dentate, or lobed. Racemes ebracteate, elongated in fruit. Fruiting pedicels straight, divaricate, or reflexed. Sepals ovate or oblong, erect, ascending, spreading, base of lateral pair saccate or not. Petals yellow, whitish-yellow or white, obovate, ovate to elliptic, apex rounded to emarginate; claw distinct, subequalling or longer than sepals. Stamens 6, tetradynamous; anthers oblong, obtuse at apex. Nectar glands 4, median and lateral, rarely 2 and lateral. Ovules 4–50 per ovary. Fruit dehiscent siliques, linear or rarely oblong, terete, 4-angled, or latiseptate, sessile or shortly stipitate, segmented; lower segment dehiscent, 4–46-seeded, longer than upper segment, smooth or torulose, valves with a prominent midvein and obscure lateral veins; distal segment seedless or 1(–3)-seeded; replum rounded; septum complete, translucent or opaque, veinless or with a distinct midvein; style obsolete or sterile terminal segment style-like; stigma capitate, entire or 2-lobed. Seeds uniseriate or rarely biseriate, wingless, globose or rarely oblong, plump or rarely slightly flattened; seed coat reticulate, mucilaginous or not when wetted; cotyledons conduplicate.

Worldwide 35 species, primarily in Mediterranean region, especially SW Europe and NW Africa. Four species in Nepal.

Key to Species

1a   Upper cauline leaves petiolate or subsessile, base not auriculate, amplexicaul, or cordate..2
   b   Upper cauline leaves sessile, base minutely auriculate to amplexicaul or deeply cordate 3

2a   Fruit strongly 4-angled, subappressed to rachis, (0.5–)1–2.5(–2.7) cm. Fruiting pedicels (2–)3–5(–6) mm..... **3. *B. nigra***
   b   Fruit terete, obscurely 4-angled, or slightly flattened, divaricate or ascending, (1.5–)2–4.5(–5) cm. Fruiting pedicels (5–)8–18(–20) mm..**4. *B. juncea***

3a   Plants glabrous throughout. Sepals erect. All filaments erect at base. Petals (1.5–)1.8–2.5(–3) cm.......... **1. *B. oleracea***
   b   Plants at least sparsely pubescent basally. Sepals ascending or rarely suberect. Filaments of lateral stamens curved at base. Petals (0.5–)0.7–1(–1.3) cm... **2. *B. rapa***

1. *Brassica oleracea* L., Sp. Pl. 2: 667 (1753).

Biennial or perennial, rarely annual herbs, (0.3–)0.6–1.5(–3) m, glabrous, glaucous. Stems erect to decumbent, branched at or above middle, sometimes fleshy at base. Basal and lowermost cauline leaves long petiolate, sometimes strongly overlapping and forming a head; petiole to 30 cm; leaf blade ovate, oblong, or lanceolate in outline, to 40 × 15 cm, margin entire, repand, or dentate, sometimes pinnatifid to pinnatisect and with a large terminal lobe and 1–13, smaller, oblong to ovate lateral lobes on each side of midvein. Upper cauline leaves sessile or subsessile in some cultivated forms, oblanceolate, ovate, to oblong, to 10 × 4 cm, base amplexicaul, auriculate, or rarely cuneate, margin entire, repand, or rarely dentate. Racemes (in cultivated forms) sometimes fleshy and condensed into a head. Sepals oblong, 0.8–1.5 cm × 1.5–2.7 mm, erect. Petals creamy yellow or rarely white, (1.5–)1.8–2.5(–3) × (0.6–)0.8–1.2 cm, ovate to elliptic, apex rounded; claw 0.7–1.5 cm. Filaments 0.8–1.2 cm, lateral pair erect at base; anthers oblong, 2.5–4 mm. Fruiting pedicels usually straight, ascending to divaricate, (0.8–)1.4–2.5(–4) cm. Fruit linear, (2.5–)4–8(–10) cm × (2.5–)3–4(–5) mm, terete, sessile or on a gynophore to 3 mm, divaricate to ascending; basal segment (2–)3–7.5(–9) cm, 20–40-seeded, valves with a prominent midvein; distal segment conical, (3–)4–10 mm, seedless or 1(or 2)-seeded; style obsolete. Seeds dark brown or blackish, globose, 1.5–2.5 mm in diameter, minutely reticulate.

Distribution: Cosmopolitan.

Ecology: Cultivated.

Flowering: March–June. **Fruiting:** April–July.

Native of Europe and cultivated throughout the world. Wild populations of the species (var. *oleracea*) are known only from the coastal cliffs of W Europe. Of the 15 varieties and 16 forms recognized by Helm (Kulturpflanze 11: 92–210. 1963), five varieties are cultivated in Nepal. Although these taxa are widely grown they are very rarely recorded nor do they persist outside cultivation, and so no distribution maps are given.

Cruciferae (Brassicaceae)

1a Young raceme fleshy, forming compact, globose to obconical head 2
 b Young raceme not fleshy, never compact, open, not forming heads 3

2a Flower buds white, densely and tightly compact. Rachis and pedicels white .... var. ***botrytis***
 b Flower buds green, somewhat loosely grouped. Rachis and pedicels green..................... var. ***italica***

3a Stem base fleshy, globose............ var. ***gongylodes***
 b Stem base not fleshy, cylindric or narrowly conical 4

4a Basal and lower cauline leaves numerous, densely or somewhat loosely grouped in heads. Axillary buds not forming small heads .... var. ***capitata***
 b Basal and lower cauline leaves few to several, widely spaced, not forming heads. Axillary buds forming small, globose or obovoid heads .... var. ***gemmifera***

Brassica oleracea var. ***botrytis*** L., Sp. Pl. 2: 667 (1753).

फूलगोभी Phulgobhi (Nepali).

Stem base elongated, not fleshy, cylindric. Basal and lower cauline leaves few to several, widely spaced, not grouped into a head. Axillary buds not forming heads. Raceme white, compact, often globose, with fleshy peduncle, rachis, pedicels and flowers.
 Widely cultivated as a vegetable. Cauliflower.

Brassica oleracea var. ***capitata*** L., Sp. Pl. 2: 667 (1753).
Brassica capitata (L.) H.Lév.

बन्दागोभी Bandagobhi (Nepali).

Stem base highly shortened, not fleshy, conical. Basal and lower cauline leaves, numerous, strongly overlapping into a compact, globose, oblong, or rarely subconical, closed, apically rounded or flattened head. Axillary buds not forming heads. Raceme neither fleshy nor condensed into a head.
 Widely cultivated as a vegetable. Cabbage.

Brassica oleracea var. ***gemmifera*** (DC.) Zenker, Fl. Thüringen.

Brassica oleracea var. *bullata* subvar. *gemmifera* DC., Syst. Nat. 2: 585 (1821); *Brassica gemmifera* (DC.) H.Lév.; *B. oleracea* subsp. *gemmifera* (DC.) Schwarz.

Stem base strongly elongated, not fleshy, cylindric. Basal and lower cauline leaves few to several, widely spaced, not overlapping into a head. Axillary buds forming small, subglobose to obovoid heads. Raceme neither fleshy nor condensed into a head.
 Cultivated occasionally as a vegetable. Brussels sprouts.

Brassica oleracea var. ***gongylodes*** L., Sp. Pl. 2: 667 (1753).

Brassica caulorapa (DC.) Pasq.; *B. oleracea* var. *caulorapa* DC.

ग्याँठगोभी Gyanthgobhi (Nepali).

Stem base highly shortened, fleshy, globose. Basal and lower cauline leaves numerous, strongly overlapping into a head. Axillary buds not forming heads. Raceme neither fleshy nor condensed into a head.
 Cultivated occasionally as a vegetable. Kohlrabi.

Brassica oleracea var. ***italica*** Plenck, Icon. Pl. Med. 6 (1794).

ब्रोकाउली Brokauli (Nepali).

Stem base elongated, not fleshy, cylindric. Basal and lower cauline leaves few to several, widely spaced, not grouped into a head. Axillary buds not forming heads. Raceme green, somewhat loose, usually obconical, with fleshy peduncle, rachis, pedicels and flowers.
 Cultivated occasionally as a vegetable. Broccoli.

2. Brassica rapa L., Sp. Pl. 1: 666 (1753).
Brassica campestris subsp. *rapa* (L.) Hook.f. & T.Anderson; *B. campestris* subsp. *rapifera* (Metzg.) Sinskaya; *B. campestris* var. *rapa* (L.) Hartm.; *B. rapa* subsp. *rapifera* Metzg.; *Raphanus rapa* (L.) Crantz.

Annual or biennial herbs, 30–120(–190) cm, glabrous or sparsely pubescent basally, rarely glaucous, sometimes with fleshy taproots. Stems erect, simple or branched above. Basal and lowermost cauline leaves petiolate, not rosulate or obscurely to strongly rosulate and forming a compact, oblong head; petiole (1–)2–10(–17) cm, slender or thickened and fleshy, sometimes strongly winged; leaf blade ovate, oblong, or lanceolate in outline, (5–)10–40(–60) × 3–10(–20) cm, margin entire, repand, dentate, or sinuate, sometimes pinnatifid to pinnatisect and with a large terminal lobe and 1–6, smaller, oblong to ovate lateral lobes on each side of midvein. Upper cauline leaves sessile, ovate, oblong, or lanceolate, 2–8(–12) × 0.8–3 cm, base amplexicaul, deeply cordate, or auriculate, margin entire or repand. Sepals oblong, (3–)4–6.5(–8) × 1.5–2 mm, ascending. Petals bright yellow, rarely pale or whitish yellow, 7–10(–13) × (2.5–)3–6(–7) mm, obovate, apex rounded. Filaments 4–6(–7) mm, lateral pair curved at base; anthers oblong, 1.5–2 mm. Fruiting pedicels straight, ascending to divaricate, (0.5–)1–2.5(–3) cm. Fruit linear, (2–)3–8(–11) cm × 2–4(–5) mm, terete, sessile, divaricate to ascending; basal segment (1.3–)2–5(–7.5) cm, 16–305-seeded, valves with a prominent midvein; distal segment conical, (0.3–)1–2.5(–3.5) cm, seedless or rarely 1-seeded; style obsolete. Seeds dark to reddish brown, globose, 1–1.8 mm in diameter, minutely reticulate.

Distribution: Asia, Europe and Africa.

Ecology: Cultivated, sometimes escaping.

Flowering: March–May. **Fruiting:** May–July.

Widely cultivated in Asia, Europe and elsewhere, sometimes escaping and semi-naturalized, but not persisting and so no distribution maps are given. Forms with three of four valves have been recognized as *Brassica trilocularis* Roxb. and *B. quadrivalvis* Hook.f. & Thomson, respectively. They were treated by Jafri (Fl. W. Pakistan 55: 24. 1973) as subspecies of *B. napus* L., but both have 2*n* = 20, and therefore they should be reduced to synonymy of or, at most, treated as a variety of *B. rapa*. Of the six varieties generally recognized in *B. rapa*, the following two occur in Nepal.

    1a Taproot fleshy, napiform; plants biennial.... var. ***rapa***
    b Taproot not fleshy, cylindric; plants annual or rarely biennial.. var. ***oleifera***

Brassica rapa var. *oleifera* DC., Syst. Nat. 2: 591 (1821).

Brassica asperifolia Lam.; *B. asperifolia* var. *sylvestris* Lam.; *B. campestris* L.; *B. campestris* subsp. *nipposinica* (L.H.Bailey) G.Olsson; *B. campestris* subsp. *oleifera* Schübl. & Mart.; *B. campestris* var. *dichotoma* (Roxb.) G.Watt; *B. campestris* var. *oleifera* DC.; *B. campestris* var. *sarson* Prain; *B. chinensis* var. *angustifolia* B.S.Sun; *B. chinensis* var. *utulis* M.Tsen & S.H.Lee; *B. dichotoma* Roxb. ex Fleming nom. nud.; *B. dubiosa* L.H.Bailey; *B. japonica* Makino; *B. napus* var. *dichotoma* (Roxb.) Prain; *B. napus* var. *quadrivalvis* (Hook.f. & Thomson) O.E.Schulz; *B. napus* var. *trilocularis* (Roxb.) O.E.Schulz; *B. perviridis* (L.H.Bailey) L.H.Bailey; *B. quadrivalvis* Hook.f. & Thomson; *B. rapa* subsp. *campestris* (L.) A.R.Clapham; *B. rapa* subsp. *nipposinica* (L.H.Bailey) Hanelt; *B. rapa* subsp. *oleifera* (DC.) Metzg.; *B. rapa* subsp. *sylvestris* (Lam.) Janch.; *B. rapa* var. *campestris* (L.) Peterm.; *B. rapa* var. *chinoleifera* Kitam.; *B. rapa* var. *dichotoma* Kitam.; *B. rapa* var. *perviridis* L.H.Bailey; *B. rapa* var. *quadrivalvis* (Hook.f. & Thomson) Kitam.; *B. rapa* var. *trilocularis* (Roxb.) Kitam.; *B. trilocularis* (Roxb.) Hook.f. & Thomson; *Sinapis dichotoma* Roxb.; *S. trilocularis* Roxb.; *S. trilocularis* Roxb. nom. nud.

तोरी Tori (Nepali).

Annual or rarely biennial herbs. Taproot not fleshy, cylindric. Basal leaves not rosulate or obscurely so; petiole slender; leaf blade subentire, sinuately lobed, pinnatifid, or incised with irregularly serrate lobes.
Fig. 18a–c.

Forms of subsp. *oleifera* with three or four-valved fruits have been recognized as var. *trilocularis* and var. *quadrivalvis*,

respectively. However, these forms are only anomalies that do not merit recognition. Although var. *oleifera* is often found as an escape from cultivation it does not truly naturalize, so a distribution map is not included here.

Widely cultivated in Asia as a source of seed oil, but also grown as a vegetable. Rapeseed.

Brassica rapa L. var. *rapa*

गान्टेमूला Gantemula (Nepali).

Biennial herbs. Taproot expanded, fleshy, napiform. Basal leaves rosulate; petiole slender; leaf blade lyrately pinnatifid or rarely sinuate-dentate.
    Cultivated for its fleshy roots. Turnip.

3. Brassica nigra (L.) W.D.J.Koch, Deutschl. Fl., ed. 3, 4: 713 (1833).
Sinapis nigra L., Sp. Pl. 2: 668 (1753); *Sisymbrium nigrum* (L.) Prantl.

कालो तोरी Kalo tori (Nepali).

Annual herbs, 0.3–2(–3.1) m, sparsely hirsute at least basally. Stems erect, branched above. Basal and lowermost cauline leaves long petiolate; petiole to 10 cm; leaf blade ovate, oblong, or lanceolate in outline, 6–30 × 1–10 cm, lyrate-pinnatifid to pinnatisect; terminal lobe ovate, dentate; lateral lobes 1–3 on each side of midvein, much smaller than terminal lobe, dentate. Upper cauline leaves petiolate, lanceolate to linear-oblong, to 5 × 1.5 cm, cuneate, margin entire or rarely dentate. Sepals oblong, 4–6(–7) cm × 1–1.5 mm, spreading to ascending. Petals yellow, (6–)7.5–11(–13) × (2.5–)3–4.5(–5.5) mm, ovate, apex rounded; claw 3–6 mm. Filaments 3.5–5 mm; anthers oblong, 1–1.5 mm. Fruiting pedicels straight, slender, erect to ascending, subappressed to rachis, (2–)3–5(–6) mm. Fruit linear or narrowly oblong-elliptic, (0.5–)1–2.5(–2.7) cm × (1.5–)2–3(–4) mm, 4-angled, sessile, subappressed to rachis; basal segment (0.4–)0.8–2(–2.5) cm, 4–10(–16)-seeded, slightly torulose, valves with a prominent midvein; distal segment style-like, sometimes narrowly conical, (1–)2–5(–6) mm, seedless. Seeds dark brown, grey, or blackish, globose, 1.2–2 mm in diameter, minutely reticulate.

Distribution: Asia, Europe and Africa.

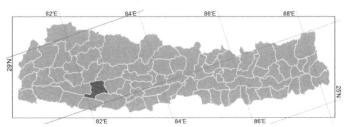

Altitudinal range: 900–2800 m.

Ecology: Slopes, steppes, field margins.

Cruciferae (Brassicaceae)

Flowering: April–July. **Fruiting:** April–July.

Black mustard. Cultivated and frequently naturalized, but under recorded in Nepal.

Tender leaves and shoots are cooked as vegetable or fermented to make gundruk.

4. _Brassica juncea_ (L.) Czern. & Coss., in Czern., Consp. Pl. Charc.: 8 (1859).

Sinapis juncea L., Sp. Pl. 2: 668 (1753); _Brassica argyri_ H.Lév.; _B. integrifolia_ (A.Stokes) Rupr. nom. nud.; _B. integrifolia_ (H.West) O.E.Schulz; _B. japonica_ (Thunb.) Siebold; _B. juncea_ subsp. _integrifolia_ (H.West) Thell.; _B. juncea_ subsp. _rugosa_ (Roxb.) Prain; _B. juncea_ var. _crispifolia_ L.H.Bailey; _B. juncea_ var. _cuneifolia_ (Roxb.) Kitam.; _B. juncea_ var. _foliosa_ L.H.Bailey; _B. juncea_ var. _gracilis_ M.Tsen & S.H.Lee; _B. juncea_ var. _integrifolia_ (A.Stokes) Sinskaya; _B. juncea_ var. _japonica_ (Thunb.) L.H.Bailey; _B. juncea_ var. _longidens_ L.H.Bailey; _B. juncea_ var. _longipes_ M.Tsen & S.H.Lee; _B. juncea_ var. _multiceps_ M.Tsen & S.H.Lee; _B. juncea_ var. _multisecta_ L.H.Bailey; _B. juncea_ var. _rugosa_ (Roxb.) Kitam.; _B. juncea_ var. _strumata_ M.Tsen & S.H.Lee; _B. juncea_ var. _subintegrifolia_ Sinskaya; _B. lanceolata_ (DC.) A.E.Lange; _B. napiformis_ var. _multisecta_ (L.H.Bailey) A.I.Baranov; _B. rugosa_ (Roxb.) L.H.Bailey; _B. rugosa_ var. _cuneifolia_ Prain; _B. taquetii_ H.Lév.; _B. wildenowii_ Boiss.; _Raphanus junceus_ (L.) Crantz; _Sinapis cernua_ Poir.; _S. chinensis_ var. _integrifolia_ A.Stokes; _S. cuneifolia_ Roxb.; _S. cuneifolia_ Roxb. nom. nud.; _S. integrifolia_ H.West; _S. japonica_ Thunb.; _S. lanceolata_ DC.; _S. patens_ Roxb.; _S. ramosa_ Roxb.; _S. rugosa_ Roxb. nom. nud.; _S. rugosa_ Roxb.

रायो Rayo (Nepali).

Annual herbs, (20–)30–100(–180) cm, pubescent or rarely glabrous, glaucous or not, sometimes with fleshy taproots. Stems erect, branched above. Basal and lowermost cauline leaves long petiolate; petiole (1–)2–8(–15) cm; leaf blade ovate, oblong, or lanceolate in outline, (4–)6–30(–80) × 1.5–15(–28) cm, lyrate-pinnatifid to pinnatisect; terminal lobe ovate, repand, dentate, or incised; lateral lobes 1–3 on each side of midvein, much smaller than terminal lobe, crisped incised, dentate, repand, or entire. Upper cauline leaves petiolate or subsessile, oblanceolate, oblong, lanceolate, or linear, to 10 × 5 cm, base cuneate to attenuate, margin entire or repand, rarely dentate. Petals yellow, (6.5–)8–11(–13) × 5–7.5 mm, ovate to obovate, apex rounded to emarginate; claw 3–6 mm. Filaments 4–7 mm; anthers oblong, 1.5–2 mm. Fruiting pedicels straight, divaricate, (0.5–)0.8–1.5(–2) cm. Sepals oblong, (3.5–)4–6(–7) × 1–1.7 mm, spreading. Fruit linear, (2–)3–5(–6) cm × 3–4(–5) mm, terete or slightly 4-angled, sessile, divaricate to ascending; basal segment (1.5–)2–4.5 cm, 12–30(–40)-seeded, slightly torulose, valves with a prominent midvein; distal segment conical, (4–)5–10(–15) mm, seedless; style obsolete. Seeds dark to light brown or grey, globose, 1–1.7 mm in diameter, minutely reticulate.

Distribution: Asia and Europe.

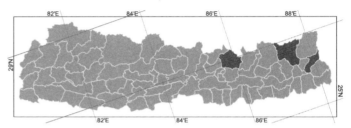

Altitudinal range: 1000–3500 m.

Ecology: Fields, waste places, roadsides. Cultivated throughout Nepal, sometimes an escape from cultivation.

Flowering: March–June. **Fruiting:** April–July.

Schulz (Pflanzen. IV. 105(Heft 70): 59. 1919) partially misapplied the name _Brassica cernua_ (Thunb.) F.B.Forbes & Hemsl. to this species.

Cultivated and widely naturalized, but under recorded in Nepal.

Mustard greens, cooked as a vegetable.

6. _Diplotaxis_ DC., Mem. Mus. Hist. Nat. 7: 243 (1821).

Perennial herbs. Trichomes absent or simple. Stems erect or ascending, rarely procumbent. Basal leaves petiolate, rosulate or not, dentate, lyrate, pinnatifid, or pinnatisect, rarely bipinnatipartite. Cauline leaves petiolate or sessile and sometimes auriculate. Racemes ebracteate or rarely lowermost flowers bracteate, elongated in fruit. Sepals oblong or linear, erect or spreading, base of lateral pair often not saccate. Petals yellow, obovate, apex obtuse or emarginate; clawed. Stamens 6, tetradynamous; filaments not dilated at base; anthers oblong, obtuse at apex. Nectar glands 4; median glands large; lateral glands smaller. Ovules 60–120 per ovary. Fruiting pedicels ascending, divaricate, or reflexed. Fruit dehiscent siliques, linear, latiseptate or terete, glabrous, sessile or long stipitate, unsegmented or segmented; proximal segment well developed, dehiscent, numerous seeded, much longer than distal segment, torulose, with a distinct midvein; distal segment indehiscent, seedless or 1- or 2-seeded; replum rounded; septum complete; style obsolete or to 2 mm; stigma capitate, entire or 2-lobed. Seeds biseriate, wingless, oblong, ovoid, or ellipsoid, slightly flattened; seed coat smooth or minutely reticulate, slightly mucilaginous or not mucilaginous when wetted; cotyledons conduplicate.

Worldwide about 25 species in NW Africa, Macaronesia, and Europe, extending into C Asia. One species in Nepal.

1. *Diplotaxis harra* (Forssk.) Boiss., Fl. Orient. 1: 388 (1867).
Sinapis harra Forssk., Fl. Aegypt.-Arab.: 118 (1775); *Diplotaxis nepalensis* H.Hara.

Herbs, (10–)20–60(–80) cm, glabrous or hirsute basally with simple trichomes. Basal leaves rosulate; petiole 0.5–2.5(–4) cm; leaf blade obovate, spatulate, oblanceolate, or ovate, (1–)2–10(–13) × (0.5–)1–3(–4) cm, base cuneate to attenuate, apex acute or obtuse, margin dentate, sinuate, or lyrate, rarely subentire. Cauline leaves short petiolate to subsessile, not auriculate, smaller than basal ones. Fruiting pedicels slender, divaricate or reflexed, arcuate or straight, (0.6–)0.8–1.5(–2.3) cm. Sepals narrowly oblong, 4–7 × 1.5–2 mm, pubescent or rarely glabrous. Petals yellow, obovate, (7–)8–13 × 3–5 mm, cuneate to a short claw. Filament 4–6 mm; anthers oblong, 1.5–2.5 mm. Ovules 66–120 per ovary. Fruit (2–)2.5–4(–4.5) cm × 2–3 mm, usually deflexed, compressed, slightly torulose; gynophore 1–4 mm; style stout, beaklike, obsolete or to 2 mm; stigma 2-lobed. Seeds ovoid or ellipsoid, 0.6–09 × 0.4–0.5 mm. Fig. 16d.

Distribution: Nepal, W Himalaya, SW Asia, Europe and Africa.

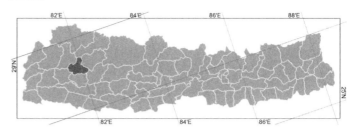

Altitudinal range: 1000–1200 m.

Ecology: Cliffs, sandy areas.

Flowering: April–August. **Fruiting:** April–August.

7. *Raphanus* L., Sp. Pl. 2: 669 (1753).

Annual or biennial herbs, glabrous, scabrous or hispid. Trichomes simple. Stems erect or prostrate, simple or branched. Basal leaves petiolate, rosulate or not, simple, dentate, lyrate, pinnatifid, or pinnatisect. Cauline leaves similar to basal, petiolate or uppermost subsessile. Racemes several-flowered, ebracteate, elongated in fruit. Sepals oblong or linear, erect, base of lateral pair saccate. Petals, white, pink, or purple, usually with darker veins, obovate or suborbicular, apex rounded or emarginate; claw subequalling or longer than sepals. Stamens 6, strongly tetradynamous; filaments not dilated at base; anthers oblong or oblong-linear, obtuse at apex. Nectar glands 4; median pair oblong; lateral pair prismatic. Ovules 2–22 per ovary. Fruiting pedicels divaricate to ascending. Fruit indehiscent siliques or silicles, often lomentaceous, breaking into 1-seeded units, linear, oblong, ovoid, ellipsoid, or lanceolate, terete or polygonal, sessile, segmented; proximal segment rudimentary, seedless, as thick as or slightly thicker than pedicel, or absent; distal segment few- to many seeded, wingless, corky, smooth to slightly torulose, not ribbed, antrorsely or retrorsely scabrous, or glabrous; replum absent; septum absent; style prominent; stigma capitate, entire or slightly 2-lobed. Seeds uniseriate, wingless, globose to ovoid; seed coat minutely reticulate, not mucilaginous when wetted; cotyledons conduplicate.

Worldwide three species native to the Mediterranean region. One species in Nepal.

1. *Raphanus sativus* L., Sp. Pl. 2: 669 (1753).
Raphanus acanthiformis J.M.Morel; *R. chinensis* Mill.; *R. macropodus* H.Lév.; *R. niger* Mill.; *R. raphanistroides* (Makino) Nakai; *R. raphinastrum* var. *sativus* (L.) Domin; *R. sativus* forma *raphanistroides* Makino; *R. sativus* var. *hortensis* Backer; *R. sativus* var. *macropodus* (H.Lév.) Makino; *R. sativus* var. *raphanistroides* (Makino) Makino.

मूला Mula (Nepali).

Herbs, 10–130 cm, Roots fleshy, white, pink, red, or black, linear, fusiform, oblong, or globose, 1–100 × 0.5–45 cm. Stems simple or branched. Basal leaves with petioles 1–30 cm; leaf blade oblong, obovate, oblanceolate, or spatulate in outline, lyrate to pinnatisect, sometimes undivided, 2–60 × 1–20 cm, apex obtuse to acute, margin dentate; lateral lobes 1–12 on each side of midvein, sometimes absent, oblong to ovate, to 10 × 5 cm. Uppermost cauline leaves subsessile, often undivided, dentate. Sepals narrowly oblong, 6–10 × 1–2 mm, glabrous or sparsely pubescent. Petals purple, pink, or sometimes white, often with darker veins, broadly obovate, 1.2–2.2 cm × 3–8 mm, apex obtuse to emarginate; claw to 1.4 cm. Filaments slender, 5–12 mm; anthers 1.5–2 mm, sagittate at base. Fruiting pedicels, straight, 0.5–4 cm. Fruits fusiform to lanceolate, sometimes ovoid or cylindric; seedless proximal segment 1–3.5 cm; seed-bearing distal segment (1–)3–15(–25) × (0.5–)0.7–1.3(–1.5) cm, corky, rounded at base, conical at apex, smooth or rarely slightly constricted between seeds, not ribbed; style 1–4 cm; stigma entire. Seeds, 2.5–4 mm in diameter. Fig. 16e.

Distribution: Cosmopolitan.

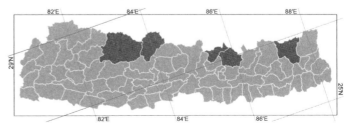

Cruciferae (Brassicaceae)

Altitudinal range: 300–3100 m.

Ecology: Fields, roadsides, waste areas.

Flowering: All year. **Fruiting:** All year.

A very variable species with regards to size, colour, and shape of fleshy roots, plant height, size and degree of division of leaves, flower colour, and fruit shape and size. Numerous infraspecific taxa have been recognized, and their taxonomy is controversial and highly confused. The interested reader should consult Pistrick (Kulturpflanze 35: 225–321. 1987).

Cultivated worldwide, primarily as a vegetable and often naturalized. Radish.

8. *Arabidopsis* (DC.) Heynh. nom cons., in Holl & Heynh., Fl. Sachsen 1: 538 (1842).

Cardaminopsis (C.A.Mey.) Hayek; *Hylandra* Á.Löve.

Annual herbs with stolons or woody caudex. Trichomes simple, mixed with stalked, 1–3-forked ones. Stems erect, often several from base, usually glabrous above. Basal leaves rosulate, entire, toothed, or pinnately lobed. Cauline leaves subsessile, entire, dentate, or rarely lyrate. Racemes few- to several-flowered, ebracteate. Sepals oblong, erect or ascending, glabrous or pubescent, base of lateral pair not saccate. Petals white, spatulate, apex obtuse or emarginate; claw distinct or not. Stamens 6, erect, slightly tetradynamous; filaments not dilated basally; anthers oblong, obtuse. Nectar glands confluent, subtending bases of filaments. Ovules 15–80 per ovary. Fruiting pedicels ascending or divaricate or slightly reflexed. Fruit dehiscent siliques, linear, terete, shortly stipitate or subsessile; valves papery, midvein distinct, glabrous, smooth or somewhat torulose; replum rounded; septum complete; style obsolete or to 1 mm; stigma capitate, entire. Seeds numerous, uniseriate, wingless or margined, ellipsoid, plump; seed coat minutely reticulate, mucilaginous or not when wetted; cotyledons incumbent.

Worldwide nine species in Eurasia and N America. One species in Nepal.

1. *Arabidopsis thaliana* (L.) Heynh., Fl. Sachsen 1: 538 (1842).
Arabis thaliana L., Sp. Pl. 2: 665 (1753); *Sisymbrium thalianum* (L.) J.Gay & Monnard.

Herbs, (2–)5–30(–50) cm. Stems erect, 1 or few from base, simple or branched above, basally with predominantly simple trichomes, apically glabrous. Basal leaves short petiolate; leaf blade obovate, spatulate, ovate, or elliptic, 0.8–3.5(–4.5) cm × (1–)2–10(–15) mm, apex obtuse, margin entire to repand or dentate, with predominantly simple and stalked 1-forked trichomes above. Cauline leaves subsessile, usually few; blade lanceolate, linear, oblong, or elliptic, (0.4–)0.6–1.8(–2.5) cm × 1–6(–10) mm, entire or rarely few-toothed. Sepals 1–2(–2.5) mm, glabrous or distally sparsely with simple hairs, inner pair not saccate. Petals white, spatulate, 2–3.5 × 0.5–1.5 mm, base attenuate to a short claw. Filaments white, 1.5–2 mm. Fruiting pedicels slender, divaricate, straight, 3–10(–15) mm. Fruit linear, terete, smooth, (0.8–)1–1.5(–1.8) cm × 0.5–0.8 mm; valves with a distinct midvein; style to 0.5 mm. Seeds ellipsoid, plump, light brown, 0.3–0.5 mm. Fig. 18d–f.

Distribution: Asia, Europe, Africa and N America.

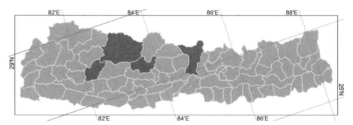

Altitudinal range: 300–4200 m.

Ecology: Plains, mountain slopes, river banks, roadsides.

Flowering: January–June(–October). **Fruiting:** January–June(–October).

The most widely studied flowering plant and considered to be the model organism for studies in genetics, development, physiology, biochemistry, and related fields. A naturalized weed throughout much of the world and under recorded in Nepal.

9. *Capsella* Medik., Pfl.-Gatt.: 85 (1792).

Annual or biennial herbs. Trichomes sessile and stellate, sometimes mixed with simple or forked ones. Stems erect. Basal leaves rosulate, simple, usually pinnately lobed, lyrate, or runcinate, rarely entire or toothed. Cauline leaves sessile, saggitate to amplexicaul or rarely auriculate, entire, dentate, or sinuate. Racemes many-flowered, ebracteate. Sepals oblong, erect or ascending, glabrous or pubescent, base of lateral pair not saccate. Petals white, rarely pinkish or yellowish, much longer or shorter than sepals, obovate, apex obtuse; claw distinct from blade. Stamens 6, erect, tetradynamous; filaments not dilated at base; anthers ovate, obtuse at apex. Median glands absent; lateral glands 1 on each side of lateral stamen. Ovules (12–)20–40 per ovary. Fruiting pedicels slender, divaricate. Fruit dehiscent silicles, obdeltoid to obdeltoid-obcordate, strongly flattened and angustiseptate, sessile;

valves papery, prominently veined, strongly keeled; replum rounded; septum complete; style less than 1 mm, included or exserted from apical notch; stigma capitate, entire. Seeds uniseriate, wingless, oblong, plump; seed coat reticulate, mucilaginous when wetted; cotyledons incumbent.

Worldwide one species, a cosmopolitan weed.

1. *Capsella bursa-pastoris* (L.) Medik., Pfl.-Gatt.: 85 (1792). *Thlaspi bursa-pastoris* L., Sp. Pl. 2: 647 (1753).

चाल्ने Chalne (Nepali).

Herbs (2–)10–50(–70) cm, sparsely to densely pubescent with sessile 3–5-rayed stellate hairs often mixed near base of plant with much longer simple trichomes. Stems erect, simple or branched. Basal leaves rosulate; petiole 0.5–4(–6) cm; leaf blade oblong to oblanceolate, (0.5–)1.5–10(–15) × 0.2–2.5(–5) cm, base cuneate to attenuate, apex acute to acuminate, margin pinnatisect, pinnatifid, runcinate, lyrate, dentate, repand, or entire. Cauline leaves sessile, narrowly oblong to lanceolate or linear, 1–5.5(–8) cm × 1–15(–20) mm, margin entire to dentate. Sepals green or reddish, 1.5–2 × 0.7–1 mm, margin membranous. Petals white, rarely pinkish or yellowish, obovate, (1.5–)2–4(–5) × 1–1.5 mm. Filaments white, 1–2 mm; anthers ovate, to 0.5 mm. Fruiting pedicels (3–)5–15(–20) mm, divaricate, usually straight, slender, glabrous. Fruit (3–)4–9(–10) × (2–)3–7(–9) mm, flat, base cuneate, apex emarginate to truncate; valves with subparallel lateral veins, glabrous; style 0.2–0.7 mm. Seeds brown, 0.9–1.1 × 0.4–0.6 mm. Fig. 16f.

Distribution: Cosmopolitan.

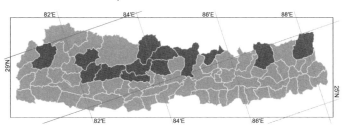

Altitudinal range: 100–4900 m.

Ecology: Roadsides, gardens, fields, waste areas, mountain slopes.

Flowering: April–July. **Fruiting:** April–July.

Cosmopolitan weed native of Europe and adjacent SW Asia. Under recorded in Nepal.

10. *Crucihimalaya* Al-Shehbaz, O'Kane & R.A.Price, Novon 9: 298 (1999).

Annual or biennial herbs, rarely perennial with a caudex. Trichomes simple and stalked, 1- or 2-forked, sometimes stellate. Stems erect or decumbent to ascending. Basal leaves rosulate or not, entire or dentate, rarely pinnately lobed. Cauline leaves sessile or subsessile, auriculate or not, sagittate or attenuate, entire, dentate, or rarely pinnately lobed, rarely absent. Racemes several- to many-flowered, ebracteate or bracteate. Sepals oblong, erect, pubescent, base of lateral pair not saccate. Petals white, lavender or purple, rarely pink, longer than sepals, spatulate to narrowly oblanceolate, rounded; claw obscurely distinct from blade. Stamens 6, erect, slightly tetradynamous; filaments not dilated at base; anthers ovate or oblong, obtuse at apex. Nectar glands confluent and subtending bases of filament. Ovules (30–)40–160 per ovary. Fruiting pedicels erect, ascending or divaricate. Fruit dehiscent siliques, linear, terete or somewhat 4-angled, rarely latiseptate, sessile or subsessile; valves with a distinct midvein, glabrous or densely pubescent, smooth or torulose; replum rounded; septum complete; style to 1 mm; stigma capitate, entire. Seeds uniseriate to sub-biseriate, wingless, oblong to subovoid, plump; seed coat minutely reticulate, mucilaginous when wetted; cotyledons incumbent.

Worldwide nine species in SW Asia, C Asia, E Asia, N Asia and the Himalayas. Five species in Nepal.

Key to Species

1a Fruit valves densely and coarsely stellate. Pedicels pubescent all around. Fruits often subappressed to rachis .... **1. *C. lasiocarpa***

b Fruit valves glabrous or very rarely puberulent. Pedicels glabrous above, rarely (*C. wallichii*) pubescent all around. Fruits not appressed to rachis 2

2a Cauline leaves distinctly auriculate or rarely saggitate at base .... **3. *C. himalaica***

b Cauline leaves petiolate or sessile, neither auriculate nor amplexicaul at base (minutely auriculate in *C. wallichii*), sometimes absent 3

3a Lowermost flowers of main raceme ebracteate. Basal leaves lyrate to pinnatifid, often canescent, persistent at time of flowering or fruiting .... **5. C. wallichii**

b Lowermost flowers of main raceme bracteate. Basal leaves entire to dentate, not canescent, withering by time of flowering or fruiting 4

4a Cauline leaves with stellate stalked trichomes above, linear-lanceolate. Plants (18–)30–85(–120) cm. Only lowermost flowers of main raceme bracteate .... **2. C. stricta**

b Cauline leaves with simple and forked trichomes above, ovate to elliptic or oblong, rarely oblanceolate. Plants 4–15(–20) cm. Main raceme bracteate nearly throughout .... **4. C. axillaris**

1. *Crucihimalaya lasiocarpa* (Hook.f. & Thomson)
Al-Shehbaz, O'Kane & R.A.Price, Novon 9: 300 (1999). *Sisymbrium lasiocarpum* Hook.f. & Thomson, J. Proc. Linn. Soc., Bot. 5: 162 (1861); *Arabidopsis lasiocarpa* (Hook.f. & Thomson) O.E.Schulz; *A. lasiocarpa* var. *micrantha* W.T.Wang; *A. monachorum* (W.W.Sm.) O.E.Schulz; *Guillenia duthiei* (O.E.Schulz) Bennet; *Hesperis lasiocarpa* (Hook.f. & Thomson) Kuntze; *Microsisymbrium duthiei* O.E.Schulz; *Sisymbrium bhutanicum* N.P.Balakr.; *S. monachorum* W.W.Sm.

Annual or biennial herbs, (10–)25–70(–120) cm. Stems erect, simple or branched at base, densely hirsute with horizontal simple trichomes to 1.3 mm, these sometimes sparser on raceme, mixed throughout with much smaller, short-stalked stellate trichomes. Basal leaves not rosulate, petiolate, often caducous by time of fruiting; petiole (0.3–)1–3 cm; blade spatulate, (0.3–)1–5(–7) cm × (2–)6–11(–20) mm, apex obtuse, margin dentate or rarely lyrate lobed, stellate pubescent and often mixed with simple or forked trichomes. Middle cauline leaves obovate to oblong or elliptic, sessile and broad at base, (0.6–)1–2(–2.5) cm × (3–)6–11 mm, stellate pubescent on both surfaces, dentate to repand or entire, reduced in size upward. Raceme bracteate throughout or only lower flowers bracteate; bracts similar to uppermost cauline leaves. Pedicel slender, stellate pubescent all around, suberect to rarely divaricate, (0.5–)1–4(–6) mm. Sepals oblong, 1.5–2(–2.5) × 0.8–1 mm, densely pubescent, sometimes narrowly membranous. Petals white to lavender, spatulate to narrowly oblanceolate, 2–3(–4) × 0.6–0.9 mm, attenuate to base; claw to 1 mm. Filaments 1.5–2.5 mm; anthers oblong-ovate, 0.3–0.4 mm. Ovules 70–160 per ovary. Fruit linear, terete, straight, often appressed to rachis, 1.5–3(–4.2) cm × 0.7–1(–1.3) mm; valves densely stellate pubescent, rounded at apex and base, with an inconspicuous midvein; style 0.3–0.6(–1) mm. Seeds light brown, oblong to subovoid, uniseriate to sub-biseriate, 0.5–0.6 × 0.3–0.4 mm.
Fig. 16g.

Distribution: Nepal, W Himalaya, E Himalaya, Tibetan Plateau and E Asia.

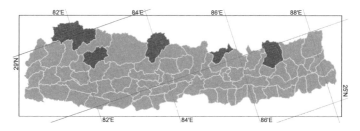

Altitudinal range: 900–4600 m.

Ecology: Fields, forest margins, grassy slopes, river banks, roadsides.

Flowering: April–August. **Fruiting:** May–September.

2. *Crucihimalaya stricta* (Cambess.) Al-Shehbaz, O'Kane & R.A.Price, Novon 9: 300 (1999).
Malcolmia stricta Cambess., in Jacquem., Voy. Inde 4(1): 16 (1841); *Arabidopsis himalaica* var. *kunawurensis* O.E.Schulz; *A. stricta* (Cambess.) N.Busch; *A. stricta* var. *bracteata* O.E.Schulz; *Hesperis stricta* (Cambess.) Kuntze; *Sisymbrium strictum* (Cambess.) Hook.f. & Thomson.

Annual or biennial herbs, (10–)22–85(–125) cm. Stems erect, simple or sometimes branched at base, densely pubescent with short-stalked stellate trichomes, often basally hirsute with horizontal simple trichomes to 1.4 mm. Basal leaves not rosulate, petiolate, often caducous by time of fruiting; petiole to 3 cm; blade spatulate, 4–9 × 1–2 cm, apex obtuse, margin coarsely dentate or rarely lyrate lobed, stellate pubescent and sometimes mixed with simple or forked trichomes. Middle cauline leaves narrowly oblong to lanceolate-linear or narrowly oblanceolate, sessile, reduced in size upward, (1–)2–5(–8) cm × (2–)3.5–10(–17) mm, base attenuate, margin dentate to repand or entire, stellate pubescent on both surfaces. Raceme basally or rarely lowermost 1 or 2 flowers bracteate; bracts similar to uppermost cauline leaves. Pedicel slender, stellate pubescent laterally and below, glabrous above, divaricate or rarely ascending, (0.5–)1–4(–6) mm. Sepals oblong, 1.7–2.5 × 0.6–0.8 mm, densely pubescent, sometimes narrowly membranous. Petals white to lavender or purple, spatulate, 2.5–3(–4) × 0.7–1 mm, attenuate to base; claw to 1 mm. Filaments 2–2.8 mm; anthers oblong, 0.4–0.6 mm. Ovules 60–120 per ovary. Fruit linear, terete, straight or curved, divaricate or rarely ascending, (1–)2–4(–5.5) cm × 0.6–1 mm; valves glabrous or very rarely puberulent, rounded at apex and base, with a conspicuous midvein; style (0.2–)0.6–1.3(–1.8) mm. Seeds brown, oblong to subovoid, uniseriate, 0.6–0.9 × 0.3–0.4 mm.

Distribution: Nepal, W Himalaya and Tibetan Plateau.

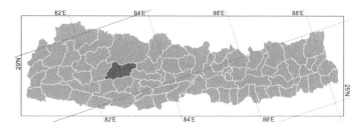

Altitudinal range: 1600–4200 m.

Ecology: Forest margins, grassy areas.

Flowering: April–August. **Fruiting:** May–September.

Schulz (Notizbl. Bot. Gart. Berlin-Dahlem 9: 1061. 1927) cited one collection (*Duthie 5352*) from Budhi village in western Nepal. *Stainton, Sykes & Williams 3365* (BM, E, G), collected from Jagat and misidentified as *Arabidopsis mollissima* (C.A.Mey.) O.E.Schulz is also *Crucihimalaya stricta*.

3. *Crucihimalaya himalaica* (Edgew.) Al-Shehbaz, O'Kane & R.A.Price, Novon 9: 301 (1999).

Arabis himalaica Edgew., Trans. Linn. Soc. London 20: 31 (1846); *Arabidopsis brevicaulis* (Jafri) Jafri; *A. himalaica* (Edgew.) O.E.Schulz; *A. himalaica* var. *harrissii* O.E.Schulz; *A. himalaica* var. *integrifolia* O.E.Schulz; *A. himalaica* var. *rupestris* (Edgew.) O.E.Schulz; *Arabis brevicaulis* Jafri; *A. rupestre* Edgew.; *Hesperis himalaica* (Edgew.) Kuntze; *Sisymbrium himalaicum* (Edgew.) Hook.f. & Thomson; *S. rupestris* (Edgew.) Hook.f. & Thomson.

Annual or biennial herbs, very rarely perennial, (3–)10–50(–70) cm. Stems erect, simple or few- to many-branched at base, densely pubescent with coarse, stalked, stellate and forked trichomes, often basally hirsute with horizontal simple or forked trichomes to 1.8 mm, rarely glabrescent above. Basal leaves rosulate, petiolate, often withering by time of fruiting; petiole (0.3–)0.7–1.5(–2) cm, often ciliate; blade spatulate, oblanceolate, ovate, or oblong, (0.4–)1–3 (–4) cm × 2–10(–14) mm, apex obtuse, margin coarsely dentate or rarely subentire, densely pubescent with coarse, stalked, stellate and forked trichomes. Middle cauline leaves oblong, rarely ovate or lanceolate, sessile, 0.5–2.5(–3.2) cm × 2–7(–11) mm, base auriculate or rarely sagittate, apex acute, margin coarsely dentate or rarely entire, pubescent as basal leaves, rarely glabrescent. Raceme bracteate throughout or only lowermost few flowers bracteate. Pedicel slender, stellate pubescent laterally and below, glabrous above, divaricate, (1–)2–7(–11) mm. Sepals often lavender, oblong, 1.5–2.5(–3) × 0.7–1 mm, densely pubescent, sometimes narrowly membranous. Petals purple to lavender or rarely white, spatulate, 2–3.5(–5) × (0.6–)1–1.5(–2) mm, attenuate to base; claw to 1 mm. Filaments 1.5–3 mm; anthers oblong, 0.3–0.5 mm. Ovules 50–110 per ovary. Fruit linear, terete, straight or rarely slightly curved, erect to divaricate-ascending, (0.8–)1.5–3.5(–4.5) cm × (0.4–)0.5–0.8(–1) mm; valves glabrous or rarely puberulent, rounded at apex and base, with an obscure or rarely prominent midvein; style (0.1–)0.4–0.6 mm. Seeds brown, oblong, uniseriate, 0.5–0.8 × 0.2–0.4 mm.

Distribution: Nepal, W Himalaya, E Himalaya, Tibetan Plateau, E Asia and SW Asia.

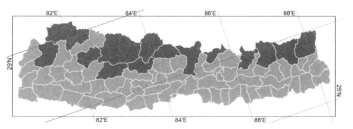

Altitudinal range: 1100–4600 m.

Ecology: Rocky hillsides, grassy meadows, sandy slopes, flood plains, scree, pastures.

Flowering: (March) April–September. **Fruiting:** June–October.

4. *Crucihimalaya axillaris* (Hook.f. & Thomson) Al-Shehbaz, O'Kane & R.A.Price, Novon 9: 301 (1999).

Sisymbrium axillare Hook.f. & Thomson, J. Proc. Linn. Soc., Bot. 5: 162 (1999); *Guillenia axillare* (Hook.f. & Thomson) Bennet; *Microsisymbrium axillare* (Hook.f. & Thomson) O.E.Schulz; *M. axillare* var. *brevipedicellatum* Jafri; *M. axillare* var. *dasycarpum* O.E.Schulz; *M. bracteosum* Jafri.

Annual or biennial herbs, (3–)10–20 cm. Stems decumbent to ascending, few- to many-branched at base, densely hispid with coarse, horizontal simple trichomes to 2 mm, these often mixed with coarse forked and much smaller stalked stellate trichomes, rarely glabrescent above or with only stellate trichomes. Basal leaves rosulate, petiolate, often withering by time of fruiting; petiole (0.2–)1–2 cm, often ciliate; blade spatulate, (0.6–)1–3.5 (–5) × (0.2–)0.5–1.5 cm, apex obtuse, margin coarsely dentate to sinuate or lyrate-pinnatifid, densely pubescent with stalked forked trichomes mixed with larger forked or simple ones below, with coarse simple and long-stalked forked trichomes above. Middle cauline leaves broadly oblong, elliptic, or obovate, sessile, 1–2.5 × (0.2–)0.7–1.3 cm, base cuneate, not auriculate, apex acute, margin dentate, pubescent as basal leaves, rarely glabrescent. Raceme bracteate throughout. Pedicel slender, stellate pubescent laterally and below, glabrous above or rarely all around, divaricate to ascending, (2.5–)4–10(–12) mm. Sepals narrowly oblong, 2–3 × 0.6–1 mm, coarsely pubescent. Petals white to lavender, spatulate, 3–4.5 × (0.8–)1–1.2 mm, attenuate to base; claw to 1 mm. Filaments 2–3 mm; anthers oblong, 0.4–0.6 mm. Ovules 60–90 per ovary. Fruit linear, terete, curved upward or sometimes straight, to divaricate, (1.4–)2.5–4(–5) cm × 0.6–0.8 mm; valves glabrous or rarely puberulent, rounded at apex and base, with an obscure midvein; style (0.1–)0.4–1 mm. Seeds brown, oblong, uniseriate, 0.8–1 × 0.3–0.4 mm.

Distribution: Nepal, W Himalaya, E Himalaya and Tibetan Plateau.

Cruciferae (Brassicaceae)

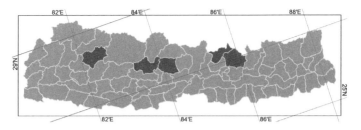

Altitudinal range: 900–3100 m.

Ecology: Rocky hillsides, shady banks, open woodlands, roadsides, rock crevices, terraces, forests.

Flowering: March–June. **Fruiting:** May–July.

The species was listed in Hara (Enum. Fl. Pl. Nepal 2: 44. 1979) in *Microsisymbrium* O. E. Schulz, an illegitimate generic name that included the type of the genus *Guillenia* Greene.

5. *Crucihimalaya wallichii* (Hook.f. & Thomson) Al-Shehbaz, O'Kane & R.A.Price, Novon 9: 301 (1999).
Sisymbrium wallichii Hook.f. & Thomson, J. Proc. Linn. Soc., Bot. 5: 158 (1861); *Arabidopsis campestre* O.E.Schulz; *A. mollissima* var. *afghanica* O.E.Schulz; *A. russelliana* Jafri; *A. taraxacifolia* (T.Anderson) Jafri; *A. wallichii* (Hook.f. & Thomson) N.Busch; *A. wallichii* var. *viridis* O.E.Schulz; *Arabis bucharica* Nevski; *A. taraxacifolia* T.Anderson; *Hesperis wallichii* (Hook.f. & Thomson) Kuntze; *Microsisymbrium angustifolium* Jafri.

Annual or biennial herbs, (5–)12–50(–80) cm. Stems erect, simple or few- to many-branched at base, densely to sparsely tomentose with finely branched stellate trichomes, often basally hirsute with horizontal simple or forked trichomes (0.5–)0.8–1.5(–2) mm, rarely glabrescent above. Basal leaves rosulate, petiolate, often lyrate to pinnatifid or pinnatisect, persistent or withering by time of fruiting; petiole 0.5–1.5(–2.5) cm, often ciliate; blade spatulate to oblanceolate or obovate in outline, 1–6 (–12) × 3–1.7(–3) cm, margin coarsely dentate to rarely entire, with up to 8 lobes on each side of midvein; lateral lobes oblong to ovate, increasing in size above, much smaller than obovate to suborbicular terminal lobe, to 1.5 × 0.8 cm, entire to dentate, densely tomentose with finely branched stellate trichomes. Middle cauline leaves oblong to oblanceolate or linear, sessile, 0.7–4(–7) cm × 1–10(–20) mm, base usually minutely auriculate, apex acute, margin entire to coarsely dentate or lyrate, pubescent as basal leaves, rarely glabrescent. Raceme ebracteate, lax. Pedicel slender, straight, stellate pubescent all around to glabrous above, divaricate or rarely ascending, (0.2–)0.5–1.6(–2.5) cm. Sepals green to lavender, oblong, 1.5–2.5(–3) × 0.7–1 mm, densely pubescent to glabrescent, sometimes narrowly membranous. Petals purple to lavender or white, spatulate, (2–)2.5–3.5(–4.5) × 0.4–0.6 mm, attenuate to base. Filaments 2.5–3(3.5) mm; anthers oblong, 0.3–0.6 mm. Ovules 70–150 per ovary. Fruit linear, terete or rarely subcompressed, arcuate to straight, divaricate to recurved or rarely ascending, (2.5–)3.5–7.5(–9.5) cm × 0.6–1 mm; valves glabrous, obtuse to acute at apex and base, with an obscure or prominent midvein; style (0.2–)0.5–1(–1.5) mm. Seeds brown, oblong, uniseriate, 0.6–1 × 0.3–0.5 mm.
Fig. 18g–i.

Distribution: Nepal, W Himalaya, E Himalaya, Tibetan Plateau, C Asia and SW Asia.

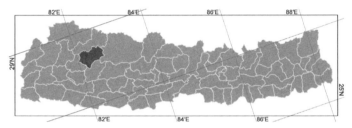

Altitudinal range: 2300–4400 m.

Ecology: Rocky slopes, hillsides, limestone crevices and ledges.

Flowering: April–August. **Fruiting:** May–September.

11. *Turritis* L., Sp. Pl. 2: 666 (1753).

Biennial, rarely short-lived perennial herbs, glaucous above. Trichomes simple and/or forked or substellate. Stems erect, simple or branched apically. Basal leaves petiolate, rosulate, simple, repand, dentate, or lobed, rarely entire. Cauline leaves sessile, auriculate, sagittate, or amplexicaul at base, entire. Racemes ebracteate, corymbose, elongated considerably in fruit. Sepals oblong, erect, base of lateral pair not saccate, margin membranous. Petals yellowish, creamy white, rarely pink; blade spatulate, oblanceolate, or rarely linear, apex obtuse; claw undifferentiated from blade. Stamens 6, erect, tetradynamous; filaments not dilated at base; anthers oblong or linear, obtuse at apex. Nectar glands confluent and subtending bases of all stamens, median glands present. Ovules 130–200 per ovary. Fruiting pedicels slender, erect. Fruit dehiscent siliques, linear, often subterete-quadrangular, sessile; valves leathery, with a prominent midvein, glabrous, smooth; replum rounded; septum complete; style short, stout; stigma capitate, subentire. Seeds biseriate, wingless, oblong to suborbicular, flattened; seed coat not mucilaginous when wetted; cotyledons accumbent.

Worldwide two species in N Africa, Eurasia and N America. One species in Nepal.

144

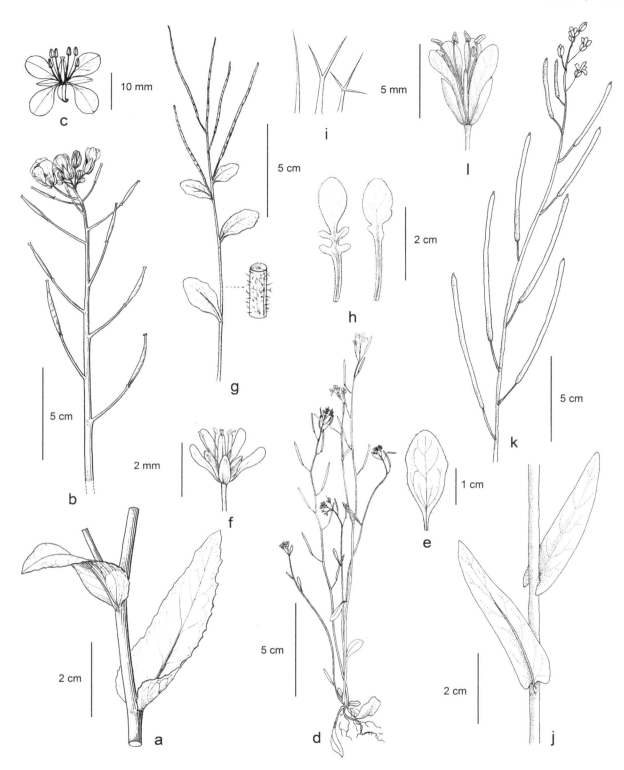

FIG. 18. CRUCIFERAE. **Brassica rapa** var. **oleifera**: a, stem with leaves; b, inflorescence with fruits; c, flower. **Arabidopsis thaliana**: d, plant with flowers and fruits; e, basal leaf; f, flower. **Crucihimalaya wallichii**: g, infructescence; h, basal leaves; i, hairs. **Turritis glabra**: j, stem with leaves; k, inflorescence with fruits; l, flower.

Cruciferae (Brassicaceae)

1. *Turritis glabra* L., Sp. Pl. 2: 666 (1753).
Arabis glabra (L.) Bernh.; *A. perfoliata* Lam.; *A. pseudoturritis* Boiss. & Heldr.; *Turritis glabra* var. *lilacina* O.E.Schulz; *T. pseudoturritis* (Boiss. & Heldr.) Velenovsky.

Herbs (30–)40–120(–150) cm, sparsely to densely pilose basally with simple and short-stalked forked trichomes, glabrous and glaucous above. Stems erect, simple basally, often branched above. Basal leaves rosulate, petiolate; leaf blade spatulate, oblanceolate, or oblong, (4–)5–12(–15) × 1–3 cm, apex obtuse, margin pinnatifid, sinuate, dentate, repand, or rarely entire, pubescent or rarely glabrous. Cauline leaves sessile, lanceolate, oblong-elliptic, or ovate, 2–9(–12) × (0.5–)1–2.5(–3.5) cm, base sagittate or auriculate, apex acute, margin dentate or entire. Sepals, (2.5–)3–5 × 0.5–1.2 mm, glabrous. Petals pale yellow, creamy white, rarely pink, linear-oblanceolate, narrowly spatulate, or rarely linear, 5–8.5 × 1.3–1.7 mm, apex obtuse. Median filament pairs 3.5–6.5 mm, lateral pair 2.5–4.5 mm; anthers narrowly oblong, 0.7–1.5 mm. Fruiting pedicels, (0.6–)0.7–1.6(–2) cm, slender, appressed to rachis, glabrous. Fruit linear, (3–)4–9(–10) cm × 0.7–1.5 mm, erect and appressed to rachis; valves glabrous, not torulose, with a prominent midvein extending full length; style 0.5–0.8(–1) mm. Seeds brown, 0.6–1.2 × 0.5–0.9 mm. Fig. 18j–l.

Distribution: Nepal, E Asia, N Asia, C Asia, SW Asia, Europe, Africa, N America and Australasia.

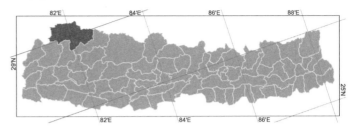

Altitudinal range: 2900–3500 m.

Ecology: Mountain slopes, forest margins, valleys, fields, meadows, woods, fields, river banks, roadsides.

Flowering: April–July. **Fruiting:** May–August.

12. *Barbarea* R.Br., in Aiton, Hortus Kew., ed. 2, 4: 109 (1812).

Biennial herbs, with rhizomes or woody caudex. Trichomes simple or absent. Stems erect, angular. Basal leaves rosulate or not, lyrate-pinnatifid or pinnatisect, rarely undivided. Cauline leaves petiolate or sessile, auriculate or amplexicaul, entire, dentate, pinnatifid, or pinnatisect. Racemes many-flowered, often ebracteate; rachis striate. Sepals oblong, erect, base of lateral pair saccate. Petals yellow, oblanceolate, apex rounded; claw obscurely differentiated. Stamens 6, tetradynamous; filaments not dilated at base; anthers oblong, obtuse at apex. Nectar glands 4; median pair tooth-like; lateral pair annular. Ovules 10–40 per ovary. Fruiting pedicels slender or thickened, erect to ascending rarely divaricate. Fruit dehiscent siliques, linear, terete to slightly compressed, 4-angled, or latiseptate, sessile or shortly stipitate; valves with a prominent midvein and distinct marginal veins, mostly glabrous, smooth or torulose; replum rounded; septum complete; style to 5 mm; stigma capitate, entire or slightly 2-lobed. Seeds uniseriate, wingless, margined, ovate to ovate-oblong, plump or slightly compressed; seed coat reticulate, rarely tuberculate, not mucilaginous when wetted; cotyledons accumbent.

Worldwide about 22 species distributed mainly in Europe and neighbouring Asia, with the centre of greatest diversity in the E Mediterranean. One species naturalized in Nepal.

1. *Barbarea intermedia* Boreau, Fl. Centre France 2: 48 (1840).
Barbarea vulgaris var. *sicula* Hook.f. & T.Anderson; *Campe intermedia* (Boreau) Rauschert.

खोले साग Khole saag (Nepali).

Herbs with stems 15–60 cm, erect, angled, glabrous. Basal and lowermost cauline leaves petiolate; petiole 0.5–3(–4.5) cm, glabrous or ciliate; leaf blade 1.5–7 cm, lyrate-pinnatifid, with 1–4(–7) lobes on each side of rachis, not fleshy; lateral lobes oblong to ovate, 3–10(–15) × 1–5(–7) mm, entire to repand; terminal lobe ovate, considerably larger than lateral ones, 1–3 × 0.7–1.5 cm. Cauline leaves pinnatifid to pinnatisect, with 1–4 lateral lobes, entire, sessile, conspicuously auriculate; auricles ovate to narrowly oblong, to 10 × 5 mm, ciliate; lateral lobes linear to linear-oblong, to 2.5 × 0.5 cm, glabrous or ciliate, entire; terminal lobe obovate-oblong, larger than lateral lobes and to 4 × 1.5 cm, entire or repand to rarely dentate. Racemes ebracteate, elongated considerably in fruit. Sepals yellow, oblong, 2–3 × 1–1.3 mm, erect, margin scarious, lateral pair slightly saccate. Petals 4–6 × 1–1.5 mm, attenuate to base. Filaments yellow, erect, 3–4 mm; anthers oblong, ca. 1 mm. Fruiting pedicels (2–)3–5 mm, terete to subquadrangular, glabrous, stout, slightly narrower than fruit. Fruit linear, (1–)1.5–3 cm × 1.5–2 mm, torulose, erect to erect-ascending; gynophore to 0.5 mm; valves with a prominent midvein and lateral veins; style 1–1.5 mm. Seeds brown, 1.2–1.7 × 1–1.2 mm. Fig. 19a.

Distribution: Nepal, W Himalaya, E Himalaya, Tibetan Plateau, E Asia, SW Asia and Europe.

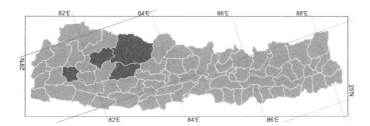

Altitudinal range: 3000–4100 m.

Ecology: Along ditches, damp slopes, marshy areas.

Flowering: June–July. **Fruiting:** July–August.

Tender shoots and leaves are eaten as a green vegetable.

13. *Cardamine* L., Sp. Pl. 2: 654 (1753).

Dentaria L.; *Loxostemon* Hook.f. & Thomson.

Annual, biennial, or rhizomatous or tuberous perennial herbs. Trichomes absent or simple. Stems erect, ascending or decumbent, leafy or rarely leafless and plant scapose. Basal leaves petiolate, rosulate or not, simple and entire, toothed, or 1–3-pinnatisect, or palmately lobed, sometimes trifoliolate, pinnately, palmately, or bipinnately compound. Cauline leaves alternate, rarely opposite or whorled, simple, or compound as basal leaves, petiolate or sessile and base cuneate, attenuate, auriculate, or sagittate. Racemes ebracteate or rarely bracteate, elongated in fruit. Sepals ovate or oblong, base of lateral pair saccate or not. Petals white, pink, purple, lilac, mauve, violet or lavender, never yellow, rarely absent, obovate, spatulate, oblong, or oblanceolate, apex obtuse or emarginate; claw absent or strongly differentiated from blade. Stamens 6 and tetradynamous, rarely 4 and equal in length; anthers ovate, oblong, or linear, obtuse at apex. Nectar glands confluent and subtending bases of all stamens; median glands 2 or rarely 4 or absent. Ovules 8–40 per ovary. Fruiting pedicels slender or thickened, ascending, erect, divaricate, or reflexed. Fruit dehiscent siliques, linear, latiseptate, sessile; valves papery, not veined, smooth or torulose, dehiscing elastically, spirally or circinately coiled; replum strongly flattened; septum complete; style distinct or rarely obsolete; stigma capitate, entire. Seeds uniseriate, wingless, rarely margined or winged, oblong, ovate or subquadrate, flattened; seed coat smooth, reticulate, colliculate, or rugose, mucilaginous or not when wetted; cotyledons accumbent or very rarely incumbent.

Worldwide about 200 species. Eleven species in Nepal.

Hara (Enum. Fl. Pl. Nepal 2: 41. 1979) noted that the type of *Cardamine nasturtioides* D.Don (Prodr. Fl. Nepal.: 201. 1825) has not been found and so taxonomic placement is uncertain. Hara suggests that this species is probably a synonym of *Cardamine scutata* subsp. *flexuosa* (With.) H.Hara (syn. *C. flexuosa* With.), but until the type is discovered this species is excluded from Flora of Nepal.

Key to Species

1a	Cauline leaves auriculate or sagittate at base, sometimes petiole absent and proximal pair of leaflets attached at or directly above node	2
b	Cauline leaves not auriculate or sagittate at base	6
2a	Cauline leaves simple, divided or not. Petals 10–17 × 4–7 mm	3
b	Cauline leaves compound. Petals 1.5–8(–9) × 1.5–3.5(–5) mm	4
3a	Cauline leaves undivided. Fruiting pedicels 0.8–3 cm	**2. *C. violacea***
b	Cauline leaves with 1–3 lateral lobes on each side, lobes much smaller than terminal one. Fruiting pedicels to 1 cm	**3. *C. nepalensis***
4a	Cauline leaves sessile. Petals purple to lavender, 6–9 mm	**7. *C. griffithii***
b	Cauline leaves petiolate. Petals white, 1.5–5(–6) mm	5
5a	Lateral leaflets of cauline leaves 1–3 on each side of rachis. Rhizomatous perennials	**8. *C. yunnanensis***
b	Lateral leaflets of cauline leaves (4–)6–11 on each side of rachis. Annuals or biennials	**10. *C. impatiens***
6a	Annuals or biennials. Petals 2.5–4(–5) mm	**11. *C. flexuosa***
b	Rhizomatous or stoloniferous perennials. Petals 5–17 mm	7
7a	Fruiting pedicels reflexed. Raceme rachis flexuous	**9. *C. elegantula***
b	Fruiting pedicels divaricate, ascending or suberect. Raceme rachis not flexuous	8

8a Rhizome without bulbils. Stem stout at attachment to rhizome, if slender then leaflets more than 1 cm wide 9
 b Rhizome with bulbils. Stems narrowed basally to a fragile slender attachment to rhizome 10

9a Petals purple or lilac, (8–)10–17 mm. Leaflets more than 2 cm ....**1. *C. macrophylla***
 b Petals white, 5–8 mm. Leaflets less than 1.5 cm...**6. *C. trifoliolata***

10a Axils of cauline leaves with bulbils. Filaments of median stamens flattened, toothed. Leaves with 1(or 2) pairs of
    lateral leaflets ....**4. *C. pulchella***
 b Axils of cauline leaves without bulbils. Filaments of median stamens slender, toothless. Leaves with 2–5 pairs of
    lateral leaflets ....**5. *C. loxostemonoides***

1. *Cardamine macrophylla* Willd., Sp. Pl., ed. 5[as 4], 3(1):
484 (1800).
Cardamine foliosa Wall. nom. nud.; *C. macrophylla* subsp.
polyphylla (D.Don) O.E.Schulz; *C. macrophylla* var. *crenata*
Trautv.; *C. macrophylla* var. *dentariaefolia* Hook.f. &
T.Anderson; *C. macrophylla* var. *diplodonta* T.Y.Cheo; *C.
macrophylla* var. *foliosa* Hook.f. & T.Anderson; *C. macrophylla*
var. *lobata* Hook.f. & T.Anderson; *C. macrophylla* var.
moupinensis Franch.; *C. macrophylla* var. *polyphylla* (D.Don)
T.Y.Cheo & R.C.Fang; *C. macrophylla* var. *sikkimensis* Hook.f.
& T.Anderson; *C. polyphylla* D.Don; *C. sachalinensis* Miyabe
& Miyake; *C. sinomanshurica* (Kitag.) Kitag.; *C. urbaniana*
O.E.Schulz; *Dentaria gmelinii* Tausch; *D. macrophylla* (Willd.)
Bunge ex Maxim.; *D. sinomanshurica* Kitag.; *D. wallichii*
G.Don; *D. willdenowii* Tausch.

तिउली Tiuli (Nepali).

Perennial herbs, (20–)30–95(–115) cm. Rhizomes creeping,
not scaly, 2–10(–30) mm in diameter, not stoloniferous. Stems
stout or slender, erect, simple or rarely branched above,
glabrous or pubescent. Basal leaves (4–)10–40(–50) cm;
petiole (1–)3–20(–25) cm; terminal leaflet sessile or petiolulate,
lanceolate, elliptic, oblong, ovate, or obovate, (1–)2–15(–25)
× (0.5–)1–3.5(–5) cm, base cuneate, apex acuminate, acute,
or subobtuse, margin serrate, crenate, dentate, or rarely
3–5-lobed; lateral leaflets (1 or)2–6 pairs, similar to terminal
but smaller. Cauline leaves compound, 3–12(–18); petiole
(1–)2–5(–6.5) cm, not auriculate at base; terminal leaflet as
those of basal leaves, sessile or with a petiolule to 1 cm,
(2–)4–12(–20) × 1–4(–5) cm, base cuneate, apex acute,
rarely acuminate, margin ciliolate and crenate, serrate,
or serrulate, rarely subentire or double serrate; lateral
leaflets 2–7(–11) pairs, sessile or petiolulate, base cuneate
or obliquely decurrent, similar to but slightly smaller than
terminal leaflet. Racemes 10–30-flowered. Sepals oblong,
3.5–6.5(–8) × 1.5–3 mm. Petals purple or lilac, obovate to
spatulate, (0.8–)1–1.7 cm × 3.5–8 mm, apex rounded or
rarely subemarginate. Median filament pairs 7–9(–11) mm,
lateral pair 6–7 mm; anthers oblong, 1–2(–2.5) mm. Ovules
8–12(–16) per ovary. Fruiting pedicels ascending or rarely
divaricate, (0.3–)0.8–2.5(–3.1) cm, straight, thickened. Fruit
linear, (2.2–)2.5–6(–7) cm × 1.5–2.5(–3) mm; valves sparsely
pubescent or glabrous; style (1–)2–6.5(–9) mm. Seeds brown,
ovoid to oblong, (1.5–)2–3(–4) × 1–1.7(–2) mm.
Fig. 19b–c.

Distribution: Nepal, W Himalaya, E Himalaya, Tibetan
Plateau, E Asia, N Asia and C Asia.

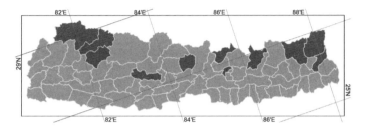

Altitudinal range: 500–4600 m.

Ecology: Damp forests, river banks, tundra, rock crevices,
meadows, damp woodlands, thickets, stream-sides, valleys,
ravines, mountain slopes, amongst boulders.

Flowering: (March–)April–October. **Fruiting:** May–October.

Cardamine macrophylla is highly variable, especially in leaflet
number, shape, size, margin, and base. A critical study of the
species from its entire range reveals that only a small number
of the variants have been recognized formally and that the
variation does not show any correlation among characters
and/or geography. Some of the variants on which varieties are
based occur within the same population, and a thorough study
at the population level is needed before the species is divided
into infraspecific taxa.

2. *Cardamine violacea* (D.Don) Wall. ex Hook.f. & Thomson,
J. Proc. Linn. Soc., Bot. 5: 144 (1861).
Erysimum violaceum D.Don, Prodr. Fl. Nepal.: 202 (1825);
Cardamine violacea subsp. *bhutanica* Grierson.

टुकी झार Tuki jhar (Nepali).

Perennial herbs, 20–100 cm, shortly pilose or subglabrous.
Rhizomes stout, to 1 cm in diameter Stems erect, simple,
stout, glabrous. Basal leaves not seen. Middle cauline leaves
simple, sessile; blade lanceolate to linearlanceolate or ovate-
lanceolate, 3.5–20 × 0.7–3.5 cm, strongly auriculate, sagittate,
to amplexicaul at base, apex acuminate to caudate, margin
ciliolate and dentate, denticulate, or entire, puberulent to
subpilose above, glabrous below; auricles oblong to ovate,
2–10 × 2–7 mm. Racemes 5–25flowered. Flowering pedicels
spreading to reflexed. Sepals oblong, 5–7 × 1.5–3 mm,
glabrous or sparsely pilose, base of lateral pair saccate.
Petals purple, spatulate to obovate, 1–1.7 cm × 4.5–7 mm.
Median filament pairs 7–9 mm, lateral pair 6–7 mm; anthers
oblong, 1.5–2 mm. Ovules 10–16 per ovary. Fruiting pedicels

148

divaricate to ascending, 0.8–3 cm, glabrous, straight. Fruit linear, 2–6 cm × 1.4–2.5 mm; valves smooth, glabrous; style 3–8 mm. Seeds brown, oblong, 2–3 × 1.4–1.8 mm, wingless.

Distribution: Nepal, E Himalaya and E Asia.

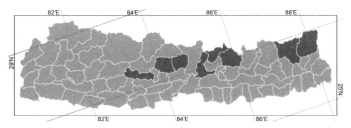

Altitudinal range: 1800–4000 m.

Ecology: Grassy slopes, stream-sides, open forests, pastures, thickets, roadside banks, sandy moist forests, forest ravines.

Flowering: May–August. **Fruiting:** July–September.

Tender shoots and leaves are cooked as a vegetable.

3. *Cardamine nepalensis* N.Kurosaki & H.Ohba, J. Jap. Bot. 64: 135 (1989).

Perennial herbs, 60–100 cm, shortly pilose or subglabrous. Rhizomes stout. Stems erect, simple, stout, glabrous, 10–18-leaved. Rhizomal leaves not seen. Middle cauline leaves simple, petiolate, minutely auriculate; blade pinnatisect, 4–15 × 1.5–5.5 cm, sparsely pilose; terminal lobe lanceolate, base cuneate, apex acuminate, margin coarsely and irregularly dentate; lateral lobes 1–3 on each side of midvein, smaller than terminal lobe, decurrent at base, apex acute, margin denticulate. Racemes 5–25-flowered. Flowering pedicels spreading to reflexed. Sepals oblong, 4–6 × 1.5–2.5 mm, glabrous or subapically sparsely pilose, base of lateral pair saccate. Petals purple, spatulate to obovate, 1–1.5 cm × 4–7 mm. Median filament pairs 7–8 mm, lateral pair 6–7 mm; anthers oblong, 1–1.5 mm. Ovules 10–14 per ovary. Fruiting pedicels to 1 cm, glabrous. Fruit linear, 2.5–4 cm × 1–2 mm; valves smooth, glabrous; style 1–1.5 mm. Seeds brown, oblong, ca. 2 × 1 mm, wingless.

Distribution: Nepal and E Himalaya.

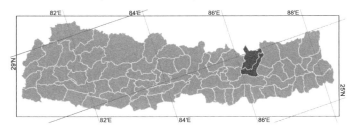

Altitudinal range: 2500–3700 m.

Ecology: Sandy forest floors and stream-sides. *Abies* forest.

Flowering: July. **Fruiting:** August.

4. *Cardamine pulchella* (Hook.f. & Thomson) Al-Shehbaz & G.Yang, Harvard Pap. Bot. 3(1): 77 (1998).
Loxostemon pulchellus Hook.f. & Thomson, J. Proc. Linn. Soc., Bot. 5: 147 (1861).

Perennial herbs, (5–)8–15(–20) cm, hirsute or glabrescent. Rhizomes to 1 cm, with stolons and numerous bulbils; bulbils whitish, fleshy scales ovoid to subglobose, with rudimentary apical appendages. Stems erect, simple; underground lower part slender, glabrous; aboveground part green to purplish, slender or stout, pilose to glabrous; bulbils of leaf axils ovoid, to 3 × 2 mm. Rhizomal leaves 1(or 2), (1.5–)3–7 cm; petiole (1–)2.2–6 cm; terminal leaflet broadly ovate to oblong, 3–10 × 2–4(–7) mm, with a petiolule to 3 mm; lateral leaflets 1(or 2) pairs, similar to terminal one. Cauline leaves compound, 1–3, 1.2–5 cm; petiole 0.5–3.5 cm, base not auriculate; terminal leaflet oblong to narrowly elliptic, (4–)6–12(–15) × (1–)1.5–4(–6) mm, base cuneate, apex mucronate, margin entire; lateral leaflets 1(or 2) pairs, similar to terminal one. Racemes 2–4(or 5)-flowered. Sepals ovate, 1.5–3 × 1–1.5 mm, margin membranous, ciliate or glabrous. Petals deep purple to mauve, broadly obovate, 5–8 × 2–4 mm, apex rounded. Median filament pairs 2–3 × 0.7–1.1 mm, flattened, extended apically into a tooth; lateral pair slender, 1–2 mm; anthers oblong, 0.5–0.6 mm. Ovules 12–16 per ovary. Fruiting pedicels ascending, 5–13 mm, straight. Fruit linear, 1–1.7 cm × 1–1.2 mm; valves smooth, glabrous; style 0.5–1 mm. Seeds brown, broadly oblong, 1.3–1.6 × 0.8–1.1 mm, wingless.

Distribution: Nepal, E Himalaya, Tibetan Plateau, Assam-Burma and E Asia.

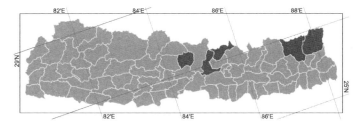

Altitudinal range: 2700–4600 m.

Ecology: Grassy marshlands, moist rocky places, stony stream banks, scree, mountain slopes.

Flowering: May–August. **Fruiting:** July–September.

5. *Cardamine loxostemonoides* O.E.Schulz, Notizbl. Bot. Gart. Berlin-Dahlem 9: 1069 (1926).
Cardamine tibetana Rashid & H.Ohba; *Loxostemon incanus* R.C.Fang ex T.Y.Cheo & Y.Z.Lan; *L. loxostemonoides* (O.E.Schulz) Y.Z.Lan & T.Y.Cheo.

चम्सुरे घाँस Chamsure ghans (Nepali).

Perennial herbs, (5–)12–30(–35) cm, glabrous or sparsely to densely pilose. Rhizomes slender, with several bulbils and stolons; bulbils with fleshy, white, scaly leaves apically with rudimentary appendages. Stems somewhat decumbent and slender below, simple. Basal leaves glabrous or pilose, compound; petiole (0.7–)2–15(–20) cm; terminal leaflet with a petiolule 3–10 mm, blade undivided and suborbicular, oblanceolate, or linear, or trifid and suborbicular to broadly obovate in outline and with obovate or oblong lobes the basal pair of which sometimes with a tiny lobule, terminal lobe 0.6–2(–3) cm × 2–5 mm; lateral leaflets 2–5 pairs, sessile or petiolulate, similar in shape and division to terminal lobe but smaller. Cauline leaves 1–4; petiole (0.3–)1–3(–4) cm, not auriculate at base; terminal and lateral leaflets similar in shape, size, and number to those of basal leaves. Racemes 2–14-flowered. Fruiting pedicels (5–)1–2.5 cm, ascending to suberect, glabrous. Sepals broadly ovate, 2.5–5 × 1.5–2.5 mm, glabrous, broadly membranous at margin and apex, base of lateral pair subsaccate. Petals purple with darker veins, broadly obovate, 0.8–1.2(–1.4) cm × 5–8(–8.5) mm, not clawed, apex rounded. Median filament pairs (3.5–)4.5–6 mm, lateral pair 2.5–4 mm; anthers narrowly oblong, 1.4–2 mm. Ovules 14–20 per ovary. Fruits linear, 2.5–3.5 cm × 1.2–1.5 mm; valves glabrous; style 1–3 mm. Seeds brown, ovate, ca. 1.5 × 1 mm, wingless.

Distribution: Nepal, W Himalaya, E Himalaya, Tibetan Plateau, Assam-Burma and E Asia.

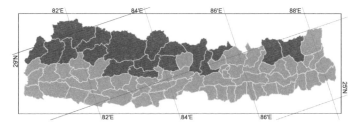

Altitudinal range: 2400–5500 m.

Ecology: Mountains slopes, along ditches, damp ground by streams, open grass and gravel, scree.

Flowering: June–July. **Fruiting:** August–September.

Hooker & Anderson (Fl. Brit. Ind. 1: 138. 1872) misapplied the name *Cardamine pratensis* L. to this species.
    Tender shoots and leaves are cooked as a vegetable.

6. Cardamine trifoliolata Hook.f. & Thomson, J. Proc. Linn. Soc., Bot. 5: 145 (1861).
Cardamine flexuosoides W.T.Wang; *C. flexuosoides* var. *glabricaulis* W.T.Wang; *Loxostemon smithii* var. *wenchuanensis* Y.Z.Lan & T.Y.Cheo.

Perennial herbs, (4–)6–18(–25) cm, often sparsely pilose at least basally. Rhizomes slender, thickened at stem base, with few stolons. Stems erect or decumbent, slender, simple or few-branched. Basal leaves 1–3, 3 or 5(or 7)-foliolate, rarely simple; petiole 1–4(–5.5) cm; terminal leaflet broadly obovate or rarely ovate, 2–12 × 3–14 mm, with a petiolule 1–6(–8) mm, base subtruncate, cordate, or rounded, apically subtruncate or obtusely 3-lobed, margin entire and obscurely 5-lobed or -crenate; lateral leaflets 1(–3) pairs, subsessile or petiolulate, resembling terminal leaflet, or not lobed and oblong or ovate, smaller. Cauline leaves 1 or 2(or 3), 3(or 5)-foliolate; petiole

0.5–1.5 cm, not auriculate at base; terminal leaflet similar to that of basal leaf, with a petiolule 0.5–3 mm; lateral leaflets similar to those of basal leaves. Racemes lax, 2–8-flowered, rachis straight. Fruiting pedicels ascending to divaricate, 0.5–1.5(–2) cm, straight, glabrous. Sepals ovate to oblong, 2–3 × 1–1.4 mm, glabrous, lateral pair subsaccate. Petals white, obovate to spatulate, 5–8 × 2.5–4 mm, not clawed, apex rounded. Median filament pairs 3–4.5 mm, slender; lateral pair 2–3 mm; anthers oblong, 0.8–1 mm. Ovules 8–12 per ovary. Young fruits glabrous. Mature fruits and seeds not seen.

Distribution: Nepal, E Himalaya, Assam-Burma and E Asia.

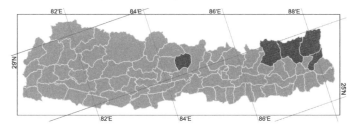

Altitudinal range: 2500–4300 m.

Ecology: Moist rocky crevices, meadows, moist forests, mossy banks, rocky areas.

Flowering: May–July. **Fruiting:** July–September.

7. Cardamine griffithii Hook.f. & Thomson, J. Proc. Linn. Soc., Bot. 5: 146 (1861).
Cardamine griffithii var. *pentaloba* W.T.Wang.

Perennial herbs, (20–)30–100(–115) cm, glabrous throughout except for leaflet margin. Rhizomes creeping, without stolons. Stems erect, simple or branched above, striate, angled, (9–)12–28(–37)-leaved. Cauline leaves compound, sessile; lower and middle ones (1–)2–9(–11) × (0.7–)1–3.5(–4.5) cm; terminal leaflet orbicular to broadly ovate or obovate, (0.5–)1–3(–3.5) × (0.3–)0.6–1.9(–2.5) cm, with a petiolule 2–10(–15) mm, base subcordate to rounded or rarely cuneate, apex obtuse to rounded, margin repand or entire and sparsely ciliate; lateral leaflets 2–4(or 5) pairs, slightly to distinctly smaller than terminal one, base obtuse or rarely slightly oblique, apex obtuse to rounded, margin entire or repand and sparsely ciliate; proximal pair of lateral leaflets auricle-like, attached at or just above node, often giving appearance of amplexicaul leaf base. Uppermost leaves smaller. Sepals ovate to oblong, 2.5–3 × 1.5–2 mm, erect. Petals purple to lavender, spatulate to obovate, 6–9 × (2.5–)3–5 mm, not clawed, apex rounded to subemarginate. Median filament pairs 3.5–4.5 mm, lateral pair 2.5–3 mm; anthers oblong, 0.9–1.1 mm. Ovules 10–22 per ovary. Fruiting pedicels ascending to divaricate, (0.4–)0.7–1.5 cm, slender, straight. Fruit linear, (1.5–)2–4 cm × 0.9–1.2 mm; valves smooth, glabrous; style 0.5–1(–2) mm; stigma 2-lobed, distinctly broader than style. Seeds brown, oblong, 1.4–1.7 × 0.8–1.1 mm, wingless.

Distribution: Nepal, E Himalaya, Tibetan Plateau, Assam-Burma and E Asia.

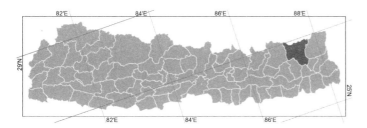

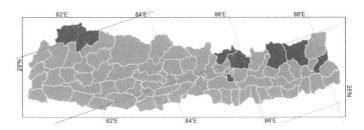

Altitudinal range: 2400–4500 m.

Ecology: Mountain slopes, valleys, stream-sides, pastures, marshy places, moist forest floors, shady rocky areas.

Flowering: May–August. **Fruiting:** June–September.

Altitudinal range: 900–4200 m.

Ecology: Moist shady places, mountain slopes, valleys, grasslands, meadows, thickets, forest openings, damp stream beds.

Flowering: March–July. **Fruiting:** April–September.

8. *Cardamine yunnanensis* Franch., Bull. Soc. Bot. France 33: 398 (1886).
Cardamine bijiangensis W.T.Wang; *C. heterophylla* T.Y.Cheo & R.C.Fang; *C. hirsuta* var. *oxycarpa* Hook.f. & T.Anderson; *C. inayatii* O.E.Schulz; *C. levicaulis* W.T.Wang; *C. longipedicellata* Z.M.Tan & G.H.Chen; *C. longistyla* W.T.Wang; *C. muliensis* W.T.Wang; *C. sikkimensis* H.Hara; *C. sinica* Rashid & H.Ohba; *C. weixiensis* W.T.Wang.

Short-lived perennial herbs with slender rhizomes, (10–)15–45(–60) cm, often pilose or puberulent, sometimes glabrescent above. Stems simple or branched from base, angled. Basal leaves petiolate, often withered by time of flowering, 3–5-foliolate, rarely simple; petiole 1–6(–8) cm; leaf blade or terminal leaflet suborbicular, ovate, to lanceolate, 0.5–3 × 0.5–2.5 cm, dentate to crenate or rarely subsinuate; lateral leaflets absent or 1 or 2 pairs. Middle cauline leaves 3–7-foliolate; petiole (1–)2–7(–8.5) cm, basally auriculate; auricles tooth-like to linear or lanceolate, (0.4–)1–3(–4) × 0.2–1(–1.5) mm; terminal leaflet lanceolate, elliptic, oblong, ovate, or suborbicular, (1–)1.3–4.5(–6) × (0.4–)0.6–2(–3) cm, base obtuse to cuneate, apex acute to acuminate, margin ciliate to ciliolate and dentate, crenate, sinuate, or rarely repand, sparsely pilose above, often glabrous below, with a petiolule (2–)4–14(–20) mm; lateral leaflets shortly petiolulate to subsessile, similar to terminal one and often oblique at base. Uppermost leaves often trifoliolate, rarely simple. Sepals oblong to nearly ovate, 2–3 × 1–1.5 mm, base not saccate. Petals white, obovate, (2.5–)3.5–5(–6) × 2–3 mm, not clawed. Median filament pairs 2.5–4 mm, lateral pair 2–3 mm; anthers oblong, 0.5–0.8 mm. Ovules 8–18 per ovary. Fruiting pedicels ascending to divaricate, 0.5–1.8(–2.3) cm, straight, slender. Fruit linear, 1.5–2.8(–3) cm × 1–1.3 mm; valves smooth, sparsely pilose; style (0.5–)1–2.5(–3.5) mm. Seeds brown, narrowly oblong, 1.3–1.8 × 0.7–1 mm, wingless. Fig. 19d–g.

Distribution: Nepal, E Himalaya, Tibetan Plateau, Assam-Burma and E Asia.

9. *Cardamine elegantula* Hook.f. & Thomson, J. Proc. Linn. Soc., Bot. 5: 146 (1861).

Perennial herbs, 8–30 cm, slender, pilose on leaves and stems with trichomes to 0.6 mm. Rhizomes slender, stoloniferous. Stems simple, erect, flexuous. Basal leaves compound, rosulate, often present at time of fruiting; petiole 0.7–2 cm; terminal leaflet broadly ovate to obovate, petiolulate, sparsely pilose, base distinctly oblique, margin entire or obscurely 1-toothed on each side; lateral leaflets 2–4 on each side, with a petiolule 0.5–1.5 mm, ovate to elliptic, 2–5(–8) × 1–3.5(–5) mm, similar to terminal one or with a lateral lobe on lower margin, distinctly mucronate. Cauline leaves compound, not auriculate at base; petiolule 0.5–3 mm; terminal leaflet elliptic to linear-elliptic, 3–6(–9) × 0.5–2 mm, base attenuate to cuneate and decurrent with adjacent lateral lobes, apex mucronate, margin entire and scabrous with trichomes 0.05–0.1 mm; lateral leaflets 5–8 on each side of rachis, petiolulate, similar to terminal leaflets. Racemes 2–5-flowered, rachis strongly flexuous. Pedicels strongly reflexed, slender, straight, 4–6 mm, glabrous. Sepals ovate, 1.5–2 × 0.8–1 mm, glabrous, membranous at margin and apex. Petals pink or white, broadly obovate, 6–7 × 3–4 mm. Median filament pairs 2.5–3 mm, filiform; lateral pair 1.5–2 mm; anthers oblong, 0.4–0.5 mm. Median nectar glands obsolete. Ovules 14–20 per ovary. Fruit linear, 1.2–1.5 cm × 0.7–1 mm; valves glabrous; style 1–2 mm. Seeds pale brown, oblong, ca. 1.8 × 0.8 mm, wingless.

Distribution: Nepal and E Himalaya.

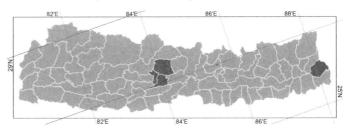

Altitudinal range: 2400–3100 m.

Ecology: Beneath wet rock overhang, damp rocks in forests, marshy areas along streams.

Flowering: May–June. **Fruiting:** May–June.

Cruciferae (Brassicaceae)

10. *Cardamine impatiens* L., Sp. Pl. 2: 655 (1753).
Cardamine basisagittata W.T.Wang; *C. dasycarpa* M.Bieb.;
C. glaphyropoda O.E.Schulz; *C. glaphyropoda* var. *crenata*
T.Y.Cheo & R.C.Fang; *C. impatiens* subsp. *elongata*
O.E.Schulz; *C. impatiens* var. *angustifolia* O.E.Schulz; *C.
impatiens* var. *dasycarpa* (M.Bieb.) T.Y.Cheo & R.C.Fang; *C.
impatiens* var. *eriocarpa* DC.; *C. impatiens* var. *fumaria* H.Lév.;
C. impatiens var. *microphylla* O.E.Schulz; *C. impatiens* var.
obtusifolia K.Knaf; *C. impatiens* var. *pilosa* O.E.Schulz; *C.
nakaiana* H.Lév.; *C. senanesis* Franch. & Sav.

Biennial or rarely annual herbs, (12–)20–65(–90) cm, glabrous
or rarely sparsely pubescent near base. Stems erect, simple
at base, usually branched, above. Basal leaves rosulate,
often withered by time of flowering; petiole 1–4 cm, not
auriculate; leaf blade pinnatisect and appearing compound.
Cauline leaves compound, up to 15 per stem; petiole
auriculate, 2–6 cm; auricles lanceolate to linear, 1–8(–10) ×
(0.1–)0.3–1.8(–2.2) mm, often ciliate; leaf blade (1–)3–18(–22)
× (0.6–)1–5.5(–7) cm, pinnatisect; terminal lobe orbicular,
obovate, ovate, or lanceolate, 1–4(–5) × 0.5–1.7 cm, with a
petiolule to 5 mm, entire or obscurely to strongly 3–5(–9)-
toothed or -lobed; lateral lobes (4–)6–11 on each side of
rachis, oblong, lanceolate, or ovate, sessile to long petiolulate,
smaller than terminal lobe, margin dentate to sublaciniate,
rarely entire; uppermost leaves with narrower segments.
Sepals oblong, 1.2–2(–2.5) × 0.7–1(–1.2) mm. Petals white,
oblanceolate, 1.5–4(–5) × 0.6–1.2 mm, rarely absent.
Filaments 2–3(–4) mm; anthers ovate, 0.3–0.5 mm. Ovules
10–30 per ovary. Fruiting pedicels divaricate to ascending,
3.5–12(–15) mm, slender. Fruit linear, (1–)1.6–3(–3.5) cm ×
0.9–1.5 mm; valves glabrous or rarely pilose, torulose; style
0.6–1.6(–2) mm. Seeds brown, oblong, 1.1–1.5 × 0.8–1 mm,
compressed, sometimes apically narrowly winged.

Distribution: Asia, Europe, Africa and N America.

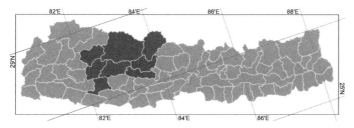

Altitudinal range: 100–4000 m.

Ecology: Shady or moist slopes, stream-sides, fields,
roadsides.

Flowering: February–July. **Fruiting:** February–July.

Tender shoots and leaves are cooked as a vegetable.

11. *Cardamine flexuosa* With., Arr. Brit. Pl., ed. 3, 3: 578
(1796).
Barbarea arisanensis (Hayata) S.S.Ying; *Cardamine
arisanensis* Hayata; *C. debilis* D.Don later homonym, non
Banks ex DC.; *C. flexuosa* subsp. *debilis* (D.Don) O.E.Schulz;
C. flexuosa var. *debilis* ((D.Don) O.E.Schulz) T.Y.Cheo &

R.C.Fang; *C. flexuosa* var. *ovatifolia* T.Y.Cheo & R.C.Fang; *C.
hirsuta* subsp. *flexuosa* (With.) Hook.f.; *C. hirsuta* var. *flaccida*
Franch.; *C. hirsuta* var. *omeiensis* T.Y.Cheo & R.C.Fang; *C.
scutata* subsp. *flexuosa* (With.) H.Hara; *C. sylvatica* Link; *C.
zollingeri* Turcz.; *Nasturtium obliquum* Zoll.

चम्सुरे Chamsure (Nepali).

Annual or biennial herbs, (6–)10–50 cm, sparsely to densely
hirsute basally or throughout, or glabrous. Stems erect,
ascending, or decumbent, 1 to several from base, simple or
branched, flexuous or straight. Basal leaves not rosulate, often
withered at anthesis, petiolate; leaf blade (2–)4–10(–14) cm,
lyrate; terminal lobe reniform, broadly ovate, or suborbicular,
repand or 3–5-lobed; lateral lobes (1 or)2–6(or 7) on each
side of midvein, petiolulate or subsessile, oblong, ovate, or
elliptic, smaller than terminal lobe, entire, repand, crenate,
or 3(–5)-lobed. Cauline leaves 3–15, including petiole
(2–)3.5–5.5(–7) cm; petiole base not auriculate; terminal
lobe 3–5-lobed; lateral lobes 2–7 on each side of midvein,
suborbicular, ovate, oblong, oblanceolate, to linear, similar to or
slightly smaller than terminal lobe, sessile or shortly petiolulate,
entire, repand, dentate, or 3(–5)-lobed. Sepals oblong, 1.5–2.5
× 0.7–1 mm. Petals white, spatulate, 2.5–4(–5) × 1–1.7 mm.
Stamens 6, rarely 4 with lateral pair absent; filaments 2–3 mm;
anthers ovate, 0.3–0.5 mm. Ovules 18–40 per ovary. Fruiting
pedicels divaricate to ascending, (5–)6–14(–17) mm, slender.
Fruit linear, (0.8–)1.2–2.8 cm × 1–1.5 mm; valves glabrous,
torulose; style 0.3–1(–1.5) mm. Seeds brown, oblong to
subquadrate, 0.9–1.5 × 0.6–1 mm, narrowly margined or not.

Distribution: Nepal, E Himalaya, Tibetan Plateau, Assam-
Burma, S Asia, E Asia, SE Asia, Europe, N America, S
America and Australasia.

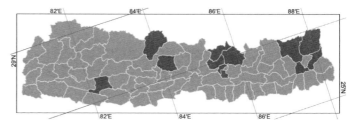

Altitudinal range: 100–4100 m.

Ecology: Fields, roadsides, grasslands, disturbed sites,
stream banks, clearings, running water, wet forests, dry sites.

Flowering: February–May. **Fruiting:** April–July.

Hooker & Anderson (Fl. Brit. Ind. 1: 138. 1872) partially
misapplied the name *Cardamine hirsuta* var. *sylvatica* Link to
this species.

Hara (Enum. Fl. Pl. Nepal 2: 41. 1979) treated *C. flexuosa*
as a variety of *C. scutata* Thunb., but the latter is an entirely
different species restricted to NE Asia, and differs from *C.
flexuosa* by having a straight instead of flexuous rachis and
terminal leaf lobe considerably larger than the laterals. Hara
also suggested that *C. nasturtioides* D.Don and *Nasturtium
sparsum* D.Don could refer to this species (see notes under
Cardamine and *Nasturtium*).

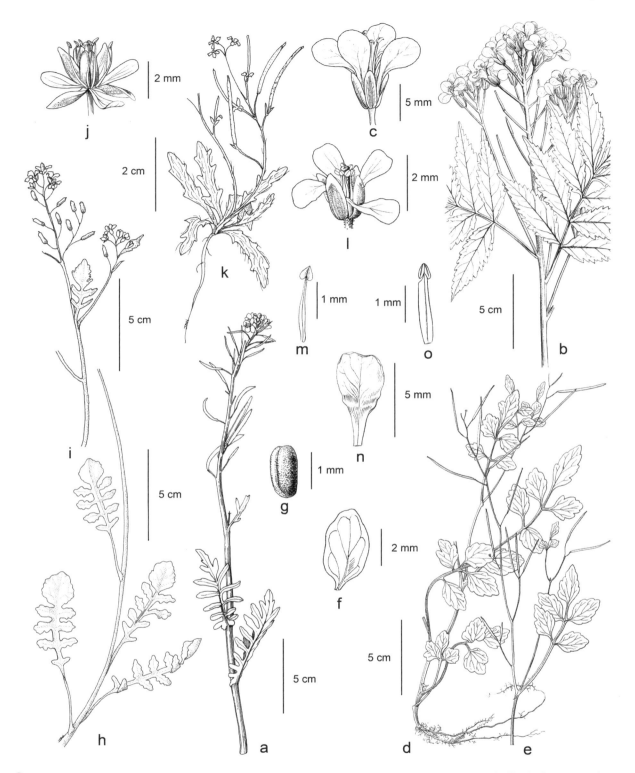

FIG. 19. CRUCIFERAE. **Barbarea intermedia**: a, shoot with flowers and fruits. **Cardamine macrophylla**: b, leaves and inflorescence; c, flower. **Cardamine yunnanensis**: d, lower part of plant with leaves; e, infructescence with leaves; f, petal; g, seed. **Rorippa palustris**: h, lower part of plant with leaves; i, inflorescence with fruits; j, flower. **Dontostemon glandulosus**: k, plant with flowers and fruits; l, flower; m, stamen. **Dontostemon pinnatifidus**: n, petal; o, stamen.

14. *Nasturtium* R.Br., in Aiton, Hortus Kew., ed. 2, 4: 110 (1812).

Perennial herbs, aquatic, rhizomatous. Trichomes absent or simple. Stems prostrate or decumbent, erect in emergent plants, rooting at lower nodes. Leaves all cauline, pinnately compound, often simple in deeply submerged plants; petiole auriculate at base; lateral leaflets 1–6(–12) pairs, petiolulate or sessile, entire, repand, or rarely dentate. Racemes many-flowered, ebracteate. Sepals oblong, erect or ascending, glabrous, base of inner pair subsaccate. Petals white to lavender, longer than sepals, obovate or narrowly spatulate, apex obtuse; clawed. Stamens 6, erect, tetradynamous; filaments base not dilated; anthers oblong, obtuse at apex. Median glands absent; lateral glands 2, annular or semiannular. Ovules 25–50 per ovary. Fruiting pedicels divaricate or recurved. Fruit dehiscent siliques, linear or rarely narrowly oblong, terete, sessile; valves obscurely veined, glabrous, smooth or slightly torulose; replum rounded; septum complete; style to 2 mm; stigma capitate, entire. Seeds biseriate, wingless, ovoid, plump; seed coat coarsely reticulate, not mucilaginous when wetted; cotyledons accumbent.

Worldwide five specie in Eurasia, N Africa and N America. One species in Nepal.

Hara (Enum. Fl. Pl. Nepal 2: 44. 1979) noted that the type of *Nasturtium sparsum* D.Don (Prodr. Fl. Nepal.: 202. 1825) has not been found and so taxonomic placement is uncertain. Hara suggests that this species is probably a synonym of *Cardamine scutata* subsp. *flexuosa* (With.) H.Hara (syn. *C. flexuosa* With.), but until the type is discovered this species is excluded from Flora of Nepal.

1. *Nasturtium officinale* R.Br., Hortus Kew., ed. 2, 4: 110 (1812).

Rorippa nasturtium-aquaticum (L.) Hayek; *Sisymbrium nasturtium-aquaticum* L.

सिम साग Sim saag (Nepali).

Herbs, 10–70(–200) cm, glabrous throughout or sparsely pubescent with simple trichomes. Leaves all cauline, pinnately compound, 5–9(–13)-foliolate; petiole auriculate at base; terminal leaflet suborbicular to oblong, 1–4 cm, base obtuse, cuneate, or subcordate, apex obtuse, margin entire or repand; lateral leaflets smaller, usually sessile. Petals white to lavender, spatulate to obovate, 2.8–4.5(–6) × 1.5–2.5 mm, apex rounded; claw ca. 1 mm. Filaments white, 2–3.5 mm; anthers oblong, ca. 0.6 mm. Sepals oblong, 2–3.5 mm. Fruiting pedicels slender, 5–12(–20) mm. Fruit cylindric, 1–1.5(–2) cm × (1.8–)2–2.5(–3) mm; valves with an obscure midvein; style 0.5–1(–1.5) mm. Seeds, ovoid, 1–1.3 × 0.7–1 mm, reddish brown, coarsely reticulate, with 25–50(–60) areolae on each side.
Fig. 16h.

Distribution: Cosmopolitan.

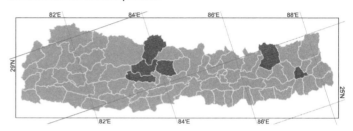

Altitudinal range: 900–3800 m.

Ecology: Streams, ditches, lakes, swamps, marshes.

Flowering: April–May. **Fruiting:** June–July.

Cultivated as a vegetable (watercress). A naturalized weed outside its native range in Eurasia.

15. *Rorippa* Scop., Fl. Carniol., ed. 2, 1: 520 (1771).

Annual, or rarely short-lived perennial herbs, usually of wet or aquatic habitats. Trichomes absent or simple. Stems erect or prostrate, simple or branched, leafy. Basal leaves petiolate, rosulate or not, simple, entire, dentate, sinuate, lyrate, pectinate, or 1–3-pinnatisect. Cauline leaves petiolate or sessile and cuneate, auriculate or sagittate at base, entire, dentate, pinnatifid, or pinnatisect. Racemes ebracteate or rarely bracteate throughout, elongated in fruit. Sepals elliptic, ovate or oblong, erect, ascending or spreading, base of lateral pair not saccate or rarely saccate, margin often membranous. Petals yellow, sometimes white or pink, rarely vestigial or absent, obovate, spatulate, oblong, oblanceolate or linear, apex obtuse or emarginate; claw sometimes distinct, often shorter than sepals. Stamens 6 and tetradynamous; anthers ovate or oblong, obtuse or rarely apiculate at apex. Nectar glands confluent, often subtending bases of all stamens. Ovules 20–170 per ovary. Fruit dehiscent siliques or silicles, linear, oblong, ovoid, or ellipsoid, terete or slightly latiseptate; valves veinless or obscurely veined, smooth or torulose; replum rounded; septum complete or rarely perforated; style obsolete or distinct; stigma capitate, entire or slightly 2-lobed. Seeds biseriate or uniseriate, wingless or rarely winged, oblong, ovoid, or ellipsoid, plump; seed coat reticulate, colliculate, rugose, tuberculate, or foveolate, mucilaginous or not when wetted; cotyledons accumbent.

Worldwide about 75 species. Four species in Nepal.

Key to Species

1a Raceme bracteate throughout or rarely along lower third.. **3. _R. benghalensis_**
  b Raceme ebracteate, rarely lowermost 1 or 2 flowers bracteate 2

2a Fruits oblong, ellipsoid, or oblong-ovoid, length less than 3 × width .... **4. _R. palustris_**
  b Fruits linear, rarely linear-oblong, length more than 4× width...3

3a Fruit often curved, (0.7–)1–2.4(–3) cm × 1–1.5(–2) mm. Seeds biseriate or nearly so. Petals longer than sepals,
     (2.5–)3–4(–4.5) × 1–1.5 mm, rarely absent ....**1. _R. indica_**
  b Fruit straight, (1.5–)2.5–4 cm × 0.7–0.9(–1) mm. Seeds uniseriate. Petals mostly absent, if present then often
     shorter than sepals and 1.5–2.5 × 0.2–0.7(–1)...**2. _R. dubia_**

1. _Rorippa indica_ (L.) Hiern, Cat. Afr. Pl. 1(1): xxvi (1896). _Sisymbrium indicum_ L., Sp. Pl., ed. 2, 2: 917 (1763); _Cardamine glandulosa_ Blanco; _Nasturtium atrovirens_ (Hornem.) DC.; _N. diffusum_ DC.; _N. heterophyllum_ D.Don later homonym, non Blume; _N. indicum_ (L.) DC.; _N. montanum_ Wall. nom. nud.; _N. montanum_ Wall. ex Hook.f. & Thomson; _N. sinapis_ (Burm.f.) O.E.Schulz; _Radicula montana_ (Wall. ex Hook.f. & Thomson) Hu ex C.Pei; _Rorippa atrovirens_ (Hornem.) Ohwi & H.Hara; _R. montana_ (Wall. ex Hook.f. & Thomson) Small; _R. sinapis_ (Burm.f.) Ohwi & H.Hara; _Sisymbrium atrovirens_ Hornem.; _S. sinapis_ Burm.f.

तोरी घाँस Tori ghans (Nepali).

Annual herbs, (6–)20–60(–75) cm, glabrous or rarely sparsely pubescent. Stems often branched basally and apically. Basal leaves withered by time of flowering. Lower and middle cauline leaves auriculate or not; petiole absent or 1–4 cm; leaf blade lyrate-pinnatipartite or undivided, obovate, oblong, or lanceolate, (2.5–)3.5–12(–16) × (0.8–)1.5–4(–5) cm,, apex obtuse to subacute margin entire or irregularly crenate to serrate; terminal lobe oblong, elliptic, to oblong-lanceolate, to 10 × 5 cm; lateral ones absent or 1–5(or 6) on each side of midvein. Uppermost leaves usually sessile, auriculate or not; leaf blade lanceolate to oblong, apex acute to acuminate, margin entire to denticulate or serrulate. Racemes ebracteate. Sepals green or pinkish, ascending, oblong-ovate, 2–3 × 0.8–1.5 mm, margin membranous. Petals yellow, obovate to spatulate, (2.5–)3–4(–4.5) × 1–1.5 mm, longer than sepals, very rarely absent. Filaments 1.5–3 mm; anthers oblong, 0.5–0.8 mm. Ovules (60–)70–110 per ovary. Fruiting pedicels slender, ascending, divaricate, or rarely slightly reflexed, straight, (2–)3–10(–15) mm. Fruit linear, (0.7–)1–2.4(–3) cm × 1–1.5(–2) mm, often curved upward; valves thin papery, not veined; style (0.5–)1–1.5(–2). Seeds reddish brown, ovate to ovate-orbicular, 0.5–0.9 × 0.4–0.6 mm, foveolate, biseriate or nearly so.

Distribution: Nepal, W Himalaya, E Himalaya, Tibetan Plateau, Assam-Burma, S Asia, E Asia, SE Asia, N America and S America.

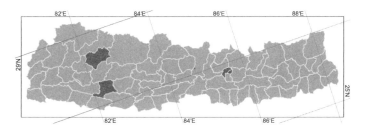

Altitudinal range: 900–2600 m.

Ecology: Roadsides, field margins, gardens, river banks.

Flowering: All year. **Fruiting:** All year.

Tender parts of the plant are cooked as a vegetable.

2. _Rorippa dubia_ (Pers.) H.Hara, J. Jap. Bot. 30: 196 (1955). _Sisymbrium dubium_ Pers., Syn. Pl. 2(2): 199 (1807); _Cardamine sublyrata_ Miq.; _Nasturtium dubium_ (Pers.) Kuntze; _N. heterophyllum_ Blume; _N. indicum_ var. _apetalum_ DC.; _N. indicum_ var. _javana_ Blume; _N. sublyratum_ (Miq.) Franch. & Sav.; _Rorippa heterophylla_ (Blume) R.O.Williams; _R. indica_ var. _apetala_ (DC.) Hochr.; _R. sublyrata_ (Miq.) H.Hara; _R. sublyrata_ (Miq.) T.Y.Cheo; _Sisymbrium apetalum_ Desf.

Annual herbs, (4–)15–33(–45) cm, glabrous or rarely sparsely pubescent. Stems often branched basally and apically. Basal leaves withered by time of flowering. Lower and middle cauline leaves auriculate or not; petiole to 4 cm, rarely absent; leaf blade lyrate-pinnatipartite or undivided, obovate, oblong, or lanceolate, (2–)3–11(–15) × (0.5–)1–3(–5) cm, apex obtuse to subacute, margin entire or irregularly crenate to serrate; terminal lobe oblong, elliptic, to oblong-lanceolate, to 14 × 4 cm; lateral ones absent or 1–4 on each side of midvein. Uppermost leaves usually sessile, auriculate or not; leaf blade lanceolate to oblong, apex acute to acuminate, margin entire to serrulate. Racemes ebracteate. Sepals often pinkish, ascending, oblong-linear, (2–)2.5–3 × 0.5–0.7 mm, margin membranous. Petals absent, if present then linear to narrowly oblanceolate, 1.5–2.5 × 0.2–0.7(–1) mm, shorter than sepals. Filaments 1.5–2.8 mm; anthers oblong, 0.5–0.8 mm. Ovules 70–90 per ovary. Fruiting pedicels slender, divaricate, straight, (2–)3–8(–10) mm Fruit linear, (1.5–)2.5–4 cm × 0.7–0.9(–1) mm, straight; valves thin, papery, not veined; style

Cruciferae (Brassicaceae)

0.2–1(–1.5) mm. Seeds reddish brown, subquadrate to ovate-orbicular, 0.5–0.8 × 0.4–0.6 mm, foveolate, uniseriate.

Distribution: Nepal, Tibetan Plateau, Assam-Burma, S Asia, E Asia, SE Asia, N America and S America.

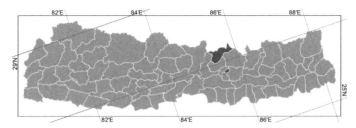

Altitudinal range: 1800–2200 m.

Ecology: Valleys, waste areas, slopes, roadsides, wet grounds, grassy places, field margins.

Flowering: All year. **Fruiting:** All year.

3. Rorippa benghalensis (DC.) H.Hara, J. Jap. Bot. 49: 132 (1974).
Nasturtium benghalense DC., Syst. Nat. 2: 198 (1821); *N. indicum* var. *benghalense* (DC.) Hook.f. & T.Anderson; *Rorippa dubia* var. *benghalensis* (DC.) Mukerjee; *R. indica* subsp. *benghalensis* (DC.) Bennet; *R. indica* var. *benghalense* (DC.) Deb; *Sinapis benghalensis* Roxb. ex DC. nom. inval.

Annual herbs, 15–65(–85) cm, glabrous or sparsely to densely hirsute with spreading to retrorse trichomes. Stems simple at base, few- to many-branched above. Basal leaves soon withering. Lowermost cauline leaves auriculate, with petioles to 3 cm; leaf blade oblong to oblong-obovate in outline, lyrate-pinnatipartite, (1.5–)2.5–12(–15) × (0.5–)1–4(–6) cm; terminal lobe broadly ovate to oblong, 1–5 × 0.7–3 cm; lateral lobes 1–4, oblong to ovate, to 2 × 1 cm, margin serrate or dentate. Upper leaves sessile, progressively reduced in size upward into bracts. Racemes bracteate throughout or rarely only along lowermost third; bracts lanceolate-linear to oblong-linear, subentire to denticulate, minutely auriculate to cuneate at base. Sepals elliptic to oblong, 1.5–2 × 0.5–0.8 mm. Petals pale yellow, spatulate to oblanceolate, 2–2.5 × 0.5–0.9 mm, apex rounded. Filaments 2–2.5 mm; anthers oblong, 0.4–0.5 mm. Ovules 100–170 per ovary. Fruiting pedicels ascending to divaricate, slender, straight, 3–6.5(–8) mm. Fruit linear, straight or curved, 0.7–1.7(–2.1) cm × 1.2–1.6 mm; style 0.3–0.8 mm. Seeds reddish brown, subglobose to broadly ovoid, biseriate, 0.5–0.6 × 0.4–0.5 mm, minutely colliculate.

Distribution: Nepal, E Himalaya, Assam-Burma, S Asia, E Asia and SE Asia.

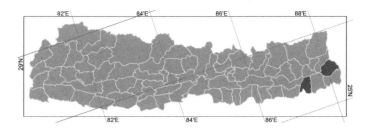

Altitudinal range: 200–300 m.

Ecology: Stream-sides, wet grounds, marsh edges.

Flowering: March–July. **Fruiting:** March–July.

4. Rorippa palustris (L.) Besser, Enum. Pl.: pl. 27 (1822).
Sisymbrium amphibium var. *palustre* L., Sp. Pl. 2: 657 (1753); *Cardamine palustre* (L.) Kuntze; *Nasturtium densiflorum* Turcz.; *N. palustre* (L.) DC.; *N. palustre* forma *longipes* Franch.; *N. palustre* forma *stoloniferum* Franch.

Annual herbs or rarely short-lived perennial, (5–)10–100(–140) cm, glabrous, rarely hirsute. Stems erect, simple or often branched above, ribbed. Basal leaves rosulate, withered early; leaf blade lyrate-pinnatisect, (4–)6–20(–30) × 1–5(–8) cm. Cauline leaves petiolate to subsessile, auriculate to amplexicaul, lyrate-pinnatisect, (1.5–)2.5–8(–12) × (0.5–)0.8–2.5(–3) cm; lateral lobes oblong to ovate, smaller than terminal lobe, (1 or)2–6(or 7) on each side of midvein, sometimes absent, margin subentire or irregularly dentate, sinuate, serrate, or crenate. Racemes ebracteate. Sepals oblong, 1.5–2.4(–2.6) × 0.5–0.8 mm. Petals yellow, spatulate, (1.5–)1.8–2.5(–3) × 0.5–1.3(–1.5) mm. Filaments 1–2.5 mm; anthers ovate, 0.3–0.5 mm. Ovules 20–90 per ovary. Fruiting pedicels divaricate or slightly to strongly reflexed, (2.5–)3–8(–12) mm, slender, straight or curved. Fruit oblong, ellipsoid, or oblong-ovoid, often slightly curved, (2.5–)4–10(–14) × (1.5–)1.7–3(–3.5) mm; valves not veined; style 0.2–1(–1.2) mm. Seeds brown to yellowish brown, ovoid to subglobose, colliculate, biseriate, 0.5–0.9 × 0.4–0.6 mm. Figs 16i, 19h–j.

Distribution: Nepal, E Himalaya, Tibetan Plateau, S Asia, E Asia, N Asia, C Asia, Europe, N America, S America and Australasia.

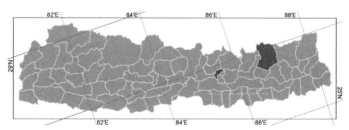

Altitudinal range: 1300–2800 m.

Ecology: Marshlands, pastures, meadows, roadsides, shores of lakes and ponds, stream banks, thickets, grasslands.

Flowering: March–October. **Fruiting:** March–October.

The first record of *Rorippa palustris* from Nepal is based on *Long & McDermott 21917* (E) from Kathmandu, and it has subsequently also been collected at Lukla. It is certainly much more widespread than the distribution map indicates.

The species is highly variable and has been divided into as many as four subspecies and seven varieties all of which are indigenous to North America. It is represented in Eurasia by subsp. *palustris*.

16. *Descurainia* Webb & Berthel. nom cons., Hist. Nat. Iles Canaries 1: 72 (1836).

Annual herbs. Trichomes short stalked, dendritic. Stems erect, simple basally, branched above. Basal leaves petiolate, withered by time of flowering, 2- or 3-pinnatisect. Lower cauline leaves similar to basal, upper sessile or short petiolate. Racemes ebracteate or rarely basally bracteate, elongated or not in fruit. Sepals oblong or linear, erect or ascending, base of lateral pair not saccate. Petals yellow, narrowly oblanceolate, apex obtuse; clawed. Stamens 6, tetradynamous; filaments not dilated at base; anthers oblong, apex obtuse. Nectar glands confluent, subtending bases of all stamens; median glands present. Ovules 5–100 per ovary. Fruiting pedicels slender, straight, divaricate to ascending. Fruit dehiscent siliques or silicles, narrowly linear, terete, sessile; valves with a prominent midvein, smooth or torulose; replum rounded; septum complete or perforated; style to 1 mm; stigma capitate, entire. Seeds uniseriate, wingless, oblong, plump; seed coat minutely reticulate, usually mucilaginous when wetted; cotyledons incumbent.

Worldwide about 40 species, mainly in N and S America and Macaronesia. One species in Nepal.

1. *Descurainia sophia* (L.) Webb ex Prantl, Nat. Pflanzenfam. 3(2): 192 (1891).
Sisymbrium sophia L., Sp. Pl. 2: 659 (1753).

मसिनो तोरीझार Masino torijhar (Nepali).

Herbs, (10–)20–70(–100) cm, eglandular, sparsely to densely pubescent with dendritic trichomes, sometimes glabrous above. Stems, simple basally, often branched above. Basal and lowermost cauline leaves 2- or 3-pinnatisect, ovate to oblong in outline, to 15 × 8 cm; petiole 0.1–2(–3) cm; leaf blade ultimate division linear to oblong, entire, acute to 10 × 2 mm. Upper cauline leaves sessile or short petiolate, smaller and often with more slender ultimate lobes, often glabrous. Sepals yellowish, oblong-linear, 2–2.5 × ca. 0.5 mm. Petals yellow, 2–2.5 × ca. 0.5 mm; claw 1.5–2 mm. Filaments 2–3 mm; anthers ca. 0.5 mm. Fruiting pedicels (0.5–)0.8–1.5(–2) cm, narrower than fruit. Fruit narrowly linear, (1–)1.5–2.7(–3) cm × 0.5–0.8(–1) mm; valves glabrous, torulose, with a prominent midvein; septum with a broad central longitudinal band appearing as 2 or 3 veins; style obsolete to 0.2 mm. Seeds uniseriate, brown, reddish, 0.7–1 × 0.3–0.6 mm.

Distribution: Nepal, E Himalaya, Tibetan Plateau, S Asia, E Asia, N Asia, C Asia, SW Asia, Europe and Africa.

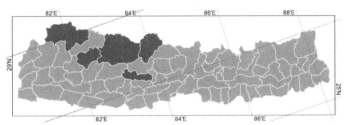

Altitudinal range: 2200–4200 m.

Ecology: Roadsides, waste places, disturbed sites, fields, pastures, deserts.

Flowering: April–June. **Fruiting:** April–June.

The seeds are sometimes used as a substitute for mustard. Young shoot and leaves are eaten as a vegetable. Preparations from the plant are taken internally to eradicate worms, and externally to treat indolent ulcers.

17. *Dontostemon* Andrz. ex C.A.Mey. nom cons., in Ledeb., Fl. Altaic. 3: 118 (1831).

Alaida Dvorák; *Dimorphostemon* Kitag.

Annual or biennial herbs. Trichomes simple, straight or crisped, mixed with multicellular, multiseriate glandular ones. Stems erect or ascending, simple or branched. Basal leaves petiolate, rosulate or not, simple, entire, dentate, pinnatifid, or pectinate-pinnatifid. Cauline leaves petiolate or sessile, similar to basal. Racemes ebracteate, elongated in fruit. Sepals oblong, not saccate. Petals white, or lavender, broadly obovate or spatulate, apex obtuse or emarginate; claw subequalling sepals or longer. Stamens 6, tetradynamous; median filament pairs free and toothed or not below anther and expanded to base, dilated at base; anthers ovate or oblong-ovate, often apiculate at apex. Nectar glands 4, lateral, 1 on each side of lateral stamen; median glands absent. Ovules 14–70 per ovary. Fruiting pedicels ascending or divaricate. Fruit dehiscent siliques, linear, terete or latiseptate; valves with a prominent midvein and often distinct marginal veins, glabrous or glandular, torulose; replum flattened; septum complete; style to 3 mm; stigma capitate, slightly lobed. Seeds uniseriate, margined or not, oblong or ovate, plump or slightly flattened; seed coat reticulate, not mucilaginous when wetted; cotyledons accumbent or incumbent.

Cruciferae (Brassicaceae)

Worldwide eleven species in E and C Asia. Two species in Nepal.

Key to Species

1a Petals obovate, (5–)6–8 mm. Median filaments abruptly expanded and toothed below anther. Seeds apically margined. Anthers 0.6–0.8 mm ....**1. *D. pinnatifidus***
b Petals spatulate, 2–4(–4.5) mm. Median filaments gradually expanded to base, toothless. Seeds not margined. Anthers 0.2–0.4 mm ....**2. *D. glandulosus***

1. *Dontostemon pinnatifidus* (Willd.) Al-Shehbaz & H.Hara, Novon 10: 96 (2000).
Cheiranthus pinnatifidus Willd., Sp. Pl., ed. 5[as 4], 3(1): 523 (1800); *Alaida pectinata* (Fisch. ex DC.) Dvorák; *Andreoskia pectinata* (Fisch. ex DC.) DC.; *Andrzeiowskia pectinata* (Fisch. ex DC.) Turcz.; *Dimorphostemon asper* Kitag.; *D. pectinatus* (Fisch. ex DC.) Golubk.; *D. pectinatus* var. *humilior* (N.Busch) Golubk.; *D. pinnatus* (Pers.) Kitag.; *D. shanxiensis* R.L.Guo & T.Y.Cheo; *Dontostemon asper* Schischk.; *D. pectinatus* (Fisch. ex DC.) Ledeb.; *D. pectinatus* var. *humilior* N.Busch; *Erysimum glandulosum* Monnet; *E. hookeri* Monnet; *Hesperidopsis pinnatifidus* (Willd.) Kuntze; *Hesperis pilosa* Poir.; *H. pinnata* Pers.; *H. punctata* Poir.; *Sisymbrium asperum* Pall.; *S. pectinatum* Fisch. ex DC.; *Torularia pectinata* (Fisch. ex DC.) Ovcz. & Junussov.

Annual or biennial herbs, (5–)10–40(–60) cm, sparsely to densely glandular. Stems erect, often simple, branched above. Basal and lowermost cauline leaves sparsely to densely pubescent with simple trichomes to 2 mm, sparsely to moderately glandular; petiole 2–10(–15) mm; leaf blade lanceolate to elliptic or oblong, (0.7–)1.5–4.5(–6) cm × (1.5–)3–10(–15) mm, base attenuate to cuneate, apex acute, margin coarsely dentate, serrate, or pinnatifid, ciliate. Middle and upper cauline leaves narrowly linear and entire, or elliptic to lanceolate and dentate to serrate. Sepals oblong, 2–3(–4) × (0.8–)1–1.5 mm, sparsely hairy apically or glabrous. Petals white, broadly obovate, (5–)6–8 × (2.5–)3–4(–5) mm, apex emarginate; claw 1–3 mm. Filaments of median stamens 2–3 mm, free, abruptly expanded and toothed below anther; filament of lateral stamens 1.5–2.5 mm, slender; anthers oblong-ovate, 0.6–0.8 mm, apiculate. Ovules 16–60 per ovary. Fruiting pedicels ascending to divaricate, often straight, (0.3–)0.5–1.5(–2.3) cm, glandular. Fruit (1.1–)1.5–4(–5) × (0.8–)1–1.3 mm, straight, erect to ascending, torulose, terete; valves sparsely to densely glandular, with prominent midvein and marginal veins; style 0.5–1.5 mm; stigma slightly lobed. Seeds brown, oblong-ovate to narrowly oblong, 1.1–2.3 × 0.7–1 mm, narrowly margined towards apex; cotyledons obliquely accumbent to incumbent.
Fig. 19n–o.

Distribution: Nepal, Tibetan Plateau, Assam-Burma, E Asia and N Asia.

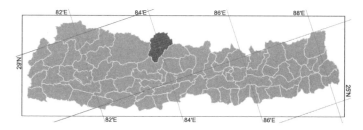

Altitudinal range: ca. 4200 m.

Ecology: Grassy plains, hillsides, rocky slopes, roadsides, sand dunes, flood plains, grasslands.

Flowering: June–August. **Fruiting:** June–August.

2. *Dontostemon glandulosus* (Karelin & Kirilov) O.E.Schulz, Notizbl. Bot. Gart. Berlin-Dahlem 10: 554 (1930).
Arabis glandulosa Karelin & Kirilov, Bull. Soc. Imp. Naturalistes Moscou 15: 146 (1842); *Alaida glandulosa* (Karelin & Kirilov) Dvorák; *Dimorphostemon glandulosus* (Karelin & Kirilov) Golubk.; *D. sergievskiana* (Polozhij) S.V.Ovchinnikova; *Neotorularia sergievskiana* (Polozhij) Czerep.; *Sisymbrium glandulosum* (Karelin & Kirilov) Maxim.; *Stenophragma glandulosa* (Karelin & Kirilov) B.Fedtsch.; *Torularia glandulosa* (Karelin & Kirilov) Vassilcz.; *T. sergievskiana* Polozhij.

Annual or biennial herbs, (1.5–)5–20(–30) cm, sparsely to densely glandular. Stems erect or ascending, simple to many-branched basally. Basal and lowermost cauline leaves sparsely to densely pubescent with simple trichomes to 1.5 mm, sparsely to moderately glandular; petiole 2–10(–25) mm; leaf blade lanceolate to oblong, (0.3–)0.5–2.5(–4) cm × 2–10(–15) mm, base attenuate to cuneate, apex acute, margin coarsely dentate to pinnatifid, ciliate. Middle and upper cauline leaves narrowly linear and entire, or elliptic to lanceolate and dentate to serrate. Sepals oblong, 1–2(–3) × 0.5–1 mm, sparsely hairy apically or glabrous. Petals lavender or white, broadly spatulate, 2–4(–4.5) × (0.5–)1–1.5(–2) mm, apex obtuse or subemarginate; claw to 1.5 mm. Filaments of median stamens 1.5–2.5 mm, free, gradually expanded to base, toothless; filament of lateral stamens 1–2 mm, slender; anthers broadly ovate, 0.2–0.4 mm, apiculate. Ovules 14–70 per ovary. Fruiting pedicels ascending to divaricate, often straight, 2–8(–12) mm, glandular. Fruit (0.7–)1.3–3(–4) × 0.8–1.3 mm, straight, erect to ascending, torulose, terete; valves sparsely to densely glandular, with prominent midvein and marginal veins; style 0.5–1 mm; stigma slightly lobed. Seeds brown, ovate to oblong, 0.8–1.7 × 0.5–0.8 mm, not

margined towards apex; cotyledons obliquely accumbent to obliquely incumbent.
Fig. 19k–m.

Distribution: Nepal, W Himalaya, E Himalaya, Tibetan Plateau, E Asia, N Asia and C Asia.

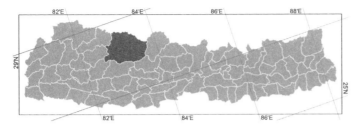

Altitudinal range: ca. 4000 m.

Ecology: Alpine meadows and steppes, sandy river banks, crevices, gravel plains, dry scrub, roadsides, scree slopes.

Flowering: June–August. **Fruiting:** June–August.

This is the first documented record of the species from Nepal, based on *Shrestha 5409* (BM) and *Grey-Wilson & Phillips 607* (K).

18. *Erysimum* L., Sp. Pl. 2: 660 (1753).

Annual or biennial herbs. Trichomes sessile, malpighiaceous or 3–5(–8)-rayed stellate. Stems simple or branched basally. Basal leaves rosulate, petiolate, simple, entire or dentate, rarely pinnatifid or pinnatisect. Cauline leaves petiolate or sessile, cuneate or attenuate at base, entire or dentate. Racemes bracteate basally, corymbose, elongated in fruit. Sepals oblong-linear, erect, pubescent, base of lateral pair saccate. Petals yellow or orange, blade oblanceolate to spatulate, apex rounded; claw differentiated from blade. Stamens 6, erect, tetradynamous; anthers linear. Nectar glands 1, 2, or 4, distinct or confluent and subtending bases of all stamens; median glands present or absent. Ovules 60–90 per ovary. Fruiting pedicels thickened and nearly as wide as fruit, erect, divaricate to ascending. Fruit dehiscent siliques or rarely silicles, linear, terete, 4-angled, latiseptate, or angustiseptate, sessile or rarely shortly stipitate; valves with a prominent midvein, pubescent on outside, keeled or not, smooth or torulose; replum rounded; septum complete; style short, often pubescent; stigma capitate, subentire. Seeds uniseriate or rarely biseriate, winged, margined, or wingless, oblong, plump or flattened; seed coat minutely reticulate, mucilaginous when wetted; cotyledons incumbent or rarely accumbent.

Worldwide about 150 species, primarily in Eurasia, but also N and C America, N Africa and Macaronesia. One species in Nepal.

Erysimum cheiri (L.) Crantz (syn. *Cheiranthus cheiri* L.) is a cultivated ornamental plant (wallflower), but it is not known if it has become naturalized in Nepal.

1. *Erysimum benthamii* Monnet, Notul. Syst. [Paris] 2: 242 (1912).
Erysimum benthamii var. *grandiflorum* Monnet; *E. dolpoense* H.Hara; *E. longisiliquum* Hook.f. & Thomson later homonym, non Schleich.; *E. pachycarpum* Hook.f. & Thomson; *E. sikkimense* Polatschek; *E. szechuanense* O.E.Schulz.

गोङ-थो-क्पा Gong-tho-kpa (Tibetan).

Herbs, (15–)30–80(–100) cm. Trichomes 3(or 4)-forked, mixed with fewer malpighiaceous ones. Stems erect, often angled, primarily with malpighiaceous trichomes. Basal leaves petiolate, rosulate, withered by time of flowering. Lower cauline leaves with petioles 1–3(–5) cm; leaf blade narrowly linear to linear-lanceolate, (2–)3–8(–11) cm × (2–)4–10(–14) mm, base attenuate to cuneate, apex acute to acuminate, margin coarsely dentate. Upper cauline leaves sessile or subsessile, denticulate to subentire. Racemes corymbose, densely flowered, elongated considerably in fruit. Sepals oblong-linear, (5–)6–8 × 1–1.5 mm, lateral pair saccate. Petals orange-yellow to yellow, (0.8–)1–1.5 cm × 2–3(–3.5) mm; claw distinct, subequalling sepals. Filaments yellow, 7–10 mm; anthers 2–3 mm. Ovules 60–90 per ovary. Fruiting pedicels divaricate to ascending, (0.5–)0.6–1.5(–2.5) cm, stout, narrower than fruit, straight. Fruit (6–)7–11(–13) cm × 1.2–1.7 mm, slightly torulose, erect to ascending, straight; valves with a prominent midvein, outside with malpighiaceous and 3- or 4-forked trichomes, inside glabrous; style slender, 1–3 mm, cylindric, narrower than fruit; stigma capitate, subentire. Seeds oblong, 1.5–2 × 0.8–1 mm.

Distribution: Nepal, E Himalaya, Tibetan Plateau, Assam-Burma and E Asia.

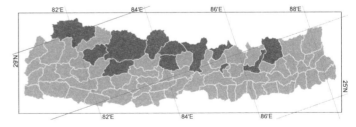

Altitudinal range: 1000–4900 m.

Ecology: Dry rocky areas, *Quercus* woods, open pastures, grassy slopes, meadows, roadsides, mountain slopes.

Flowering: May–July. **Fruiting:** July–September.

Cruciferae (Brassicaceae)

The reports of *Erysimum hieraciifolium* L. from Nepal, Bhutan, and Sikkim are based on misidentified plants of *E. benthamii*. Hara (Enum. Fl. Pl. Nepal 2: 43. 1979) recognized three species of *Erysimum* in Nepal, but the collections on which these species were based clearly belong to the variable *E. benthamii*.

19. *Braya* Sternb. & Hoppe, Denkschr. Königl.-Baier. Bot. Ges. Regensburg 1(1): 65 (1815).

Platypetalum R.Br.

Perennial herbs. Caudex simple or many-branched. Trichomes simple or forked. Plants rarely glabrous or glabrescent. Basal leaves petiolate, rosulate, simple, entire or dentate, rarely pinnately lobed, petiolar base persistent. Cauline leaves absent, or few, sessile or nearly so, not auriculate, entire or rarely dentate. Racemes bracteate, ebracteate or only lowermost flowers bracteate, elongated or not in fruit. Sepals oblong, caducous or persistent, erect, base of lateral pair not saccate, margin membranous. Petals white, pink, or purple, spatulate to obovate, apex obtuse or rounded; claw shorter than sepals. Stamens 6, tetradynamous; filaments dilated at base; anthers ovate or oblong, obtuse at apex. Nectar glands 4, 1 on each side of lateral stamen; median glands absent. Ovules 8–40 per ovary. Fruiting pedicels erect, divaricate to ascending. Fruit dehiscent siliques or silicles, linear, oblong, ovoid, terete or slightly latiseptate, sessile; valves with a distinct midvein, glabrous or pubescent, smooth or torulose; replum rounded; septum complete; style obsolete or to 1 mm; stigma capitate, entire to 2-lobed. Seeds uniseriate (rarely biseriate), wingless, oblong or ovoid, plump; seed coat minutely reticulate, not mucilaginous when wetted; cotyledons incumbent.

Worldwide about 20 polymorphic species in the alpine, subarctic, or northern temperate regions of N America, Europe, and Asia. Two species in Nepal.

Key to Species

1a   Stems leafy. Racemes bracteate at least basally. Fruit linear siliques .... **1. *B. humilis***
b   Stems leafless. Racemes ebracteate. Fruit oblong or ovoid silicles .... **2. *B. rosea***

1. *Braya humilis* (C.A.Mey.) B.L.Rob., Syn. Fl. N. Amer. 1(1): 141 (1895).
Sisymbrium humile C.A.Mey., in Ledeb., Icon. Pl. 2: 16 (1830); *Arabidopsis tuemurnica* K.C.Kuan & Z.X.An; *Arabis piasezkii* Maxim.; *Dichasianthus humilis* (C.A.Mey.) Soják; *Erysimum alyssoides* Franch.; *E. stigmatosum* Franch.; *Hesperis hygrophila* Kuntze; *H. piasezkii* (Maxim.) Kuntze; *Malcolmia perennans* Maxim.; *Neotorularia humilis* (C.A.Mey.) Hedge & J.Léonard; *N. maximowiczii* (Botsch.) Botsch.; *N. piasezkii* (Maxim.) Botsch.; *Sisymbrium humile* var. *piasezkii* (Maxim.) Maxim.; *S. nanum* Bunge; *S. piasezkii* Maxim.; *Torularia humilis* (C.A.Mey.) O.E.Schulz; *T. humilis* var. *maximowiczii* (Botsch.) H.L.Yang; *T. humilis* var. *piasezkii* (Maxim.) Jafri; *T. maximowiczii* Botsch.; *T. piasezkii* (Maxim.) Botsch.

Herbs, (4–)8–25(–35) cm, sparsely to densely covered with short-stalked to subsessile, submalpighiaceous or rarely 2-forked trichomes often mixed along petioles and stem bases with simple trichomes, rarely plants glabrescent. Stems leafy, usually few to many from the base, ascending to erect, rarely subdecumbent. Basal leaves rosulate; petiole 2–16(–35) mm; leaf blade obovate, spatulate, oblanceolate, oblong, to sublinear, (0.3–)0.5–2(–3.5) cm × 1–8(–10) mm, base attenuate to cuneate, apex acute to obtuse, margin entire, repand, dentate, to pinnatifid, sparsely to densely pubescent or rarely glabrous. Cauline leaves similar to basal leaves but progressively smaller upward, uppermost sessile to subsessile. Racemes bracteate at least basally. Sepals, 2–3 × 0.8–1.2 mm, subsaccate or not saccate at base. Petals white, pink, or purple, spatulate to broadly obovate,

3–5(–8) × (1–)1.5–2.5(–4) mm, apex rounded. Filaments 2–3(–4) mm; anthers oblong, 0.4–0.7 mm. Ovules 20–40 per ovary. Fruiting pedicels erect and subappressed to rachis to ascending or divaricate, slender, much narrower than fruit, (2.5–)3–8(–12) mm. Fruit linear, 1.2–2.5(–3.2) cm × 0.5–0.9(–1) mm, mostly straight, terete, torulose, pubescent with submalpighiaceous trichomes rarely mixed with fewer straight ones, rarely glabrescent; style slender, 0.3–0.8(–1) mm; stigma entire to strongly 2-lobed. Seeds oblong 0.6–0.9 × 0.4–0.5 mm.
Figs 16j, 20a–c.

Distribution: Nepal, W Himalaya, E Himalaya, Tibetan Plateau, Assam-Burma, N Asia, C Asia, SW Asia and N America.

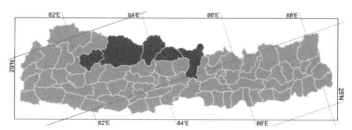

Altitudinal range: 1000–5300 m.

Ecology: Sandy areas, river terraces, open stony slopes, scree, limestone ledges.

Flowering: May–August. **Fruiting:** June–August.

160

A highly variable species in leaf shape and margin, flower size and colour, pubescence, fruit length and orientation, cotyledonary position, chromosome numbers, and length of the bracteate portion of racemes. Numerous taxa have been described in China, Russia, and North America, but the variation is continuous and does not support the recognition of infraspecific taxa. Fully bracteate to only basally bracteate racemes are often found within the same population and throughout the range of the species. The above synonymy all pertains to E and C Asia, and it represents about half of the total synonymy on the species.

2. Braya rosea (Turcz.) Bunge, Del. Sem. Hort. Dorpater: 7 (1841).
Platypetalum roseum Turcz., Bull. Soc. Imp. Naturalistes Moscou 11: 87 (1838); *Braya aenea* Bunge; *B. aenea* subsp. *pseudoaenia* V.V.Petrovsky; *B. angustifolia* (N.Busch) Vassilcz.; *B. brachycarpa* Vassilcz.; *B. brevicaulis* Schmid; *B. limosella* Bunge; *B. limoselloides* Bunge ex Ledeb.; *B. marinellii* Pamp.; *B. rosea* var. *aenea* (Bunge) Malyschev; *B. rosea* var. *brachycarpa* (Vassilcz.) Malyschev; *B. rosea* var. *glabra* Regel & Schmalh.; *B. rosea* var. *leiocarpa* O.E.Schulz; *B. rosea* var. *multicaulis* B.Fedtsch.; *B. rosea* var. *simplicior* B.Fedtsch.; *B. sinuata* Maxim.; *B. thomsonii* Hook.f.; *B. tibetica* Hook.f. & Thomson; *B. tibetica* forma *linearifolia* Z.X.An; *B. tibetica* forma *sinuata* (Maxim.) O.E.Schulz; *B. tibetica* var. *breviscapa* Pamp.; *B. tinkleri* Schmid; *Hesperis limosella* (Bunge) Kuntze; *H. limoselloides* (Bunge ex Ledeb.) Kuntze; *H. rosea* (Turcz.) Kuntze; *Sisymbrium alpinum* var. *aeneum* (Bunge) Trautv.; *S. alpinum* var. *roseum* (Turcz.) Trautv.; *S. limosella* (Bunge) E.Fourn.

Herbs (1–)3–10(–16) cm, scapose, densely to sparsely pilose with short-stalked forked trichomes sometimes mixed with simple ones, rarely glabrous throughout except for petiole margin. Caudex slender, sometimes with petiolar remains of previous years, simple or few-branched. Basal leaves rosulate; petiole (0.2–)0.4–1.6(–3) cm, slender or slightly expanded at base, ciliate; leaf blade linear, oblong, oblanceolate, obovate, (0.4–)1–3(–4) cm × 0.5–3.5(–6) mm, base attenuate, apex obtuse to acute, margin entire to dentate or sinuate, densely to sparsely pilose or glabrous. Scapes densely pilose to glabrous, leafless. Raceme ebracteate, capitate or rarely considerably elongated in fruit. Sepals subapically purple or greenish, oblong, 1.5–2.5(–3) × 1–1.2 mm, glabrous to densely pubescent, broadly white margined. Petals purple, pink, to white, spatulate to obovate, (2.5–)3–4(–4.5) × (0.7–)1–1.5 mm, apex obtuse. Filaments 1.3–1.8 mm; anthers ovate, ca. 0.3 mm. Ovules 8–12 per ovary. Fruiting pedicels divaricate to ascending, 1.5–5(–7) mm. Fruit ovoid to oblong, (2–)3–6.5(–8) × 1–2 mm, glabrous to densely pubescent; style 0.2–0.7(–1) mm. Seeds ca. 0.7–1 × 0.4–0.5 mm.

Distribution: Nepal, W Himalaya, E Himalaya, Tibetan Plateau, E Asia, N Asia and C Asia.

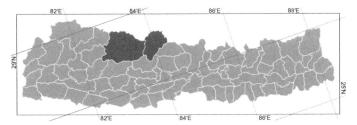

Altitudinal range: 2500–5300 m.

Ecology: Mountain slopes, river banks, scree, weathered marble rock and debris, steppe, alpine cushions.

Flowering: June–July. **Fruiting:** July–September.

This is the first documented record of this species for Nepal, and it is based on *Wald 24* (BM), *Stainton 4305* (BM) and *Grey-Wilson & Phillips 540* (K). Critical examination of numerous collections, including types and authentic material, of *Braya rosea, B. aenea, B. tibetica,* and *B. thomsonii* reveal that there is not a single character that can be used to reliably distinguish between them. There is continuous variation in the amount of indumentum, fruit length and shape, leaf shape and margin, duration of sepals, relative length of scape to basal rosette, and petal colour and shape.

20. *Christolea* Cambess., in Jacquem., Voy. Inde 4(1): 17 (1841).

Koelzia Rech.

Perennial herbs with a woody caudex or herbaceous base. Trichomes simple. Stems branched from caudex and above, sometimes woody at base. Basal leaves absent. Cauline leaves petiolate, dentate, uppermost sometimes entire. Racemes ebracteate, elongated in fruit. Sepals oblong, caducous, base of lateral pair not saccate. Petals cream to yellow with purple base, spatulate or oblong-obovate, apex rounded; claw purple, subequalling sepals. Stamens 6, tetradynamous; filaments not dilated at base; anthers oblong, apiculate at apex. Nectar glands confluent and subtending bases of all stamens; median glands present. Ovules 10–20 per ovary. Fruiting pedicels slender, ascending, suberect, or reflexed. Fruit dehiscent siliques, linear to oblong or lanceolate, latiseptate, sessile; valves papery, with a distinct midvein, glabrous or puberulent, torulose; replum rounded, covered by connate valve margin; septum complete; style obsolete, rarely to 0.5 mm; stigma capitate, entire or slightly 2-lobed. Seeds uniseriate, wingless, sometimes appendaged towards apex, oblong, slightly flattened, often transversely oriented in locule; seed coat minutely reticulate, not mucilaginous when wetted; cotyledons incumbent.

Worldwide two species in the Himalaya and C Asia. One species in Nepal.

Cruciferae (Brassicaceae)

1. *Christolea crassifolia* Cambess., in Jacquem., Voy. Inde 4(1): 17 (1841).
Christolea afghanica (Rech.f.) Rech.; *C. crassifolia* var. *pamirica* (Korsh.) Korsh.; *C. incisa* O.E.Schulz; *C. pamirica* Korsh.; *Ermania pamirica* (Korsh.) Ovcz. & Junussov; *Koelzia afghanica* Rech.f.; *Parrya ramosissima* Franch.

Herbs (8–)15–40(–50) cm. Caudex woody, compactly branched, to 2.5 cm in diameter Stems branched, sometimes woody at base, hirsute with simple trichomes to 1.2 mm, rarely glabrous. Leaves all cauline, several to many; petiole (2–)5–12(–15) mm; leaf blade highly variable in shape, pubescence, or margin, usually obovate, spatulate, rhomboid, oblong, or elliptic, (0.8–)1.3–3.5(–5) × 0.4–1.6(–2.5) cm, base cuneate to attenuate, apex acute to obtuse, margin dentate to rarely incised along upper half, sometimes entire, leathery, densely to sparsely pubescent, rarely glabrous. Raceme several- to many-flowered. Sepals oblong, 3–4 × 1.5–2 mm, glabrous or sparsely puberulent. Petals 5–6.5 × 2–3 mm; claw 2.5–3.5 mm. Filaments white, median pairs 2.5–3 mm, lateral pair 1.5–2 mm; anthers oblong, 0.9–1.2 mm. Fruiting pedicels 5–9(–12) mm, glabrous or sparsely puberulent. Fruit (0.9–)1.5–3(–3.5) cm × (2–)2.7–4(–4.5) mm, subappressed to rachis, flattened; valves constricted between laterally oriented seeds, glabrous or puberulent, base obtuse, apex acute to subacuminate; style obsolete or to 0.4 mm. Seeds brown, oblong, 1.8–2.3 × 0.9–1.5 mm.
Figs 16p, 20d–e.

Distribution: Nepal, W Himalaya, Tibetan Plateau, C Asia and SW Asia.

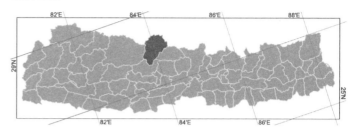

Altitudinal range: 3500–4700 m.

Ecology: Alpine steppe, rocky slopes, bare slopes.

Flowering: June–September. **Fruiting:** July–October.

---

21. *Dilophia* Thomson, Hooker's J. Bot. Kew Gard. Misc. 5: 19 (1853).

Perennial herbs. Trichomes absent or simple. Roots fleshy, conical. Stems erect, simple to ground level, branched above. Basal leaves sessile, fleshy, rosulate, simple, entire or sinuate or dentate. Cauline leaves sessile, attenuate and not auriculate, entire. Corymbs several- to many-flowered, bracteate or ebracteate, simple or compound, not elongated in fruit. Sepals broadly ovate, persistent, erect to ascending, glabrous or puberulent, base of lateral pair not saccate. Petals white to lavender or purple, longer than sepals; blade spatulate to spatulate-linear, apex obtuse to subemarginate; claw obscure to distinct. Stamens 6, slightly spreading, slightly tetradynamous; filaments not dilated at base; anthers ovate, apiculate at apex. Nectar glands confluent and subtending bases of all filaments. Ovules 4–12 per ovary. Fruiting pedicels ascending or divaricate. Fruit dehiscent silicles, obcordate, angustiseptate, sessile; valves membranous or thin papery, obscurely veined, glabrous or puberulent, rounded, smooth, apically gibbous and cristate; replum strongly flattened, base much broader; septum complete or perforated; style to 0.5 mm, included in apical notch of fruit; stigma capitate, entire. Seeds wingless, broadly oblong, slightly flattened; seed coat minutely reticulate, not mucilaginous when wetted; cotyledons incumbent or oblique.

Worldwide two species in the Himalayas. One species in Nepal.

---

1. *Dilophia salsa* Thomson, Hooker's J. Bot. Kew Gard. Misc. 5: 20 (1853).
Dilophia dutreuili Franch.; *D. kashgarica* P.Rupr.; *D. salsa* var. *hirticalyx* Pamp.

Herbs glabrous throughout, or sparsely to densely pubescent on fruit valves and/or sepals with simple flattened trichomes to 0.5 mm. Stems erect, somewhat fleshy, 1.5–10 cm, often arranged in a flat mat 1–12 cm in diameter. Leaves spatulate to linear-spatulate or oblong to linear, (3–)6–20(–30) × (1–)2–3(–5) mm, fleshy, base attenuate, apex obtuse, margin sinuate to dentate or repand, sometimes entire. Corymbs few- to many-flowered, lowermost or most flowers bracteate, sometimes flowers solitary from centre of rosette. Sepals (1–)1.5–2.5 × 1–1.5 mm, glabrous to densely pubescent, margin broadly membranous, apex denticulate or entire. Petals white to lavender, drying purplish, 1.8–2.5(–3.2) × 0.5–1(–1.5) mm, apex obtuse to subemarginate; claw to 2 mm. Filaments white, 1.1–1.6 mm; anthers ovate, (0.2–)0.3–0.5 mm, apicula triangular, 0.05–0.1 mm. Fruiting pedicels 3–10 mm. Fruit valves oblong to ovate, glabrous to densely pubescent, 1–2.5 mm, apical cristae 3–14; replum 1.2–2.5 mm wide at base; style 0.2–0.5 mm. Seeds brown to blackish, broadly oblong, (2 or)4–8(–12) per fruit, 0.7–1.1(–1.5) × 0.5–6 mm.
Fig. 20f–g.

Distribution: Nepal, W Himalaya, E Himalaya, Tibetan Plateau, E Asia, N Asia and C Asia.

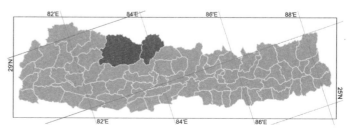

162

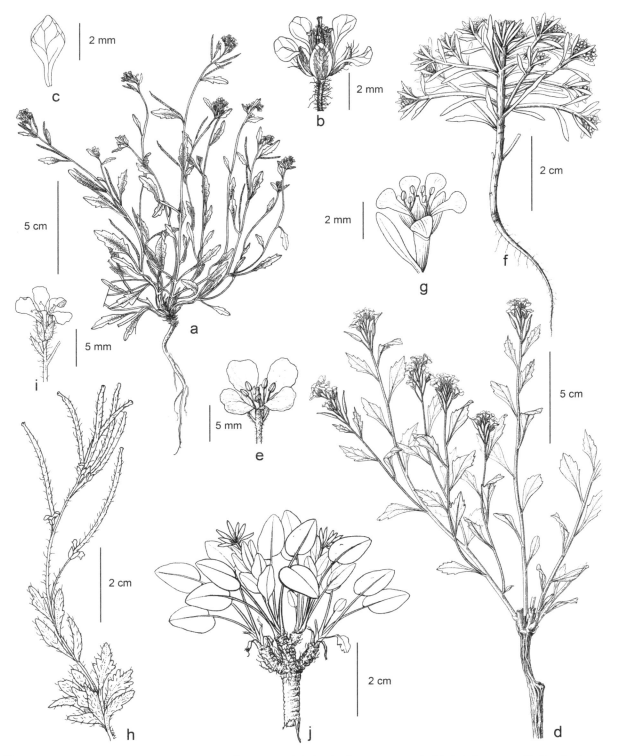

FIG. 20. CRUCIFERAE. **Braya humilis**: a, plant with flowers and fruits; b, flower; c, petal. **Christolea crassifolia**: d, plant with flowers and fruits; e, flower. **Dilophia salsa**: f, flowering plant; g, flower. **Solms-laubachia linearis**: h, fruiting plant; i, flower. **Eutrema heterophyllum**: j, flowering plant.

163

Cruciferae (Brassicaceae)

Altitudinal range: 2200–5500 m.

Ecology: Among sand in river bottoms or by marshes, damp mud by stream-sides, debris covered glacial moraine, scree, grassland in valley, steep slopes with semiconsolidated scree, turf cushions, sand plains, permafrost, salty pastures, dunes, alpine steppe.

Flowering: June–August. **Fruiting:** July–September.

22. *Lepidostemon* Hook.f. & Thomson, J. Proc. Linn. Soc., Bot. 5: 161 (1861).

Chrysobraya H.Hara.

Annual or caespitose perennial herbs. Trichomes dendritic, forked, submalpighiaceous, or simple, often more than 1 kind present. Stems erect, simple from rosettes, leafy or leafless. Basal leaves petiolate, rosulate or not, simple, entire or dentate. Cauline leaves absent or similar to basal, sometimes pinnatifid. Racemes few- to many-flowered, ebracteate or bracteate throughout, elongated in fruit, sometimes flowers solitary on long pedicels originating from axils of rosette leaves. Sepals oblong, persistent or caducous, base of lateral pair not saccate, margin membranous. Petals yellow, white, lavender, or purple, broadly obovate, apex rounded or emarginate; claw subequalling sepals. Stamens 6, tetradynamous; filaments winged or wingless, toothed or rarely toothless; anthers reniform, obtuse at apex. Nectar glands 4, lateral; median glands absent. Ovules 8–28 per ovary. Fruiting pedicels slender, ascending to divaricate or recurved. Fruit dehiscent siliques, linear-oblong to linear, terete or latiseptate, sessile; valves papery, obscurely veined, rarely marginal veins prominent, pubescent, torulose or rarely smooth; replum rounded; septum complete or rarely perforate; style distinct, to 2 mm; stigma capitate, entire. Seeds uniseriate, wingless, oblong or ovate, plump; seed coat minutely reticulate, not mucilaginous when wetted; cotyledons accumbent.

Worldwide five species in Bhutan, China and Sikkim. Three species in Nepal.

Hara (Enum. Fl. Pl. Nepal 2: 38. 1979) did not record the genus for Nepal. Instead, he listed one species (*Lepidostemon glaricola* (H.Hara) Al-Shehbaz) as *Chrysobraya* and another (*L. williamsii* (H.Hara) Al-Shehbaz) as *Draba*.

Key to Species

1a Annuals. Filaments of median stamens flattened, often appendaged. Cotyledons accumbent ....**1. *L. gouldii***
 b Perennials. Filaments of median stamens slender, unappendaged. Cotyledons incumbent...2

2a Racemes ebracteate. Stems leafless. Trichomes exclusively dendritic. Nectar glands clavate, 0.5–1 mm. Petals bright yellow ....**2. *L. glaricola***
 b Racemes bracteate throughout. Stems leafy. Most trichomes submalpighiaceous. Nectar glands to 0.2 mm. Petals white .... **3. *L. williamsii***

1. *Lepidostemon gouldii* Al-Shehbaz, Novon 10: 331 (2000).

Annual herbs, 2–15 cm. Trichomes simple, stalked forked, or subsessile and submalpighiaceous on stems and pedicels, almost exclusively simple and to 1.5 mm long on leaves. Basal leaves rosulate, smaller than and somewhat similar to cauline ones, dry by flowering time. Cauline leaves few to many, well-spaced below raceme; petiole obsolete or rarely to 2 mm; leaf blade spatulate to oblanceolate, 2–15 × 1–3(–4) mm, base attenuate, apex obtuse, margin entire or obscurely dentate with 1 or 2 teeth on each side, sparsely pubescent. Racemes 3–12-flowered, ebracteate, elongated considerably in fruit. Sepals obong, 1.5–2 × 1–1.5 mm, spreading, caducous. Petals purple, lavender, or rarely white, broadly obovate, 4.5–6 × 3–4 mm, caducous, apex emarginate; claw 1–2 mm. Filaments 1.5–2.5 mm, caducous; median pairs with a linear wing 0.3–0.5 mm wide, unexpanded portion of filament to 0.3 mm; lateral pair entire; anthers reniform, 0.3–0.5 mm. Nectar glands ovate, 0.2–0.3(–0.4) mm. Ovules 8–12 per ovary. Fruiting pedicels divaricate, recurved, 5–12 mm. Fruit (immature) narrowly oblong, 5–10 × ca. 1 mm, terete; valves densely pubescent with stalked or dendritic trichomes; style ca. 1 mm; stigma wider than style, entire. Immature seeds oblong.

Distribution: Nepal and E Himalaya.

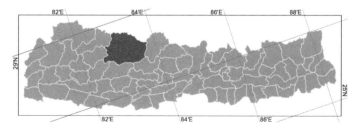

Altitudinal range: ca. 3600 m.

Ecology: Sandy soil near streams, moist *Abies* forest.

Flowering: May–July. **Fruiting:** July–August.

2. *Lepidostemon glaricola* (H.Hara) Al-Shehbaz, Novon 10: 333 (2000).
Chrysobraya glaricola H.Hara, J. Jap. Bot. 49: 195 (1974).

Perennial caespitose herbs, 1–10 cm. Trichomes crisped, dendritic or rarely few forked. Basal leaves rosulate, many; petiole 3–12 mm; leaf blade spatulate to obovate or oblanceolate, 3–15 × 1.5–5 mm, base attenuate, apex subacute to obtuse, margin 1–4-toothed on each side, rarely subentire. Cauline leaves absent. Racemes numerous-flowered, ebracteate, elongated in fruit. Sepals oblong, 2–2.5 × 1–1.5 mm, slightly spreading, hairy, persistent. Petals bright yellow, broadly obovate, 3–6 × 2–4 mm, persistent, apex emarginate to rounded; claw 1–2 mm. Filaments 1.5–2.5 mm, persistent, neither winged nor dentate; anthers reniform, 0.5–0.6 mm. Nectar glands clavate, 0.5–1 mm. Ovules 8–12 per ovary. Fruiting pedicels ascending to divaricate, slender, straight or slightly curved, (0.5–)1–3 cm. Fruit linear-oblong, 4–11 × 1.2–2 mm, terete; valves torulose, densely pubescent with dendritic trichomes, obscurely veined; septum complete or perforate; style 0.3–1 mm; stigma wider than style, entire. Seeds oblong, 1–1.2 × 0.5–0.6 mm; cotyledons incumbent.

Distribution: Nepal and E Himalaya.

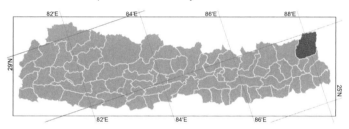

Altitudinal range: ca. 4100 m.

Ecology: Rocks, mossy boulder scree, wet stream beds, scree with running water, meadows, rock crevices.

Flowering: May–July. **Fruiting:** July–October.

3. *Lepidostemon williamsii* (H.Hara) Al-Shehbaz, Edinburgh J. Bot. 59(3): 446 (2000).
Draba williamsii H.Hara, J. Jap. Bot. 52: 353 (1977).

Perennial herbs, 1.5–12 cm. Trichomes simple and subsetose on sepals and along leaf margin and upper surface, to 1.2 mm, those on stems, pedicels, lower leaf surface, and fruits stalked-forked or subsessile and submalpighiaceous. Stems simple at base. Basal leaves not rosulate, soon withered. Cauline leaves subsessile or with petioles to 5 mm; leaf blade spatulate, oblanceolate, or lanceolate, 3–25 × 1–5 mm, base attenuate, apex acute to obtuse, margin dentate with 1–4 teeth on each side. Racemes 3–20-flowered, bracteate throughout, elongated considerably in fruit. Sepals oblong, 2–3 × 1–1.5 mm, ascending, caducous. Petals white, broadly obovate, 4–6 × 2.5–4 mm, caducous, apex emarginate; claw 2–3 mm. Filaments 1.5–2.5 mm, caducous, slender, unappendaged; anthers reniform, 0.3–0.5 mm. Nectar glands ca. 0.1 mm, subconfluent. Ovule number unknown. Fruiting pedicels divaricate, recurved, 5–15 mm. Fruit linear, 8–13 × ca. 1 mm, terete, not torulose; valves densely pubescent; style 0.4–0.8 mm; stigma as wide as style, entire. Seeds ovate, brown, 0.7–0.8 × ca. 0.5 mm; cotyledons incumbent.

Distribution: Endemic to Nepal.

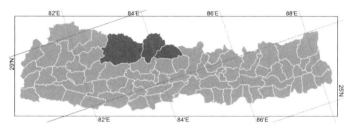

Altitudinal range: 3100–4600 m.

Ecology: Grassy slopes, among rocks in river shingles, edge of pastures, wet stony slopes, rock crevices, exposed soil near snow patches.

Flowering: June–July. **Fruiting:** June–July.

One of the paratypes cited by Hara (1977) as *Draba williamsii*, *Ludlow & Sherriff 16839* (BM), is the closely related *Lepidostemon gouldii* Al-Shebaz. *Lepidostemon williamsii* differs from *L. gouldii* by having fully bracteate instead of ebracteate racemes and unappendaged instead of apically appendaged median staminal filaments.

23. *Pycnoplinthopsis* Jafri, Pakistan J. Bot. 4: 73 (1972).

Perennial herbs, caespitose, scapose. Caudex simple to many-branched. Trichomes eglandular, dendritic or forked, occasionally with few simple ones. Stems reduced to tiny portions added annually to the apex of caudex or its branches. Basal leaves petiolate, not fleshy, forming well-defined rosette; petiole caducous, thin; leaf blade simple, coarsely dentate to incised towards apex. Cauline leaves absent. Racemes solitary flowers borne on long pedicels originating from centre of rosette. Sepals united into campanulate calyx, persistent, not saccate; lobes ovate to deltoid. Petals white, longer than sepals, broadly obovate, shallowly emarginate at apex; claw obscurely differentiated from blade, much shorter than calyx. Stamens 6, included, erect, tetradynamous; filaments filiform, wingless, unappendaged, not dilated at base, glabrous, free; anthers blackish, ovate, sagittate at base, apiculate at apex. Nectar glands 2, lateral, semiannular and intrastaminal; median glands absent. Ovules 8–20 per ovary. Fruiting pedicels slender, terete, strongly recurved, much shorter than leaves. Fruit dehiscent capsular siliques or silicles, linear to oblong, terete, sessile, unsegmented, geocarpic; valves papery, not navicular, not keeled, torulose, wingless, unappendaged, obscurely veined; replum rounded; septum complete; gynophore absent; style to 1.5 mm, cylindric, persistent, glabrous; stigma capitate, entire. Seeds uniseriate, wingless, oblong, plump; seed coat obscurely reticulate, not mucilaginous when wetted; cotyledons incumbent.

Worldwide one species in Nepal, Bhutan, Sikkim, India and China.

1. *Pycnoplinthopsis bhutanica* Jafri, Pakistan J. Bot. 4: 74 (1972).
Pegaeophyton bhutanicum H.Hara; *Pycnoplinthopsis minor* Jafri.

Herbs with slender to stout, apically branched or rarely unbranched caudex to 2 cm in diameter. Petiole (0.2–)0.5–3(–5) cm, thin, flattened, not persistent; leaf blade spatulate to oblanceolate, rarely obovate, (0.5–)1–4(–4.5) × 0.4–1.6 cm, base subattenuate to cuneate, apex acute, margin subincised to deeply dentate towards apex and with (3–)4–8(–12) teeth on each side, thin, glabrous to densely pubescent above in upper half with forked to dendritic, often crisped trichomes to 0.6 mm. Calyx (2.5–)3.5–5 mm, membranous, persisting with fruit, sometimes splitting as fruit develops; lobes ovate-deltoid, 1.5–3.5 × 1–2.5 mm. Petals white, broadly obovate, (5.5–)8–13 × (4–)5.5–8 mm, apex subemarginate; claw obscure, rarely to 1 mm. Filaments slender, median pairs 2–3.5 mm, lateral pair 1.5–2 mm; anthers ca. 0.4 mm. Fruiting pedicels, (0.5–)1–2.5(–4) cm. Fruit linear to oblong, 5–11 × 2–3 mm; valves papery; style 0.5–1 mm. Seeds oblong, light brown, 1–1.4 × 0.6–8 mm.

Distribution: Nepal, E Himalaya, Tibetan Plateau and Assam-Burma.

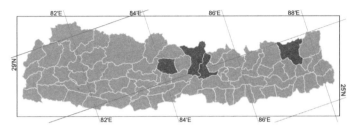

Altitudinal range: 3000–4500 m.

Ecology: Stony areas at stream edge, crevices of steep wet rocks, open scree, mossy areas below waterfalls, wet rock ledges.

Flowering: May–July. **Fruiting:** August–September.

24. *Solms-laubachia* Muschl., Notes Roy. Bot. Gard. Edinburgh 5: 205 (1912).

Desideria Pamp.; *Ermaniopsis* H.Hara; *Oreoblastus* Suslova.

Perennial herbs, sometimes pulvinate. Caudex branched covered with petioles of previous years. Trichomes simple, rarely shortly stalked, branched. Stems minute or present and simple, leafy or leafless. Basal leaves petiolate, rosulate, simple, entire or 3–9(–11)-toothed, pinnately or palmately veined. Cauline leaves absent or similar to basal ones and short petiolate to subsessile, not auriculate. Flowers in (3–)6–35-flowered, bracteate or ebracteate corymbose racemes elongated or not in fruit. Sepals oblong to ovate, free or united, persistent or caducous, erect, equal, base of inner pair not saccate, margin membranous. Petals purple, blue to lilac, greenish-blue, pink, or rarely white, suborbicular, obovate, to spatulate, apex obtuse to emarginate; claw subequalling or longer than sepals. Stamens 6; filaments free, dilated or not at base; anthers oblong-linear to ovate, not apiculate at apex; nectar glands 2 and lateral, or confluent and subtending bases of all filaments; median nectaries absent or present. Ovules 6–30 per ovary. Fruit dehiscent silique or silicle, linear, oblong, ovate, or lanceolate, latiseptate, sessile, readily detached at maturity from pedicel, rectangular in cross section; valves papery, reticulate veined, with a prominent midvein and marginal veins, glabrous or pubescent, smooth, adnate with replum at fruit apex, margin angled; replum rounded, concealed by connate valve margins; septum complete or rarely perforate or reduced to a rim, membranous, translucent, rarely absent; style obsolete, rarely to 1 mm long; stigma capitate, entire or 2-lobed, lobes not decurrent. Seeds uniseriate or biseriate, wingless, broadly ovate to suborbicular or oblong, flattened; seed coat reticulate, rugose, or papillate, not mucilaginous when wetted; cotyledons accumbent.

Worldwide 26 species in the Himalayas, E and C Asia. Five species in Nepal.

Key to Species

1a	Leaves entire	**4. *S. jafrii***
b	Leaves dentate	2
2a	Sepals united, persistent	**2. *S. nepalensis***
b	Sepals free, caducous	3

3a   Racemes ebracteate. Filaments flattened, subapically toothed..**1. *S. haranensis***

b   Racemes bracteate throughout. Filaments not flattened or toothed 4

4a   Fruit linear, (0.8–)1–1.7(–2) mm wide. Petals 4–5(–5.5) × 1.5–2.5 mm. Seeds uniseriate, 0.8–1.1 × 0.5–0.8 mm
     ....**3. *S. linearis***

b   Fruit lanceolate to linear-lanceolate, (3–)4–6 mm wide. Petals (6–)6.5–8 × 3–4 mm. Seeds biseriate,
     (1.5–)1.8–2(–2.3) × 1–1.4 mm ....**5. *S. himalayensis***

1. *Solms-laubachia haranensis* (Al-Shehbaz) J.P.Yue, Al-Shehbaz & H.Sun, Ann. Missouri Bot. Gard. 95: 534 (2008). *Desideria haranensis* Al-Shehbaz, Ann. Missouri Bot. Gard. 87: 559 (2001); *Ermaniopsis pumila* H.Hara.

Herbs 2–6 cm. Trichomes simple, straight, to 0.5 mm, mixed on leaves with short-stalked unequally branched forked ones. Stems erect, simple, pilose to hirsute. Basal leaves fleshy, persistent; petiole 2–12 mm, sparsely to densely pilose with simple trichomes, ciliate at base, not expanded or papery at base; leaf blade broadly ovate, suborbicular, to obovate, 3–13 × 3–11 mm, base cuneate or obtuse, apex obtuse, margin 1–5-toothed, sparsely to densely pubescent. Cauline leaves absent. Racemes 3–8-flowered, ebracteate. Pedicel divaricate, straight, 4–12 mm, pilose. Sepals free, oblong, 3.5–4.5 × 1.7–2 mm, caducous, pilose, base not saccate, margin membranous. Petals white tinged with greenish blue, obovate, 6.5–8 × 3–4 mm, apex obtuse; claw 3–4 mm. Filaments white, flattened and subapically toothed, median pairs 3–4 mm, lateral pair 2–3 mm; anthers oblong, 0.9–1.1 mm. Ovules 10–14 per ovary. Immature fruit linear, flattened, sessile, straight, retrorsely pilose; septum complete; style-like glabrous apex to 1.5 mm; stigma capitate, subentire. Seeds not seen.

Distribution: Endemic to Nepal.

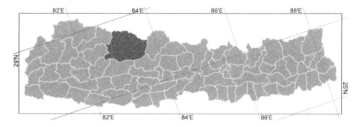

Altitudinal range: 5000–5900 m.

Ecology: Scree slopes.

Flowering: June.

Known from two collections from Dolpa District.

2. *Solms-laubachia nepalensis* (H.Hara) J.P.Yue, Al-Shehbaz & H.Sun, Ann. Missouri Bot. Gard. 95: 535 (2008). *Desideria nepalensis* H.Hara, J. Jap. Bot. 50: 264 (1975).

Herbs 2–3 cm. Trichomes simple, straight, to 1 mm. Stems minute, simple, glabrous. Basal leaves slightly fleshy; petiole 2–5 mm, sparsely pilose with simple trichomes, ciliate at base, not expanded or papery at base; leaf blade broadly obovate to subflabellate, 2–3 × 1–3 mm, base cuneate, apex acute, margin 3–5-toothed, densely pubescent. Cauline leaves absent. Flowers 2–4, ebracteate. Pedicel ascending, straight, 3–5 mm, solitary from basal rosette, spreading, pilose. Sepals united, 5–6 × 3–4 mm, densely pilose, base not saccate; calyx lobes ovate, 1.5–2 mm, margin membranous. Petals purplish, obovate, 11–13 × 5–6 mm, apex obtuse; claw 6–7 mm. Filaments 4–5.5 mm, slightly dilated at base, median pairs 4.5–5.5 mm, lateral pair 3–4 mm; anthers oblong, 0.9–1.1 mm. Ovules number, fruits, and seeds unknown.

Distribution: Endemic to Nepal.

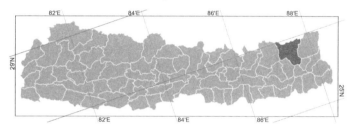

Altitudinal range: ca. 5400 m.

Ecology: Unknown.

Flowering: May.

Solms-laubachia nepalensis is currently known only from the type collection, Swan 71-72 (BM), collected in the Barun Valley, Sankhuwasabha District.

3. *Solms-laubachia linearis* (N.Busch) J.P.Yue, Al-Shehbaz & H.Sun, Ann. Missouri Bot. Gard. 95: 535 (2008). *Christolea linearis* N.Busch, Fl. URSS 8: 636 (1939); *Desideria linearis* (N.Busch) Al-Shehbaz; *Ermania kachrooi* Dar & Naqshi; *E. kashmiriana* Dar & Naqshi; *E. linearis* (N.Busch) Botsch.; *E. parkeri* O.E.Schulz; *Oreoblastus linearis* (N.Busch) Suslova; *O. parkeri* (O.E.Schulz) Suslova.

Herbs 4–15 cm, densely pilose throughout to subglabrous. Trichomes simple, to 1.5 mm. Stems simple, pilose to glabrous. Basal leaves not fleshy, pilose to glabrous, persistent; petiole 2–7(–12) mm, not ciliate; leaf blade broadly obovate to spatulate, 4–15 × 2–12 mm, base cuneate to attenuate, apex acute, margin 3–5-toothed or rarely entire. Cauline leaves similar to basal or linear to lanceolate, 5–10 × 1–3 mm, often entire, short petiolate to subsessile. Racemes 8–20-flowered, bracteate throughout; bracts similar to cauline leaves but smaller, often adnate to pedicel. Sepals free, oblong to ovate, 2–3 × 1–1.5 mm, caducous, pilose or with a terminal tuft of hairs, base not saccate, margin membranous. Petals purple or lavender with paler base, narrowly spatulate, 4–5(–5.5) × 1.5–2.5 mm, apex rounded;

Cruciferae (Brassicaceae)

claw 2–2.5 mm. Filaments white, slightly dilated at base, median pairs 2.5–3.5 mm, lateral pair 1.8–2.5 mm; anthers ovate, 0.4–0.5 mm. Ovules 16–26 per ovary. Fruiting pedicels ascending, straight, 2–8(–12) mm, pilose or glabrous. Fruit linear, (1.5–)2–3.5(–4.2) cm × (0.8–)1–1.7(–2) mm, flattened; valves pilose or glabrous, distinctly veined; septum complete, membranous; style obsolete; stigma 2-lobed. Seeds brown, ovate, 0.8–1.1 × 0.5–0.8 mm, uniseriate, minutely reticulate. Fig. 20h–j.

Distribution: Nepal, W Himalaya, N Asia and C Asia.

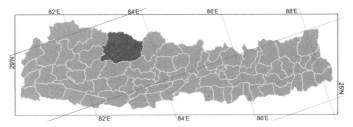

Altitudinal range: 4800–5800 m.

Ecology: Gravelly or sandy slopes, scree, gravelly moraine below glacier.

Flowering: June–August. **Fruiting:** July–September.

Hedge (Fl. Iranica Cruciferae: 211. 1968) misapplied the name *Ermania himalayensis* (Cambess.) O.E.Schultes to this species.

4. *Solms-laubachia jafrii* (Al-Shehbaz) J.P.Yue, Al-Shehbaz & H.Sun, Ann. Missouri Bot. Gard. 95: 534 (2008). *Phaeonychium jafrii* Al-Shehbaz, Nord. J. Bot. 20: 160 (2000).

Herbs 8–30 cm. Caudex stout, woody, few- to many-branched, to 3 cm in diameter, covered with petiolar bases of previous years. Trichomes simple and forked stalked or subsessile, crisped, flattened. Stems erect, simple and to 25 from caudex, tomentose. Basal leaves rosulate; petiole 1–5(–7) cm, ciliate with simple trichomes to 2 mm, becoming thickened and corky, to 6 mm wide at base; leaf blade green, broadly ovate to oblong, occasionally lanceolate to oblanceolate, 1–5.5(–7) × 1–2.5 cm, base cuneate to attenuate, apex obtuse to acute, margin entire, subtomentose. Cauline leaves absent. Racemes (8–)12–35-flowered, ebracteate. Sepals oblong, 3–4 × 1.5–2 mm, tomentose, persistent, margin membranous. Petals lavender to white flushed basally with purplish, obovate, 6.5–10 × 4–5 mm, apex obtuse; claw 3–4 mm. Filaments purple; median pairs 3–4 mm; lateral pair 2–2.5 mm; anthers oblong, 0.9–1.1 mm. Ovules 10–16 per ovary. Fruiting pedicels suberect to ascending, straight, (0.3–)0.6–1.5(–2.5) cm, tomentose. Fruit linear, (0.9–)1.5–2.5(–3.5) cm × 1.5–2 mm uniseriate, slightly flattened, sessile, straight; valves finely tomentose, with a distinct midvein and marginal veins; style 0.3–0.7 mm; stigma capitate, entire. Seeds brown to blackish, oblong, 2–2.5 × 0.9–1.1 m, slightly flattened, margined towards apex.

Distribution: Nepal, E Himalaya and Tibetan Plateau.

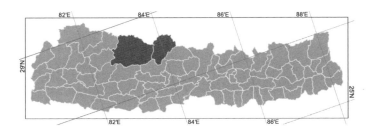

Altitudinal range: 4000–4900 m.

Ecology: Scrub, cliff ledges, steep rocky hillsides.

Flowering: July–September. **Fruiting:** July–September.

Hara (Enum. Fl. Pl. Nepal 2: 44. 1979) misinterpreted the limits of *Phaeonychium parryoides* and recorded the species in Nepal based on *Polunin, Sykes & Williams 1106* (BM), but this collection is now assigned to *Solms-laubachia jafrii*.

5. *Solms-laubachia himalayensis* (Cambess.) J.P.Yue, Al-Shehbaz & H.Sun, Ann. Missouri Bot. Gard. 95: 534 (2008). *Cheiranthus himalayensis* Cambess., in Jacquem., Voy. Inde 4(3): 14 (1844); *C. himalaicus* Hook.f. & Thomson; *Christolea himalayensis* (Cambess.) Jafri; *Desideria himalayensis* (Cambess.) Al-Shehbaz; *Ermania himalayensis* (Cambess.) O.E.Schulz nom. illegit.; *Oreoblastus himalayensis* (Cambess.) Suslova.

Herbs 4–20 cm, densely pilose throughout to subglabrous. Trichomes simple, to 1.5 mm. Stems simple, pilose or glabrous. Basal leaves not fleshy, pilose or glabrous, persistent; petiole 0.4–1.6(–3) cm, not ciliate; leaf blade broadly obovate to spatulate, 4–14 × 3–9 mm, base cuneate to attenuate, apex acute, margin (3–)5-toothed. Cauline leaves similar to basal or linear to lanceolate, 5–17 × 1–4 mm, often entire, short petiolate to subsessile. Racemes 6–25-flowered, bracteate throughout; bracts similar to cauline leaves but smaller, sometimes adnate to pedicel. Sepals free, oblong, 3–4 × 1.2–1.5 mm, caducous, pilose or with a terminal tuft of hairs, base not saccate, margin membranous. Petals purple or lilac with yellowish centre, broadly spatulate, (6–)6.5–8 × 3–4 mm, apex subemarginate; claw 3–4 mm. Filaments white, slightly dilated at base, median pairs 3–4 mm, lateral pair 2–4 mm; anthers ovate, ca. 0.6 mm. Ovules 14–24 per ovary. Fruiting pedicels ascending, straight or curved, 3–10 mm, pilose or glabrous. Fruit lanceolate to linear-lanceolate, (1.7–)2–3.5(–4) cm × (3–)4–6 mm, strongly flattened; valves pilose or glabrous, distinctly veined; septum complete, membranous; style obsolete; stigma 2-lobed. Seeds brown, ovate, (1.5–)1.8–2(–2.3) × 1–1.4 mm, biseriate, minutely reticulate. Fig. 16k.

Distribution: Nepal, W Himalaya and Tibetan Plateau.

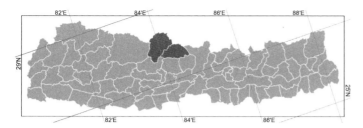

Altitudinal range: 4200–5600 m.

Ecology: Alpine tundra, open hills, sandstone scree.

Flowering: June–August. **Fruiting:** July–October.

Hara (Enum. Fl. Pl. Nepal 2: 43. 1979) based his record of *Solms-laubachia himalayensis* (using the illegitimate *Ermania himalayensis*) on *Polunin, Sykes & Williams 37* (BM). However, this is *S. linearis* (N.Busch) J.P.Yue, Al-Shehbaz & H.Sun, a species that he correctly recorded from Nepal, as *E. linearis* (N.Busch) Botsch), based on *Lowndes 1159* (BM).

The present record of *S. himalayensis* is based on *Wald 65* (BM) from Dhaulagiri Himal, *McBeath 1406* (E) and *McBeath 1486* (E) from Marsyandi, and *Komarkova 18* (GH) from Annapurna Himal.

25. *Eutrema* R.Br., Chlor. Melvill.: 193 (1823).

Wasabia Matsum.

Annual, biennial, or perennial herbs, glaucous or not above, with slender or fleshy and fusiform roots, rhizomes, or caudex. Trichomes absent or simple. Stems leafy or rarely leafless, erect, ascending, decumbent, or prostrate, simple or branched at base and/or apically. Basal leaves petiolate, rosulate or not and often withered by time of flowering, oblong, ovate, lanceolate, or reniform to cordate, simple, entire, dentate, or pinnately or palmately lobed, pinnately or palmately veined. Cauline leaves petiolate or sessile and cuneate or auriculate, sagittate, or amplexicaul at base, pinnately or palmately veined, entire, dentate, or crenate, lowermost alternate or whorled; ultimate veins ending or not with apiculate callosites. Racemes bracteate throughout or basally, or ebracteate, elongated considerably or not elongated in fruit. Sepals ovate or oblong, base of lateral pair rarely saccate. Petals white or rarely lavender, longer or shorter than sepals; blade spatulate, obovate, oblong, or obcordate, apex obtuse, rounded to emarginate; claw undifferentiated from blade. Stamens 6, tetradynamous, equal or subequal in length; filaments dilated or not at base, terete or rarely flattened and laterally toothed; anthers ovate or oblong, obtuse or apiculate at apex. Nectar glands lateral or confluent, subtending bases of all stamens; median glands present or absent. Ovules 4–12 per ovary. Fruiting pedicels erect and subappressed to stem, ascending, or divaricate, slender, straight or reflexed. Fruit dehiscent, siliques or silicles, linear, oblong, or ovoid, terete, slightly 4-angled or latiseptate, sessile or on a short gynophore; valves with an obscure or prominent midvein, smooth or torulose, glabrous or rarely papillate; replum rounded, expanded or flattened; septum complete, absent or perforate and reduced to a rim, veinless; style obsolete or distinct and up to 3 mm long, slender or clavate; stigma capitate, entire or rarely 2-lobed. Seeds uniseriate or biseriate, wingless, oblong, plump or compressed; seed coat obscurely reticulate to foveolate or papillate, slightly mucilaginous or not when wetted; cotyledons incumbent, accumbent or obliquely so.

Worldwide 26 species, primarily in the Himalaya and C and E Asia, with one extending into N America. Four species in Nepal.

Key to Species

1a	Raceme ebracteate, subumbellate in fruit	**1. *E. heterophyllum***
b	Raceme bracteate, elongated or not in fruit	2
2a	All leaves alternate	**2. *E. hookeri***
b	At least some of the lowermost cauline leaves whorled, other leaves opposite or alternate	3
3a	Petals 3–5 × 1.5–2 mm. Sepals persistent. Septum present. Seeds papillate	**3. *E. lowndesii***
b	Petals (7–)8–9(–10) × (3.5–)4.5–6 mm. Sepals caducous. Septum absent. Seeds foveolate	**4. *E. verticillatum***

1. *Eutrema heterophyllum* (W.W.Sm.) H.Hara, J. Jap. Bot. 48: 97 (1973).
Braya heterophylla W.W.Sm., Notes Roy. Bot. Gard. Edinburgh 11: 201 (1920); *Eutrema compactum* O.E.Schulz; *E. edwardsii* var. *heterophyllum* (W.W.Sm.) W.T.Wang; *E. obliquum* K.C.Kuan & Z.X.An.

Perennial herbs, 2–15(–25) cm, glabrous throughout or rarely puberulent, with a caudex and fleshy root. Stems erect, simple, often few from caudex. Basal leaves rosulate, somewhat fleshy; petiole (0.5–)1–5.5(–7.5) cm; leaf blade ovate, suborbicular, lanceolate or rhombic, (0.3–)0.5–2(–2.5) cm × (2–)4–10(–15) mm, base cordate, truncate, to cuneate, sometimes distinctly oblique, apex obtuse to subacute, margin entire. Middle cauline leaves lanceolate, ovate, or linear-lanceolate, 0.7–2.5(–3) cm × 2–9 mm, pinnately veined, sessile, base cuneate, apex subacute, margin entire. Fruiting raceme compact, subumbellate or short racemes, little or not elongated in fruit, 0.2–2(–3.5) cm. Sepals ovate, 1.5–2 × ca. 1 mm, persistent through fruit maturity. Petals white, spatulate, 2–3.5 × 1–1.7 mm. Filaments white, 1–2 mm, slightly dilated at base; anthers ovate, 0.2–0.3 mm. Ovules 4–10 per ovary. Fruiting pedicels divaricate to slightly reflexed, 1–4(–5) mm. Fruit linear to oblong, (4–)5–10(–12) × 1.5–2 mm, slightly 4-angled, not torulose; valves cuneate at both ends, with a prominent midvein; gynophore to 0.5 mm; septum mostly perforate; style obsolete or to 0.6 mm. Seeds oblong, plump, 1.5–2 × ca. 1 mm. Cotyledons incumbent. Figs 16l, 20j.

Distribution: Nepal, E Himalaya, Tibetan Plateau, E Asia, N Asia and C Asia.

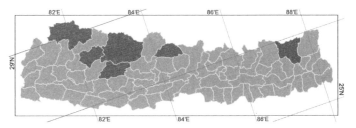

Altitudinal range: 2500–5400 m.

Ecology: Alpine meadows, scree, grassy slopes, near glaciers, *Kobresia* turf, alpine mats, sandstone ridges.

Flowering: June–July. **Fruiting:** July–August.

2. *Eutrema hookeri* Al-Shehbaz & S.I.Warwick, Harvard Pap. Bot. 10: 133 (2005).
Cochlearia himalaica Hook.f. & Thomson, J. Proc. Linn. Soc., Bot. 5: 154 (1861); *Taphrospermum himalaicum* (Hook.f. & Thomson) Al-Shehbaz & Kats.Arai.

Short-lived perennial herbs, glabrous throughout except for sparsely puberulent fruit and upper portion of calyx. Root narrowly fusiform-linear, fleshy, apex with minute scale-like leaves. Stems solitary from fleshy root then producing a rosette with prostrate or rarely ascending to erect branches, (2–)5–10(–15) cm. Leaves not rosulate; petioles of basal and lowermost leaves (0.5–)1–2.5(–4) cm, gradually shorter upward; leaf blade broadly ovate to oblong, 4–10(–15) × 2–6(–10) mm, gradually reduced in size upward, base obtuse to subcordate, apex obtuse or rounded, margin lobed or toothed to repand or entire. Racemes densely flowered, elongated considerably or not elongated in fruit, bracteate throughout; bracts leafy and representing all cauline leaves. Sepals oblong, 1.2–2 × 0.8–1 mm, persistent, membranous at margin, sparsely hairy distally. Petals white, obovate to spatulate, 2.5–4(–6) × 1.5–2.5(–3.5) mm, attenuate to base, apex slightly emarginate. Filaments white, 1.5–2 mm, median dilated at base; anthers ovate, 0.2–0.4 mm. Ovules 4–12 per ovary. Fruiting pedicels glabrous, slender, straight or strongly recurved and fruits appearing geocarpic, 4–9(–13) mm. Fruit ovoid to oblong, latiseptate, not torulose, (2–)4–8(–10)

× (1.5–)2–3.4(–4) mm, obtuse at both ends; valves membranous, sparsely puberulent with papillae to 0.5 mm, rarely subglabrous, obscurely veined; septum absent; style slender, 0.5–1(–1.2) mm. Seeds (2–)6–8(–12), brown, oblong, 1–1.5 × 0.8–1 mm, foveolate; cotyledons obliquely accumbent.

Distribution: Nepal, E Himalaya, Tibetan Plateau and Assam-Burma.

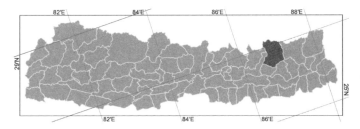

Altitudinal range: ca. 4600 m.

Ecology: Rocky ground on exposed slopes, scree, deep rich soil, muddy slopes, stream banks, sandy beds, moist granite ledges, alpine pastures and dwarf scrub, *Kobresia* turf, under *Juniperus* trees.

Flowering: June–August. **Fruiting:** July–September.

3. *Eutrema lowndesii* (H.Hara) Al-Shehbaz & S.I.Warwick, Harvard Pap. Bot. 10: 133 (2005).
Glaribraya lowndesii H.Hara, J. Jap. Bot. 53: 136 (1978); *Taphrospermum lowndesii* (H.Hara) Al-Shehbaz.

Short-lived perennial herbs, 5–12 cm, glabrous throughout or puberulent with papillae 0.05–0.2 mm. Root slender, slightly fleshy, base with a whorl of oblong scale-like leaves ca. 2–5 × 0.5–1 mm. Stems erect, 1–3 from root, lowermost leafless part 2–9 cm. Leaves glabrous, fleshy, lowermost cauline ones whorled, others alternate; petioles 2–5 mm, gradually shorter upward; leaf blade spatulate, 3–15 × 1–5 mm, gradually reduced in size upward, base attenuate, apex retuse or rounded, margin entire. Racemes densely flowered, elongated considerably in fruit, bracteate throughout; bracts leafy, smaller than cauline leaves; rachis glabrous or papillate. Sepals oblong, 1.5–2.5 × ca. 1 mm, persistent well after fruit dehiscence, membranous at distal margin and apex, not ciliolate, glabrous or papillate. Petals white, narrowly obovate, 3–5 × 1.5–2 mm, attenuate to claw-like base to 1.5 mm, apex obtuse. Filaments white, 1.5–2.5 mm; anthers broadly ovate, 0.2–0.3 mm. Ovules 4–6 per ovary. Fruiting pedicels glabrous, slender, straight or curved, 4–8 mm. Fruit oblong or narrowly so, strongly latiseptate, not torulose, 7–15 × 3–5 mm, obtuse to cuneate at base; valves membranous, glabrous, often distinctly veined, smooth; septum complete, membranous; style slender, (1–)2–3 mm. Seeds 3–8, brown, oblong, compressed, 1.9–2.1 × 1.2–1.4 mm, densely papillate with papillae 0.1–0.2 mm; cotyledons accumbent.

Distribution: Nepal and Tibetan Plateau.

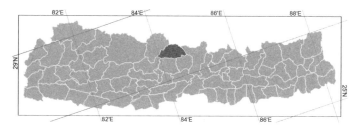

Altitudinal range: 5000–5200 m.

Ecology: Scree slopes.

Flowering: July. **Fruiting:** July–August.

The species record from Nepal is based on the type collection, *Lowndes L1287* (BM), from Khangsar, Manang.

4. *Eutrema verticillatum* (Jeffrey & W.W.Sm.) Al-Shehbaz & S.I.Warwick, Harvard Pap. Bot. 10: 134 (2005).
Cardamine verticillata Jeffrey & W.W.Sm., Notes Roy. Bot. Gard. Edinburgh 8: 120 (1913); *Braya verticillata* (Jeffrey & W.W.Sm.) W.W.Sm.; *Staintoniella verticillata* (Jeffrey & W.W.Sm.) H.Hara; *Taphrospermum verticillatum* (Jeffrey & W.W.Sm.) Al-Shehbaz.

Herbs (4–)6–15(–23) cm tall, sparsely to moderately pubescent, rarely glabrous. Root narrowly fusiform-linear, fleshy, base with oblong or ovate scale-like leaves 3–7 × 1–2.5 mm. Stems erect, often solitary from fleshy root, lowermost leafless part (0.5–)4–10(–15) cm. Leaves glabrous, lowermost cauline leaves verticillate, others opposite or alternate; petioles (0.4–)0.6–2 cm, gradually shorter upward; leaf blade oblong, rarely oblanceolate, 5–15(–25) × 2–7(–9) mm, gradually reduced in size upward, base obtuse or rarely attenuate, margin entire, apex rounded. Racemes densely flowered, elongated considerably in fruit, bracteate throughout; bracts leafy, smaller than cauline leaves; rachis sparsely pubescent with retrorse trichomes 0.4–0.6 mm. Sepals oblong, 2.5–3.5(–4) × 1.5–2 mm, caducous, membranous at margin, sparsely ciliolate near apex with trichomes 0.05–0.1 mm, glabrous or distally sparsely pubescent with trichomes to 0.5 mm. Petals white or rarely lavender, broadly obovate, (7–)8–9(–10) × (3.5–)4.5–6 mm, cuneate or claw-like base to 4 mm, apex emarginate. Filaments white or lavender, (3–)3.5–4.5 mm, dilated at base; anthers ovate, 0.6–0.8 mm. Ovules 4–8 per ovary. Fruiting pedicels glabrous, slender, straight or curved, (4–)6–12(–18) mm. Fruit ovate or oblong, strongly latiseptate, not torulose, 7–13 × 4–7 mm, obtuse or cuneate at base; valves membranous, glabrous, often distinctly veined, smooth; septum absent; style slender, (1–)2–3 mm. Seeds 3–8, brown, oblong, foveolate, compressed, 1.8–2.2 × 1.2–1.4 mm; cotyledons accumbent.

Distribution: Nepal, Tibetan Plateau and E Asia.

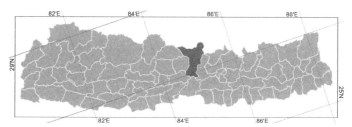

Altitudinal range: ca. 4900 m.

Ecology: Scree, cliff ledges, glaciers, open stony moorland, siliceous scree slopes.

Flowering: June–July. **Fruiting:** July–August.

26. *Pegaeophyton* Hayek & Hand.-Mazz., Anz. Akad. Wiss. Wien, Math.-Naturwiss. Kl. 59: 245 (1922).

Perennial herbs. Caudex simple or branched. Trichomes absent or simple. Stems reduced to tiny portions added annually to apex of caudex or its branches. Basal leaves fleshy or not, rosulate, simple, entire or toothed. Cauline leaves absent. Flowers solitary, borne on pedicels from axils of rosette leaves. Sepals free, caducous at time of fruiting, broadly ovate or oblong, ascending or spreading, glabrous or pubescent, base not saccate. Petals white, pink, blue, lilac or violet, longer than sepals, broadly obovate or suborbicular, rarely spatulate, apex rounded or subemarginate; clawed or not. Stamens 6, erect or spreading, subequal; filaments dilated at base; anthers ovate suborbicular or oblong, obtuse at apex. Nectar glands confluent, subtending bases of all stamens. Ovules 2–15 per ovary. Fruiting pedicels ascending, straight. Fruit dehiscent silicles, rarely siliques, oblong, orbicular, ovoid, or globose, terete, latiseptate, not geocarpic, sessile or stipitate; valves membranous or papery, not veined or obscurely veined, smooth, glabrous or pubescent; replum flattened; septum absent; style obsolete or to 3 mm, subconical; stigma discoid, entire. Seeds uniseriate, wingless, oblong or broadly-ovate, plump or flattened; seed coat obscurely reticulate, not mucilaginous when wetted; cotyledons accumbent.

Worldwide six species in S and E Asia. Three species in Nepal.

Key to Species

1a Fruiting pedicels pubescent along one side, persisting for more than one season. Fruit narrowly oblong, length:width ratio 3–5. Sepals apically ciliate ....**1. *P. minutum***

  b Fruiting pedicels glabrous or distally pubescent on all sides, not persistent. Fruit oblong, ovate, orbicular, ovoid, to subglobose, length to width ratio 1–1.5(–2). Sepals not ciliate...2

2a   Fruit oblong, orbicular to ovate, flattened, valves papery, glabrous. Sepals (2–)2.5–6(–10) mm. Petals (3.5–)5–12(–15) mm. Seeds flattened, 1.5–3.5(–4) × 1–2.5(–3) mm. .... **2. *P. scapiflorum***

b   Fruit ovoid to subglobose, valves membranous, puberulent apically. Sepals 1.1–1.3 mm. Petals 1.6–2(–2.5) mm. Seeds plump, 1–1.1(–1.3) × 0.5–0.6(–0.8) mm .... **3. *P. nepalense***

1. *Pegaeophyton minutum* H.Hara, J. Jap. Bot. 47: 270, fig. 2 (1972).
Pegaeophyton garhwalense H.J.Chowdhery & Surendra Singh.

Herbs with slender, branched caudex 0.75–2 mm in diameter. Petiole (3–)5–10(–15) mm; leaf blade obovate, spatulate, ovate, oblong, or oblanceolate, 1–4(–5) × (0.5–)1–1.5(–2) mm, somewhat fleshy, sparsely puberulent above with trichomes 0.02–0.08 mm, rarely glabrous, glabrous below, base cuneate or subattenuate, apex obtuse to subrounded, margin entire. Sepals broadly ovate, 1.2–2 × 0.8–1.2 mm, free, glabrous, distally ciliate with flattened trichomes to 0.05 mm. Petals white, lilac or violet, broadly obovate or suborbicular, (1.5–)2–3(–4.5) × 1.5–2.5(–3) mm, tapering to claw-like base 0.7–1.5 mm. Filaments dilated at base, (0.9–)1–1.5 mm; anthers broadly ovate or suborbicular, 0.3–0.4 mm. Fruiting pedicels slender, (1–)1.5–2.5(–4) cm, persistent, puberulent along 1 side of entire length with trichomes 0.02–0.08 mm. Fruit narrowly oblong, 3–5 × 1–1.2 mm; valves nearly flat, extending along part of fruit length; gynophore 0.25–0.5 mm; style 0.3–0.4 mm. Seeds 3–5, broadly ovate, brown, plump, 1–1.2 × 0.6–0.7 mm.

Distribution: Nepal, W Himalaya, E Himalaya and Tibetan Plateau.

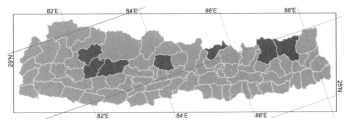

Altitudinal range: 3600–5200 m.

Ecology: Mossy wet ledges, hillsides, streams, mossy granite or boulders, steep grassy slopes.

Flowering: May–July.

2. *Pegaeophyton scapiflorum* (Hook.f. & Thomson) C.Marquand & Airy Shaw, J. Linn. Soc., Bot. 48: 229 (1929).
Cochlearia scapiflora Hook.f. & Thomson, J. Proc. Linn. Soc., Bot. 5: 154 (1861); *Pegaeophyton scapiflorum* var. *pilosicalyx* R.L.Guo & T.Y.Cheo.

सो-लो-कार्पो So-lo-karpo (Tibetan).

Herbs with slender, apically branched caudex 1–8(12) mm in diameter. Petiole 1–8(–13) cm; leaf blade ovate, oblong, elliptic, obovate, spatulate, oblanceolate, or narrowly linear, (1–)1.5–8(–10) × 0.2–1.2(–2.5) cm, base cuneate or

subattenuate, apex obtuse or acute, margin entire or dentate, sometimes minutely ciliate, somewhat fleshy or not, glabrous or sparsely pubescent above with trichomes 0.2–0.5 mm, glabrous below. Sepals ovate or oblong, (2–)2.5–6(–10) × 1.5–3.5(–4.5) mm, glabrous or sparsely pubescent, not ciliate. Petals white, pink, or blue, sometimes white with greenish or bluish centre, broadly obovate, spatulate, or suborbicular, (3.5–)5–7 × (1.5–)2–3(–3.5) mm, tapering to claw-like base (0.5–)1–3(–3.5) mm. Filaments dilated at base, (2.5–)3–5(–7) mm; anthers oblong to narrowly so, (0.5–)1–1.5(–2) mm, sagittate at base. Fruiting pedicels slender or stout, glabrous or rarely sparsely pubescent apically all around with trichomes 0.2–0.5 mm, (1.2–)2.5–15(–20) cm, not persistent. Fruit, oblong or orbicular to ovate, (4–)5–13(–20) × (2–)4–8(–10) mm; valves nearly flat, extending along part of fruit length, papery, glabrous; gynophore (0.5–)2–5 mm; style 1–2(–3) mm. Seeds (1–)3–10(–12), broadly ovate, brown, flattened, 1.5–2(–2.5) × 1–1.6(–1.8) mm. Fig. 21a–b.

Distribution: Nepal, W Himalaya, E Himalaya, Tibetan Plateau, Assam-Burma, E Asia and N Asia.

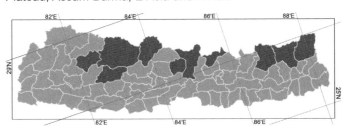

Altitudinal range: 4000–5600 m.

Ecology: Alpine tundra, alpine meadows, muddy gravelly slopes, gravel near glaciers, grassy slopes, water at lake shores, moist pastures, stony slopes with unconsolidated scree, seepage areas in scree, in moss by streams, rock crevices, boggy ground by lakes, sandy soil at edge of streams, in melting snow or running water.

Flowering: May–August. **Fruiting:** July–September.

Pegaeophyton scapiflorum includes two subspecies, of which only subsp. *scapiflorum* occurs in Nepal. Subspecies *robustum* (O.E.Schulz) Al-Shebaz, T.Y.Cheo, L.L.Lu & G.Yang has a stouter caudex (up to 30 mm diameter), larger petals (8–12 mm,) and seeds (2.5–3.5 mm), but is restricted to China and adjacent Bhutan.

3. *Pegaeophyton nepalense* Al-Shehbaz, Kats.Arai & H.Ohba, Novon 8: 327 (1998).

Herbs with slender caudex ca. 1 mm in diameter. Leaves 5–12 per caudex; petiole (2–)6–10(–14) mm, slender at base, glabrous or with few trichomes; blade suborbicular to broadly obovate, 2–4(–5) × 1.5–3.5(–4.5) mm, base obtuse, apex rounded or subrounded, margin entire, somewhat fleshy, moderately pubescent above with trichomes 0.3–0.5 mm, glabrous below. Flowers 3–8 per plant. Pedicels slender, distally pubescent with trichomes 0.3–0.5 mm all around, 2–5 mm at anthesis, not elongated in fruit. Sepals oblong, 1.1–1.3(–1.5) × 0.5–0.7 mm, free, spreading to ascending, not saccate, pubescent on distal half with trichomes to 0.3 mm, membranous margin 0.05–0.1 mm wide, obtuse. Petals white, broadly obovate to suborbicular, slightly emarginate, 1.6–2(–2.5) mm; blade 0.8–1.5 × 0.8–1.5 mm; claw 0.8–1.2 mm. Filaments erect, white, slightly dilated at base, 1.2–1.5 mm, persistent to fruit maturity; anthers suborbicular, 0.25–0.3 mm, slightly sagittate at base. Fruit, broadly ovoid to subglobose, 2–3 × 1.8–2 mm; valves membranous, rounded, extending along part of fruit length, glabrous or minutely puberulent distally; gynophore 0.1–0.2 mm; style 0.5–0.7 mm. Seeds oblong, brown, plump, 2–4 per fruit, 1–1.1(–1.3) × 0.5–0.6(–0.8) mm.

Distribution: Nepal and E Himalaya.

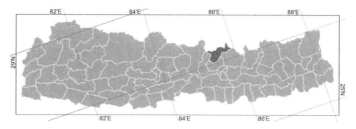

Altitudinal range: 4200–5200 m.

Ecology: Not known.

Flowering: June–August. **Fruiting:** August–September.

One collection, *Polunin 1439* (BM), included plants of both *Pegaeophyton nepalense* and *P. minutum* H.Hara, but no intermediates have yet been found and they are maintained as distinct species.

27. *Lepidium* L., Sp. Pl. 2: 643 (1753).

Annual or biennial herbs. Trichomes absent or simple. Stems erect or prostrate, simple or branched basally and/or apically. Basal leaves rosulate or not, simple, entire or pinnately dissected. Cauline leaves petiolate or sessile, base cuneate, attenuate, auriculate, sagittate, or amplexicaul, margin entire, dentate, or dissected. Racemes ebracteate, corymbose, elongated or not in fruit. Sepals oblong, base of lateral pair not saccate. Petals white to lavender, erect or spreading, sometimes rudimentary or absent, obovate or spatulate, apex obtuse, rounded, or emarginated, base attenuate or cuneate. Stamens 2 and median, 6 and tetradynamous or subequal in length, or 4 and all median or 2 median and 2 lateral; anthers ovate or oblong. Nectar glands 4 or 6, distinct; median glands present. Ovules 2 per ovary; placentation apical. Fruiting pedicels terete, slender, erect, ascending or divaricate, recurved or straight. Fruit dehiscent silicles, oblong, ovate to elliptic, strongly angustiseptate; valves veinless or prominently veined, keeled or not, apically winged or wingless; replum rounded; septum complete or perforated; style absent, obsolete, or distinct, included or exserted from apical notch of fruit; stigma capitate, entire or rarely 2-lobed. Seeds winged, margined, or wingless, oblong or ovate, plump or flattened; seed coat smooth, minutely reticulate, or papillate, usually copiously mucilaginous when wetted; cotyledons incumbent.

Worldwide about 180 species on all continents except Antarctica. Three species in Nepal.

Key to Species

1a Stamens 6. Fruit (4–)5–6 mm, winged all around, broadly so apically. Plants glabrous or pilose with, slender trichomes. Cotyledons 3-lobed ....**1. *L. sativum***

  b Stamens 2 or 4. Fruit 1.8–3.1 mm, obscurely winged only apically. Plants puberulent with clavate or capitate trichomes. Cotyledons not lobed 2

2a Middle and upper cauline leaves usually pinnatifid. Infructescence capitate. Petals 0.5–1 mm. Stamens 4 .... **2. *L. capitatum***

  b Middle and upper cauline leaves dentate, serrate, or entire. Infructescence elongate, racemose. Petals absent or rudimentary. Stamens 2 ....**3. *L. apetalum***

1. *Lepidium sativum* L., Sp. Pl. 2: 644 (1753).

चम्सुर Chamsur (Nepali).

Annual herbs, (10–)20–80(–100) cm. Stems erect, simple or branched, sparsely crisped pilose above. Basal leaves not rosulate; petiole 1–4 cm; leaf blade 1- or 2-pinnatifid to -pinnatisect, 2–8(–10) × 1–3(–5) cm, ultimate lobes ovate to oblong, apex acute, margin dentate. Cauline leaves petiolate; leaf blade similar to basal ones but less divided, with 1–4 lateral lobes on each side of midvein; uppermost leaves subsessile, linear, margin entire. Sepals oblong, 1–1.8 × 0.5–0.8 mm, glabrous or pubescent below. Petals white to lavender, spatulate or obovate, 2.5–3.5(–4) × 0.7–1.4 mm,

base attenuate. Stamens 6; filaments 1.5–2 mm; anthers oblong, 0.4–0.5 mm. Fruiting pedicels suberect and appressed to rachis, or ascending, straight, 1.5–4(–6) mm, terete to slightly flattened, glabrous. Fruit oblong-ovate to elliptic, (4–)5–6(–7) × 3–4.5(–5.5) mm, base rounded, margin and apex winged, apex emarginate; wings all around fruit, 1–1.5 mm at apex; apical notch 0.2–0.5 mm; style 0.2–0.5(–0.8) mm, free from wings, included or rarely exserted from apical notch. Seeds reddish brown, oblong, 2–2.6 × 1–1.3 mm, wingless; cotyledons incumbent, 3-lobed.
Figs 16m, 21c.

Distribution: Cosmopolitan.

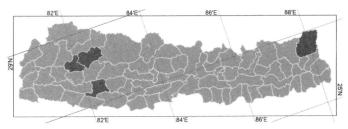

Altitudinal range: 200–3000 m.

Ecology: Cultivated or naturalized.

Flowering: June–July. **Fruiting:** August–September.

Young leaves are cooked as a vegetable.

2. Lepidium capitatum Hook.f. & Thomson, J. Proc. Linn. Soc., Bot. 5: 175 (1861).
Lepidium incisum Edgew.; *L. kunlunshanicum* G.L.Zhou & Z.X.An.

Annual or biennial herbs, (5–)10–35(–50) cm, densely covered with clavate or capitate papillate trichomes. Stems prostrate or rarely suberect, branched basally and above. Basal and lower cauline leaves with petioles 0.5–4 cm; leaf blade oblong to spatulate or lanceolate, (0.5–)1–4(–7) × 0.2–2 cm, usually glabrous, pinnatifid, base attenuate; lobes oblong to lanceolate or linear, 2–8(–15) × 0.5–3(–5) mm, 1- or 2-pinnatifid or -partite, serrate, to entire. Upper cauline leaves short petiolate to subsessile, similar lowermost leaves, progressively smaller upward. Racemes capitate, elongated slightly or not elongated in fruit. Sepals oblong, 0.8–1 × 0.3–0.4(–0.5) mm, glabrous or with crisped trichomes. Petals white, narrowly obovate, (0.5–)0.6–0.9(–1) × 0.2–0.3 mm, base cuneate, apex subtruncate to emarginate. Stamens 4; filaments 0.8–1 mm; anthers ovate, 0.1–0.2 mm. Fruiting pedicels slender, divaricate, slightly recurved or straight, (1–)2–3(–4) mm, puberulent above with clavate or capitate trichomes. Fruit broadly ovate, 1.8–2.5 × 1.7–1.8 mm, glabrous; wing apical, 0.1–0.3 mm; apical notch 0.05–0.3 mm; style obsolete, rarely to 0.15 mm, included in apical notch of fruit. Seeds brown, oblong-ovate, 1–1.1 × 0.6–0.7 mm; cotyledons incumbent, unlobed.

Distribution: Nepal, W Himalaya, E Himalaya, Tibetan Plateau and E Asia.

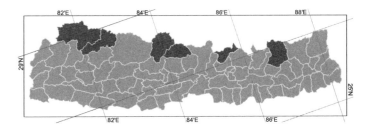

Altitudinal range: 2700–5300 m.

Ecology: Mountain slopes, disturbed areas, plains.

Flowering: June–September. **Fruiting:** June–September.

Hara (Enum. Fl. Pl Nepal 2: 44. 1979) reported *Polunin, Sykes & Williams 2268* (BM) as *Lepidium capitatum*, but that collection is *L. apetalum* Willd.

3. Lepidium apetalum Willd., Sp. Pl., ed. 5[as 4], 3(1): 439 (1800).
Lepidium chitungense Jacot Guill.

चक्मा Chakma (Nepali).

Annual or biennial herbs, (5–)10–25(–40) cm, puberulent with clavate or capitate trichomes. Stems erect, branched basally and above. Petiole of basal leaves (0.5–)1–3 cm; leaf blade oblong, lanceolate, or oblanceolate, 1.5–4(–5) × 0.7–1.2(–1.5) cm, pinnatifid, sinuate, or dentate. Upper cauline leaves sessile; leaf blade narrowly lanceolate, linear-oblong, to linear, 6–30(–40) × 1–3(–5) mm, base subauriculate or sometimes cuneate, apex acute to subobtuse, margin remotely serrate to entire. Sepals caducous, oblong, 0.7–0.8 × 0.3–0.35 mm, glabrous or puberulent, white at margin and apex. Petals absent or rudimentary and ca. 0.3 mm. Stamens 2; filaments 0.7–0.8 mm; anthers broadly ovate, 0.1–0.2 mm. Fruiting pedicels slender, often recurved, 2–4(–5) mm, puberulent only above with clavate or capitate trichomes, slightly flattened or narrowly winged. Fruit broadly elliptic, 2.2–3.1 × 1.7–2.3 mm, widest at middle, apex with narrow winged; apical notch 0.1–0.3 mm; style 0.05–0.15 mm, included in apical notch. Seeds reddish brown, oblong-ovate, 1.1–1.3 × 0.6–0.8 mm, wingless, finely papillate; cotyledons incumbent, unlobed.

Distribution: Nepal, Tibetan Plateau, S Asia, E Asia, N Asia and C Asia.

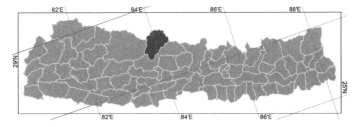

Altitudinal range: 2500–4100 m.

Ecology: Roadsides, slopes, waste places, ravines, plains, fields.

Flowering: May–August. **Fruiting:** May–August.

Hooker & Anderson (Fl. Brit. Ind. 1: 160. 1872) and Kitamura (Fauna Fl. E. Himalaya: 139. 1955) and other authors have

misapplied the name *Lepidium ruderale* L. to this species. Hara (Enum. Fl. Pl Nepal 2: 44. 1979) misapplied the name *L. capitatum* Hook.f. & Thomson to this species.

28. *Megacarpaea* DC., Syst. Nat. 2: 417 (1821).

Perennial herbs. Caudex terminated by petiolar remains of previous years. Trichomes simple. Stems erect, simple basally, branched apically. Basal leaves petiolate, rosulate, simple, pinnately lobed, or 1–3-pinnatisect. Cauline leaves petiolate or sessile, auriculate at base, sinuate, pinnately lobed, or pinnatisect. Racemes ebracteate, in panicles, elongated in fruit. Sepals oblong, deciduous, base of lateral pair not saccate, margin membranous. Petals yellow, obovate, oblong, entire; claw much shorter than sepals. Stamens (8–)12–16(–24), slightly tetradynamous or equal in length; filaments dilated at base; anthers oblong or linear, obtuse at apex. Nectar glands confluent around bases of all stamens; median glands present. Ovules 2 per ovary. Fruiting pedicels slender, ascending, divaricate, or strongly recurved. Fruit indehiscent, schizocarpic, didymous, angustiseptate silicles, sessile or shortly stipitate; valves (mericarps) oblong, obovate or orbicular, 1-seeded, leathery, smooth, broadly winged, keeled, glabrous; replum rounded; style obsolete, rarely to 0.5 mm; stigma capitate, entire. Seeds wingless, broadly ovate, strongly flattened; seed coat smooth, not mucilaginous when wetted; cotyledons accumbent.

Worldwide nine species in C Asia and the Himalayas. One species in Nepal.

1. *Megacarpaea polyandra* Benth., Hooker's J. Bot. Kew Gard. Misc. 7: 356 (1855).

रुबिको साग Rubiko saag (Nepali).

Herbs, (0.5–)1–2 m, tomentose above. Caudex to 15 cm diameter. Trichomes flattened, crisped. Stems erect, branched above. Basal and lower cauline leaves oblong to oblanceolate in outline, pinnatisect, 15–60 cm, glabrous to sparsely pubescent; lateral lobes 6–12 on each side, oblong to lanceolate, 6–20 × 1–5 cm, base decurrent, apex acuminate, margin serrulate to sinuate-dentate or subentire. Upper cauline leaves auriculate, similar to lower ones but gradually reduced in size. Sepals yellowish, 3.5–5(–6) × 1.5–2.5 mm, sparsely pubescent to glabrous. Petals yellow, oblong to obovate, entire, 4–6(–7) × 2–4(–6) mm, base cuneate, apex rounded to subacute; claw 1–2 mm. Stamens (8–)12–16(–24); filaments 3.5–5 mm; anthers narrowly oblong, 1–1.3 mm. Fruiting pedicels slender, 1.5–4 cm, strongly recurved, pubescent. Fruit halves suborbicular to obovate-orbicular, 3–4 × 2.5–3 cm; wings 5–10(–18) mm wide; replum 7–16 mm; locule 1.5–2(–2.8) × 1.2–1.7 cm. Seed brown, broadly ovate, 1.4–1.7 × 0.7–1.1 cm.

Distribution: Nepal, W Himalaya and Tibetan Plateau.

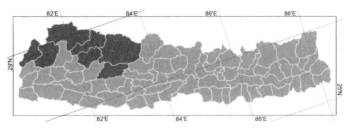

Altitudinal range: 2700–4600 m.

Ecology: Slopes, rocky areas, along streams.

Flowering: May–August. **Fruiting:** August–October.

Megacarpaea bifida Benth., a little collected and poorly known species endemic to Kashmir, is probably conspecific (or at best infraspecific) with *M. polyandra*, but differs in its entire leaf lobes and slightly elongated fruit halves.

Young leaves and shoots are cooked as a vegetable.

29. *Noccaea* Moench, Suppl. Meth.: 89 (1802).

Perennial herbs, often glabrous and glaucous. Trichomes absent or simple. Stems erect, ascending or decumbent, simple. Basal leaves petiolate, rosulate or not, simple, entire or dentate. Cauline leaves sessile, auriculate, sagittate, or amplexicaul at base, entire or rarely denticulate. Racemes ebracteate, elongated or not in fruit. Sepals oblong, erect or ascending, base of lateral pair not saccate, margin membranous. Petals white, pale mauve to lavender, apex rounded; claw differentiated or not from blade. Stamens 6, tetradynamous; filaments dilated or not at base; anthers ovate or oblong, obtuse at apex. Nectar glands 1 on each side of lateral stamens; median glands absent. Ovules 4–10 per ovary. Fruiting pedicels slender, divaricate or slightly reflexed. Fruit dehiscent siliques or silicles, oblong, elliptic or ob-lanceolate, often apically notched, strongly angustiseptate, sessile; valves keeled, winged or wingless; replum rounded; septum complete; style usually prominent, exserted or included in apical notch of fruit; stigma capitate, entire. Seeds uniseriate, wingless, oblong, ovoid, or ellipsoid, plump; seed coat smooth or obscurely reticulate; cotyledons accumbent.

Cruciferae (Brassicaceae)

Worldwide about 65–80 species in temperate Eurasia, N America, and southern S America. Three species (one endemic) in Nepal.

Key to Species

1a   Fruit oblong, contorted, apex narrowly winged, emarginate .... **3. *N. cochlearioides***

b   Fruit elliptic, straight, apex neither winged nor emarginate 2

2a   Style 0.7–1.5 mm. Fruit elliptic 5–8 × 3–4 mm. Petals (5–)6–7 × 2.5–3.5 mm .... **1. *N. andersonii***

b   Style 3–4 mm. Fruit narrowly oblong-oblanceolate, 6–10 × 1.5–2 mm. Petals 4.5–5 × 1.5–2 mm......... **2. *N. nepalensis***

1. *Noccaea andersonii* (Hook.f. & Thomson) Al-Shehbaz, Adansonia, Ser. 3 24(1): 91 (2002).
Iberidella andersonii Hook.f. & Thomson, J. Proc. Linn. Soc., Bot. 5: 177 (1861); *I. tibetica* C.Marquand & Airy Shaw; *Thlaspi andersonii* (Hook.f. & Thomson) O.E.Schulz.

Herbs (4–)6–15(–20) cm, glabrous throughout. Stems erect to decumbent, simple. Basal leaves rosulate; petiole (2–)5–10(–15) mm; leaf blade oblong, spatulate, or suborbicular, (0.3–)0.5–1.5(–1.8) cm × 2–6(–10) mm, base obtuse, apex rounded, margin entire. Middle cauline leaves sessile, oblong, ovate, or suborbicular, 5–10(–15) × 2–5(–7) mm, base auriculate, apex rounded, obtuse, to subacute, margin repand or entire. Sepals oblong, 2–2.5 × 1.2–1.5 mm, not saccate, margin white. Petals white with lavender tinge, spatulate, (5–)6–7 × 2.5–3.5 mm, apex rounded. Filaments 2.5–3.5 mm; anthers oblong, 0.6–0.8 mm. Ovules 4–8(–10) per ovary. Fruiting pedicels (2–)4–7 mm, slender, divaricate, straight. Fruit elliptic, 5–8 × 3–4 mm, base obtuse, apex obtuse to subacute, without apical notch; apical wings absent; style 0.7–1.5 mm. Seeds dark brown, oblong, 1.6–1.9 × 1–1.2 mm, smooth.

Distribution: Nepal, W Himalaya, E Himalaya and Tibetan Plateau.

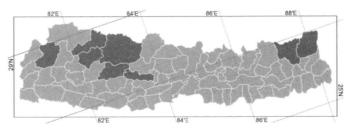

Altitudinal range: 3000–5200 m.

Ecology: Rocky crevices, scree, moist grounds, steep hillsides, grassy river banks.

Flowering: (March–)May–July. **Fruiting:** June–August.

2. *Noccaea nepalensis* Al-Shehbaz, Adansonia, Ser. 3 24(1): 89 (2002).

Herbs 30–40 cm, glaucous, glabrous throughout. Stems erect or ascending, simple and several from base, main stem 1-branched below raceme. Basal leaves rosulate; petiole 5–25 mm; leaf blade spatulate to suborbicular, 7–15

× 4–7 mm, base cuneate, apex rounded, margin entire or repand. Cauline leaves 5–7, widely spaced, sessile, oblong to ovate, middle ones 10–25 × 3–10 mm, base auriculate, apex obtuse, magin repand or entire; auricles 4 × 3 mm. Racemes ebracteate, corymbose, elongated considerably in fruit and to 15 cm. Sepals oblong, 2.5–3 × 1–1.2 mm, not saccate, margin white. Petals pale mauve, spatulate, 4.5–5 × 1.5–2 mm, apex rounded. Filaments 2.5–3 mm; anthers oblong, 0.7–0.8 mm. Ovules 8–10 per ovary. Fruiting pedicels 5–12 mm, slender, divaricate, straight. Fruit narrowly oblong-oblanceolate, strongly angustiseptate, carinate, widest slightly above middle 6–10 × 1.5–2 mm, wingless, base cuneate, apex obtuse, apical notch absent or obsolete; style filiform, 3–4 mm. Seeds dark brown, ovoid, ca. 1.2 × 0.6 mm.

Distribution: Endemic to Nepal.

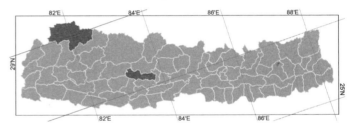

Altitudinal range: 2700–3900 m.

Ecology: Partial shade in mixed forests.

Flowering: May–June. **Fruiting:** May–June.

3. *Noccaea cochlearioides* (Hook.f. & Thomson) Al-Shehbaz, Adansonia, Ser. 3 24(1): 91 (2002).
Thlaspi cochlearioides Hook.f. & Thomson, J. Proc. Linn. Soc., Bot. 5: 177 (1861).

Herbs 5–10 cm, glabrous throughout. Stems erect to ascending, simple. Basal leaves rosulate; petiole (0.5–)1–3.5 cm; leaf blade broadly ovate to orbicular, 4–12 × 3–10 mm, base obtuse, apex rounded, margin entire, repand or obscurely denticulate. Middle cauline leaves sessile, oblong to ovate, 0.6–2.2 cm × 3–8(–10) mm, base auriculate, apex obtuse to acute, margin repand or entire, rarely denticulate. Sepals oblong, 1.8–2.5 × 1–1.5 mm, not saccate, margin white. Petals white, spatulate, 3.5–5.5 × 1.8–2.5 mm, apex rounded. Filaments 1.5–2.5 mm; anthers ovate, 0.4–0.6 mm. Ovules 6–10 per ovary. Fruiting pedicels 3–12 mm, slender, divaricate or slightly reflexed, straight. Fruit oblong, 5–8(–11)

× 3–4 mm, usually curved and somewhat contorted, base obtuse, apex emarginate and apical notch 0.5–1 mm deep; apical wings 0.5–1 mm wide; style 0.5–1 mm. Seeds brown, ovate-oblong, 1–1.5 × 0.6–0.8 mm, obscurely reticulate.

Distribution: Nepal, W Himalaya and E Himalaya.

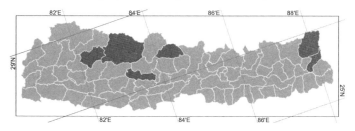

Altitudinal range: 2500–5800 m.

Ecology: Pasture, among rocks, stream-sides, river banks and peaty soils on consolidated scree.

Flowering: June–July. **Fruiting:** June–July.

30. *Sisymbrium* L., Sp. Pl. 2: 657 (1753).

Annual herbs. Trichomes absent or simple. Stems erect, branched. Basal leaves petiolate, rosulate or not, simple, entire or variously pinnately dissected. Cauline leaves petiolate or sessile. Racemes ebracteate or rarely bracteate, often elongated considerably in fruit. Sepals ovate or oblong to linear, erect or spreading, base of lateral pair sometimes subsaccate. Petals yellow, spatulate, oblong, or oblanceolate, apex obtuse or emarginate; claw subequalling sepals. Stamens 6, tetradynamous; filaments not dilated at base; anthers oblong, obtuse at apex. Nectar glands confluent and subtending bases of all stamens; median glands present. Ovules 40–150 per ovary. Fruiting pedicels slender or thickened. Fruit dehiscent siliques, narrowly linear, terete; valves papery to subleathery, with a prominent midvein and 2 conspicuous marginal veins, smooth or torulose; replum rounded; septum complete; style distinct; stigma capitate, 2-lobed, lobes not decurrent. Seeds uniseriate, wingless, oblong, plump; seed coat reticulate or papillate, not mucilaginous when wetted; cotyledons incumbent or obliquely so.

Worldwide about 40 species distributed primarily in Eurasia and N Africa, with a few in S Africa and the New World. Two species in Nepal.

Key to Species

1a   Fruit divaricate to recurved, (5–)7–10(–12) cm. Style 0.7–2(–2.5) mm. Sepals 4–6 mm. Petals 6–10(–12) mm ....**1. *S. brassiciforme***
  b   Fruit erect to ascending, (2.5–)3–4(–5) cm. Style 0.2–0.5 mm. Sepals 2–2.5 mm. Petals 2.5–3.5(–4) mm.........**2. *S. irio***

1. *Sisymbrium brassiciforme* C.A.Mey., in Ledeb., Fl. Altaic. 3: 129 (1831).
Sisymbrium ferganense Korsh.; *S. iscandericum* Kom.

Herbs (25–)35–100(–125) cm. Stems erect, branched above, glabrous or sparsely to densely soft hairy at least near base, usually glabrous above. Basal leaves rosulate, somewhat fleshy; petiole (1–)2–5(–8) cm; blade broadly oblanceolate to oblong-oblanceolate in outline, lyrate-pinnatipartite to lyrate, (1.5–)3–15(–26) × (0.5–)1–6(–9) cm, dentate to rarely subentire; lateral lobes 1 or 2(or 3) on each side of midvein, much smaller than the oblong, dentate terminal lobe. Uppermost cauline leaves narrowly lanceolate to linear, often entire, rarely lobed. Sepals oblong-linear, spreading, 4–6 × 1–1.5 mm. Petals yellow, spatulate, 6–10(–12) × 1.5–2.5 mm. Filaments yellowish, erect, 3–6 mm; anthers oblong, 1.5–2 mm. Ovules 90–150 per ovary. Fruiting pedicels horizontal to divaricate-ascending, stout, narrower than fruit, 5–10(–12) mm. Fruits narrowly linear, stout, terete, (5–)7–10(–12) cm × 1–1.5 mm, usually recurved; valves glabrous or pubescent, subtorulose; style stout, 0.7–2(–2.5) mm; stigma

2-lobed; septum slightly thickened. Seeds oblong, 0.9–1.3 × 0.5–0.7 mm, inserted in depressions of septum.

Distribution: Nepal, W Himalaya, Tibetan Plateau, N Asia, C Asia and SW Asia.

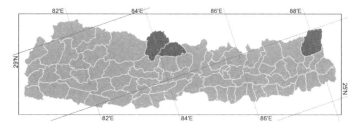

Altitudinal range: 2700–3700 m.

Ecology: Roadsides, rocky places, fields.

Flowering: June–August. **Fruiting:** August–September.

The report in Hara (Enum. Fl. Pl. Nepal 2: 45. 1979) of

Cruciferae (Brassicaceae)

Sisymbrium heteromallum C.A.Mey. from Nepal is based on plants of *S. brassiciforme* misidentified by Jafri as *S. heteromallum*. The latter species does not occur in Nepal.

Although Hara (1979) cited only *Nakao s.n.* from Mustang in his account, there have been numerous other collections including *Stainton, Sykes & Williams 870* (BM, E), *Stainton, Sykes & Williams 7235* (BM, E), *Stainton, Sykes & Williams 1219* (BM, E) and *Stainton, Sykes & Williams 1483* (BM, E).

2. *Sisymbrium irio* L., Sp. Pl. 2: 659 (1753).
Arabis charbonnelii H.Lév.

Herbs (10–)20–60(–75) cm. Stems erect, branched below and above, glabrous to sparsely pubescent at least basally. Basal leaves not rosulate; petiole (0.5–)1–4.5(–6) cm; blade oblanceolate to oblong in outline, runcinate-pinnatisect, (1.5–)3–12(–15) × (0.5–)1–6(–9) cm; lateral lobes (1 or)2–6(–8) on each side of midvein, smaller than terminal, oblong to lanceolate, entire or dentate to lobed. Uppermost cauline leaves smaller than basal, entire or 1–3-lobed. Sepals oblong, erect, 2–2.5 × 1–1.5 mm. Petals yellow, oblong-oblanceolate, 2.5–3.5(–4) × 1–1.5 mm. Filaments yellowish, erect, 2.5–4 mm; anthers oblong, 0.5–0.9 mm. Ovules 40–90 per ovary. Fruiting pedicels divaricate to ascending, slender and much narrower than fruit, (5–)7–12(–20) mm. Fruits narrowly linear, slender, terete, (2.5–)3–4(–5) cm ×

0.9–1.1 mm, erect to ascending, younger ones overtopping flowers; valves glabrous, slightly torulose; style 0.2–0.5 mm; stigma prominently 2-lobed; septum membranous. Seeds oblong, 0.8–1 × 0.5–0.6 mm, inserted in depressions of septum.
Fig. 21d.

Distribution: Nepal, W Himalaya, E Asia, N Asia, C Asia, SW Asia and Europe.

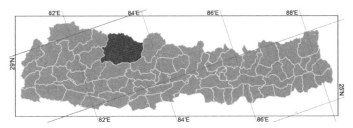

Altitudinal range: 60–1700 m.

Ecology: Rocky slopes, orchards, roadsides, fields, pastures, waste grounds, prairies, disturbed sites.

Flowering: May–August. **Fruiting:** June–September.

Native of Europe and western and central Asia. This is the first report of the species from Nepal, based on *Dobremez et al. 2722* (G).

31. *Smelowskia* C.A.Mey. nom cons., in Ledeb., Icon. Pl. 2: 17 (1830).

Perennial or annual herbs. Caudex well-developed, often branched. Trichomes dendritic, mixed with fine, simple and stalked forked ones. Stems procumbent to ascending, simple or branched above. Basal leaves petiolate, rosulate, 1- or 2-pinnatisect, sometimes flabellate. Cauline leaves petiolate or sessile, not auriculate, entire to pinnatisect. Racemes corymbose, many-flowered or bracteate, often elongated considerably in fruit. Sepals oblong, suberect to spreading, caducous or persistent, base of lateral pair not saccate Petals white, longer than sepals; blade spatulate to obovate or suborbicular, apex rounded; claw differentiated from blade. Stamens 6, slightly tetradynamous; filaments often dilated at base; anthers ovate or oblong, obtuse at apex; nectar glands usually confluent, subtending bases of all stamens, median glands present or absent. Ovules 20–46 per ovary. Fruiting pedicels slender, erect to ascending. Fruits broadly oblong, oblong-linear or suborbicular, terete, 4-angled, latiseptate, or angustiseptate, sessile or rarely short stipitate, unsegmented; valves with a prominent or obscure midvein, glabrous to pubescent, smooth; replum rounded; septum complete or perforated; style obsolete or distinct; stigma capitate, entire or rarely slightly 2-lobed. Seeds uniseriate, wingless, oblong, plump; seed coat minutely reticulate, not mucilaginous when wetted; cotyledons incumbent.

Worldwide 25 species in the Himalayas, C Asia, Mongolia, and the Russian Far East. One species in Nepal.

1. *Smelowskia tibetica* (Thomson) Lipsky, Trudy Imp. S.-Peterburgsk. Bot. Sada 23: 76 (1904).
Hutchinsia tibetica Thomson, Icon. Pl. 9: pl. 900 (1862); *Capsella thomsonii* Hook.f.; *Hedinia elata* C.L.He & Z.X.An; *H. rotundata* Z.X.An; *H. taxkargannica* G.L.Zhou & Z.X.An; *H. taxkargannica* var. *hejigensis* G.L.Zhou & Z.X.An; *H. tibetica* (Thomson) Ostenf.

Herbs (1–)5–30(–45) cm, densely to sparsely pubescent, canescent or green. Stems procumbent to ascending, densely hirsute basally with primarily simple trichomes to 1.3 mm. Basal leaves sparsely to densely pubescent; petiole

(0.2–)0.5–2(–3.5) cm, often ciliate basally; leaf blade ovate to narrowly oblong in outline, 1- or 2-pinnatisect, (0.3–)1–4(–7) × (0.2–)0.7–2(–2.5) cm; ultimate lobes ovate to oblong-linear, 1–12 × 0.5–3 mm. Cauline leaves similar to basal, reduced in size and divisions toward stem apex. Raceme bracteate throughout or rarely only basally; upper bracts subsessile, sometimes adnate to pedicel. Sepals oblong, 1.3–2 × 0.7–0.9 mm, sparsely pubescent. Petals obovate, 2–3.2 × (0.6–)0.9–1.4 mm; claw ca. 1.5 mm. Filaments 1.5–2 mm; anthers 0.3–0.4 mm. Ovules 20–46 per ovary. Fruiting pedicels straight, erect to ascending, 1.5–3.5(–5) mm, pubescent, subappressed to rachis. Fruit broadly oblong, rarely

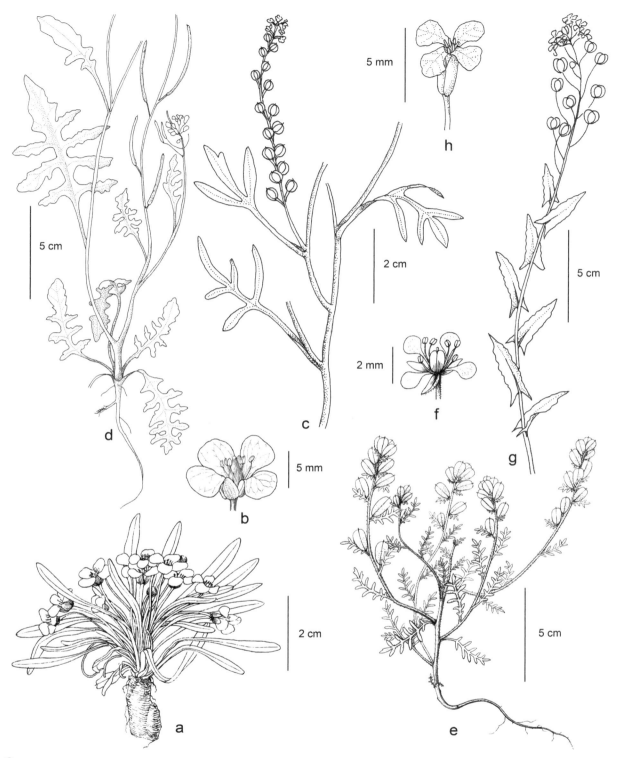

FIG. 21. CRUCIFERAE. **Pegaeophyton scapiflorum**: a, flowering plant; b, flower. **Lepidium sativum**: c, infructescence with leaves. **Sisymbrium irio**: d, plant with flowers and fruits. **Smelowskia tibetica**: e, fruiting plant; f, flower. **Thlaspi arvense**: g, stem with inflorescence and fruits; h, flower.

oblong-linear or suborbicular, (4.5–)5–10(–14) × 3–5 mm, flat or slightly twisted, obtuse to slightly retuse or rarely subacute at both ends, appressed to rachis; valves glabrous to pubescent; style 0.3–0.8 mm, glabrous or rarely pubescent. Seeds light to dark brown, oblong, 0.8–1.1 × 0.4–0.6 mm. Figs 16n, 21e–f.

Distribution: Nepal, E Himalaya, Tibetan Plateau, Assam-Burma, E Asia, N Asia and C Asia.

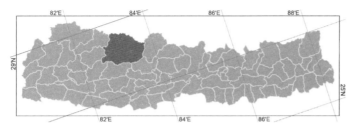

Altitudinal range: ca. 4500 m.

Ecology: Sandy or sandstone gravel, alpine meadows and steppes, scree, sandy slopes.

Flowering: June–August. **Fruiting:** July–September.

32. *Alliaria* Heist. ex Fabr., Enum.: 161 (1759).

Biennial herbs. Trichomes simple. Stems erect. Basal leaves long petiolate, rosulate, simple, crenate or dentate. Cauline leaves petiolate, dentate. Racemes several-flowered, ebracteate, bracteate throughout, or only basally bracteate. Sepals oblong, erect, glabrous, base of lateral pair not saccate. Petals white, longer than sepals; blade oblanceolate, apex obtuse; claw obscurely differentiated from blade. Stamens 6, slightly tetradynamous; filaments not dilated at base; anthers oblong, obtuse at apex. Nectar glands confluent, subtending bases of all stamens. Ovules 14–20 per ovary. Fruiting pedicels slender or thickened, narrower than or as thick as fruit, terete. Fruit dehiscent siliques, linear, terete or 4-angled, sessile; valves with a prominent midvein and distinct marginal veins, glabrous, torulose; replum rounded; septum complete; style distinct and to 6 mm; stigma capitate. Seeds uniseriate, wingless, oblong, plump; seed coat longitudinally striate, not mucilaginous when wetted; cotyledons incumbent.

Worldwide two species, one endemic to the Caucasus and the other a Eurasian weed now cosmopolitan. One species in Nepal.

1. *Alliaria petiolata* (M.Bieb.) Cavara & Grande, Bull. Orto Bot. Regia Univ. Napoli 3: 418 (1913).
Arabis petiolata M.Bieb., Fl. Taur.-Caucas. 2: 126 (1808); *Alliaria officinalis* Andrz. ex DC.; *A. officinalis* Andrz. ex M.Bieb. nom. inval.; *Erysimum alliaria* L.; *Sisymbrium alliaria* (L.) Scop.

लसुने साग Lasune saag (Nepali).

Herbs with garlic smell when crushed. Stems erect, (15–)30–90(–130) cm, simple or branched above, glabrous or pilose basally with trichomes to 1.5 mm. Basal leaves rosulate, withered by time of fruiting; petiole 3–10(–16) cm; leaf blade reniform or cordate, (0.6–)1.5–5(–7) cm wide, shorter in length, base cordate, margin crenate or dentate, glabrous or pilose. Cauline leaves with much shorter petioles, ovate, cordate, or deltoid, to 15 × 15 cm, base cordate or truncate, apex acute, margin acutely to obtusely toothed. Racemes ebracteate or rarely lowermost flowers bracteate. Sepals oblong, (2–)2.5–3.5(–4.5) × 0.7–1.5 mm. Petals white, oblanceolate, (2.5–)4–8(–9) × (1.5–)2–3(–3.5) mm, attenuate to claw-like base. Filaments 2–3.5(–4.5) mm; anthers oblong, 0.7–1 mm. Ovules 14–20 per ovary. Fruiting pedicels divaricate or ascending, (2–)3–10(–15) mm, nearly as thick as

fruit. Fruit linear, (2–)3–7(–8) cm × 1.2–2.5 mm, subtorulose, quadrangular or subterete, divaricate-ascending; valves glabrous; style (0.2–)1–2(–3) mm. Seeds brown or black, narrowly oblong, 2–4.5 × 0.7–2 mm.

Distribution: Nepal, W Himalaya, Tibetan Plateau, N Asia, C Asia, SW Asia and Europe.

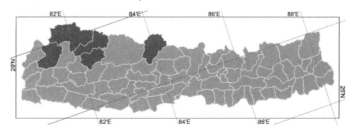

Altitudinal range: 2100–3200 m.

Ecology: Waste areas, roadsides, fields, river banks, woodlands.

Flowering: April–June. **Fruiting:** May–July.

Naturalized in many countries and under recorded in Nepal.

33. *Thlaspi* L., Sp. Pl. 2: 645 (1753).

Annual herbs, often glabrous and glaucous. Trichomes absent or simple. Stems erect, simple or branched. Basal leaves petiolate or subsessile, subrosulate, simple, entire or dentate. Cauline leaves sessile, often auriculate, sagittate, or amplexicaul at base, entire or dentate. Racemes ebracteate, elongated in fruit. Sepals ovate or oblong, erect or ascending, base of lateral pair not saccate, margin membranous. Petals white, spatulate, apex obtuse or emarginate; claw differentiated from blade. Stamens 6, tetradynamous; anthers ovate, obtuse at apex. Nectar glands 1 on each side of lateral stamen; median glands absent. Ovules 6–16 per ovary. Fruiting pedicels slender, divaricate. Fruit dehiscent silicles, obovate or suborbicular, apically notched, strongly angustiseptate, sessile; valves keeled, winged all around or rarely only distally; replum rounded; septum complete; style absent or very short and included in apical notch of fruit; stigma capitate, entire. Seeds uniseriate, wingless, ovoid, plump; seed coat concentrically striate; cotyledons accumbent.

Worldwide six species in Europe and SW Asia. One species in Nepal.

1. *Thlaspi arvense* L., Sp. Pl. 2: 645 (1753).

तिते Tite (Nepali).

Herbs (9–)15–55(–80) cm, glabrous throughout, often glaucous, foetid when crushed. Stems erect, simple or branched above. Petiole of basal leaves 0.5–3 cm; leaf blade oblanceolate to spatulate or obovate, 1–5 × 0.4–2.3 cm, base attenuate to cuneate, margin entire, repand, or coarsely toothed, apex rounded. Middle cauline leaves sessile, oblong, (0.5–)1.5–4(–8) × (0.2–)0.5–1.5(–2.5) cm, base sagittate or auriculate, apex rounded, obtuse, to subacute, margin dentate, repand, or entire. Sepals, (1.5–)2–3(–3.3) × 1–1.5 mm, not saccate, margin white. Petals, (2.4–)3–4.5(–5) × (0.8–)1.1–1.6 mm, narrowed to a claw-like base ca. 1 mm, apex obtuse or emarginate. Filaments (1–)1.5–2 mm; anthers ovate, 0.3–0.5 mm. Fruiting pedicels (0.5–)0.9–1.3(–1.5) cm, slender, divaricate, straight or slightly curved upward. Fruit, (0.6–)0.9–2 × (0.5–)0.7–2 cm, base obtuse to rounded, apex deeply emarginate and apical notch ca. 5 mm deep; wings 1–1.5 mm wide at base, 3.5–5 mm wide apically; style absent or 0.1–0.3 mm. Seeds blackish brown, ovoid, (1.2–)1.6–2(–2.3) × 1.1–1.3 mm.
Figs 16o, 21g–h.

Distribution: Cosmopolitan.

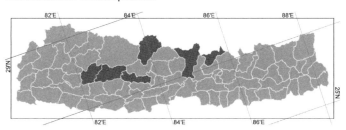

Altitudinal range: 2000–4600 m.

Ecology: Roadsides, grassy slopes, fields, waste places.

Flowering: March–October. **Fruiting:** March–October.

A cosmopolitan weed of Eurasian origin.

Used as a medicinal plant and source of oil. Tender shoots are cooked as a vegetable. The seeds are febrifuge and anti-inflammatory, and are used against pus in the lungs and renal inflammation. Cattle feeding on the plant produce tainted milk.

Moringaceae

Stephen Blackmore & Mark F. Watson

Deciduous trees or shrubs with soft wood and resinous bark; bark and pith with gum canals. Stipules absent. Leaves alternate, 2–4-odd pinnate, leaflets opposite, entire, glabrous to puberulent, base of petioles and pinnae jointed and with conspicuous stalked glands. Flowers in large, axillary, bracteate panicles, bisexual, zygomorphic. Sepals 5, petal-like, base fused in to a short tube, lobes often reflexed, unequal, imbricate in bud. Petals, free, unequal, imbricate in bud, the outer 3 larger, inner 2 smaller. Stamens 5, fused to petals, alternate with 5 staminodes, hairy at base, anthers dehiscing longitudinally. Ovary superior, 1-celled, on a short gynophore, pubescent, ovules 10–20 on 3 parietal placentas. Style filiform, stigma minute. Fruit a 3-angled, 3-valved, dehiscent capsule. Seeds embedded in spongy valves of the capsule, usually 3-winged but sometimes lacking wings, oily, endosperm almost absent.

Worldwide one genus and about 13 species of semi-arid environments of Africa and Asia. One species in Nepal.

1. *Moringa* Adans., Fam. Pl. 2: 318 (1763).

Guilandina L.

Generic description as for Moringaceae.

1. *Moringa oleifera* Lam., Encycl. 1(2): 398 (1785).
Guilandina moringa L.; *Moringa pterygosperma* Gaertn.

सइजान Saijan (Nepali).

Tree to 10–12 m, bark pale, smooth to rugose. Leaves 15–30(–60) cm, leaflets ovate, obovate or elliptic, 1–2 × 0.5–1.5 cm, of two distinct sizes with those closer to the rachis about half the size of the more terminal leaflets, base rounded to cuneate, apex rounded, obtuse or emarginate, puberulous when young, glabrescent, petiolules slender, 1–2 mm. Panicles erect, 20–30 cm, widely spreading. Bracts linear, ca. 1 mm. Flowers white to cream, fragrant, resembling an inverted pea flower with 2 dorsal sepals and 1 dorsal petal usually remaining unreflexed and 'keel-like', each flower borne on a false pedicel 7–15 mm, true pedicel 1–2 mm. Sepals narrowly elliptic to linear-lanceolate, 0.7–1.4 mm, usually puberluent. Petals spatulate to narrowly elliptic, 1–1.4(–2) cm, glabrous or puberulent only at base. Capsules pendent, 15–30(–50) × 1–3 cm. Seeds ovoid to globose, 3-angled, 1-1.5 cm (excluding wings), wings 3, up to 7 mm broad. Fig. 22a–e.

Distribution: Nepal, W Himalaya, E Himalaya, E Asia and Africa.

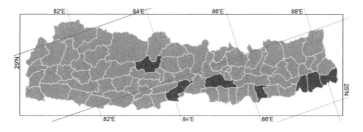

Altitudinal range: 100–1100 m.

Ecology: Widely cultivated as an ornamental and sometimes naturalizing.

Flowering: (January–)February–May(–December). **Fruiting:** May–December.

Known as the horse-radish tree on account of the mustard oil glycosides. Native to NW India, widely grown as an ornamental and useful tree in tropical areas in the Himalayan region, into S China and beyond. Seldom collected and poorly represented in herbarium collections. Planted trees probably occur above the recorded upper altitude limit.

Young fruits are eaten as a vegetable, and young roots make a good substitute for true horse-radish (*Armoracia rusticana* G.Gaertn., B.Mey. & Scherb.). Seeds contain a fine oil (Ben Oil) used in artists' paints and by watchmakers, and they also contain a powerful flocculent used to clarify turbid water. Leaves, bark, gum, roots, flowers and seeds are used medicinally to treat a very wide range of conditions. In recent years they have been widely planted in poverty alleviation projects, especially in Africa, because of tolerance to aridity and the fact that much of the plant is edible to humans and livestock.

Droseraceae

Anjana Giri & Mark F. Watson

Annual, biennial or perennial insectivorous herbs, fibrous-rooted with short rhizomes or with underground tubers. Stipules present or absent. Leaves in a basal rosette or alternate on erect stems, densely covered with sticky, reddish-green stalked insect-trapping glands, juvenile leaf blade circinate. Flowers in subterminal or terminal racemes on leafy stems or leafless scapes, bisexual, actinomorphic, bracts and bracteoles sometimes present, simple. Sepals (4–)5, mostly free, fused at base, persistent. Petals white or flushed rose pink, (4–)5(–8), free, mostly free or fused only at base, glabrous, closing and contorted after anthesis, persistent in fruit. Stamens 5, mostly free, alternate with petals, anthers 2-celled, introrse, dehiscing by longitudinal slits. Ovary superior, 1-locular, globose or ovoid, ovules numerous, placentation parietal. Styles 3–5, free or fused at base, persistent in fruit, stigma capitate, fimbriate or tooth-like. Fruit a 3–5-valved, loculicidal capsule, dehiscing at apex. Seeds numerous, ellipsoid or linear, endosperm fleshy.

Worldwide three or four genera and 105 species, distributed in nearly all temperate and tropical regions except the Pacific Islands. One genus and two species in Nepal.

1. *Drosera* L., Sp. Pl. 1: 281 (1753).

Description as for Droseraceae.

Worldwide about 100 species, cosmopolitan, especially diverse in Australia. Two species in Nepal.

Key to Species

1a   Stem elongate. Leaves alternate. Stipules absent..**1. *D. peltata***
  b   Stem very short (flowers on scapes). Leaves in a basal rosette. Stipules present ....**2. *D. burmanni***

1. *Drosera peltata* Thunb., Drosera: 7 (1797).
Drosera lunata Buch.-Ham. ex DC.; *D. peltata* Sm. later homonym, non Thunb.; *D. peltata* var. *lunata* (Buch.-Ham. ex DC.) C.B.Clarke.

झिंगा जाले Jhinga jaale (Nepali).

Perennial herbs, (6–)9–25(–38) cm. Tuber ca. 2 cm below ground, 5–8(–10) mm in diameter, glabrous or with black, papillose glands. Stems erect, short lived, pale green, usually simple and sparingly branched above, or with a few short branches arising from the tuber. Stipules absent. Basal leaves yellow-green, densely whorled or more usually absent, petiole 2–8 mm, blade peltate, orbicular to suborbicular, 2–4 × 6–8 mm, surface covered with red gland-tipped cilia. Cauline leaves smaller, yellow-green, numerous, widely spaced, petiole 5–11(–13) mm, blade peltate, semi-circular to lunate with attenuate tails at the basal angles, 1–2(–3) × 3–4(–5) mm, margin with spreading red gland-tipped cilia ('tentacles') 0.5–4(–5) mm, upper surface with short-stalked glands. Inflorescence terminal or subterminal, 1.7–4(–7) cm, (3–)4–5(–8)-flowered. Bracts and bracteoles few, mostly absent, linear, 1–3 mm. Pedicels slender, (2–)4–8 mm. Sepals yellowish green, ovate to narrowly so, (1.5–)2–3(–4) × (1.1–)1.4–2 mm, margin and apex fimbriation very variable, usually 5(–7)-fid, glabrous to glandular hairy. Petals white, occasionally flushed pink, oblong to obovate, (2.5–)4–5(–6) × (1.8–)2.5–3 mm. Stamens 2–4 mm. Ovary subglobose, 1.5–2 mm, styles 3, ca. 0.8 mm, each deeply divided into filiform segments, glandular, stigmas (2–)3-fid, minutely fringed. Capsule subglobose, 2–4 × 2–2.5 mm, enclosed by persistent sepals and petals.
Fig. 22f–g.

Distribution: Nepal, W Himalaya, E Himalaya, Tibetan Plateau, Assam-Burma, S Asia, E Asia, SE Asia and Australasia.

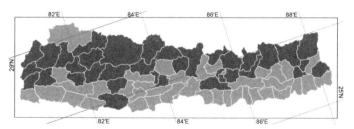

Altitudinal range: 900–4000 m.

Ecology: A common plant of wet or moist ground, often in disturbed open or rocky areas, meadows, wet cliff faces, sometimes under *Pinus*.

Flowering: (June–)July–August(–September). **Fruiting:** August–September.

Past literature recognise the Nepalese material as *Drosera peltata* var. *lunata*, however, *D. peltata* is a highly variable species and recent treatments include *lunata* within the

Droseraceae

general variation of the species and so do not recognize it at any rank.

Used in Ayruvedic medicine as a tonic and anti-syphilitic drug.

2. *Drosera burmanni* Vahl, Symb. Bot. 3: 50 (1794).

Annual or biennial rosette herbs to 22 cm. Roots fibrous, tuber absent. Stem unbranched, extremely short, sometimes to 1 cm when growing in shade. Stipule pale brown, 3–7 mm, fused to petiole at base, deeply 3-fid, lobes laciniate, papery, persistent. Leaves forming persistent flat rosette 1.5–3 cm diameter, subsessile or shortly petiolate. Leaf blade yellowish-green to reddish-violet, obovate-spatulate, 6–10 × 5–7 mm, base attenuate, with glandular hairs to 3 mm, apex fimbriate. Inflorescence terminal on 1 or 2 unbranched scapes, 6–22 cm, 2–19-flowered, glabrous or with white or to reddish-violet glands. Bracts and bracteoles usually present, linear, 1–3 mm. Pedicels slender, 1–7 mm. Sepals yellowish-green to reddish-violet, narrowly oblong, 2–3.5 × 0.5–1 mm, apex truncate to obtuse, margin and apex entire, not ciliate, outer surface with short glandular hairs and white glands. Petals white to reddish-violet, obovate, 3–4 × 2–3 mm. Stamens 2–3 mm. Ovary subglobose, glabrous, styles 5(–6), filiform, 2–3 mm, undivided, incurved, stigmas tooth-like. Capsule 2–3 mm, enclosed by persistent floral remains.
Fig. 22h–i.

Distribution: Nepal, Assam-Burma, S Asia, SE Asia, Africa and Australasia.

Altitudinal range: 60–200 m.

Ecology: Predominantly a tropical species of wet places and marshy ground at very low altitudes, sometimes in areas of cultivation.

Flowering: All year. **Fruiting:** All year.

No Nepalese herbarium specimens of this species have been found and the only reference to it occurring in Nepal is the literature citation by Don (Prodr. Fl. Nepal.: 212. 1825) of a Wallich specimen said to be collected in Nepal. This would normally refer to to material collected by Edward Gardener in Nepal between 1817 and 1819, and sent to Lambert by Wallich. The specimen Don used should now be in the Natural History Museum, London (BM), but it has not been traced, and so Don's citation cannot be verified. However, this predominantly more southern latitude species has been found at 200 m in Darjeeling district and is likely to occur in Nepal and so it is included here as a full entry without distribution map.

Used medicinally as a powerful rubifacient to ease muscle pain.

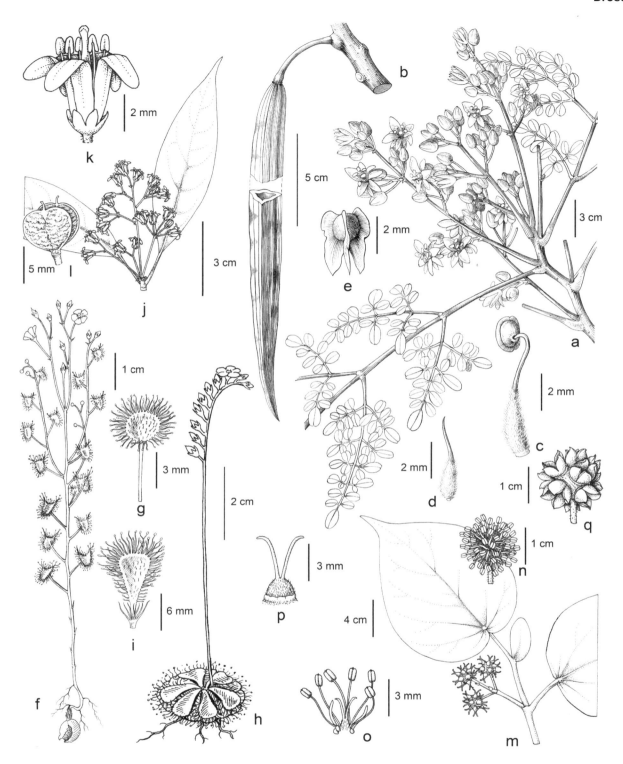

FIG. 22. Moringaceae. **Moringa oleifera**: a, flowering branch; b, fruit; c, stamen; d, staminode; e, seed. Droseraceae. **Drosera peltata**: f, flowering plant; g, leaf. **Drosera burmanni**: h, flowering plant; i, leaf and stipule. Pittosporaceae. **Pittosporum napaulense**: j, inflorescence and leaves; k, flower; l, dehisced capsule. Hamamelidaceae. **Exbucklandia populnea**: m, shoot with female flowers; n, male head; o, male flower in section; p, female flower; q, fruiting head.

Crassulaceae

Hideaki Ohba & Keshab R. Rajbhandari

Succulent herbs, subshrubs or shrubs. Stipules absent. Leaves alternate, opposite or verticillate, petiolate or sessile, usually simple, pinnately veined, entire or slightly incised, rarely lobed or imparipinnate. Inflorescences terminal or axillary, cymose, corymbiform, spiculate, racemose, paniculate, or sometimes reduced to a solitary flower. Flowers usually bisexual, sometimes unisexual in *Rhodiola* (plants then dioecious or rarely gynodioecious), actinomorphic, (3–)5(or 6)-merous. Sepals almost free or basally connate, persistent. Petals free or more or less connate. Stamens as many or twice as many as petals. Nectar scales at or near base of carpels. Fruit a cluster of follicles often surrounded by persistent perianth parts. Seeds few or numerous.

Worldwide about 53 genera and over 1500 species in Africa, America, Asia, Europe. Eight genera and 45 species in Nepal.

Key to Genera

1a	Petals united at least to middle	2
b	Petals free or up to one third united	3
2a	Calyx united, tube longer than lobes. Flowers nodding. Stamens inserted at base of corolla	**1. Bryophyllum**
b	Calyx free or slightly connate at base. Flowers erect. Stamens inserted at middle of corolla	**2. Kalanchoe**
3a	Leaves opposite, united at base to form sheath around stem	**3. Tillaea**
b	Leaves alternate or opposite, not united at base	4
4a	Scale leaves present at least at apex of rhizomes. Plants dioecious or with bisexual flowers	**4. Rhodiola**
b	Scale leaves absent. Flowers bisexual	5
5a	Plants with conspicuous rosette throughout year	6
b	Plants without rosette or with small, seasonal rosette	7
6a	Leaves flat, wider than 8 mm, with serrate or crenulate margins	**7. Phedimus**
b	Leaves flat or often terete, up to 8 mm wide, with entire margin (in *Sedum filipes* leaves up to 4 cm wide, but margins entire)	**8. Sedum**
7a	Plants hairy, often with glandular hairs. Petals basally connate. Flowering stem from axil of rosulate leaves	**5. Rosularia**
b	Plants glabrous. Petals free. Flowering stem from centre of rosette	**6. Sinocrassula**

1. *Bryophyllum* Salisb., Parad. Lond. 1(1): pl. 3 (1805).

Herbs, rarely subshrubs or shrubs. Roots fibrous. Stems usually erect. Leaves opposite, rarely 3-verticillate, petiolate, pinnately compound, rarely simple or pinnately lobed (or simple and bearing bulbils along margin), ± fleshy. Inflorescences terminal, cymose, many-flowered. Flowers bisexual, usually pendulous, 4-merous. Calyx tubular or rarely campanulate; tube sometimes basally dilated. Corolla tubular to salverform, equalling or longer than calyx; lobes shorter than or scarcely longer than tube. Stamens twice as many as petals, inserted below middle of corolla tube, usually near base; filaments equalling corolla tube. Nectar scales entire or emarginate. Carpels erect, free. Styles long. Follicles many-seeded.

Worldwide about 20 species in Asia and Africa, widely introduced throughout the tropics. One species in Nepal.

1. *Bryophyllum pinnatum* (Lam.) Oken, Allg. Naturgesch. 3(3): 1966 (1841).
Cotyledon pinnata Lam., Encycl. 2(1): 141 (1786); *Bryophyllum calycinum* Salisb.; *Cotyledon rhizophylla* Roxb.; *Kalanchoe pinnata* (Lam.) Pers.

अजम्बरी Ajambari (Nepali).

Perennial herb, entirely glabrous, with stout erect stem to 1.5 m. Petioles 1–7.5 cm, with broadened base. Earliest leaves simple, later leaves pinnate with 3 or 5 leaflets, base cuneate to truncate, apex obtuse or rounded, margin crenate, with bulbils at sinuses; blade of simple leaves ovate to oblong, to 10 × 4 cm; blade of pinnate leaves 6–24 × 4–12 cm, leaflets ovate to oblong or oblong-ovate. Inflorescence lax, paniculate, many-flowered. Pedicels 1–1.2 cm. Calyx green with red or violet stripes, campanulate; tube 2–3 cm; lobes triangular-ovate, 7–11 mm, apex acute to acuminate. Corolla pale green or greenish red-purple, ± cylindrical, connate; tube 2.5–4 cm; lobes oblong-ovate or triangular-ovate, 0.9–1.4 × 0.4–0.6 cm, apex acute. Stamens inserted below middle of corolla-tube; anthers deep yellow. Nectar scales pale green, rectangular, 1.8–2.6 mm. Carpels1–1.7 cm, with styles 2–3 cm. Fig. 23a–e.

Distribution: Asia, Africa, N America, S America and Australasia.

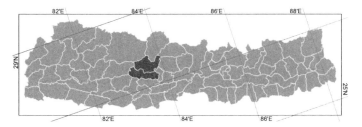

Altitudinal range: 800–1400 m.

Ecology: Rocky ground, way-sides.

Flowering: January–March.

Currently only known from two collections in Nepal, but presumably much more widely distributed.
    The plant is poisonous to cattle. Burnt leaves are externally applied to wounds, boils, snake bites, burns, corns and opthalmia.

2. *Kalanchoe* Adans., Fam. Pl. 2: 248 (1763).

Herbs sometimes biennial, subshrubs or shrubs. Roots usually fibrous. Leaves opposite, petiolate or sessile, usually amplexicaul; leaf blade margin entire, dentate, crenate, or leaves pinnate. Inflorescences terminal, cymose, sometimes also with subterminal cymes and thus paniculate, many-flowered; bracts small. Flowers bisexual, erect, 4-merous. Sepals free or basally subconnate, triangular to lanceolate, usually shorter than corolla tube. Corolla yellow (or white, red, pink or orange), salverform; tube subquadrangular or basally inflated and urn-shaped, base slightly narrowed; lobes longer than tube. Stamens twice as many as petals, inserted near middle of corolla tube; filaments unequal in length, usually very short. Nectar scales linear to suborbicular. Carpels erect. Styles short or long. Follicles many-seeded. Seeds ellipsoid.

Worldwide about 125 species in Africa and Asia. One species in Nepal.

1. *Kalanchoe spathulata* DC., Pl. Hist. Succ.: pl. 65 (1801). *Kalanchoe varians* Haw.

Kalanchoe spathulata var. **_staintonii_** H.Ohba, J. Jap. Bot. 78: 253 (2003).

हाथी काने Haathi kane (Nepali).

Stems to 1.2 m, glabrous, smooth, to 15 mm thick. Leaves glabrous, smooth, conspicuously petiolate; petioles of lower leaves 4–6 cm. Lamina narrowly oblong-ovate to narrowly triangular-ovate or widely lanceolate, to 20 × 8 cm, base rounded or widely cuneate, apex rounded, margins rather regularly crenate to crenulate on distal two thirds. Leaves in upper part narrowly ovate to widely lanceolate, conspicuously petiolate, margins crenate to crenulate. Inflorescence elongate lax few-flowered panicles, to 40 cm, glabrous. Pedicel 5–10 cm, glabrous, smooth. Flowers erect glabrous, smooth. Sepal lanceolate, 6–7 cm. Corolla bright to pale yellow; tube cylindrical, shallowly 4-angled, ca. 1.5 cm; lobes broadly lanceolate ca. 1 × 0.3 cm, apex acute to acuminate, spreading. Stamens inserted in upper part of corolla tube; filaments green. Carpels green.

Fig. 23f–i.

Distribution: Nepal, W Himalaya, E Himalaya and Assam-Burma.

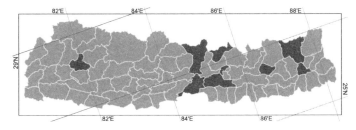

Altitudinal range: 900–1700 m.

Ecology: On exposed waste soil but sometimes rocky slopes or banks, especially path-sides in subtropical to warm temperate zones.

Flowering: May–August.

Kalanchoe spathulata is found from northern India to Sri Lanka and eastwards to China and SE Asia and though it has often been known under the name *K. integra* (Medik.) Kuntze, the latter is a synonym of *K. deficiens* (Forssk.) Asch. & Schweinf., which is endemic to the Arabian Peninsula (Ohba, J. Jap. Bot. 78: 247. 2003).

Kalanchoe spathulata includes seven varieties with rather restricted distributions in addition to the more widespread var. *spathulata*. In Nepal var. *staintonii* is distinguished from var.

spathulata by its large, long-petiolate leaves with regularly crenulate margins and rounded or widely cuneate bases.

Leaf paste and juice are applied to sprains, burns, eczema, and boils. Leaf juice is given for diarrhoea and dysentery. The leaves are also used to treat dysuria.

3. *Tillaea* L., Sp. Pl. 1: 129 (1753).

Small herbs, often annual, usually glabrous. Roots fibrous. Stems erect or ascending, sometimes ± stoloniferous at base. Leaves opposite, fused at base to form short sheath, flat or terete, margin entire. Inflorescences axillary, cymose, often shorter than subtending leaf, 1- to few-flowered. Flowers bisexual, (3 or)4- or 5-merous, inconspicuous. Sepals free, spurless. Petals inconspicuous, free. Stamens as many as petals; filaments filiform. Carpels free. Stigma terminal. Follicles dehiscent along adaxial suture. Seeds smooth, papillate, or striate.

Worldwide about 16 species in the Old World. One species in Nepal.

Although *Tillaea* has been treated as a synonym of *Crassula* (e.g. Grierson, Fl. Bhutan 1: 474. 1987), molecular evidence indicates that they should remain distinct (Gilbert *et al.*, Novon 10: 366. 2000).

1. *Tillaea schimperi* (Fisch. & C.A.Mey.) M.G.Gilbert, H.Ohba & K.T.Fu, Novon 10: 366 (2000).
Crassula schimperi Fisch. & C.A.Mey., Index Seminum Hort. Petrop. 8: 56 (1842); *Crassula pentandra* (Royle ex Edgew.) Schönland; *Tillaea pentandra* Royle ex Edgew.

Perennial herb to 10 cm. Stems sparsely branched. Leaves opposite, sessile, narrowly lanceolate to linear, to 12 × 3 mm, base slightly spurred, apex acute. Flowers usually 5-merous, shortly pedicellate. Sepals narrowly triangular, 1–1.5 mm. Petals shorter than sepals, 0.7–1 mm, lanceolate, whitish. Follicles erect, purplish red, with 2 seeds in each locule. Fig. 23j–n.

Distribution: Nepal, W Himalaya, E Himalaya, Tibetan Plateau, S Asia, SW Asia and Africa.

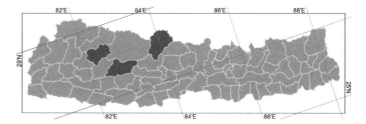

Altitudinal range: 1900–2500 m.

Ecology: On shady banks.

Flowering: July–August. **Fruiting:** September–October.

4. *Rhodiola* L., Sp. Pl. 1: 1035 (1753).

Chamaerhodiola Nakai; *Rhodia* Adans.; *Sedum* sect. *Rhodiola* (L.) Scop.; *Tetradium* Dulac.

Perennial herbs with slender or thick, suberect rhizomes bearing scales at apex. Flowering stems 1 to many, arising from axils of scales, annual, unbranched, leafy; dried remains of previous years' stems sometimes persistent. Radical leaves sometimes present. Stem leaves usually alternate, occasionally verticillate or opposite, simple, fleshy. Inflorescences terminal, flowers several or numerous in dense or spreading cymes, sometimes solitary. Flowers bisexual or unisexual (plants dioecious or rarely gynodioecious). Calyx (3 or)4- or 5(or 6)-parted, shortly tubular at base, lobes more or less erect. Petals free, as many as sepals. Stamens twice as many as petals, alternipetalous and oppositepetalous; oppositepetalous stamens adnate to basal part of petals. Nectar scales linear, oblong, suborbicular or quadrangular. Ovary superior to half-inferior; carpels as many as petals. Follicles few- to many-seeded.

Worldwide about 90 species, in mountains and high latitudes of the N Hemisphere. 22 species in Nepal.

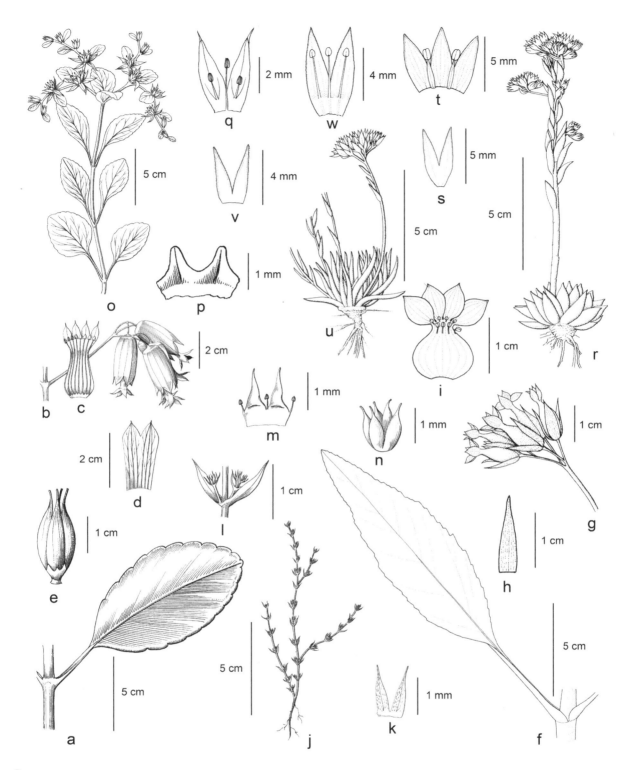

FIG. 23. CRASSULACEAE. **Bryophyllum pinnatum**: a, leaf; b, inflorescence; c, opened corolla showing stamens; d, calyx portion; e, fruit. **Kalanchoe spathulata** var. **staintonii**: f, leaf; g, inflorescence; h, sepal; i, opened corolla showing stamens. **Tillaea schimperi**: j, flowering plant; k, sepals; l, flowers and leaves; m, petals and stamens; n, fruit. **Phedimus odontophyllus**: o, stem and inflorescence; p, sepals; q, petals and stamens. **Sinocrassula indica**: r, flowering plant; s, sepals; t, petals and stamens. **Rosularia marnieri**: u, flowering plant; v, calyx portion, w, petals and stamens.

189

Crassulaceae

Key to Species

1a Flowers bisexual. Ovaries ± superior 2
 b Flowers unisexual (plants dioecious or rarely gynodioecious). Ovaries ± half-inferior (subgen. *Rhodiola*) 8

2a Foliage or dimorphic scaly leaves present at apex of rhizomes 3
 b Foliage and dimorphic scaly radical leaves absent, i.e. radical leaves all scaly (subgen. *Crassipedes*) 4

3a Petals oblong-ovate or ovate, white (sect. *Primuloides*) .... **13. *R. humilis***
 b Petals narrowly oblong or narrowly lanceolate, pink (sect. *Smithia*) .... **20. *R. smithii***

4a Leaves ovate to oblong-ovate or elliptic to (ob)lanceolate, wider than 5 mm 5
 b Leaves linear to lanceolate, less than 5 mm wide 6

5a Leaves ovate to oblong-ovate. Flowering stem smooth without papillae .... **4. *R. chrysanthemifolia***
 b Leaves elliptic to (ob)lanceolate. Flowering stem with dense papillae .... **16. *R. nepalica***

6a Leaves ± irregularly lobulate in upper half of margins .... **19. *R. sinuata***
 b Leaves entire to nearly entire or remotely minutely crenulate 7

7a Flowering stems 4–10 cm long. Leaves 5–12 mm long, entire to nearly entire .... **1. *R. amabilis***
 b Flowering stems 15–40 cm long. Leaves 1.2–3 cm long, remotely minutely crenulate .... **22. *R. wallichiana***

8a Flowering stems annual but marcescent (sect. *Chamaerhodiola*) 9
 b Flowering stems annual, deciduous 12

9a Stamens apparently shorter than petals. Flowering stems and leaves densely papillate .... **11. *R. himalensis***
 b Stamens nearly as long as petals or longer than petals. Flowering stems and leaves smooth 10

10a Leaves (1.6–)2–4 mm wide, narrowly oblong or oblong-ovate. Stamens 2.7–3 mm long. Ovules ca. 16 in
    each locule .... **21. *R. tibetica***
 b Leaves 0.6–1.5 mm wide, linear to linear-elliptic. Stamens 3–6 mm long. Ovules 6–16 in each locule 11

11a Ovaries of female flowers 3–5 mm long with ca. 6 ovules in each locule. Male flowers with petals 2.5–3.5 mm
    and stamens 2.3–4 mm .... **5. *R. coccinea***
 b Ovaries of female flowers 7–9 mm long with 14–16 ovules in each locule. Male flowers with petals 3.5–6 mm
    and stamens 4–6 mm .... **9. *R. fastigiata***

12a Leaves 4–6, nearly verticillate, long petiolate (sect. *Prainia*) .... **17. *R. prainii***
 b Leaves more than 10, alternate, sessile (shortly petiolate in *R. hookeri*) (sect. *Rhodiola*) 13

13a Leaves subopposite or alternate, rhombic-ovate to elliptic .... **3. *R. calliantha***
 b Leaves always alternate, shape variable but never rhombic-ovate 14

14a Rhizomes slender, 2–4 mm thick, surculose .... **7. *R. cretinii***
 b Rhizomes usually thicker than 7 mm, without suckers 15

15a Leaves triangular- or pentagonal-ovate, 1.2–2.5 × 1–1.5 cm .... **10. *R. heterodonta***
 b Leaves linear to narrowly oblong, elliptic or lanceolate to narrowly ovate (rarely ovate, in that case [*R. bupleuroides*
    var. *parva*] leaves less than 8 mm long, and flowers dark reddish purple) 16

16a Flowers white or yellowish 17
 b Flowers dark purple to purplish red, bright crimson to violet-blue (in dried specimens usually dull yellow) or green
    with deep purplish red stripes 18

17a Leaves with entire or nearly entire margins, 0.3–0.7 cm wide .... **14. *R. imbricata***
 b Leaves with remotely and roughly crenulate margins, 0.8–1.5 cm wide .... **15. *R. lobulata***

18a Flowering stems 7–9 mm thick. Styles of female flowers 2–3 mm long .... **6. *R. crenulata***
 b Flowering stems less than 6 mm thick. Styles of female flowers shorter than 1 mm 19

19a Leaves with truncate or rounded base 20
  b Leaves with cordate or attenuate base 21

20a Plants ± glaucous. Leaves petiolate (petiole ca. 1 mm long), linear- or oblong-lanceolate to lanceolate-ovate, apex acute to acuminate ....**8. *R. discolor***
  b Plants not glaucous. Leaves sessile, elliptic to narrowly ovate, apex rounded to obtuse ....**18. *R. purpureoviridis***

21a Leaves with cordate (or rarely cordate-rounded) base, sessile, oblong, oblong-ovate or ovate, margin entire...**2. *R. bupleuroides***
  b Leaves with attenuate base, usually petiolate, rectangular-oblong or -obovate, margin shallowly crenulate ....**12. *R. hookeri***

1. *Rhodiola amabilis* (H.Ohba) H.Ohba, J. Jap. Bot. 51: 386 (1976).
Sedum amabile H.Ohba, J. Jap. Bot. 51: 295 (1976).

Plants 4–10 cm, bisexual. Rhizomes dauciform, usually sparsely branched. Flowering stems ca. 1 mm thick, pale green, glabrous, smooth. Leaves alternate, widely spreading, linear to linear-lanceolate, 5–12 × 0.8–1.5 mm, base attenuate, apex rounded or obtuse, margin entire, glabrous, smooth. Inflorescences 1–3(–5)-flowered, 3–5 × 0.5–1 mm. Pedicels ca. 1 mm long. Calyx tube ca. 0.5 mm, glabrous; lobes triangular-lanceolate, 4–5 mm, apex rounded or obtuse, tinged with red. Petals white or pink near apex, spreading at flowering, narrowly elliptic to elliptic, 5–7.5 mm, apex obtuse, often with short mucro. Stamens distinctly shorter than petals; anthers deep purplish red. Nectar scales creamy white, square, 0.6–1 mm. Carpels 5–6 mm with styles 1.5 mm. Ovules 12–18 in each locule.

Distribution: Nepal and E Himalaya.

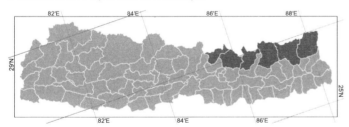

Altitudinal range: 2300–5200 m.

Ecology: On exposed moss-covered rocks.

Flowering: August–September.

Raymond-Hamet (Acta Horti Gothob. 2: 394. 1926) and some later authors have partially misapplied the name *Sedum linearifolium* Royle to this species
 A putative hybrid between *Rhodiola amabilis* and *R. wallichiana* (Hook.) S.H.Fu was recorded from C Nepal (Beding in Rolwaling Himal) by Ohba *et al.* (Acta Phytotax. Geobot. 37: 123: 1986).

2. *Rhodiola bupleuroides* (Wall. ex Hook.f. & Thomson) S.H.Fu, Acta Phytotax. Sin., Addit. 1: 124 (1965).
Sedum bupleuroides Wall. ex Hook.f. & Thomson, J. Proc. Linn. Soc., Bot. 2: 98 (1858); *S. bupleuroides* Wall. nom. nud.; *S. cooperi* Praeger later homonym, non Clemenc.

सोहलो-मार्पो Sohlo-marpo (Tibetan).

Plants dioecious. Rhizomes cylindrical, 1–2 cm thick. Flowering stems pale green, glabrous, smooth. Leaves sessile, margin entire, glabrous, smooth. Inflorescences glabrous, smooth. Flowers 4–7 mm across. Pedicels 0.2–1 cm. Calyx purplish red, glabrous; tube 1.6–8 mm; lobes narrowly oblong or oblong-ovate, 2–5 mm, apex rounded. Male flowers with petals linear or oblanceolate, 2.8–4 mm long, spreading to somewhat reflexed. Stamens usually as long as petals, filaments purplish red; anthers dark purplish red. Nectar scales oblong, shining, blackish purple-red, 0.6–1.2 mm long. Female flowers with petals oblong or oblong-ovate, 2.8–4 mm long, ascending to spreading. Carpels 3.5–9 mm, upper part abruptly out-curved with styles less than 0.3 mm long. Ovules 10–16 in each locule. Follicles 5–12 mm long. Seeds nearly cylindrical with round apex, 1.8–2.5 mm.

Distribution: Nepal, W Himalaya, E Himalaya, Tibetan Plateau, Assam-Burma and E Asia.

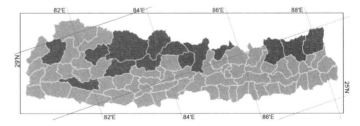

Praeger (Notes Roy. Bot. Gard. Edinburgh, 13: 78. 1921) partially misapplied the name *Sedum discolor* Franch. to this species.

1a Flowering stems 15–50 cm. Leaves 1–6 × 0.5–4.5 cm. Inflorescence with 7–40 flowers .... var. ***bupleuroides***
  b Flowering stems 2–5 cm. Leaves 4–8 × 3–5 mm. Flowers 1 to few...................................... var. ***parva***

191

Crassulaceae

Rhodiola bupleuroides (Wall. ex Hook.f. & Thomson) S.H.Fu var. *bupleuroides*

Plants 15–30 cm, dioecious. Leaves narrowly oblong-ovate to ovate or obovate, 1–6 × 0.5–4 cm, base cordate or cordate-rounded, apex rounded. Inflorescence 7–40-flowered. Flowers dark purplish red.

Distribution: Nepal, W Himalaya, E Himalaya, Tibetan Plateau, Assam-Burma and E Asia.

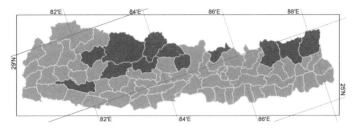

Altitudinal range: 3200–4500 m.

Ecology: On scree, stony or rocky slopes or banks, and sometimes among rocks, and on cliffs.

Flowering: June–August. **Fruiting:** August–September.

Rhodiola bupleuroides var. *parva* (Fröd.) H.Ohba, Alp. Fl. Jaljale Himal: 26 (1992).
Sedum bupleuroides var. *parvum* Fröd., Ark. Bot. 30A(9): 5 (1943).

Flowering stems 2–5 cm long. Leaves ovate to oblong-ovate, 4–8 × 3–5 mm, base cordate or rarely cordate-truncate, apex rounded to obtuse. Flowers purple, 1 to few.

Distribution: Nepal and Tibetan Plateau.

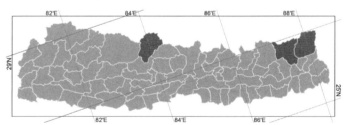

Altitudinal range: 3300–4500 m.

Ecology: On windswept alpine scree or among rocks.

Flowering: June–August.

3. *Rhodiola calliantha* (H.Ohba) H.Ohba, J. Jap. Bot. 51: 386 (1976).
Sedum callianthum H.Ohba, J. Jap. Bot. 49: 325 (1974).

Plants 12–18 cm, dioecious. Rhizomes cylindrical, ca. 1.5 cm thick. Flowering stems pale purplish red, smooth. Petioles ca. 2 mm. Leaves subopposite or alternate, narrowly to normally rhombic-ovate or elliptic, 4–5.5 × 1–2.2 cm, base attenuate,

apex acute, margin apically roughly serrate-crenate, almost smooth. Inflorescences 20–35-flowered. Pedicels 1.5–2.5 mm long, scarcely papillate. Calyx lobes triangular to triangular-ovate, 1.2–1.6 mm, apex obtuse. Petals pinkish to purple, narrowly obovate, 3–4.5 × 0.8–1.5 mm, apex obtuse, margin entire or erose towards apex. Stamens shorter than petals; anthers deep purplish red. Nectar scales deep purplish red, narrowly oblong, ca. 1 mm, apex rounded-truncate. Carpels tapering, styles indistinct. Ovules 2 in each locule.

Distribution: Nepal and Tibetan Plateau.

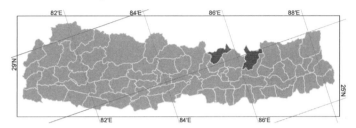

Altitudinal range: 3700–4300 m.

Ecology: On rocks on shady slopes.

Flowering: June.

4. *Rhodiola chrysanthemifolia* (H.Lév.) S.H.Fu, Acta Phytotax. Sin., Addit. 1: 127 (1965).
Sedum chrysanthemifolium H.Lév., Repert. Spec. Nov. Regni Veg. 12: 283 (1913); *Rhodiola ovatisepala* (Raym.-Hamet) S.H.Fu; *Sedum linearifolium* var. *ovatisepalum* Raym.-Hamet; *S. ovatisepalum* (Raym.-Hamet) H.Ohba.

Plants 5–25 cm, bisexual. Main root branched, thick with slender upper parts. Rhizomes short, ascending or erect. Flowering stems pale green, long, smooth. Petioles 0.5–1.5 cm. Leaves aggregated in upper part of flowering stem or evenly distributed, margin 3- or 4-lobed or cleft, papillate. Inflorescences rather compact, with few flowers. Pedicels often papillate. Petals white or greenish white, 5–7 mm. Stamens slightly shorter than or as long as petals; anthers purplish red. Carpels erect, 4–5 mm; styles ca. 1 mm.

Distribution: Nepal, E Himalaya, Tibetan Plateau, Assam-Burma and E Asia.

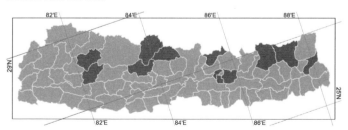

Weibel (Candollea 16: 144. 1958) partially misapplied the name *Sedum linearifolium* Royle to this species.

1a Leaves ± aggregate at apical part of flowering stem. Calyx lobes ovate to triangular-ovate, 1.5–2.5 mm. Petals ca. 5 mm .... subsp. *chrysanthemifolia* b Leaves distributed throughout flowering stem. Calyx lobes oblong to oblong-ovate, 3.5–5 mm long. Petals 5–7 mm .... subsp. *sacra*

Rhodiola chrysanthemifolia (H.Lév.) S.H.Fu subsp. *chrysanthemifolia*

Plants 5–25 cm. Flowering stems smooth. Leaves aggregated in upper part of flowering stem, ovate, oblong-ovate or oblong, 3–5 × 1.1–2.5 cm, base attenuate and decurrent, margin 3- or 4-lobed or -cleft, papillate. Pedicels 4–10 mm, often papillate. Calyx lobes ovate or triangular-ovate, 1.5–2.5 mm, apex rounded. Petals greenish white, narrowly oblong-ovate, ca. 5 mm. Anthers purplish red. Nectar scales deep to pale yellow, oblong, apex truncate to emarginate. Carpels erect, 4–5 mm; styles ca. 1 mm. Follicles erect, ca. 6 mm. Seeds oblong-lanceolate.

Distribution: Nepal, E Himalaya, Tibetan Plateau, Assam-Burma and E Asia.

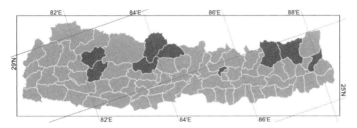

Altitudinal range: 2700–4600 m.

Ecology: On forested slopes, moss-covered tree trunks, rocks, rocky cliffs.

Flowering: June–September. **Fruiting:** August–September.

Rhodiola chrysanthemifolia subsp. *sacra* (Prain ex Raym.-Hamet) H.Ohba, J. Fac. Sci. Univ. Tokyo, Sect. 3, Bot. 12: 384 (1980).
Sedum linearifolium var. *sacrum* Prain ex Raym.-Hamet, Acta Horti Gothob. 2: 395 (1926); *Rhodiola sacra* (Prain ex Raym.-Hamet) S.H.Fu; *Sedum sacrum* (Prain ex Raym.-Hamet) H.Ohba.

Plants to to 20 cm. Flowering stems densely papillate. Leaves evenly distributed, shortly petiolate, obovate-oblong to obovate, 1.5–2.5 × 0.7–1.7 cm, base attenuate, apex rounded, margin with 4 or 5 crenate or dentate lobes on each side, lower surface smooth. Pedicels 1–2 mm, densely papillate. Calyx lobes oblong to oblong-ovate, 3.5–5 mm. Petals white, narrowly oblong, 5–7 mm. Stamens slightly shorter than or as long as petals; anthers deep purplish red. Nectar scales dark orange-red, rectangular-ovate, 0.6–1 mm. Carpels 4.5–6 mm with ca. 1.5 mm style. Ovules 14–20 in each locule.

Distribution: Nepal, E Himalaya and Tibetan Plateau.

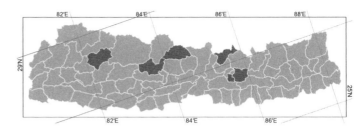

Altitudinal range: 2200–3300 m.

Ecology: On rocky grassland slopes, rock crevices.

Flowering: July.

5. *Rhodiola coccinea* (Royle) Boriss., in Kom., Fl. URSS 9: 41 (1939).
Sedum coccineum Royle, Ill. Bot. Himal. Mts. [7]: 223, pl. 48 fig. 3 (1835); *Rhodiola asiatica* D.Don; *Sedum asiaticum* (D.Don) DC.; *S. juparense* Fröd.

Plants 1–5 cm, dioecious. Main root 10–30 cm or longer. Rhizomes thick, with a large number of persistent old flowering stems. Flowering stems erect or curved, 1–2 mm thick. Leaves sessile, linear to linear-elliptic, 3–7 × 0.6–1.2 mm, base attenuate, apex rounded or obtuse, margin entire, glabrous, smooth. Inflorescences 0.8–1 cm wide, with 1 to few flowers. Flowers 4- (rarely 5-)merous, pedicels less than 2 mm. Calyx 2.5–3.5 mm, reddish, lobes linear-ovate to subulate, 1–2.5 mm, apex rounded or obtuse. Petals red or yellow, oblong-obovate or narrowly oblong-ovate, 2.5–3.5 mm, apex rounded to obtuse. Stamens longer than petals; anthers red. Nectar scales yellow, narrowly oblong, 0.6–1.3 mm. Carpels of female flowers 4–6 mm, with reflexed short apical beak; ovaries 3–5 mm. Ovules ca. 6 in each locule. Follicles reddish, 6–7 mm. Seeds oblong, 1–1.5 mm, both ends winged.

Distribution: Nepal, W Himalaya, E Himalaya, Tibetan Plateau and SW Asia.

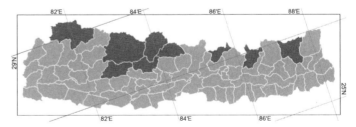

Altitudinal range: 3200–5800 m.

Ecology: On stony soils, rocks, rock crevices on slopes.

Flowering: June–July.

6. *Rhodiola crenulata* (Hook.f. & Thomson) H.Ohba, J. Jap. Bot. 51: 386 (1976).
Sedum crenulatum Hook.f. & Thomson, J. Proc. Linn. Soc., Bot. 2: 96 (1858); *Rhodiola euryphylla* (Fröd.) S.H.Fu;

R. rotundata (Hemsl.) S.H.Fu; *Sedum bupleuroides* var. *rotundatum* (Hemsl.) Fröd.; *S. euryphyllum* Fröd.; *S. megalanthum* Fröd.; *S. megalophyllum* Fröd.; *S. rotundatum* Hemsl.; *S. rotundatum* var. *oblongatum* C.Marquand & Airy Shaw.

सोहलो-मार्पो Sohlo-marpo (Tibetan).

Plants 15–21 cm, dioecious. Caudex stout, with many persistent blackish old flowering stems. Flowering stems straw-coloured or reddish, 0.7–0.9 cm thick, nearly smooth. Petioles 0.8–2 mm. Leaves alternate, elliptic-oblong to suborbicular, 1.2–3.3 × 0.6–1.7 cm, base rounded or widely cuneate, apex obtuse or rounded, margin usually crenulate and often undulate, smooth. Inflorescences ca. 2 × 2–3 cm, 20–40-flowered. Pedicels 3–5 mm, smooth. Calyx lobes linear to linear-elliptic or lanceolate, 2.6–4 mm, apex obtuse. Petals red to purplish red, oblong-ovate or obovate to angularly obovate, 4.5–7 mm, apex rounded to obtuse. Stamens slightly longer than petals; anthers deep purplish red. Nectar scales deep purplish red, oblong, 0.8–1.2 mm. Carpels of female flowers 7–10 mm, styles 2–3 mm. Ovules 6–12 in each locule. Follicles reddish, 8–10 mm. Seeds obovoid to ovoid, 1.5–2 mm, both ends winged.
Fig. 24a–c.

Distribution: Nepal, E Himalaya, Tibetan Plateau and E Asia.

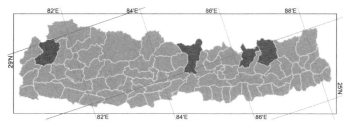

Altitudinal range: 3700–5800 m.

Ecology: On scree, moraines of high elevations.

Flowering: June–September.

7. *Rhodiola cretinii* (Raym.-Hamet) H.Ohba, J. Jap. Bot. 51: 386 (1976).
Sedum cretini Raym.-Hamet, J. Bot. 54(suppl.1): 16 (1916); *Chamaerhodiola cretinii* (Raym.-Hamet) Nakai; *Rhodiola crassipes* var. *cretinii* (Raym.-Hamet) H.Jacobsen; *Sedum crassipes* var. *cretinii* (Raym.-Hamet) Fröd.; *S. wallichianum* var. *cretinii* (Raym.-Hamet) H.Hara.

Plants 2–12 cm, dioecious. Rhizomes slender, erect, 2–4 mm thick, producing suckers. Flowering stems ca. 1.2 mm thick, ascending to suberect, papillate. Leaves linear to narrowly elliptic or linear-obovate, 7–10 × 1.5–2.5 mm, base long attenuate, apex rounded, obtuse or acute, margin entire or shallowly 3–5-crenate, smooth. Inflorescences compact, ca. 1 cm wide. Pedicels 1–5 mm, papillate. Calyx lobes linear to subulate, 3–4.5 mm, apex obtuse. Petals greenish to yellowish white, linear-oblanceolate or narrowly elliptic, 3.5–6 × 1–1.5 mm, apex obtuse. Stamens slightly longer than petals,

anthers dark purplish. Nectar scales pale yellow or orange-yellow, oblong, 0.6–0.9 mm. Carpels 5–7 mm, styles ca. 1.5 mm. Ovules 14–24 in each locule.

Distribution: Nepal, E Himalaya and Tibetan Plateau.

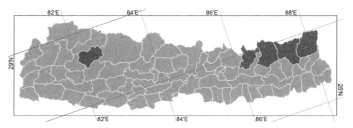

Altitudinal range: 3700–4800 m.

Ecology: On scree, gravel slopes, or among rock crevices.

Flowering: June–August.

8. *Rhodiola discolor* (Franch.) S.H.Fu, Acta Phytotax. Sin., Addit. 1: 124 (1965).
Sedum discolor Franch., J. Bot. [Morot] 10: 285 (1896); *S. bupleuroides* var. *discolor* (Franch.) Fröd.

Plants 12–40 cm, dioecious. Rhizomes subprostrate with ascending or erect apical part, 3–5 mm thick. Flowering stems glaucous, pale green, glabrous, smooth. Petioles ca. 1 mm. Leaves linear to oblong-lanceolate or lanceolate-ovate, 9–25 × 3–5(–7) mm, base truncate to rounded (sometimes with small auricles), apex acute to acuminate, margin obscurely dentate to subentire and usually revolute, upper surface light green, lower surface ± glaucescent. Inflorescences 3–5 × 5–10 cm. Pedicels 3–7 mm. Calyx lobes narrowly triangular, 1–2.5 × 0.5–1 mm. Petals purplish red, oblong to oblong-obovate, 5–6 mm in male flowers, 3–4 mm in female flowers. Stamens slightly shorter than petals, 4.5–5.5 mm; anthers reddish purple. Nectar scales deep reddish purple, oblong-quadrangular, 0.7–1.2 mm, apex emarginate. Carpels 7–9 mm, with indistinct styles. Ovules 10–14 in each locule. Follicles 0.8–1 cm.
Fig. 24d–f.

Distribution: Nepal, E Himalaya and E Asia.

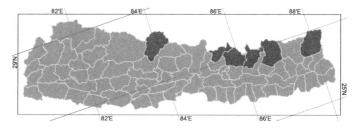

Altitudinal range: 3600–4600 m.

Ecology: On floor of alpine *Rhododendron* thickets and rocky cliffs.

Flowering: June–July.

Praeger (J. Roy. Hort. Soc. 46: 49. 1921) partially misapplied the name *Sedum tibeticum* Hook.f. & Thomson to this species.

9. *Rhodiola fastigiata* (Hook.f. & Thomson) S.H.Fu, Acta Phytotax. Sin., Addit. 1: 122 (1965).
Sedum fastigiatum Hook.f. & Thomson, J. Proc. Linn. Soc., Bot. 2: 98 (1858); *Chamaerhodiola fastigiata* (Hook.f. & Thomson) Nakai; *Sedum quadrifidum* var. *fastigiatum* (Hook.f. & Thomson) Fröd.

Plants 8–20 cm, dioecious. Rhizomes erect, often branched, 1–1.5 cm thick, with persistent old flowering stems. Flowering stems 1.2–2 mm thick, densely leafy. Leaves linear, linear-lanceolate, elliptic or oblanceolate, 8–12 × 1–4 mm, base attenuate, apex obtuse, margin entire, finely mammillate. Inflorescences dense, ca. 1 × 2 cm. Pedicels 1–4 mm, papillate. Calyx lobes linear to narrowly triangular, ca. 3 mm, apex obtuse. Petals creamy white or yellow, often reddish outside, oblong-lanceolate, ca. 5 × 1.3 mm, apex obtuse. Stamens slightly longer than petals, 5–6.5 mm; anthers deep red. Nectar scales bright yellow, transversely oblong, ca. 0.7 mm, apex emarginate. Carpels 6–8 mm long, styles indistinct. Ovules 14–16 in each locule. Follicles 7–9 mm. Seeds linear-ellipsoidal, 1.5–2.5 mm.

Distribution: Nepal, W Himalaya, E Himalaya, Tibetan Plateau and E Asia.

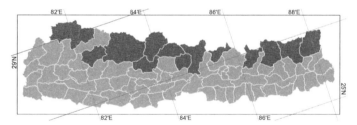

Altitudinal range: 3600–5700 m.

Ecology: On scree and boulders, often on rocks.

Flowering: June–August.

10. *Rhodiola heterodonta* (Hook.f. & Thomson) Boriss., Fl. URSS 9: 32 (1939).
Sedum heterodontum Hook.f. & Thomson, J. Proc. Linn. Soc., Bot. 2: 95 (1858); *S. roseum* var. *heterodontum* (Hook.f. & Thomson) Fröd.

Plants 30–40 cm, dioecious. Roots vertical, stout. Rhizomes branched. Flowering stems 4–5 mm thick. Leaves sessile, usually triangular-ovate or pentagonal-ovate, 1.2–2.5 × 1–1.5 cm, base cordate and amplexicaul, apex acute, margin coarsely serrate. Inflorescences compact, 1–1.5 × 1.5–2 cm, with 80–120 bractless flowers. Flowers sessile. Calyx-lobes linear, ca. 3 mm long, apex obtuse. Petals greenish yellow, linear, to 7 × 1.3 mm, apex subobtuse. Stamens nearly twice as long as petals, 6–9 mm; anthers dull purplish red. Nectar scales reddish, oblong, 0.8–1.1 mm, apex shallowly concave.

Carpels ca. 6 mm long with indistinct styles. Ovules 8–10 in each locule. Follicles 0.8–1.2 cm long. Seeds ellipsoid, ca. 1.5 mm.

Distribution: Nepal, W Himalaya, Tibetan Plateau and SW Asia.

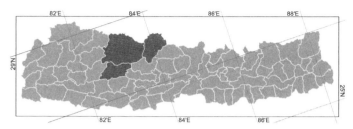

Altitudinal range: 3900–5100 m.

Ecology: On rocky slopes and among boulders.

Flowering: May–June.

11. *Rhodiola himalensis* (D.Don) S.H.Fu, Acta Phytotax. Sin., Addit. 1: 121 (1965).
Sedum himalense D.Don, Prodr. Fl. Nepal.: 212 (1825); *Chamaerhodiola asiatica* (D.Don) Nakai; *C. himalensis* (D.Don) Nakai; *C. quadrifida* (Pall.) Nakai; *Rhodiola quadrifida* (Pall.) Fisch. & C.A.Mey.; *Sedum himalayanum* Wall. nom. nud.; *S. hypericifolium* Wall. nom. nud.; *S. quadrifidum* Pall.; *S. quadrifidum* var. *acuminatum* Pall.; *S. quadrifidum* var. *coccineum* (Royle) Pall.; *S. quadrifidum* var. *himalense* (D.Don) Fröd.; *S. quadrifidum* var. *scoparium* Pall.

सेन-चूड़बा Tsen-chungba (Tibetan).

Plants 20–30 cm, dioecious. Rhizomes elongate, cylindrical, creeping to ascending, up to 40 × 2–3(–4) cm, apical part erect with persistent old flowering stems. Flowering stems purplish red, densely papillate. Leaves sessile 0.7–2.7 × 0.2–1 cm. Inflorescences compact. Pedicels to 3 mm long, papillate. Calyx lobes narrowly triangular, 1.5–2.7 mm, apex rounded, smooth. Petals deep reddish purple, oblong-lanceolate, 2.5–4.5 mm, apex obtuse or rounded. Stamens distinctly shorter than petals, 2–3 mm long; anthers deep reddish purple. Nectar scales blackish purple, oblong, 0.7–0.9 mm, apex emarginate. Carpels 4.5–5.5 mm, tapering, styles indistinct. Ovules 10–14 in each locule. Follicles 7–8 mm. Seeds ellipsoidal, 2.5–3 mm.

Distribution: Nepal, E Himalaya, Tibetan Plateau and E Asia.

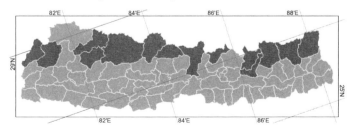

Praeger (Notes Roy. Bot. Gard. Edinburgh 13: 94. 1921) partially misapplied *Sedum scabridum* Franch. to this species.

Crassulaceae

The plant is considered diuretic and effective in the treatment of kidney problems, lung infections and asthma. The sap is used to treat skin diseases and oral infections.

1a Leaves papillate on lower surface, margins entire but often shallowly repand. Calyx papillate.. subsp. ***bouvieri***

b Leaves without papillae on lower surface, margins entire or remotely crenulate-denticulate. Calyx smooth .... subsp. ***himalensis***

Rhodiola himalensis subsp. ***bouvieri*** (Raym.-Hamet) H.Ohba, J. Fac. Sci. Univ. Tokyo, Sect. 3, Bot. 13: 137 (1982).
Sedum bouvieri Raym.-Hamet, J. Bot. 54(suppl.1): 11 (1916); *Chamaerhodiola bouvieri* (Raym.-Hamet) Nakai; *Rhodiola bouvieri* (Raym.-Hamet) H.Ohba; *Sedum quadrifidum* var. *bouvieri* (Raym.-Hamet) Fröd.

Leaves ovate, base shortly attenuate, apex attenuate or acute, margin entire, often shallowly repand, lower surface with papillae 0.2–0.4 mm long. Calyx papillate.

Distribution: Nepal and W Himalaya.

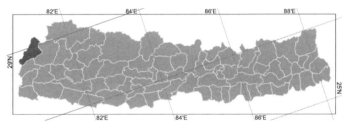

Altitudinal range: ca. 3900 m.

Ecology: On mossy rock.

Flowering: July.

A near endemic to Nepal currently known only from the type collection, *Duthie 5565* (BM) from Nampa Gadh, and *Anon. s.n.* (E), Chahlek, Byans in Kumaon.

Rhodiola himalensis (D.Don) S.H.Fu subsp. ***himalensis***

Leaves narrowly lanceolate to lanceolate, oblanceolate, oblong-oblanceolate or obovate, base rounded, apex acute to apiculate, margin entire or remotely crenulate-denticulate, lower surface smooth. Calyx smooth.

Distribution: Nepal, E Himalaya, Tibetan Plateau and E Asia.

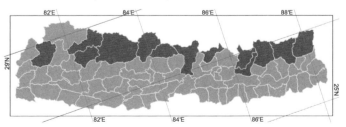

Altitudinal range: 3300–4800 m.

Ecology: On rocky slopes, on rocks, among rock crevices in alpine scree and boulders.

Flowering: May–June.

12. *Rhodiola hookeri* S.H.Fu, Acta Phytotax. Sin., Addit. 1: 124 (1965).
Rhodiola bhutanica (Praeger) S.H.Fu; *Sedum bhutanense* Praeger; *S. bhutanicum* Praeger; *S. elongatum* Wall. ex Hook.f. & Thomson nom. illegit.; *S. elongatum* Wall. nom. nud.; *S. hookeri* (S.H.Fu) N.P.Balakr.; *S. thomsonianum* H.Ohba.

Plants mostly 30–70 cm, dioecious. Rhizomes cylindrical, 1–2 cm thick. Flowering stems to 8 mm thick, pale green, glabrous, smooth. Petioles to 1 cm. Leaves usually rectangular-oblong or -obovate, 3–10 × 1.8–3.5 cm, base attenuate, apex obtuse or rounded, margin irregularly and shallowly crenulate, glabrous, smooth. Inflorescences 40–90-flowered, glabrous, smooth. Flowers ca. 7 mm across. Pedicels 0.5–1 cm. Calyx lobes narrowly oblong or oblong-ovate, 3–5 mm. Petals dark purplish red, sometimes with green shade, narrowly oblong or narrowly oblong-ovate, 3–4 mm. Stamens usually as long as petals; anthers dark purplish red. Nectar scales oblong, blackish purple-red, ca. 1 mm. Carpels to 6 mm, upper part extremely out-curved with short styles less than 0.5 mm. Ovules 10–16 in each locule.

Distribution: Nepal, E Himalaya and Tibetan Plateau.

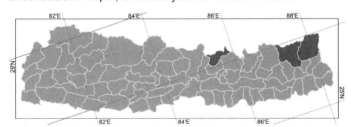

Altitudinal range: 2900–4200 m.

Ecology: On humid, rocky slopes or scree.

Flowering: June–July.

Rhodiola hookeri differs from *R. bupleuroides* (Wall. ex Hook.f. & Thomson) S.H.Fu by its rectangular-oblong or -obovate leaves with sparse and irregular serration, and the greenish, erect, long and thick (to 8 mm across) flowering stems.

13. *Rhodiola humilis* (Hook.f. & Thomson) S.H.Fu, Acta Phytotax. Sin., Addit. 1: 119 (1965).
Sedum humile Hook.f. & Thomson, J. Proc. Linn. Soc., Bot. 2: 99 (1858); *Chamaerhodiola humilis* (Hook.f. & Thomson) Nakai; *Sedum barnesianum* Praeger; *S. levii* Raym.-Hamet.

Plants 1.5–4 cm, bisexual. Main root thick. Caudex simple, erect, short, with persistent leafy scales with petioles 8–18 mm. Blade green, linear-elliptic to rhombic-lanceolate

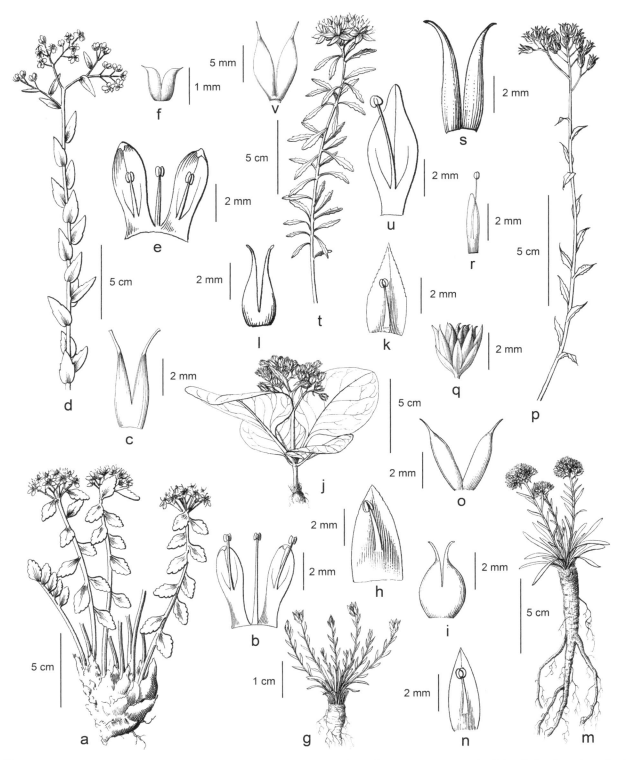

FIG. 24. CRASSULACEAE. **Rhodiola crenulata**: a, flowering plant; b, petals and stamens; c, carpels. **Rhodiola discolor**: d, stem with leaves and inflorescence; e, petals and stamens; f, carpels. **Rhodiola humilis**: g, flowering plant; h, petal and stamen; i, carpels. **Rhodiola prainii**: j, flowering plant; k, petal and stamen; l, carpels. **Rhodiola smithii**: m, flowering plant; n, petal and stamen; o, carpels. **Rhodiola tibetica**: p, stem with leaves and infructescence; q, female flower; r, petal and stamen; s, carpels. **Rhodiola wallichiana**: t, stem with leaves and inflorescence; u, petal and stamen; v, carpels.

or oblanceolate, 3–8 mm, apex rounded or obtuse, base long-attenuate. Flowering stems few, smooth. Leaves linear-elliptic or linear-lanceolate, 4.5–9.5 × 1–2.3 mm, base attenuate, apex rounded, margin entire. Inflorescences solitary or few flowered. Pedicels 1.2–2 mm, smooth. Calyx lobes ovate, ca. 2.6–3.5 mm, apex obtuse to rounded. Petals white at first becoming pink, ovate or oblong-ovate, 5–6.5 mm, apex attenuate. Stamens always shorter than petals; anthers reddish. Nectar scales transversely oblong, 0.5–0.9 mm, apex truncate. Carpels 4–5 mm; tapering. Ovules 8–10 in each locule.
Fig. 24g–i.

Distribution: Nepal, E Himalaya and Tibetan Plateau.

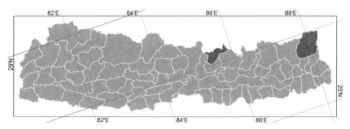

Altitudinal range: 2100–5000 m.

Ecology: Alpine meadows.

Flowering: September. **Fruiting:** September.

14. *Rhodiola imbricata* Edgew., Trans. Linn. Soc. London, Bot. 20: 47 (1846).
Sedum imbricatum (Edgew.) Walp.; *S. imbricatum* Hook.f. & Thomson later homonym, non (Edgew.) Walp.

Plants to 30 cm, dioecious. Caudex subcylindrical to long-ovoid, 2–2.5 cm wide, sparsely branched. Flowering stems 10–20 cm × 4–6 mm, after flowering 20–30 cm × 6–9 mm, glabrous, smooth. Leaves evenly distributed, oblanceolate to narrowly elliptic, 20–30 × 3–7 mm, base rounded or cuneate, apex acute, margin nearly entire to remotely denticulate in upper half, smooth. Inflorescence compact, 1–1.5 × 2.5–3 cm, 20–40-flowered, pedicels 3–5 mm, smooth. Calyx tube 1.2–2 mm; lobes linear to subulate, 3–4 mm. Petals angular-oblanceolate or oblong-obovate, 5–6.5 mm. Stamens distinctly longer than petals, 5.5–8 mm; anthers dark purplish red. Nectar scales reddish, widely oblong, 0.7–1.2 mm, dark red. Carpels 3–5 mm long, tapering, styles indistinct. Ovules 8–10 in each locule. Follicles 8–12 mm long. Seeds 1.5 mm long, ellipsoidal.

Distribution: Nepal, W Himalaya and Tibetan Plateau.

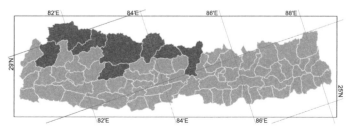

Altitudinal range: 4100–5700 m.

Ecology: On rocky or gravelly slopes in alpine zones.

Flowering: June–August.

Clarke (Fl. Brit. Ind. 2: 417. 1878) partially misapplied the name *Sedum roseum* DC. to this species, and Fröderström (Acta Horti Gothob. 5: app. 37. 1930) similarly partially misapplied the name *S. roseum* Scop.

15. *Rhodiola lobulata* (N.B.Singh & U.C.Bhattach.) H.Ohba, J. Jap. Bot. 61: 205 (1986).
Rhodiola imbricata var. *lobulata* N.B.Singh & U.C.Bhattach., Bull. Bot. Surv. India 25: 246 (1985).

Plants 10–30 cm, dioecious. Caudex robust. Flowering stems suberect ca. 5 mm thick, glabrous. Leaves remote, sessile, narrowly elliptic or narrowly (ob-)lanceolate, 2–4 × 0.8–1.5 cm, base narrowly cuneate, apex subacute, margin remotely crenulate, glabrous. Flowers (only female known) 5-merous. Calyx 5–6 mm including tube 1.5–2 mm; lobes subulate. Petals free, whitish, suberect, linear, 4–5 × 1–1.2 mm, apex obtuse. Stamens absent. Nectar scales widely oblong, 0.7–1.2 mm, apex rounded or truncate. Carpels basally connate with calyx tube, free part erect, 4.5–7 mm, tapering into indistinct style. Ovules 20–22 in each locule.

Distribution: Nepal and W Himalaya.

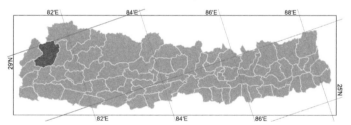

Altitudinal range: ca. 4100 m.

Ecology: On rocks in scree.

Flowering: July.

16. *Rhodiola nepalica* (H.Ohba) H.Ohba, J. Jap. Bot. 51: 386 (1976).
Sedum nepalicum H.Ohba, J. Jap. Bot. 49: 322 (1974).

Plants 10–30 cm, bisexual. Rhizomes subcylindrical, 6–10 mm wide, upper part branched. Flowering stems papillate. Leaves spreading, narrowly elliptic, ovate or lanceolate, 1–4.5 × 0.6–1.5 cm, base long cuneate to attenuate, apex rounded, 5- or 7-lobulate or remotely crenate-serrate, glabrous. Inflorescence 6–15-flowered. Pedicels 1–6.5 mm, papillate. Calyx 5–7 mm, glabrous; lobes linear, linear-subulate to narrowly triangular, 4–5.5 mm. Petals white, narrowly elliptic, narrowly lanceolate or narrowly obovate, 8–15 mm. Stamens as long as petals; anthers dark purplish red. Nectar scales yellow, oblong, 0.9–1.2 mm. Carpels ca. 12–18 mm, tapering. Ovules 14–24 in each locule.

Distribution: Endemic to Nepal.

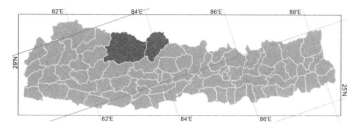

Altitudinal range: 3500–4600 m.

Ecology: On gravel slopes or on rocks.

Flowering: August–September.

17. *Rhodiola prainii* (Raym.-Hamet) H.Ohba, J. Jap. Bot. 51: 386 (1976).
Sedum prainii Raym.-Hamet, Bull. Soc. Bot. France 56: 566 (1909); *S. apiculatum* Craib ex Raym.-Hamet nom. inval.; *S. stewartii* Craib ex Raym.-Hamet nom. inval.

Plants to 8 cm, dioecious. Caudex erect, thick, to 2 cm in diameter. Petioles 1–3 cm. Leaves 4–6, verticillate, oblong-elliptic to ovate or reniform-orbicular, (1.5-)2–6 × 2.5–4 cm, base abruptly narrowed to long attenuate, apex rounded or obtuse, margin entire, glabrous or minutely few mammillate. Inflorescences 1–4 cm in diameter, 13–18-flowered, pedicel 1.5–2.5 cm. Calyx lobes narrowly triangular-ovate to oblong-triangular, 2.5–4 mm, apex acute to acuminate. Petals white, pink or pale red, oblong-ovate, ovate or orbicular, 4–7 × 2–3 mm, apex acute or short acuminate, margin erose or apically minutely dentate. Stamens shorter than petals; anthers deep purplish red. Nectar scales deep purplish red, oblong, 0.6–1 mm. Carpels 4–7 mm, tapering, styles indistinct. Ovules 30–40 in each locule.
Fig. 24j–l.

Distribution: Nepal, E Himalaya and Tibetan Plateau.

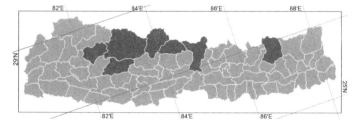

Altitudinal range: 3300–5300 m.

Ecology: On rocks or rock crevices in subalpine or alpine zones.

Flowering: July.

18. *Rhodiola purpureoviridis* (Praeger) S.H.Fu, Acta Phytotax. Sin., Addit. 1: 125 (1965).

Rhodiola purpureoviridis* subsp. *phariensis (H.Ohba) H.Ohba, J. Jap. Bot. 61: 206 (1986).
Sedum phariense H.Ohba, J. Jap. Bot. 48: 328 (1973); *Rhodiola phariensis* (H.Ohba) S.H.Fu.

Plants 10–25 cm, dioecious. Caudex thick, erect, to 2 cm in diameter. Flowering stems erect, pale green, smooth. Leaves sessile, elliptic to narrowly ovate, 1–2.5 × 0.4–0.9 cm, glabrous, base rounded, apex rounded to obtuse, margin remotely serrate. Pedicels 2–3 mm, smooth. Calyx lobes green, ascending or suberect, 1.2–2.4 mm. Petals pale green with deep purplish red stripes, linear-oblanceolate to narrowly obovate, ca. 3.2 × 0.9–1.1 mm, apex rounded. Stamens as long as petals; filaments purplish; anthers greenish red. Nectar scales blackish purple, subrectangular, ca. 1.2 mm long, apex emarginate. Carpels green, later tinged with red, erect, lanceolate, ca. 2 mm. Follicles ca. 6 mm.

Distribution: Nepal and Tibetan Plateau.

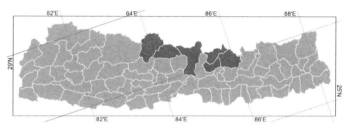

Altitudinal range: 3100–4800 m.

Ecology: On rocks or on scree.

Flowering: August.

Subsp. *phariensis* differs from subsp. *purpureoviridis* by its smaller stature, smaller, glabrous leaves and reddish, bisexual flowers. Subsp. *purpureoviridis* is endemic to China where it is more widely distributed than subsp. *phariensis*.

19. *Rhodiola sinuata* (Royle ex Edgew.) S.H.Fu, Acta Phytotax. Sin., Addit. 1: 127 (1965).
Sedum sinuatum Royle ex Edgew., Trans. Linn. Soc. London, Bot. 20: 47 (1846); *Rhodiola fui* Boriss.; *R. linearifolia* (Royle) S.H.Fu later homonym, non Boriss.; *R. trifida* (Wall. ex C.B.Clarke) H.Jacobsen; *Sedum linearifolium* Royle; *S. linearifolium* var. *pauciflorum* (Edgew.) C.B.Clarke; *S. linearifolium* var. *sinuatum* (Royle ex Edgew.) Raym.-Hamet; *S. mucronatum* Edgew.; *S. pauciflorum* Edgew.; *S. sinuatum* Royle nom. nud.; *S. trifidum* Wall. ex C.B.Clarke; *S. trifidum* nom. illegit.; *S. trifidum* Wall. nom. nud.

Plants to 15 cm, bisexual. Rhizomes short, ascending or erect, often with long roots. Flowering stems erect or ascending, finely papillate. Leaves linear-oblanceolate to narrowly oblanceolate, 1.5–3.5 × 0.2–1 cm, base long-attenuate, apex acute, glabrous, lobulate or pinnately parted or divided in upper third; lobes 2 or 3, linear, 2–10 × 1.5–3 mm.

Crassulaceae

Inflorescences compact, ca. 1.5 × 2.5 cm. Pedicels 3–6 mm, finely papillate. Calyx lobes ovate or triangular-lanceolate, 2–3.3 mm. Petals white inside, greenish white outside, lanceolate to ovate or elliptic, 7–11 mm long, apex rounded to obtuse. Stamens shorter than petals; anthers deep purplish red. Nectar scales deep or pale yellow, subquadrangular. Carpels 5–10 mm; styles 1.5–3.5 mm long. Ovules 12–18 in each locule. Follicles erect, ca. 6 mm long. Seeds brown, ovoid-oblong, slightly winged at both ends.

Distribution: Nepal, W Himalaya and Tibetan Plateau.

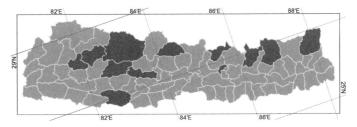

Altitudinal range: 1200–4400 m.

Ecology: On rocks, among rock crevices, on rocky slopes or scree.

Flowering: August.

20. *Rhodiola smithii* (Raym.-Hamet) S.H.Fu, Acta Phytotax. Sin., Addit. 1: 122 (1965).
Sedum smithii Raym.-Hamet, Bot. Jahrb. Syst. 50(Beibl. 112): 8 (1913); *S. chumbicum* Prain ex Raym.-Hamet nom. inval.

Plants 1.7–7 cm, bisexual. Caudex simple, erect, thick. Outer scales brownish, triangular-ovate to triangular-semicircular, 5–8.5 × 3–5 mm, apex long acuminate; inner scales with 3.5–6 mm long linear green apex. Flowering stems ca. 1.7 mm thick, smooth. Leaves lanceolate to linear-elliptic, 5–14 × 1.1–2.2 mm, base shortly attenuate, apex rounded or rarely obtuse, margin entire, smooth. Inflorescences lax, 0.5–2 × 1–3.5 cm, with 5–10 flowers. Pedicels less than 1.5 mm, smooth. Calyx lobes lanceolate or linear-lanceolate, 2.5–4 mm. Petals pinkish, narrowly oblong or narrowly lanceolate, 3.7–6.5 × 1.4–2 mm, apex rounded or obtuse. Stamens distinctly shorter than petals; anthers red. Nectar scales reddish, quadrangular, 0.5–0.9 mm, apex emarginate. Carpels 3.6–7.4 mm, with styles 1.4–2 mm. Ovules 4–8 in each locule. Follicles erect. Seeds oblong-oblanceolate, ca. 1.3 mm.
Fig. 24m–o.

Distribution: Nepal, E Himalaya and Tibetan Plateau.

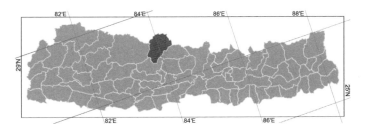

Altitudinal range: 3500–4600 m.

Ecology: On sandy grasslands, gravelly places in glacial valleys or among rock crevices.

Flowering: July–September.

21. *Rhodiola tibetica* (Hook.f. & Thomson) S.H.Fu, Acta Phytotax. Sin., Addit. 1: 121 (1965).
Sedum tibeticum Hook.f. & Thomson, J. Proc. Linn. Soc., Bot. 2: 96 (1858); *S. quadrifidum* var. *tibeticum* (Hook.f. & Thomson) Fröd.; *S. stracheyi* Hook.f. & Thomson; *S. tibeticum* var. *stracheyi* (Hook.f. & Thomson) C.B.Clarke.

Plants 7–25 cm, dioecious. Rhizomes obconical, erect, 1.5–2.5 cm thick, with persistent old flowering stems. Flowering stems 1.5–2 mm thick, glabrous, smooth. Leaves narrowly oblong to oblong-ovate, 5–12 × 2–4 mm, base truncate to rounded, apex acute to obtuse, margin entire, smooth. Inflorescences 2–3 × 2–3 cm, with 20–40 flowers, axes smooth. Pedicels to 4 mm, smooth. Male flowers with calyx lobes linear-subulate to linear-lanceolate, 1.4–1.7 mm, apex rounded or rarely truncate, petals purplish red, oblong, 2.6–3 mm, apex rounded to obtuse. Female flowers with calyx lobes linear-subulate to linear-lanceolate, 1.7–2.1 mm, apex rounded or rarely truncate, petals purplish red, narrowly oblong to oblong 2.3–3.2 mm, apex rounded to obtuse. Stamens slightly longer than petals; anthers reddish. Nectar scales yellow, oblong, 0.8–1 mm, apex praemorse or emarginate to truncate. Carpels 6–9 mm, with conspicuously out-curved apex, styles indistinct. Ovules ca. 16 in each locule. Follicles 8–10 mm. Seeds linear to oblong, 1.7–2.7 mm.
Fig. 24p–s.

Distribution: Nepal, W Himalaya and Tibetan Plateau.

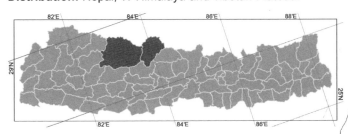

Altitudinal range: 4000–4500 m.

Ecology: Stony slopes.

Flowering: July–August.

22. *Rhodiola wallichiana* (Hook.) S.H.Fu, Acta Phytotax. Sin., Addit.: 125 (1965).
Sedum wallichianum Hook., Icon. Pl. 7: pl. 604 (1844); *Chamaerhodiola crassipes* (Wall. ex Hook.f. & Thomson) Nakai; *Rhodiola crassipes* (Wall. ex Hook.f. & Thomson) Boriss.; *Sedum crassipes* Wall. nom. nud.; *S. crassipes* Wall. ex Hook.f. & Thomson; *S. crassipes* var. *cholaense* Praeger.

अम्बासिंह Ambasingh (Nepali).

Plants 15–40 cm, bisexual. Caudex prostrate, slender, ca. 1 cm thick. Flowering stems 3–6 mm thick, greenish, glabrous, usually smooth. Leaves linear or narrowly lanceolate to linear-oblanceolate, 1.2–3 × 0.2–0.6 cm, base long attenuate, apex obtuse, margin minutely serrate in apical third, glabrous. Inflorescences 2–3.5 cm across, usually with 6–20 flowers. Pedicels 1–4 mm, mostly densely papillate. Calyx lobes subulate, 5.5–8 mm, apex rounded. Petals greenish, pale yellowish green, or rarely pinkish, narrowly elliptic or linear, 7–11 × 1.8–2.5 mm, apex rounded. Stamens nearly as long as petals; anthers deep purplish red. Nectar scales pale or orange-yellow, oblong, 1–1.2 mm. Carpels 0.9–1.4 cm, tapering, with 2.5–4 mm style. Ovules 28–36 in each locule. Follicles 1–1.5 cm. Seeds ellipsoidal, both ends winged. Fig. 24t–v.

Distribution: Nepal, W Himalaya, E Himalaya and Tibetan Plateau.

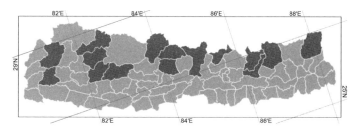

Altitudinal range: 3000–5500 m.

Ecology: On rocky slopes and river beds.

Flowering: August–September. **Fruiting:** October.

Clarke (Fl. Brit. Ind. 2: 419. 1878) and Smith (Rec. Bot. Surv. Ind. 4: 259. 1911) misapplied the name *Sedum asiaticum* (D.Don) DC. to this species.
Juice of the plant is applied to burns and wounds.

5. *Rosularia* (DC.) Stapf, Bot. Mag. 149: pl. 8985 (1923).

Dwarf perennial herbs, usually hairy. Rootstock usually fleshy. Leaves mostly in dense, basal rosettes, usually with several rosettes per plant; leaves on flowering stems alternate, sessile, flat. Flowering stems often several, arising from axils of rosette leaves or solitary and arising from centre of rosette; cauline leaves alternate. Inflorescence cymose-corymbiform, or spicate-paniculate, lax to dense. Flowers bisexual, 5–9-merous. Sepals connate at base. Petals connate at least at base, pink or white, sometimes with red or purple markings, urceolate, tubular to funnel-shaped, campanulate or stellate; lobes partly connate at base, limb erect to spreading, membranous. Stamens twice as many as petals. Nectar scales cuneate to cuneate-spathulate-quadrate. Carpels erect, free, often hairy. Follicles erect, free, many-seeded. Seeds striate.

Worldwide 36 species in the Mediterranean, C Asia and SW Asia. Three species in Nepal.

Key to Species

1a Rosulate leaves sword-shaped with acuminate to acute apex ....**2. *R. marnieri***
b Rosulate leaves spathulate to obovate or oblong-spathulate with obtuse or rounded apex 2

2a Plants without stolons. Calyx tube cup-shaped. Petals 6–7.3 mm long ....**1. *R. adenotricha***
b Plants stoloniferous. Calyx tube saucer-shaped. Petals 5–5.5 mm long...**3. *R. rosulata***

1. *Rosularia adenotricha* (Wall. ex Edgew.) C.-A.Jansson, Fl. Iranica 72: 29 (1970).
Sedum adenotrichum Wall. ex Edgew., Trans. Linn. Soc. London 20: 48 (1846); *Cotyledon papillosa* Aitch. & Hemsl.; *C. tenuicaulis* Aitch. & Hemsl.; *Sedum adenotrichum* Wall. nom. nud.; *S. anoicum* Praeger; *S. cuneatum* Wall. ex Raym.-Hamet nom. inval.; *S. talichiense* Werderm.; *Umbilicus papillosus* (Aitch. & Hemsl.) Boiss.; *U. tenuicaulis* (Aitch. & Hemsl.) Boiss.

Plants 8–15 cm. Rosettes 1.5–2 cm across, with 10–20 leaves. Rosulate leaves sessile, spathulate or obovate, 8–14

× 4–5 mm, apex rounded to obtuse, margin entire, glabrous, smooth, costa not prominent. Flowering stem glandular-pubescent. Cauline leaves sessile, spathulate to narrowly oblong, 4–6 × 2–3 mm, base truncate, apex rounded to obtuse, margin entire, both surfaces densely glandular-pubescent. Inflorescence lax, 10–20-flowered, 4–5 cm long and wide; axis densely glandular-pubescent. Pedicels 4–7 mm, densely glandular-pubescent. Calyx glandular-pubescent outside; tube cup-shaped, ca. 0.5 mm; lobes ovate to narrowly ovate or narrowly oblong, 1.7–4.2 mm, apex acute to rounded. Petals usually white, connate ca. 2.5 mm from base, lanceolate-oblong to oblong, 6–7.3 × 1.4–2.2 mm, apex acuminate to acute, erect at flowering. Stamens shorter than petals; anthers yellow. Nectar scales yellowish green, oblong, ca. 0.8 mm. Carpels 3.7–4.2 mm, upper ventral side glandular-pubescent, with slender style ca. 1 mm. Ovules 30–40 in each locule.

Distribution: Nepal, W Himalaya and SW Asia.

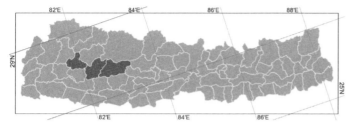

Altitudinal range: 1300–2300 m.

Ecology: On rock crevices.

Flowering: July–August.

2. *Rosularia marnieri* (Raym.-Hamet ex H.Ohba) H.Ohba, J. Jap. Bot. 52: 7 (1977).
Sedum marnieri Raym.-Hamet ex H.Ohba, J. Jap. Bot. 49: 260 (1974); *Rosularia alpestris* subsp. *marnieri* (Raym.-Hamet ex H.Ohba) Eggli.

Plants 7–19 cm. Conspicuously papillate throughout except for stamens and nectar scales. Roots thick, to 13 cm long. Rosettes 2–4 cm across, with 20–40 leaves. Rosulate leaves sessile, gladiate, 15–22 x 2.2–4.5 mm, apex acuminate to acute, margin entire, glabrous, smooth, costa not prominent. Cauline leaves sessile, narrowly lanceolate to narrowly elliptic, 7–14 × 1.5–3 mm, base truncate, apex acuminate, margin entire, both surfaces densely papillate. Inflorescence compact, 10–20-flowered, 1.5–5 cm long and wide; axis densely papillate. Pedicels 2–3 mm long, densely papillate. Calyx papillate; tube ca. 0.5 mm; lobes ovate or triangular-ovate, 3.5–4.2 mm, apex acute. Petals pinkish, connate ca. 1.5 mm from base, ovate, 7–7.7 × 3.5–3.8 mm, apex obtuse or rarely acute, margin minutely erose, ascending at flowering. Stamens shorter than petals; anthers deep red. Nectar scales whitish, square, 0.5–0.8 mm long. Carpels 6–7 mm, sparsely papillate, with slender style ca. 1 mm. Ovules 15–20 in each locule. Fig. 23u–w.

Distribution: Endemic to Nepal.

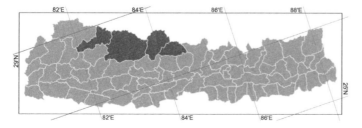

Altitudinal range: 3000–4400 m.

Ecology: On exposed or arid, stony or sandy slopes, or among rock crevices.

Flowering: July–September.

3. *Rosularia rosulata* (Edgew.) H.Ohba, J. Jap. Bot. 52: 9 (1977).
Sedum rosulatum Edgew., Trans. Linn. Soc. London 20: 48 (1846); *S. pyriforme* Royle ex Hook.f. & Thomson nom. inval.; *Umbilicus radicans* Klotzsch.

Plants 8–15 cm with long stolons arising from axils of former year's rosulate leaves. Rosettes 1.5–1.8 cm across, with 6–10 leaves. Rosulate leaves sessile, spathulate with petiole-like base, 6–21 × 2–8 mm, apex rounded, margin entire, glabrous, smooth, costa not prominent. Flowering stem glandular-pubescent. Cauline leaves sessile, oblanceolate or linear-oblanceolate to linear, 7–18 × 3–7 mm, base shortly attenuate, apex rounded, margin entire, both surfaces densely glandular-pubescent. Inflorescence lax, 3–5-flowered; axis densely glandular-pubescent. Pedicels 3–15 mm, densely glandular-pubescent. Calyx glandular-pubescent outside; tube saucer-shaped, ca. 0.3 mm long; lobes ovate to lanceolate, 1.2–2.2 mm, apex obtuse. Petals white, connate ca. 0.3 mm from base, lanceolate or oblong-lanceolate, 5–5.5 × 1.5–1.7 mm, apex acute, erect or ascending at flowering. Stamens shorter than petals; anthers yellow. Nectar scales whitish, narrowly oblong, ca. 0.8 mm. Carpels 3.6–4.5 mm, ventral side sparsely glandular-pubescent, with slender style ca. 1 mm. Ovules 12–16 in each locule.

Distribution: Nepal, W Himalaya and SW Asia.

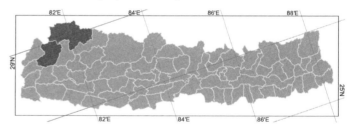

Altitudinal range: 2000–3500 m.

Ecology: Among rock crevices.

Flowering: July–August.

6. *Sinocrassula* A.Berger, Nat. Pflanzenfam., ed. 2, 18a: 462 (1930).

Biennial herbs, glabrous, ± xerophytic, papillose or rarely pubescent. Sterile stems usually present. Leaves mostly in basal rosettes, often with several rosettes per plant, often caducous and lost by anthesis, alternate, apex obtuse or acuminate, sometimes with spine. Inflorescence terminal or lateral, paniculate-corymbiform with long, basally subopposite branches, rarely simple and raceme-like; bracts leaf-like, laxly arranged. Flowers erect, pedicellate, bisexual, 5-merous, campanulate-urceolate. Calyx subglobose, sepals erect, triangular or triangular-lanceolate, base connate. Petals subglobose-urceolate, free or almost so, yellow to red or purplish red. Stamens 5, alternating with and slightly shorter than petals. Nectar scales entire, apex emarginate or dentate. Carpels somewhat wide, base abruptly narrowed. Styles short, stigmas capitate. Follicles many-seeded.

Worldwide four species in the Himalaya and China. One species in Nepal.

1. *Sinocrassula indica* (Decne.) A.Berger, Nat. Pflanzenfam., ed. 2, 18a: 463 (1930).
Crassula indica Decne., in Jacquem., Voy. Inde 4(1): 61, pl. 73 f.1 (1841); *Sedum cavaleriei* H.Lév.; *S. indicum* (Decne.) Raym.-Hamet; *S. indicum* var. *ohtusifolium* Fröd.; *S. martini* H.Lév.; *S. paniculatum* Wall. nom. nud.; *S. paniculatum* var. *indicum* (Decne.) Fröd. nom. inval.; *S. scallanii* Diels.

Plants 5–20 cm. Basal leaves not spurred, linear, 1–4 × 0.5–1.5 cm, apex acuminate, margin entire, elliptic in cross section, fleshy, often glaucous. Flowering stems erect, arising from the centre of rosette, ± glaucous. Cauline leaves sessile, not spurred, linear, apex long apiculate, margin entire, elliptic in cross section. Inflorescence corymbose, with great number of flowers. Pedicels smooth. Sepals ± fleshy, not spurred, basally connate; lobes subulate to linear-triangular, 1.5–2 mm, apex acute or shortly acuminate, erect at flowering. Petals orange red or greenish-yellow, triangular or triangular-lanceolate, 3–5 mm, erect at flowering. Stamens shorter than petals; anthers blackish red-purple. Nectar scales square to oblong. Follicles erect.
Fig. 23r–t.

Distribution: Nepal, W Himalaya, E Himalaya, Tibetan Plateau and E Asia.

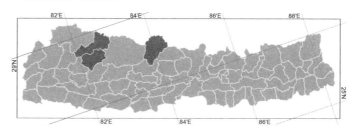

Altitudinal range: 2300–4100 m.

Ecology: On rocks, among rocks or rock crevices on boulders or river beds.

Flowering: July–August.

7. *Phedimus* Raf., Amer. Monthly Mag. & Crit. Rev. 1: 438 (1817).

Perennial or rarely annual herbs, usually glabrous, sometimes stems woody at base. Leaves opposite, petiolate, flat, margin serrate or crenate. Inflorescences terminal, cymose with 3 main branches, bractless, many-flowered. Flowers sessile or nearly so, bisexual, mostly 5-merous. Sepals basally connate, fleshy, without spur. Petals free, usually spreading. Stamens twice as many as petals. Nectar scales entire or apex emarginate. Ovaries and follicles with adaxial outgrowth. Styles short, oblique or spreading at flowering. Follicles many-seeded. Seeds striate.

Worldwide about 12 species in Asia and Europe. One species in Nepal.

1. *Phedimus odontophyllus* (Fröd.) 't Hart, in 't Hart & Eggli, Evol. Syst. Crassulaceae: 169 (1995).
Sedum odontophyllum Fröd., Acta Horti Gothob. 7: append. 117 (1932); *Aizopsis odontophyllum* (Fröd.) Grulich.

Perennial herb, 13–20 cm. Stems with procumbent, sparsely branched, terete, 3–4 mm thick, with 1 or 2 sterile branches, glabrous, smooth. Petioles 2–10 mm. Leaves broadly ovate or rarely broadly lanceolate, 3–6 × 0.7–2.5 cm, base attenuate, apex rounded to obtuse, margin loosely crenulate, smooth, thick herbaceous. Inflorescence a cyme with 3 or rarely 4 primary axes, 20–30-flowered. Flowers almost sessile, 6–8 mm across. Sepals green, ± fleshy, connate ca. 1 mm from base, broadly ovate to ovate, 1.5–1.7 mm, apex cuspidate with rounded tip, smooth. Petals long-lanceolate to linear-ovate, 4.3–5 × 1.3–1.6 mm, apex long acuminate, deep yellow, often with red shade, ascending at flowering. Stamens shorter than petals; anthers deep yellow. Nectar scales transversely oblong, 0.7 mm. Carpels ca. 3 mm, with style 0.8–1 mm. Ovules 25–30 in each locule. Seeds ellipsoidal, with longitudinally striate testa.
Fig. 23o–q.

Distribution: Nepal and E Asia.

Crassulaceae

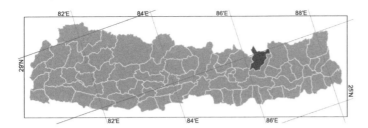

Altitudinal range: ca. 3900 m.

Ecology: Rocky places.

Flowering: April–June.

Phedimus odontophyllus predominantly occurs in SW Sichuan, but is also known from one collection in Nepal, *Zimmermann 1375* (G), from Beding, Dolkha District. It is a good example of the Sichuan/Yunnan to Central Nepal disjunction in distribution that is occasionally seen for Nepalese species.

8. *Sedum* L., Sp. Pl. 1: 430 (1753).

Glabrous or pubescent, perennial or annual herbs to subshrubs with usually much-branched non-flowering shoots, rarely rosulate. Roots fibrous or with tubers or taproots. Stems erect or decumbent, sometimes fasciculate or moss-like, fleshy, glabrous or hairy, base rarely woody. Leaves alternate, opposite or verticillate, base often spurred, margin normally entire. Inflorescence terminal or axillary, cymose, often corymbiform, 1- to many-flowered. Flowers usually bisexual, rarely unisexual, (3–)5(–9)-merous; bracts usually present, often leaf-like. Sepals free or basally connate. Petals free or almost so. Stamens usually twice as many as petals, rarely equal in number to petals; oppositipetalous stamens adnate to petals. Nectar scales entire or apex emarginate. Carpels usually as many as petals, occasionally fewer, free or basally widened and connate. Styles short or long. Follicles many- or few-seeded. Seeds ovoid to ellipsoid, smooth or papillate, less often striate.

Worldwide about 470 species, mainly in the N hemisphere, but extending to the S hemisphere in Africa and S America. 15 species in Nepal.

Key to Species

1a   Flowers white ....**2. *S. filipes***
  b   Flowers yellow...2

2a   Ovaries kyphocarpic, i.e. swollen in ventral (adaxial) side with lip along suture 3
  b   Ovaries orthocarpic, i.e. not swollen in ventral side and without lip along suture 5

3a   Ovaries usually 3. Leaves spathulate with retuse apex, flat.....................................**14. *S. triactina***
  b   Ovaries usually 5, sometimes 4. Leaves linear-lanceolate to narrowly oblanceolate, ± fleshy with elliptic cross section 4

4a   Calyx basally connate, lobes with acuminate apex. Petals free ....**7. *S. multicaule***
  b   Calyx free, lobes with rounded-obtuse apex. Petals basally connate........................**12. *S. pseudomulticaule***

5a   Basal part of flowering stems long creeping. Leaves spathulate, usually ternate ....**1. *S. chauveaudii***
  b   Flowering stems hardly or shortly creeping. Leaves not spathulate, opposite or alternate................6

6a   Flowering stems with numerous, densely tufted sterile branches at base 7
  b   Flowering stems without or with very few sterile branches at flowering time 8

7a   Petals ovate-lanceolate or narrowly rhombic with attenuate base.....................................**4. *S. gagei***
  b   Petals widely trullate (angular-ovate) with narrow claw ....**15. *S. trullipetalum***

8a   Petals apparently connate at base...9
  b   Petals free or nearly so 10

9a   Petals lanceolate or elliptic-oblong with entire margins. Stamens more than two thirds as long as petals
      ....**8. *S. obtusipetalum***
  b   Petals widely oblanceolate to obovate, usually with erose margins. Stamens less than half as long as petals ....**9. *S. oreades***

10a Petals narrowly obtrullate (angular-ovate) .... **6. _S. magae_**
  b Petals lanceolate to ovate or linear to oblong 11

11a Calyx connate at base, without spur. Placenta basal .... **11. _S. przewalskii_**
  b Calyx free or nearly so, with spur. Placenta marginal 12

12a Flowering stems 1.5–3 cm 13
  b Flowering stems (3–)5–25 cm 14

13a Ovaries two thirds connate .... **5. _S. henrici-roberti_**
  b Ovaries free almost from base .... **10. _S. perpusillum_**

14a Leaves linear, (4–)8–17 mm, with rounded apex. Petals narrowly oblong to narrowly oblong-ovate, 4.5–6 mm
       .... **3. _S. forrestii_**
  b Leaves widely lanceolate to ovate, 4–8 mm, with obtuse or abruptly apiculate apex. Petals lanceolate,
       3–4.5 mm .... **13. _S. roborowskii_**

1. _Sedum chauveaudii_ Raym.-Hamet, Notul. Syst. [Paris] 1: 137 (1910).
Sedum triphyllum Praeger.

Perennial herbs, 6–16 cm. Flowering stems glabrous, papillate, erect with long creeping base. Leaves ternate but partly alternate, narrowly spathulate or narrowly obovate, 6–20 × 3–6 mm, base long attenuate or petiole-like, apex rounded, margin entire, flat, glabrous, papillate. Inflorescence corymbiform cyme, lax, many-flowered, bracts leaf-like. Flowers shortly pedicellate, 5-merous. Sepals free, shortly spurred, oblanceolate or linear-spathulate, unequal in size, 6.5–8 mm, apex rounded, papillate. Petals yellow, connate ca. 1.5 mm from base, lanceolate to linear-lanceolate, 8–12 mm, apex rounded with small mucro, suberect at flowering. Stamens 10. Nectar scales square, 0.5–0.6 mm. Carpels basally connate for ca. 1.5 mm, with short style. Follicles divergent, ca. 7.5 mm, with small lip along suture. Seeds ovoid.
Fig. 25a–d.

Distribution: Nepal and E Asia.

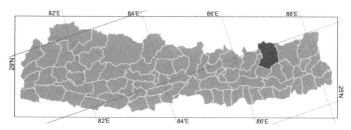

Altitudinal range: ca. 2900 m.

Ecology: On rocks in forests.

Flowering: August–November. **Fruiting:** October–December.

Disjunctly distributed between SW China (Yunnan and Sichuan) and Solu Khumbu District. Only known from one collection in Nepal (_Zimmermann 1679, G_).

2. _Sedum filipes_ Hemsl., in F.B.Forbes & Hemsl., J. Linn. Soc., Bot. 23: 284 (1887).
Sedum filipes var. _major_ Hemsl.; _S. filipes_ var. _pseudostapfii_ (Praeger) Fröd.; _S. pseudo-stapfii_ Praeger; _S. trientaloides_ Praeger.

Annual or biennial, tender herb to 16 cm. Flowering stems glabrous, papillate, erect or ascending with short creeping base. Leaves opposite, in 1–4 pairs, petiolate or petiole-like base, spathulate or (ob)ovate to broadly (ob)ovate, 1–4 × 0.8–2.5 cm, apex rounded, margin entire, flat, glabrous, papillate. Inflorescence large, terminal, but sometimes from axils of uppermost leaves. Pedicels 2 or 3 times as long as flowers, papillate. Sepals free, shortly spurred, triangular-ovate or -lanceolate, 1.5–2.5 mm, apex obtuse to rounded, papillate. Petals white, basally connate, oblong-lanceolate to linear-ovate, 3–5 mm, apex obtusely acute, suberect at flowering. Stamens 10. Nectar scales spathulate, obovate. Carpels with short style. Follicles suberect, without lips along suture. Seeds ellipsoid.

Distribution: Nepal, E Himalaya, Assam-Burma and E Asia.

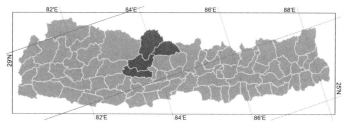

Altitudinal range: 2200–2700 m.

Ecology: On rocks in forests.

Flowering: August–October. **Fruiting:** October.

Fröderström (Acta Horti Gothob. 6: app. 36. 1931) partially misapplied the name _Sedum elatinoides_ DC. to this species.

3. Sedum forrestii Raym.-Hamet, Notes Roy. Bot. Gard. Edinburgh 5: 118, pl. 86 (1912).
Sedum holei Raym.-Hamet.

Annual herbs, 3–10 cm. Flowering stems glabrous, smooth, erect from base, usually branched from base; branches suberect. Leaves alternate, sessile, spur 0.5–1 mm with rounded or truncate apex; blade linear, (4–)8–17 × 1–3 mm, apex rounded, margin entire, flat, glabrous, smooth. Inflorescence a cyme with 3 primary axes, 0.5–2 × 1–4 cm, with 10–70 flowers; primary axis dichasially branched, smooth. Flowers 6–9 mm across. Pedicels 1–3 mm, smooth. Sepals free, spurred, narrowly oblong or narrowly lanceolate to lanceolate, 3–4.5(–5.5) mm, apex obtuse or rounded, smooth, suberect. Petals yellow, free or nearly so, narrowly oblong to narrowly oblong-ovate, 4.5–6 mm, apex obtuse, suberect or ascending. Stamens shorter than or as long as the petals; anthers deep yellow. Nectar scales translucent, linear, 0.5–1.4 mm, apex rounded-truncate. Carpels erect, 3–6 mm, almost free or connate to 1 mm from base; style 0.6–1.2 mm long. Ovules 4–8 in each locule.
Fig. 25e–g.

Distribution: Nepal, Tibetan Plateau and E Asia.

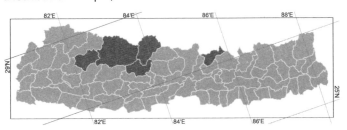

Altitudinal range: 2800–4600 m.

Ecology: On mossy rock or rock crevices.

Flowering: July–August. **Fruiting:** August–September.

Sedum forrestii is distinguished from the similar *S. henrici-robertii* Raym.-Hamet by the combination of the characters of leaves, flowers, petals and carpels (see note under *S. henrici-robertii*).

4. Sedum gagei Raym.-Hamet, Repert. Spec. Nov. Regni Veg. 8: 263 (1910).

Perennial herbs, 1.5–10 cm, with numerous sterile branches. Flowering stems glabrous, smooth, erect from creeping, branched base. Leaves alternate, imbricate in sterile branches, sessile, spur 0.5–1.1 mm, with shallowly 3-lobate apex; blade subulate or linear-triangular to narrowly triangular-lanceolate, 4–6 × 1–1.5 (–1.8) mm, apex acuminate, margin entire and papillate, flat, thick herbaceous, smooth. Inflorescence a compact cyme with 3 primary axes, 1–2 × 2–3 cm, primary axis ascending, 1–2 cm, smooth. Flowers 6–8 mm across. Pedicel 0.5–3.5 mm, smooth. Sepals free, spurred, lanceolate or triangular-lanceolate, 5.5–7 mm long, apex acuminate, smooth or rarely sparsely papillate. Petals yellow, connate 0.7–1 mm from base, ovate-lanceolate or narrowly rhombic,

(6–)8–9.5 × 1.3–1.7 mm, apex acuminate to acute, suberect at flowering. Stamens shorter than petals; anthers yellow. Nectar scales square, ca. 0.4 mm, apex retuse. Carpels 7.5–9 mm, connate 2.5–3.5 mm from base; style 2.5–3 mm. Ovules 40–50 in each locule.

Distribution: Nepal, W Himalaya, E Himalaya, Tibetan Plateau and E Asia.

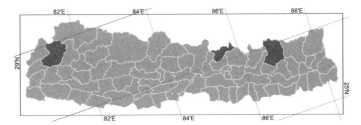

Altitudinal range: 3900–4600 m.

Ecology: On mossy rocks in alpine zone.

Flowering: August–September. **Fruiting:** August–September.

Sedum gagei resembles *S. trullipetalum* Hook.f. & Thomson, but differs in having free, spurred sepals, and basally connate petals with acuminate or acute apex, and higher connate ovaries.

5. Sedum henrici-roberti Raym.-Hamet, Repert. Spec. Nov. Regni Veg. 12: 407 (1913).

Annual herbs, 1.5–3(–5) cm. Stems several times branched near base, erect, glabrous, smooth. Leaves alternate, sessile, spur ca. 0.5 mm with rounded or truncate apex; blade linear to narrowly oblong, 3.8–5 × 0.9–1.1 mm, apex obtuse to acute, margin entire, flat, glabrous, smooth. Inflorescences a cyme, 0.5–2 × 0.5–1.5 cm, with 10–20 flowers, conspicuously bracteate, primary axes short, ascending, smooth. Flowers often haplostemonous, 2–3 mm across at flowering. Pedicels 1–1.5 mm, smooth. Sepals free, spurred, narrowly oblong or oblong-lanceolate, 3–4 mm, apex obtuse or acute, smooth. Petals yellow, free or nearly so, narrowly triangular or triangular-lanceolate, 2.5–3.2 mm, apex rounded or rarely obtuse, suberect at flowering. Stamens 10 or 5, shorter than petals; anthers deep yellow. Nectar scales translucent, linear, ca. 1.2 mm, apex rounded. Carpels 2.2–2.8 mm, connate 1.5–1.8 mm from base, erect; style ca. 0.3 mm. Ovules 4–8 in each locule, placenta marginal. Follicles ± divergent, 2.5–3 mm. Seeds narrowly ovoid or ellipsoid, ca. 0.7 mm.

Distribution: Nepal and Tibetan Plateau.

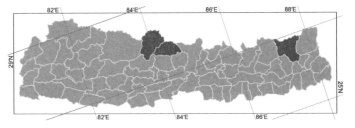

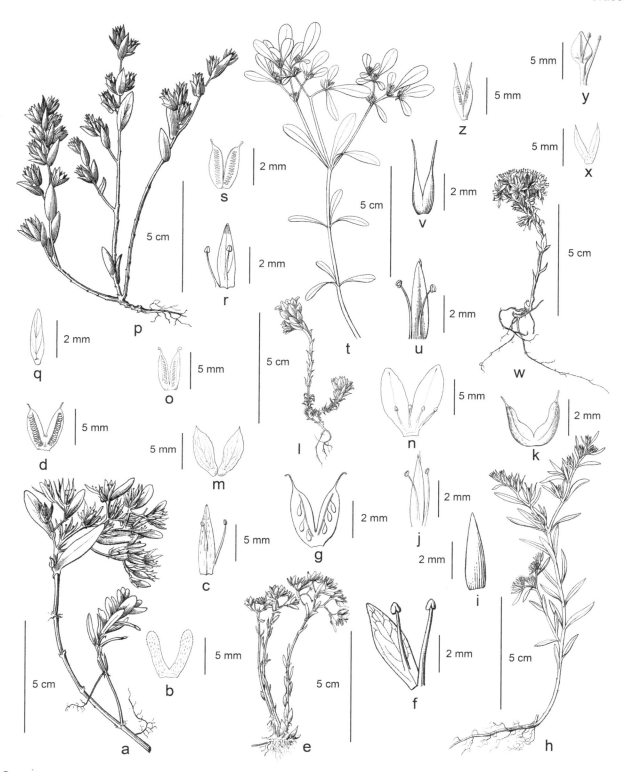

FIG. 25. CRASSULACEAE. **Sedum chauveaudii**: a, flowering plant; b, sepals; c, petal and stamens; d, longitudinal section of carpels. **Sedum forrestii**: e, flowering plant; f, petal and stamens; g, longitudinal section of carpels. **Sedum multicaule**: h, flowering plant; i, sepal; j, petal and stamens; k, carpels. **Sedum oreades**: l, flowering plant; m, sepals; n, petals and stamens; o, longitudinal section of carpels. **Sedum roborowskii**: p, flowering plant; q, sepal; r, petal and stamens; s, longitudinal section of carpels. **Sedum triactina**: t, stem with leaves and inflorescence; u, petal and stamens; v, carpels. **Sedum trullipetalum**: w, flowering plant; x, sepals; y, petal and stamens; z, longitudinal section of carpels.

Crassulaceae

Altitudinal range: 4200–4700 m.

Ecology: On rocks or sandy soils in scree or sandy stream side banks.

Flowering: July. **Fruiting:** August.

Sedum henrici-robertii resembles *S. magae* Raym.-Hamet or *S. perpusillum* Hook.f. & Thomson. *Sedum henrici-robertii* is distinguished from *S. magae* by its narrowly triangular or triangular-lanceolate petals and the highly connate ovaries with 4–8 ovules. *Sedum perpusillum* differs from *S. henrici-robertii* in having oblong-ovate or triangular-oblong petals and less connate ovaries. *Sedum henrici-robertii* is distinguished from *S. forrestii* Raym.-Hamet by the small leaves 3.8–5 × 0.9–1.1 mm with obtuse or acute apex, the flowers 2–3 mm across with lanceolate petals 2.5–3.2 mm long and carpels 2.2–2.8 mm long and connate 1.5–1.8 mm from base.

6. *Sedum magae* Raym.-Hamet, Repert. Spec. Nov. Regni Veg. 13: 350 (1914).
Sedum tillaeoides Duthie ex C.E.C.Fisch.; *S. tillaeoides* Duthie nom. nud.

Annual herb, 1.5–3 cm. Stems several-times branched near base, glabrous, smooth, branches erect to ascending. Leaves alternate, sessile, spur 0.3–0.6 mm with rounded or truncate apex; blade linear to linear-oblanceolate, 4–5.2 × 1–1.2 mm, apex acute to obtuse, margin entire, rather flat, glabrous, smooth. Inflorescence a compact cyme, 0.5–2 × 0.5–1.5 cm, flowers usually 10–20, primary axis short, ascending, smooth. Flowers haplostemonous, 2–3 mm across. Pedicels to 2.5 mm, smooth. Sepals free, spurred, narrowly obovate to oblong, 3–4 mm, apex obtuse or rounded, smooth. Petals yellow, free, narrowly obtrullate to obtrullate, 2–2.6 mm, apex rounded or obtuse, suberect at flowering. Stamens 5, nearly as long as petals; anthers deep yellow. Nectar scales translucent, linear, 1.2–1.7 mm, apex truncate or emarginate. Carpels erect, 2.3–2.5 mm, connate 0.3–0.8 mm from base; style ca. 0.3 mm. Ovules 12–18 in each locule. Follicles 2.5–2.7 mm, suberect. Seeds narrowly ovoid, ca. 0.6 mm.

Distribution: Nepal and W Himalaya.

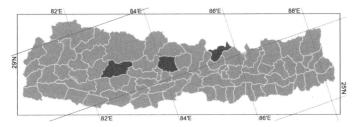

Altitudinal range: 3900–4600 m.

Ecology: On rocks or sandy soils on scree.

Flowering: July.

Sedum magae is distinguished from *S. perpusillum* Hook.f. & Thomson by the obtrullate petals and the ovaries with 12–18 ovulate locules. *Sedum magae* differs from *S. henrici-robertii*

Raym.-Hamet in having obtrullate petals and less connate ovaries with 12–18 ovules.

7. *Sedum multicaule* Wall. ex Lindl., Edward's Bot. Reg. 26: 58 (1840).
Sedum multicaule Wall. nom. nud.

Perennial herb to 15 cm, with numerous sterile branches with creeping base. Flowering stems ascending, much-branched, dull yellow, often tinged purplish red . Leaves linear-lanceolate to narrowly lanceolate or linear, 10–15 x 2–4 mm, base rounded-attenuate, apex acuminate, flat, dull red to purplish green, smooth. Inflorescence with 3 branches, with 10–40 flowers. Flowers sessile. Sepals basally connate, unequal in size, linear-lanceolate, 4.7–7.5 mm, apex acuminate or mucronate, often with short mucro, smooth, suberect to ascending at flowering. Petals yellow, shorter than sepals, free, narrowly oblong or linear-lanceolate, 4.5–6 mm, apex acuminate, often with mucro, ascending at flowering. Stamens shorter than petals; anthers deep yellow. Nectar scales pale green, rectangular, ca. 1 mm. Follicles stellately patent with distinct lips along suture. Seeds oblong, ca. 1 mm. Fig. 25h–k.

Distribution: Nepal, W Himalaya, E Himalaya, Tibetan Plateau, Assam-Burma and E Asia.

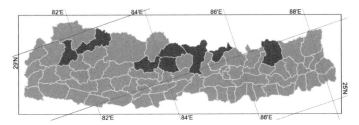

Altitudinal range: 1300–3200 m.

Ecology: On and among rocks and rock crevices.

Flowering: July–August. **Fruiting:** August–September.

8. *Sedum obtusipetalum* Franch., J. Bot. [Morot] 10: 289 (1896).

Sedum obtusipetalum subsp. *dandyanum* Raym.-Hamet ex H.Ohba, J. Jap. Bot. 61: 274 (1986).

Sedum costantinii Raym.-Hamet.

Annual herbs, 4–15 cm. Stems branched several times throughout, glabrous, smooth. Leaves ± remote, sessile, spur 0.5–1 mm, with truncate apex; blade lanceolate to linear, 4–11 × 1.2–2.5 mm, apex obtuse to acute, margin entire, flat, smooth. Inflorescence a cyme, 1.5–2.5 × 2–4 cm, with 10–40 flowers. Pedicels 1–3 mm, smooth. Sepals free, spurred basally, linear to lanceolate, 3.5–6 mm, apex rounded to obtuse, smooth. Petals yellow, basally slightly connate, lanceolate or oblong-elliptic, 6–8.5 × 2.5 mm, margin entire, apex obtuse to acute, suberect at flowering. Stamens more than two thirds as long as petals; anthers deep yellow. Nectar scales translucent, linear, ca. 1.1 mm, apex emarginate or rounded. Carpels erect, 1–1.1 cm, connate 0.9–1.8 mm from base; style 1–2 mm. Ovules 8–20 in each locule. Follicles erect or suberect.

Distribution: Endemic to Nepal.

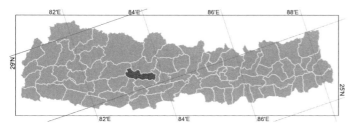

Altitudinal range: ca. 3300 m.

Ecology: On soil banks on south-facing slopes.

Flowering: August.

This subspecies is only known from the type, *Stainton, Sykes and Williams 9011* (BM, TI), from Phagune Dhuri, Baglung. It is distinguished from subsp. *obtusipetalum* by the sparsely flowered inflorescences and the broader petals. The typical subspecies is restricted to SW Sichuan and NW Yunnan, and they are a good example of the interesting SW China–Nepal disjunction seen in some groups.

9. *Sedum oreades* (Decne.) Raym.-Hamet, Bull. Soc. Bot. France 56: 571 (1909).
Umbilicus oreades Decne., in Jacquem., Voy. Inde 4(1): 62 (1841); *Cotyledon oreades* (Decne.) C.B.Clarke; *C. spathulata* (Hook.f. & Thomson) C.B.Clarke; *Sedum filicaule* Duthie ex Raym.-Hamet nom. inval.; *S. jaeschkei* Kurz; *S. squarrosum* Royle ex Raym.-Hamet; *Umbilicus luteus* Decne.; *U. spathulatus* Hook.f. & Thomson.

Perennial herbs, 2–10(–20) cm. Flowering stems erect, glabrous, smooth, with few sterile branches. Leaves alternate, sessile, spur 0.3–0.6 mm, with rounded or very shallowly trilobulate to nearly truncate apex; blade oblong to broadly oblong, (3-) 5–8 × (1-)2–3 mm, apex obtuse or rarely acute, margin entire, flat, glabrous, smooth. Inflorescence a cyme with (1–)4–10 flowers, with smooth, dichasially branched 3 primary axes. Flowers usually 5- or 6-, rarely 7-merous, 4–8 mm across. Pedicels to 5 mm. Sepals free, spurred, oblong or rarely oblanceolate, 4.5–7 mm, apex obtuse, smooth, ascending. Petals yellow, connate 1–2 mm from base, membranous, oblanceolate or obovate, 7–10 × 2.5–3.8 mm, margins usually erose, apex obtuse or rarely rounded, suberect at flowering. Stamens less than half as long as petals, oppositipetalous stamens often reduced; anthers deep yellow. Nectar scales green with reddish base, linear, ca. 1.2 mm. Carpels 4.5–6 mm, connate 0.5–1 mm from base, straight; style ca. 0.8 mm. Ovules 10–22 in each locule.
Fig. 25l–o.

Distribution: Nepal, W Himalaya, E Himalaya, Tibetan Plateau, Assam-Burma and E Asia.

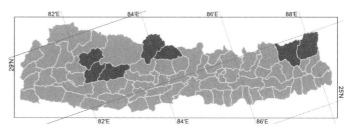

Altitudinal range: 3200–5200 m.

Ecology: On mossy rocks, scree, boulders, meadows in alpine zone.

Flowering: July–August. **Fruiting:** September–October.

10. *Sedum perpusillum* Hook.f. & Thomson, J. Proc. Linn. Soc., Bot. 2: 103 (1858).

Annual herbs, 1.5–3(–4) cm. Flowering stems branched several times from base, glabrous, smooth, branches ascending or erect. Leaves alternate, sessile, spur ca. 0.5 mm, with truncate or rounded-truncate apex; blade linear-lanceolate or narrowly oblong, 3.5–5 × 1–1.3 mm, apex acute or rarely obtuse, margin entire, flat, glabrous, smooth. Inflorescence a cyme, 0.5–1 × 0.5–1.5 cm, usually with 3–10 flowers, primary axes short, ascending or suberect, smooth. Flowers 2–3 mm across, often haplostemonous. Pedicels to 1.5 mm, smooth. Sepals free, spurred, narrowly ovate to ovate or oblong-ovate, 3–4 mm, apex obtuse or acute, smooth. Petals yellow, free, triangular with oblong base or oblong-ovate, 2.8–3.2 mm, apex acute, suberect at flowering. Stamens shorter than petals; anthers deep yellow. Nectar scales translucent, linear, 1.2–1.5 mm, apex rounded. Carpels 2–3 mm, nearly free, erect; style ca. 0.5 mm. Ovules 6–8 in each locule. Follicles 3–3.5 mm, carpels not divergent. Seeds ellipsoid, ca. 0.7 mm, apex rounded.

Distribution: Nepal, E Himalaya and Tibetan Plateau.

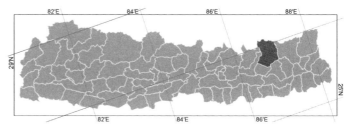

Altitudinal range: 3800–4500 m.

Ecology: On mossy rocks in scree or meadows in alpine zone.

Flowering: May.

Sedum perpusillum is similar to *S. magae* Raym.-Hamet from which it differs by having locules with fewer ovules (6–8) and triangular petals with an oblong or oblong-ovate base.

Crassulaceae

11. *Sedum przewalskii* Maxim., Bull. Acad. Imp. Sci. Saint-Pétersbourg 29: 156 (1883).

Annual herbs, 3–5 cm. Flowering stems glabrous, smooth, often branched from base, erect or ascending from base. Leaves alternate, sessile, spur 0.3–0.5 mm, with truncate or emarginate (rarely rounded) apex; blade oblong to oblong-ovate or rarely lanceolate, 2–4 (–7) × 0.6–1.1 mm, apex rounded to obtuse or rarely acute, margin entire, flat, glabrous, smooth. Inflorescence a lax, few-flowered cyme, 1.5–3 cm long and wide, axes smooth. Flowers often haplostemonous. Pedicels 1–3 mm, smooth. Sepals without spur, connate 0.8–1.4 mm from base, oblong or broadly ovate to ovate, 2–3 mm, apex rounded to obtuse or rarely acute, smooth, suberect. Petals yellow, free, oblong-lanceolate to narrowly ovate or narrowly triangular, 2.5–4.5 mm, apex rounded to obtuse or rarely acute, suberect or erect at flowering. Stamens shorter than petals; anthers deep yellow. Nectar scales translucent, linear or linear-obovate, 0.8–1.2 mm, apex rounded. Carpels 2–3.2 mm, carpels erect, free to connate 0.5–0.7 mm from base; style 0.7–1.2 mm long. Ovules 2–6 in each locule; placenta basal. Follicles 3–4 mm, not divergent. Seeds oblongoid, 0.7–0.8 mm, apex rounded.

Distribution: Nepal, E Himalaya, Tibetan Plateau and E Asia.

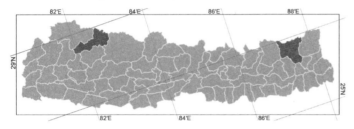

Altitudinal range: 4200–4500 m.

Ecology: On scree or gravelly slope in alpine zone.

Flowering: August. **Fruiting:** September.

12. *Sedum pseudomulticaule* H.Ohba, J. Jap. Bot. 53: 328 (1978).

Perennial herbs to 20 cm. Flowering stem simple, with long creeping base, 2.5–3.5 mm thick, glabrous, smooth. Leaves 3-verticillate to opposite or alternate, sessile, shortly spurred, narrowly oblanceolate to linear-oblanceolate, 25–40 × 3–5 mm, base attenuate, apex obtuse to acute, margin entire, flat, glabrous, smooth. Inflorescence with 10–30 flowers. Pedicels 2–4 mm, ± papillate. Sepals free, without spur, linear, 4.5–7 mm, apex rounded or obtuse, smooth, ascending at flowering. Petals yellow, connate ca. 1.5 mm from base, 7.5–9 mm, suberect at flowering. Stamens shorter than petals; anthers orange yellow. Nectar scales 1–1.2 mm. Carpels 6.5–8.2 mm long, connate 1.5–1.8 mm from base, with ca. 1.7 mm style. Ovules ca. 30 in each locule.

Distribution: Endemic to Nepal.

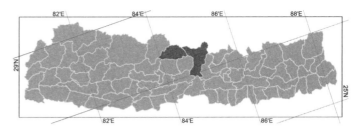

Altitudinal range: ca. 2100 m.

Ecology: On mossy rocks.

Flowering: July.

This differs from *Sedum multicaule* Wall. ex Lindl. in having free sepals with rounded or obtuse apex, slightly connate petals, ovaries connate 1.5–1.8 mm from base and longer leaves.

13. *Sedum roborowskii* Maxim., Bull. Acad. Imp. Sci. Saint-Pétersbourg 29: 154 (1883).

Annual herbs, 5–25 cm long. Flowering stems glabrous, smooth or mammillate, simple or branched, erect from ascending base. Leaves alternate, sessile, spur 0.5–1 mm, with truncate apex; blade oblong-obovate or obovate to narrowly obovate, 4–16 × 2–9 mm, apex rounded, margin entire, flat, upper surface and margin usually smooth, rarely mammillate. Inflorescences a cyme, sparsely to densely flowered, 2–7 × 1–4 cm; primary axes 3, ascending to suberect, zigzag, usually smooth. Flowers usually haplostemonous, 4–7 mm across. Pedicels 0.5–1 mm, smooth. Sepals free, spurred, narrowly oblong or linear-obovate, 2.2–3.7 mm, apex rounded to ovate, smooth, rarely sparsely mammillate, suberect. Petals yellow, free, linear-subulate to narrowly triangular or linear-lanceolate, 3–4.5 mm, apex acute or obtuse, suberect or ascending at flowering. Stamens shorter than petals; anthers deep yellow. Nectar scales translucent, oblong-ovate, 0.5–0.7 mm, apex truncate. Carpels erect, 2.5–3.5 mm, connate 0.6–1.2 mm from base, style 0.3–0.5 mm. Ovules 20–40 in each locule. Follicles 4–6 mm, not divergent. Seeds oblong-ellipsoid, ca. 0.7 mm. Fig. 25p–s.

Distribution: Nepal, Tibetan Plateau and E Asia.

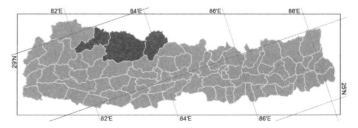

Altitudinal range: 2000–4600 m.

Ecology: Among stones, on rocky or soil grounds in scree or meadow.

Flowering: August–September. **Fruiting:** September.

14. *Sedum triactina* A.Berger, in Engl. & Prantl, Nat. Pflanzenfam., ed. 2, 18a: 460 (1930).
Sedum verticillatum (Hook.f. & Thomson) Raym.-Hamet later homonym, non L.; *Triactina verticillata* Hook.f. & Thomson.

Perennial herbs, 7–25 cm. Flowering stems annual, tender, often with short sterile branches, glabrous, smooth. Leaves verticillate but partly opposite or rarely alternate, almost sessile, with ca. 0.6 mm round spur; blade herbaceous, oblong-spathulate or oblanceolate, 10–25 × 3–8 mm, apex retuse to rounded, base long attenuate, flat, glabrous, smooth. Inflorescence a cyme with 3 main axes, 4–13 × 1–5 mm, conspicuously bracteate, with 12–30 flowers, primary axes long, smooth. Flowers sessile, 7–9 mm across. Sepals connate ca. 0.5 mm from base, without spur, oblong or triangular-ovate, ca. 1 mm, apex rounded. Petals yellow, linear-lanceolate to subulate, 4–6.5 mm, apex acute, widely spreading at flowering. Stamens shorter than petals; anthers deep orange yellow. Nectar scales yellow, narrowly oblong, apex rounded. Carpels 3, rarely 2, 4–5.5 mm, swollen on ventral side; style 1.2–1.5 mm. Ovules 2 in each locule. Follicles ± divergent with ventrally swollen carpels, 6–7 mm. Seeds ca. 1 mm.
Fig. 25t–v.

Distribution: Nepal, E Himalaya, Tibetan Plateau, Assam-Burma and E Asia.

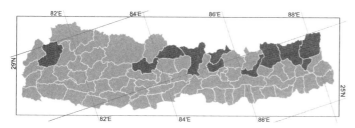

Altitudinal range: 2700–4300 m.

Ecology: On mossy rocks in forests.

Flowering: August–September.

15. *Sedum trullipetalum* Hook.f. & Thomson, J. Proc. Linn. Soc., Bot. 2: 102 (1858).

Perennial herbs, 6–10 cm. Flowering stems erect from procumbent base, much-branched, with many (often tufted) sterile branchlets, glabrous, smooth. Leaves imbricate, sessile, spur 0.4–0.6 mm, with rounded or shallowly trilobulate apex; blade narrowly oblong to oblong or lanceolate, 5–8 × 1.1–1.8 mm, apex long-acuminate, margin entire, flat, glabrous, smooth. Inflorescence a rather large cyme, nearly hemispherical, 1.5–2.5 × 2–3 cm, with 20–50 flowers. Flowers 5–7 mm across. Pedicels 0.8–2.5 mm, smooth. Sepals not spurred, connate ca. 0.8 mm from base, oblong or oblong-lanceolate, 4–6.3 mm, apex acuminate, smooth, suberect to ascending. Petals yellow, free, ovate, 6–8.5 mm, with 2.5–3.5 mm claw, apex obtuse or rounded, suberect at flowering. Stamens shorter than petals; anthers dark purplish red. Nectar scales pale green, narrowly oblong to oblong, 0.6–0.7 mm, apex emarginate. Carpels erect, 6–7 mm, connate 0.6–2.5 mm from base; style tapering, 1.2–1.5 mm. Ovules ca. 30 in each locule.
Fig. 25w–z.

Distribution: Nepal, W Himalaya, E Himalaya, Tibetan Plateau, Assam-Burma and E Asia.

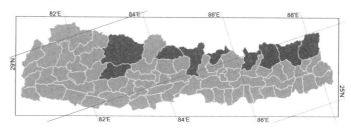

Altitudinal range: 2400–5500 m.

Ecology: On mossy rocks in scree and boulders in alpine zone.

Flowering: August–October. **Fruiting:** October–November.

Sedum trullipetalum is easily distinguished by its clawed, ovate petals. *Sedum gagei* Raym.-Hamet resembles *S. trullipetalum* and is distinguished from the latter by the spurred sepals and the petals with narrowly trullate lamina.

Saxifragaceae

Shinobu Akiyama, Richard J. Gornall, Bhaskar Adhikari, Colin A. Pendry & Mark F. Watson

Herbs. Stipules present or absent. Leaves simple or compound; basal leaves sometimes rosetted; cauline leaves alternate, rarely opposite. Flowers usually in cymes, panicles or racemes, or solitary, actinomorphic, bisexual or sometimes unisexual. Hypanthium usually present. Sepals 4 or 5. Petals usually 5, free, sometimes absent and then the sepals petaloid. Stamens 5–10, free. Ovary superior or half-inferior, carpels 2 or rarely 3, more or less connate, placentation axile, parietal or marginal in free part, styles 2. Fruit a capsule. Seeds numerous.

Worldwide 30 genera and 625 species with a subcosmopolitan distribution; particularly common in temperate and cold regions of N hemisphere. Seven genera and 106 species in Nepal.

Recent analyses have clearly shown that the traditional, broad circumscription of the Saxifragaceae (e.g. Engler, Pflanzenfam. 18a: 74-226. 1928) is highly polyphyletic. It includes many lineages which are only distantly related and can no longer be maintained (Morgan & Soltis, Ann. Miss. Bot. Gard. 80: 631-660. 1993; Soltis & Soltis, Am. J. Bot. 84: 504-522. 1997).

Key to Genera

1a	Leaves compound	2
b	Leaves simple	3
2a	Leaves pinnate. Flowers ca. 5 mm across	**1. Rodgersia**
b	Leaves 2- or 3-ternate, the leaflets sometimes pinnate. Flowers 2–3 mm across	**2. Astilbe**
3a	Petals absent (very rarely present in *Tiarella*)	4
b	Petals present (rarely absent in *Saxifraga*)	5
4a	Flowers in racemes. Carpels unequal in size	**3. Tiarella**
b	Flowers in cymes, sometimes ± crowded. Carpels of equal size	**4. Chrysosplenium**
5a	Plants with thick rhizomes. Leaves medium-sized or large, 4–25 cm	**5. Bergenia**
b	Plants often tufted or caespitose, sometimes creeping at base but then with slender rhizomes. Leaves small, seldom attaining 4 cm	6
6a	Flowering stem leafy. Leaves lacking crystals, distributed along stem, sometimes aggregated toward base, or sometimes forming columnar rosettes	**6. Saxifraga**
b	Flowering stem leafless. Leaves containing crystals, all arranged in a compact, basal rosette	**7. Micranthes**

1. *Rodgersia* A.Gray, Mem. Amer. Acad. Arts, Ser. 2 6(1): 389 (1858).

Shinobu Akiyama & Mark F. Watson

Robust perennial herbs, rhizomes horizontal, thick and elongate. Stipules linear-lanceolate, acuminate, fused at base to petiole, persistent. Leaves basal and cauline, petiolate, odd-pinnate, alternate, leaflets subsessile to shortly petiolulate. Inflorescence a terminal, many-flowered, lax, paniculate, much-branched cyme, bracts and bracteoles absent. Flowers greenish white to yellowish green, bisexual. Hypanthium obconic, short. Sepals 5, conspicuous, petaloid, spreading. Petals absent. Stamens 10, opposite and between sepals, exserted. Ovary semi-inferior. Carpels 2–3, ovules numerous, placentation axile, styles 2. Capsules 2–3-valved, seeds numerous.

Worldwide five species in E Himalaya and E Asia. One species in Nepal.

1. *Rodgersia nepalensis* Cope ex Cullen, Notes Roy. Bot. Gard. Edinburgh 34: 116 (1975).

Flowering stems 1–1.5(–2) m, villous and glandular pubescent especially above. Rhizome 3–4 cm thick. Stipules brown, 30–40 × 7–10 mm, papery. Basal leaves to 1.5 m, petioles long, 30–40 cm. Leaflets 7–14, oblong-elliptic, 12–40 × 4–12 cm, base cuneate or rounded often asymmetric, apex shortly acuminate, margin double-serrate upper surface glabrous, lower surface (and rachis) with long (to 2 cm) brown

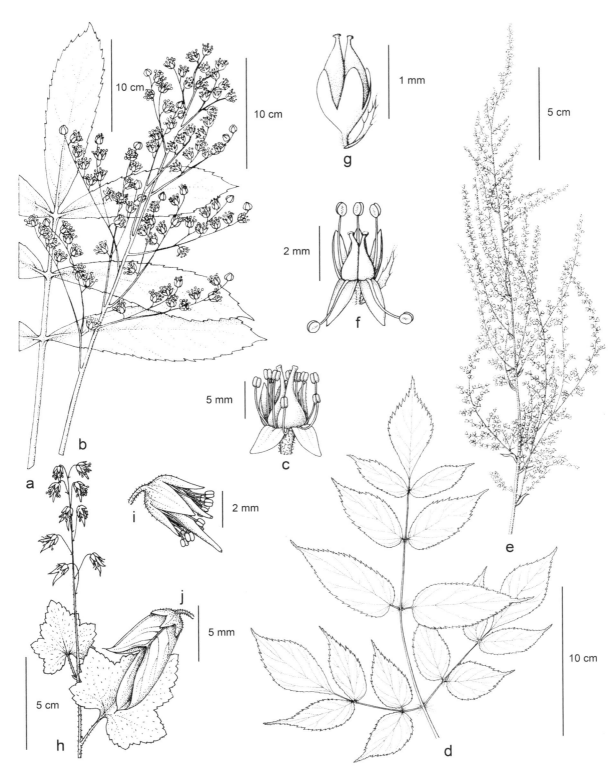

FIG. 26. SAXIFRAGACEAE. **Rodgersia nepalensis**: a, leaf; b, inflorescence; c, flower with two calyx lobes bent forward. **Astilbe rivularis**: d, leaf; e, inflorescence; f, flower with two calyx lobes bent forward; g, fruit. **Tiarella polyphylla**: h, stem with leaves and inflorescence; i, flower; j, fruit.

scale-like hairs on veins, petiolules 0–1 cm. Cauline leaves similar but reducing up the stem, the uppermost almost sessile and often 3-foliolate. Inflorescence 30–40 × 30–40 cm, floccose, villous and glandular pubescent. Pedicels 2–10 mm, glandular. Hypanthium 3–4 mm, glandular. Sepals triangular, 3.5–5 × 2–2.5 mm, apex acute or subacuminate, glandular. Stamens greenish white, 5–8 mm, unequal Capsules 7–10 mm, tapering above into the divergent persistent styles. Fig. 26a–c.

Distribution: Nepal and E Himalaya.

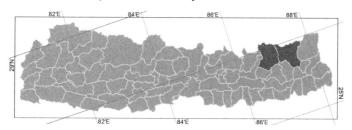

Altitudinal range: 2500–3400 m.

Ecology: Mixed temperate broadleaf and coniferous woodland with *Abies*, *Tsuga*, *Rhododendron*, *Pieris* and bamboos, often in moist, humus-rich soil growing amongst shrubs and other tall herbs.

Flowering: June–August. **Fruiting:** September–October.

Near endemic to Nepal. Predominantly found in north eastern districts of Nepal, but also one collection from central Sikkim (Toong). Morphologically distinct and widely separated geographically from the other species in *Rodgersia* which are found in N Myanmar and SW China (especially Sichuan, Yunnan, SE Xizang) and eastwards to Korea and Japan. The long, brown, scale-like hairs on the stem and leaf rachis, floccose inflorescence branches, glandular sepals and long stamens are quite unlike other members of the genus.

2. *Astilbe* Buch.-Ham. ex D.Don, Prodr. Fl. Nepal.: 210 (1825).

Shinobu Akiyama & Colin A. Pendry

Perennial herbs, rhizomatous, without stolons or bulbils. Stipules membranous, adnate to petiole base and sheathing stem. Leaves cauline, not in a basal rosette, long petiolate, alternate, bi- or tri-ternate, the leaflets sometimes pinnate, margins biserrate. Inflorescence a terminal, very many-flowered, bracteate panicle. Flowers white or reddish, bisexual, actinomorphic. Hypanthium very small, adnate to base of ovary. Sepals (4–)5, rather petaloid. Petals absent. Stamens 5, opposite sepals. Ovary half-inferior. Carpels 2 or rarely 3, equal, connate at base, ovules many, placentation axile. Styles 2 (rarely 3), free. Fruit a capsule with 2 (rarely 3) equal, ventrally dehiscent valves. Seeds numerous, small, narrowly fusiform.

Worldwide about 20 species in Asia and N America. One species in Nepal.

1. *Astilbe rivularis* Buch.-Ham. ex D.Don, Prodr. Fl. Nepal.: 211 (1825).
Spiraea barbata Wall. ex Cambess.

ठूलो औषधी Thulo aushadhi (Nepali).

Plants 0.6–2.5 m. Rhizome thick. Stems brown, long glandular-hairy. Leaves 15–50 cm including petiole. Petiole and petiolules glabrescent to sparsely long brown-pilose with denser tufts at base of petiolules and axes of rachis. Leaflets ovate to elliptic, 4–14 × 2–8 cm, base rounded to cordate, apex acuminate, sparsely brown glandular-hairy above, long brown-pilose and glandular-hairy on veins below. Panicles to 60 cm, rachis pale brown hairy. Bracts 1.5–2 mm, to 3 mm at base of panicle branches. Bracteoles 2, 1 at base of pedicel, 1 towards apex, filiform, ca. 1 mm. Pedicels ca. 2 mm, pale brown hairy. Sepals white to reddish, narrowly ovate, 1.5–2 mm. Stamens ca. 3 mm. Carpels 1–1.5 mm, glabrous. Valves ovoid, 3–4 mm. Seeds ca. 1 mm. Fig. 26d–g.

Distribution: Nepal, W Himalaya, E Himalaya, Tibetan Plateau, Assam-Burma and SE Asia.

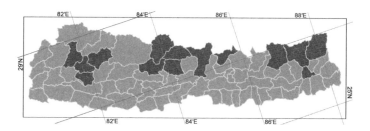

Altitudinal range: 1400–3600 m.

Ecology: *Pinus wallichiana* forests and mixed evergreen and deciduous forests, under shade and in the open.

Flowering: July–September. **Fruiting:** September–October.

Applied externally to sprains and muscular swellings. The juice is used to treat diarrhoea, dysentery, prolapse of the uterus and haemorrhage.

3. *Tiarella* L., Sp. Pl. 1: 405 (1753).

Shinobu Akiyama & Colin A. Pendry

Perennial herbs, rhizomatous, without stolons and bulbils. Stipules membranous, adnate to base of petiole. Leaves basal and cauline, petiolate, simple, alternate, palmately 3–5-lobed, the basal leaves larger with longer petioles, margin irregularly dentate. Inflorescence few- to many-flowered, terminal, a raceme or sparsely-branched panicle. Bracts minute, filiform. Flowers white or pale pink, bisexual, zygomorphic. Hypanthium very small, adnate to base of ovary. Sepals 5, petaloid. Petals absent, rarely 5. Stamens 10. Ovary superior. Carpels 2, unequal, connate at base, ovules many, placentation parietal. Styles 2, free. Fruit a capsule with 2 very unequal, ventrally dehiscent valves. Seeds ellipsoid.

Worldwide three species in the Himalaya, E Asia and N America. One species in Nepal.

1. *Tiarella polyphylla* D.Don, Prodr. Fl. Nepal.: 210 (1825).

Plants 20–45 cm. Stems unbranched, glandular-hairy. Stipules 2–6 mm, margin fimbriate. Petioles 2–12 cm, glandular-hairy. Leaves broadly ovate to almost orbicular, 2–6 × 2.5–6 cm, shallowly lobed, base cordate, apex acute, coarsely hirsute above, glandular-hairy below, especially on veins. Cauline leaves 2 or 3, petioles ca. 1 cm. Raceme 5–20 cm, rachis, bracts and pedicels glandular-hairy. Flowers 6–40. Pedicel ca. 1 cm. Sepals erect, ovate, ca. 3 × 1 mm. Stamens and styles shortly exserted. Stamens 2.5–3 mm. Larger carpel 3.5–4 mm, the smaller 2.5–3 mm. Capsule with larger valve ca. 10 mm, smaller valve ca. 6 mm. Seeds ca. 1 mm.
Fig. 26h–j.

Distribution: Nepal, E Himalaya, Tibetan Plateau and E Asia.

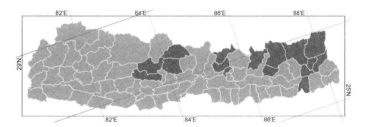

Altitudinal range: (2000–)2300–3300(–4000) m.

Ecology: In mixed deciduous and coniferous forest, often with *Tsuga dumosa* and *Pinus wallichiana*.

Flowering: May–July. **Fruiting:** May–October.

Root juice is used to treat fevers, and leaf juice is applied as hot compress to treat muscular swelling.

4. *Chrysosplenium* L., Sp. Pl. 1: 398 (1753).

Shinobu Akiyama, Bhaskar Adhikari & Colin A. Pendry

Perennial herbs, rhizomatous or caespitose, often with stolons. Rhizomes either rather stout, densely set with roots and clothed with scales, or more slender, long-creeping and without scales. Stipules absent. Leaves simple, petiolate, opposite or alternate, toothed or lobed. Basal leaves sometimes present, long-petiolate, arising from rhizome, similar to cauline leaves. Lower cauline leaves occasionally scale-like. Flowers small, bisexual, actinomorphic, green, yellow, purplish or brownish, usually in cymes surrounded with bracteal leaves, rarely flowers solitary. Hypanthium funnel- or cup-shaped, more or less adnate to ovary. Sepals 4, one pair overlapping the other in bud, persistent. Petals absent. Disk around style, ± developed, nectariferous. Stamens 8, inserted on margin of hypanthium. Filaments needle-shaped. Anthers 2-locular, laterally dehiscent. Ovary semi-inferior, 1-locular. Carpels 2, connate, with 2 parietal placentae. Styles 2, free. Stigmas punctate. Capsules semi-inferior to semi-superior, 2-lobed, flattened, dehiscent along inner suture. Seeds numerous, oblong to globose with a carina on one side, thick-walled, smooth, sometimes hairy, ridged or minutely papillose.

Worldwide about 55 species in temperate and arctic regions of Eurasia, N Africa, Greenland, and N and S America. Ten species in Nepal.

Key to Species

1a   Leaves opposite, at least on sterile branches..2
  b   Leaves all alternate 3

2a   Leaves ovate-rounded to orbicular, up to 1 cm. Flowers 2.5–3.5 mm across. Seeds smooth or minutely papillose under a microscope .... **1. *C. nepalense***

b   Leaves elliptic to broadly ovate, 1–2 cm. Flowers ca. 4 mm across. Seeds conspicuously patently papillose-pilose..**2. *C. trichospermum***

3a   Stem and leaves villous 4

b   Stem and leaves glabrous..5

4a   Flowering stems 7–20 cm. Leaves with 14–20 teeth...**5. *C. lanuginosum***

b   Flowering stems 2–7 cm. Leaves with 3–7 teeth .... **6. *C. singalilense***

5a   Basal leaves absent, lower cauline leaves sometimes present, scale-like or leaf-like with petioles 1–3 cm 6

b   Basal leaves arising from rhizome, petioles (1–)3–19 cm 8

6a   Lower cauline leaves scale-like, sessile...**4. *C. carnosum***

b   Lower cauline leaves leaf-like, petiolate...7

7a   Bracteal leaves 10–13 × 8–15 mm. Rhizome without scales .... **3. *C. uniflorum***

b   Bracteal leaves 3–5 × 5–7 mm. Rhizome scaly..**10. *C. tenellum***

8a   Stamens 0.6–0.8 mm ....**7. *C. griffithii***

b   Stamens 1–2 mm 9

9a   Plants 3–12 cm. Basal leaves with 7–15 lobes..**8. *C. nudicaule***

b   Plants 13–25 cm. Basal leaves with 17–27 lobes..**9. *C. forrestii***

1. *Chrysosplenium nepalense* D.Don, Prodr. Fl. Nepal.: 210 (1825).

Herbs 6–20 cm, glabrous throughout. Rhizome long-creeping, without scales. Stolons present. Sterile branches well developed. Leaves opposite. Basal leaves absent. Scale leaves absent. Cauline leaves numerous, evenly spaced. Petioles 3–18 mm. Leaves ovate to orbicular, 2–10 × 2–8 mm, base broadly cuneate to truncate, apex obtuse to retuse, margin crenate with 3–12 short teeth. Cyme somewhat lax, 1–2 cm across, 2–7-flowered. Bracteal leaves with 2–10 mm petioles, broadly ovate to orbicular, 4–10 × 3–11 mm, margin crenate. Flowers yellowish green, 2.5–3.5 mm across. Sepals ovate-triangular, ca. 1 mm. Stamens shorter than sepals, 0.6–0.8 mm. Capsule lobes ascending to divergent, ca. 2 mm. Seeds brown, ellipsoid to ovoid, 0.8–1 mm, smooth, lustrous or sometimes with obscure flat ridges.

Distribution: Nepal, E Himalaya, Tibetan Plateau and Assam-Burma.

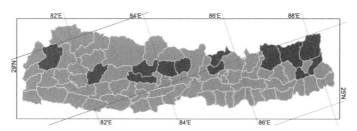

Altitudinal range: 1900–4000 m.

Ecology: Damp ground in shady forests, near streams, on moist rocks.

Flowering: April–June. **Fruiting:** April–June.

2. *Chrysosplenium trichospermum* Edgew. ex Hook.f. & Thomson, J. Proc. Linn. Soc., Bot. 2: 73 (1857).

Herbs 10–26 cm, glabrous throughout. Rhizome long-creeping, without scales. Stolons poorly developed. Sterile branches well developed. Leaves opposite. Basal leaves absent. Scale leaves absent. Petioles 5–10 mm. Leaves elliptic to ovate, 6–20 × 5–13 mm, base broadly cuneate to attenuate, apex obtuse, margin serrate with 7–11(–15) teeth. Cyme somewhat lax, 1–3 cm across. Bracteal leaves with 2–10 mm petioles, obovate, 8–10 × 5–8 mm. Flowers yellowish green, ca. 4 mm across. Sepals ovate, ca. 2 mm. Stamens ca. 1 mm, slightly shorter than sepals. Anthers yellow. Capsule lobes ca. 5 mm. Seeds brown, ovoid, ca. 1 mm, conspicuously patently papillose-pilose.

Distribution: Nepal and W Himalaya.

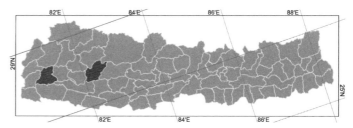

Altitudinal range: 2100–3000 m.

Ecology: On stream-sides and other wet places.

Flowering: April–June. **Fruiting:** April–June.

3. *Chrysosplenium uniflorum* Maxim., Bull. Acad. Imp. Sci. Saint-Pétersbourg 27: 468 & 472 (1881).

Herbs 3–10 cm, glabrous throughout. Rhizome creeping, without scales. Stolons mainly from axils of lower leaves, very long, filiform, often below ground. Sterile branches absent. Leaves alternate. Basal leaves absent. Scale leaves absent or occasionally present at base of stem, whitish, fleshy, ovate. Cauline leaves several, evenly spaced. Petioles 1–4 cm, longer on lower stem. Leaves reniform to orbicular 5–15 × 8–18 mm, base cordate to broadly cuneate, apex retuse, margin crenate with (6–)8–10(–12) teeth on each side. Upper leaves somewhat congested and smaller with fewer marginal teeth. Cyme dense, 1–3-flowered. Bracteal leaves with ca. 5 mm petioles, ovate, 10–13 × 8–15 mm. Flowers greenish, ca. 5 mm across. Sepals ascending, broadly ovate, ca. 3 mm. Stamens 1.3–1.6 mm. Capsule lobes ca. 1mm. Seeds elliptic, smooth, to 0.9 mm.

Distribution: Nepal, Tibetan Plateau and E Asia.

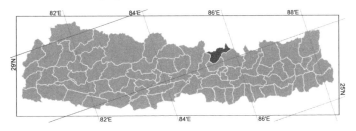

Altitudinal range: ca. 4000 m.

Ecology: In wet places in *Abies* or *Juniperus* forest or alpine areas.

Flowering: May–July. **Fruiting:** May–July.

4. *Chrysosplenium carnosum* Hook.f. & Thomson, J. Proc. Linn. Soc., Bot. 2: 73 (1857).
Chrysosplenium carnosulum Hook.f. & Thomson ex Maxim.

यहकी-मह Yahkee-mah (Tibetan).

Herbs 7–15 cm, glabrous throughout, often caespitose, with branches arising from leaf axils. Rhizome thick, scaly. Stolons absent. Leaves alternate. Basal leaves absent. Lower cauline leaves scale-like, 3–7 × 2–3 mm. Upper leaves with petioles ca. 1 mm, obovate to spathulate, 5–12 × 3–7 mm, base cuneate, apex obtuse, sometimes retuse, margin crenate with 5–9 teeth. Cyme dense, 1–3 cm across, 2–3-flowered. Bracteal leaves often yellowish, with 1–5 mm petioles, ovate, 5–12 × 3–9 mm, base cuneate, apex obtuse to slightly retuse, margin crenate with 5–9 teeth. Flowers greenish yellow, often brownish or dark purplish, 3–5 mm across. Sepals reniform to rounded, ca. 2 mm. Stamens ca. 1 mm, slightly shorter than sepals. Stamens ca. 1 mm, often purplish. Capsule lobes ca. 3 mm, ovoid. Seeds brown, ovoid, ca. 1 mm, smooth, lustrous, glabrous.

Distribution: Nepal, W Himalaya, E Himalaya and Tibetan Plateau.

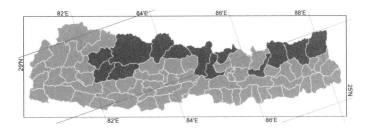

Altitudinal range: 3700–5500 m.

Ecology: Alpine grasslands, moist places, scree slopes, earth pockets among boulders.

Flowering: July–August. **Fruiting:** July–August.

The plant is anti-inflammatory, febrifuge and cholagogue, and used to cure headache and inflammation of the gall bladder.

5. *Chrysosplenium lanuginosum* Hook.f. & Thomson, J. Proc. Linn. Soc., Bot. 2: 74 (1857).
Chrysosplenium adoxoides Hook.f. & Thomson ex Maxim.;
Saxifraga adoxoides Griff.

Herbs, 7–20 cm, villous with long, soft, rusty-brown (when dry), spreading hairs. Rhizome long-creeping, sometimes with filiform stolons. Leaves alternate. Basal leaves absent. Scale leaves absent. Sterile branches well-developed, densely villous above, with leaves crowded in upper part, petioles1–5 cm, blade ovate to orbicular, 1–3 × 1–2 cm, base broadly cuneate to rounded, apex obtuse, margin crenate with 14–16 teeth, margins crenate with coarse, flattish, emarginate teeth, pubescent above and on margins. Cauline leaves 1–3, with petioles 5–20 mm, broadly ovate to reniform, 10–20 × 4–20 mm, base cuneate, sometimes truncate, margin minutely crenate with ca. 20 teeth, both surfaces sparsely pilose. Cyme lax, 3–5 cm, branched, ca. 10-flowered. Bracteal leaves green, with 2–12 mm petioles, ovate 2–10 × 2–12 mm, base broadly cuneate, apex obtuse, margin crenate with 4–6 teeth, both surfaces pilose to glabrous. Flowers 3.5–5 mm across, lower ones long pedicellate. Sepals green, ovate, ca. 1 × 2 mm, glabrous. Stamens ca. 0.5 mm. Capsule lobes ovoid, 3–4 mm. Seeds brown, ovoid, less than 1 mm, smooth (very minutely papillose under microscope).

Distribution: Nepal, E Himalaya, Tibetan Plateau and Assam-Burma.

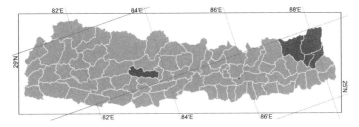

Altitudinal range: 2700–3500 m.

Ecology: Shady wet forest, on rocks.

Flowering: April–June. **Fruiting:** April–June.

Chrysosplenium lanuginosum is unlikely to be confused with any of the other Nepalese species because of its rusty-brown villous indumentum and sterile branches with leaves which are very much larger than the leaves on fertile stems.

6. *Chrysosplenium singalilense* H.Hara, J. Jap. Bot. 36: 77 (1961).

Slender herbs, 2–7 cm, more or less rusty-brown pubescent throughout. Rhizome short creeping. Leaves alternate. Scale leaves absent. Basal leaves with petioles 3–5(–10) mm, blade ovate to orbicular, 2.5–5(–6) × 3–6(–8) mm, base truncate to rounded, margin crenate with 3–7 teeth, sparsely pubescent. Cauline leaves 2 or 3, similar to basal leaves but smaller. Cyme lax, ca. 1 cm across, (1–)2–6-flowered. Bracteal leaves with 2–8 mm petioles, depressed ovate, 2–5 mm, with 3–5 crenate teeth. Flowers yellow, 3–4 mm across. Sepals patent, ovate, 1–1.6(–2) mm. Stamens 0.5–0.7 mm long. Capsules subinferior. Seed brown, oval, ca. 0.6 × 0.4 mm, very minutely papillate under microscope.

Distribution: Nepal and E Himalaya.

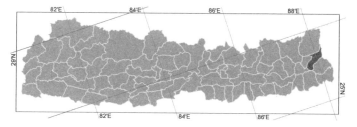

Altitudinal range: ca. 3900 m.

Ecology: Mossy stony slopes.

Flowering: May–June. **Fruiting:** May–June.

A near endemic to Nepal, known only from the type specimen *Kanai et al. 764 p.p.* (TI) from Singalila and *Shakya & Bhattacharya 2387* (KATH) probably from Dhading. The latter is a little larger in its leaves and flowers and its identity is uncertain.

7. *Chrysosplenium griffithii* Hook.f. & Thomson, J. Proc. Linn. Soc., Bot. 2: 74 (1857).

Herbs 10–17 cm, almost glabrous, sometimes with scattered rusty-brown hairs. Rhizome stout, scaly. Often caespitose, without stolons. Leaves alternate. Scale leaves absent. Basal leaf often present, with 8–10 cm petiole, reniform, 1.5–2 × 2–3 cm, base deeply cordate, margin 7–13-lobed, lobe apex usually retuse, both surfaces glabrous. Cauline leaves alternate, 1–3, with 1–3 cm petioles, similar to basal leaf. Cyme lax, branched, 1–2 cm across, 5–13-flowered. Bracteal leaves greenish, smaller than cauline leaves, with 2–5 mm petioles, ovate, 5–15 × 5–20 mm, base cordate to broadly cuneate, margin 3–9-lobed, lobe apex obtuse, glabrous. Flowers greenish, 4–6 mm across. Sepals patent,

rhombic-ovate, 2–5 mm, glabrous. Stamens 0.6–0.8 mm. Capsules semi-superior. Seeds oval, 0.7–0.9 mm, smooth, lustrous.
Fig. 27a–c.

Distribution: Nepal, E Himalaya and Tibetan Plateau.

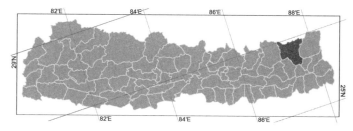

Altitudinal range: ca. 3600 m.

Ecology: Wet rocks in ravines or woods, or in alpine meadows.

Flowering: May–June. **Fruiting:** May–June.

In Nepal *Chrysosplenium griffithii* is known only from *Swan 475-474* (BM) from Sankhuwasaba. It is close to *C. forrestii* Diels, but the leaves of the latter have more lobes which have squarish rather than rounded apices. *Chrysosplenium nudicaule* Bunge is similar but smaller and its cyme and cauline leaf are much more congested at the apex of the flowering stem.

8. *Chrysosplenium nudicaule* Bunge, Fl. Altaic. 2: 114 (1830).

Chrysosplenium nudicaule var. ***intermedium*** H.Hara, J. Fac. Sci. Univ. Tokyo, Sect. 3, Bot. 7: 65, pl. 13A (1957).

Herbs, 3–13 cm, almost glabrous. Rhizome thick, scaly. Stolons long filiform. Sterile branches absent. Leaves alternate. Scale leaves absent. Basal leaves 1 or 2, with 1–7 cm petioles, reniform, 3–8 × 8–15 mm, base deeply cordate, apex rounded, margin with 7–15 lobes, lobes rectangular, apex obtuse, sometimes overlapping. Cauline leaf absent or 1, very close to cyme, with 2–5 cm petiole, broadly ovate to reniform, 5–10 × 3–10 mm, base cuneate, margin crenate with 3–7 lobes, lobes rounded, glabrous. Cyme very dense, 1–3 cm, 2–6-flowered. Bracteal leaves greenish, similar to cauline leaf. Flowers 4–5 mm across. Sepals obovate to rhombic, 1–2 mm, glabrous. Stamens ca. 1 mm; anthers yellow. Capsule subsuperior, ca. 1.5 mm. Seed brown, oval, ca. 0.9 mm, shiny.

Distribution: Nepal, E Himalaya and Tibetan Plateau.

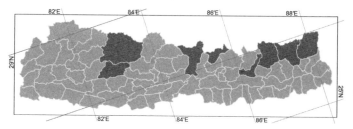

Altitudinal range: 3700–4600 m.

Ecology: Along streams in shady forests or alpine areas.

Flowering: June–October. **Fruiting:** June–October.

Chrysosplenium nudicaule var. *intermedium* is similar to *C. forrestii* Diels and *C. griffithii* Hook.f. & Thomson but smaller and its cyme and cauline leaf are much more congested at the apex of the flowering stem. Var. *intermedium* differs from the N Asian var. *nudicaule* in having a cauline leaf, larger flowers and longer stamens

9. *Chrysosplenium forrestii* Diels, Notes Roy. Bot. Gard. Edinburgh 5: 282 (1912).

Herbs 13–23 cm, glabrous. Rhizome creeping, thick, scaly. Stolons absent. Leaves alternate. Scale leaves absent. Basal leaves 0–2, with 5–19 cm petioles, reniform, 1–3 × 1–6 cm, base cordate, apex rounded, margin crenate with 15–23 lobes, apex of lobe usually retuse, both surfaces sparsely pubescent to glabrous. Cauline leaf 1, with 5–15 mm petiole, blade reniform to broadly ovate, 1.3–2.5 × 2–4.5 cm, base subcordate to cuneate, margin crenate with 7–15 lobes, apex of lobe retuse, sometimes obtuse, both surfaces sparsely pubescent. Cyme dense, 1–4 cm 8–14-flowered. Bracteal leaves with 2–5 mm petiole, reniform to orbicular, 1–2.3 × 1–3.5 cm, base cuneate, margin crenate with 7–17 lobes, apex obtuse, sometimes retuse, sparsely pubescent. Inner leaves large often bright yellow at anthesis, reniform. Flowers yellow, ca. 4 mm across. Sepals reniform to rounded, ca. 3 × 3 mm, apex minutely crenate, glabrous. Stamens 8, ca. 2 mm. Style ca. 8 mm. Capsule lobes 1.5 mm. Seed brown, ovoid, ca. 0.9 mm, shiny.

Distribution: Nepal, E Himalaya, Tibetan Plateau and Assam-Burma.

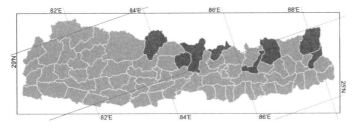

Altitudinal range: 3600–4800 m.

Ecology: Moist alpine grasslands, small ravines, among boulders.

Flowering: June–October. **Fruiting:** June–October.

Chrysosplenium forrestii is close to *C. griffithii* Hook.f. & Thomson but the leaves of the latter have fewer lobes which have rounded rather than squarish apices and are more spreading. *Chrysosplenium nudicaule* is similar but smaller and its cyme and cauline leaf are much more congested at the apex of the flowering stem.

10. *Chrysosplenium tenellum* Hook.f. & Thomson, J. Proc. Linn. Soc., Bot. 2: 73 (1857).

Slender herbs, 4–9 cm, glabrous, often matted. Rhizome short, slender, scaly. Stolons from lower axils, filiform, with small leaves. Leaves alternate. Basal leaves absent. Scale leaves absent. Cauline leaves several, some clustered around lower stem, with 0.5–3 cm petioles, longest on lower leaves, reniform to orbicular, 3–10 × 3–15 mm, base cordate, margin irregularly 3–7-lobed, lobe apex retuse, glabrous. Cyme very lax, 5–10 mm across, 1–5(–6)-flowered. Bracteal leaves with 1–5 mm petioles, broadly ovate to reniform, 3–5 × 5–7 mm, base broadly cuneate, margin 3–5-lobed, lobe apex obtuse. Flowers yellowish green, 3.5–5 mm across. Sepals patent, reniform, ca. 2 mm, base narrowed, glabrous. Stamens ca. 0.5 mm. Seeds red-brown, oval, ca. 6 mm, smooth, glabrous.

Distribution: Nepal, W Himalaya and E Himalaya.

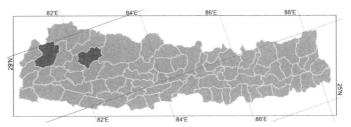

Altitudinal range: 2700–4300 m.

Ecology: Moist places.

Flowering: May–July. **Fruiting:** May–July.

This diminutive species is probably very under collected and is likely to occur in E Nepal.

5. *Bergenia* Moench, Methodus: 664 (1794).

Shinobu Akiyama & Colin A. Pendry

Perennial herbs, without stolons or bulbils, with thick rhizomes and forming large clumps. Stipule membranous, adnate to base of petiole, sheathing stem. Leaves all basal, simple, alternate, entire, petiolate, margin entire or dentate. Inflorescence a few-flowered panicle with cymose branches and a single large basal bract. Individual flowers with or without bracts. Flowers white to reddish purple, bisexual, actinomorphic. Hypanthium cup-shaped, free from the ovary. Sepals 5, not petaloid. Petals 5(–6), white, pink or purplish red. Stamens 10, filaments needle-shaped. Ovary half-inferior. Carpels 2, equal, connate at base, ovules many, placentation axile below, marginal above. Styles 2, free. Fruit a 2-valved capsule dehiscing ventrally. Seeds numerous, flattened, ovoid.

Saxifragaceae

Ten species in Asia. Two species in Nepal.

Key to Species

1a Leaf blade elliptic to ovate-elliptic, margins glabrous or ciliate near base. Petals deep purplish red to bright pink .... **1. *B. purpurascens***
b Leaf orbicular to broadly obovate, margins ciliate. Petals white tinged pink .... **2. *B. ciliata***

1. *Bergenia purpurascens* (Hook.f. & Thomson) Engl., Bot. Zeitung [Berlin] 26: 841 (1868).
Saxifraga purpurascens Hook.f. & Thomson, J. Proc. Linn. Soc., Bot. 2: 61 (1857).

लि-गा-दुर Li-ga-dur (Tibetan).

Plants 7–55 cm. Rhizome scales with entire margins. Flowering stems reddish brown, thick, glandular-hairy. Stipules 3.5–5 cm, margin entire. Petioles 1.5–10 cm. Leaves elliptic to ovate-elliptic, 7–25 × 5–17 cm, base cuneate to rounded, apex rounded, margins entire or shallowly sinuate or glandular-denticulate, glabrous or ciliate near base, glandular punctate. Peduncle, pedicel, hypanthium and calyx deep purplish or brownish red, glandular or glandular-hairy. Basal bract 1–3 cm, inflorescence bracts (3–)6–11 mm. Flowers 1–8, nodding. Pedicels 8–20 mm. Hypanthium 3–6 mm. Sepals oblong, 5–8 mm, apex rounded, entire. Petals deep purplish red to bright pink, broadly spathulate to obovate, 15–25 × 7–9 mm, tapering into a basal claw 2–3 mm. Stamens 9–14 mm. Ovary 6–7 mm. Styles 5–6 mm. Capsule 14–20 mm. Seeds 1.5–2 mm.
Fig. 27d–g.

Distribution: Nepal, E Himalaya, Tibetan Plateau and Assam-Burma.

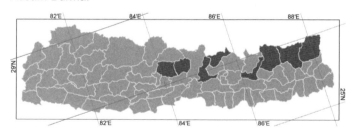

Altitudinal range: 3200–4700 m.

Ecology: Alpine meadows and among dwarf shrubs.

Flowering: May–July. **Fruiting:** August–September.

2. *Bergenia ciliata* (Haw.) Sternb., Revis. Saxifrag. Suppl.2: 2 (1831).
Megasea ciliata Haw., Saxifrag. Enum.: 7 (1821); *Saxifraga ciliata* Royle; *S. pacumbis* Buch.-Ham. ex D.Don nom. inval.

पाषनभेद Pakhanbhed (Nepali).

Plants 5–25 cm. Rhizome scales with ciliate margins. Flowering stems green to reddish, thick, sparsely glandular.

Stipules 0.5–2.5 cm, margin ciliate. Petioles 1–6 cm. Leaves orbicular to broadly obovate, 4–15(–30) × 4–14(–25) cm, base rounded to cordate, apex rounded, margins finely denticulate and ciliate, glabrous or sparsely pubescent on both surfaces. Peduncle, pedicel, hypanthium and calyx green to reddish, glabrous to sparsely glandular. Basal bract 7–20 mm, inflorescence bracts absent. Flowers 1–20, erect or spreading. Pedicels 2–18 mm. Hypanthium 3–7 mm. Sepals oblong, 3–6 mm, apex rounded, margin denticulate or ciliate. Petals pink or white tinged pink, spreading, obovate 11–15 × 7–13 mm, base with a claw 1.5–2 mm. Stamens 6–12 mm. Ovary 5–8 mm. Styles 4–6 mm. Capsule 9–15 mm. Seeds 1.5–2 mm.

Distribution: Nepal, W Himalaya, E Himalaya, Tibetan Plateau, Assam-Burma and SW Asia.

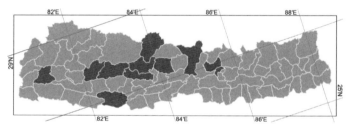

Juice or powder of the whole plant is taken to treat urinary problems. Juice of the rhizome is taken in cases of haemorrhoids, asthma and urinary problems and the rhizome is also used as an anthelmintic.

1a Leaves pubescent at least below......... forma *ciliata*
b Leaves glabrous on both surfaces.... forma *ligulata*

Bergenia (Haw.) Sternb. *ciliata* forma *ciliata*.

Leaves sometimes pubescent above, always below.

Distribution: Nepal and W Himalaya.

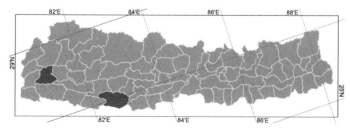

Altitudinal range: 900–2500 m.

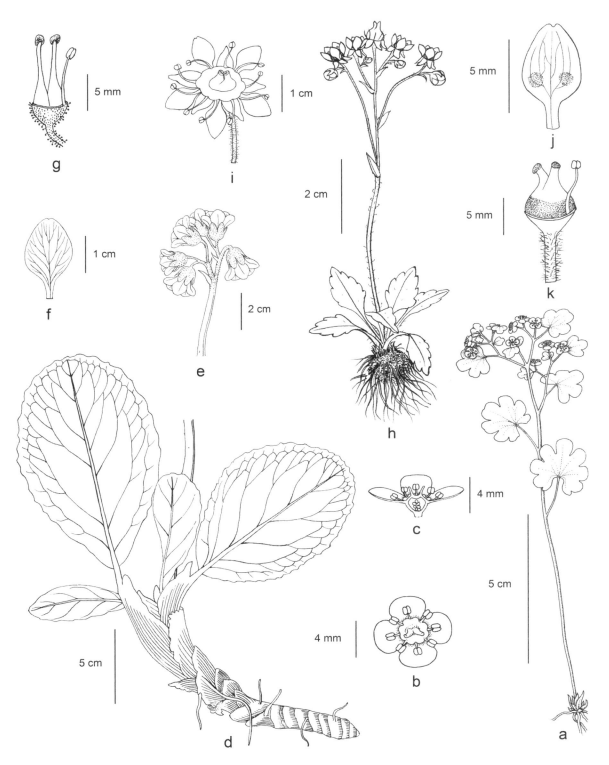

FIG. 27. Saxifragaceae. **Chrysosplenium griffithii**: a, flowering plant; b, flower from above; c, longitudinal section of flower. **Bergenia purpurascens**: d, leaves and rhizome; e, inflorescence; f, petal; g, ovary and stamen. **Micranthes melanocentra**: h, flowering plant; i, flower; j, petal; k, ovary and stamen.

Saxifragaceae

Ecology: *Quercus* forest.

Flowering: March–May.

This forma is apparently undercollected in Nepal.

Bergenia ciliata forma *ligulata* (Wall.) Yeo, Kew Bull. 20: 134 (1966).
Saxifraga ligulata Wall., Asiat. Res. 13: 398, pl. s.n.[11] (1820).

Leaves glabrous on both surfaces.

Distribution: Nepal, W Himalaya, E Himalaya, Tibetan Plateau, Assam-Burma and SW Asia.

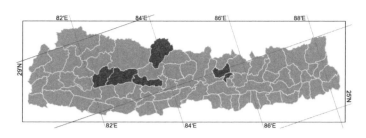

Altitudinal range: 1300–3200 m.

Ecology: *Quercus* and *Pinus* forests.

Flowering: March–May.

6. *Saxifraga* L., Sp. Pl. 1: 398 (1753).

Shinobu Akiyama & Richard J. Gornall

Perennial herbs. Stems caespitose or simple, solitary or grouped into tufts, mats or cushions. Stolons sometimes present. Stipules absent. Leaves usually alternate, rarely opposite, in basal rosettes and/or on stems, petiolate or not, blade simple, margin entire, lobed or dentate, variously pubescent with glandular or eglandular hairs, chalk glands sometimes present. Flowering stems variously pubescent with glandular or eglandular hairs, usually leafy. Inflorescence a solitary flower or a few- to many-flowered bracteate cyme. Flowers usually bisexual, sometimes unisexual and plants then dioecious, actinomorphic, rarely zygomorphic. Hypanthium cup-shaped to saucer-shaped. Sepals 5, free, sometimes with chalk glands. Petals 5, free, often yellow or white, sometimes pink or red to purple, margin usually entire, callose or not. Stamens 8 or 10, free, filaments subulate to linear. Nectary sometimes well developed as an annular or semiannular disk. Ovary superior to inferior, 2-carpellate, carpels fused at base or along the entire placental region, 2-locular, ovules many, placentation axile, integuments 2. Styles 2, free, stigma capitate. Fruit a 2-valved dehiscent capsule. Seeds many, small.

Worldwide about 460 species, in temperate regions of Asia, Europe, N America and S America (Andes). 87 species (21 endemic) in Nepal.

Few collections are made of fruiting material, so it is not possible to give fruiting times for most species. The following taxa have been recorded from Nepal but are probably not present. *Saxifraga cernua* L. was recorded by Baehni (Candollea 16: 215. 1958) based on *Zimmerman 1368, 1506* (G). We have not seen these specimens but given the localities (Beding and Menlungtse) we suspect a misidentification or misapplication of the name. *Saxifraga flagellaris* var. *stenosepala* (Royle) Hultén was recorded by Kitamura (Fauna Fl. Nepal Himalaya: 144. 1955) based on a collection by Nakao, 30 June 1953 (KYO), which we have not seen. The name, however, is a synonym of *S. setigera* Pursh, an American species, and we strongly suspect a misidentification. *Saxifraga duthiei* Gand. was originally collected from W Himalaya (Karakorum) and although this species has been recorded from Nepal (Bull. Dept. Med. Pl. Nepal 7: 83. 1976), we have seen no specimens. *Saxifraga ramulosa* Wall. ex Ser. was published based on a specimen from 'Nepal Bhuddringth, Kamroop', but the locality was actually Surinagar (Garhwal) and not Nepal (Hara, Enum. Fl. Pl. Nepal 2: 154. 1979), and it is yet to be found in Nepal.

Key to Sections

1a Leaves with sunken submarginal or subapical chalk glands on upper surface, bearing calcareous deposits ....**3. *S.* sect. *Porphyrion*** (p 255)
 b Leaves without chalk glands 2

2a Leaves usually entire. Petals usually yellow, occasionally white or pink to red or purple .... **1. *S.* sect. *Ciliatae*** (p 223)
 b Leaves usually conspicuously toothed or lobed. Petals usually white or pink to red or purple, rarely yellow..................3

3a Leaves with petiole at least 2× length of lamina. Ovary superior ....**2. *S.* sect. *Mesogyne*** (p 254)
 b Leaves with petiole equalling or shorter than lamina. Ovary inferior ....**4. *S.* sect. *Saxifraga*** (p 264)

1. *Saxifraga* sect. *Ciliatae* Haw. Misc. Nat.: 160(1803).

Plants caespitose, many branched, forming mats or cushions, or with stems erect. Rhizome sometimes with leafy buds or stoloniferous. Leaves alternate (except in *S. contraria* Harry Sm.), without chalk glands. Petals usually yellow, rarely white or pink to red or purple, often with calloses. Stamens 10. Ovary superior to semi-inferior.

Key to Species

1a	Stem nodes glabrous or only with straight, glandular hairs, glands brown or black	2
b	At least lower stem nodes and petiole bases with brown, crisped, villous hairs (mostly eglandular)	36
2a	Basal leaf axils producing filiform (not leafy) stolons	3
b	Basal leaf axils rarely stoloniferous, but if so then stolons leafy, not filiform	12
3a	Petals shorter than or only slightly exceeding sepals	4
b	Petals at least 1.5× as long as sepals	7
4a	Stolons arising only from axils of basal leaves	5
b	Stolons arising from axils of median leaves	6
5a	Upper surface of basal leaves glabrous	**52. *S. consanguinea***
b	Upper surface of basal leaves densely glandular-pubescent	**53. *S. pilifera***
6a	Flowers bisexual. Petals 3–6 mm	**50. *S. tentaculata***
b	Flowers dioecious. Petals 2.5–3.5 mm	**51. *S. neopropagulifera***
7a	Pedicels at least 3× as long as cauline leaves. Cyme lax, many-flowered	**49. *S. brunonis***
b	Pedicels mostly less than 2× as long as cauline leaves. Cyme corymbose, compact or flower solitary	8
8a	Stolons arising from axils of median leaves	9
b	Stolons arising only from axils of basal leaves	10
9a	Flowers bisexual. Petals 3–6 mm long	**50. *S. tentaculata***
b	Flowers dioecious. Petals 2.5–3.5 mm long	**51. *S. neopropagulifera***
10a	Margin of basal leaves slender-ciliate, often glandular, longest hairs less than 0.5 mm	**46. *S. stenophylla***
b	Margin of basal leaves coarsely eglandular setose-ciliate, longest bristles 0.5–1 mm	11
11a	Inflorescence of 1–3 flowers, not obviously umbellate. Stem leaves shorter than or equalling the basal leaves	**47. *S. mucronulata***
b	Inflorescence of more than 3 flowers, umbellate. Stem leaves longer than or equalling the basal leaves	**48. *S. mucronulatoides***
12a	Leaves often shiny, leathery. Leafy buds produced in axils of cauline leaves, or rhizome scales, sometimes developing into short, sterile shoots. Petals white or yellow	13
b	Leaves not shiny, carnose. Long sterile shoots sometimes arising from axils of basal leaves. Petals yellow	21
13a	Leaf margin coarsely toothed or lobed	14
b	Leaf margin entire	16
14a	Leafy buds conspicuous in axils of upper cauline leaves or bracts. Sepals reflexed in fruit. Petals white	**40. *S. strigosa***
b	Leafy buds conspicuous or inconspicuous in axils of lower and basal leaves. Sepals erect or spreading in fruit. Petals yellow	15
15a	Median stem leaves distributed evenly, each usually with 3–5 apical lobes	**38. *S. hispidula***
b	Median leaves aggregated, often into a rosette, each with several teeth	**39. *S. substrigosa***
16a	Upper surface of leaves pubescent	17
b	Upper surface of leaves glabrous	18

| 17a | Upper surface of leaves strigose. Petals clawed | **38. *S. hispidula*** |
| b | Upper surface of leaves sparingly pubescent. Petals tapered at base | **45. *S. jaljalensis*** |

| 18a | Stem often prostrate, many-branched from the middle | **41. *S. filicaulis*** |
| b | Stem usually erect, simple or branched only at base | 19 |

| 19a | Broadest leaves at least 3 mm wide. Petals 2-callose | **43. *S. wallichiana*** |
| b | Broadest leaves to 3 mm wide. Petals not callose | 20 |

| 20a | Stem branched at base. At least some leaves recurved | **42. *S. serrula*** |
| b | Stem simple. Leaves straight | **44. *S. brachypoda*** |

| 21a | Plants forming well-defined basal leaf rosette at anthesis. Inflorescence usually several-flowered | 22 |
| b | Plants without a well-defined basal leaf rosette at anthesis (stem with axillary shoots forming mats or cushions, or simple). Inflorescence 1(or 2–5)-flowered | 23 |

| 22a | Margin of basal leaves setose-ciliate. Inflorescence (1 or)2- or 3-flowered. Petals spotted | **64. *S. punctulata*** |
| b | Margin of basal leaves entire. Inflorescence 2–10-flowered. Petals without spots | **65. *S. umbellulata*** |

| 23a | Margin of lower leaves glabrous | 24 |
| b | Margin of lower leaves glandular- or eglandular-ciliate | 26 |

| 24a | Leaves opposite, occasionally alternate on young shoots | **61. *S. contraria*** |
| b | Leaves alternate | 25 |

| 25a | Petals yellow, elliptic ca. 4 × 1.5 mm | **60. *S. engleriana*** |
| b | Petals green, obovate, ca. 2.3 × 1.1 mm | **63. *S. microphylla*** |

| 26a | Margin of leaves glandular-ciliate | 27 |
| b | Margin of leaves eglandular-ciliate | 29 |

| 27a | Flowering stem submerged in foliage, even in fruit | **57. *S. jacquemontiana*** |
| b | Flowering stem visible, flowers clearly overtopping foliage | 28 |

| 28a | At least some leaves with short, glandular hairs. Sepals reflexed | **58. *S. stella-aurea*** |
| b | Lower leaves with long, curly, glandular and eglandular hairs at least at margin. Sepals erect | **59. *S. llonakhensis*** |

| 29a | Nectary a conspicuous disk surrounding ovary | 30 |
| b | Nectary tissue obscure | 32 |

| 30a | Leaves opposite, occasionally some alternate on young shoots | **61. *S. contraria*** |
| b | Leaves alternate | 31 |

| 31a | Petals yellow, elliptic, ca. 4 × 1.5 mm | **60. *S. engleriana*** |
| b | Petals green, obovate, ca. 2.3 × 1.1 mm | **63. *S. microphylla*** |

| 32a | Petals white, unspotted | **54. *S. williamsii*** |
| b | Petals yellow, often spotted orange | 33 |

| 33a | Sepals spreading to reflexed | 34 |
| b | Sepals erect | 35 |

| 34a | Sepals to 3 × 1.5 mm | **58. *S. stella-aurea*** |
| b | Sepals to 1.5 × 1 mm | **62. *S. nanella*** |

| 35a | Margin of leaves setose-ciliate. Pedicels to 10 mm | **55. *S. perpusilla*** |
| b | Apical margin of leaves fimbriate. Flowers sessile | **56. *S. hemisphaerica*** |

36a Pedicels with brown, crisped, villous hairs only 37
b Pedicels with black-tipped glandular hairs, occasionally with brown, crisped villous hairs or glabrous 48

37a Basal leaves absent at anthesis...**12. S. sikkimensis**
b Basal leaves present at anthesis 38

38a Upper surface of basal leaves brown crisped villous... 39
b Upper surface of basal leaves glabrous.. 41

39a Sepals pubescent on outer surface...**37. S. sinomontana**
b Sepals glabrous on outer surface 40

40a Petals 6.5–7.5 mm, longer than sepals, yellow..**13. S. lamninamensis**
b Petals 3.5–4 mm, not or scarcely exceeding sepals, greenish.............................**33. S. hirculoides**

41a Basal leaves without a well-defined petiole, blade linear-oblong, to 1 mm wide................................. 42
b Basal leaves clearly petiolate, blade narrowly elliptic to ovate or lanceolate, more than 1 mm wide 43

42a Leaves broadened at base, clasping stem ....**29. S. matta-viridis**
b Leaves tapered at base, not clasping stem .... **30. S. saginoides**

43a Ovary with a conspicuous nectary disk..**32. S. tangutica**
b Ovary without a conspicuous nectary disk... 44

44a Margin of sepals glabrous or glandular-pubescent 45
b Margin of sepals brown crisped villous 46

45a Margin of sepals densely glandular-pubescent. Petals to 6.4 mm ....**31. S. parva**
b Margin of sepals glabrous or sparsely glandular-pubescent. Petals more than 6 mm..............**34. S. elliptica**

46a Lower outer surface and lower margin of petals brown crisped villous ....**35. S. montanella**
b Petals glabrous.. 47

47a Petals not callose ....**36. S. namdoensis**
b Petals callose ....**37. S. sinomontana**

48a Leaves often glaucous, with prominent submarginal vein running from proximal to distal ends. Basal or proximal
leaves glandular-pubescent or glabrous below, ± glabrous above.. 49
b Leaves not glaucous, without prominent submarginal vein running from proximal to distal ends. Leaf pubescence
variable 62

49a Basal leaves persistent, mostly present at flowering time 50
b Basal leaves caducous, mostly absent at flowering time.. 55

50a Basal leaf blade oblanceolate, attenuate at base ....**7. S. rolwalingensis**
b Basal leaf blade ovate to broadly ovate, cuneate to rounded or cordate at base 51

51a Lower median cauline leaves petiolate ....**5. S. hookeri**
b Lower median cauline leaves sessile 52

52a Cauline leaves rounded at base, not amplexicaul ....**1. S. diversifolia**
b At least some cauline leaves cordate at base, amplexicaul 53

53a Median cauline leaves 2–3.5 cm. Stems densely brown crisped villous proximally............**2. S. parnassifolia**
b Median cauline leaves 0.8–2 cm. Stems glabrous proximally 54

54a Median cauline leaves glabrous...**3. S. sphaeradena**
b Median cauline leaves pubescent ....**4. S. dhwojii**

1. _Saxifraga diversifolia_ Wall. ex Ser., in DC., Prodr. 4: 44 (1830).

Stems caespitose or simple, erect, sometimes forming clumps. Stolons absent. Leaves alternate. Basal leaves petiolate. Petiole 2.5–10(–13) cm, brown crisped eglandular or glandular-villous. Blade ovate, 1.5–7 × 1–4.6 cm, base rounded to cordate, apex acute, margin brown crisped villous, both surfaces glabrous or brown crisped villous. Median cauline leaves sessile, ovate to narrowly ovate, 1–6.3 × 0.4–4 cm, base rounded, apex acute, margin brown crisped eglandular-villous at base, both surfaces glabrous or pubescent. Upper cauline leaves smaller, margin shortly brown glandular-pubescent, both surfaces glabrous. Flowering stems 10–45 cm, basal parts brown crisped eglandular or glandular-villous or glabrous, most densely shortly brown glandular-pubescent above, bulbils absent. Flowers 3–10 (or more), in a corymbose cyme, bisexual. Pedicels 6–15 mm, shortly brown glandular-pubescent. Sepals reflexed, elliptic to ovate, 3–6 × 1.3–4 mm, outer surface and margin shortly brown glandular-pubescent or glabrous, veins 3–5, free or partly confluent. Petals yellow, obovate, 5–8 × 2–6 mm, base narrowed, apex obtuse, not callose. Stamens 4–5.5 mm. Nectary band obscure. Ovary almost superior, carpels fused for more than half their length, tapered into conical, 1–1.5 mm styles.

Distribution: Nepal, W Himalaya, E Himalaya and Tibetan Plateau.

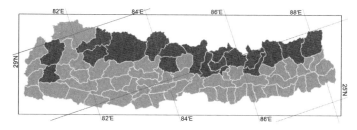

Altitudinal range: 2400–4800 m.

Ecology: Mid hills to alpine.

Flowering: July–September.

2. _Saxifraga parnassifolia_ D.Don, Trans. Linn. Soc. London 13(2): 405 (1821).
Saxifraga diversifolia var. _parnassifolia_ (D.Don) Ser.

Stems caespitose or simple, erect, often forming clumps. Stolons absent. Leaves alternate. Basal leaves petiolate. Petiole 1.5–8 cm, brown crisped eglandular- or glandular-villous. Blade ovate, 2–6.5 × 1.5–4.5 cm, base rounded to cordate, apex acute to obtuse, both surfaces and margin brown crisped villous. Median cauline leaves sessile, ovate, (1–)2–3.5 × (0.8–)1.5–2.5 cm, base cordate, amplexicaul, apex acute to obtuse, both surfaces and margin brown glandular-pubescent, or both surfaces glabrous or nearly so. Upper cauline leaves smaller, both surfaces and margin shortly brown glandular-pubescent or lower surface glabrous. Flowering stems (8–)15–50 cm, brown crisped eglandular- or glandular-villous at base, brown crisped glandular-villous or glabrous in median part, densely short brown glandular-pubescent above, bulbils absent. Flowers 3–10 (or more) in a corymbose cyme, bisexual. Pedicels 5–30 mm, shortly brown glandular-pubescent. Sepals erect to spreading, narrowly ovate, 3–4 × 1.5–2 mm, apex obtuse to subacute, outer surface and margin shortly brown glandular-pubescent, veins 3, confluent near apex. Petals yellow, obovate, 6–8 × 3.5–4 mm, base contracted into a claw ca. 1 mm, apex obtuse, 4-callose. Stamens 2–3 mm. Nectary band obscure. Ovary semi-inferior, carpels fused for about half their length, contracted into slender, ca. 3 mm styles.

Distribution: Nepal, W Himalaya, E Himalaya and Tibetan Plateau.

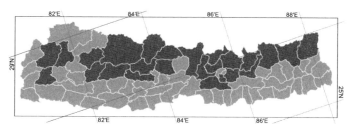

Altitudinal range: 1400–4900 m.

Ecology: Mid hills to alpine.

Flowering: July–September. **Fruiting:** September.

3. _Saxifraga sphaeradena_ Harry Sm., Bull. Brit. Mus. (Nat. Hist.), Bot. 2: 235, pl. 15B fig. 4a (1960).

Stems caespitose or simple, erect, often forming clumps. Stolons absent. Leaves alternate. Basal leaves petiolate. Petiole 1–7 cm, glabrous or brown crisped villous. Blade ovate, 10–20(–30) × 6–15(–20) mm, base rounded to cordate, apex acute, both surfaces glabrous. Cauline leaves sessile, cordate to ovate, 8–20(–30) × 5–15(–20) mm, base cordate, amplexicaul, apex acute, margin brown crisped villous at base or glabrous, both surfaces glabrous. Uppermost cauline leaves margin shortly blackish glandular-pubescent. Flowering stems 8–20(–30) cm, lower parts glabrous, shortly blackish glandular-pubescent above, bulbils absent. Flowers solitary or to 3(–5) in a cyme, dioecious. Pedicels 5–30 mm, shortly blackish glandular-pubescent. Sepals spreading, ovate, ca. 4 ×

2 mm, apex obtuse, margin blackish glandular-pubescent, both surfaces glabrous, veins 3–6, confluent. Petals yellow, obovate, ca. 7 × 5 mm, base without a claw, apex obtuse, not callose. Stamens ca. 7 mm in male flowers. Nectary band obscure. Ovary (female flowers only) almost superior, carpels fused for more than half their length, tapered into ca. 1.5 mm styles.

Distribution: Nepal, E Himalaya and Tibetan Plateau.

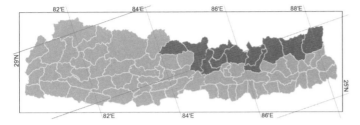

Altitudinal range: 3300–4800 m.

Ecology: Mid hills to alpine, in masses on rocks, on stony hill slopes.

Flowering: July–September.

4. *Saxifraga dhwojii* (Harry Sm.) S.Akiyama & H.Ohba, Bull. Natl. Sci. Mus., Tokyo, B. 33(3/4): 97 (2007).
Saxifraga sphaeradena subsp. *dhwojii* Harry Sm., Bull. Brit. Mus. (Nat. Hist.), Bot. 2: 236, pl. 4 fig. e (1960).

Stems caespitose or simple, erect, often forming clumps. Stolons absent. Leaves alternate. Basal leaves petiolate. Petiole ca. 10 mm, glabrous or brown crisped villous at least near the base. Blade ovate, 8–10 × 5–7 mm, base rounded to cordate, apex acute, adaxial surface eglandular-pubescent. Cauline leaves sessile, cordate to ovate, 5–12 × 3–6 mm, base cordate, amplexicaul, apex acute, margin brown crisped villous at base or glabrous, upper surface eglandular-pubescent or glabrous. Uppermost cauline leaves margin shortly blackish glandular-pubescent. Flowering stems 5–12 cm, lower parts glabrous or sparingly eglandular-brown-pubescent, shortly black glandular-pubescent above, bulbils absent. Flowers solitary. Pedicels 12–35 mm, shortly blackish glandular-pubescent. Sepals spreading, ovate, 3–3.5 × 1.5–2 mm, apex obtuse, margin blackish glandular-pubescent, adaxial surface black glandular-pubescent proximally, veins 3–6, confluent. Petals yellow, ovate to obovate, 4–7 × 4–5 mm, base with a claw 1–1.5 mm, apex rounded, 2–4 callose. Stamens 3–4 mm. Nectary band obscure. Ovary semi-inferior, carpels fused for more than half their length, tapered into ca. 1.5 mm styles.

Distribution: Nepal, E Himalaya and Tibetan Plateau.

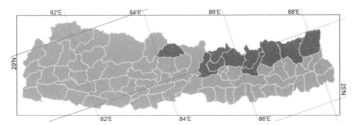

Altitudinal range: 3600–4500 m.

Ecology: Alpine.

Flowering: September.

An imperfectly known species that appears also to be closely related to *Saxifraga parnassiifolia* D.Don.

5. *Saxifraga hookeri* Engl. & Irmsch., Bot. Jahrb. Syst. 48: 582 (1912).
Saxifraga corymbosa Hook.f. & Thomson.

Stems caespitose or simple, erect, forming clumps. Stolons absent. Leaves alternate. Basal and lower cauline leaves sometimes absent at anthesis, petiolate. Petiole to 4 cm, brown crisped villous. Blade elliptic to ovate, 15–20 × 8–10 mm, base cuneate to rounded, apex acute, margin brown crisped villous, both surfaces pubescent. Median cauline leaves shortly petiolate to sessile, smaller than lower leaves, blade ovate to narrowly ovate, base brown crisped villous, apex acute, both surfaces pubescent or sometimes glabrous above. Upper cauline leaves sessile, lanceolate to narrowly elliptic, upper surface and margin short blackish glandular-pubescent. Flowering stems 12–30(–45) cm, brown crisped villous, mostly shortly blackish glandular-pubescent in upper parts, sometimes glabrous, bulbils absent. Flowers (1–)3–12 in a corymbose cyme, bisexual. Pedicels 3–20 mm, shortly blackish glandular-pubescent. Sepals spreading, lanceolate to ovate, 3–4 × 1–2 mm, apex acute, outer surface and sometimes margin short blackish glandular-pubescent, veins 3, free. Petals yellow, minutely spotted orange, obovate, 6–8 × 3–4 mm, base contracted to a claw, apex obtuse, 3–4-callose. Stamens 3–5 mm. Nectary band obscure. Ovary almost superior, carpels fused for more than half their length, tapered into conical ca. 1 mm styles.

Distribution: Nepal, E Himalaya and Tibetan Plateau.

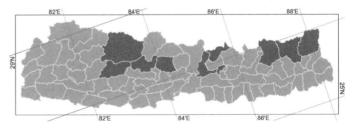

Altitudinal range: 3000–4600 m.

Ecology: Alpine, open slopes, stony ground.

Flowering: July–September.

6. *Saxifraga latiflora* Hook.f. & Thomson, J. Proc. Linn. Soc., Bot. 2: 71 (1857).

Stems caespitose or simple, erect, forming clumps. Stolons absent. Leaves alternate. Basal leaves often absent at flowering. Lower cauline leaves petiolate Petiole 1–4 cm, brown crisped villous, sheathing in basal leaves. Blade lanceolate, 2–4 × 0.6–1.2 cm, base attenuate, apex acute, margin brown crisped glandular-villous, texture herbaceous, both surfaces brown crisped glandular-villous. Middle and upper cauline leaves sessile, similar to but smaller than the lower leaves. Flowering stems 6–15 cm, glabrous at base, mostly brown crisped glandular-villous above, bulbils absent. Flowers usually solitary, rarely 2–4, bisexual. Pedicels 3–10 mm, brown crisped glandular-villous. Sepals erect, ovate, ca. 7 × 4 mm, apex obtuse, outer surface and margin brown glandular-ciliate, veins 3–5, confluent near apex. Petals yellow, ovate, 10–15 × 6–10 mm, base with a short claw, apex subacute to obtuse, many-callose. Stamens 3.5–5 mm. Nectary band obscure. Ovary almost superior, carpels fused for more than half their length, narrowed into ca. 1 mm styles.

Distribution: Nepal and E Himalaya.

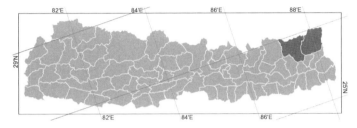

Altitudinal range: 4100–4600 m.

Ecology: Alpine.

Flowering: July–August.

7. *Saxifraga rolwalingensis* H.Ohba, J. Jap. Bot. 59(12): 360 (1984).

Stems caespitose or simple, erect, forming clumps. Stolons absent. Leaves alternate. Basal leaves somewhat persistent. Lower cauline leaves petiolate or almost sessile. Blade oblanceolate, 3.5–6 × 0.7–0.9 cm, base attenuate, apex acute, lower surface brown crisped villous, upper surface glabrous. Median and upper cauline leaves sessile, ovate to lanceolate, smaller than lower leaves, base rounded, truncate to cordate, lower surface brown crisped villous or glabrous, upper surface glabrous. Flowering stems (8–)14–25(–35) cm, brown crisped villous towards base, glabrous in median part, glandular-pubescent above, bulbils absent. Flowers (1–)3–5, in a cyme, bisexual. Pedicels to 2 cm, long glandular-pubescent. Sepals spreading, ovate to broadly oblong, 4–5 × 2–3.5(–4) mm, apex obtuse, margin short blackish glandular-pubescent, veins 3–5, confluent. Petals yellow, spotted orange, narrowly obovate, 7.5–8.5 × 5–6.2 mm, base attenuate, apex rounded, not callose. Stamens ca. 3 mm. Nectary band obscure. Ovary almost superior, carpels fused for more than half their length, narrowed into ca. 1 mm styles.

Distribution: Endemic to Nepal.

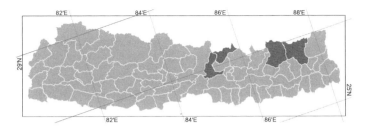

Altitudinal range: 4000–4600 m.

Ecology: Open, stony slopes.

Flowering: September.

8. *Saxifraga moorcroftiana* (Ser.) Wall. ex Sternb., Revis. Saxifrag. Suppl.2: 28, pl. 24 (1831).
Saxifraga diversifolia var. *moorcroftiana* Ser., Prodr. 4: 44 (1830); *Hirculus moorcroftianus* (Ser.) Losinsk.; *Saxifraga lysimachioides* Klotzsch.

Stems caespitose or simple, erect, forming clumps. Stolons absent. Leaves alternate. Basal leaves usually absent at flowering time. Cauline leaves usually sessile, reducing upwards, the basal-most ones caducous. Lower cauline leaves oblanceolate to oblong, 2–6 × 0.6–1.5 cm, base cuneate, apex acute, lower surface pubescent, upper surface and margin glabrous. Median cauline leaves lanceolate, base rounded to cordate or amplexicaul, apex acute, glabrous, uppermost leaves marginally sparsely glandular-pubescent. Flowering stems 12–50 cm, brown crisped villous towards base, nearly glabrous in median part, brown glandular-pubescent above, bulbils absent. Flowers (1 or)2–6(–10) in a corymbose cyme, bisexual. Pedicel 7–25 mm, glandular-pubescent. Sepals erect to spreading, ovate, ca. 4 × 3 mm, apex rounded to obtuse, outer surface glabrous, margin shortly black glandular-pubescent, veins 3–5, median 3 confluent near apex. Petals yellow, obovate, 7–10 × 3.5–6 mm, base without a claw, apex rounded, 2-callose. Stamens 5–6 mm. Nectary band obscure. Ovary almost superior, carpels fused for more than half their length, narrowed into ca. 1 mm styles.

Distribution: Nepal, W Himalaya, E Himalaya, Tibetan Plateau and E Asia.

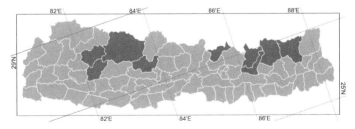

Altitudinal range: 3500–4600(–4900) m.

Ecology: Alpine, open slopes.

Flowering: July–September.

9. *Saxifraga kingiana* Engl. & Irmsch., Bot. Jahrb. Syst. 48: 610 (1912).
Saxifraga gageana Engl. & Irmsch. later homonym, non W.W.Sm.

Stems caespitose or simple, erect, forming clumps. Stolons absent. Leaves alternate. Basal leaves absent at flowering time. Cauline leaves sessile, lanceolate to ovate, to 3.5–5.5 × 1.8–3.2 cm, base cordate, amplexicaul, apex acute to acuminate, margin entire, both surfaces and margin villous (glandular in upper leaves). Basal and upper leaves smaller than the median, most lower leaves caducous. Flowering stems to 70 cm, densely pale brown crisped villous towards base, pale brown glandular-villous above, bulbils absent. Flowers numerous in a pryamidal cyme, terminal and in axils of upper leaves, bisexual. Pedicels 8–20 mm, pale brown glandular-villous. Sepals spreading, ovate, 5–6.5 × 3–4.5 mm, apex acute, pale brown slightly glandular crisped villous, on outer surface, apex and margin, veins 3, confluent near apex. Petals yellow, obovate to suborbicular, 7–10 × 4–8 mm, base with a short claw, apex obtuse, very finely several-callose. Stamens 4–5 mm. Nectary band obscure. Ovary almost superior, carpels fused for more than half their length, narrowed into conical ca. 1.5 mm styles.

Distribution: Nepal, E Himalaya and Tibetan Plateau.

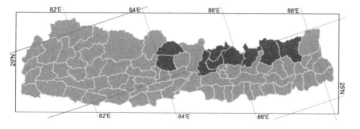

Altitudinal range: 2900–4400 m.

Ecology: Alpine, rock slopes, scree slopes, stony places.

Flowering: July–August(–September).

10. *Saxifraga kingdonii* C.Marquand, J. Linn. Soc., Bot. 48: 179 (1929).
Saxifraga nayari Wadhwa.

Stems caespitose or simple, erect, forming clumps. Stolons absent. Leaves alternate. Basal leaves absent at flowering time. Basal leaves mostly caducous, with petiole 3–5 mm, blade elliptic to broadly ovate, 5–12 × 4–5 mm, both surfaces and margin crisped-rufous hairy. Cauline leaves sessile, elliptic, 4.5 × 2.5 mm, reducing in size upwards, apex obtuse, both surfaces eglandular-pilose, sometimes densely so, margin pilose, sometimes sparingly. Flowering stems ca. 25 cm, rufous-pubescent, proximally eglandular, distally glandular. Flower solitary, or cyme 2-flowered. Pedicels glandular-villous. Sepals erect to spreading, oblong to broadly ovate, 6.5–10 × 2.5–7 mm, abaxially and marginally glandular- or eglandular-villous; veins 3–5, partly confluent at apex. Petals yellow to orange, elliptic, (6–)8–12 × 5–9 mm, base truncate to a short claw, several-callose, 5–7-veined. Stamens

ca. 6 mm. Ovary subsuperior, carpels fused for more than half their length, encircled at base by a conspicuous nectary band, tapered into conical styles.

Distribution: Nepal, Tibetan Plateau and Assam-Burma.

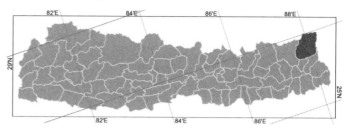

Altitudinal range: ca. 4000 m.

Ecology: Mountain slopes.

Flowering: September.

Known in Nepal from only a single collection (*Binns et al. BMW126*, Yalung. KATH). Specimens that are less pilose and with leaves and petals less broadly elliptic (as here) were distinguished as *Saxifraga nayarii* Wadhwa, but the distinction is not convincing.

11. *Saxifraga excellens* Harry Sm., Bull. Brit. Mus. (Nat. Hist.), Bot. 2: 230, pl. 13B fig. 1t (1960).

Stems caespitose or simple, erect, sometimes forming clumps. Stolons absent. Leaves alternate. Basal leaves absent or sometimes present, similar to lower cauline leaves but smaller. Lower cauline leaves petiolate. Petiole 1–7 cm, long brown crisped pubescent, sometimes winged and amplexicaul. Blade obovate-elliptic, 7–9.5 × 4–5.3 cm, margin and lower surface long brown crisped pubescent, upper surface sparsely pubescent, Median and upper cauline leaves sessile, broadly elliptic, 7–13 × 4–6.5 cm, base rounded to amplexicaul, apex obtuse, margin and lower surface long brown crisped pubescent, upper surface sparsely pubescent. Flowering stems 15–40 cm, basal parts glabrous, brown crisped eglandular-villous above, bulbils absent. Flowers 2–10 in a cyme, bisexual. Pedicel 3–20 mm, glandular- or eglandular-pubescent. Sepals erect, broadly ovate, ca. 4.5 × 3.5 mm, apex obtuse, very minutely laciniate, margin glandular-pubescent or glabrescent at base, outer surface glabrous, veins 5, median 3 confluent. Petals reddish purple, obovate, 9–12 × 6–10 mm, base gradually narrowed, apex rounded, not callose. Stamens ca. 6 mm. Ovary almost superior, carpels fused for more than half their length, tapered into conical 1–2 mm styles.

Distribution: Endemic to Nepal.

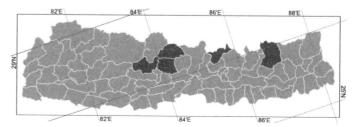

Saxifragaceae

Altitudinal range: 3600–4700 m.

Ecology: In *Abies* forest to alpine habitats, under overhanging boulders, on mossy rocks.

Flowering: August–September.

Although endemic it is known from at least eight localities.

12. *Saxifraga sikkimensis* Engl., Bot. Jahrb. Syst. 48: 573 (1912).
Saxifraga bertholdii Baehni.

Stems caespitose, simple, erect, forming clumps. Stolons absent. Leaves alternate. Basal leaves absent at flowering time. Lower leaves petiolate, upper leaves sessile. Petiole 3–12(–15) mm, brown crisped villous. Blade lanceolate to ovate, 6–14 × 2–5 mm, base rounded to cuneate, attenuate into petiole, apex acute, margin entire, sometimes with brown hairs, texture herbaceous, upper surface glabrous or pubescent, lower surface usually glabrous or rarely sparsely pubescent. Flowering stems 6–20 cm, brown crisped villous, leaves 6–12, bulbils absent. Flowers usually solitary, rarely 2, bisexual. Pedicels 5–20 mm, brown crisped villous, sometimes sparingly so. Sepals spreading, oblong-elliptic, 3–4 × 2–2.5 mm, apex rounded, margin and adaxial surface glabrous, veins 3, free. Petals yellow, obovate, 6–10 × 3–7 mm, base with a short claw, apex obtuse to slightly retuse, finely 2-callose or not. Stamens 3–5 mm. Nectary band obscure. Ovary almost superior, carpels fused for more than half their length, narrowed into ca. 1 mm styles.

Distribution: Nepal, E Himalaya and Assam-Burma.

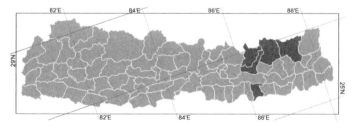

Altitudinal range: 3200–4600(–5200) m.

Ecology: Alpine, shrubby banks, sandy slopes.

Flowering: August–October.

13. *Saxifraga lamninamensis* H.Ohba, J. Jap. Bot. 59: 362 (1984).

Stems caespitose or simple, erect, often producing stoloniferous leafy shoots from axillary buds on rhizome. Leaves alternate. Basal and lower cauline leaves petiolate. Petiole 10–20 mm, brown crisped villous. Blade lanceolate, 15–20 × 5–10 mm, base attenuate, apex acute, margin brown crisped villous, both surfaces pubescent. Median and upper cauline leaves sessile, similar to but smaller than lower leaves, margin brown crisped villous at base, both surfaces

glabrous to sparsely pubescent. Flowering stems 10–25 cm, sparsely crisped brown eglandular villous, bulbils absent. Flowers (1–)4–5(–10) in a cyme, bisexual. Pedicels ca. 4 cm, glandular-pubescent. Sepals erect to spreading, linear-oblong to oblong, 4–4.5 × 1.8–2 mm, apex obtuse, glabrous, veins 3, confluent. Petals yellow, obovate-oblong to narrowly oblong, 6.5–7.5 × 3.5–4 mm, base attenuate, apex rounded, not callose. Stamens 4–5 mm. Ovary almost superior, carpels fused for more than half their length, tapered into ca. 1mm styles.

Distribution: Nepal and E Himalaya.

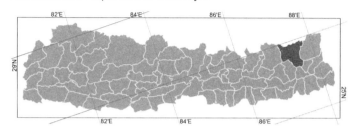

Altitudinal range: 3200–4200 m.

Ecology: Alpine.

Flowering: August.

14. *Saxifraga amabilis* H.Ohba & Wakab., J. Jap. Bot. 62: 162 (1987).

Stems caespitose, many-branched, forming mats. Stolons absent. Leaves alternate. Basal leaves absent at time of flowering. Cauline leaves sessile, oblanceolate to narrowly ovate, 4–5 × 1.5–2.2(–2.5) mm, base cuneate, apex aristate in lower leaves, obtuse to acute with or without glandular hairs in median and upper leaves, margin hispid in lower leaves, glandular-pubescent in upper leaves, both surfaces glabrous. Flowering stems 3–5 cm, glabrous at base, blackish-glandular-pubescent above, bulbils absent. Flower solitary, bisexual. Pedicels glandular-pubescent. Sepals erect or ascending, oblong to oblong-spatulate, 4–5 × 2–2.5 mm, apex apiculate to acute with or without a glandular hair, outer surface and margin blackish glandular-pubescent, veins usually 3, confluent. Petals yellow, spotted orange at base, obovate, 5–6.5 × 3–4 mm, base narrowed into a claw, apex rounded. Stamens 3–3.5 mm. Nectary band obscure. Ovary superior, carpels fused for more than half their length, tapered into slender ca. 1 mm styles.

Distribution: Endemic to Nepal.

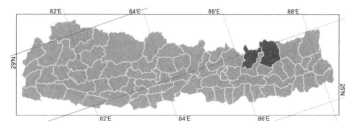

Saxifragaceae

Altitudinal range: 4000–4600 m.

Ecology: Alpine.

Flowering: July–August(–September).

15. *Saxifraga lepida* Harry Sm., Bull. Brit. Mus. (Nat. Hist.), Bot. 2: 239, pl. 17A (1960).

Stems caespitose or simple, erect, forming clumps. Stolons absent. Leaves alternate. Basal leaves petiolate. Petiole to 15 mm, glabrous or sparsely pubescent. Blade lanceolate-linear to lanceolate, 5–15 × 0.6–2 mm, base attenuate, apex obtuse to subacute, both surfaces glabrous. Lower cauline leaves petiolate, median and upper leaves sessile. Cauline leaves linear, 5–13 × ca. 1 mm, smaller above, both surfaces glabrous or pubescent. Flowering stems 3.5–8 cm, glabrous except brown crisped hairs in axils, leaves 3–8, bulbils absent. Flowers solitary, unisexual. Pedicels 15–30 mm, glabrous. Sepals spreading or reflexed at anthesis, elliptic, 2.5–3.5 × 1.5–1.6 mm, apex obtuse, margin glabrous or sparsely brown crisped pubescent, outer surface glabrous, veins 3, confluent. Petals yellow, rotund-obovate to obovate, 6–7.5 × 4–4.2 mm, base with a short claw, apex rounded, not callose. Stamens ca. 3.5 mm. Nectary band obscure. Ovary in female flowers almost superior, carpels fused for more than half their length, tapered into ca. 1 mm styles.

Distribution: Nepal and E Himalaya.

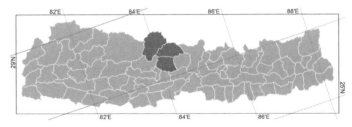

Altitudinal range: 3100–4300 m.

Ecology: Alpine, grassy ravines.

Flowering: August.

16. *Saxifraga mallae* H.Ohba & Wakab., J. Jap. Bot. 62: 165 (1987).

Stems caespitose, simple, erect, sometimes in clumps. Stolons absent. Leaves alternate. Basal leaves petiolate. Petiole 1–2.5 cm, glabrous or very rarely brown crisped villous. Blade linear to linear-lanceolate, 10–25 × 0.6–2 mm, glabrous. Lower cauline leaves petiolate, upper leaves sessile. Leaves decreasing in size upwards, apex obtuse, glabrous or very rarely brown crisped villous at base. Flowering stems 3–8 cm, lower parts glabrous, shortly black glandular-pubescent above, bulbils absent. Flowers solitary, sometimes unisexual. Pedicels shortly blackish glandular-pubescent. Sepals spreading at anthesis, narrowly oblong to broadly lanceolate, 3–4 × 1–2 mm, apex obtuse, glabrous or just margin blackish

glandular-pubescent, veins 3, confluent. Petals yellow, often spotted yellow, obovate, 7–9 × 4–5 mm, base cuneate to attenuate, apex rounded, not callose. Stamens 3–4 mm (male flowers). Nectary band obscure. Ovary superior, carpels fused for more than half their length, tapered into ca. 1 mm styles.

Distribution: Endemic to Nepal.

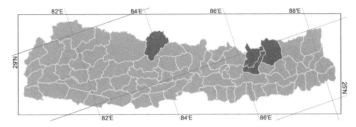

Altitudinal range: 3900–5000 m.

Ecology: Alpine.

Flowering: July–August(–September).

17. *Saxifraga nakaoi* Kitam., in H.Kihara, Fauna Fl. Nepal Himalaya: 144, fig. 25 (1955).
Saxifraga breviglandulosa Harry Sm. nom. nud.

Stems caespitose, branched, forming clumps. Stolons absent. Leaves alternate. Basal leaves petiolate. Petiole 1–2 cm, pale brown crisped pubescent. Blade narrowly ovate to ovate-lanceolate, 3–8 × 1.5–2.5 mm, base attenuate into petiole, apex acute, surfaces glabrous. Lower cauline leaves petiolate, upper leaves sessile, blade similar to basal leaves, lanceolate 6–12 × 1.5–2 mm, somewhat decreasing in size upwards. Flowering stems 5–12 cm, lower parts glabrate except nodes with brown crisped hairs, dark brown glandular-pubescent above, bulbils absent. Flowers solitary, functionally unisexual. Pedicels 5–15 mm, densely black glandular-pubescent. Sepals spreading at anthesis, ovate, 3–3.5 × 1–2 mm, apex acute, proximal margin black glandular-pubescent, outer surface glabrous, veins 3–5, confluent. Petals yellow, broadly obovate, 6.5–8 × 4–5 mm, base cuneate, apex rounded, 2–4 callose. Stamens 3.5–4 mm in male flowers. Nectary band obscure. Ovary superior, carpels fused for more than half their length, abruptly tapered into styles. Styles 1–1.5 mm in female flowers.

Distribution: Nepal, W Himalaya and E Himalaya.

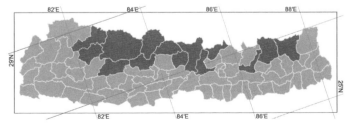

Altitudinal range: 3500–5200 m.

Ecology: Alpine, open grass slopes, open stony cliffs.

Flowering: July–August(–September).

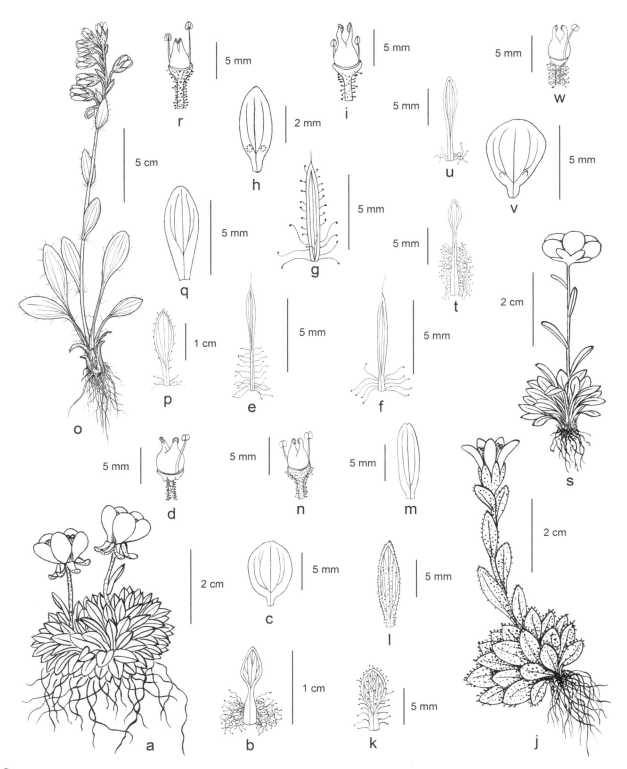

FIG. 28. SAXIFRAGACEAE. **Saxifraga caveana**: a, flowering plant; b, basal leaf; c, petal; d, ovary and stamen. **Saxifraga aristulata**: e, basal leaf; f, lower cauline leaf; g, upper cauline leaf; h, petal; i, ovary and stamens. **Saxifraga lychnitis**: j, flowering plant; k, basal leaf; l, cauline leaf; m, petal; n, ovary and stamens. **Saxifraga nigroglandulifera**: o, flowering plant; p, cauline leaf; q, petal; r, ovary and stamens. **Saxifraga parva**: s, flowering plant; t, basal leaf; u, cauline leaf; v, petal; w, ovary and stamens.

18. *Saxifraga glabricaulis* Harry Sm., Bull. Brit. Mus. (Nat. Hist.), Bot. 2: 241, pl. 18A fig. 7 (1960).
Saxifraga palpebrata var. *parceciliata* Engl. & Irmsch.

Stems caespitose, many-branched, forming cushions. Stolons absent. Leaves alternate. Basal leaves petiolate. Petiole to 10 mm, margin eglandular or glandular hispid. Blade ovate to elliptic, 2–3 × 1–2 mm, base attenuate, apex obtuse to acute, margin eglandular hispid, upper surface eglandular hispid, lower surface glabrous. Lower cauline leaves shortly petiolate, upper leaves sessile. Blade elliptic, 4–5 × 1–2 mm, base cuneate, apex obtuse, margin eglandular hispid, upper surface eglandular hispid. Flowering stems 2–4 cm, glabrous, bulbils absent. Flowers solitary, unisexual. Pedicels 1–5 mm, glabrous. Sepals erect to spreading, ovate, 3–4 × 1–3 mm, apex obtuse, glabrous, veins 3–6, confluent. Petals yellow, elliptic, ca. 8 × 5.5 mm, base narrowed into a claw ca. 1.5 mm, apex obtuse, 2–4 callose. Stamens 2–3.5 mm in male flowers (smaller and sterile in female flowers). Nectary a prominent annular disk. Ovary superior, carpels fused for more than half their length, narrowed into divergent styles. Styles ca. 1.5 mm in female flowers.

Distribution: Nepal, E Himalaya and Tibetan Plateau.

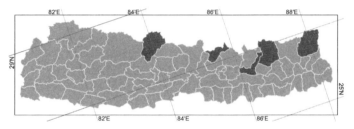

Altitudinal range: 4100–4500 m.

Ecology: Alpine, sandy ground, on rocks.

Flowering: July–August(–September).

19. *Saxifraga zimmermannii* Baehni, Candollea 16: 226 (1958).

Stems caespitose, branched, forming clumps. Stolons absent. Leaves alternate. Basal leaves petiolate. Petiole 4–6 mm, glabrous or sparsely brown crisped villous. Blade ovate to narrowly ovate, 10–20 × 0.6–5 mm, base rounded to tapered, apex obtuse, margin long-ciliate, both surfaces long eglandular villous. Cauline leaves sessile, oblong-lanceolate, apex obtuse, glabrous, brown crisped villous at nodes. Flowering stems 7–12 cm, lower parts glabrate, shortly black glandular-pubescent above, bulbils absent. Flowers solitary or 2 in a cyme. Pedicels 5–13 mm, shortly black glandular-pubescent. Sepals erect at anthesis, elliptic-ovate, ca. 4 × 2.5 mm, apex obtuse, glabrous, veins 3. Petals yellow, obovate, 7.5–8 × 4–4.5 mm, base cuneate, apex obtuse, not callose. Stamens ca. 4 mm. Nectary obscure. Ovary superior, carpels fused for more than half their length, narrowed into ca. 1.5 mm styles.

Distribution: Endemic to Nepal.

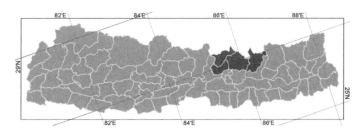

Altitudinal range: (3600–)4000–4500 m.

Ecology: Alpine.

Flowering: August–September.

A narrow endemic known with certainty only from the type collection (*Zimmerman 1436*, G, BM).

20. *Saxifraga ganeshii* H.Ohba & S.Akiyama, J. Jap. Bot. 74: 223 (1999).

Stems caespitose. Stolons absent. Leaves alternate. Basal leaves petiolate, Petiole to 15 mm, sparsely pubescent. Blade ovate-lanceolate to lanceolate, 5–15 × 2.5–5 mm, base narrowed into petiole, apex obtuse, upper surface and margin pale brown crisped pubescent, lower surface glabrous. Lower cauline leaves shortly petiolate, upper leaves sessile. Blade narrowly ovate-elliptic to elliptic, 3–8 × 2–3 mm, reducing in size upwards, pale brown crisped pubescent and sometimes pale brown glandular-pubescent. Flowering stems 4–11 cm, sparsely pale brown pubescent throughout, densely brown crisped villous in leaf axils, pale brown glandular-pubescent above, bulbils absent. Flowers solitary, bisexual. Pedicels up to 2 mm, pale brown glandular and sparsely pale brown crisped pubescent. Sepals oblong, ca. 3 × 2.5–3 mm, apex obtuse, margin pale brown crisped eglandular or glandular-pubescent, both surfaces glabrous. Petals yellow, unspotted, broadly elliptic, 5–6 × 3.5–4.5 mm, base contracted into a very short claw, apex slightly retuse to obtuse, (3 or)4-callose. Ovary superior, carpels fused for more than half their length, narrowed into styles.

Distribution: Endemic to Nepal.

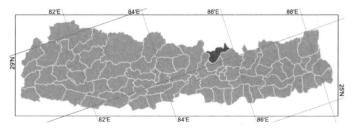

Altitudinal range: ca. 4250 m.

Ecology: Alpine.

Flowering: August.

Known only from the type collection from Rasuwa (*Miyamoto et al. 9420122*, A, TNS).

21. *Saxifraga cordigera* Hook.f. & Thomson, J. Proc. Linn. Soc., Bot. 2: 68 (1857).

Stems caespitose, simple or many-branched, forming cushions or mats. Stolons absent. Leaves alternate. Basal leaves petiolate. Petiole ca. 5 mm, margin brown eglandular crisped villous. Blade ovate to obovate, ca. 3 × 2 mm, base attenuate, apex obtuse, margin brown eglandular crisped villous, upper surface villous or glabrous, lower surface glabrous. Cauline leaves sessile, lowest leaves elliptic, median and upper leaves ovate, 4–7 × 3–4 mm, base cordate, apex obtuse to subacute, margin long brownish villous (hairs longer than 1 mm), both surfaces glabrous or lower surface sparsely villous. Flowering stems 2–6 cm, brown crisped villous, bulbils absent. Flower solitary, bisexual. Pedicels 2–8 mm, eglandular crisped villous. Sepals erect, elliptic to obovate, 2.5–4 × 1.5–2 mm, apex obtuse, outer surface glabrous, margin villous-pubescent, veins 3–5, confluent near apex. Petals yellow, orange spotted, obovate, 5–8 × 2.5–4.5 mm, base contracted to a claw, apex obtuse, not callose. Stamens ca. 1.5 mm. Nectary band obscure. Ovary superior, carpels fused for more than half their length, contracted into 0.5–0.75 mm styles.

Distribution: Nepal, E Himalaya and Tibetan Plateau.

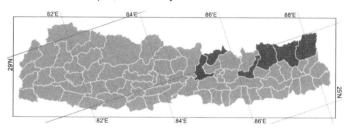

Altitudinal range: 3900–5100 m.

Ecology: Alpine, on rocks, among rocks, rocky slopes.

Flowering: July–August(–September).

22. *Saxifraga palpebrata* Hook.f. & Thomson, J. Proc. Linn. Soc., Bot. 2: 67 (1857).

Stems caespitose, simple or sometimes many-branched, forming cushions. Stolons absent. Leaves alternate. Basal leaves petiolate. Petiole 5–15 mm, villous-ciliate. Blade spatulate, 4–10 × 2–5 mm, base tapered, apex obtuse, margin and upper surface villous, lower surface glabrous. Lower cauline leaves shortly petiolate, upper leaves sessile, similar to basal leaves, oblong to ovate, ca. 4.5 mm. Flowering stems 3–10 cm, glandular-villous, bulbils absent. Flowers solitary, bisexual or perhaps sometimes female only, subsessile. Pedicels 1–3 mm, villous. Sepals erect or spreading, broadly oblong to ovate, 4–5 × 2–3 mm, apex obtuse, surfaces and margin glabrous, veins 3, confluent. Petals yellow, obovate, 6–7 × 4–5 mm in male flowers, ca. 5.5 mm in female flowers, base cuneate or with a short claw, apex rounded, not callose. Stamens 3–4 mm. Nectary disk prominent, annular. Ovary superior, carpels fused for more than half their length, tapered into conical, ca. 1 mm styles.

Distribution: Nepal, W Himalaya and E Himalaya.

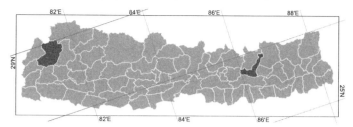

Altitudinal range: 4100–4200 m.

Ecology: Alpine.

Flowering: August.

23. *Saxifraga caveana* W.W.Sm., Rec. Bot. Surv. India 4(5): 193 (1911).
Saxifraga diapensia Harry Sm.

Stems caespitose, many-branched, forming mats or cushions. Stolons absent. Leaves alternate. Basal leaves petiolate. Petiole 3–10 mm, margin crisped glandular-villous towards base. Blade oblong to ovate, 4–10 × 1.1–5 mm, base cuneate, apex subacute, glabrous. Flowering stems 2.5–4.5 cm, dark brown glandular-pubescent, leafless but with bracts, bulbils absent. Flowers solitary, bisexual. Pedicels 1–2(–7) cm, dark brown glandular-pubescent. Sepals spreading to reflexed, ovate to lanceolate, gibbous, 4–6 × 2–4 mm, apex subacute, outer surface and margin dark brown glandular-pubescent or glabrous, veins 3, confluent near apex. Petals yellow, elliptic to obovate-elliptic, 7–10 × 4–6 mm, base with a claw, 0.8–1 mm, apex retuse, sometimes 2-callose. Stamens 5–6 mm. Nectary band obscure. Ovary superior, carpels fused for more than half their length, tapered into conical, 1–2 mm styles.
Fig. 28a–d.

Distribution: Nepal and E Himalaya.

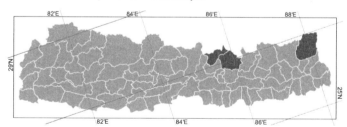

Altitudinal range: 4200–5000 m.

Ecology: Alpine.

Flowering: July–August.

24. *Saxifraga harae* H.Ohba & Wakab., J. Jap. Bot. 62: 163 (1987).

Stems caespitose, many-branched, forming mats. Stolons absent. Leaves alternate. Basal leaves petiolate. Petiole ca.

8 mm, margin eglandular brown crisped villous. Blade linear-spatulate to linear-rhombic, 3–8 × 1.5 mm, base attenuate, apex acute, both surfaces and margin shortly glandular-pubescent. Median and upper cauline leaves sessile, rhombic to linear, 4–8 × 0.7–1.5 mm, both surfaces and margin shortly glandular-pubescent, margin brown crisped villous at base. Flowering stems 4–5 cm, brown glandular-pubescent, bulbils absent. Flowers solitary, bisexual. Pedicels shortly glandular-pubescent. Sepals suberect, oblong-ovate, 3–4 × 2–2.5 mm, apex obtuse, margin shortly glandular-pubescent, outer surface sparsely glandular-pubescent, veins 3, usually confluent. Petals yellow, spotted orange, broadly oblong to oblong, 6–8 × 3–5 mm, base constricted into a claw 0.5–1 mm, apex rounded, not callose. Stamens 2.5–4 mm. Nectary band obscure. Ovary superior, carpels fused for more than half their length, narrowed into styles.

Distribution: Endemic to Nepal.

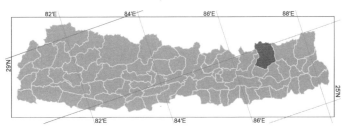

Altitudinal range: 3900–5000 m.

Ecology: Alpine, on rocks, rock slopes.

Flowering: August–September.

Apparently restricted to Solu Khumbu district, but collected there at least eight times.

25. *Saxifraga aristulata* Hook.f. & Thomson, J. Proc. Linn. Soc., Bot. 2: 68 (1857).
Hirculus aristulata (Hook.f. & Thomson) Losinsk.

Stems caespitose, usually many-branched, forming clumps, mats or cushions, sometimes simple. Stolons absent. Leaves alternate. Basal leaves petiolate. Petiole ca. 3 mm, margin brown crisped glandular-pubescent. Blade linear, 3–8 × 0.5–1 mm, base attenuate into petiole, apex acute, aristate, both surfaces glabrous. Lower cauline leaves petiolate, median and upper leaves sessile. Petiole 1–2 mm. Blade linear-lanceolate 5–8 × 0.5–1 mm, apex aristate, margin brown glandular-pubescent at base, both surfaces glabrous. Flowering stems 2–5(–10) cm, crisped brown glandular-pubescent at base and the middle, dark brown glandular-pubescent above, bulbils absent. Flowers solitary, bisexual. Pedicels 4–15 mm, black glandular-pubescent or glabrous. Sepals erect, then spreading to reflexed, elliptic to ovate, 2–2.5 × 1–1.2 mm, apex obtuse or acute, glabrous, veins 3, confluent near apex. Petals yellow, elliptic, 3.5–5 × 1.5–2 mm, base with a claw ca. 0.5 mm, apex obtuse to acute, 2-callose. Stamens 2–3 mm. Nectary band obscure. Ovary almost superior, carpels fused for more than half their length, tapered into ca. 1.5 mm styles.

Fig. 28e–i.

Distribution: Nepal, W Himalaya, E Himalaya and Tibetan Plateau.

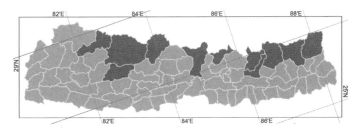

Altitudinal range: 3500–5200 m.

Ecology: Alpine, scree slopes, stony slopes, grassy slopes.

Flowering: August–September.

26. *Saxifraga lychnitis* Hook.f. & Thomson, J. Proc. Linn. Soc., Bot. 2: 68 (1857).

Stems caespitose or simple, erect, sometimes forming clumps. Stolons absent. Leaves alternate. Basal leaves aggregated into a rosette, sessile, elliptic, 5–10 × 3–5 mm, base attenuate, apex acute, both surfaces and margin eglandular or glandular brown villous. Cauline leaves sessile, elliptic to narrowly elliptic, 7–15 × 2–7 mm, base cuneate, apex acute, surfaces and margin brown glandular-pubescent. Flowering stems 3–8 cm, lower parts eglandular brown villous, glandular brown villous above, bulbils absent. Flowers solitary, usually nodding, bisexual. Pedicels 2–5 mm, densely glandular-pubescent. Sepals erect, oblong-lanceolate, 4–7 × 1.5–3 mm, apex acute to obtuse, outer surface and margin densely glandular-pubescent, veins 3, confluent near apex. Petals yellow, oblong, 8–15 × 2–3 mm, base with a claw, apex obtuse, not callose. Stamens 3–4 mm. Nectary band obscure. Ovary almost superior, carpels fused for more than half their length, tapered into 1.5–2 mm styles.
Fig. 28j–n.

Distribution: Nepal, W Himalaya, E Himalaya and Tibetan Plateau.

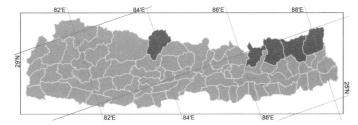

Altitudinal range: 3900–5300 m.

Ecology: Alpine, on rocks, open mossy places.

Flowering: June–September. **Fruiting:** September.

27. *Saxifraga viscidula* Hook.f. & Thomson, J. Proc. Linn. Soc., Bot. 2: 69 (1857).

Stems caespitose or simple, erect, sometimes forming clumps. Stolons absent. Leaves alternate. Basal leaves aggregated into a rosette, sometimes absent at anthesis. Petiole 5–10 mm. Blade oblong to elliptic, 5–10 × 3–5 mm, base attenuate, apex acute, both surfaces and margin eglandular- or brown glandular-pubescent. Cauline leaves sessile, ascending, oblong-elliptic, 3–6 × 1.5–2.5 mm, base tapered, apex acute or obtuse, both surfaces and margin brown glandular-pubescent. Flowering stems 5–12 cm, lower parts and nodes eglandular brown villous, black glandular-pubescent above, bulbils absent. Flowers 1–3, in a cyme, bisexual. Pedicels 10–20 mm, densely glandular-pubescent. Sepals erect or spreading, oblong-ovate, 5–6 × 1.5–2.5 mm, apex obtuse, margin and outer surface black glandular-pubescent, veins 3. Petals yellow, linear-oblong, 10–15 × 2–3 mm, base contracted to a short claw, apex obtuse, margin minutely glandular-pubescent, 2-callose. Stamens 3–4 mm. Nectary band obscure. Ovary almost superior, carpels fused for more than half their length, tapered into 1.5–2 mm styles.

Distribution: Nepal, W Himalaya and E Himalaya.

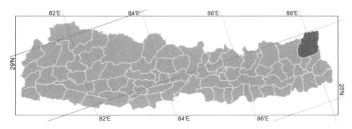

Altitudinal range: 4000–4900 m.

Ecology: Alpine, on rocks and in meadows.

Flowering: July–September. **Fruiting:** September–October.

28. *Saxifraga nigroglandulifera* N.P.Balakr., J. Bombay Nat. Hist. Soc. 67: 59 (1970).
Hirculus nutans (Hook.f. & Thomson) Losinsk.; *Saxifraga nutans* Hook.f. & Thomson.

Stems caespitose or simple, erect, sometimes forming clumps. Stolons absent. Leaves alternate. Basal leaves petiolate. Petiole 1–5 cm, margin crisped villous. Blade elliptic to ovate, 10–35 × 5–15 mm, base tapered or truncate to rounded, apex acute to obtuse, margin crisped villous, upper surface glabrous or sometimes sparsely pubescent, lower surface glabrous. Lower cauline leaves petiolate, upper leaves sessile. Cauline leaves similar to basal leaves but smaller, margin crisped villous. Flowering stems 5–24 cm, sparsely brown crisped villous at lower leaf axils, densely black glandular-pubescent above, bulbils absent. Flowers 2–7, somewhat secund in a raceme-like cyme, usually nodding, bisexual. Pedicels 3–8 mm, densely dark brown glandular-pubescent. Sepals erect, ovate to ovate-lanceolate, 3–5 × 1.5–2 mm, apex obtuse or acute, margin and outer surface dark brown glandular-pubescent, veins 3–6, free. Petals yellow, oblanceolate, 7–8 ×

2.5–3 mm, base tapered, apex rounded, not callose. Stamens 4–7 mm. Ovary semi-inferior, carpels fused for more than half their length, tapered into 1–1.5 mm styles.
Fig. 28o–r.

Distribution: Nepal, E Himalaya and Tibetan Plateau.

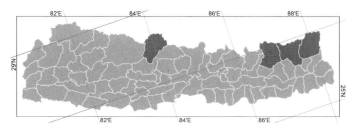

Altitudinal range: 4000–4800 m.

Ecology: Alpine, wet soil slopes, grassy places on rocky slopes.

Flowering: July–September. **Fruiting:** September–October.

29. *Saxifraga matta-viridis* Harry Sm., Bull. Brit. Mus. (Nat. Hist.), Bot. 2: 243, 8 fig. e (1960).

Stems caespitose, many-branched, forming moss-like mats scarcely more than 1 cm. Stolons absent. Leaves alternate. Basal leaves petiolate. Petiole ca. 1 mm, margin membranous, broadening and clasping stem, fimbriate. Blade linear, subcylindric, 4–5 × 0.5–0.75 mm, base narrowed to petiole, apex obtuse to subacute or acute, margin and surfaces glabrous. Cauline leaves similar to basal but wider, up to 1 mm across. Flowering stems 3 mm, glabrous, leaves 2 or 3, bulbils absent. Flowers solitary, bisexual. Pedicels 1–2 mm, glabrous or very sparingly brown-crisped villous. Sepals erect, lanceolate to ovate-lanceolate, 2–3.5 × 0.75–1 mm, apex subacute, margin and surfaces glabrous, veins 3, confluent. Petals yellow, narrowly obovate, 2.5–3.5 × 1.5 mm, base tapered, apex obtuse, not callose. Stamens 1.5–2 mm. Nectary a somewhat conspicuous fleshy disk. Ovary almost superior, carpels fused for more than half their length, tapered into 0.75 mm styles.

Distribution: Nepal and E Himalaya.

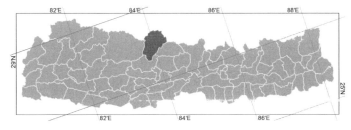

Altitudinal range: ca. 3500 m.

Ecology: Alpine, on open hillside.

Flowering: July–August.

30. *Saxifraga saginoides* Hook.f. & Thomson, J. Proc. Linn. Soc., Bot. 2: 68 (1857).
Hirculus saginoides (Hook.f. & Thomson) Losinsk.

Stems densely caespitose, many-branched, forming dense cushions. Stolons absent. Leaves alternate. Basal leaves slightly dilated at base forming a petiole. Petiole 2–4 mm, margin brown crisped glandular-villous. Blade suboblong, 3–4 × ca. 1 mm, apex subobtuse, glabrous. Cauline leaves linear, ca. 6 × 1 mm, base narrowed into petiole, apex obtuse, margin brown crisped villous at base, both surfaces glabrous. Flowering stems to 15 mm, mostly embedded in the cushion, crisped brown eglandular villous, bulbils absent. Flowers solitary, bisexual or sometimes unisexual. Pedicels up to 1 mm, densely crisped brown eglandular villous. Sepals erect, elliptic, ca. 3.5 × 1–1.7 mm, apex obtuse, glabrous, veins 3–4, free. Petals yellow, ovate to elliptic, 3–4 × 1–1.5 mm, base contracted into a claw 0.5–1 mm, apex obtuse, not callose. Stamens 1–2 mm. Nectary band obscure. Ovary almost superior, carpels fused for half their length, narrowed into ca. 0.75 mm styles.
Fig. 29o–s.

Distribution: Nepal, W Himalaya, E Himalaya and Tibetan Plateau.

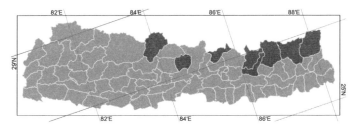

Altitudinal range: 4100–5300 m.

Ecology: Alpine, rock slopes, scree slopes, stony places.

Flowering: July–September. **Fruiting:** September.

At least some populations appear to be dioecious.

31. *Saxifraga parva* Hemsl., J. Linn. Soc., Bot. 30: 112 (1895).

Stems caespitose or simple, erect. Stolons absent. Leaves alternate. Basal leaves petiolate. Petiole 4–9 mm, margin sparsely crisped glandular-villous. Blade ovate-elliptic to oblong, 4–4.5 × 1.5–2 mm, base attenuate, apex obtuse, glabrous or margin brown crisped glandular-villous. Cauline leaves 3–10, lowermost ones petiolate. Petiole 2–4.5 mm, margin crisped glandular-villous at base. Blade similar to basal leaves, ovate to oblong, 3.5–7 × 1.5–2 mm, glabrous or margin sparsely brown crisped pubescent. Upper cauline leaves sessile, oblong to lanceolate to linear-oblanceolate, 5–8 × 1.2–3.2 mm, margin crisped glandular-villous. Flowering stems 0.7–4.5 cm, brown crisped glandular-villous, bulbils absent. Flowers solitary, bisexual. Pedicels 1–3 mm, brown crisped glandular-villous. Sepals erect, elliptic to ovate, 2–3.6 × 1–2.3 mm, apex obtuse, margin glandular-pubescent,

glabrous, veins 3, free. Petals reddish outside, yellow inside, obovate to elliptic, 2.3–6.4 × 1.3–4.7 mm, base with 0.2–1 mm claw, apex obtuse, 2-callose. Stamens 3–5 mm. Nectary band obscure. Ovary almost superior, carpels fused for more than half their length, tapered into 1–1.5 mm styles.
Fig. 28s–w.

Distribution: Nepal, E Himalaya and Tibetan Plateau.

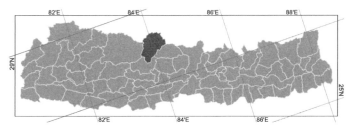

Altitudinal range: 4700–5000 m.

Ecology: Alpine.

Flowering: July–August.

Saxifraga parva is known in Nepal only from *Stainton, Sykes & Williams 2358* (BM), but there is some doubt about the identity of this specimen and whether it may actually be *S. tibetica* Losink.

32. *Saxifraga tangutica* Engl., Bull. Acad. Imp. Sci. Saint-Pétersbourg 29: 115 (1883).
Hirculus tangutica (Engl.) Losinsk.; *Saxifraga hirculus* var. *subdioica* C.B.Clarke; *S. montana* var. *subdioica* (C.B.Clarke) C.Marquand; *S. subdioica* (C.B.Clarke) Engl. ex W.W.Sm. & Cave.

Stems caespitose or simple, erect, sometimes forming clumps. Stolons absent. Leaves alternate. Basal leaves petiolate. Petiole 8–30 mm, base brown crisped villous. Blade lanceolate, 8–15 × 3–5 mm, apex obtuse or acute, margin sparsely brown crisped villous, both surfaces glabrous, upper surface rarely brown crisped villous. Cauline leaves almost sessile, lanceolate, 5–8 × 1–3 mm, margin sparsely brown crisped villous, both surfaces glabrous. Flowering stems 3–12 cm, crisped brown eglandular villous, bulbils absent. Flowers in 3–8-flowered cymes, apparently unisexual. Pedicels 3–7 mm, densely brown crisped eglandular villous. Sepals erect, eventually reflexed, ovate, 2–3 × 1–2 mm, apex obtuse, margin brown crisped villous, outer surface brown crisped villous, veins 3–5, free. Petals purple on outside, yellow within, elliptic, 2–3 × ca. 1 mm, base with a 0.3-0.8 mm claw, apex obtuse, 2-callose. Stamens 1–2 mm. Nectary a conspicuous annular disk. Ovary nearly inferior, carpels fused for more than half their length, tapered into conical, ca. 1 mm styles.

Distribution: W Himalaya, E Himalaya and Tibetan Plateau.

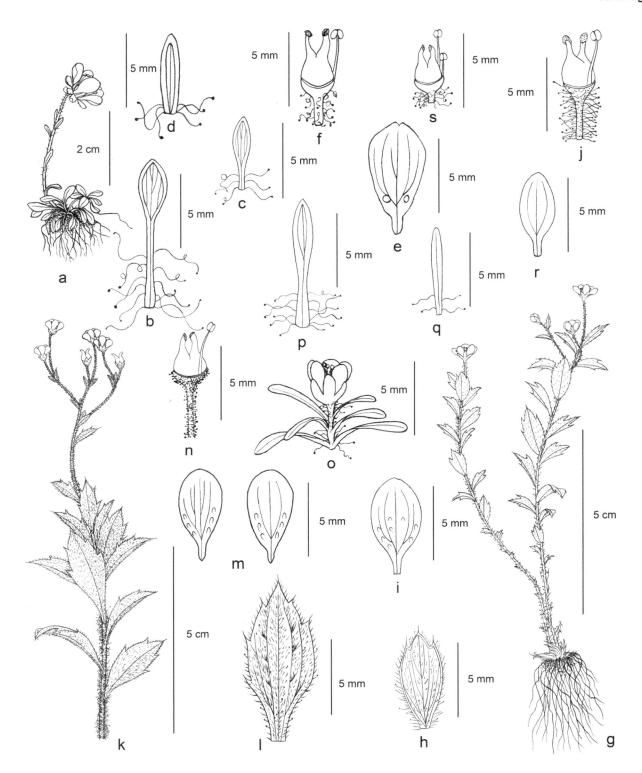

FIG. 29. Sᴀxɪꜰʀᴀɢᴀᴄᴇᴀᴇ. **Saxifraga elliptica**: a, flowering plant; b, basal leaf; c, lower cauline leaf; d, upper cauline leaf; e, petal; f, ovary and stamen. **Saxifraga hispidula**: g, flowering plant; h, leaf; i, petal; j, ovary and stamen. **Saxifraga substrigosa**: k, flowering part of plant; l, leaf; m, petals; n, ovary and stamen. **Saxifraga saginoides**: o, flowering part of plant; p, basal leaf; q, cauline leaf; r, petal; s, ovary and stamens.

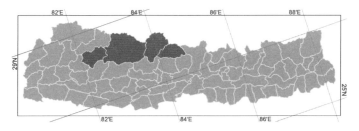

Altitudinal range: 4300–5400 m.

Ecology: Alpine, on stony slopes.

Flowering: July–August(–September). **Fruiting:** September.

33. *Saxifraga hirculoides* Decne., in Jacquem., Voy. Inde 4(1): 67, pl. 78.1 (1841).
Saxifraga hirculus var. *hirculoides* (Decne.) C.B.Clarke.

Stems caespitose or simple, erect, sometimes forming clumps. Stolons absent. Leaves alternate. Basal leaves petiolate. Petiole 7–30 mm, brown crisped villous. Blade ovate to narrowly elliptic, 7.5–9 × 4–8 mm, base attenuate, apex subacute, margin sparingly brown crisped villous. Cauline leaves 3–6, similar to basal leaves, lowermost petiolate, otherwise sessile. Blade oblong, 5.5–20 × 1.3–5 mm, margin brown crisped villous. Flowering stems 1.3–17 cm, brown crisped villous, bulbils absent. Flowers solitary or 2–4 in a cyme, bisexual. Pedicels 3–10 mm, densely crisped brown eglandular villous. Sepals erect, broadly ovate to broadly elliptic, 2.4–3 × 2.2–2.5 mm, apex obtuse, margin crisped brown eglandular villous, outer surface glabrous or sparingly crisped brown eglandular villous, veins 3, free. Petals pale yellow-green or green, elliptic or ovate to obovate, 3.5–4 × 2–3.3 mm, base with a 0.2–0.6 mm claw, apex obtuse, margin crisped brown eglandular villous, not callose. Stamens 2–3 mm. Nectary a conspicuous annular disk. Ovary semi-inferior, carpels fused for more than half their length, narrowed into 0.5 mm styles.

Distribution: Nepal, W Himalaya and Tibetan Plateau.

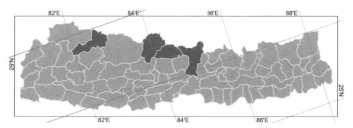

Altitudinal range: 4200–5100 m.

Ecology: Alpine, among boulders on barren slopes, in scree, moist banks.

Flowering: July–August.

34. *Saxifraga elliptica* Engl. & Irmsch., Bot. Jahrb. Syst. 48: 585 (1912).

Stems caespitose, simple or many-branched, forming mats. Stolons absent. Leaves alternate. Basal leaves petiolate. Petiole 3–7.5 mm, margin brown crisped glandular-villous. Blade elliptic to ovate-lanceolate, 5–9 × 2–4 mm, base tapered, apex obtuse to acute, lower surface tinged purple, glabrous or eglandular villous. Lower cauline leaves petiolate, upper leaves sessile. Blade linear-oblong to ovate, (2–)4–5 × 1–2 mm, base tapered, apex obtuse to acute, margin eglandular villous, surfaces glabrous or glabrate. Flowering stems 3–7 cm, brown glandular crisped villous, bulbils absent. Flowers solitary, bisexual. Pedicels 3–14 mm, brown crisped glandular-villous. Sepals erect, then spreading to reflexed, broadly ovate to elliptic or oblong, 2–3.2 × 1–2.2 mm, leathery, glabrous or sometimes glandular-pubescent at margin, veins 3, free. Petals yellow, obovate to broadly obovate to broadly elliptic, (4–)6–8.6 × (2.5–)4–5.3 mm, base with a 1.1–1.2 mm claw, apex rounded to retuse, 2-callose. Stamens 3–5 mm. Ovary superior, carpels fused for more than half their length, narrowed into ca. 0.5 mm styles.
Fig. 29a–f.

Distribution: Nepal, E Himalaya and Tibetan Plateau.

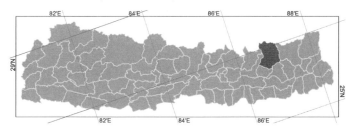

Altitudinal range: 4200–4300 m.

Ecology: Alpine.

Flowering: July–September.

35. *Saxifraga montanella* Harry Sm., Bull. Brit. Mus. (Nat. Hist.), Bot. 2: 238, pl. 16B fig. 6 (1960).

Stems caespitose or simple, erect, often forming clumps. Stolons absent. Leaves alternate. Basal leaves petiolate. Petiole 3–12 mm, margin brown crisped villous. Blade ovate to lanceolate, 4–10 × 2–5 mm, base cuneate to rounded, margin brown crisped villous, apex acute, lower surface glabrous, upper surface brown crisped villous. Lowermost cauline leaves petiolate or sessile. Blade lanceolate to elliptic to oblong, 4–11 × 1.3–5 mm, base cuneate to rounded, apex obtuse to acute, margin brown crisped villous, lower surface glabrous, upper surface brown crisped villous in lower leaves, glabrate in upper leaves. Flowering stems 2–10 cm, brown crisped villous, bulbils absent. Flower solitary, bisexual. Pedicels 5–30 mm, brown eglandular or glandular crisped villous. Sepals spreading, ovate to elliptic, 2.8–3.5 × 2.3 mm, apex obtuse, outer surface glabrous, margin brown crisped villous, veins 3–9, free. Petals yellow, ovate to obovate, 5–6 × 2.5–4 mm, base with a 0.5–0.7 mm claw, apex obtuse or retuse, 2-callose. Stamens 2–5 mm. Ovary almost superior, carpels fused for more than half their length, tapered into 1–1.5 mm styles.

Distribution: Nepal, E Himalaya and Tibetan Plateau.

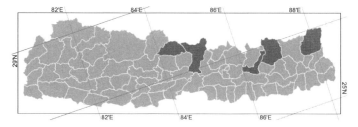

Altitudinal range: 4100–4800 m.

Ecology: Alpine, on cliffs.

Flowering: July–August.

At least some collections are functionally dioecious.

36. *Saxifraga namdoensis* Harry Sm., Bull. Brit. Mus. (Nat. Hist.), Bot. 2: 237, pl. 16A fig. 5 (1960).

Stems caespitose, erect, forming clumps. Stolons absent. Leaves alternate. Basal leaves petiolate. Petiole 1–5 cm, base sparsely brown crisped pubescent. Blade lanceolate, 10–30 × 5–9 mm, base tapered, apex subacute, margin and both surfaces glabrous. Lowermost cauline leaves petiolate, otherwise sessile. Blade lanceolate to narrowly elliptic,15–30 × 4–9 mm, decreasing in size upwards, margin sparsely brown crisped pubescent, surfaces glabrous. Flowering stems 8–25 cm, lower parts glabrate except in axils, densely brown crisped villous above, leaves 6–9, bulbils absent. Flowers 2–4 in a cyme, bisexual. Pedicels 5–20 mm, densely brown crisped villous or glabrescent. Sepals more or less spreading at anthesis, ovate to oblong, 4–5 × 2.5–4 mm, apex rounded, margin densely brown crisped villous, outer surface glabrous or very sparsely brown crisped villous, veins 5–7, confluent. Petals yellow, obovate-elliptic to broadly elliptic, 5.5–11 × 4–7 mm, base without a claw or with a very short claw (0.1–0.2 mm), apex rounded, not callose. Stamens ca. 5 mm. Nectary band inconspicuous. Ovary subsuperior, carpels fused for more than half their length, tapered abruptly into ca. 1 mm styles.

Distribution: Endemic to Nepal.

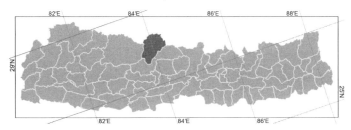

Altitudinal range: 4500–4700 m.

Ecology: Alpine, on grass banks of streams, meadows.

Flowering: August.

Very close to *Saxifraga sinomontana* J.T.Pan & Gornall, but appears to differ in its more laxly branched inflorescence with more numerous flowers.

37. *Saxifraga sinomontana* J.T.Pan & Gornall, Novon 10: 377 (2000).
Hirculus montanus (Harry Sm.) Losinsk.; *Saxifraga hirculus* var. *indica* C.B.Clarke; *S. montana* Harry Sm.; *S. montana* forma *rubra* Harry Sm.

Stems caespitose or simple, erect, sometimes forming clumps. Stolons absent. Leaves alternate. Basal leaves petiolate. Petiole 7–10 mm, margin brown crisped pubescent. Blade elliptic-lanceolate to elliptic, 5–20 × 2–6 mm, base attenuate, apex acute to subacute, margin glabrous or sparingly brown crisped pubescent, both surfaces glabrous. Lower cauline leaves petiolate, upper leaves sessile. Blade lanceolate to linear, 5–8 × 1–1.5 mm, base, apex obtuse or acute, margin sparsely brown crisped pubescent, both surfaces glabrous. Flowering stems 3–25 cm, lower parts sparsely brown crisped pubescent, densely so above, bulbils absent. Flowers 1–3, in a cyme, bisexual. Pedicels 5–15 mm, densely crisped brown eglandular villous. Sepals erect, oblong-obovate, 4–5 × 2–3 mm, apex obtuse, margin densely crisped brown eglandular villous, outer surface usually glabrous, veins 5–8, free. Petals yellow (rarely red), orange spotted, broadly obovate, 6–12 × 4–7 mm, base truncate or with a short claw up to 0.5 mm, apex obtuse to rounded, 2-callose. Stamens 4–5 mm. Nectary band obscure. Ovary almost superior, carpels fused for more than half their length, tapered into 1–2 mm styles.

Distribution: Nepal, W Himalaya, E Himalaya and Tibetan Plateau.

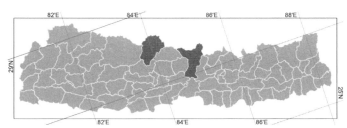

Altitudinal range: 4100–5000 m.

Ecology: Alpine, open stony slopes, on soil ground.

Flowering: July–August(–September).

Saxifraga montana forma *rubra* refers to plants with red petals. However the petals of these plants are smaller (ca. 5 × 2 mm) than is usually the case in *S. sinomontana* and dry greenish-tinged purple. Such plants may be allied more with *S. hirculoides* Decne.

38. *Saxifraga hispidula* D.Don, Trans. Linn. Soc. London 13(2): 380 (1822).
Hirculus hispidulus (D.Don) Losinsk.; *Saxifraga evolvuloides* Wall. ex Ser.; *S. hispidula* var. *dentata* Franch.

Stems caespitose or simple, branched towards the base, leafy throughout. Stolons absent. Leaves alternate. Leafy buds usually present in at least some axils. Lower cauline leaves smaller than median ones. Median cauline leaves sessile, ovate to elliptic, 3–12 × 1–6 mm, base rounded, apex acute, aristate, margin with 1 or 2 bristle-pointed teeth on each side, sometimes entire, eglandular-ciliate, both surfaces eglandular- or glandular-strigose. Flowering stems, 3–15 cm, eglandular- or glandular-pubescent, distally glandular-pubescent, leafy. Flowers usually solitary, occasionally up to 4, in a cyme, bisexual. Pedicels 5–10 mm, densely glandular-pubescent. Sepals erect to spreading, ovate, 2.5–4.5 × 1.5–2.5 mm, apex acute, bristle-pointed, outer surface and margin eglandular- or glandular-pubescent, veins 3–8, confluent. Petals yellow, 5–7 × 3–4 mm, base contracted into a ca. 0.6 mm claw, apex obtuse, 2-callose. Stamens 3–4 mm. Nectary band obscure. Ovary almost superior, carpels fused for more than half their length, narrowed into conical, 0.5–1 mm styles.
Fig. 29g–j.

Distribution: Nepal, W Himalaya, E Himalaya, Tibetan Plateau, Assam-Burma and E Asia.

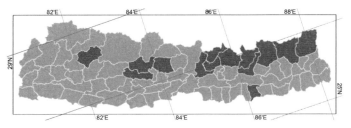

Altitudinal range: 3000–4500 m.

Ecology: Alpine, on cliffs, in forests.

Flowering: July–September.

In Nepal plants with toothed leaves (var. *dentata*) appear to be more common than those with entire leaves (var. *hispidula*).

39. *Saxifraga substrigosa* J.T.Pan, in C.Y.Wu, Fl. Xizang. 2: 463 (1985).
Saxifraga bumthangensis Wadhwa ex Grierson nom. nud.

Stems caespitose, branched or simple, erect. Stolons absent. Leaves alternate. Leafy buds inconspicuous, in axils of at least lower leaves, sometimes developing into sterile, leafy shoots by anthesis. Lowest leaves less than half the size of median leaves, caducous. Median cauline leaves ovate or obovate to oblong, 10–30 × 5–10 mm, base cuneate, margin 5- or 6-serrate, both surfaces strigose, apex acute. Upper cauline leaves reduced. Flowering stems usually simple, sometimes branched proximally, 10–30 cm tall, eglandular villous below, glandular pubescent towards apex. Cyme ca.

5 cm, 2–10-flowered. Pedicel ca. 10 mm, glandular pubescent. Sepals erect, ovate, 2.6–3 × 1.4–1.8 mm, abaxially glandular pubescent, margin glabrous; veins 5–8, confluent at apex. Petals yellow, obovate, 6.5–7 × 3.5–4 mm, ca. 6-callose, base gradually narrowed into a ca. 1.5 mm claw, apex obtuse, 3–8-veined. Stamens ca. 3.5 mm. Nectary band obscure. Ovary almost superior, carpels fused for more than half their length, narrowed into ca. 1.5 mm styles.
Fig. 29k–n.

Distribution: Nepal and E Asia.

Altitudinal range: 2700–4200 m.

Ecology: Mixed forests, forest margins, alpine meadows and rock crevices.

Flowering: July–September.

The record for Nepal is based on *Skully s.n.* (K) which lacks all collection details and so cannot be mapped. The ecological information has been taken from Pan *et al.* (Fl. China 8: 317. 2001).

40. *Saxifraga strigosa* Wall. ex Ser., in DC., Prodr. 4: 41 (1830).
Hirculus strigosus (Wall. ex Ser.) Losinsk.

Stems caespitose, branched or simple, erect. Stolons absent. Leaves alternate. Leafy buds present in axils of rosette leaves and bracts, often replacing flowers. Lower cauline leaves of flowering stem less than half the size of median ones. Median cauline leaves aggregated into a rosette, subsessile or short-petiolate, elliptic to obovate, 10–20 × 5–10 mm, base cuneate to attenuate, apex acute, aristate, margin 2–3-aristate-dentate, both surfaces strigose. Upper cauline leaves smaller than median ones. Flowering stems 6–25 cm, eglandular crisped-villous towards base, blackish glandular-pubescent above. Flowers solitary or 3–10, in a cyme, bisexual. Pedicels 5–15 mm, black glandular-pubescent. Sepals spreading to reflexed, ovate to narrowly ovate, 1.5–2 × 1–1.5 mm, apex acute, outer surface strigose, margin somewhat glabrous, veins 3–7, confluent. Petals white, spotted reddish brown and yellow, ovate to elliptic, 4–5.5 × 1.5–2.5 mm, base contracted to a claw, apex acute to obtuse, 2–4 callose. Stamens 2–4 mm. Nectary band obscure. Ovary almost superior, carpels fused for more than half their length, narrowed into 0.5–2 mm styles.
Fig. 30a–d.

Distribution: Nepal, W Himalaya, E Himalaya, Tibetan Plateau, Assam-Burma and E Asia.

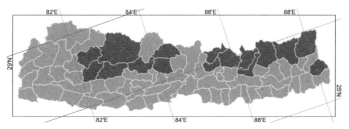

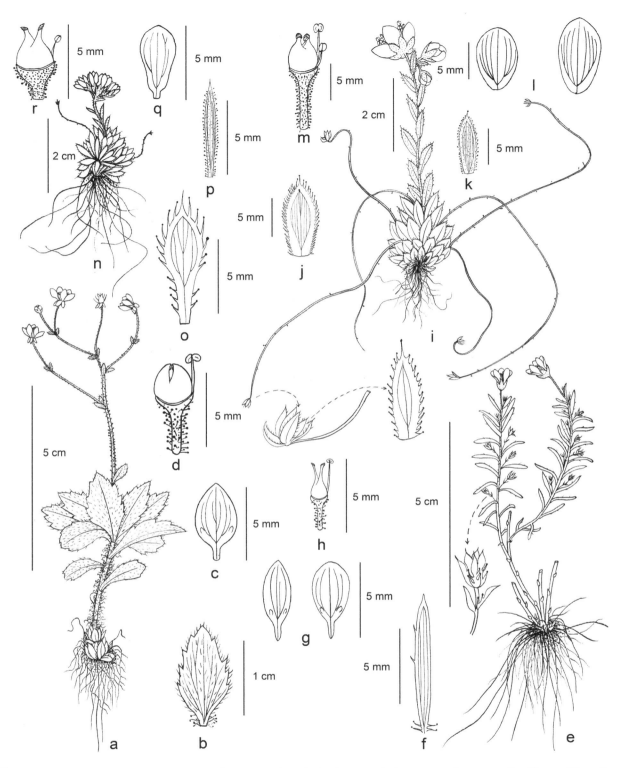

FIG. 30. SAXIFRAGACEAE. **Saxifraga strigosa**: a, flowering plant; b, leaf; c, petal; d, ovary and stamen. **Saxifraga filicaulis**: e, flowering plant; f, leaf; g, petals; h, ovary and stamen. **Saxifraga stenophylla**: i, flowering plant; j, basal leaf; k, cauline leaf; l, petals; m, ovary and stamens. **Saxifraga mucronulata**: n, flowering plant; o, basal leaf; p, cauline leaf; q, petal; r, ovary and stamen.

Saxifragaceae

Altitudinal range: 2100–3700(–4200) m.

Ecology: Forest margin, meadows, rock crevices.

Flowering: July–September. **Fruiting:** September–October.

41. *Saxifraga filicaulis* Wall. ex Ser., in DC., Prodr. 4: 46 (1830).
Hirculus filicaulis (Wall. ex Ser.) Losinsk.

Stems caespitose, many-branched, erect or trailing, forming clumps or mats, leafy throughout. Stolons absent. Leaves alternate. Leafy buds present in axils, often developing into short shoots. Lower cauline leaves scale-like. Median and upper cauline leaves sessile, linear, 3–10 × 1–2 mm, apex acute, aristate, margin recurved, eglandular- or glandular-ciliate, both surfaces usually glabrous. Flowering stems 8–12 cm, eglandular-pubescent or basal parts nearly glabrous, glandular-pubescent above. Flowers usually solitary, or up to 3, in a cyme, bisexual. Pedicels 3–5(–7) mm, glandular-pubescent. Sepals erect to spreading, ovate, 2–2.5 × 1–2 mm, apex acute to obtuse, outer surface glandular-pubescent, margin often glabrous, veins 3–5, confluent. Petals yellow, elliptic to obovate, 4–5.5 × 2–3 mm, base contracted into a ca. 1mm claw, apex obtuse, 2-callose. Stamens 3–4 mm. Nectary band obscure. Ovary almost superior, carpels fused for more than half their length, tapered into slender, ca. 2 mm styles. Fig. 30e–h.

Distribution: Nepal, W Himalaya, E Himalaya, Tibetan Plateau, Assam-Burma and E Asia.

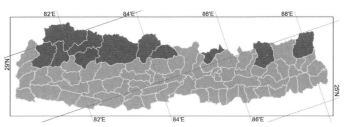

Altitudinal range: (2300–)3000–4500 m.

Ecology: On cliffs, grassy and rocky slopes.

Flowering: July–September. **Fruiting:** September–October.

42. *Saxifraga serrula* Harry Sm., Bull. Brit. Mus. (Nat. Hist.), Bot. 2: 252, pl. 21A fig. 13 f (1960).

Stems caespitose, many-branched, erect or trailing, forming clumps or mats, leafy throughout. Stolons absent. Leaves alternate. Leaf buds present in leaf axils. Median cauline leaves larger than basal and upper ones, linear, 6–10 × 1–1.5 mm, apex acute, bristle-pointed, margin eglandular (rarely glandular) ciliate, both surfaces glabrous. Flowering stems 4–20 cm, densely glandular-pubescent above. Flowers solitary, bisexual. Pedicels 5–15 mm, densely glandular-pubescent. Sepals ovate, 3–4 × 2–3 mm, apex acute, bristle-pointed, margin glandular or eglandular-pubescent, both

surfaces glabrous, veins 3, free. Petals yellow, elliptic to ovate, 5–7.5 × 3–4 mm, base constricted into a ca. 1 mm claw, apex obtuse, margin entire, not callose. Stamens 4–5 mm. Nectary obscure. Ovary superior, carpels fused for more than half their length, tapered into conical, ca. 1.5 mm styles.

Distribution: Nepal, E Himalaya and E Asia.

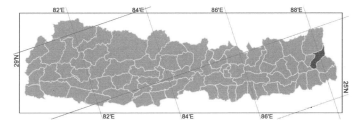

Altitudinal range: 2800–3500 m.

Ecology: Alpine, on dry soil.

Flowering: July–September.

43. *Saxifraga wallichiana* Sternb., Revis. Saxifrag. Suppl.2: 21, pl. 22 (1831).
Saxifraga brachypoda var. *fimbriata* (Wall. ex Ser.) Engl. & Irmsch.; *S. fimbriata* Wall. ex Ser. later homonym, non D.Don.

Stems caespitose or simple, erect, leafy throughout. Stolons absent. Leaves alternate. Leafy buds present in axils of leaves and bracts. Median cauline leaves larger than lower and upper leaves, ovate to lanceolate, (6–)8–15 × 2.5–4.5(–5) mm, base slightly cordate to truncate, apex acute, margin cartilaginous, shiny, eglandular- or glandular-ciliate, both surfaces glabrous. Flowering stems 8–25 cm, upper parts densely glandular-pubescent. Flowers usually solitary, bisexual. Pedicels 3–6 mm, densely black glandular-pubescent. Sepals erect, ovate to narrowly ovate, (3–)3.5–4.5 × 2–3 mm, apex acute, outer surface glabrous or sparsely glandular-pubescent, margin sparsely glandular-ciliate, veins 3–7, partly or fully confluent. Petals yellow, elliptic to obovate, 6.5–8 × 2.5–4 mm, base contracted into a 1 mm claw or attenuate, apex obtuse, margin very finely laciniate or entire, 2-callose. Stamens 4–4.5 mm. Nectary band obscure. Ovary almost superior, carpels fused for half their length, tapered into slender, 2–2.5 mm styles.

Distribution: Nepal, W Himalaya, E Himalaya, Tibetan Plateau, Assam-Burma and E Asia.

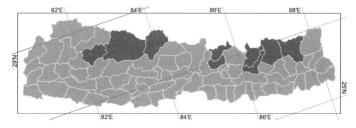

Altitudinal range: 3300–4600(–5000) m.

Ecology: Alpine, on rocky slopes.

Flowering: July–September. **Fruiting:** September–October.

44. *Saxifraga brachypoda* D.Don, Trans. Linn. Soc. London 13(2): 378 (1822).
Hirculus brachypodus (D.Don) Losinsk.; *Saxifraga glandulosa* Wall. ex DC.

Stems simple or caespitose, often forming clumps, leafy throughout. Stolons absent. Leaves alternate. Leafy buds present in leaf axils. Median cauline leaves larger than basal and upper leaves, lanceolate, 7–12(–15) × (1–)1.5–2.5(–3) mm, base subcordate to truncate, apex acute, margin eglandular- or glandular-ciliate, both surfaces glabrous, shiny. Flowering stems 4–19(–26) cm, glandular-pubescent towards apex. Flowers solitary, bisexual. Pedicels 8–19(–25) mm, densely glandular-pubescent. Sepals erect to spreading, ovate, (3–)4–5 × (1.5–)2.5–3.5 mm, apex acute, margin sparingly glandular-pubescent, outer surface glabrous, or glandular-pubescent, veins 3–7, free or partly or fully confluent. Petals yellow, elliptic to ovate, 5–8 × 3.5–5 mm, base contracted into a 0.6–1 mm claw, apex obtuse, margin very finely laciniate or entire, not callose. Stamens 4–5 mm. Nectary band obscure. Ovary almost superior, carpels fused for more than half their length, tapered into conical, ca. 1.5 mm styles.

Distribution: Nepal, W Himalaya, E Himalaya, Tibetan Plateau, Assam-Burma and E Asia.

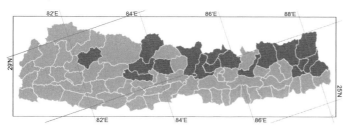

Altitudinal range: (2500–)3300–5000 m.

Ecology: Subalpine, alpine, rocky slopes.

Flowering: (July–)August–October. **Fruiting:** September–October.

45. *Saxifraga jaljalensis* H.Ohba & S.Akiyama, Univ. Mus. Univ. Tokyo Nat. Cult. 4: 31 (1992).

Stems simple. Stolons absent. Leaves alternate. Bulbils present in leaf and bract axils. Basal rosette leaves narrowly spatulate to linear-oblanceolate, 7–10 mm, base gradually attenuate, apex rounded, upper surface sparingly pubescent, lower surface glabrous. Cauline leaves linear to linear-oblanceolate, to 3 mm long. Flowering stems to 2 cm. Flowers solitary, bisexual. Sepals spreading, narrowly oblong, 2–3 × ca. 1 mm, apex obtuse. Petals yellow with orange spots towards the base, obovate, 4–5 × 3.5–4 mm, base attenuate, apex rounded. Stamens ca. 2mm. Nectary band obscure. Carpels fused for more than half their length, styles ca. 1.5 mm.

Distribution: Endemic to Nepal.

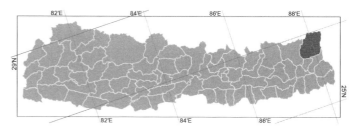

Altitudinal range: ca. 4300 m.

Ecology: Alpine, steep rocky slopes.

Flowering: August.

Apparently known only from the type collection from the Jaljale Himal (*Ohba et al. 9110380*, TI).

46. *Saxifraga stenophylla* Royle, Ill. Bot. Himal. Mts. [7]: 227, pl. 50 fig. 1 (1835).
Saxifraga flagellaris subsp. *hoffmeisteri* (Klotzsch) Hultén; *S. hoffmeisteri* Klotzsch; *S. stenophylla* subsp. *hoffmeisteri* (Klotzsch) H.Hara.

Stems simple, erect or caespitose, sometimes forming small clumps. Stolons arising from axils of basal leaves, sparsely glandular-pubescent, apex usually gemmiferous. Leaves alternate. Basal leaves aggregated into a rosette, sessile, narrowly elliptic to subspatulate, 8–13 × 2–4.5 mm, apex acute, glandular, margin slender glandular-ciliate, leathery, both surfaces somewhat glandular-pubescent. Cauline leaves on flowering stem similar to basal leaves, remote or overlapping, 5.5–11 × 1.5–3 mm. Flowering stems 5–18 cm, densely glandular-pubescent. Flowers solitary or 2 or 3 in a cyme, bisexual. Pedicels 6–14 mm, densely glandular-pubescent. Sepals erect, ovate to lanceolate, 4–6 × 1.2–3 mm, apex usually mucronate, somewhat fleshy, outer surface and margin glandular-pubescent, veins 5–9, partly or fully confluent. Petals red to yellow, obovate to broadly obovate to elliptic, 8–12 × 4.5–7.5 mm, base without a claw, apex obtuse, not callose. Stamens 4–5.7 mm. Nectary band obscure. Ovary almost superior, carpels fused for more than half their length, narrowed into ca. 1.5 mm styles.
Fig. 30i–m.

Distribution: Nepal, W Himalaya, Tibetan Plateau and SW Asia.

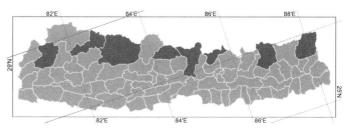

Altitudinal range: 4300–5400(–5700) m.

Ecology: Alpine, scree, grassland.

Saxifragaceae

Flowering: July–August.

Plants with crowded, overlapping cauline leaves that overtop the solitary flower may be recognized as subsp. *hoffmeisteri*.

47. *Saxifraga mucronulata* Royle, Ill. Bot. Himal. Mts. [7]: 227 (1835).
Saxifraga flagellaris subsp. *mucronulata* (Royle) Engl. & Irmsch.; *S. flagellaris* var. *mucronulata* (Royle) C.B.Clarke; *S. spinulosa* Royle.

Stems simple, erect or caespitose, sometimes forming small clumps. Stolons arising from axils of basal leaves, sometimes from axils of median cauline leaves, glandular-pubescent. Leaves alternate. Basal leaves aggregated into a rosette, spatulate to linear-spatulate, 8–9.5 × 1.6–2 mm, base attenuate, apex mucronate, margin denticulate-ciliate, both surfaces glabrous. Cauline leaves of flowering stem similar to basal leaves, often decreasing in size upwards, margin eglandular- or glandular-ciliate. Flowering stems 2–9 cm, glandular-pubescent. Flowers 1–6(–10) in a cyme, bisexual. Pedicels to 5 mm, glandular-pubescent. Sepals erect, ovate, ca. 2.5 × 1 mm, apex subacuminate, outer surface and margin glandular-pubescent, veins 3, free or partly confluent. Petals yellow to orange, obovate, 4–5 × 2.5–3 mm, base contracted to a claw, apex obtuse, not callose. Stamens 1–3 mm. Nectary band obscure. Ovary semi-inferior, carpels fused for more than half their length, narrowed into 0.8–1 mm styles.
Fig. 30n–r.

Distribution: Nepal, W Himalaya, E Himalaya and Tibetan Plateau.

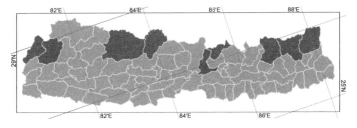

Altitudinal range: 3700–4800 m.

Ecology: Alpine, scree slopes.

Flowering: July–August.

48. *Saxifraga mucronulatoides* J.T.Pan, Acta Phytotax. Sin. 29: 223 (1991).
Saxifraga flagellaris subsp. *sikkimensis* Hultén; *S. mucronulata* subsp. *sikkimensis* (Hultén) H.Hara.

Stems simple, erect or caespitose, sometimes forming small clumps. Stolons arising from axils of basal leaves, glandular-pubescent. Leaves alternate. Basal leaves aggregated into a rosette, spatulate to linear-spatulate, 8–15 × 1.5–3 mm, base attenuate, apex aristate, both surfaces glabrous, margin denticulate-ciliate. Cauline leaves of flowering stem linear to linear-spatulate, sometimes longer than basal leaves, both surfaces and margin eglandular- or glandular-pubescent. Flowering stems 2.5–6 cm, glandular-pubescent. Flowers (4 or)5–10 in an umbellate cyme, bisexual. Pedicels 3–4 mm, glandular-pubescent. Sepals erect, ovate to lanceolate, 2.5–3.5 × 0.75–1 mm, apex acute, often aristate, outer surface and margin glandular-pubescent, veins 4 or 5, confluent. Petals yellow to orange, obovate to spatulate, 4.5–7 × 2.5–3 mm, base contracted to a claw, apex obtuse, 2-callose. Stamens 1.5–2.5 mm. Nectary band obscure. Ovary semi-inferior, carpels fused for more than half their length, narrowed into 1–1.5 mm styles.
Fig. 31a–d.

Distribution: Nepal, E Himalaya and Tibetan Plateau.

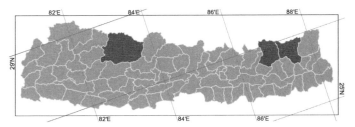

Altitudinal range: 4500–5100 m.

Ecology: Alpine, scree slopes.

Flowering: July–August.

49. *Saxifraga brunonis* Wall. ex Ser., in DC., Prodr. 4: 45 (1830).
Hirculus brunonianus Losinsk.; *Saxifraga brunoniana* Wall. ex Sternb. nom. illegit.

Stems caespitose or simple, sometimes forming mats. Stolons arising from axils of basal leaves, sparsely glandular-pubescent. Leaves alternate. Basal leaves aggregated into a rosette, grey-green, shiny, oblong-ensiform, 6–8 × 1–1.5 mm, apex cartilaginous aristate, hairs sometimes gland-tipped, margin cartilaginous setose-ciliate, both surfaces glabrous. Cauline leaves of flowering stem remote, similar to basal leaves, decreasing in size upwards. Flowering stems 6–9 cm, glabrous at base, sparsely glandular-pubescent above. Flowers 1–3, in a cyme, bisexual. Pedicels 1.2–3 cm, sparsely glandular-pubescent. Sepals spreading, ovate, 1.2–2 × 1–1.5 mm, apex obtuse, outer surface and margin glabrous, veins 3–5, free or partly or fully confluent. Petals yellow, sometimes spotted orange, elliptic, 6–7 × 2–3.5 mm, base cuneate or tapered to a short claw, apex rounded, obscurely 2-callose. Stamens 3–4 mm. Nectary band obscure. Ovary almost superior, carpels fused for more than half their length, narrowed into 1–1.2 mm styles.
Fig. 31e–i.

Distribution: Nepal, W Himalaya, E Himalaya and Tibetan Plateau.

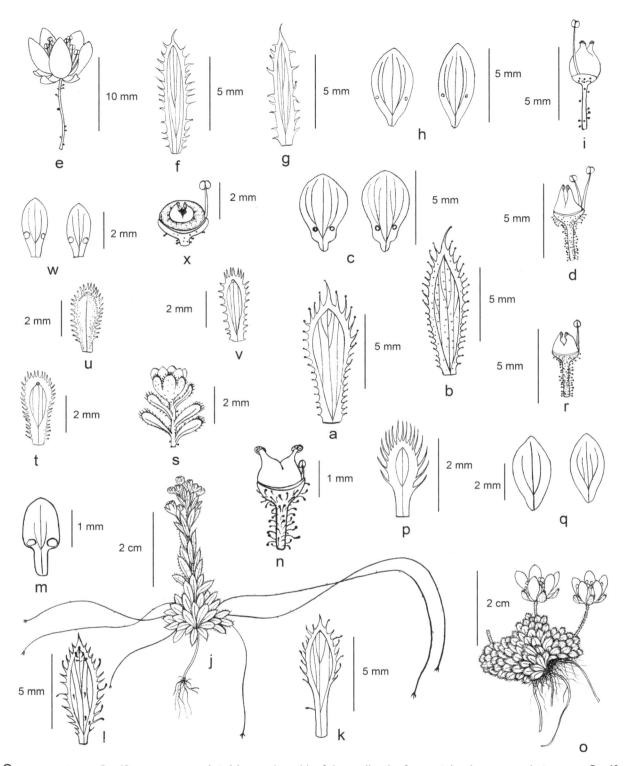

FIG. 31. SAXIFRAGACEAE. **Saxifraga mucronulatoides**: a, basal leaf; b, cauline leaf; c, petals; d, ovary and stamens. **Saxifraga brunonis**: e, flower; f, basal leaf; g, cauline leaf; h, petals; i, ovary and stamen. **Saxifraga consanguinea**: j, flowering plant; k, basal leaf; l, cauline leaf; m, petal; n, ovary and stamen. **Saxifraga perpusilla**: o, flowering plant; p, leaf; q, petals; r, ovary and stamen. **Saxifraga hemisphaerica**: s, flower with cauline leaf; t, shoot leaf adaxial view; u, shoot leaf abaxial view; v, cauline leaf; w, petals; x, ovary and stamen.

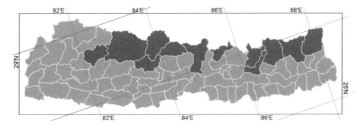

Altitudinal range: (1400–)2400–5600 m.

Ecology: Mainly subalpine and alpine, sandy places near streams.

Flowering: July–August.

50. *Saxifraga tentaculata* C.E.C.Fisch., Bull. Misc. Inform. Kew 1940: 295 (1941).

Stems simple, erect or caespitose. Stolons arising from axils of median cauline leaves, glabrous or sparsely glandular-pubescent. Leaves alternate, sessile, tending to be aggregated into loose rosettes in lower and upper parts of flowering stem, upper surface green, lower surface red, oblanceolate, 5–10 × 2–3 mm, base attenuate, apex obtuse, margin glandular-pubescent or glabrous, both surfaces glabrous. Flowering stems 2.5–5 cm, glabrous towards base, glandular-pubescent towards apex. Flowers 1–2, in a cyme, bisexual. Pedicel 5–10 mm, glandular-pubescent. Sepals erect, ovate, 2–3 × 1.5–3 mm, apex obtuse, outer surface glandular-pubescent or glabrous, margin glandular or eglandular-pubescent; veins 3. Petals yellow, spotted red, elliptic to obovate or ovate, 3–6 × 1–2.5 mm, base clawed, apex obtuse, not callose. Stamens 2.5–4 mm. Nectary band obscure. Ovary almost inferior, carpels fused more than half the length, narrowed into 1 mm styles.

Distribution: Nepal, E Himalaya and Tibetan Plateau.

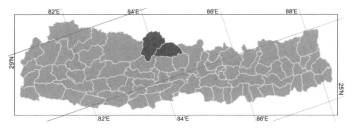

Altitudinal range: 3500–4800 m.

Ecology: Alpine, rock slopes, scree, stony places.

Flowering: July–August(–September). **Fruiting:** September.

51. *Saxifraga neopropagulifera* H.Hara, J. Jap. Bot. 51(1): 7 (1976).

Stems simple or caespitose, erect, sometimes forming mats. Stolons arising from axils of median cauline leaves, glabrous. Leaves alternate, tending to be aggregated into loose rosettes in lower and upper parts of flowering stem. Rosulate leaves oblong-spatulate, 3–8 × 1.3–3 mm, margin ciliolate. Cauline leaves obovate or oblong-spatulate, 3–10 × 1–3 mm, base attenuate, apex subobtuse, margin shortly eglandular-ciliate, both surfaces glabrous. Flowering stems 1–12 cm, glabrate. Flowers 1–3 in a cyme, dioecious. Pedicel 2–6 mm, glandular-pubescent. Male flowers 6–8 mm diameter, female ones smaller. Sepals erect, oblong-ovate, 2.2–3 × 1–1.5 mm, apex obtuse, glabrous or minutely glandular-pubescent, veins 3, confluent. Petals yellow, oblong-ovate, 2.5–3.5 × 1–1.5 mm, base distinctly clawed, apex subobtuse, 2-callose. Stamens ca. 2.5 mm in male flowers and shorter than stigma and sterile, in female ones. Nectary band obscure. Ovary (female flowers only) almost inferior, carpels fused for more than half their length, narrowed to ca. 1 mm styles.

Distribution: Endemic to Nepal.

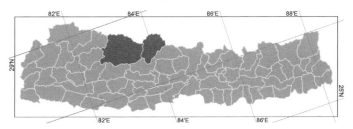

Altitudinal range: 4500–5600 m.

Ecology: Alpine areas on loose scree.

Flowering: July–August.

A narrow endemic restricted to Dolpo and Mustang.

52. *Saxifraga consanguinea* W.W.Sm., Notes Roy. Bot. Gard. Edinburgh 8: 132 (1913).

Stems simple, erect or caespitose, sometimes forming small clumps. Stolons arising from axils of basal leaves, sparsely glandular-pubescent. Leaves alternate, sessile. Basal leaves aggregated into a rosette, narrowly elliptic to oblanceolate, 4.5–9 × 1–2 mm, apex mucronate, margin eglandular- or glandular-ciliate, both surfaces glabrous. Cauline leaves of flowering stem elliptic to oblanceolate, 5–12 × 1–2 mm, apex mucronate, margin eglandular- or glandular-pubescent, both surfaces glabrous. Flowering stems 1.5–2.5 cm, glandular-pubescent. Flowers 1–10, in a cyme, unisexual (at least in some populations). Pedicels to 5 mm, glandular-pubescent. Sepals erect, ovate, 1.8–2.5 × 1–1.5 mm, apex obtuse to acute, outer surface glabrous or glandular-pubescent, margin sparsely glandular-pubescent, veins 3–5 confluent or not. Petals red, pink or yellow, elliptic, 1.5–2.5 × 1–1.5 mm, base contracted to a claw, apex obtuse, 2-callose. Stamens 1.5–2 mm in male flowers, shorter and sterile in females. Nectary a conspicuous, annular disk. Ovary (in females) semi-inferior, carpels fused for more than half their length, narrowed into ca. 1 mm styles.
Fig. 31j–n.

Distribution: Nepal and Tibetan Plateau.

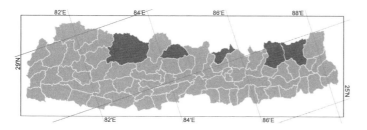

Altitudinal range: 4400–5200 m.

Ecology: Alpine, on ground.

Flowering: (June–)July–August(–September).

53. *Saxifraga pilifera* Hook.f. & Thomson, J. Proc. Linn. Soc., Bot. 2: 66 (1857).

Stems simple, erect or caespitose, sometimes forming small clumps. Stolons arising from axils of basal leaves, sparsely glandular-pubescent or glabrous. Leaves alternate. Basal leaves aggregated into a rosette, elliptic to oblanceolate, 5–8 × 1.5–3 mm, base attenuate, apex acute, both surfaces and margin shortly glandular-pubescent. Cauline leaves of flowering stem similar to basal. Flowering stems 2–8 cm, densely glandular-pubescent. Flowers 2–12, in a cyme, dioecious (at least in some populations). Pedicels 1–2(–5) mm, glandular-pubescent. Sepals erect, ovate, 2–2.5 × 1–1.2 mm, apex acute, outer surface and margin glandular-pubescent, veins 3–5 free or partly confluent. Petals pale green and red, or red, obovate, 2–3 × 1–1.5 mm, usually overlapping, base narrowed, apex obtuse, 2-callose. Stamens 1.5–2 mm. Nectary band obscure. Ovary semi-inferior, carpels fused for more than half their length, narrowed into ca. 1 mm styles.

Distribution: Nepal and E Himalaya.

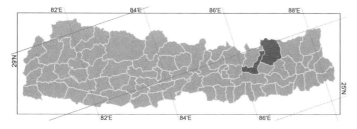

Altitudinal range: 4500–4800 m.

Ecology: Alpine.

Flowering: July–August.

54. *Saxifraga williamsii* Harry Sm., Bull. Brit. Mus. (Nat. Hist.), Bot. 2: 100, pl. 3 fig. i (1958).

Stems caespitose, many-branched, crowded leafy shoots forming cushions. Stolons absent. Leaves alternate. Cauline leaves loosely arranged or in some parts dense, linear to obovate-linear, 4–5 × 1.7 mm, apex obtuse, margin

setose-ciliate, somewhat fleshy, surfaces glabrous. Flowering stem to 3 mm, embedded in vegetative cushion, bulbils absent. Flowers solitary, bisexual. Pedicels up to 1 mm, glabrous. Sepals erect, spreading in fruit, rounded to ovate or broadly ovate, 2.5–4 × 2.5–3 mm, apex obtuse to subacute, margin patchily setose-ciliate, outer surface glabrous, veins 3, partly confluent. Petals white, orbicular, 6–8.5 × 5–5.5 mm, base abruptly narrowed to a ca. 2 mm claw, apex rounded, occasionally emarginate, not callose, persistent. Stamens 3.5–4 mm, anthers brown. Nectary band obscure. Ovary almost superior, carpels fused for half their length, narrowed into ca. 1.5 mm styles.

Distribution: Endemic to Nepal.

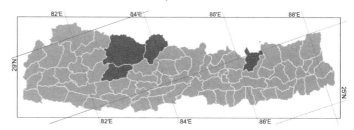

Altitudinal range: 4000–5100 m.

Ecology: Alpine, open grass slopes.

Flowering: June–August.

Sufficiently morphologically similar at first glance to be confused with members of section *Porphyrion*. However, *Saxifraga williamsii* lacks the chalk glands characteristic of that section and moreover, has the finely striate pollen surface typical of section *Ciliatae*.

55. *Saxifraga perpusilla* Hook.f. & Thomson, J. Proc. Linn. Soc., Bot. 2: 72 (1857).

Stems caespitose, many-branched, crowded leafy shoots forming cushions. Stolons absent. Leaves alternate sessile. Cauline leaves aggregated towards base into a rosette, obovate, ca. 2.6 × 1 mm, apex obtuse, margin white fimbriate, herbaceous, lower surface white pubescent towards apex, upper surface glabrous. Flowering stems 10–20 mm, embedded in cushion, bulbils absent. Flowers solitary, bisexual. Pedicels 5–8 mm, densely short glandular-pubescent. Sepals erect or ascending, broadly ovate to oblong, ca. 1.5 × 1.1 mm, apex scarious fimbriate, margin white glandular or eglandular-pubescent, outer surface white glandular-pubescent, veins 3, confluent. Petals yellow, elliptic to oblong, ca. 3–4 × 1–1.5 mm, base with a claw, apex obtuse, not callose. Stamens 2–2.5 mm. Nectary band obscure. Ovary semi-inferior, carpels fused for more than half their length, tapered into conical ca. 0.75 mm styles.
Fig. 31o–r.

Distribution: Nepal, E Himalaya and Tibetan Plateau.

Saxifragaceae

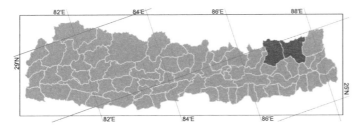

Altitudinal range: 4700–5300(–5600) m.

Ecology: Alpine, scree, rocky bare places.

Flowering: July–September.

56. *Saxifraga hemisphaerica* Hook.f. & Thomson, J. Proc. Linn. Soc., Bot. 2: 62 (1857).

Stems caespitose, many-branched, crowded leafy shoots forming cushions. Stolons absent. Leaves alternate. Cauline leaves imbricate, aggregated into a rosette, subspatulate, concave above, ca. 3 × 1 mm, apex obtuse, margin setose ciliate at base, fimbriate at the apex, somewhat fleshy, both surfaces glabrous. Flowering stems embedded among rosette leaves, not visible, 2–5 cm, bulbils absent. Flower solitary, unisexual, sessile. Pedicels less than 0.5 mm, glandular-pubescent. Sepals erect, linear-oblong, ca. 1 × 0.7 mm, apex scarious fimbriate, margin glandular-ciliate at base, somewhat fleshy, outer surface glabrate, veins 3–7, confluent. Petals yellow, 2.5–3.5 × ca. 1 mm, base with a claw, apex obtuse to acute, 2-callose or absent. Stamens ca. 2 mm in male flowers, female flowers with staminodes shorter than sepals. Nectary a prominent, annular disk. Ovary inferior, carpels fused for more than half their length, tapered into conical, up to 1 mm styles. Fig. 31s–x.

Distribution: Nepal, W Himalaya, E Himalaya and Tibetan Plateau.

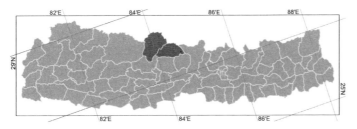

Altitudinal range: 4500–5100 m.

Ecology: Alpine.

Flowering: July–August.

57. *Saxifraga jacquemontiana* Decne., in Jacquem., Voy. Inde 4(1): 68, pl. 78 fig. 2 (1841).

Stems caespitose, many-branched, crowded leafy shoots forming cushions. Stolons absent. Leaves alternate. Cauline leaves persistent, aggregated distally into a terminal

rosette, elliptic, ca. 3.3 × 1.6 mm, apex obtuse, margin glandular-pubescent, somewhat fleshy, lower surface glandular-pubescent, upper surface sparingly so. Flowering stems 10–15 mm, embedded among rosette leaves, not or only slightly visible, glandular-pubescent, bulbils absent. Flowers solitary, bisexual. Pedicels 0.5–5 mm, glandular-pubescent. Sepals erect to spreading, ovate, ca. 2.8 × 1.8 mm, apex obtuse, margin glandular-pubescent, outer surface glandular-pubescent, veins 3, confluent. Petals yellow, obovate, ca. 5 × 2.5–3 mm, base with a ca. 0.7 mm claw, apex rounded to subacute, 2-callose. Stamens 2–3.5 mm. Nectary band obscure. Ovary semi-inferior, carpels fused for about half their length, tapered into slender, 1–1.5 mm styles. Fig. 32a–e.

Distribution: Nepal, W Himalaya, E Himalaya and Tibetan Plateau.

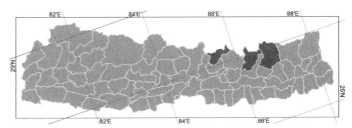

Altitudinal range: 4600–5300 m.

Ecology: Alpine.

Flowering: July–August.

58. *Saxifraga stella-aurea* Hook.f. & Thomson, J. Proc. Linn. Soc., Bot. 2: 72 (1857).
Saxifraga jacquemontiana var. *stella-aurea* (Hook.f. & Thomson) C.B.Clarke; *S. stella-aurea* var. *polyadena* Harry Sm.

Stems caespitose, many-branched, crowded leafy shoots forming cushions. Stolons absent. Leaves alternate. Cauline leaves aggregated below into a columnar cluster, sessile, spatulate to elliptic, 3–5 × 1.5–1.8 mm, apex obtuse, margin glandular- or eglandular-ciliate or glabrous, somewhat fleshy, both surfaces glabrous. Flowering stems 2–4 cm, dark brown glandular-pubescent, bulbils absent. Flowers solitary, bisexual. Pedicels 1–3 cm, glandular-pubescent. Sepals reflexed, elliptic to ovate, ca. 3 × 1.5 mm, apex obtuse to acute, sparsely glandular-pubescent or glabrous below, veins 3–6, free or partly or fully confluent. Petals yellow, spotted orange at base, ovate to elliptic, 3.5–4.5 × ca. 2 mm, base with a claw, apex obtuse, obscurely 2-callose. Stamens 4–4.3 mm. Nectary band obscure. Ovary almost superior, carpels fused for more than half their length, tapered into ca. 1 mm styles. Fig. 32f–i.

Distribution: E Himalaya, Tibetan Plateau and Assam-Burma.

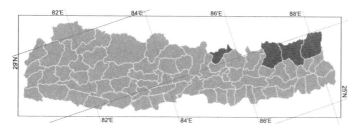

Altitudinal range: 4000–4700(–5300) m.

Ecology: Alpine.

Flowering: July–August.

This species is somewhat variable, and plants with larger petals (usually greater than 4 × 2 mm) and glandular-pubescent sepals have been referred to var. *polyadena*. However, since mixed populations and intermediates occur, we have refrained from recognizing any infraspecific taxa.

59. *Saxifraga llonakhensis* W.W.Sm., Rec. Bot. Surv. India 4: 192 (1911).

Stems caespitose, leafy shoots branched forming mats. Stolons absent. Leaves alternate. Basal leaves oblanceolate-linear, 5.2–8 × ca. 1 mm, apex mucronate, margin setose-villous, somewhat fleshy, both surfaces glabrous or eglandular- or glandular-villous. Cauline leaves oblong to linear, 2.8–5 × 0.6–1.6 mm, margin dark brown glandular-ciliate, apex obtuse, fleshy, both surfaces glabrous or dark brown glandular-pubescent. Flowering stems 4.2–9 cm, dark brown glandular-pubescent, bulbils absent. Flower solitary, rarely 2 or 3 in a cyme, bisexual. Pedicels slender, 8–15 mm, dark brown glandular-villous. Sepals erect, broadly ovate to triangular-ovate, 2–2.5 × 1.2–2.3 mm, apex obtuse, margin dark brown glandular-pubescent, leathery, outer surface dark brown glandular-pubescent, veins 3–5, free. Petals yellow, obscurely orange-yellow spotted, pandurate-oblong, 6.2–9.2 × 2–3.1 mm, base usually rounded to cuneate, with a 0.2–1 mm claw, apex obtuse, 4–6-callose. Stamens ca. 4.5 mm. Nectary band obscure. Ovary almost superior, carpels fused for about half their length, tapered into slender, up to 2 mm styles. Fig. 32j–n.

Distribution: Nepal, E Himalaya, Tibetan Plateau, Assam-Burma and E Asia.

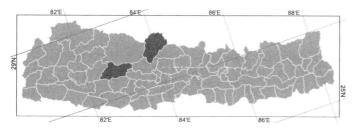

Altitudinal range: 4200–5200 m.

Ecology: Alpine, open stony slopes.

Flowering: July–August.

60. *Saxifraga engleriana* Harry Sm., Repert. Spec. Nov. Regni Veg. 20: 16 (1924).

Stems creeping, many-branched, forming small clumps or mats. Stolons absent. Leaves alternate, sessile. Cauline leaves clustered into a rosette near middle of stem, remote towards base and apex, obovate, ca. 3 × 1.5 mm, base cuneate, apex obtuse, fleshy, both surfaces usually glabrous. Flowering stems 1.5–8 cm, glabrous towards base, eglandular pale brown crisped pubescent above, without long crisped hairs in axils. Flower solitary, bisexual. Pedicels 4–7 mm, eglandular crisped pubescent. Sepals erect at first, then spreading, ovate to elliptic, 2–2.5 × 1.2–1.5 mm, apex obtuse, fleshy, glabrous, veins 3, confluent. Petals yellow, elliptic 3–4 × 1.5–2 mm, base narrowed into a claw, apex obtuse to subacute, 2-callose. Stamens 2.5–4 mm. Nectary a conspicuous annular disk. Ovary almost superior, carpels fused for more than half their length, tapered into slender, 1–2 mm styles.

Distribution: Nepal, E Himalaya and Tibetan Plateau.

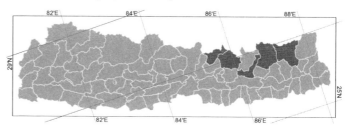

Altitudinal range: 4400–5000 m.

Ecology: Alpine.

Flowering: July–August.

61. *Saxifraga contraria* Harry Sm., Bull. Brit. Mus. (Nat. Hist.), Bot. 2: 252, 14 fig. a (1960).

Stems caespitose, shoots many-branched, sometimes forming loose mats. Stolons absent. Cauline leaves opposite, occasionally some alternate on young shoots, aggregated into a rosette. Blade elliptic, ca. 2 × 1 mm, apex obtuse, margin glabrous, occasionally setose-ciliate, fleshy, both surfaces glabrous. Flowering stems 1–5 cm, glabrous towards base, white pubescent above. Flowers solitary, bisexual. Pedicels 1–5 mm, white pubescent. Sepals spreading, subovate, 1.5–1.8 × 1–1.5 mm, apex obtuse, fleshy, glabrous, veins 3, free. Petals yellow, narrowly ovate to elliptic 2–3.5 × 1–1.3 mm, base contracted into a ca. 0.5 mm claw, apex retuse, 2-callose. Stamens ca. 2 mm. Nectary a conspicuous annular disk. Ovary semi-inferior, carpels fused for more than half their length, tapered into conical, ca. 0.5 mm styles. Fig. 32o–r.

Distribution: Nepal, E Himalaya and Tibetan Plateau.

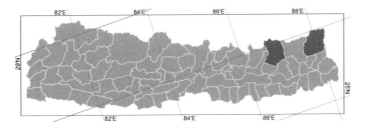

Altitudinal range: 4500–5300 m.

Ecology: Alpine, among stones beside streams.

Flowering: June–August. **Fruiting:** August–October.

62. *Saxifraga nanella* Engl. & Irmsch., Bot. Jahrb. Syst. 50: 44 (1914).

Stems caespitose, with branched leafy shoots, sometimes forming loose cushions. Stolons absent. Leaves alternate. Cauline leaves mostly caducous towards base, somewhat aggregated above, obovate, 3–4.5 × 1–1.5 mm, apex obtuse, margin eglandular-ciliate, somewhat fleshy, both surfaces glabrous. Flowering stems 1.5–4 cm, brown glandular-pubescent. Flowers solitary (or rarely 2), bisexual. Pedicels 6–12 mm, sparsely glandular-pubescent. Sepals spreading to reflexed, ovate, 1.3–1.5 × 0.8–1 mm, apex obtuse, margin glabrous, somewhat fleshy, outer surface glandular or eglandular-pubescent, veins 3–5, partly or fully confluent. Petals yellow, spotted orange, oblong, 4–6 × 1.5–2 mm, base with a short claw, apex obtuse to acute, not or 2-callose. Stamens ca. 3–4 mm. Nectary band obscure. Ovary almost superior, carpels fused for more than half their length, tapered into conical, ca. 1 mm styles.

Distribution: Nepal, E Himalaya, Tibetan Plateau and E Asia.

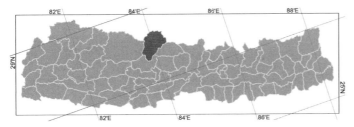

Altitudinal range: 4500–5000 m.

Ecology: Alpine, scree.

Flowering: July–August.

63. *Saxifraga microphylla* Royle ex Hook.f. & Thomson, J. Proc. Linn. Soc., Bot. 2: 72 (1857).
Saxifraga microviridis H.Hara.

Stems creeping, many-branched, forming loose mats. Stolons absent. Leaves alternate, sessile. Cauline leaves often clustered into a rosette near the middle of stem, remote towards ends, or remote throughout, oblong, ca. 3 × 1.5 mm,

base cuneate, apex obtuse, fleshy, both surfaces glabrous. Flowering stems to 2.5 cm, glabrous towards base, eglandular pale brown glandular-pubescent above, Flowers solitary, bisexual. Pedicels 1–3 mm, densely eglandular-crisped pubescent. Sepals erect, elliptic, ca. 1.8 × 1.2–1.5 mm, apex obtuse, fleshy, glabrous, veins 3, confluent. Petals green, obovate, 2–2.5 × 1–1.5 mm, base cuneate, apex obtuse, not callose. Stamens 2–3 mm. Nectary a conspicuous, annular disk. Ovary semi-inferior, carpels fused for more than half their length, tapered into 0.5 mm styles.

Distribution: Nepal and E Himalaya.

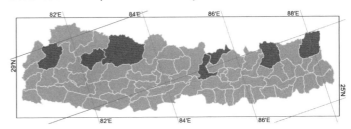

Altitudinal range: 3900–5600 m.

Ecology: Alpine.

Flowering: July–August.

64. *Saxifraga punctulata* Engl., Bot. Jahrb. Syst. 48: 601 (1912).

Stems caespitose or simple, erect. Stolons absent. Leaves alternate. Basal leaves forming a rosette, blade spatulate, 4–7 × 1.5–2 mm, fleshy, apex obtuse, margin cartilaginous, eglandular setose ciliate, upper surface glabrous, pustulate towards apex. Cauline leaves aggregated at stem tip, oblanceolate, 4–6 × 1–1.2 mm, apex obtuse to acute, margin blackish glandular-pubescent, both surfaces blackish glandular-pubescent, upper surface pustulate at apex. Flowering stems 1.5–4 cm, blackish glandular-pubescent. Flowers solitary or 2–4 in an umbellate cyme, bisexual. Pedicels 1–2.5 cm, blackish glandular-pubescent. Sepals erect to spreading, ovate to triangular, 2–2.5 × 2–2.5 mm, apex acute, outer surface and margin blackish glandular-pubescent, veins 5, confluent. Petals pale creamy white, with yellow and purple spots near base, elliptic, 7–9 × 3–5 mm, base abruptly narrowed into a ca. 0.6 mm claw, apex obtuse, not or obscurely callose. Stamens 3–4 mm. Nectary band obscure. Ovary almost superior, carpels fused for more than half their length, tapered into conical, ca. 0.5 mm styles.

Distribution: Nepal, E Himalaya and Tibetan Plateau.

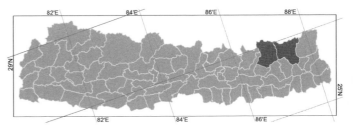

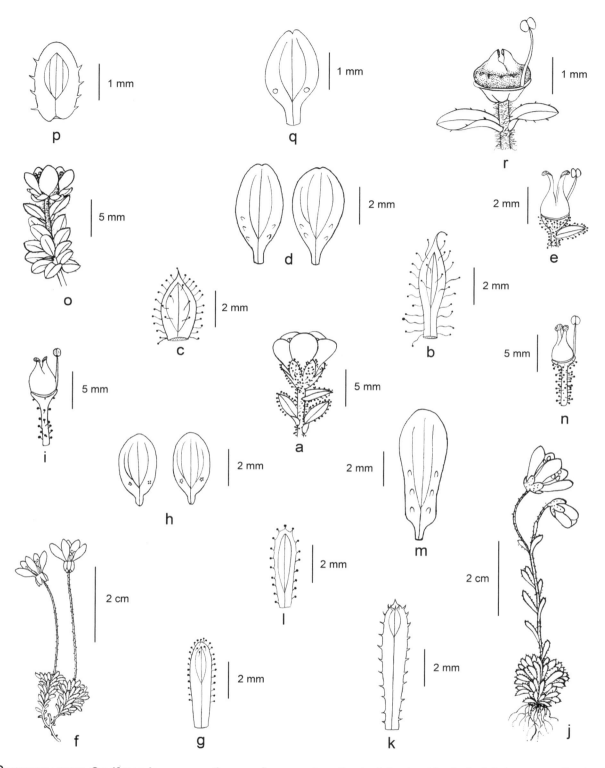

FIG. 32. SAXIFRAGACEAE. **Saxifraga jacquemontiana**: a, flower and cauline leaf; b, shoot leaf adaxial view; c, cauline leaf; d, petals; e, ovary and stamen. **Saxifraga stella-aurea**: f, flowering plant; g, leaf; h, petals; i, ovary and stamen. **Saxifraga llonakhensis**: j, flowering plant; k, basal leaf; l, cauline leaf; m, petal; n, ovary and stamen. **Saxifraga contraria**: o, stem with flower and cauline leaf; p, leaf; q, petal; r, ovary and stamen.

Saxifragaceae

Altitudinal range: 4500–4900(–5800) m.

Ecology: Alpine.

Flowering: July–August.

65. *Saxifraga umbellulata* Hook.f. & Thomson, J. Proc. Linn. Soc., Bot. 2: 71 (1857).

सून-चू-टेग Sun-chu-teg (Tibetan).

Stems simple, erect. Leaves alternate. Basal leaves aggregated into a rosette, spatulate, 6–14 × 1.5–3 mm, apex obtuse, margin glabrous, upper surface glabrous, somewhat pustulate towards apex. Cauline leaves oblong to subspatulate, 4.5–6.6 × 1.5–2 mm, margin brown glandular-pubescent, both surfaces brown glandular-pubescent or upper surface glabrous. Flowering stems up to 6 cm, brown glandular-pubescent. Flowers 2–23 in an umbellate cyme, bisexual. Pedicels 5–17 mm, brown glandular-pubescent. Sepals usually erect, ovate to narrowly triangular-ovate, (1.5–)2.2–3.5 × 1–1.3 mm, apex subobtuse to acute, outer surface and margin brown glandular-pubescent, veins 3, free. Petals yellow, sometimes with orange streaks, spathulate, 6.5–9 × 2–3 mm, base with a claw 0.4–0.5 mm, apex rounded to acute, 2-callose. Stamens 3–3.5 mm. Nectary band obscure. Ovary almost superior, carpels fused for more than half their length, tapered into slender, ca. 2 mm styles.

Distribution: Nepal, E Himalaya and Tibetan Plateau.

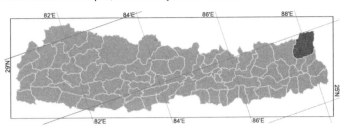

Altitudinal range: ca. 3700 m.

Ecology: Alpine, on rocks.

Flowering: July–August.

Pan *et al.* (Fl. China 8: 331. 2001) included *Saxifraga pasumensis* Marquand & Airy-Shaw, *S. muricola* Marquand & Airy-Shaw and *S. lhasana* H. Sm. within *S. umbellulata*. The differences between these species are summarised in that work. Here, we prefer to take a narrower view of the variation and the taxon described above is *S. umbellulata s.s.* We have seen no Nepalese specimens of the other taxa.

---

2. *Saxifraga* sect. *Mesogyne* Sternb., Revis Saxifrag. Suppl. 2:29 (1831).

Plants caespitose, with erect stems arising from a bulbiliferous rhizome, not forming cushions or mats and not stoloniferous. Leaves alternate. Chalk glands absent. Nectary a band of tissue encircling the base of the ovary, not disc-shaped. Petals white, without calloses. Stamens 10. Ovary superior.

Key to Species

1a	Leaf teeth or lobes mostly blunt	**66.** *S. asarifolia*
b	Leaf teeth acute	2
2a	Bulbils absent from axils of cauline leaves	**67.** *S. sibirica*
b	Bulbils (minute, less than 0.5 mm) present in axils of cauline leaves	**68.** *S. granulifera*

66. *Saxifraga asarifolia* Sternb., Revis. Saxifrag. Suppl.2: 33, pl. 24 (1831).
Saxifraga odontophylla Wall. ex Hook.f. & Thomson.

Stems caespitose, erect, arising from a bulbiliferous rhizome. Leaves alternate. Petiole of basal leaf 3–12(–15) cm. Blade suborbicular to reniform, 1.5–4.5 × 3–6 cm, base deeply cordate, apex obtuse, margin crenate with (7–)9–17 rounded teeth, ciliate, upper surface sparsely eglandular-pubescent, lower surface usually purplish. Cauline leaves similar to basal leaves, decreasing in size upwards, teeth more acute, axillary bulbils absent. Flowering stems 7–20(–30) cm, sparsely brownish or whitish villous. Flowers 2–10 in a lax cyme, bisexual. Pedicels 3–25 mm, glandular-pubescent. Sepals erect to spreading, ovate to oblong-ovate, 3.5–5 × ca. 2 mm, apex acute, outer surface and margin glandular-pubescent, veins 3–5, confluent near apex. Petals white, sometimes spotted or flushed reddish within, obovate, 8–12 × 6–8 mm, narrowed at the base, apex rounded. Stamens 3–4 mm, anthers reddish or orange. Ovary superior, carpels fused for more than half their length, narrowed into 2 mm styles.

Distribution: Nepal, W Himalaya and E Himalaya.

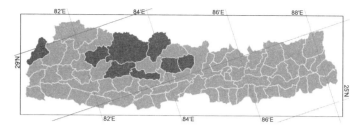

Altitudinal range: 3600–4600 m.

Ecology: Alpine meadows.

Flowering: July–August. **Fruiting:** September–October.

67. *Saxifraga sibirica* L., Syst. Nat., ed. 10, 2: 1027 (1759).
Saxifraga odontophylla Wall. ex Sternb.

Stems caespitose, erect, arising from a bulbiliferous rhizome. Leaves alternate. Petiole of basal leaf 1–4 cm, glandular-pubescent. Blade reniform, 6–8 × 8–12 mm, base cordate to truncate, apex obtuse, margin 5–7-lobed, upper surface and margin glandular-pubescent, lower surface glabrous. Cauline leaves similar to basal leaves, decreasing in size upwards, ultimately 3-lobed to entire and sessile, both surfaces glandular-pubescent or glabrous, margin glandular-pubescent; axillary bulbils absent. Flowering stems 6–15 cm, densely glandular-pubescent. Flowers 2–5, in a lax cyme, bisexual. Pedicels slender, 8–20 mm, glandular-pubescent. Sepals erect, lanceolate, 4–5 × 1–2 mm, apex obtuse to subacute, outer surface and margin glandular-pubescent, veins 3, confluent. Petals white, obovate-cuneate, 8–12 × 3–4 mm, base attenuate, apex rounded. Stamens 3–5 mm, anthers cream or white. Ovary superior, carpels fused up to half their length, narrowed into 1–2 mm styles.
Fig. 33a–c.

Distribution: Nepal, W Himalaya, E Himalaya, Tibetan Plateau, E Asia, N Asia and Europe.

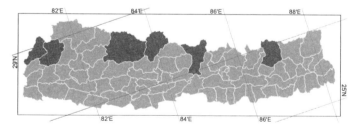

Altitudinal range: 3200–5000(–5800) m.

Ecology: Alpine meadows and pastures.

Flowering: July–August. **Fruiting:** ?September.

68. *Saxifraga granulifera* Harry Sm., Bull. Brit. Mus. (Nat. Hist.), Bot. 2: 259, pl. 21B (1960).
Saxifraga sibirica var. *bulbifera* Harry Sm.

Stems caespitose, erect, arising from a bulbiliferous rhizome. Leaves alternate. Petiole of basal leaf 1.5–4 cm, glandular-pubescent. Blade reniform, 5–12 × 8–15 mm, base cordate or truncate, apex acute, 5–9-lobed, lobes broadly ovate to triangular-ovate, both surfaces nearly glabrous, margin glandular-pubescent. Cauline leaves similar to basal leaves, decreasing in size upwards, ultimately sessile, axillary bulbils present, 0.2–0.5 mm, reddish brown. Flowering stems 10–15 cm, sparsely glandular-pubescent. Flowers 1(–5) in a lax cyme, bisexual. Pedicels slender, 1–7 cm, glandular-pubescent. Sepals erect, ovate or narrowly ovate to lanceolate, 1.2–2.5 × ca. 1 mm, acute to subacute, outer surface and margin red-glandular pubescent, veins 3, confluent. Petals white, obovate-cuneate to oblanceolate, (5–)7–12 × 2–2.5(–4) mm, base attenuate, apex obtuse to rounded. Stamens 3.5–5 mm, anthers purple. Ovary superior, carpels fused for more than half their length, narrowed into slender, ca. 2.5 mm styles.

Distribution: Nepal, W Himalaya, E Himalaya, Tibetan Plateau and E Asia.

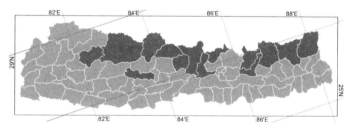

Altitudinal range: (2700–)3100–4350(–5000) m.

Ecology: Alpine, mossy boulders, cliffs, among rocks.

Flowering: July–September. **Fruiting:** September–October.

3. *Saxifraga* sect. *Porphyrion* Tausch, Hort. Canal. I (1823).

Plants caespitose, many branched, forming mats or cushions. Rhizome neither bulbiliferous nor stoloniferous. Leaves alternate or opposite, with chalk glands on the upper surface near the margin and apex, glands surrounded by white crystalline deposits. Petals white or pink to red or purple, without calloses. Stamens 8 or 10. Ovary superior to semi-inferior.

Key to Species

1a	Leaves opposite. Flowers white, solitary (1–3-flowered in *Saxifraga roylei*) 2	
b	Leaves alternate. Flowers white to pink or red, 1–several..5	

| 2a | Bases of opposite pair of leaves joining at an acute angle, margin denticulate-ciliate....................................**72. *S. roylei*** |
| b | Bases of opposite pair of leaves confluent, margin glabrous ....3 |

| 3a | Flowers sessile or subsessile. Stems forming large usually loose mat-like tufts. Leaves hardly imbricate or imbricate only at top of stems, 3–3.5 mm ....**70. *S. georgei*** |
| b | Pedicels longer than 5 mm. Shoots columnar, forming small sparse tufts. Leaves densely imbricate, 2–2.5 mm ....4 |

| 4a | Sepal veins confluent. Sepal chalk gland absent...**69. *S. alpigena*** |
| b | Sepal veins free. Sepal chalk gland present...**71. *S. quadrifaria*** |

| 5a | Flowers 2–several (solitary only in dwarfed specimens)...6 |
| b | Flowers solitary ....11 |

| 6a | Sepals with 1–3 chalk glands..7 |
| b | Sepals without chalk glands...8 |

| 7a | Petals 3–5.5 mm, white or pale pink. Leaf apex obtuse to subacute, subtruncate, often appearing horseshoe-like ....**81. *S. andersonii*** |
| b | Petals ca. 6.5 mm, deep rose. Leaf apex acute ....**82. *S. rhodopetala*** |

| 8a | Leaf chalk glands ca. 7. Flowering stems up to ca. 3 cm. Plant forming a tight cushion ....9 |
| b | Leaf chalk glands more than 9. Flowering stems at least 5 cm. Plant forming a loose cushion ....10 |

| 9a | Sepal veins confluent. Leaves 5–6 mm, apical quarter much thickened...........................**83. *S. afghanica*** |
| b | Sepal veins free. Leaves 7–8 mm, apex not thickened ....**85. *S. micans*** |

| 10a | Leaves at least 10 mm ....**84. *S. cinerea*** |
| b | Leaves up to 7 mm..**86. *S. stolitzkae*** |

| 11a | Leaves with several chalk glands..12 |
| b | Leaves with 1 chalk gland (sometimes obscure in *Saxifraga subsessiliflora*) ....14 |

| 12a | Leaves imbricate only at top of stems. Shoots forming a loose mat ....**78. *S. poluninana*** |
| b | Leaves imbricate throughout. Shoots forming tight cushions ....13 |

| 13a | Sepals ovate, obtuse, up to 3mm, sparingly glandular-pubescent ....**79. *S. mira*** |
| b | Sepals narrowly triangular, acute, longer than 4 mm, densely glandular pubsecent...........**80. *S. staintonii*** |

| 14a | Leaf apex obtuse, not thickened, chalk gland subapical ....15 |
| b | Leaf apex truncate-triquetrous with chalk gland set in the tip..16 |

| 15a | Petals pink or lilac ....**76. *S. lowndesii*** |
| b | Petals white ....**77. *S. subsessiliflora*** |

| 16a | Leaf apex cartilaginous-fimbriate ....**73. *S. hypostoma*** |
| b | Leaf apex glabrous ....17 |

| 17a | Leaf with base slightly dilated, glandular-pubescent and apex minutely apiculate...........**74. *S. kumaunensis*** |
| b | Leafwith base tapered, denticulate-ciliate and apex obtuse ....**75. *S. pulvinaria*** |

69. *Saxifraga alpigena* Harry Sm., Bull. Brit. Mus. (Nat. Hist.), Bot. 2: 97, 1 fig. i (1958).

Plants caespitose, many-branched, crowded leafy shoots 4 mm diameter, forming loose cushions. Leaves on vegetative shoots opposite, tightly imbricate, sessile, ovate to obovate, 1.5–2.5 × 1–2 mm, apex obtuse, thickened, margin of leaf-pairs confluent, glabrous, cartilaginous, both surfaces glabrous, chalk gland solitary. Flowering stems 5–15 mm above cushion surface, glandular-pubescent, leaves 1 or 2. Flowers solitary, bisexual. Pedicels 2–5 mm, glandular-pubescent. Sepals erect, ovate, ca. 2 mm, apex obtuse, outer surface and margin glandular-pubescent, veins 3, confluent near apex. Petals white, ca. 6 × 4 mm, limb orbicular, base narrowed to a claw, apex rounded. Stamens 8, 1.7–2.7 mm, anthers red-brown. Nectary band obscure. Ovary semi-inferior, carpels fused only at base, narrowed into ca. 1 mm styles.

Distribution: Endemic to Nepal.

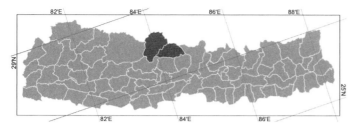

Altitudinal range: 3400–4200 m.

Ecology: Alpine shingle and moraine, amongst rocks.

Flowering: June.

An endemic only known from a few collections.

70. *Saxifraga georgei* J.Anthony, Notes Roy. Bot. Gard. Edinburgh 18: 33 (1933).

Plants caespitose, many-branched, crowded leafy shoots forming hemispherical cushions. Leaves on vegetative shoots opposite, spreading, sessile, broadly ovate, 3–4 × 1.5–2 mm, apex obtuse, recurved, margin of opposite leaf-pairs confluent, quite thick, both surfaces glabrous, chalk gland solitary. Flower stems 5–10 mm, glandular- and eglandular-pubescent. Flowers solitary, bisexual. Pedicels 1–3 mm. Sepals erect, broadly ovate, 1.5–2 × 1.5–2 mm, apex subobtuse, glabrous or sparingly glandular-pubescent on outer surface and basal margin, veins 3, confluent near apex or not, chalk gland solitary. Petals white to very pale pink, obovate, 3–5 × 2–3 mm, base contracted into a claw, apex obtuse. Stamens 8, equalling or longer than sepals, anthers brownish to blackish. Nectary band obscure. Ovary semi-inferior, carpels fused for about half thier length, tapering into erect, 2.5–3 mm styles.

Distribution: Nepal, E Himalaya, Tibetan Plateau and E Asia.

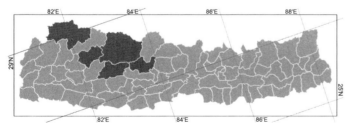

Altitudinal range: 3900–5400 m.

Ecology: Alpine, on wet rocks, under overhanging rocks, on boulders.

Flowering: May–July.

71. *Saxifraga quadrifaria* Engl. & Irmsch., in Engl., Pflanzenr. IV. 117(Heft 69): 575 (1919).

Plants caespitose, many-branched, crowded leafy shoots

forming pulvinate cushions. Leaves on vegetative shoots opposite, tightly imbricate, sessile, ovate, 2–2.5 × 1–1.5 mm, apex subtruncate-triquetrous, margin of leaf-pairs forming an acute angle, denticulate-ciliate, both surfaces glabrous, chalk gland solitary. Flowering stems 5–6 mm above cushion surface, glandular-pubescent, leaves 0 or 1. Flowers solitary, bisexual. Pedicels 5–6 mm, glandular-pubescent. Sepals erect, ovate, 1.5–2 × ca. 1 mm, apex subacute, outer surface and margin glandular-pubescent, veins 3, free. Petals white, 4–6 × 2 mm, limb narrowly obovate, base attenuate, apex rounded. Stamens 8, 2–3 mm, anthers red-brown. Nectary band obscure. Ovary semi-inferior, carpels fused only at base, narrowed into ca. 1 mm styles.

Distribution: Nepal, E Himalaya and Tibetan Plateau.

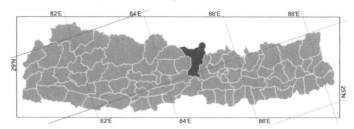

Altitudinal range: ca. 4100 m.

Ecology: Alpine.

Flowering: June.

Described from Tibet and recorded from Nepal by Kitamura (Fauna Fl. Nepal Himalaya: 146. 1955), based on material collected by Nakao (3 June 1953) from Manaslu (probably KYO), but not seen by the authors. Alpine plant enthusiasts, however, report seeing either it or putative hybrids involving it in Nepal.

72. *Saxifraga roylei* Harry Sm., Bull. Brit. Mus. (Nat. Hist.), Bot. 2: 95, 1 fig. o (1958).

Plants caespitose, many-branched, crowded leafy shoots forming cushions. Leaves on vegetative stems opposite, loosely arranged or imbricate above, ascending, not or slightly recurved towards apex, sessile, ovate- or obovate-linear, to 3.5 mm, apex subacute, usually minutely mucronulate, margin of opposite leaf-pairs meeting at an acute angle, not confluent, somewhat leathery, both surfaces glabrous, margin denticulate-ciliate at base, chalk gland solitary. Flowering stems 10–15 mm, much shorter than vegetative shoots, with long, fine glandular hairs, leaves 3 or 4, distant. Flower solitary or 2–3 in a cyme, bisexual. Pedicels 2–3 mm, glandular-pubescent. Sepals erect, ovate, 2–2.5 × 1.5 mm, apex subacute, margin and outer surface sparingly glandular pubescent, veins 3, confluent near apex, minute solitary chalk gland near apex. Petals white, ca. 4.6 × 3.2 mm, limb suborbicular, base abruptly narrowed to a claw, apex rounded. Stamens 10, ca. 3 mm, anthers reddish. Nectary obscure. Ovary semi-inferior, carpels fused only at base, tapering into erect, 2.5–3 mm styles.

Distribution: Endemic to Nepal.

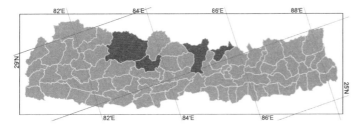

Altitudinal range: 3200–3800 m.

Ecology: Alpine, streamsides and ledges on rock faces.

Flowering: June.

73. *Saxifraga hypostoma* Harry Sm., Bull. Brit. Mus. (Nat. Hist.), Bot. 2: 103, 4 fig. d (1958).

Plants caespitose, many-branched, crowded leafy shoots forming pale green cushions. Leaves on vegetative shoots alternate, imbricate, sessile, ovate to oblong, 1.5–3 × 1 mm, apex truncate, triquetrous, margin eglandular-ciliate, cartilaginous-laciniate towards apex, both surfaces glabrous, chalk gland solitary, subapical. Flowering stems very short, embedded among vegetative shoots, glabrous, leaves 3 or 4. Flowers solitary, bisexual. Pedicels 0.5–3 mm, sparingly glandular-pubescent. Sepals erect, ovate, ca. 1.5 × 1.5 mm, apex obtuse, outer surface glabrous, margin glandular-villous-pubescent, veins 3, confluent near apex, chalk gland absent. Petals white, orbicular-obtriangular, 3–4 × 2.5–4 mm, base tapered into a short claw, apex rounded. Stamens 10, 1–1.5 mm, anthers red. Nectary band obscure. Ovary semi-inferior, carpels fused for about half their length, abruptly tapered into erect, ca. 1 mm styles.

Distribution: Endemic to Nepal.

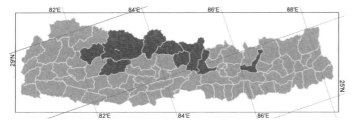

Altitudinal range: 3900–5300 m.

Ecology: Alpine regions on or amongst rocks, rock ledges, scree.

Flowering: (May–)June–September.

74. *Saxifraga kumaunensis* Engl., in Engl., Pflanzenr. IV. 117(Heft 69): 576, fig. 119 (1919).

Plants caespitose, many-branched, crowded obconical leafy shoots, to 10 mm diameter, forming cushions. Leaves on vegetative shoots alternate, densely imbricate, appressed to the stem, sessile, oblong, 2.5–4 × 1–1.2 mm, apex obtuse, mucronulate, leathery, both surfaces glabrous, margin glabrous towards apex, ciliate on dilated leaf-base, chalk gland solitary. Flowering stems to 10 mm, glandular and eglandular-pubescent, leaves 4 or 5. Flowers solitary, bisexual. Pedicels 1–2 mm. Sepals erect to spreading, broadly ovate, 1.5–2 × ca. 1.2 mm, apex obtuse, outer surface glabrous, margin glandular-pubescent, veins 3, free, chalk gland solitary. Petals white, obovate, ca. 4 × 2.5 mm, base cuneate or tapered into a short claw, apex rounded. Stamens 10, ca. 2 mm, anthers red-brown. Nectary band obscure. Ovary semi-inferior, carpels fused only at base, tapered into erect, ca. 2 mm styles.

Distribution: Nepal and W Himalaya.

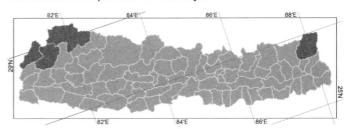

Altitudinal range: 3300–4800 m.

Ecology: Alpine, rock crevices.

Flowering: May–July.

75. *Saxifraga pulvinaria* Harry Sm., Bull. Brit. Mus. (Nat. Hist.), Bot. 2: 105, pl. 4 fig. m (1958). *Saxifraga imbricata* Royle.

Plants caespitose, many-branched, crowded columnar leafy shoots, 5–6 mm diameter, forming cushions. Leaves on vegetative shoots alternate, imbricate, sessile, appressed to stem, linear-oblong, 3.5–5 × 1–2 mm, apex truncate, triquetrous, somewhat leathery, both surfaces glabrous, margin denticulate-ciliate, chalk gland solitary. Flowering stems embedded among vegetative shoots, ca. 15 mm, sparingly glandular-pubescent. Flowers solitary, bisexual, sessile. Sepals erect, subtriangular-ovate to broadly ovate, 1.5–2 × 1.3–1.4 mm, apex obtuse to acute, outer surface glabrous, margin glandular-pubescent, veins 3, confluent or not near apex, chalk gland solitary. Petals white, obovate to oblanceolate or oblong, 3.5–5.3 × 2–3 mm, base gradually narrowed to a claw, apex obtuse or rounded to retuse. Stamens 10, ca. 2 mm, anthers yellow. Nectary obscure. Ovary semi-inferior, carpels fused only at base, tapered into erect, 1–2 mm styles. Fig. 33d–h.

Distribution: Nepal and W Himalaya.

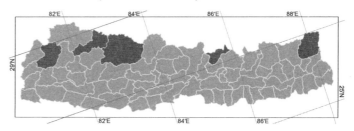

Altitudinal range: 3800–5900 m.

Ecology: Alpine, stony slopes, rock crevices, cliffs.

Flowering: April–July(–August).

76. *Saxifraga lowndesii* Harry Sm., Bull. Brit. Mus. (Nat. Hist.), Bot. 2: 106, 5 fig. d (1958).

Plants caespitose, many-branched, crowded, decumbent, leafy shoots 7–9 mm diameter, forming mats or loose cushions. Leaves on vegetative shoots alternate, loosely imbricate, sessile, obovate-linear, 5–7 × 2–2.5 mm, apex obtuse or retuse, margin cartilaginous, both surfaces glabrous, margin sparingly glandular-pubescent at base, chalk gland solitary. Flowering stems 2 mm, scarcely exceeding subtending vegetative shoots, glabrous. Flowers solitary, bisexual. Pedicels ca. 1 mm. Sepals erect to spreading, broadly ovate, ca. 2.5 × 3 mm, apex rounded, margin glandular-ciliate, both surfaces glabrous, veins 3–5, confluent, chalk glands 1 or 2. Petals rose-lilac or white, suborbicular, 6–8 × 4–7 mm, base narrowed to a claw, apex truncate. Stamens 10, 3–4.5 mm, anthers red-brown. Nectary band obscure. Ovary semi-inferior, carpels fused only at base, tapered into erect, 2–2.5 mm styles.

Distribution: Endemic to Nepal.

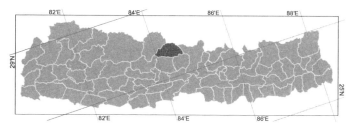

Altitudinal range: 3800–4200 m.

Ecology: Alpine, among rocks.

Flowering: June. **Fruiting:** July.

A poorly recorded endemic known from Sabche Khola and an unlocalised collection from E Nepal (Swan 421, BM).

77. *Saxifraga subsessiliflora* Engl. & Irmsch., in Engl., Pflanzenr. IV. 117(Heft 69): 573 (1919).
Saxifraga lolaensis Harry Sm.; *S. matta-florida* Harry Sm.

Plants caespitose, many-branched, crowded cylindrical leafy shoots forming cushions. Leaves on vegetative shoots alternate, densely imbricate, sessile, appressed to the stem, oblong-obovate, 3–6 × 1–2.5 mm, slightly keeled towards base on lower surface, apex obtuse to acute, not thickened, both surfaces glabrous, margin minutely glandular-ciliate or glabrous, chalk gland solitary. Flowering stems embedded among vegetative shoots, 1–1.5 mm, subglabrous. Flowers solitary, bisexual, sessile. Sepals erect or ascending, ovate, 2–2.5 × ca. 1.5 mm, apex obtuse, both surfaces glabrous,

margin glandular-pubescent, veins 3, confluent near apex, chalk gland solitary. Petals white, oblong-obovate, 3–5 × ca. 1.5 mm, base cuneate, apex rounded. Stamens 10, 2–2.5 mm, anthers reddish brown. Nectary inconspicuous. Ovary semi-inferior, carpels fused for half their length, tapered into free, thick, ca. 1 mm styles.
Fig. 33i–m.

Distribution: Nepal and E Himalaya.

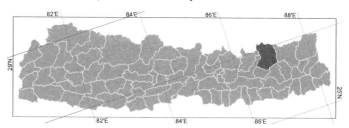

Altitudinal range: 4400–4600 m.

Ecology: Alpine.

Flowering: June.

78. *Saxifraga poluninana* Harry Sm., Bull. Brit. Mus. (Nat. Hist.), Bot. 2: 114, pl. 8 fig. h (1958).

Plants caespitose, many-branched, woody at base, decumbent leafy shoots, 10–13 mm diameter, forming cushions. Leaves on vegetative shoots alternate, densely packed, long-persistent, spreading, recurved, sessile, linear, 5–6.2 × ca. 1.5 mm, apex subacute, somewhat thickened, both surfaces glabrous, margin denticulate-ciliate in basal third, chalk glands 5–7, submarginal from apex downwards. Flowering stems 5–25 mm, glandular-pubescent, leaves 3–4. Flowers solitary, bisexual. Pedicels 1–2 mm, glandular-pubescent. Sepals erect, ovate, 3–4 × 1.5–2.5 mm, apex obtuse to acute, recurved, outer surface and margin densely glandular-pubescent except at the apex, veins 3, confluent, chalk gland 1, inconspicuous. Petals white to pink, obovate to suborbicular, 8–10 × 4–5 mm, base gradually narrowed, apex rounded. Stamens 10, 4–5 mm, anthers dark red. Nectary obscure. Ovary semi-inferior, carpels fused only at base, tapered into erect, 3–4 mm styles.

Distribution: Endemic to Nepal.

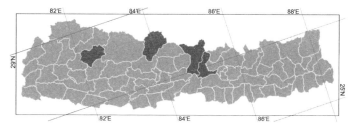

Altitudinal range: 3200–3500 m.

Ecology: Alpine, crevices on rocks beside streams, usually in shade.

Saxifragaceae

Flowering: May.

Believed to hybridize with *Saxifraga cinerea* Harry Sm. where their respective geographical and altitudinal ranges overlap. Records at lower altitude (2000 m) require confirmation.

79. *Saxifraga mira* Harry Sm., Bull. Brit. Mus. (Nat. Hist.), Bot. 2: 114, 8 fig. d (1958).

Plants caespitose, many-branched, woody at base, crowded, decumbent leafy shoots to 10 mm diameter, forming cushions. Leaves on vegetative shoots alternate, densely packed, long-persistent, spreading, sessile, linear-oblong, 5–6.5 × 2–2.5 mm, apex obtuse or acute, thick, both surfaces glabrous, margin denticulate-ciliate in upper third, chalk glands 5–7, submarginal from apex downwards. Flowering stems 6–8 mm, densely glandular-pubescent, leaves 2–3. Flowers solitary, bisexual. Pedicels 1–2 mm. Sepals erect, broadly ovate, ca. 3 × 3 mm, apex obtuse, outer surface minutely glandular-pubescent towards base, margin minutely glandular-pubescent, veins 3, free, chalk glands absent. Petals white or rose pink, orbicular, 7–7.5 × 5–6 mm, base abruptly tapered into a short claw ca. 1.3 mm, apex rounded. Stamens 10, 3–4.5 mm, anthers yellow. Nectary band inconspicuous. Ovary semi-inferior, carpels fused only at base, tapered into erect, 3–4 mm styles.

Distribution: Endemic to Nepal.

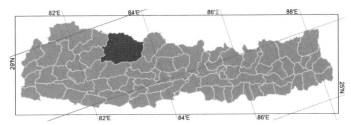

Altitudinal range: ca. 4400 m.

Ecology: Alpine, on cliffs, often on shaded north-facing rock faces.

Flowering: June.

An endemic only known from the type collection, *Polunin, Sykes and Williams 1094* (BM, E) on the Barbung Khola.

80. *Saxifraga staintonii* Harry Sm., Bull. Brit. Mus. (Nat. Hist.), Bot. 2: 118, pl. 13 fig. j (1958).

Plants caespitose, many-branched, somewhat woody at base, crowded leafy shoots up to 13 mm diameter, forming cushions. Leaves on vegetative shoots alternate, imbricate, ascending or spreading to recurved, sessile, linear-oblong, ca. 9 × 2 mm, apex acute to acuminate, leathery, both surfaces glabrous, margin denticulate-ciliate on slightly dilated leaf-base, chalk glands 9–13, submarginal from apex downwards. Flowering stems ca. 5 cm, long glandular-hairy, leaves 7–9. Flowers solitary, bisexual. Pedicels 10–15 mm, glandular-pubescent.

Sepals erect, apex recurved, ovate to narrowly triangular, 4–4.5 × ca. 2 mm, apex acute, narrowly membranous, outer surface and margin glandular-pubescent, veins 3, confluent near apex, chalk gland subapical, solitary. Petals white, narrowly obovate, ca. 10 × 4 mm, base gradually narrowed, apex rounded. Stamens 10, ca. 4 mm, anthers yellow. Nectary obscure. Ovary semi-inferior, carpels fused only at base, tapered into ca. 2 mm styles.

Distribution: Endemic to Nepal.

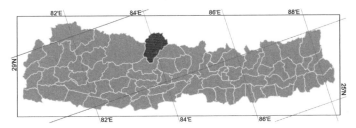

Altitudinal range: ca. 4800 m.

Ecology: Alpine, on steep rocks.

Flowering: August.

Apparently known only from the type collection, *Stainton, Sykes & Williams 7276* (BM) from Samargaon.

81. *Saxifraga andersonii* Engl., Bot. Jahrb. Syst. 48: 609 (1912).

Plants caespitose, many-branched, somewhat woody at base, crowded leafy shoots 8–12 mm diameter, forming compact cushions. Leaves on vegetative shoots alternate, densely packed, aggregated into an apical rosette, sessile, obovate to oblong to oblanceolate-linear or oblong-spatulate, 5–10 × 1.2–2.6(–3) mm, apex obtuse to subacute, recurved, leathery, both surfaces glabrous, margin ciliate at base, chalk glands 3–7. Flowering stems ca. 3 cm, glandular-pubescent, leaves 4–8. Flowering stem leaves oblanceolate-linear to oblong-spatulate, 5–6 × 1.5–1.8 mm, apex somewhat recurved, chalk glands 3–7, both surfaces glabrous, margin ciliate at base. Flowers (1 or)2–7 in a cyme, bisexual. Pedicels to 4 mm in flower, elongating to 10 mm in fruit, minutely glandular-pubescent. Sepals reflexed, ovate, 1.5–2 × 1–1.5 mm, apex obtuse, outer surface and margin sparsely glandular-pubescent, veins 3, confluent at apex, chalk gland solitary near apex. Petals white, oblong-obovate, 3–5 × 1.5–3.5 mm, base narrowed to a claw, apex obtuse. Stamens 10, 2–3 mm, anthers reddish to blackish. Nectary band obscure. Ovary semi-inferior, carpels fused only at base, tapered into erect, ca. 3 mm styles.
Fig. 33n–r.

Distribution: Nepal, E Himalaya and Tibetan Plateau.

260

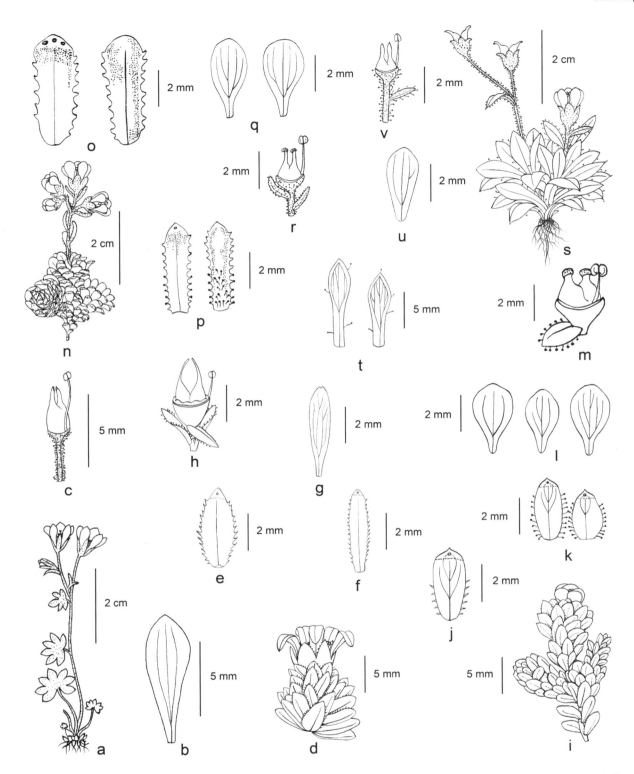

FIG. 33. SAXIFRAGACEAE. **Saxifraga sibirica**: a, flowering plant; b, petal; c, ovary and stamen. **Saxifraga pulvinaria**: d, flowering plant; e, basal leaf; f, cauline leaf; g, petal; h, ovary and stamen. **Saxifraga subsessiliflora**: i, flowering plant; j, lower shoot leaf; k, upper shoot leaf; l, petals; m, ovary and stamen. **Saxifraga andersonii**: n, flowering plant; o, shoot leaves; p, cauline leaves; r, ovary and stamen. **Saxifraga coarctata**: s, plant with flowers and fruits; t, basal leaf; u, petal; v, ovary and stamen.

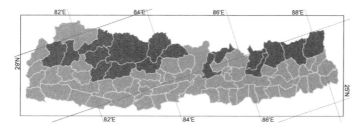

Altitudinal range: 3400–5500 m.

Ecology: Alpine, on soil slopes, rocky slopes, rocks, crevices, gravels, moraine.

Flowering: May–July. **Fruiting:** August–October.

Putative hybrids with *Saxifraga hypostoma* Harry Sm. (syn. *S.* x *tukuchensis* J. Bürgel) are common in central Nepal, where the parents meet. *Saxifraga* x *hetenbeliana* J. Bürgel is believed to be the result of crossing with *S. pulvinaria* Harry Sm. Field observations suggest also that hybridization with *S. lowndesii* Harry Sm. may occur where their respective geographical and altitudinal ranges overlap. Putative hybrids with *S. quadrifaria* Engl. & Irmsch. have also been observed, and one of the products of this hybrid combination is said to be *S. alpigena* Harry Sm. (q.v.). Experimental evidence is needed in support of all these suggestions.

82. *Saxifraga rhodopetala* Harry Sm., Bull. Brit. Mus. (Nat. Hist.), Bot. 2: 124, 12 fig. e (1958).

Plants caespitose, many-branched, woody at base, leafy shoots, ca. 13 mm diameter, forming cushions. Leaves on vegetative shoots alternate, densely packed, spreading, recurved, sessile, linear to obovate-linear, ca. 5 mm, apex subacute to acute, thick, both surfaces glabrous, margin denticulate-ciliate in basal third, long-persistent, chalk glands 5–11. Flowering stems 3–4 cm, densely glandular-pubescent, leaves 5 or 6. Flowers 5–9 in a cyme, bisexual. Pedicels 1–8 mm. Sepals erect, ovate, ca. 3 × 1.5–2 mm, apex obtuse, outer surface and margin densely glandular-pubescent except at apex, veins 3, confluent, chalk glands 1–3. Petals deep rose, narrowly obovate, ca. 6.5 × 2.5 mm, apex obtuse, base gradually narrowed. Stamens 10, 5 mm. Nectary obscure. Ovary semi-inferior, carpels fused only at base, tapered into erect, ca. 3 mm styles.

Distribution: Endemic to Nepal.

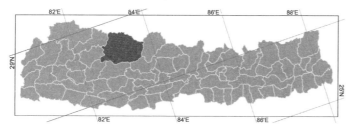

Altitudinal range: 3900–4600 m.

Ecology: Alpine, cliff faces, rock ledges and stony slopes.

Flowering: June. **Fruiting:** September.

A narrow endemic known from the type collection near Phoksundo Tal (*Stainton, Sykes and Williams 2196*, BM, E) and from Lulo Khola (*Polunin, Sykes & Williams 3472*, BM, E)

83. *Saxifraga afghanica* Aitch. & Hemsl., J. Linn. Soc., Bot. 18: 56 (1880).

Plants caespitose, many-branched, crowded leafy shoots, ca. 10 mm diameter, forming cushions. Leaves on vegetative shoots alternate, loosely imbricate, sessile, oblong to subspatulate, 4–7 × 1.3–2 mm, apex obtuse, slightly recurved, leathery, margin narrowly cartilaginous, both surfaces glabrous, margin denticulate-ciliate towards base, chalk glands 3–8. Flowering stems 8–25 mm, glandular-pubescent, leaves several. Flowers (2 or)3–4 in a cyme, rarely solitary, bisexual. Pedicels to 2 mm, glandular and eglandular-pubescent. Sepals erect, oblong, 2–3 × 1–1.4 mm, apex obtuse, outer surface and margin sparsely glandular-pubescent, veins 3, free, chalk gland 1. Petals rose, obovate, ca. 5 × 2.5–3.3 mm, base gradually narrowed into a long claw, apex obtuse. Stamens 10, 3–3.5 mm, anthers red. Nectary band obscure. Ovary semi-inferior, carpels fused for half their length, narrowed into the erect, 1.5–2 mm styles.

Distribution: Nepal, W Himalaya, Tibetan Plateau and SW Asia.

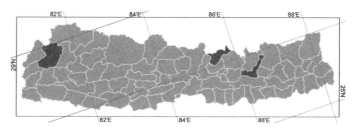

Altitudinal range: 4600–4800 m.

Ecology: Alpine, on rocks.

Flowering: May–June.

84. *Saxifraga cinerea* Harry Sm., Bull. Brit. Mus. (Nat. Hist.), Bot. 2: 128, 14 fig. k (1958).

Plants caespitose, many-branched, woody at base, crowded leafy shoots, ca. 20 mm diameter, forming loose cushions. Leaves on vegetative shoots alternate, sessile, linear, 10–12 × 1.5–2 mm, apex acute, recurved, leathery, margin with a narrow cartilaginous border, both surfaces glabrous, margin denticulate-ciliate in the basal part, chalk glands 13–18. Flowering stems up to 8 cm, densely glandular-pubescent, leaves ca. 5, shorter than internodes. Flowers 2–6 in a cyme, bisexual. Pedicels 2–10 mm, dark brown glandular-pubescent. Sepals erect, ovate, ca. 3 × 1.5–2 mm, apex subacute, outer surface and margin glandular-pubescent, veins 3, free, chalk gland absent. Petals pure white, obovate, 8–10 × ca. 6 mm, base gradually narrowed, apex obtuse. Stamens 10, ca. 3.5 mm. Nectary band obscure. Ovary semi-inferior, carpels fused only at base, tapered into 1–1.5 mm styles.

Distribution: Endemic to Nepal.

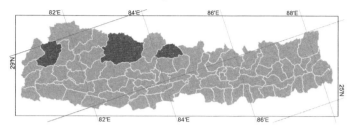

Altitudinal range: 2700–3300 m.

Ecology: Stony banks and among rocks.

Flowering: April–June. **Fruiting:** July.

Hybridizes with *Saxifraga poluninana* Harry Sm. in areas where their geographical and altitudinal ranges overlap. It has been claimed that both *S. micans* Harry Sm. and *S. staintonii* Harry Sm. are of hybrid origin with this parentage.

85. *Saxifraga micans* Harry Sm., Bull. Brit. Mus. (Nat. Hist.), Bot. 2: 126, pl. 12 fig. i (1958).

Plants caespitose, many-branched, woody at base, crowded leafy shoots, 11–13 mm diameter, forming cushions. Leaves on vegetative shoots alternate, densely imbricate towards stem tips, silvery, sessile, linear, 6–9 × 1.5–2 mm, apex acute, thickened, both surfaces glabrous, margin denticulate-ciliate in lower quarter, chalk glands 7. Flowering stems 2–3 cm, densely shortly brown glandular-pubescent, leaves 4–6. Flowers 3 or 4 in a cyme, bisexual. Pedicels 2–4 mm, dark brown glandular-pubescent. Sepals erect, ovate to broadly ovate, 2–3 × 1.5–2.5 mm, apex obtuse, outer surface and margin short glandular-pubescent, veins 4 or 5, free, the median ones terminating in 1–3 chalk glands. Petals white to tinged with pink, obovate to suborbicular, 9–12.5 × 6–8.5 mm, base gradually narrowed, apex rounded. Stamens 10, 3–4 mm, anthers red-brown. Nectary band inconspicuous. Ovary semi-inferior, carpels fused only at base, tapered into 2–2.5 mm styles.

Distribution: Endemic to Nepal.

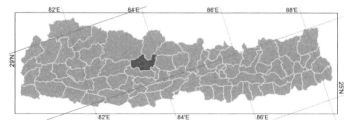

Altitudinal range: ca. 3800 m.

Ecology: Alpine, on rock faces.

Flowering: June.

An endemic only known from the type collection, *Stainton, Sykes & Williams 3074* (BM, E) near Gurjakhani, Myagdi District.

86. *Saxifraga stolitzkae* Duthie ex Engl. & Irmsch., in Engl., Pflanzenr. IV. 117(Heft 69): 569, fig. 116D (1919).

Plants caespitose, many-branched, somewhat woody at base, crowded, ascending, leafy shoots 12–15 mm diameter, forming cushions. Leaves on vegetative shoots alternate, imbricate, aggregated tips into rosettes, sessile, almost linear, 4–7 × 1.3–2 mm, apex obtuse to subacute, recurved, leathery, both surfaces glabrous, margin denticulate-ciliate at base, chalk glands 7–13 in the upper half to three quarters. Flowering stems 5–8 cm, densely glandular-pubescent, leaves up to 12. Flowers 3–8 in a subumbellate cyme, bisexual. Pedicels 2–10 mm, dark-brown glandular-pubescent. Sepals erect, ovate, ca. 3 × 1 mm, outer surface and margin glandular-pubescent, veins 3, free, chalk gland solitary. Petals pink, obovate, 5–6 × 2.5–3.5 mm, base gradually narrowed into a claw, apex obtuse. Stamens 10, 3–3.5 mm, anthers reddish. Nectary band obscure. Ovary semi-inferior, carpels fused for half their length, tapered into erect, 1.5–2.5 mm styles.

Distribution: Nepal and W Himalaya.

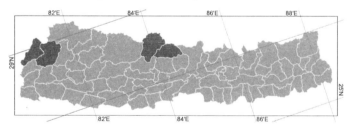

Altitudinal range: 3000–4000(–4300) m.

Ecology: Alpine, on rocks.

Flowering: May–September.

4. *Saxifraga* sect. *Saxifraga*

Plants caespitose, many branched, forming mats or cushions. Rhizome neither bulbiliferous nor stoloniferous. Leaves alternate, without chalk glands. Petals white, without calloses. Stamens 10. Ovary inferior.

87. *Saxifraga coarctata* W.W.Sm., Rec. Bot. Surv. India 4: 194 (1911).
Saxifraga humilis Engl. & Irmsch.

Plants caespitose, many-branched, forming dense cushions. Basal leaves aggregated into rosettes, subspatulate, 7–15 × 1.5–4 mm, apex acute, margin entire or 2- or 3-dentate, glandular-pubescent, both surfaces usually glabrous. Cauline leaves oblanceolate, 5–15 × 1.5–3 mm, base attenuate, apex acute, margin entire, lanate-ciliate, both surfaces glabrous. Flowering stems 1.5–4 cm, embedded in cushion at anthesis, elongating in fruit, densely glandular-pubescent, bulbils absent. Flowers solitary, bisexual. Pedicel to 5 mm, glandular-pubescent. Sepals erect, ovate, 2–2.5 × ca. 1.5 mm, apex obtuse, outer surface and margin glandular-pubescent, veins 3, more or less confluent near apex. Petals white, obovate to oblong, 3–4 × 2–2.5 mm, base cuneate, apex obtuse. Stamens 2–2.5 mm. Nectary band obscure. Ovary subinferior, carpels fused for more than half their length, styles erect, ca. 1.5 mm.
Fig. 33s–v.

Distribution: Nepal, W Himalaya and Tibetan Plateau.

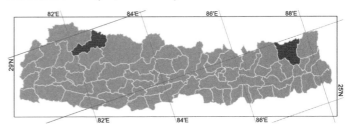

Altitudinal range: 4400–4900 m.

Ecology: Alpine.

Flowering: July–August.

7. *Micranthes* Haw., Syn. Pl. Succ.: 320 (1812).

Shinobu Akiyama & Richard J. Gornall

Perennial herbs. Stems caespitose or simple. Stolons absent. Stipules absent. Leaves alternate, in basal rosettes, petiolate or not, containing crystals (druses), chalk glands absent; blade simple, margin entire or dentate, glabrous or variously pubescent. Flowering stems glandular- or eglandular-pubescent, usually leafless, rarely with leaf-like bracts. Inflorescence a solitary flower or a few- to many-flowered bracteate cyme. Flowers bisexual, actinomorphic, rarely slightly zygomorphic. Hypanthium cup-shaped to saucer-shaped. Sepals 5, free. Petals 5, free, white, pink or red to purple, margin entire, not callose. Stamens 10, free, filaments subulate, linear or clavate. Nectary sometimes well developed as an annular disk. Ovary superior to semi-inferior, 2-carpellate, carpels fused for half or less of their length, 2-locular, ovules many, placentation axile, integuments 1. Styles 2, free, stigma capitate. Fruit a 2-valved dehiscent capsule. Seeds many, small.

Worldwide about 80 species in montane and arctic regions of Asia, Europe and N America. Four species in Nepal.

DNA sequence data have shown that *Micranthes* is not part of *Saxifraga* but instead is closer to the monotypic American genera *Cascadia* and *Saxifragodes* (Soltis *et al*. Ann. Miss. Bot. Gard. 88: 669-693. 2001). It can be distinguished morphologically from *Saxifraga* by its leafless flowering stems, carpels that are united to below the middle, reticulate pollen exine and only a single integument.

Key to Species

1a Petals entirely red. Nectary disk annular, conspicuous..**1. *M. gageana***
 b Petals mainly or entirely white. Nectary disk not conspicuous 2

2a Filaments clavate ....**2. *M. pallida***
 b Filaments linear...3

3a Inflorescences not flat-topped at flowering. Petals 3–5 × 2–3 mm.....................**3. *M. pseudopallida***
 b Inflorescences flat-topped at flowering. Petals 6.5–10 × 4–6 mm ....**4. *M. melanocentra***

1. Micranthes gageana (W.W.Sm.) Gornall & H.Ohba, Fl. Nepal 3: xviii (2011).
Saxifraga gageana W.W.Sm., Rec. Bot. Surv. India 4: 265 (1911).

Stems caespitose or simple, erect, solitary or forming clumps, 2–5(–8) cm, eglandular or glandular crisped-villous. Leaves petiolate. Petiole 5–15 mm. Blade spatulate to elliptic to ovate, 3–12 × 1.5–5 mm, base cuneate to truncate, apex acute to obtuse, margin entire or serrate, crisped-villous, both surfaces glabrous or sparsely pubescent. Flowers solitary or up to 3. Bracts ovate to linear, 1.5–3 mm, apex acute, margin entire to crenate, glandular-pubescent. Pedicels 6–15 mm, eglandular or glandular crisped-villous. Sepals spreading, purplish, oblong to ovate, 1.5–3 × 1–2 mm, apex rounded, glabrate, veins 3, confluent near apex. Petals red, obovate to ovate, 2–3.5 × 1–2.5 mm, base cuneate, apex rounded. Stamen filaments linear, 3–4 mm, anthers yellow. Nectary a conspicuous annular disk, margin lobed. Ovary up to one third inferior, carpels dark red, fused for half their length, narrowed into styles, styles ca. 1 mm.

Distribution: Nepal, E Himalaya and Tibetan Plateau.

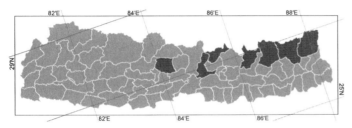

Altitudinal range: 3600–5400 m.

Ecology: Alpine meadows.

Flowering: July–August.

Although *Micranthes gageana* was treated as a synonym of *Saxifraga melanocentra* Franch. by Pan *et al.* (Fl. China 8: 283. 2001), we prefer to maintain these as distinct species. The distribution of *M. gageana* in China is uncertain.

2. Micranthes pallida (Wall. ex Ser.) Losinsk., Izv. Glavn. Bot. Sada S.S.S.R. 27: 601 (1928).
Saxifraga pallida Wall. ex Ser., in DC., Prodr. 4: 38 (1830); *S. himalaica* M.S.Balakr.; *S. micrantha* Edgew.

Stems caespitose or simple, erect, solitary or forming clumps, (1.5–)7–25 cm, simple, eglandular or glandular crisped-villous. Leaves petiolate. Petiole 1–9 cm, crisped-villous. Blade ovate, 1–3 × 1–2.5 cm, base cuneate to truncate, apex obtuse, margin crenate, crisped pubescent, both surfaces glabrous or adaxially sparsely pubescent. Inflorescences not flat-topped at flowering, 3–many-flowered. Bracts narrowly ovate to ovate, 0.5–3 cm, apex obtuse to acute, margin entire or crenate, pubescent. Pedicels 3–20 mm, eglandular or glandular-pubescent. Sepals spreading to reflexed, ovate, 1–2 × 0.8–1.6 mm, apex obtuse to subacute, glabrous, veins 3–5, confluent near apex. Petals white, with 2 yellow spots

near base, ovate, 1.5–4 × 1.2–3 mm, base with a claw, apex acute, obtuse or retuse. Stamen filaments clavate, 1.5–4 mm, anthers violet. Nectary obscure. Ovary up to one third inferior, carpels green, dark purple in fruit, fused for up to half their length, narrowed into conical styles, styles ca. 1 mm.

Distribution: Nepal, E Himalaya and Tibetan Plateau.

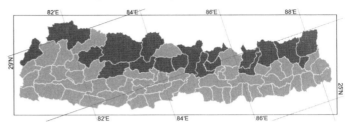

Altitudinal range: 3200–4900 m.

Ecology: Alpine boggy places.

Flowering: July–August.

3. Micranthes pseudopallida (Engl. & Irmsch.) Losinsk., Izv. Glavn. Bot. Sada S.S.S.R. 27: 601 (1928).
Saxifraga pseudopallida Engl. & Irmsch., Bot. Jahrb. Syst. 50, Beibl. 114: 40 (1914).

Stems caespitose or simple, erect, solitary or forming clumps, 5–15 cm, eglandular or glandular crisped-villous. Leaves petiolate. Petiole 1–3.5 cm, blade rhombic-ovate to ovate to oblong-ovate, 0.8–3 × 0.5–1.8 cm, base cuneate, apex acute to obtuse, margin crenate-serrate to serrate, sparsely eglandular or glandular-pubescent, both surfaces glabrous or pubescent. Inflorescences not flat-topped at flowering, (1–)2–8-flowered. Bracts ovate to narrowly elliptic, 5–15 × 0.5–10 mm, apex acute, margin serrate or entire, surfaces glabrous or pubescent. Pedicels 3–12 mm, crisped eglandular or glandular-villous. Sepals spreading to reflexed, triangular-ovate, 2.5–3.5 × 1.5–2.5 mm, apex acute, glabrous, veins 3–7, confluent. Petals white, 2 yellow spots near base, ovate to elliptic to obovate, 3–5 × 2–3 mm, base contracted or narrowed into a claw, apex obtuse. Stamen filaments linear, 3–5 mm, anthers dark red. Nectary an obscurely lobed annular band. Ovary semi-inferior, carpels reddish to dark red, fused for half their length, tapered into conical styles, styles 1.5–2 mm.

Distribution: Nepal, W Himalaya, E Himalaya and Tibetan Plateau.

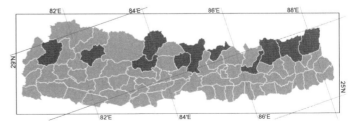

Altitudinal range: 3400–4800 m.

Ecology: Alpine, among boulders near streams.

Flowering: July–August. **Fruiting:** August.

4. *Micranthes melanocentra* (Franch.) Losinsk., Izv. Glavn. Bot. Sada S.S.S.R. 27: 601 (1928).
Saxifraga melanocentra Franch., J. Bot. [Morot] 10: 263 (1896); *S. atrata* var. *subcorymbosa* Engl.

वे-धेन-कार्पो Woe-dhen-karpo (Tibetan).

Stems caespitose or simple, erect, solitary or forming clumps, 3–9 cm, eglandular or glandular crisped-villous. Leaves petiolate. Petiole 1–2 cm. Blade rhombic-ovate to ovate to elliptic, 0.8–2.5 × 0.5–1.8 cm, base cuneate, apex obtuse to acute, margin serrate, sparsely eglandular or glandular-pubescent, both surfaces glabrous or pubescent. Inflorescence (1–)2–6-flowered in a corymbose cyme. Bracts narrowly elliptic, 5–15 × 0.5–1.5 mm, apex acute, margin entire or serrate, surfaces glabrous or pubescent. Pedicels 5–25 mm, eglandular or glandular crisped-villous. Sepals spreading to reflexed, triangular-ovate, 4–6 × 2.5–3 mm, apex acute, glabrous, veins 3–8, confluent. Petals white, 2 yellow spots near base, ovate to elliptic to obovate, 6.5–10 × 4–6 mm, base contracted or narrowed into a claw, apex obtuse to acute. Stamen filaments linear, 6.5–9.5 mm, anthers dark red or black. Nectary an obscurely lobed proximal band. Ovary up to one third inferior, carpels dark red or black, fused for half their length, narrowed into conical styles, styles 1–3 mm. Fig. 27h–k.

Distribution: Nepal, W Himalaya, E Himalaya, Tibetan Plateau and E Asia.

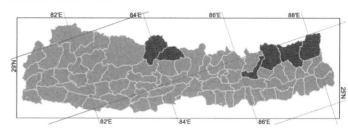

Altitudinal range: 3900–5300 m.

Ecology: Alpine meadows, open shrubland, amongst boulders near streams.

Flowering: July–August. **Fruiting:** August.

Parnassiaceae

Bhaskar Adhikari & Shinobu Akiyama

Perennial herbs, glabrous. Flowering stems erect, solitary to several. Radical leaves one to many, petiolate, palmate-veined. Cauline leaf solitary, sessile or semi-amplexicaul, usually with a few brown hairs at the base, palmate-veined. Flowers solitary, terminal, actinomorphic, bisexual. Hypanthium campanulate, free or adnate to lower half of ovary. Sepals 5. Petals 5, white or green, margin entire, erose or fimbriate. Stamens 5, opposite to sepals, equal or unequal. Staminodes 5, opposite to petals, discoid or divided into lobes or finger-like segments, apical glands distinct or not. Ovary semi-inferior or superior, 1-locular. Ovules many, placentation parietal. Style short with 3 or 4 lobes. Fruit 3 or 4-valved capsule.

Worldwide one genus and about 70 species in temperate regions of the N hemisphere, especially E Asia. Seven species in Nepal.

1. *Parnassia* L., Sp. Pl. 1: 273 (1753).

Description as for Parnassiaceae.

Key to Species

1a	Flowers green. Staminodes entire, discoid	**1. *P. tenella***
b	Flowers white. Staminodes divided into 3–6 lobes or segments	2
2a	Apex of staminodes divided into 5–6 linear segments	**2. *P. wightiana***
b	Apex of staminodes divided into 3 lobes or segments	3
3a	Stems 15–45 cm tall. Petals 10–15 × 5–10 mm	**3. *P. nubicola***
b	Stems 2–13 cm tall. Petals 3–10 × 1.5–4 mm	4
4a	Petals up to 1.5× longer than sepals. Cauline leaf usually near the flower or in upper half of stem	**4. *P. kumaonica***
b	Petals at least 2× longer than sepals. Cauline leaf usually in the middle of lower half of stem	5
5a	Petal margin distinctly fimbriate	**5. *P. chinensis***
b	Petal margin entire or erose	6
6a	Radical leaves usually cordate at base	**6. *P. pusilla***
b	Radical leaves cuneate to rounded at base	**7. *P. trinervis***

1. *Parnassia tenella* Hook.f. & Thomson, J. Proc. Linn. Soc., Bot. 2: 80 (1857).

Flowering stems usually solitary, rarely 2, 6–15 cm. Radical leaves 1 or 2, petioles 4–9 cm. Leaves reniform, 1–2 × 1.5–2.5 cm, base deeply cordate, apex obtuse or rounded, margin entire. Cauline leaf near middle or in upper half of stem, cordate, 0.5–1.0 × 0.5–1 cm, base cordate, apex obtuse or rounded, margin entire. Flowers green, 0.5–1 cm across. Hypanthium 1–2 mm. Sepals ovate to oblong-ovate, 2–3 × 1–2 mm, base rarely with a few brown hairs, apex acute or obtuse, margin entire. Petals broadly obovate, 3–5 × 3–4 mm, base cuneate or narrowed into a claw, apex obtuse, margin entire. Filaments usually equal, 3–4 mm; anthers ca. 1 mm. Staminodes 2–3 mm, discoid ca. 1 mm in diameter, glands absent. Ovary superior or slightly inferior, 2–3 mm. Style ca. 0.5 mm, stigma 3 or 4-lobed.
Fig. 34l–n.

Distribution: Nepal, E Himalaya, Tibetan Plateau and E Asia.

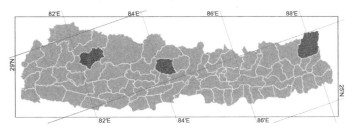

Altitudinal range: 3000–3500 m.

Ecology: Alpine meadows.

Flowering: August–September. **Fruiting:** September–October.

2. *Parnassia wightiana* Wall. ex Wight & Arn., Prodr. Fl. Ind. Orient. 1: 35 (1834).
Parnassia ornata Wall. ex Arn.; *P. ornata* Wall. nom. nud.; *P. wightiana* Wall. nom. nud.

Flowering stems 1–2, 12–30 cm. Radical leaves many, petioles 5–20 cm. Leaves usually cordate, rarely ovate, 1.5–4 × 1–4 cm, base cuneate, cordate or truncate, apex acute or obtuse, margin entire. Cauline leaf near middle or in lower half of stem, ovate or cordate, 1.5–3 × 1–3.5 cm, base semi-amplexicaul, apex acute or obtuse, margin entire. Flowers white, 1.5–3 cm across. Hypanthium ca. 2 mm. Sepals oblong-ovate, 5–10 × 3–5 mm, base usually with a few brown hairs, apex bluntly pointed, margin entire. Petals obovate, 10–15 × 3–8 mm, base abruptly narrowed into a claw, apex obtuse, margin entire or undulate above, densely fimbriate in lower half. Filaments sometimes unequal, 3–7 mm; anthers 2–2.5 mm. Staminodes 4–6 mm, segments 5 or 6 finger-like, 2–2.5 mm, glands distinct. Ovary semi-inferior, 5–5.5 mm. Style indistinct or ca. 1 mm, stigma 3-lobed.
Fig. 34r–t.

Distribution: Nepal, W Himalaya, E Himalaya, Tibetan Plateau, Assam-Burma, E Asia and SE Asia.

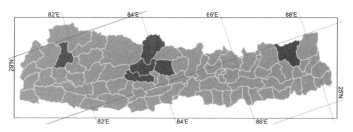

Altitudinal range: 2200–3500 m.

Ecology: Moist places.

Flowering: July–September. **Fruiting:** August–November.

3. *Parnassia nubicola* Wall. ex Royle, Ill. Bot. Himal. Mts. [7]: 227, pl. 50 fig. 3 (1835).
Parnassia nubicola Wall. nom. nud.

निरविशी Nirbishi (Nepali).

Flowering stems 1–3, 15–45 cm. Radical leaves many, petioles 3–10 cm. Leaves ovate or narrowly ovate to elliptic, 2–8 × 1.5–3 cm, base cuneate to truncate or attenuate to petiole, rarely cordate, apex acute, rarely obtuse, margin entire. Cauline leaf near middle or in lower half of stem, ovate to elliptic, 1.5–5 × 1–2.5 cm, base semi-amplexicaul, apex acute or obtuse, margin entire. Flowers white, 1–3 cm across. Hypanthium 2.5–4 mm. Sepals ovate to oblong-ovate, 5–10 × 3–5 mm, base usually with a few brown hairs, apex bluntly pointed, margin entire. Petals obovate, 10–15 × 5–10 mm, base cuneate or narrowed into a claw, apex obtuse or retuse, margin entire or minutely fimbriate in lower half. Filaments usually unequal, 1.5–6 mm; anthers 1.5–3 mm. Staminodes 4–6 mm, segments 3, 1–1.5 mm,

glands usually distinct. Ovary semi-inferior, 5–8 mm. Style ca. 2 mm, stigma 3-lobed.
Fig. 34f–h.

Distribution: Nepal, W Himalaya, E Himalaya, Tibetan Plateau, E Asia and SW Asia.

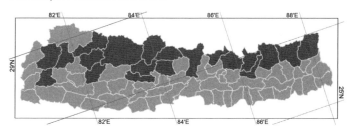

Altitudinal range: 2100–4600 m.

Ecology: Moist, open grasslands.

Flowering: July–September. **Fruiting:** August–December.

Juice of the roots is applied to wounds.

4. *Parnassia kumaonica* Nekr., Bull. Soc. Bot. France 74: 646 (1927).

Flowering stems 1–6, 2–6 cm. Radical leaves 1–6, petioles 1–3 cm. Leaves cordate to ovate, 4–10 × 3–8 mm, base cordate or truncate, apex obtuse, rarely acute, margin entire. Cauline leaf usually near flower or in the upper half of stem, cordate to ovate, 3–10 × 2–8 mm, base semi-amplexicaul, apex acute or obtuse, margin entire. Flowers white, 0.5–1 cm across. Hypanthium 1.5–2 mm. Sepals ovate to oblong-ovate, 3–6 × 2–4 mm, base rarely with a few brown hairs, apex obtuse, margin entire. Petals obovate, 3–8 × 2–4 mm, base cuneate, apex obtuse or retuse, margin entire, undulate or rarely minutely fimbriate at lower half. Filaments usually unequal, 1.5–5 mm; anthers 1–1.5 mm. Staminodes 2.5–3.5 mm, 3-lobed, lobes ca. 0.5 mm, glands absent or indistinct. Ovary semi-inferior, 4–5 mm. Style 0.5–1.5 mm, stigma 3-lobed.
Fig. 34c–e.

Distribution: Nepal and W Himalaya.

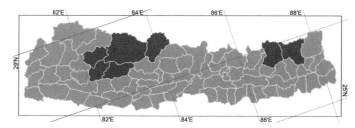

Altitudinal range: 3500–4700 m.

Ecology: Open, moist places and on mossy rocks.

Flowering: June–September. **Fruiting:** August–October.

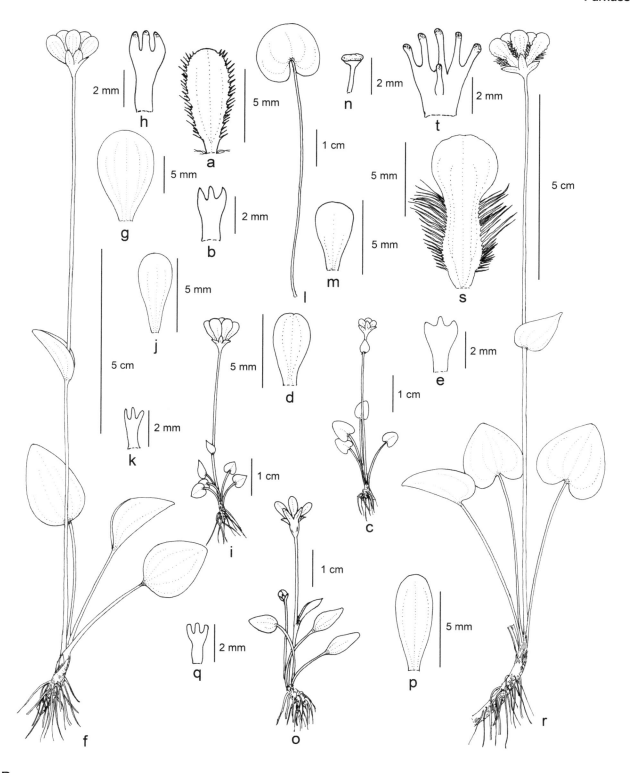

FIG. 34. PARNASSIACEAE. **Parnassia chinensis** var. **chinensis**: a, petal; b, staminode. **Parnassia kumaonica**: c, flowering plant; d, petal; e, staminode. **Parnassia nubicola**: f, flowering plant; g, petal; h, staminode. **Parnassia pusilla**: i, flowering plant; j, petal; k, staminode. **Parnassia tenella**: l, leaf; m, petal; n, staminode. **Parnassia trinervis**: o, flowering plant; p, petal; q, staminode. **Parnassia wightiana**: r, flowering plant; s, petal; t, staminode.

5. *Parnassia chinensis* Franch., Bull. Soc. Bot. France 44: 252 (1897).

Flowering stem one to many, 2.5–7 cm. Radical leaves 1 to many, petioles 1–4 cm. Leaves broadly ovate to cordate, 3–10 × 2–10 mm, base usually cordate, rarely truncate, apex obtuse, margin entire. Cauline leaf near middle or in lower half of stem, rarely in upper half, ovate to cordate, 3–7 × 2–6 mm, base semi-amplexicaul, apex obtuse, margin entire. Flowers white, 0.5–1 cm across. Hypanthium 2–3 mm. Sepals ovate to oblong-ovate, 2.5–5 × 1.5–4 mm, base rarely with brown hairs, apex obtuse, margin entire. Petals obovate to narrowly obovate, 5–10 × 2–3.5 mm, base cuneate or sometimes narrowed into a claw, usually with few hairs at base, apex obtuse, margin fimbriate except at apex and base. Filaments 1.5–7 mm, sometimes unequal; anthers 3–4 mm. Staminodes 2–3 mm, 3-lobed, lobes 0.5–0.7 mm, glands absent or indistinct. Ovary semi-inferior, 3–4 mm. Style 0.5–2 mm, stigma 3-lobed.

Distribution: Nepal, E Himalaya, Tibetan Plateau, Assam-Burma and E Asia.

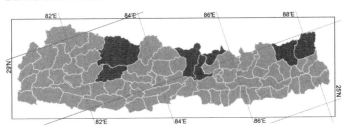

1a Margin of petals densely fimbriate except at base and apex, usually with a few hairs at base .... var. ***chinensis***
  b Margin of petals slightly fimbriate except at apex and base, base without hairs .... var. ***ganeshii***

Parnassia chinensis Franch. var. ***chinensis***

Margin of petals densely fimbriate except at base and apex, usually with a few hairs at base.
Fig. 34a–b.

Distribution: Nepal, E Himalaya, Tibetan Plateau, Assam-Burma and E Asia.

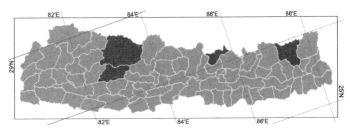

Altitudinal range: 3300–4600 m.

Ecology: Moist and rocky places.

Flowering: July–August. **Fruiting:** August–September.

Parnassia chinensis var. ***ganeshii*** S.Akiyama & H.Ohba, Bull. Natl. Sci. Mus., Tokyo, B. 27(4): 129 (2001).

Margin of petals slightly fimbriate except at apex and base, base without hairs.

Distribution: Endemic to Nepal.

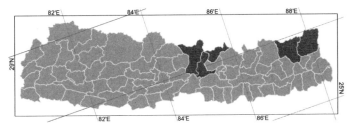

Altitudinal range: 3900–5200 m.

Ecology: Alpine meadows.

Flowering: July–August. **Fruiting:** August–September.

6. *Parnassia pusilla* Wall. ex Arn., Companion Bot. Mag. 2: 315 (1837).
Parnassia affinis Hook.f. & Thomson; *P. pusilla* Wall. nom. nud.

Flowering stem usually solitary, 2–10 cm. Radical leaves many, petioles 1–3 cm. Leaves usually cordate, rarely ovate, 3–10 × 2–8 mm, base usually cordate, rarely truncate, apex acute or obtuse, margin entire. Cauline leaf near middle or in lower half of stem, rarely in upper half, cordate or ovate, 3–8 × 1–7 mm, base semi-amplexicaul, apex acute or obtuse, margin entire. Flowers white, 5–8 mm across. Hypanthium ca. 2 mm. Sepals ovate to oblong-ovate, 2–4 × 1.5–3 mm, base rarely with brown hairs, apex acute, margin entire. Petals obovate or narrowly obovate, 4–8 × 1.5–3.5 mm, base cuneate sometimes narrowed into a claw, apex obtuse or retuse, margin entire or slightly erose. Filaments sometimes unequal, 1.5–4 mm; anthers 0.5–1 mm. Staminodes 2–2.5 mm, 3-lobed, lobes 0.5–1 mm, glands absent or indistinct. Ovary semi-inferior, 1.5–3 mm. Style 0.5–0.7 mm, stigma 3-lobed. Fig. 34 i–k.

Distribution: Nepal, W Himalaya, E Himalaya and Tibetan Plateau.

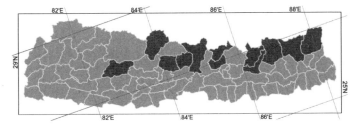

Altitudinal range: 2200–4400 m.

Ecology: Moist places.

Flowering: July–September. **Fruiting:** August–November.

Clarke misapplied the names *Parnassia mysorensis* Heyne (Fl. Brit. Ind. 2: 402. 1878) and *P. ovata* Ledeb. (Fl. Brit. Ind. 2: 403. 1878) to this species.

7. *Parnassia trinervis* Drude, Linnaea 39: 322 (1875).

Flowering stem usually 1–3, rarely more, 2.5–13 cm. Radical leaves 3–9 (to many), petioles (0.5–)1–1.5(–3) cm. Leaves ovate to narrowly ovate, 0.5–1.5 × 0.3–1 cm, base rounded to cuneate, apex acute, margin entire. Cauline leaf usually in lower half of stem, sometimes in the middle, ovate to ovate-lanceolate, 0.5–1 × 0.5–0.8 cm, base semi-amplexicaul, apex acute or obtuse, margin entire. Flowers white, ca. 1 cm across. Hypanthium ca. 2 mm. Sepals oblong-ovate, 2.5–4.5 × 1.5–3 mm, base rarely with brown hairs, apex rounded to obtuse, margin entire. Petals obovate or narrowly obovate, 6.5–10 × 2–4 mm, base gradually narrowed, apex obtuse or rounded, margin entire or sometimes slightly erose. Filaments usually equal, 3–5 mm; anthers 0.5–1 mm. Staminodes 2–2.5 mm, 3-lobed, lobes ca. 0.5 mm, glands absent or indistinct. Ovary semi-inferior, ca. 2 mm. Style ca. 0.5 mm, stigma 3-lobed.
Fig. 34 o–q.

Distribution: Nepal, Tibetan Plateau, Assam-Burma and E Asia.

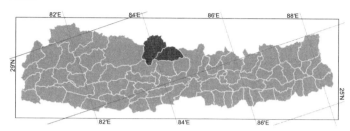

Altitudinal range: 3500–4900 m.

Ecology: Open, moist places.

Flowering: July–August. **Fruiting:** August–September.

271

Hydrangeaceae

Sirjana Shrestha, Krishna K. Shrestha & Colin A. Pendry

Small trees, shrubs, subshrubs or climbing shrubs. Leaves simple, opposite (rarely subopposite), subsessile to petiolate, exstipulate, margins more or less serrate, glabrous or with indumentum of simple or stellate hairs, pinnately veined or 3- or 5-veined at base. Inflorescence a cyme, corymb or raceme, sometimes with larger, sterile flowers on margins. Bracts and bracteoles present. Fertile flowers bisexual, small or medium sized, actinomorphic. Sepals 4 or 5, connate into a tube more or less adnate to ovary, lobes distinct, sepaloid or tooth-like. Petals 4 or 5, free. Stamens 10–30, diplostemonous, filaments linear, subulate or dilated, sometimes 2-dentate at apex. Ovary 3–5-lobed, inferior to partially superior, styles 2–6, free to partially connate. Fruit a many-seeded, loculicidal capsule or berry.

Worldwide about 17 genera and 250 species, particularly in temperate and subtropical regions of the N hemisphere. Four genera and eight species in Nepal.

Key to Genera

1a	Sterile flowers present	**1. Hydrangea**
b	Sterile flowers absent	2
2a	Inflorescence a terminal panicle. Fruit a berry	**2. Dichroa**
b	Inflorescence cymose corymbs or racemes borne on leafy lateral branchlets. Fruit a capsule	3
3a	Flowers in cymose corymbs. Petals 5. Leaves with stellate hairs. Stamens 10, with 2-dentate filaments	**3. Deutzia**
b	Flowers in racemes. Petals 4. Leaves with simple hairs. Stamens 20–30, with simple filaments	**4. Philadelphus**

1. *Hydrangea* L., Sp. Pl. 1: 397 (1753).

Shrubs, subshrubs or small trees, erect or climbing. Leaves opposite, petiolate, pinnately veined. Indumentum glabrous to variously hairy, hairs simple. Leaves pinnately veined. Inflorescence a terminal, occasionally axillary, corymbose cyme. Sterile flowers present, more numerous in cultivated taxa, occasionally absent in wild plants, borne at the margin of inflorescence on long pedicels, sepals 4 or 5, petaloid. Fertile flowers usually numerous, bisexual, smaller than sterile flowers. Pedicels short. Calyx tube adnate to ovary, 4- or 5-toothed. Petals 4 or 5, free, rarely connate and forming a calyptra, ovate, valvate. Stamens 10–14, inserted on disk, filaments linear, anthers oblong. Ovary inferior with flat apex to semi-superior with the apex projecting beyond the calyx tube, 2–4(or 5)-locular, placentation axile or parietal, ovules numerous. Styles 2–4, free or basally connate, stigma terminal or decurrent. Fruit a loculicidal capsule.

Worldwide about 30 species, from the Himalaya to E Asia, SE Asia, N and S America. Four species in Nepal with a further three cultivated.

Hydrangea macrophylla (Thunb.) Ser. (syn. *Viburnum macrophyllum* Thunb.), *H. paniculata* Siebold and *H. stylosa* Hook.f. & Thomson (syn. *H. macrophylla* subsp. *stylosa* (Hook.f. & Thomson) E.M.McClintock) are cultivated in the Kathmandu Valley and may be found more widely in the rest of Nepal. All are included in the key. *Hydrangea macrophylla*, native to Japan, is the most commonly grown species throughout the world and includes numerous cultivars. It is notable for the range of colours in its flowers which respond to soil pH, with blue flowers in acid soils and pink flowers in base-rich soils. *Hydrangea paniculata* from China, Japan and the far east of Russia has white flowers in elongate, pyramidal panicles which are quite distinct from all other species in Nepal. *Hydrangea stylosa* from Bhutan, Sikkim, Burma and Yunnan may yet be found growing wild in E Nepal. It is has blue flowers and is easily separable from the other species by the key characters.

Key to Species

1a	Leaves pubescent along veins and in vein axils only, otherwise glabrous	2
b	Leaves more or less pubescent on both surfaces	3
2a	Climbing shrub. Petals connate to form a calyptra	**1. H. anomala**
b	Erect shrub. Petals free	**H. stylosa**

3a   Inflorescence a pyramidal, paniculate cyme. Leaves ovate... *H. paniculata*
b   Inflorescence a corymbose cyme. Leaves elliptic, oblong or broadly or narrowly ovate 4

4a   Ovary partially superior, apex conical 5
b   Ovary completely inferior, apex flat 6

5a   Sterile and fertile flowers white or yellowish ....**2. *H. heteromalla***
b   Sterile and fertile flowers pink, blue or purple..*H. macrophylla*

6a   Leaves narrowly ovate to elliptic or oblong, 1.5–9 cm wide. Inflorescence axis slender. Petals blue............**3. *H. aspera***
b   Leaves broadly ovate, 5–15 cm wide. Inflorescence axis stout. Petals purple..**4. *H. robusta***

1. *Hydrangea anomala* D.Don, Prodr. Fl. Nepal.: 211 (1825).
Hydrangea altissima Wall.

Climbing shrub to 2–6 m. Branchlets brown, glabrous. Petioles 2–8 cm. Leaves ovate, elliptic, oblong or oblong-ovate, 4–17 × 2.5–12 cm, base rounded to cuneate, apex acuminate, margin serrate, glabrous above, pubescent below along the veins and in their axils, secondary veins 6–8 pairs. Inflorescence 10–30 cm, loose, spreading, about 25 cm across. Sterile flowers white, sometimes absent, sepals 4, obovate to suborbicular, base cuneate to rounded, apex obtuse to slightly obcordate, 7–9 × 7–10 mm, margin entire. Fertile flowers creamy white, calyx tube campanulate, 1.5–2 mm in diameter, teeth triangular, ca. 0.5 mm. Petals 5, connate apically, forming a 2–2.5 × 1–1.5 mm calyptra. Stamens 10 or 12, subequal, filaments 2–3 mm, anthers subglobose ca. 0.5 × 0.5 mm. Ovary inferior, placentation axile, styles 2 or 3, free, recurved, ca.1.5 mm in fruit. Capsule urceolate, 3–3.5 mm in diameter.

Distribution: Nepal, W Himalaya, E Himalaya, Tibetan Plateau, Assam-Burma and E Asia.

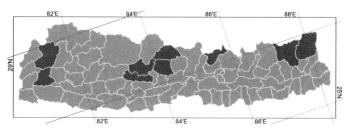

Altitudinal range: 1700–2900 m.

Ecology: Broad-leaved forests, stream-sides and rocky mountain slopes.

Flowering: April–July. **Fruiting:** August–October.

2. *Hydrangea heteromalla* D.Don, Prodr. Fl. Nepal.: 211 (1825).
Hydrangea vestita Wall.

फुस्रे काठ Phusre kath (Nepali).

Shrub to 3–10 m. Branchlets red-brown, with a few elliptical lenticels, pubescent with simple hairs. Petioles 2–5 cm,

with pilose hairs. Leaves obovate to elliptic or oblong-ovate, 10–25 × 4–11 cm, base cuneate, obtuse or truncate, apex acuminate, margin serrulate or doubly serrate, the teeth acuminate, sparsely pubescent above, densely strigose-pubescent below, secondary veins 8–11 pairs. Inflorescence 12–20 × 12–20 cm, ca. 27 cm wide in fruit. Sterile flowers white or yellowish. Sepals 4 or 5, obovate, ovate or orbicular, 0.5–2 × 0.5–2 cm, base cuneate to rounded, apex acute to acuminate, margin entire, sparsely ciliate. Fertile flowers white or yellowish, ca. 6 mm across. Pedicels 2–3 mm, pubescent. Calyx tube funneliform to campanulate, ca. 1 mm in diameter, teeth triangular. Petals 5, oblong, 2.2–3 × 1–1.5 mm, apex acute to shortly acuminate. Stamens 10, unequal, longer stamens opposite petals and shorter stamens alternate with petals, 2–4 mm, filaments broad at the base. Ovary up to half superior, placentation axile, styles 3 or 4, erect, subulate, 1.5–2 mm, stigma thickened. Capsule subglobose, 3–5 × 2–3 mm.

Distribution: Nepal, E Himalaya, Tibetan Plateau, Assam-Burma and E Asia.

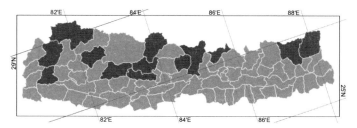

Altitudinal range: 1600–3200 m.

Ecology: Mountain forests, especially with *Tsuga* and *Picea*.

Flowering: June–August. **Fruiting:** September–October.

Hara (Fauna Fl. Nepal Himalaya: 142. 1955) misspelt this species as '*heterophylla*'.
    Juice of the bark is used to treat coughs and colds.

Hydrangeaceae

3. *Hydrangea aspera* Buch.-Ham. ex D.Don, Prodr. Fl. Nepal.: 211 (1825).
Hydrangea vestita var. *fimbriata* Wall. ex DC.; *H. vestita* var. *fimbriata* Wall. nom. nud.

Shrub or small trees to 3–5 m. Branchlets brown, terete or obscurely 4-angled, strigose or pubescent. Petioles 1–4.5 cm. Leaves narrowly ovate to elliptic or oblong, 4–28 × 1.5–9 cm, base rounded to cuneate, apex acuminate, margin serrate with teeth alternating in size, sparsely to densely strigose above, pubescent to villous below, secondary veins 6–10 pairs. Inflorescence 6–25 cm across. Sterile flowers greenish white, pinkish or red, sepals 4 or 5, ovate, obovate or orbicular, 1.5–2.5 × 0.8–2.5 cm, base cuneate, apex mucronulate, margin entire to serrate. Fertile flowers purple blue, calyx tube campanulate to cupular, 1–1.5 mm in diameter, teeth triangular to ovate-triangular, 0.5–1 mm. Petals 5, free, ovate, 2–2.5 × 1–1.5 mm. Stamens 10, unequal, 2.5–4 mm, anther globose, ca. 1 × 1 mm. Ovary inferior, placentation axile, styles 2 or 3, free, recurved, ca. 1–2 mm in fruit. Capsule urceolate, 3–3.5 mm in diameter.
Fig. 35a.

Distribution: Nepal, W Himalaya, E Himalaya, Tibetan Plateau, Assam-Burma and E Asia.

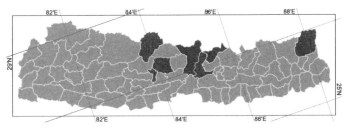

Altitudinal range: 1200–2700 m.

Ecology: Dense forests or thickets in valleys or on mountain slopes.

Flowering: July–September. **Fruiting:** October–November.

4. *Hydrangea robusta* Hook.f. & Thomson, J. Proc. Linn. Soc., Bot. 2: 76 (1857).
Hydrangea aspera subsp. *robusta* (Hook.f. & Thomson) E.M.McClint.

फिरफिरे घाँस Phirphire ghans (Nepali).

Shrubs to 4 m. Branchlets terete, pubescent to hispid. Petioles 3–8 cm. Leaves broadly ovate to slightly obovate, 9–22 × 5–15 cm, base rounded, obtuse or truncate, apex acuminate, margin doubly serrate with teeth alternating in size, acuminate, sparsely pubescent above, densely pubescent to strigose below, secondary veins 8–13 pairs. Inflorescence 10–12 × 10–30 cm. Sterile flowers white, sepals 4, ovate to orbicular, 1.5–2 × 1–2 cm, base rounded to cuneate, apex acute to obtuse, margin serrate to dentate in the upper half. Fertile flowers purple, calyx tube cupular, 1.5–2 mm in diameter, teeth ovate to triangular, 0.5–1 mm. Petals 5, free, ovate to lanceolate, 2–3 × 1.5 mm. Stamens 10–14, unequal, 6–10 mm. Ovary inferior, placentation axile, styles 2, spreading to recurved, 1–2 mm in fruit. Capsule cupular, 3–4 × 3–5 mm.

Distribution: Nepal, E Himalaya, Tibetan Plateau, Assam-Burma, S Asia and E Asia.

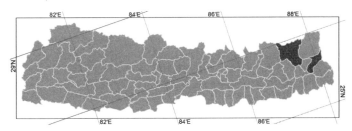

Altitudinal range: 1900–2500 m.

Ecology: Forests or thickets in valleys, along stream banks or on mountain slopes.

Flowering: July–August. **Fruiting:** September–February.

2. *Dichroa* Lour., Fl. Cochinch. 1: 301 (1790).

Adamia Wall.

Shrubs or subshrubs. Indumentum of simple hairs. Leaves opposite or rarely subopposite, petiolate, opposite, rarely alternate, pinnately veined. Inflorescence a terminal panicle. Flowers all fertile. Calyx tube adnate to ovary, obconical, 5(or 6)-toothed. Petals 5 or 6, free, induplicate-valvate. Stamens 10–20, filaments filiform to subulate, anthers ellipsoid to ovoid. Ovary inferior, incompletely 3–5 locular, placentation parietal, ovules numerous, styles 4(–6), connate only at base, divergent, stigma oblong to subglobose. Fruit a fleshy berry; seeds ovoid, minute.

One species in Nepal.

1. *Dichroa febrifuga* Lour., Fl. Cochinch. 1: 301 (1790).
Adamia cyanea Wall.

बँसुली Bansuli (Nepali).

Shrub, 1–2 m. Stems glabrous, flowering branchelets pubescent. Petioles 1–3.5 cm. Leaves elliptic, obovate, elliptic oblong 6–25 × 1.5–5.5 cm, base cuneate, apex acute to acuminate, margin serrate, glabrous above except on veins, slightly pubescent below, denser on veins, secondary veins

274

FIG. 35. HYDRANGEACEAE. **Hydrangea aspera**: a, inflorescence and leaves. **Dichroa febrifuga**: b, inflorescence and leaves; c, fruit. **Deutzia compacta**: d, inflorescence and leaves; e, opened flower; f, fruit. **Philadelphus tomentosus**: g, flowering branch; h, opened flower, i, fruit.

Hydrangeaceae

8–10 pairs, Inflorescence 3–15 cm, pubescent. Calyx tube adnate to ovary, 5 mm in diameter, teeth 0.5 × 1 mm. Petals blue, elliptic to oblong-lanceolate, 4–6 × 1.5–2.5 mm, reflexed at maturity. Stamens 5–6 mm, filaments 3–4 mm, anthers ca. 2 × 1 mm. Styles 2–3 mm, stigmas subglobose to oblong, ca. 1 × 1 mm. Berry intense metallic blue, subglobose, 3–8 mm in diameter.
Fig. 35b–c.

Distribution: Nepal, E Himalaya, Assam-Burma, S Asia, E Asia and SE Asia.

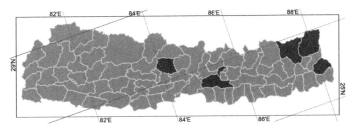

Altitudinal range: 1100–2500 m.

Ecology: Mixed forests on mountain slopes or in valleys.

Flowering: February–May. **Fruiting:** June–August.

The plant is emetic and febrifuge. Roots are used in the treatment of malaria and coughs.

3. *Deutzia* Thunb., Nov. Gen. Pl. 1: 19 (1781).

Erect shrubs. Indumentum of stellate hairs. Leaves opposite, subsessile to shortly petiolate, pinnately veined. Inflorescence a cymose corymb, terminal on lateral branchlets from buds enclosed by imbricate scales persisting at base. Flowers all fertile. Calyx tube adnate to ovary, campanulate, 5-toothed. Petals 5, induplicate or imbricate. Stamens 10, 2-seriate, filaments flat, filaments of outer stamens 2-dentate at apex, filaments of inner stamens truncate to 2-dentate at apex, anthers shortly stalked, subglobose. Ovary inferior, 3–5-loculed, placentation axile, ovules numerous, styles 3–5, free, stigmas terminal or decurrent. Fruit a capsule, 3(–5)-valved, dehiscing loculicidally or between styles; seeds numerous, oblong, compressed.

Worldwide about 60 species in the warm temperate regions of the N hemisphere. Two species in Nepal.

Key to Species

1a   Petals oblong to elliptic, induplicate. Leaves ovate.. **1. *D. staminea***
  b   Petals obovate to ovate or orbicular, imbricate. Leaves elliptic to narrowly ovate .... **2. *D. compacta***

1. *Deutzia staminea* R.Br. ex Wall., Pl. Asiat. Rar. 2[8]: 82, pl. 191 (1831).
Deutzia bhutanensis Zaik.; *D. brunoniana* Wall. nom. nud.; *D. brunoniana* Wall. ex G.Don; *D. staminea* Wall. nom. nud.; *D. staminea* var. *brunoniana* (Wall. ex G.Don) Hook.f. & Thomson.

सुन्तले Suntale (Nepali).

Shrubs 2–4 m. Stems 9–14-rayed stellate hairy, soon glabrescent. Petioles 1–2.5(–5) mm. Leaves ovate, 2–5(–7.5) × 1–3.5 cm, base rounded or broadly cuneate, apex acute to acuminate, margin serrulate, regularly 4–8-rayed stellate hairy above, densely 9–14-rayed stellate hairy below, secondary veins 3–6 pairs. Inflorescences 2–5(–9) cm, 9–25-flowered, terminal or on 2–6-leaved flowering branchlets, axes stellate hairy. Pedicels 4–6 mm. Flowers ca. 1 cm across. Calyx tube cupular, densely yellowish stellate hairy, teeth triangular to narrowly ovate, 2 × 1 mm. Petals white, induplicate, oblong to elliptic, 6–10 mm × 3–6 mm, stellate hairy outside, glabrous within. Outer stamens 6–8 mm, inner stamens 4–6 mm, anthers globose, on 1–1.5 mm stalk. Styles 7–8 mm. Capsule hemispheric, 3–4 mm in diameter.

Distribution: Nepal, W Himalaya, E Himalaya, Tibetan Plateau and E Asia.

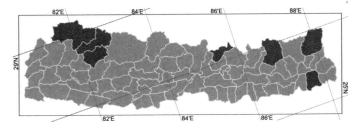

Altitudinal range: 1100–3500 m.

Ecology: In thickets on mountain slopes.

Flowering: April–July. **Fruiting:** August–November.

Juice of the plant is used to cure fever. The plant is used as fodder.

2. Deutzia compacta Craib, Kew Bull. 1913: 264 (1913).
Deutzia corymbosa var. *hookeriana* C.K.Schneid.; *D. hookeriana* (C.K.Schneid.) Airy Shaw.

Shrubs 2–2.5 m. Stem 6–8-rayed stellate hairy, glabrescent. Petioles 2–3 mm. Leaves narrowly ovate to elliptic, 2–7.5 × 0.8–2.5 cm, base rounded to cuneate, apex acute to acuminate, margin serrulate, regularly 4–6-rayed stellate hairy above, hairs sometimes with erect central ray, more densely stellate hairy below, secondary veins 3–4 pairs. Inflorescences 2–8 cm, 8–15 flowered, terminal or on 4–6-leaved flowering branchlets, axes rather sparsely stellate hairy. Pedicels 3–15 cm. Flowers ca. 1 cm across. Calyx tube cupular, stellate hairy, teeth broadly ovate, 1.5–2 × 1.5–2.5 mm. Petals white to pink or purplish, imbricate, obovate to ovate or suborbicular, 6–7 × 5–6 mm, stellate hairy outside, glabrous within. Outer stamens 4–7 mm, inner stamens 3–4 mm, anthers globose, on ca. 1 mm stalks. Styles 3 or 4(or 5), free, 4–5 mm. Capsule hemispheric, 3–4 mm in diameter.
Fig. 35d–f.

Distribution: Nepal, E Himalaya, Tibetan Plateau, Assam-Burma and E Asia.

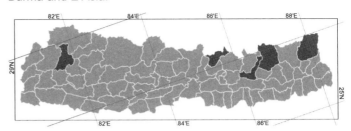

Altitudinal range: 2100–3900 m.

Ecology: Riversides and alpine scrubland.

Flowering: April–June. **Fruiting:** June–September.

Clarke (Fl. Brit. Ind. 2: 406. 1878) partially misapplied *Deutzia corymbosa* R.Br ex G.Don to this species.
   Noshiro et al. 9840077 (E), collected in fruit from Sankhuwasaba, was previously determined as *D. compacta*, but its flowers have 5 or 6 styles and it has larger leaves (to 9 × 4 cm) whose underside have an even indumentum of stellate hairs with long, erect central rays. It appears not to match any Himalayan species, but the material is not complete enough to be certain that it is a new species.

4. *Philadelphus* L., Sp. Pl. 1: 470 (1753).

Erect shrubs. Indumentum of simple hairs. Leaves opposite, petiolate, 3–5-veined above base. Inflorescence racemose, terminal on lateral branchlets from buds enclosed by imbricate scales persisting at base. Flowers all fertile. Calyx tube adnate to ovary, 4(or 5)-lobed. Petals 4, contorted. Stamens 20–30, filaments subulate, free, anthers ovoid or oblong. Ovary inferior to slightly superior, 4 (or 5)-locular, placentation axile, style solitary (3 or)4-lobed, stigmas clavate. Fruit a 4–5-valved loculicidal capsule.

Worldwide about 80 species in temperate regions of N hemisphere. One species in Nepal.

1. Philadelphus tomentosus Wall. ex G.Don, Gen. Hist. 2: 807 (1832).
Philadelphus coronarius var. *tomentosus* (Wall. ex G.Don) Hook.f. & Thomson; *P. nepalensis* Koehne; *P. nepalensis* Wall. ex Loudon nom. nud.; *P. tomentosus* Wall. nom. nud.; *P. tomentosus* forma *nepalensis* (Koehne) H.Hara; *P. tomentosus* Wall. ex G.Don forma *tomentosus*; *P. triflorus* Wall. nom. nud.; *P. triflorus* Wall. ex S.Y.Hu.

Shrubs to 2–3 m. Branchlets red-brown, glabrous or sparsely villous. Petioles 0.3–1 cm, villous. Leaves ovate, 2–10 × 0.5–5 cm, base rounded or cuneate, apex acuminate, margin minutely and distantly serrate, glabrous above, below pubescent along veins and margin or densely pubescent throughout. Inflorescence 5–7-flowered, 3–10 cm, on 2–6-leaved lateral branchlets, axes glabrous. Flowers 1.5–2 cm across. Pedicels 0.5–1.5 cm. Calyx tube adnate to ovary, 1–1.7 cm in diameter, sepals ovate, 5–6.5 × 2–4 mm, acuminate, glabrous or sparsely pubescent outside, pubescent within towards apex. Petals 4, white, obovate, 10–15 × 7–12 mm. Stamens 20–30, filaments 4–7 mm, anthers ca. 1 × 1 mm. Style 7–8 mm, apically divided for one third of its length. Capsule, ellipsoid, ca. 1 × 0.5 cm, sepals persistent, erect to reflexed; seeds flattened, ca. 2 mm.
Fig. 35g–i.

Distribution: Nepal, W Himalaya, E Himalaya, Tibetan Plateau and Assam-Burma.

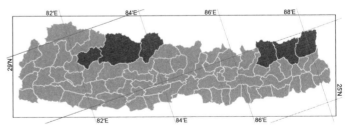

Altitudinal range: 1800–3600 m.

Ecology: Open slopes, tracksides.

Flowering: April–July. **Fruiting:** August–October.

Schilling (J. Roy. Hort. Soc. 94: 227. 1969) misapplied the name *Philadelphus coronarius* L. to this species.

Grossulariaceae

J. Crinan M. Alexander & Bandana Shakya

Deciduous, sometimes spiny, usually glandular, shrubs, rarely small trees, often dioecious, sometimes epiphytic. Buds with several scarious or papery scales. Stipules absent. Leaves petiolate, alternate, rarely fascicled, simple, palmately 3 or 5-lobed, rarely entire. Flowers bisexual, or unisexual in dioecious plants, in many- or few-flowered racemes, sometimes clustered or solitary. Bracts conspicuous, paired or solitary, narrowly ovate to ligulate, caducous or persistent. Hypanthium shallowly cup-shaped to short-tubular, basally adnate to ovary in bisexual or female flowers, bearing sepals, petals and stamens in alternate whorls on rim; sepals 5, petal-like, erect or reflexed at anthesis, sometimes changing in fruit. Petals 5, usually inconspicuous and much smaller than sepals. Stamens 5, vestigial or with undeveloped pollen in female flowers. Style 2-lobed or bifid, often appearing unreduced in male flowers (even when ovary vestigial). Ovary inferior, short-stalked, 1-celled, with many ovules, vestigial or lacking in male flowers; placentation parietal. Fruit a juicy berry, with persistent apical calyx. Seeds many.

Worldwide one genus and about 145 species from the temperate and mountainous regions of the N hemisphere, especially in E Asia. Seven species in Nepal.

1. *Ribes* L., Sp. Pl. 1: 200 (1753).

Description as for Grossulariaceae.

The name *Ribes vilmorinii* Jancz. appears in Hara (Enum. Fl. Pl. Nepal 2: 159. 1979) based on *Zimmerman 616* (G), quoting Baehni (Candollea 16: 219. 1958). As no specimens of this species have been seen, and the nearest record is in Yunnan, it has not been included in this account.

Key to Species

1a	Flowers bisexual	2
b	Flowers unisexual, plants dioecious	4
2a	Plants spiny	**1. *R. alpestre***
b	Plants unarmed	3
3a	Sepals erect or incurved at or soon after anthesis. Racemes 5–10 cm. Petioles usually lacking straggling glandular hairs near base. Bracts shorter than pedicels	**2. *R. himalense***
b	Sepals reflexed at or soon after anthesis. Racemes to 20 cm. Petioles with straggling glandular hairs near base. Bracts longer than pedicels	**3. *R. griffithii***
4a	Leaves broadly ovate or circular to reniform, lobes obtuse or rounded, crenate or obtusely toothed	**7. *R. orientale***
b	Leaves ovate, lobes acute or acuminate, acutely toothed	5
5a	Leaves unlobed or lobes 3 or 5, to 10 × 10 cm, mid lobe often curved to one side	**6. *R. acuminatum***
b	Leaves 3- or 5-lobed, to 5 × 4.5 cm, mid-lobe not curved	6
6a	Mid-lobe more than half the length of the leaf, usually long-acuminate	**4. *R. glaciale***
b	Mid-lobe less than half the length of the leaf, obtuse to acute	**5. *R. luridum***

1. *Ribes alpestre* Wall. ex Decne., in Jacquem., Voy. Inde 4(1): 64, pl. 75 (1841).
Ribes himalensis Royle nom. nud.

मसिनो किम्बु Masino kimbu (Nepali).

Shrubs to 2.5 m. Nodal spines 3, whorled, stout, 7–25 mm. Branches sparsely and minutely prickly or with stalked glands; older branches unarmed, black-scurfy. Petioles to 3 cm, sparsely hairy or glandular-hairy. Leaves broadly ovate, 1.5–2.5 × 2–3.5 cm, base broadly rounded or truncate to cordate; lobes 3 or 5, obtuse; terminal about equal to or slightly longer than laterals; margin coarsely and irregularly crenate to obtuse-serrate or doubly so; glabrous or sparsely hairy, especially on veins below, ciliate on margins, with occasional stalked glands above and sessile glands below. Flowers bisexual, axillary, 1–3 in short racemes. Bracts paired, broadly ovate to ovate-triangular, 2–3 mm. Pedicels 5–8 mm, sparsely glandular-hairy. Calyx pale green to brown, darker at base, hypanthium cylindrical to narrowly campanulate,

4–5.5 mm, usually with sparse hair and stalked glands; sepals oblong to ligulate, 5–7 mm, pale green to pink, margin purplish, erect in fruit. Petals elliptic to oblong, 2.5–3.5 mm, white, not exceeding sepals. Stamens pale, ca. 4–5 mm, exceeding petals. Ovary with dense stalked glands; style split to nearly half-way. Fruit green to reddish purple, spherical to ellipsoid, to 15 mm, usually with stalked glands.
Fig. 36a.

Distribution: Nepal, W Himalaya, E Himalaya, Tibetan Plateau, E Asia and SW Asia.

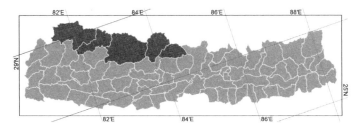

Altitudinal range: 3100–3800 m.

Ecology: Open and scrubby areas on mountain slopes, open *Picea* forest.

Flowering: April–June. **Fruiting:** June–October.

Roxburgh (Fl. Ind. 2: 515. 1824) and Clarke (Fl. Brit. Ind. 2: 410. 1878) misapplied the name *Ribes grossularia* L. to this species.

Ripe fruits are edible and also preserved as jams, jellies, and sauces.

2. Ribes himalense Royle ex Decne., in Jacquem., Voy. Inde 4(1): 66, pl. 77 (1841).
Ribes emodense Rehder; *R. emodense* var. *urceolatum* (Jancz.) Rehder; *R. glaciale* Wall. nom. nud.; *R. himalense* var. *appendiculatum* Jancz.; *R. himalense* var. *decaisnei* Jancz.; *R. himalense* var. *urceolatum* Jancz.; *R. melanocarpum* K.Koch.

खमार Khamaar (Nepali).

Shrubs to ca. 4.5 m, unarmed. Branches glabrous. Buds purplish brown, 2.5–4 mm, acute. Petioles green to red, 3–6 cm, glabrous or pubescent, especially at apex, glandular-hairy near base and apex. Leaves broadly ovate to suborbicular, 3.5–5 × 4.5–6.5 cm, base cordate; lobes 3 or 5, ovate-triangular, acute to shortly acuminate; terminal equalling or slightly longer than laterals; margin sharply double-serrate, some teeth simple; sparsely glandular-hairy to glabrous above, ± glabrous below though hairy between vein bases. Racemes dense, to ca. 10 cm, with 8–25 flowers; rachis and pedicels pubescent to densely hairy, sometimes with sparse stalked glands. Bracts solitary, narrowly triangular to ovate, 1–2.5 mm, usually shorter than pedicels, sometimes ciliate. Pedicels 1.5–3 mm. Flowers bisexual, 4–6 mm across. Calyx pale green tinged with pink or purple; hypanthium funnel-shaped to narrowly campanulate, 1.5–2 mm; sepals obovate-spathulate to circular, 2–3.5 mm, erect. Petals yellowish or reddish green, tinged with purplish, circular to flabellate, 1–1.5 mm,

not exceeding sepals. Stamens about equalling petals. Ovary glabrous; style equalling or slightly exceeding stamens, 2-fid apically. Fruit red, becoming purplish black, spherical, 6–7 mm, glabrous.
Fig. 36b–d.

Distribution: Nepal, W Himalaya, E Himalaya, Tibetan Plateau, Assam-Burma and E Asia.

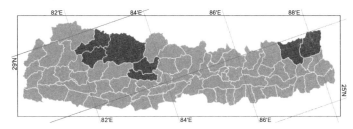

Altitudinal range: 1800–3400 m.

Ecology: Shrubby hillsides and ravines; open or dense woodland.

Flowering: April–May. **Fruiting:** July–August.

Hooker & Thomson (J. Proc. Linn. Soc., Bot. 2: 89. 1857) and Clarke (Fl. Brit. Ind. 2: 411. 1878) misapplied the name *Ribes rubrum* L. to this species.

This species has sometimes been misspelt '*himalayense*' (e.g. Jancz., Bot. Jahrb. 36 Beibl. 82: 51. 1905).

Leaf juice is given for diarrhoea and dysentery, and a paste is applied to cuts and wounds.

3. Ribes griffithii Hook.f. & Thomson, J. Proc. Linn. Soc., Bot. 2: 88 (1857).

Shrubs to ca. 4.5 m, unarmed. Branches glabrous. Buds brown, to ca. 6 mm, acute. Petioles green to reddish, to ca. 11 cm, glabrous to sparsely pubescent, especially at apex, with straggling, marginal, glandular hairs to ca. 5 mm near base. Leaves circular to elliptic, wider than long, 5–9 × 6–12 cm, base truncate to deeply cordate; lobes (3 or)5, ovate to triangular, acuminate; terminal equalling or slightly longer than laterals; margin sharply double-serrate, some teeth simple; above and often below with sparse short-stalked glandular hairs,. Racemes dense, to ca. 20 cm, with 10–40 flowers; rachis and pedicels densely crisped-hairy. Bracts solitary, narrowly triangular to ligulate, to 10 mm, exceeding pedicel, ciliate, pubescent near base. Pedicels to ca. 3 mm. Flowers bisexual, 5–7 mm across. Calyx pale green to dark reddish purple; hypanthium broadly campanulate, 2–3 mm; sepals narrowly elliptic to oblong, 2–3 mm, erect. Petals pale yellowish or reddish green, tinged purplish, spathulate to flabellate, 1–2 mm, not exceeding sepals. Stamens about equalling petals. Ovary glabrous; style exserted, 2-fid apically. Fruit ovoid to spherical, to ca. 1 cm, red, glabrous.
Fig. 36e.

Grossulariaceae

Distribution: Nepal, E Himalaya, Tibetan Plateau, Assam-Burma and E Asia.

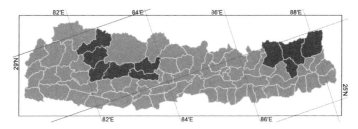

Altitudinal range: 2800–4000 m.

Ecology: Shady areas in evergreen, deciduous and mixed forest; ravines.

Flowering: April–June. **Fruiting:** (April–)July–October.

Ribes griffithii is notable for having particularly prominent venation on the undersides of its leaves compared with the other Nepalese species.

4. *Ribes glaciale* Wall. ex Roxb., Fl. Ind. 2: 513 (1824).
Ribes glaciale var. *laciniatum* (Hook.f. & Thomson) C.B.Clarke;
R. laciniatum Hook.f. & Thomson.

Dioecious, arching or pendent shrubs to 8(–10) m, unarmed. Branches grey to black, with dense, short, stiff hairs, later glabrous. Buds reddish brown, narrowly ovoid, 4–6 mm, acute, scales scarious. Petioles 1.5–2.5 cm, pale reddish brown, glabrous to sparsely crisped-hairy, sometimes sparsely glandular-hairy. Leaves broadly ovate, 3–5 × 2–4.5 cm, base widely cordate to subtruncate; lobes 3 or 5; terminal triangular-ovate, more than half leaf length, acuminate; laterals asymmetrically ovate, acute; margin coarsely serrate, some teeth double; above and below with sparse glands and/or glandular hairs, sometimes also with short minute hairs above especially on veins. Male racemes 3–6.5 cm, 10–30-flowered, female racemes 1–3 cm (to ca. 4.5 cm in fruit), 4–10-flowered; rachis and pedicels crisped-hairy and with short-stalked glands. Bracts solitary, narrowly ovate to ellipsoid, 5–7.5 mm, ciliate, 1-veined. Pedicels 1.5–4 mm (female longer than male). Calyx dark red to reddish brown, glabrous; hypanthium shallowly cup-shaped, 1–2 mm; sepals oblong-ligulate, rounded, ca. 2 mm, erect. Petals spathulate to flabellate, much smaller than sepals. Stamens exceeding petals, not exceeding sepals. Ovary obovoid-spherical, glabrous; style 2-lobed at apex. Fruit red, ± spherical, 5–7 mm, glabrous.
Fig. 36f.

Distribution: Nepal, W Himalaya, E Himalaya, Tibetan Plateau, Assam-Burma and E Asia.

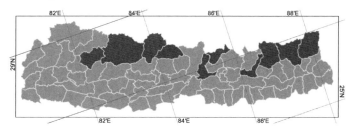

Altitudinal range: 2700–4000 m.

Ecology: Scrub, open coniferous and mixed woodland, secondary forest (with *Quercus*, *Berberis*, *Rhododendron*, *Abies*, *Juniperus*), often hanging from rocks or boulders.

Flowering: May–June. **Fruiting:** July–September.

Ribes glaciale is close to *R. luridum* Hook.f. & Thomson but can usually be distinguished by the acuminate terminal leaf-lobe being clearly more than half the length of the leaf. *Ribes laciniatum* has been treated variously as a separate species or as a variety of *R. glaciale*, differentiated on its more laciniate leaves and lanceolate sepals, but as Hooker and Thomson themselves say (Fl. Br. Ind. 2: 410. 1879), the leaf character is not reliable. As sepal shape also varies, it seems best not to recognise *R. laciniatum* formally.
Ripe fruits are edible.

5. *Ribes luridum* Hook.f. & Thomson, J. Proc. Linn. Soc., Bot. 2: 87 (1857).

Dioecious, arching, shrub to ca. 3 m, unarmed. Branches glabrous, exfoliating, red becoming intensely black. Buds brown, ovoid to oblong-ovoid, 4–6 mm, acute, glabrous. Petioles to 3 cm, glabrous or slightly puberulent, with sparse short-stalked glands. Leaves broadly ovate to orbicular, 2–4 × 3.5–4.5 cm base truncate to shallowly cordate; lobes 3(or 5), terminal triangular to almost rhombic, acute; laterals slightly shorter, asymmetrically ovate, acute; margin coarsely dentate, some teeth double; above and below sparsely to densely glandular-hairy, often also sparsely to densely strigose above. Male racemes 2.5–5 cm, 8–20-flowered; female racemes shorter, with fewer flowers; rachis densely crisped-hairy and sparsely to densely glandular-hairy. Bracts solitary, elliptic-lanceolate to oblong, attenuate or slightly erose, 5–7 mm, pubescent near margin and towards tip, a few hairs glandular, 1-veined. Pedicels 5–15 mm, crisped and glandular-hairy. Calyx dull red to purple, glabrous to sparsely pubescent; hypanthium shallowly cup-shaped, 1.5–2 mm; sepals ovate to ligulate 2–2.5 mm, erect, sometimes with central paler or greenish stripe. Petals ca. 0.5–0.7 mm, spathulate to circular. Stamens not exceeding sepals. Ovary glabrous; style exceeding stamens (even when non-functional in male flowers), 2-lobed. Fruit black, subglobose, 5–7 mm, glabrous.
Fig. 36g–h.

Distribution: Nepal, E Himalaya, Assam-Burma and E Asia.

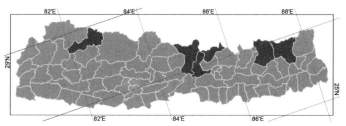

Altitudinal range: 3400–4100 m.

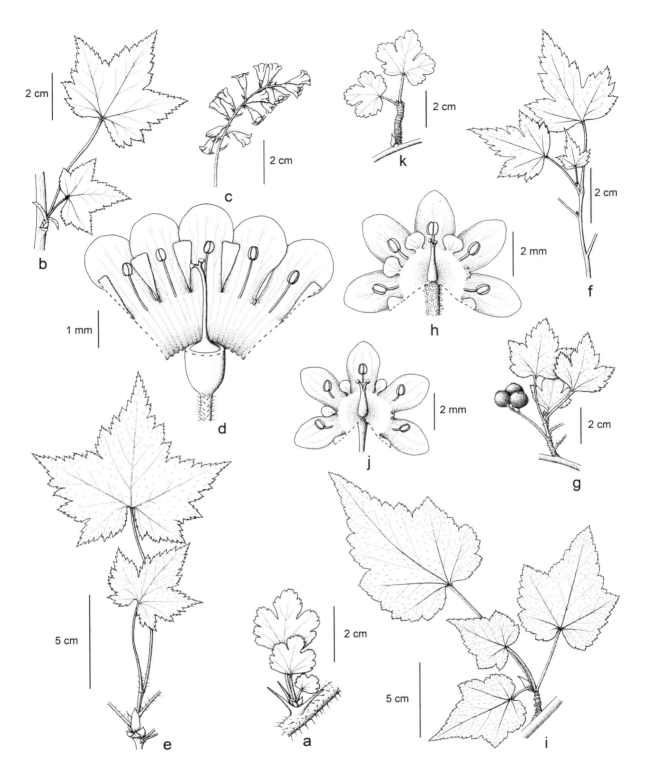

FIG. 36. GROSSULARIACEAE. **Ribes alpestre**: a, leaves and spines. **Ribes himalense**: b, leaves; c, inflorescence; d, opened flower. **Ribes griffithii**: e, leaves. **Ribes glaciale**: f, leaves. **Ribes luridum**: g, leaves and infructescence; h, opened flower. **Ribes acuminatum**: i, leaves; j, opened flower. **Ribes orientale**: k, leaves.

Ecology: Shrubby areas and open forest among *Abies*, *Juniperus* and *Rhododendron*.

Flowering: May–July. **Fruiting:** July–September.

Ribes luridum is close to *R. glaciale* Wall. ex Roxb. but can usually be distinguished by the obtuse to acute terminal leaf lobe being less than half the overall length of the leaf.

6. *Ribes acuminatum* Wall. ex G.Don, Gen. Hist. 3: 187 (1834).
Ribes acuminatum Wall. nom. nud.; *R. acuminatum* var. *desmocarpum* (Hook.f. & Thomson) Jancz.; *R. desmocarpum* Hook.f. & Thomson; *R. takare* D.Don; *R. takare* forma *desmocarpum* (Hook.f. & Thomson) H.Hara.

टाँफू Tanfu (Nepali)

Dioecious shrub to ca. 4.5 m, unarmed. Branches stout, red, later grey to black, glabrous or slightly glandular-hairy. Buds reddish brown, ovoid, 4–6 mm. Petioles to 6 cm, finely puberulent and often with glandular hairs. Leaves broadly-ovate to circular, 5–10 × 4–10 cm, base widely to narrowly cordate, rarely subtruncate; unlobed or lobes 3 or 5; terminal ovate-triangular, acuminate, longer than the shortly acuminate upper laterals; margin irregularly and deeply double-serrate; glabrous to sparsely glandular-hairy, rarely pubescent. Racemes to 15 cm, erect to spreading; female racemes robust, often shorter but lengthening in fruit; rachis and pedicels pubescent and shortly glandular-hairy. Bracts solitary, narrowly elliptic to ligulate, to ca. 8 mm, margin (at least apically) coarsely glandular-hairy or ciliate, 1-veined. Pedicels 3–5 mm. Calyx green to dark red or reddish brown, sparsely pubescent; hypanthium shallowly cup-shaped, 1–2 mm; sepals oblong, 2–3 mm (shorter in female flowers), 3-veined, erect or spreading in fruit. Petals smaller, orbicular. Stamens not exceeding sepals. Ovary obovoid, glabrous or finely pubescent; style 2-lobed at apex. Fruit yellowish green turning reddish brown, ovoid to globose, 5–7 mm, glabrous or pubescent, sometimes with glandular hairs. Fig. 36i–j.

Distribution: Nepal, W Himalaya, E Himalaya, Tibetan Plateau, Assam-Burma and E Asia.

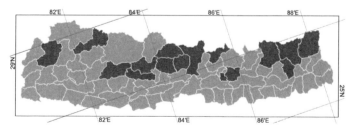

Altitudinal range: 1800–3600 m.

Ecology: Shrubby areas, evergreen and mixed forest, sometimes epiphytic.

Flowering: April–June. **Fruiting:** (May–)July–August.

Baehni (Candollea 16: 220. 1958) misapplied the name *Ribes orientale* Desf. to this species.

The name *R. acuminatum* Wall ex G.Don was applied to this species by Grierson (Fl. Bhutan 1: 524. 1987) stating that 'the original description of *R. takare* D.Don could apply to several species and as the type specimen appears to be lost, it certainly cannot be accepted as an earlier name for *R. acuminatum*'. In this Grierson follows Hooker & Thomson (J. Proc. Linn. Soc. 2: 89. 1857) who considered the name *R. takare* to be 'indeterminable'.

Plants having branches, leaves, inflorescences and fruit with both glandular and non-glandular hairs have been distinguished as forma *desmocarpum*, but the distribution of indumentum types appears too random for formal taxonomic recognition.

Ripe fruits are edible.

7. *Ribes orientale* Desf., Hist. Arbr. France 2: 88 (1809).
Ribes leptostachyum Decne.; *R. villosum* Wall.

Dioecious shrub to ca. 2 m, unarmed, most parts pubescent, sticky-glandular and usually with short-stalked glands. Young branches reddish brown; older branches stout, grey, glabrous. Buds reddish brown, ovoid to oblong, 5–6 mm. Petioles to ca. 2 cm. Leaves broadly ovate or circular to reniform, 1–3.5 × 2.5–3.5 cm, base broadly cuneate to truncate or shallowly cordate; lobes 3(or 5), obtuse or rounded; terminal about equalling laterals; margin irregularly and coarsely crenate to obtuse-dentate. Male racemes 2–5 cm, 8–30-flowered, erect; female racemes 2–3 cm, 5–15-flowered, to 4 cm in fruit. Flowers rarely bisexual. Bracts solitary, narrowly triangular or lanceolate to elliptic, acute, to ca. 8 mm. Pedicels 2–3 mm, longer in fruit. Calyx pink to dark red or purplish brown; hypanthium broadly cup-shaped, 1.5–2 mm; sepals ovate to oblong, 2–2.5 mm, erect. Petals darker than sepals, spathulate, 0.5–1 mm. Stamens slightly exceeding petals. Ovary ovoid; style 2-lobed. Fruit red to purple, globose, 7–9 mm, glabrous to pubescent, sometimes with sessile and/or short-stalked glands. Fig. 36k.

Distribution: Nepal, W Himalaya, E Himalaya, Tibetan Plateau, E Asia, C Asia, SW Asia and Europe.

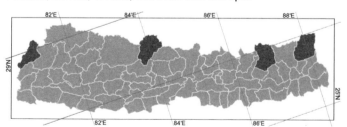

Altitudinal range: 3600–4200 m.

Ecology: Shrubby and open areas, roadsides and cliffs, often on sandy soils.

Flowering: May–June. **Fruiting:** July–September.

Pittosporaceae

Martin R. Pullan & Mark F. Watson

Evergreen trees or shrubs, sometimes climbing, branchlets often whorled. Stipules absent. Leaves alternate, clustered at ends of branches appearing opposite or almost whorled, simple, entire, pinnately veined, leathery. Inflorescences terminal or axillary umbellate panicles. Flowers bisexual, actinomorphic. Bracts and bracteoles present, simple. Sepals 5, small, fused at base, imbricate in bud. Petals 5, free and erect. Stamens 5, opposite sepals, filaments filiform, anthers 2-celled, elongate, sagittate, dorsifixed, dehiscing by longitudinal slits. Ovary superior, 2-celled, ovules numerous, placentation parietal. Style 1, short, columnar, stigma minute. Fruit a 2-valved capsule, dehiscing by 2 valves, pericarp leathery or somewhat woody, striate within. Seeds numerous, testa thin, endosperm well developed.

Worldwide seven to nine genera and about 250 species in tropical and warm temperate regions of Africa, Asia, the Pacific Islands, and particularly diverse in Australia. One genus and one species in Nepal.

1. *Pittosporum* Banks ex Gaertn. nom cons., Fruct. Sem. Pl. 1: 286 (1788).

Tobira Adans. nom. rej.

Description as for Pittosporaceae.

Worldwide 150 to 200 species in tropical and warm temperate regions of Africa, Asia, Australia and the Pacific Islands. One species in Nepal.

1. *Pittosporum napaulense* (DC.) Rehder & E.H.Wilson, Pl. Wilson. 3(2): 326 (1916).
Senacia napaulensis DC., Prodr. 1: 347 (1824); *Celastrus verticillatus* Roxb. later homonym, non Ruiz & Pav.; *Pittosporum napaulense* var. *rawalpindiense* Gowda; *P. verticillatum* Wall. nom. nud.

टिबिल्टी Tibilti (Nepali).

Shrubs or trees to 6–10(–20) m, sometimes epiphytic or climbing, glabrous. Branchlets brown, lenticellate. Petioles stout, 0.8–2.1 cm. Leaves biennial, elliptic to oblong-elliptic, 7–20 × 2.5–5.5 cm, base cuneate, apex acuminate or acute, margin flat but reflexed on drying, lateral veins convex on underside, glabrous. Panicles terminal, few to many, 4–5 cm, densely pubescent with short white (or brownish) hairs, flowers sweet scented. Pedicels 3–5 mm, densely pubescent in flower, glabrescent. Sepals elliptic to oblong-elliptic, 2–3 mm, fused slightly at base, apex obtuse to acute, glabrous or margin ciliate. Petals yellow, obovate-oblong to spatulate, 4–6 × 1–2 mm, glabrous. Stamens ca. 4 mm, glabrous. Ovary 2.0–2.3 mm, densely pubescent, style 1–2 mm, glabrous. Capsules globose, compressed, broader than long, ca. 5–6 × 5–7 mm, style persistent, seeds red, 4–6, 2–3 mm diameter, glutinous-sticky.
Fig. 22j–l.

Distribution: Nepal, W Himalaya, E Himalaya, Tibetan Plateau, Assam-Burma, S Asia and E Asia.

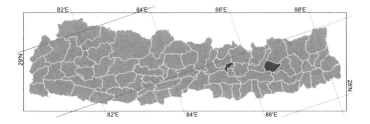

Altitudinal range: 1300–1500(–2000) m.

Ecology: Warm evergreen broad-leaved forests.

Flowering: (March–)April–July(–September). **Fruiting:** August–October(–March).

Hooker & Thomson (Fl. Brit. Ind. 1: 199. 1872) partly misapplied the name *Pittosporum floribundum* Wight & Arn. to this species.

Under represented in herbaria and only known from a few collections in Nepal. It has sometimes been misspelt '*nepalensis*'.

Hamamelidaceae

Mark F. Watson

Large evergreen trees, branches stout with distinct nodes, glabrous or with a few simple hairs. Stipules solitary or fused in pairs, folded against one another, enclosing axillary buds and young inflorescences, large and leathery, leaving scars at nodes. Petioles long and slender. Leaves alternate, simple, entire (or 3–5-lobed on juvenile growth), palmately 5–7-veined at base, thick and leathery. Inflorescence axillary, capitate, pedunculate, initially enclosed by a pair of stipules, with either bisexual or only female flowers. Flowers fused in globular heads, actinomorphic. Calyx tube adnate to base of ovary, lobes inconspicuous or absent. Petals 4–6 or absent, free, straight in bud. Stamens 10–14 or absent, free, born on a lobed disk, filaments of variable length, subulate, anther sacs opening by a single valve. Ovary semi-inferior, 2-celled, styles 2, free, elongate and spreading, stigmas decurrent, ovules 4–9 per cell, placentation axile. Bisexual flowers lacking calyx lobes, petals white, linear, ovary with 6–8 ovules per cell. Female flowers with calyx of 5, truncate rim-like lobes, petals and stamens absent. Fruit a woody capsule, aggregated into heads, each 2-celled, dehiscing by 4 valves, 4–8 seeds per cell, upper 4–5 seeds sterile, unwinged, lower 1–2 seeds fertile with oblong wing.

Worldwide about 30 genera and 100–140 species distributed across all continents, especially in tropical regions. One genus and one species in Nepal.

A family which exhibits many morphological features considered primitive. Morphological diversity is wide and unusual, the above description referring only to Nepalese material.

1. *Exbucklandia* R.W.Br., J. Wash. Acad. Sci. 36: 348 (1946).

Description as for Hamamelidaceae.

Worldwide two to four species distributed from E Himalaya to S China and SE Asia. One species in Nepal.

1. *Exbucklandia populnea* (R.Br. ex Griff.) R.W.Br., J. Wash. Acad. Sci. 36: 348 (1946).
Bucklandia populnea R.Br. ex Griff., Asiat. Res. 19: 95, pl. 13 – 14 (1836); *B. populifolia* Hook.f. & Thomson; *Symingtonia populnea* (R.Br. ex Griff.) Steenis.

पिप्ली Pipli (Nepali).

Tree, 15–30(–50) m. Stipules green or flushed red, obliquely obovate-oblong, 2–6 × 1–2.5 cm, apex obtuse or rounded. Petioles 2–8 cm, longest on juvenile growth, glabrous. Leaves broadly ovate, 10–27 × 7–20 cm, base truncate or shallowly cordate, apex acuminate, margin entire, usually glabrous. Peduncles 1–2.5 cm. Flower heads 2–4 per axil, (5–)8–12-flowered, female heads 8–15 mm diameter, bisexual heads 15–20 mm in diameter. Bisexual flowers with petals 2–3 mm. Stamens creamy white, ca. 5 mm, ovary yellowish brown, pubescent, styles 3–4 mm, erect. Fruiting heads 1.7–2.2 cm in diameter. Capsules 7–9 × 5–6 mm, smooth, dehiscing above middle. Fertile seeds with wings ca. 9 × 2 mm.
Fig. 22m–q.

Distribution: Nepal, E Himalaya, Assam-Burma, E Asia and SE Asia.

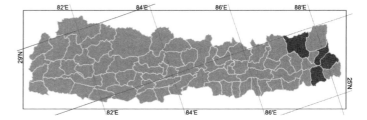

Altitudinal range: 1300–2300 m.

Ecology: Warm temperate mixed evergreen hill forests with *Quercus*, Lauraceae and *Magnolia*, sometimes in disturbed areas with *Alnus nepalensis*.

Flowering: May–July. **Fruiting:** September–February.

As most Nepalese specimens are sterile phenological data has been extrapolated from Bhutanese material and Zhang *et al.* (Fl. China 9: 24. 2003). The Nepalese local name Pipli is similar to Pipal (*Ficus religiosa* L.), perhaps on account of the similarly shaped leaves.

An ornamental tree with valuable timber used in building construction and bark used medicinally for muscle inflammation. Regenerates freely on disturbed soil of road cuttings and land slips and so is useful in stabilizing hillsides against erosion.

Rosaceae

Hiroshi Ikeda, Colin A. Pendry, Hideaki Ohba, Mark F. Watson, David E. Boufford, Mohan Siwakoti, Anthony R. Brach, J. Crinan M. Alexander, Vidya K. Manandhar, Sangeeta Rajbhandary, Nirmala Joshi, Cathy King & Cliodhna D. Ní Bhroin

Trees, shrubs or herbs, deciduous or evergreen. Herbs unarmed, trees and shrubs sometimes with thorns, spines or prickles. Stems erect, scandent, arching, prostrate or creeping. Stipules paired, free or adnate to petiole, rarely absent, persistent or deciduous. Petiole usually 2-glandular apically. Leaves alternate, simple or compound, margins often variously serrate, rarely entire. Inflorescences various, from solitary flowers to umbels, corymbs, racemes or cymose panicles. Flowers actinomorphic, bisexual, rarely unisexual and then plants dioecious. Hypanthium composed of basal parts of sepals, petals and stamens, free from or adnate to ovary, short or elongate. Sepals (technically calyx lobes) usually 5, rarely fewer or more, imbricate. Epicalyx of 5 episepals sometimes present, alternate with and usually smaller than sepals. Petals as many as sepals, free, imbricate, sometimes absent or similar to calyx lobes. Stamens usually numerous, rarely few, always in a complete ring. Filaments free. Anthers small, didymous, rarely elongate, 2-locular. Ovary inferior, semi-inferior or superior. Carpels 1 to many, free to connate and then adnate to inner surface of hypanthium. Ovules usually 2 in each carpel, sometimes 1–many. Styles as many as carpels, terminal, lateral or basal, free or sometimes connate. Fruit a follicle, pome, achene, drupe or aggregate of drupelets, exposed or enclosed in persistent hypanthium and sometimes also by sepals.

Worldwide about 85 genera and 3000 species, cosmopolitan but most diverse in the temperate and warm regions of the N hemisphere. 26 genera and 155 species in Nepal.

A family of huge agricultural and horticultural importance in which the taxonomy of many groups is complicated by apomixis. For pragmatic reasons the arrangement of genera in this account mainly follows Lu *et al.* (Fl. China 9: 46. 2003), and so does not necessarily reflect the current understanding of phylogenetic relationships within the family.

Key to Genera

1a	Leaves pinnately or palmately divided to midrib	2
b	Leaves simple or shallowly lobed	18
2a	Evergreen or deciduous trees or shrubs	3
b	Deciduous perennial herbs or plants with woody stock and annual flowering shoots	7
3a	Ovary inferior	**9. Sorbus**
b	Ovary superior	4
4a	Stems without prickles	5
b	Stems prickly	6
5a	Leaves with 7–12 pairs of leaflets	**3. Sorbaria**
b	Leaves with 1 or 2 pairs of leaflets	**16. Potentilla**
6a	Carpels on convex receptacle	**14. Rubus**
b	Carpels enclosed in fleshy hypanthium	**19. Rosa**
7a	Leaves with 3 leaflets	8
b	Leaves pinnately divided or palmately divided with more than 3 leaflets	11
8a	Plants without stolons or rhizomes	9
b	Plants with spreading stolons or rhizomes	10
9a	Stamens 10–30	**16. Potentilla**
b	Stamens 4 or 5	**17. Sibbaldia**
10a	Flowers yellow or creamy-yellow	**16. Potentilla**
b	Flowers white or pink	**18. Fragaria**
11a	Petals absent	**21. Sanguisorba**
b	Petals present	12

| 12a | Epicalyx absent | 13 |
| b | Epicalyx present | 15 |

| 13a | Flowers yellow | **20. Agrimonia** |
| b | Flowers cream, white, pink or purplish | 14 |

| 14a | Flowers unisexual, in spike-like racemes | **2. Aruncus** |
| b | Flowers bisexual, in large panicles | **13. Filipendula** |

| 15a | Rosette herb with long, slender stolons | **18. Fragaria** |
| b | Rosette or prostrate herbs without long, slender stolons | 16 |

| 16a | Leaves pinnate, with terminal leaflet much larger than lateral leaflets | **15. Geum** |
| b | Leaves palmate or if pinnate then all leaflets similar in size | 17 |

| 17a | Stamens 10–30 | **16. Potentilla** |
| b | Stamens 4 or 5 | **17. Sibbaldia** |

| 18a | Deciduous perennial herbs, or plants with woody stock and annual leafy and flowering shoots | 19 |
| b | Evergreen or deciduous trees and shrubs | 20 |

| 19a | Leaves entire. Petals white | **14. Rubus** |
| b | Leaves palmately divided. Petals absent | **22. Alchemilla** |

| 20a | Ovary superior | 21 |
| b | Ovary inferior or semi-inferior | 27 |

| 21a | Carpel(s) enclosed in hypanthium | 22 |
| b | Carpel(s) clearly visible | 23 |

| 22a | Stems without prickles. Carpel 1, enclosed in leathery hypanthium | **4. Neillia** |
| b | Stems prickly. Carpels many, enclosed in fleshy hypanthium | **19. Rosa** |

| 23a | Carpels 5 | **1. Spiraea** |
| b | Carpel 1 | 24 |

| 24a | Branches conspicuously thorny | **23. Prinsepia** |
| b | Branches thornless | 25 |

| 25a | Petals distinct from sepals | **24. Prunus** |
| b | Petals not clearly distinguishable from sepals | 26 |

| 26a | Leaves entire | **25. Pygeum** |
| b | Leaves toothed | **26. Maddenia** |

| 27a | Ovary semi-inferior, carpels partially free above | 28 |
| b | Ovary inferior, carpels completely united | 30 |

| 28a | Branches conspicuously thorny | **6. Pyracantha** |
| b | Branches thornless | 29 |

| 29a | Leaves acute to rounded, if acuminate to 7 cm or rarely 8 cm | **5. Cotoneaster** |
| b | Leaves acuminate, more than 7 cm long | **7. Photinia** |

| 30a | Carpels with 3 to many ovules | 31 |
| b | Carpels with 1 or 2 ovules | 32 |

| 31a | Leaves toothed, at least at apex, sometimes slightly lobed | **10. Docynia** |
| b | Leaves entire | **Cydonia** |

32a	Leaves evergreen, leathery	33
b	Leaves deciduous, papery	34

33a	Inflorescence a corymb	**7. Photinia**
b	Inflorescence a panicle	**8. Eriobotrya**

34a	Petals 3–8 mm. Inflorescence many-flowered. Leaves not clustered on short side shoots	**9. Sorbus**
b	Petals 8–20 mm. Inflorescence few-flowered. Leaves clustered on short side shoots	35

35a	Styles free	**11. Pyrus**
b	Styles connate at base	**12. Malus**

Key to Genera (fruiting material)

1a	Fruit a dehiscent follicle, rarely a capsule. Carpels 1–5. Stipules present or absent	2
b	Fruit an indehiscent achene, pome or drupe. Carpels 1–many. Stipules present	5

2a	Leaves pinnately or palmately divided to mibrib. Leaflets stalked or stalkless	3
b	Leaves simple, entire or shallowly lobed, if deeply lobed then at least part of the blade visible on either side of midrib	4

3a	Perennial herb with woody stock and annual flowering shoots	**2. Aruncus**
b	Trees or shrubs	**3. Sorbaria**

4a	Stipules absent. Carpels 3–5. Follicles exposed	**1. Spiraea**
b	Stipules present. Carpel 1. Follicle enclosed within hypanthium	**4. Neillia**

5a	Ovary inferior or semi-inferior. Carpels (1 or)2–5, ± adnate to inner side of cupular hypanthium. Fruit a pome	6
b	Ovary superior. Carpels 1 to many, free from hypanthium. Fruit a drupe, achene or aggregate of several achenes or drupelets	16

6a	Leaves pinnate	**9. Sorbus**
b	Leaves simple, entire or lobed	7

7a	Fruit with 1–5 bony nutlets	8
b	Fruit with 1–5 leathery or papery cells, each containing 1 or more seeds	9

8a	Branches thornless	**5. Cotoneaster**
b	Branches conspicuously thorny	**6. Pyracantha**

9a	Leaves often deeply lobed to more than half-way towards midrib	**10. Docynia**
b	Leaves entire or shallowly toothed	10

10a	Leaf-margin toothed	11
b	Leaf-margin entire	12

11a	Inflorescence corymbose	**7. Photinia**
b	Inflorescence paniculate	**8. Eriobotrya**

12a	Inflorescence compound corymbose	13
b	Inflorescence umbellate, racemose, fasciculate or flowers solitary	14

13a	Leaves leathery, evergreen	**7. Photinia**
b	Leaves papery, deciduous	**9. Sorbus**

14a	Carpels with numerous ovules	**Cydonia**
b	Carpels with 1 or 2 ovules	15

15a	Styles free. Inflorescence corymbose-racemose. Fruit with numerous grit cells	**11. Pyrus**
b	Styles connate basally. Inflorescence a fascicle. Fruit without or with few grit cells	**12. Malus**

287

16a Carpel 1(–2). Fruit a drupe, sepals often deciduous. Leaves simple 17
  b Carpels usually numerous, rarely few. Fruit 1 or more achenes, or an aggregate of fleshy drupelets, sepals persistent. Leaves compound, very rarely simple 20

17a Branches conspicuously thorny...**23. Prinsepia**
  b Branches thornless..18

18a Fruit obscurely 2-lobed..**25. Pygeum**
  b Fruit unlobed 19

19a Leaf margins eglandular or glands sessile. Stipules up to 15 mm, usually caducous....................................**24. Prunus**
  b Leaf margins at base with gland tipped hairs. Stipules longer than 15 mm, often persistent ....**26. Maddenia**

20a Leaves simple or shallowly lobed, if deeply lobed then at least part of the blade visible on either side of midrib..........21
  b Leaves pinnately or palmately divided to mibrib 22

21a Fruit a berry of fleshy drupelets ....**14. Rubus**
  b Fruit a cluster of achenes enclosed in persistent, dry hypanthium...**22. Alchemilla**

22a Fruit a berry of fleshy drupelets ....**14. Rubus**
  b Fruit of 1 or more achenes, exposed or enclosed in persistent hypanthium 23

23a Evergreen or deciduous trees or shrubs..24
  b Deciduous perennial herbs or plants with woody stock and annual flowering-shoots......................................25

24a Stems without prickles. Hypanthium dry and hard..**16. Potentilla**
  b Stems prickly. Fruit of achenes enclosed in fleshy hypanthium (hip) ....**19. Rosa**

25a Leaves with 3 leaflets..26
  b Leaves pinnately divided or palmately divided with more than 3 leaflets 29

26a Plants without stolons or rhizomes...27
  b Plants with spreading stolons or rhizomes...28

27a Stamens 10–30 ....**16. Potentilla**
  b Stamens 4 or 5...**17. Sibbaldia**

28a Fruiting receptacle dry...**16. Potentilla**
  b Fruiting receptacle juicy, strawberry-like ....**18. Fragaria**

29a Fruit of 1 or more achenes enclosed in persistent hypanthium 30
  b Fruit of achenes exposed on a flat or convex receptacle 32

30a Leaves palmately divided or lobed. Epicalyx present ....**22. Alchemilla**
  b Leaves pinnately divided. Epicalyx absent...31

31a Fruiting hypanthium with hook-like spines outside..**20. Agrimonia**
  b Fruiting hypanthium winged, without hook-like spines..**21. Sanguisorba**

32a Fruiting receptacle juicy, strawberry-like ....**18. Fragaria**
  b Fruiting receptacle dry..33

33a Achenes up to 12, borne on flat receptacle ....**13. Filipendula**
  b Achenes numerous, borne on convex receptacle..34

34a Styles lengthening in fruit, often with feathery hairs, or jointed ....**15. Geum**
  b Styles not lengthening in fruit...35

35a Stamens 10–30 ....**16. Potentilla**
  b Stamens 4 or 5...**17. Sibbaldia**

1. *Spiraea* L., Sp. Pl. 1: 489 (1753).

J. Crinan M. Alexander & Nirmala Joshi

Deciduous branched shrubs, rarely unbranched subshrubs, bisexual or dioecious. Twigs unarmed. Stipules absent. Leaves simple, sessile or shortly petiolate, serrate or incised to lobed, rarely entire, pinnately veined, rarely with 3–5 veins from the base. Inflorescences umbels, umbel-like racemes, corymbs or panicles. Flowers bisexual or unisexual, in umbels, umbel-like racemes, corymbs or panicles. Bracts linear to lanceolate, at base of pedicel, persistent or caducous. Hypanthium campanulate to cup-shaped, bearing sepals, petals, anthers and nectary-disk on rim. Sepals 5, valvate or slightly overlapping, usually slightly shorter than calyx-tube. Petals 5, overlapping or contorted, broadly obovate to orbicular, short-clawed, usually longer than sepals, glabrous. Stamens 15–40. Ovary superior, carpels (3–)5(–8), free; ovules pendulous, several (rarely 2 or 3) per carpel. Styles terminal or subterminal, not exceeding stamens; stigmas capitate or disciform. Fruit of (3–)5(–8) follicles, usually dehiscing along inner suture. Seeds minute, linear to oblong.

Worldwide about 100 species in the N hemisphere, from temperate regions to subtropical mountains. Nine species in Nepal, one of which is introduced.

Material should be collected with both fertile and sterile branches as leaves are often larger on the latter and this can be taxonomically informative. It is also important to note the growth form in order to distingiush *Spiraea hemicryptophyta* Grierson from similar species.

Key to Species

1a Leaves on flowering shoots ± sessile, narrowly attenuate at base...2
b Leaves on flowering shoots with distinct petioles 3

2a Leaves obovate to spathulate, rounded, not toothed or lobed, to 6 mm wide...................... **6. *S. hypericifolia***
b Leaves obovate or elliptic to almost circular in outline, some with teeth or lobes, some more than 7 mm
wide .... **7. *S. hypoleuca***

3a Main stems arched or spreading. Inflorescences lateral, on closely and regularly arranged erect side branches to
5(–7) cm. Leaves on fertile branches to 1.5 cm...4
b Main stems erect to ascending. Inflorescences terminal, or on side branches, the lower more than 7 cm. Some leaves
on fertile branches longer than 1.5 cm...5

4a Carpels glabrous. Pedicels and calyx tube moderately to densely hairy, sometimes reddish. Main stems arched,
often black. Usually above ca. 3500 m .... **4. *S. arcuata***
b Carpels, pedicels and calyx tube densely white-hairy. Main stems straight or arched, red to black. Usually
below ca. 3500 m .... **5. *S. canescens***

5a Plant entirely glabrous (in Nepal). Inflorescence a dense umbel with many pedicels arising from a common
peduncle. Peduncles and pedicels reddish.. **9. *S. cantoniensis***
b Plant hairy, at least in inflorescence. Inflorescences corymbose, peduncle dividing into smaller rays which further
divide into pedicels. Peduncles and pedicels not reddish...6

6a Plant almost always unbranched above ground. Stems annual, arising from an underground rhizome.
Inflorescences terminal .... **2. *S. hemicryptophyta***
b Plant branched above ground. Stems perennial. Inflorescences terminal or lateral 7

7a Inflorescences mostly terminal, to 30 cm across. Flowers to 5 mm across. Leaf length:width ratio > 2,
tips attenuate. Carpels hairy .... **3. *S. micrantha***
b Inflorescences terminal and lateral, to ca. 5.5 cm across. Flowers more than 5 mm across. Leaf length:width
ratio < 2, tips not or scarcely attenuate. Carpels glabrous or slightly pubescent on ventral suture 8

8a Leaves with pointed teeth. Flowers usually pink, sometimes white. Carpels 2.5–3.5 mm .... **1. *S. bella***
b Leaves crenate, obtusely toothed to entire. Flowers white. Carpels 1.5–2 mm .... **8. *S. vacciniifolia***

Rosaceae

1. *Spiraea bella* Sims, Bot. Mag. 50: pl. 2426 (1823).
Spiraea amoena Spae; *S. expansa* Wall. nom. nud.

सेतो खरेटो Seto khareto (Nepali).

Branched, usually dioecious shrubs to ca. 2.5 m. Branches yellowish to reddish brown, slightly angled, ± pubescent; buds pale brown to reddish brown, obtuse, glabrous. Petioles 2–5 mm, sparsely pubescent. Leaves grey-green below, brighter green above, thinly leathery, ovate to ovate-elliptic, 2–4 × 1–2 cm, base truncate or rounded to broadly cuneate, apex acute, sharply serrate or doubly serrate from below middle, glabrous or rarely puberulous above, almost glabrous to pubescent on veins below, pinnately veined. Corymbs terminal and lateral, compound, pubescent to densely hairy, 2–5.5 cm across, many-flowered. Pedicels 5–8 mm. Flowers mostly unisexual, 5–7 mm across, female slightly smaller. Hypanthium campanulate, slightly pubescent outside, more densely so inside. Sepals triangular, acute, gland-tipped, reflexed in fruit. Petals pink, rarely white, ca. 2.5 mm. Stamens ca. 20, reduced and shorter than petals in female flowers, slightly longer than petals in male flowers. Disk 10-lobed. Carpels pubescent, reduced in male flowers. Follicles spreading, well exserted, glabrous or very slightly puberulous on inner suture; styles spreading.
Fig. 37a.

Distribution: Nepal, E Himalaya, Tibetan Plateau, Assam-Burma and E Asia.

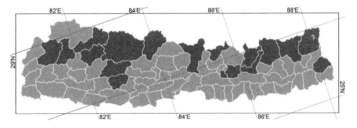

Altitudinal range: 1800–4300 m.

Ecology: Broad-leaved, coniferous or mixed forests, thickets on slopes, open rocky hillsides.

Flowering: May–August. **Fruiting:** August–September.

Intermediates with *Spiraea micrantha* Hook.f. are sometimes found and the two species are not always clearly distinct. The length:width ratio of the leaves is usually less than 2.0, while in *S. micrantha* it is usually greater than 2.0.

2. *Spiraea hemicryptophyta* Grierson, Notes Roy. Bot. Gard. Edinburgh 44: 262 (1987).

Perennial, rhizomatous, dioecious subshrubs to ca. 45 cm. Rhizome horizontal, at or below ground level, thin, stem-like. Stems annual, unbranched or with a few weak ascending branches below, pale yellowish to reddish brown, terete, finely striate, softly pubescent above, glabrous below. Buds pale to dark brown, acute, slightly hairy. Petioles 2–5 mm. Leaves olive-green above, paler below, broadly ovate-elliptic

to orbicular, 2.5–5.0 × 1.8–3 cm, base rounded to broadly truncate, apex acute, coarsely double-serrate, glabrous, sparsely pubescent on veins beneath, pinnately veined. Corymbs terminal, compound, to 6 cm wide, many-flowered; moderately to densely crisped-hairy; primary stalks to ca. 15 mm. Pedicels to 6 mm. Flowers unisexual, to 7 mm across. Hypanthium cupular, 1.5–2 × 2.5–3 mm, pubescent outside. Sepals triangular, 1.5–2 mm, persistent, reflexed in fruit, slightly acuminate, gland-tipped. Petals white to pink, 2.5–3 mm. Disk irregularly 8–12-lobed. Male flowers: stamens ca. 25; filaments ca. 3 mm; carpels reduced. Female flowers: staminodes 0.5–1 mm; carpels 5 or 6, ellipsoid, 1.5–2 mm; style ca. 1.5 mm, ± exserted from hypanthium. Follicles ca.3 mm, shiny, brown to red or purple, hairy near ventral suture ca. half exserted.
Fig. 37b.

Distribution: Nepal, E Himalaya, Tibetan Plateau, Assam-Burma and E Asia.

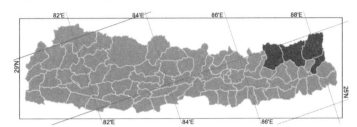

Altitudinal range: 3500–4300 m.

Ecology: Open areas in *Rhododendron* forests, rocky slopes.

Flowering: June–August. **Fruiting:** August–October.

This species is related to *Spiraea bella* Sims though of very distinctive habit, consisting of tufted, upright, usually unbranched stems, arising from an underground rhizome. It was not recognized as a distinct species in Flora of China (9: 53. 2003).

3. *Spiraea micrantha* Hook.f., Fl. Brit. Ind. 2[5]: 325 (1878). *Spiraea japonica* subsp. *micrantha* (Hook.f.) Kitam.; *S. japonica* var. *himalaica* Kitam.

Branched shrubs to 2.5 m. Branches reddish brown, finely ridged, pubescent. Buds brown to reddish brown, acute, pubescent. Petioles to ca. 1 cm. Leaves mid-green, slightly paler below, ovate to elliptic, 2.5–7 × 0.7–3 cm, base rounded to broadly cuneate, sharply serrate or doubly serrate from below middle, apex attenuate, acute, pubescent on veins below, less so above, otherwise ± glabrous, pinnately veined. Corymbs terminal or on long, leafy lateral branches, compound, pubescent, spreading, to 30 cm across, much-branched and many-flowered. Pedicels to ca. 4 mm. Flowers usually bisexual, 4–5 mm across. Hypanthium campanulate, pubescent outside. Sepals ca. 1.8–2.0 mm, reflexed in fruit. Petals white, rarely pinkish; blade ca. 1.5 mm. Stamens ca. 20, much longer than petals. Disk 10-lobed. Carpels hairy. Follicles ± erect, moderately to densely pubescent, 1–2 mm; styles spreading, slightly abaxial.
Fig. 37c.

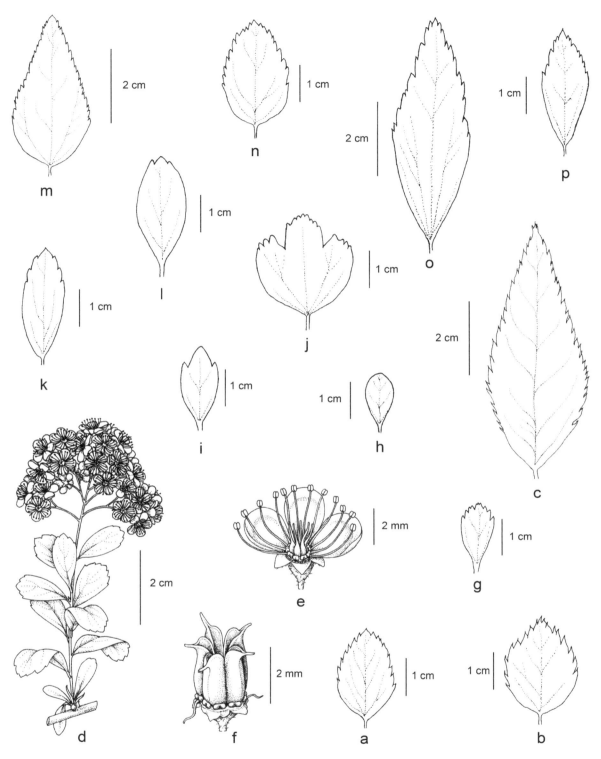

FIG. 37. RosACEAE. **Spiraea bella**: a, leaf. **Spiraea hemicryptophyta**: b, leaf. **Spiraea micrantha**: c, leaf. **Spiraea arcuata**: d, inflorescence leaves; e, opened flower with two petals removed. **Spiraea canescens**: g, leaf. **Spiraea hypericifolia**: h, leaf. **Spiraea hypoleuca**: i–j, leaves (vegetative shoot); k–l, leaves (fertile shoot). **Spiraea vacciniifolia**: m–n, leaves. **Spiraea cantoniensis**: o–p, leaves.

Distribution: Nepal, E Himalaya, Tibetan Plateau and Assam-Burma.

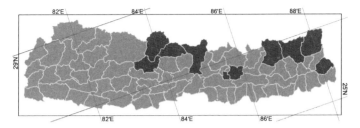

Altitudinal range: 1500–3000 m.

Ecology: Deciduous and mixed primary and secondary forest, open and shrubby mountain slopes.

Flowering: May–August. **Fruiting:** (May–)July–November.

Spiraea micrantha is not clearly distinct from *S. bella* Sims, and intermediates between the two species are sometimes found.

4. Spiraea arcuata Hook.f., Fl. Brit. Ind. 2[5]: 325 (1878).
Spiraea zabeliana C.K.Schneid.

पण्डा Panda (Nepali).

Arched shrubs to ca. 2 m. Branches robust, dark reddish brown to black, shiny, strongly ridged, glabrescent. Buds dark reddish brown, acuminate, glabrous. Petioles to ca. 4 mm, glabrous or subglabrous. Leaves olive-green above, paler below, narrowly elliptic to obovate, 8–12 × 3–5 mm on flowering shoots, up to ca. 15 × 10 mm on vegetative shoots, base cuneate, apex obtuse, rarely subacute, entire to obtusely 3–8-serrate or shallowly lobed apically, glabrous or subglabrous on both surfaces, pinnately veined. Corymbs terminal on short, lateral branchlets, compound, densely many-flowered, puberulous, to ca. 2.5 × 2.5 cm. Pedicels to ca. 8 mm (< 13 mm in fruit). Flowers bisexual, 6–8 mm across. Hypanthium turbinate, pubescent outside. Sepals triangular, reflexed in fruit, usually acute. Petals pink, ca. 3 mm. Stamens ca. 20, slightly shorter than or about equalling petals. Disk broadly crenate. Carpels glabrous. Follicles spreading, wholly exserted, shining, glabrous, rarely puberulous on inner suture only; styles terminal, slightly abaxial, divergent.
Fig. 37d–f.

Distribution: Nepal, W Himalaya, E Himalaya, Tibetan Plateau, Assam-Burma and E Asia.

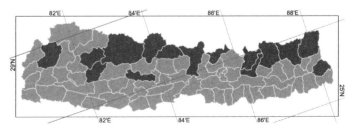

Altitudinal range: 2100–4900 m.

Ecology: Thickets and open areas on mountain slopes and river banks, subalpine rocky places.

Flowering: June–September. **Fruiting:** July–October.

Similar to, and sometimes hard to distinguish from, *Spiraea canescens* D.Don, though the carpels are almost always glabrous and it generally occurs above 3500 m.

5. Spiraea canescens D.Don, Prodr. Fl. Nepal.: 227 (1825).
Spiraea cuneifolia Wall. ex Cambess.; *S. cuneifolia* Wall. nom. nud.

झिल्लेटी Jhilleti (Nepali).

Shrubs to ca. 3.2 m. Branches spreading or slightly arched, brown or grey-brown, angled, pubescent, later glabrescent. Buds reddish brown, obtuse, pubescent. Petioles ca. 2 mm (up to 4 mm on vegetative shoots), pubescent. Leaves olive-green above, paler below, ovate or elliptic to obovate, 1–1.5 × 0.5–1.0 cm, base cuneate, apex obtuse, sometimes inconspicuously 3-lobed, margin entire or obtusely 3–5-dentate above middle, pubescent to glabrous above and below, pinnately veined. Corymbs terminal on short, lateral branchlets, compound, 2–4.5 × 3–5 cm, many-flowered, usually densely hairy. Pedicels 4–8 mm. Flowers bisexual, 5–7 mm across. Hypanthium campanulate, pubescent outside. Sepals triangular, erect or spreading in fruit, acute. Petals white to pink, ca. 2 mm. Stamens ca. 20, about equalling or slightly longer than petals. Disk with 10 broad emarginate lobes. Carpels moderately to densely hairy; styles shorter than stamens. Follicles slightly spreading, usually densely hairy, sometimes glabrous on outer suture; styles on outer side, divergent, ca. half exserted.
Fig. 37g.

Distribution: Nepal, E Himalaya, Tibetan Plateau and E Asia.

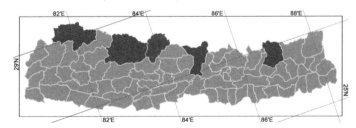

Altitudinal range: 1500–3000 m.

Ecology: *Pinus wallichiana* forest, open shrubby vegetation of degraded areas and streamsides.

Flowering: April–June. **Fruiting:** September–October.

Similar to, and sometimes hard to distinguish from, *Spiraea arcuata* Hook.f., though the carpels are densely hairy and it generally occurs below 3500 m.

6. *Spiraea hypericifolia* L., Sp. Pl. 1: 489 (1753).

Shrubs to 1.6 m. Branches erect to spreading, greyish brown, becoming dark red as bark splits, slender, terete, glabrous or puberulous when young. Buds brown, acute, glabrous or subglabrous. Leaves ± sessile, greyish olive-green above, paler below, obovate to spathulate, 10–15 × 5–7 mm, base narrowly attenuate, apex rounded, margin entire, glabrous above, with long sparse hairs on margin and midrib below. Umbels ± sessile, to 1.5 cm across, (1–)5–8-flowered, with cluster of leaves at base; pedicels 4–10 mm, glabrous, pinnately veined. Flowers bisexual, 5–7 mm across. Hypanthium campanulate, glabrous outside. Sepals greenish brown, triangular, ca. 1 mm, erect to spreading in fruit, acute. Petals white, ca. 4 mm. Stamens ca. 20, slightly shorter than or about equalling petals. Disk 10-lobed; lobes almost separate. Carpels and follicles glabrous though with fringe of hairs on ventral suture; styles not exceeding stamens. Follicles erect, ca. half exserted; styles terminal on outer side, ± erect. Fig. 37h.

Distribution: Nepal, W Himalaya, E Asia, N Asia, SW Asia and Europe.

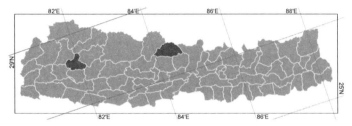

Altitudinal range: ca. 1700 m.

Ecology: Sparse forests and thickets, open dry areas.

Flowering: March–April. **Fruiting:** April–June.

It is uncertain whether *Spiraea hypericifolia* occurs in Nepal as the name has been widely misinterpreted. The lectotype (*LINN 651/5*) has small, ± spathulate leaves without any teeth or lobes, even on vegetative shoots, and glabrous inflorescences. The protologue says "Canada", probably in error. No Nepalese material seen matches this, though *Polunin, Sykes & Williams 1903* (BM) from Tila Valley (Western Nepal) is similar in leaf-shape but has hairy inflorescences. Plants assigned to this species from Turkey have toothed or divided leaves, and Lu & Alexander (Fl. China 9: 72. 2003) describe the leaves on sterile shoots as obtusely 2- or 3-dentate. Morphological data for this description were mostly taken from *Lambert, s.n. 10.3.1926 & 15.6.1926*, Kashmir (HUH), which closely match the Linnaean type. The record from Manang relates to *Nakao s.n. 26 May 1953* (KYO) (Kitamura, Fauna Fl. Nepal Himalaya: 161. 1955) but it has not been possible to check the specimen.

7. *Spiraea hypoleuca* Dunn, Bull. Misc. Inform. Kew 1921: 119 (1921).
Spiraea diversifolia Dunn.

Shrubs to ca. 2.5 m. Branches erect to spreading, grey to black, becoming dark or reddish brown, finely ridged, pubescent, becoming glabrous or black-scurfy. Buds brown, acute, densely grey- or red-silky. Leaves on flowering shoots ± sessile, olive-green above, much paler beneath, obovate to elliptic, 10–20 × 4–10 mm, base narrowly cuneate-attenuate, apex obtuse to rounded, margin entire below middle, bluntly and irregularly toothed towards apex, glabrous above and below, though young unfurling leaves sometimes silky near base, pinnately veined. Leaves on vegetative shoots larger, with petiole to ca. 7 mm, broadly ovate-elliptic to orbicular, to ca. 2.5 × 2 cm, often 3-palmatifid at apex. Umbels on short, lateral, few-leaved shoots to 1.5(–2) cm, 7–12-flowered. Pedicels 8–12 mm, glabrous or pubescent. Flowers bisexual, ca. 7 mm across. Hypanthium campanulate, glabrous or pubescent outside. Sepals triangular, green to brown, 1.5–2 mm, spreading to recurved in fruit. Petals white, ca. 4 mm. Stamens ca. 25, slightly shorter than or about equalling petals. Disk entire to 10-lobed. Carpels glabrous or pubescent; styles slightly shorter than stamens. Follicles ± erect, slightly exserted, glabrous; styles divergent. Fig. 37i–l.

Distribution: Nepal and W Himalaya.

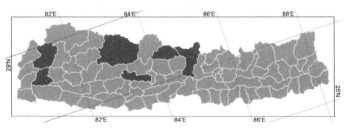

Altitudinal range: 2100–4400 m.

Ecology: Evergreen and mixed forest, Juniper scrub, thickets, shady banks.

Flowering: April–June. **Fruiting:** April–August.

Dunn described *Spiraea diversifolia* as differing from *S. hypoleuca* in its glabrous inflorescences and sterile branches with circular, palmatifid leaves. Very often herbarium specimens do not include sterile branches. Most of those that do have palmatifid leaves and as some of these also have hairy inflorescences, it is not possible to distinguish the two species. It is suspected that most plants assigned to *S. hypoleuca* would have circular, palmatifid leaves on their sterile branches

8. *Spiraea vacciniifolia* D.Don, Prodr. Fl. Nepal.: 227 (1825).
Spiraea laxiflora Lindl.; *S. rhamnifolia* Wall. nom. nud.

Branched shrubs to ca. 2 m. Branches reddish brown, smooth to slightly striate, densely appressed-hairy when young, glabrescent. Buds reddish brown, densely hairy, glabrescent. Petioles to ca. 8 mm, densely hairy, glabrescent. Leaves grey-green to olive-green above, ± glaucous below, broadly ovate-elliptic to rhombic, 1.5–4(–5) × 0.5–2(–2.5) cm, base rounded to broadly cuneate, apex obtuse to rounded, regularly crenate to blunt-toothed above middle, glabrous or with scattered hairs above and below, pubescent on veins especially below, pinnately veined; on finer branches leaves sometimes elliptic, minutely crenate or toothed, 1–2 cm × 3–6 mm. Corymbs compound, densely crisped-hairy, many-flowered, terminal corymbs to 10 cm across, lateral corymbs to ca. 5 cm across, on branches to 25 cm. Pedicels 5–8 mm. Flowers bisexual, ca. 4 mm across. Hypanthium campanulate, pubescent outside. Sepals triangular, green to brown, ca. 1 mm, crisped-hairy, erect in fruit. Petals white, 1.5–2.0 mm. Stamens 20–25. Disk 10-lobed. Carpels and follicles glabrous, ± erect; styles spreading, slightly abaxial. Fig. 37m–n.

Distribution: Nepal and W Himalaya.

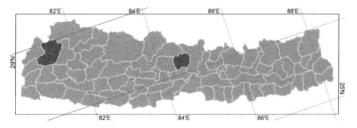

Altitudinal range: 1500–2300 m.

Ecology: Among scrub on dry stony hillsides, *Castanopsis* forest.

Flowering: May–June. **Fruiting:** July–October.

9. Spiraea cantoniensis Lour., Fl. Cochinch. 1: 322 (1790).

Glabrous shrubs to ca. 1.5 m. Branches ascending, dark reddish brown, later greyish brown, slender, terete. Buds grey to dark brown, obtuse, hairy. Petioles 4–7 mm. Leaves grey-blue below, dark green above, rhombic-lanceolate to rhombic-oblong, 2–8 × 0.8–2 cm, base cuneate, apex obtuse, margin irregularly and coarsely crenate to obtusely toothed above middle, with 3 main veins from near base. Umbels terminal on leafy lateral branches, well clear of leaves, to 4 cm across, many-flowered, simple with pedicels all from one point, to 14 mm. Flowers bisexual, to 7 mm across. Hypanthium campanulate. Sepals triangular or ovate-triangular, erect in fruit, apex acute or shortly acuminate. Petals white, to 4 × 4.5 mm. Stamens 20–30, slightly shorter than to nearly equalling petals. Disk with irregular emarginate lobes. Styles shorter than stamens. Follicles spreading, glabrous; styles terminal, usually divergent. Fig. 37o–p.

Distribution: Nepal and E Asia.

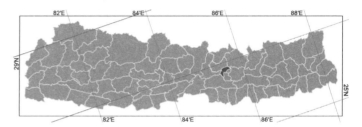

Altitudinal range: (200–)500–2200 m.

Ecology: Possibly cultivated.

Flowering: November–May. **Fruiting:** Not known.

It is possible that the single Nepalese specimen, *Pande 72*, Sundarijal (BM), is cultivated, though this is not mentioned on the label. According to Lu & Alexander (Fl. China 9: 62. 2003) *Spiraea cantoniensis* is "Native at least in N Jiangxi; widely cultivated elsewhere in China [Japan], 200–300 m". *Spiraea cantoniensis* was not included in Flora of Bhutan.

2. Aruncus L., Opera Var.: 259 (1758).

Hiroshi Ikeda

Dioecious perennial herbs. Twigs unarmed. Stipules obscure. Leaves radical and cauline, 2- or 3-ternate, petiolate; leaflets petiolulate, thinly charactaceous, margin doubly serrate, pinnately veined. Inflorescence a panicle with many flowers, terminal. Flowers unisexual, pedicellate. Bracts absent, bracteole solitary, linear, at apex of pedicel, persistent or caducous. Hypanthium shallowly cupulate. Sepals 5, valvate in bud. Petals 5, white, spathulate, inserted on the throat of hypanthium. Male flowers: stamens ca. 20; carpels degenerate. Female flowers: stamens degenerate; ovary superior, carpels 3–5(–7), free, erect. Style terminal. Fruit a follicle, pendulous with deflexed pedicel, coriaceous, ventrally dehiscent. Seeds several, narrowly fusiform, without endosperm.

Worldwide three to six poorly defined species in temperate regions of the N hemisphere. One species in Nepal.

1. Aruncus dioicus (Walter) Fernald, Rhodora 41: 423 (1939).

Actaea dioica Walter, Fl. Carol.: 152 (1788); *Aruncus silvester* Kostel. ex Maxim. orth. var.; *A. silvester* Kostel. nom. nud.; *A. sylvester* Kostel. ex Maxim.; *Spiraea aruncus* L.

Aruncus dioicus subsp. **triternatus** (Maxim.) H.Hara, in H.Ohashi, Fl. E. Himalaya, Sec. Rep., Bull.: 50 (1971).

Aruncus dioicus var. *triternatus* (Maxim.) H.Hara; *A. sylvester* var. *triternatus* Maxim.; *Spiraea triternata* Wall. nom. nud.

Plants 30–100 cm. Rhizomes short-creeping, stout. Stems

erect, with a few scales at base, densely hirsute in lower part, sparsely pilose and pubescent above. Petioles 5–14 cm, glabrescent. Leaves to 20 cm; leaflets narrowly ovate to ovate-orbicular, 3–10 cm × 1–6 cm, base cuneate to truncate, apex caudate to acuminate, glabrescent above, glabrescent below except pubescent on veins. Inflorescence 10–30 cm; peduncles and pedicels densely short pubescent. Hypanthium nearly glabrous outside; sepals ovate, ca. 1 mm, apex obtuse. Male flowers: petals 1–1.2 × 0.3–0.6 mm. Stamens ca. 20; filaments 2–3 mm, glabrous; anthers globose, 0.3–0.5 mm across, apex rounded. Female flowers: petals 0.6–0.8 × 0.2–0.3 mm. Ovaries ca. 1 mm, glabrous; style ca. 0.5 mm, glabrous. Follicles oblong to ellipsoidal, 2–2.5 × 1–1.2 mm, greyish brown, glabrous. Seeds fusiform, ca. 1.5 × ca. 0.5 mm. Fig. 38a–d.

Distribution: Nepal, W Himalaya, E Himalaya, Tibetan Plateau and E Asia.

Altitudinal range: 3000–4200 m.

Ecology: Open meadows and within forests.

Flowering: June–August. **Fruiting:** July–September.

Don (Prodr. Fl. Nepal.: 228. 1825) and Hooker (Fl. Brit. Ind. 2: 323. 1878) partially misapplied the name *Spiraea aruncus* L. to this subspecies, and Kitamura (Fauna Fl. Nepal Himalaya: 147. 1955) misapplied the name *Aruncus sylvester* Kostel. ex Maxim. (as '*silvester*').

This subspecies is characterized by smaller, long acuminate leaflets with villous veins beneath, smaller flowers and smaller follicles.

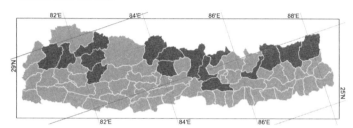

3. *Sorbaria* (Ser.) A.Braun nom cons., Fl. Brandenburg 1(1): 177 (1860).

Spiraea sect. *Sorbaria* Ser., in DC., Prodr. 2: 545 (1825).

Colin A. Pendry

Deciduous shrubs or small trees. Twigs unarmed. Stipules small, free, persistent. Leaves alternate, imparipinnate; leaflets opposite, sessile, biserrate, pinnately veined. Inflorescence a large, bracteate, terminal panicle. Flowers small, numerous, bisexual. Bracts linear, at base of pedicel, persistent. Hypanthium shallowly cup-shaped. Sepals 5, short, broad, persistent. Petals 5, imbricate, white. Stamens 20–25, equalling or longer than petals. Ovary superior, carpels 5, opposite sepals, basally connate, glabrous or subglabrous. Style terminal. Fruit of 5 basally connate follicles, glabrous, dehiscent along inner suture, remains of hypanthium persistent at base. Seeds several, linear, flattened.

Worldwide about nine species in temperate Asia. One species in Nepal.

1. *Sorbaria tomentosa* (Lindl.) Rehder, J. Arnold Arbor. 19: 74 (1938).
Schizonotus tomentosus Lindl., Bot. Reg. 26: Misc. 71 (1840); *Spiraea lindleyana* Wall. ex Lindl.; *S. lindleyana* Wall. ex Royle nom. nud.; *S. lindleyana* Wall. nom. nud.; *S. sorbifolia* L.

Erect, much-branched, arching shrub to 4 m. Twigs smooth, glabrous. Stipules linear, 4–8 mm. Leaves 13–30 mm, rachis glabrous to minutely pubescent, especially around insertion of leaflets. Leaflets (4–)6–11 pairs, narrowly ovate, 3–8 × 1–2 cm, base rounded to cuneate, sometimes oblique, apex acuminate, villous on veins below, secondary veins 12–18(–24) pairs. Inflorescence to 25 × 20 cm, rachis glabrous to minutely pubescent. Bracts 2–3 mm. Pedicels 1–3 mm, glabrous. Flowers 5–6 mm across. Hypanthium ca. 1 × 2 mm, glabrous to minutely pubescent. Sepals rounded,

to ca. 1 mm, membranous. Petals suborbicular, 1.5–2 mm. Stamens 2–2.5 mm. Carpels ca. 0.5 mm, ovoid, glabrous or basally pubescent. Ovules pendulous, 8–10. Styles 1.5 mm, spreading. Follicles 3 mm. Seeds linear, flattened. Fig. 38e–g.

Distribution: Nepal, W Himalaya, C Asia and SW Asia.

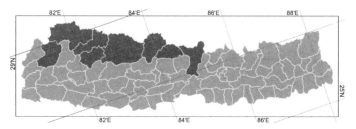

Rosaceae

Altitudinal range: 1800–2900 m.

Ecology: *Pinus wallichiana* forest and degraded scrub.

Flowering: June–July. **Fruiting:** July–August.

Hooker (Fl. Brit. Ind. 2: 324. 1878) misapplied the name *Spiraea sorbifolia* (L.) Br. to this species.

The reddish brown infructescence persists on the plant until the following season.

Juice from the seed is used to treat liver problems.

4. *Neillia* D.Don, Prodr. Fl. Nepal.: 228 (1825).

Colin A. Pendry & Cliodhna D. Ní Bhroin

Deciduous shrubs. Twigs unarmed. Stipules conspicuous, entire or toothed, caducous or persistent. Leaves simple, petiolate, more or less 3-lobed, rarely entire, margins biserrate, pinnately veined. Inflorescences terminal, racemes or loose panicles. Flowers pedicellate, bisexual. Bracts linear to elliptic or narrowly ovate, at base of pedicel, persistent. Hypanthium campanulate or campanulate-urceolate. Sepals 5, persistent in fruit. Petals 5, equal to sepals. Stamens 8–30, in 1 or 2 whorls, those of the inner whorl shorter. Ovary superior, carpel 1. Style terminal. Fruit an ovoid follicle, dehiscing along one suture, surrounded by the persistent, often glandular-hairy hypanthium. Seeds 6–10, obovoid.

Worldwide about 20 species from the Himalaya, E Asia and SE Asia. Three species in Nepal.

Key to Species

1a Inflorescence an open panicle, 6–14 cm. Stipule margin toothed .... **1. *N. thrysiflora***
 b Inflorescence a compact raceme, rarely with a few short branches at base, 2–6 cm. Stipule margin simple 2

2a Ovary glabrous. Leaves glabrous or sparsely villous above and below, not velvety to the touch.............. **2. *N. rubiflora***
 b Ovary villous. Leaves villous above and below, velvety to the touch .... **3. *N. velutina***

1. *Neillia thrysiflora* D.Don, Prodr. Fl. Nepal.: 228 (1825). *Neillia virgata* Wall. nom. nud.

भोग्ला Bhogla (Gurung).

Shrubs to 2 m. Twigs very sparsely pilose, glabrescent. Buds usually in threes, with two supra-axillary. Stipules ovate-lanceolate to 6–8 mm, margins toothed. Petioles 9–14 mm, pilose, soon glabrescent. Leaves ovate, more or less 3-lobed, sometimes with the lobes indistinct or absent, 4–9 × 2.5–5 cm, base cordate to truncate or rounded, apex acuminate but occasionally acute, margins biserrate, occasionally serrulate, glabrous or with a few stiff, appressed hairs towards the margin above, sparsely pilose only on veins below. Inflorescence an open, spreading panicle to 14 cm, peduncle tomentose. Bracts 3–4 mm. Pedicels 2–3 mm, tomentose. Hypanthium campanulate, 2–3 × 3 mm, tomentose. Sepals triangular, 3–4 mm, acuminate, tomentose along middle outside, almost glabrous within. Petals white, broadly ovate or triangular, 2–3 mm. Stamens 8–20, to 1.5 mm. Ovary ovoid, ca. 2 mm, glabrous. Style ca. 2 mm. Capsule ovoid, 6–7 mm, sparsely pubescent around suture, enclosed by persistent hypanthium with 1–2 mm stalked glandular hairs. Fig. 38h–i.

Distribution: Nepal, E Himalaya, Assam-Burma and E Asia.

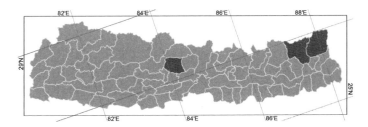

Altitudinal range: 1600–2400 m.

Ecology: *Castanopsis* forest and in scrub.

Flowering: August–September. **Fruiting:** September–November.

Ripe fruits are said to be edible.

2. *Neillia rubiflora* D.Don, Prodr. Fl. Nepal.: 229 (1825).

Shrubs to 4 m. Twigs tomentose, glabrescent. Buds usually solitary, axillary, rarely with 1 or 2 supra-axillary buds. Stipules ovate-lanceolate to linear-lanceolate, to 10 mm, margins entire. Petioles 10–12(–15) mm, glabrous to tomentose. Leaves ovate, usually 3-lobed, occasionally with the lobes indistinct or absent, 3–10 × 2–5 cm, base cordate, rarely truncate or rounded, apex acute to acuminate, margins serrate to biserrate, almost glabrous to densely hairy below, indumentum denser on veins. Inflorescence dense, racemose,

296

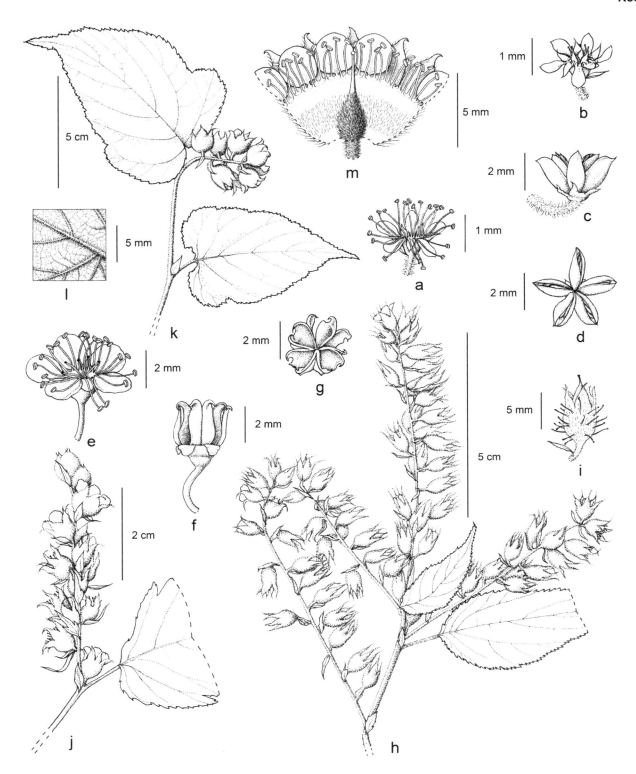

FIG. 38. Rosaceae. **Aruncus dioicus**: a, male flower; b, female flower; c, fruit side view; d, fruit vertical view. **Sorbaria tomentosa**: e, flower; f, fruit side view; g, fruit vertical view. **Neillia thrysiflora**: h, inflorescence with fruits; i, fruit. **Neillia rubiflora**: j, inflorescence. **Neillia velutina**: k, inflorescence and leaves; l, lower surface of leaf; m, opened flower.

rarely with up to 4 short branches at base, to 6 cm, peduncle tomentose. Bracts 4–6(–9) mm. Pedicels 2–3 mm, tomentose. Hypanthium campanulate-urceolate, 3–4 × 4 mm, tomentose. Sepals triangular, 2.5–4 mm, acute to acuminate, tomentose on inner and outer surfaces. Petals white or pink, orbicular, ca. 3 mm. Stamens 1.5–2 mm. Ovary subglobose, ca. 2 mm, slightly hairy at apex. Style ca. 3 mm. Capsule ovoid, ca. 6 mm, glabrous apart from a few sparse hairs around suture, enclosed by hypanthium with at least a few 2 mm long stalked glandular hairs.
Fig. 38j.

Distribution: Nepal, E Himalaya, Tibetan Plateau and E Asia.

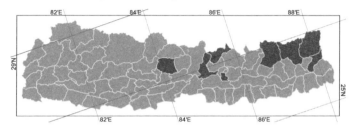

Altitudinal range: 1500–3500 m.

Ecology: In coniferous or broad-leaved evergreen forest with *Abies*, *Tsuga*, *Quercus* and *Rhododendron* and in scrub or grassland.

Flowering: June–July. **Fruiting:** August–September.

3. *Neillia velutina* Pendry, Phytotaxa 10: 38 (2010).

Shrub to 2 m. Twigs villous, glabrescent. Buds usually solitary. Stipules narrowly ovate, 4–7 mm, margin entire, rarely obscurely toothed, sparsely villous. Petioles 6–10 mm, densely villous. Leaves ovate, entire or somewhat 3-lobed, 2.5–7 × 1.5–4.5 mm, base cordate, apex acute to acuminate, margin doubly serrate, sparsely villous above, denser below, especially on the veins, secondary veins 5–6 pairs. Inflorescence a terminal raceme, usually simple or occasionally branched at the base, 1.5–4 cm, peduncle densely villous. Bracts 3–4 mm. Pedicels 1.5–4 mm. Hypanthium urceolate-campanulate, 3–4 × 4–5 mm, densely villous outside, glabrous or with a few hairs basally within. Sepals triangular, 2.5–3 mm, apex acute to acuminate, densely villous outside, sparsely villous within. Petals suborbicular, shortly clawed, 2.5–3 mm. Stamens 20–30, irregularly 2-whorled, filaments 1–2 mm, anthers 0.5 mm. Ovary 2 mm, densely villous. Style 2.5 mm, glabrous except at base. Fruits not seen.
Fig. 38k–m.

Distribution: Nepal and E Himalaya.

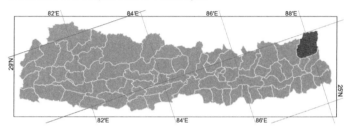

Altitudinal range: 2000–2600 m.

Ecology: In grassland and scrub.

Flowering: June–July. **Fruiting:** August–September.

The velvety leaves of *Neillia velutina* readily distinguish it from *N. rubiflora* D.Don and *N. thyrsiflora* D.Don.

5. *Cotoneaster* Medik., Philos. Bot. 1: 154 (1789).

Anthony R. Brach

Shrubs, rarely small trees, erect, decumbent or prostrate, deciduous, semi-evergreen or evergreen. Branchlets mostly terete, rarely slightly angulate, unarmed. Stipules small, subulate or lanceolate, usually caducous. Leaves alternate, simple, entire, shortly petiolate, papery or leathery. Inflorescences terminal or axillary, corymbose or cymose, sometimes flowers solitary or in fascicles. Flowers bisexual. Bracts at base of pedicel; bracteoles present or absent. Hypanthium turbinate or campanulate, rarely cylindric, adnate to ovary. Sepals 5, persistent, short. Petals 5, erect or spreading, imbricate in bud, white, pink or red. Stamens 10–20(–22), inserted in mouth of hypanthium. Ovary inferior or semi-inferior, 2–5-loculed; carpels 2–5, connate below, free above; ovules 2 per carpel, erect. Styles 2(–4), free; stigmas dilated. Fruit a red, brownish red or orange to black pome, with persistent, incurved, fleshy sepals, containing nutlets. Nutlets (1 or)2–5, bony, 1-seeded; seeds compressed.

Worldwide about 50–70 species widespread in temperate and subtropical N Africa, Asia and Europe, most abundant in SW China. 13 species in Nepal.

The taxonomy of the genus is complicated by hybridization and apomixis and while this account follows a wide view of species concepts (Dickoré & Kasperek, Willdenowia 40: 28. 2010) other authors recognise up to 400 species worldwide (Fryer & Hylmö, Cotoneasters. 2009). The 39 species accepted in the Ann. Checkl. Fl. Pl. Nepal (2000) are here reduced to 13 more clearly defined species.

Key to Species

1a Inflorescences compact compound corymbs, with more than 20 flowers or fruits. Petals white, spreading. Leaves large, more than 2.5 cm long 2

b Inflorescences lax corymbs, with fewer than 20 flowers or fruits, sometimes reduced to a solitary flower. Petals either erect and pink or red, or spreading and white. Leaves often less than 2.5 cm long 3

2a Leaves oblong or elliptic to ovate-lanceolate, to 8(–12) cm. Branchlets purplish brown or greyish brown. Fruit bright red, ellipsoid, 4–5 mm in diameter ....**1. C. frigidus**

b Leaves ovate or elliptic-ovate to obovate, to 5(–10) cm. Branchlets reddish brown. Fruit purplish brown to black, ovoid to globose, 6–8 mm in diameter ....**2. C. affinis**

3a Leaves longer than 2 cm, rarely slightly shorter 4

b Leaves often less than 2 cm, rarely slightly longer 7

4a Leaves subleathery. Flowers 9–10 mm across ....**4. C. hebephyllus**

b Leaves papery. Flowers 6–9 mm across..5

5a Inflorescences with 3–20 flowers or fruits. Leaves oblong to elliptic or ovate, glabrous or slightly pilose above, white tomentose below. Petals spreading ....**3. C. racemiflorus**

b Inflorescences with 1–5 flowers or fruits. Leaves elliptic to ovate-lanceolate, both surfaces pubescent, or sparsely so above. Petals erect..6

6a Flowers or fruits 3–6. Petioles 3–5 mm. Leaves broadly elliptic to suborbicular, to 2.5(–3) cm, both surfaces strigose-tomentose, sparsely so or subglabrous above, apex acute to acuminate. Petals dark red ....**10. C. symondsii**

b Flowers or fruits 1 or 2. Petioles 5–7 mm. Leaves ovate to lanceolate, to 6.5(–8) cm, both surfaces villous or pilose, apex long acuminate to acute. Petals white or pink, rarely red................................**12. C. acuminatus**

7a Petals erect, red, rarely pink or white. Prostrate, spreading or erect, deciduous or semi-evergreen shrubs...................8

b Petals spreading, white or pinkish or tinged red. Prostrate or low evergreen shrubs 12

8a Prostrate dwarf shrubs ....**11. C. adpressus**

b Erect shrubs 9

9a Stems ± distichously branched. Leaves ovate or obovate to suborbicular, both surfaces appressed pubescent. Petals white, pink, or stained reddish. Fruit obovoid or globose 10

b Stems irregularly branched. Leaves ovate to elliptic, both surfaces pubescent, sometimes sparsely so above. Petals red or pink. Fruit subglobose or turbinate 11

10a Branchlets strigose, glabrescent, not verruculose. Leaves apically acute and mucronate or obtuse. Flowers or fruit 2–6 ....**9. C. nitidus**

b Branchlets conspicuously verruculose. Leaves apically emarginate or partly mucronulate. Flowers or fruit solitary ....**13. C. verruculosus**

11a Flowers or fruit 3–6. Petioles 3–5 mm. Leaves broadly elliptic to suborbicular, both surfaces strigose-pilose, or becoming subglabrous above, apex acute to acuminate. Petals red. Leaves deciduous or semi-evergreen ....**10. C. symondsii**

b Flowers or fruit 1 or 2. Petioles 5–7 mm. Leaves ovate to lanceolate, both surfaces villous or pilose, apex long acuminate to acute. Petals white or pink, rarely red. Leaves deciduous.............................**12. C. acuminatus**

12a Leaves suborbicular or broadly ovate, base broadly cuneate to rounded. Flowers 1–3. Fruit red, obovoid to turbinate, with 2 or 3 nutlets..**8. C. rotundifolius**

b Leaves elliptic, oblong or obovate to oblong-obovate, rarely oblanceolate, base cuneate to obtuse. Flowers 1–12. Fruit red or scarlet, globose, with 1 or 2 nutlets..13

13a Leaves to 30 mm long. Flowers or fruit usually 3–12 ....**4. C. hebephyllus**

b Leaves to 10 mm long. Flowers or fruit usually 1–5, except *C. buxifolius* and *C. integrifolius* up to 10-flowered..........14

14a   Small shrubs, rigid, semi-evergreen. Branchlets greyish brown or brown. Leaves not shiny above, papery or thinly leathery. Flowers or fruit 3–5(–9) .... **5. *C. buxifolius***

   b   Prostrate or carpet-forming shrubs, flexuous, evergreen. Branchlets black, brown, red, green or purple. Leaves shiny above, leathery. Flowers or fruit 1(–3) 15

15a   Prostrate shrubs with irregular branching. Leaf margin flat, apex obtuse to retuse, rarely acute, glabrous or glabrescent below ....**6. *C. microphyllus***

   b   Low or dwarf shrubs, divaricate or squarrose, with regular branching. Leaf margins ± revolute, apex obtuse, acute or acuminate, tomentose, pubescent or glabrescent below... **7. *C. integrifolius***

1. *Cotoneaster frigidus* Wall. ex Lindl., Edward's Bot. Reg. 15: pl. 1229 (1829).
Cotoneaster bacillaris Wall. nom. nud.; *C. frigidus* Wall. nom. nud.; *C. gamblei* G.Klotz; *C. himalaiensis* Zabel; *C. nepalensis* K.Koch.

छार्वा Chharwa (Tibetan).

Shrubs or small trees, deciduous, to 10 m or more, scandent, with spreading branches. Branchlets spiralled, purplish brown or greyish brown, angular, initially tomentose, glabrescent. Stipules linear-lanceolate, tomentulose, caducous. Petioles 4–7(–10) mm, tomentose. Leaves distichous, shiny, dark green above, oblong or elliptic to ovate-lanceolate, (3–)3.5–8(–12) × 1.5–3(–6) cm, base cuneate or obtuse, margin flat or revolute, apex acute or obtuse, sometimes mucronate with spiniform point, thickly papery to subleathery, glabrous above, initially tomentose below, gradually glabrescent, veins slightly impressed above and prominent below. Compound corymbs dense, 4–6 × 3–5 cm, 20–40-flowered or more; rachis and pedicels densely tomentose. Bracts linear-lanceolate. Pedicels 2–4 mm. Flowers 6–7 mm across. Hypanthium campanulate or broadly subcylindric, densely tomentose outside. Sepals triangular, apex obtuse or acute, tomentose or villous. Petals spreading, white, broadly ovate or suborbicular, 3–3.5 mm and nearly as broad, apex obtuse, glabrousor puberulous inside near base, rarely somewhat emarginate. Stamens 18–20, slightly shorter than petals; anthers pink or red to black. Styles 2. Fruit bright to dark red, ellipsoid or globose to obovoid, 4–5 mm in diameter; nutlets (1 or)2.
Fig. 39a–c.

Distribution: Nepal, W Himalaya, E Himalaya and Tibetan Plateau.

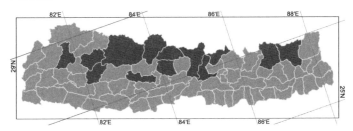

Altitudinal range: (900–)2200–3400 m.

Ecology: River valleys, broad-leaved deciduous forests on slopes.

Flowering: March–May(–June). **Fruiting:** September–October(–November).

2. *Cotoneaster affinis* Lindl., Trans. Linn. Soc. London 13(1): 101 (1822).
Cotoneaster affinis forma *bacillaris* (Wall. ex Lindl.) Samjatnin; *C. affinis* var. *bacillaris* (Wall. ex Lindl.) C.K.Schneid.; *C. bacillaris* Wall. ex Lindl.; *C. bacillaris* var. *affinis* (Lindl.) Hook.f.; *C. confusus* G.Klotz ex Arv.Kumar & Panigrahi; *C. cooperi* C.Marquand; *C. frigidus* var. *affinis* (Lindl.) Wenz.; *C. hedegaardii* J.Fryer & B.Hylmö; *C. ignotus* G.Klotz; *C. obovatus* Wall. ex Dunn; *C. obtusus* Wall. ex Lindl.; *C. virgatus* G.Klotz; *C. wattii* G.Klotz; *Mespilus affinis* (Lindl.) D.Don.

राइन्स Raeens (Nepali).

Shrubs, deciduous, to 5(–8) m, with spreading branches. Branchlets spiralled, dark reddish brown, terete, verruculose, initially pilose or strigose. Stipules linear or lanceolate, tomentulose, caducous. Petioles 4–5(–11) mm, densely yellow tomentose. Leaves distichous, dark green and dull to shiny above, broadly ovate or elliptic-ovate to obovate, 2.5–5(–10) × 1.4–2(–6) cm, base broadly cuneate, rounded or obtuse, margin flat or revolute, apex obtuse or acute, sometimes truncate, rounded or apiculate, papery, glabrous or puberulous only along midvein above, densely yellow tomentose below, gradually glabrescent, midvein impressed above and prominent below. Compound corymbs (2.5–)3–4 cm across, densely (3–)15–30-flowered or more; rachis and pedicels densely yellow tomentose. Bracts linear-lanceolate. Pedicels 2–3(–5) mm. Flowers ca. 1 cm across. Hypanthium campanulate, tomentose outside. Sepals broadly triangular, apex acute, pilose or strigose. Petals spreading, white, suborbicular or ovate, (2.5–)3–4 mm, apex obtuse, pubescent within near base. Stamens 20, shorter than sepals; anthers white or pink to purple. Styles 2. Fruit purplish brown to dark violet or black, ovoid to globose, 6–8(–10) mm in diameter; nutlets (1 or)2.
Fig. 39d–f.

Distribution: Nepal, W Himalaya, E Himalaya, Tibetan Plateau, Assam-Burma, S Asia, E Asia and C Asia.

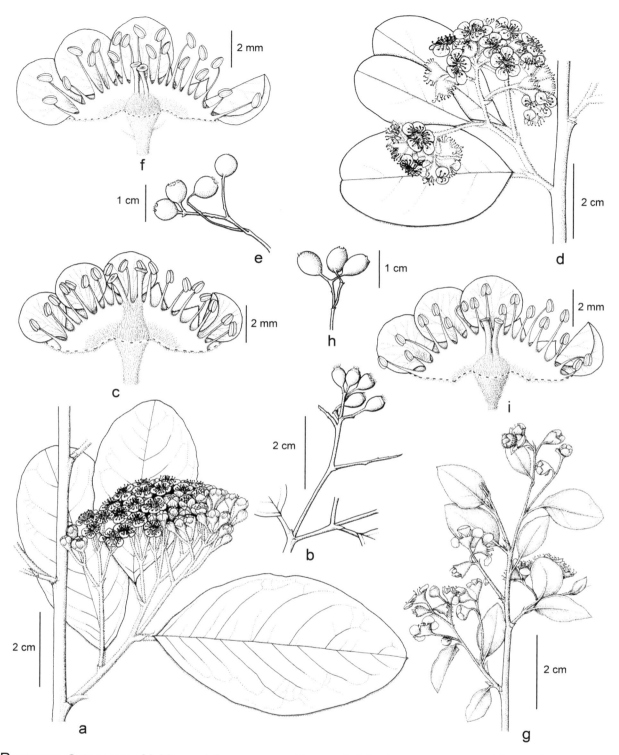

FIG. 39. Rosaceae. **Cotoneaster frigidus**: a, inflorescence and leaves; b, infructescence; c, opened flower. **Cotoneaster affinis**: d, inflorescence and leaves; e, infructescence; f, opened flower. **Cotoneaster racemiflorus**: g, inflorescence and leaves; h, infructescence; i, opened flower.

Rosaceae

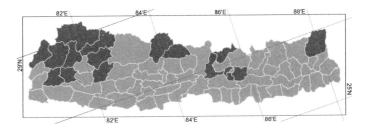

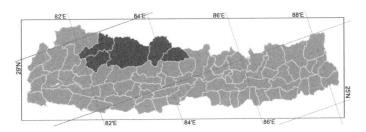

Altitudinal range: (1100–)2200–3000(–3900) m.

Ecology: Slopes, mixed forests, thickets of river valleys, among shrubs on open ground.

Flowering: April–June. **Fruiting:** (June–) September–October(–November).

Further research is needed on the distribution of this taxon.

The wood is used for making walking sticks and tent pegs. A paste of the root is applied to treat headaches and the juice is given in cases of indigestion.

3. *Cotoneaster racemiflorus* (Desf.) K.Koch, Dendrologie 1: 170 (1869).
Mespilus racemiflorus Desf., Cat. Pl. Horti Paris., ed. 3: 409 (1829); *Cotoneaster inexpectatus* G.Klotz; *C. racemiflorus* var. *royleanus* Dippel; *C. tibeticus* G.Klotz; *C. zayulensis* G.Klotz.

Shrubs, deciduous, (0.3–)1–2.5 m, with spreading or ascending, arched branches. Branchlets spiralled, greyish brown or reddish brown, terete, slender, initially densely grey tomentose, glabrous when old. Stipules lanceolate, pilose or villous, caducous. Petioles (1.4–)2–5 mm, tomentose. Leaves spiralled or distichous, dull olive-green above, oblong or ovate to elliptic, rarely obovate, (1–)1.5–5 × (0.5–)1–2 cm, base rounded or broadly cuneate, margin revolute, apex obtuse and mucronate, often emarginate, sometimes acute or shortly acuminate, papery, glabrous or slightly pilose above, white tomentose below, veins slightly impressed above and prominent below. Corymbs (0.5–)1.5–2 cm, (3–)5–9(–20)-flowered; rachis and pedicels white tomentose. Pedicels 2–3 mm. Flowers 8–9 mm across. Hypanthium campanulate or turbinate, white tomentose outside. Sepals broadly triangular, 1–2 mm, apex acute, subglabrous or glabrous. Petals spreading, white, suborbicular or ovate to obovate, 2.5–3.5 × 2–3 mm, apex obtuse, rarely emarginate, white puberulous near base within. Stamens 18–20, slightly shorter than petals; anthers yellow, white or purple. Styles 2. Fruit red to bright red, broadly ellipsoid to subglobose or pyriform, (5–)8–10 mm; nutlets (1 or)2.
Fig. 39g–i.

Distribution: Nepal, W Himalaya, E Himalaya, Tibetan Plateau, Assam-Burma, E Asia, N Asia, C Asia, SW Asia and Africa.

Altitudinal range: (1400–)2700–4000 m.

Ecology: Shrub thickets on dry, stony slopes, forests, forest margins, river and mountain valleys.

Flowering: (April–)May–June. **Fruiting:** July–September(–October).

Some individuals have been observed with flowers possessing a single style and 1-locular ovary. In *Dobremez 2787* (BM), the type of *Cotoneaster inexpectatus* G.Klotz, some flowers have 2 styles and others 1 style. Further research is needed on the distributional range of this taxon.

4. *Cotoneaster hebephyllus* Diels, Notes Roy. Bot. Gard. Edinburgh 5: 273 (1912).
Cotoneaster ludlowii G.Klotz; *C. rotundifolius* T.T.Yu later homonym, non Wall. ex Lindl.; *C. schlechtendalii* G.Klotz; *C. sherriffii* G.Klotz.

Shrubs, semi-evergreen or deciduous, erect, to 1.5 m tall or more, with dense, ascending to irregularly spreading branches. Branchlets spiralled, greyish brown to reddish, terete, initially sparsely accumbent, strigose, glabrate, densely grey lenticellate when old. Stipules lanceolate or linear-lanceolate, villous, caducous. Petioles 1–4 mm, villous. Leaves distichous or spiralled, dark or pale green above, broadly elliptic or oblong-obovate to lanceolate-elliptic, rarely oblanceolate, 0.6–4 cm × 4–25 mm, base cuneate, rounded, or obtuse, margin flat or revolute, apex rounded, obtuse, or acute, sometimes mucronate, papery or subleathery, villous or glabrous above, densely yellowish or grey appressed villous, later subglabrous below, midvein impressed above, prominent below. Corymbs erect, 1–2.5 cm, 3–9(–15)-flowered; rachis and pedicels initially villous, glabrescent. Pedicels 3–6 mm. Flowers 9–10 mm across. Hypanthium campanulate, ± villous outside. Sepals triangular, apex acute, subglabrous. Petals spreading, white or pinkish, suborbicular, 3–4 mm wide, apex obtuse, puberulous inside near base. Stamens 16–20, shorter than petals; anthers purple-black. Styles 1 or 2. Fruit red or orange-red to bright red, globose or depressed-globose, 5–10 mm in diameter; nutlets (1 or)2.

Distribution: Nepal, E Himalaya, Tibetan Plateau, Assam-Burma and E Asia.

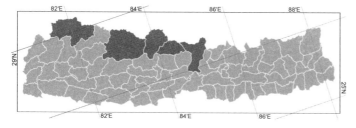

Altitudinal range: 2000–4000(–4100) m.

Ecology: River valleys, forests, slopes, shrubby areas.

Flowering: May–June. **Fruiting:** September–October.

Grierson (Fl. Bhutan 1: 589. 1987) included *Cotoneaster ludlowii* within the synonymy of *C. sherriffii*, and together they are treated here within *C. hebephyllus*.
   Ripe fruits are eaten to relieve coughs and colds.

5. *Cotoneaster buxifolius* Wall. ex Lindl., Edward's Bot. Reg. 15: pl. 1229 (1829).

Shrubs, small, semi-evergreen to evergreen, erect, rigid, to 1.5(–3) m tall. Branchlets spiralled, dark greyish brown or brownish, terete, initially densely whitish tomentose, glabrate. Stipules subulate, tomentulose, caducous. Petioles 1–3 mm, tomentose. Leaves spiralled, green above, elliptic to oblong or elliptic-obovate, 5–10(–16) × (2–)4–9 mm, base narrowly to broadly cuneate, margin revolute, apex acuminate or acute and mucronate, subleathery or thickly papery, initially appressed pubescent above, glabrescent, densely grey tomentose below, midvein impressed above, prominent below. Inflorescences solitary flowers or cymose, 5–10 mm, (1 or)3–5(–9)-flowered. Bracts lanceolate. Pedicels 3–5 mm, tomentose. Flowers 7–9 mm across. Hypanthium campanulate, tomentose outside. Sepals ovate-triangular, apex acute, tomentose, glabrescent. Petals spreading, white, suborbicular or broadly ovate, 3–4 mm wide, apex obtuse, glabrous. Stamens 20, slightly shorter than petals; anthers pink to blackish. Styles 2, nearly as long as stamens. Fruit red, subglobose or depressed globose, 5–6 mm in diameter; nutlets usually 2.
Fig. 40a.

Distribution: Nepal, W Himalaya, E Himalaya, Tibetan Plateau, Assam-Burma and E Asia.

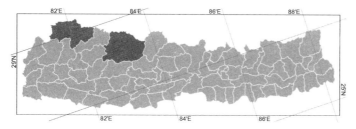

Altitudinal range: 2200–2900 m.

Ecology: Mountain regions, rocky mountain slopes, thickets, roadsides, river valleys.

Flowering: April–June. **Fruiting:** September–October.

6. *Cotoneaster microphyllus* Wall. ex Lindl., Bot. Reg. 13: pl. 1114 (1827).
Cotoneaster buxifolius forma *cochleatus* Franch.; *C. cochleatus* (Franch.) G.Klotz; *C. glacialis* (Hook.f. ex Wenz.) Panigrahi & Arv.Kumar; *C. microphyllus* var. *cochleatus* (Franch.) Rehder & E.H.Wilson; *C. microphyllus* var. *glacialis* Hook.f. ex Wenz.; *C. microphyllus* var. *nivalis* G.Klotz; *C. nivalis* (G.Klotz) Panigrahi & Arv.Kumar; *C. thymifolius* var. *cochleatus* (Franch.) Franch. ex H.Lév.

छालेप Chhalep (Tibetan).

Shrubs, evergreen, dwarf, mostly prostrate or mound-forming, flexuous, to 1 m tall, with irregular, spreading branches. Branchlets spiralled, reddish brown to blackish brown, terete, initially yellow pubescent, gradually glabrate. Stipules lanceolate or linear, slightly pubescent, caducous. Petioles 1–2 mm or more, pubescent. Leaves spiralled or distichous, shiny, pale to dark green, obovate to obcordate or lanceolate, 4–10(–16) × (1.5–)4–8(–9) mm, base broadly cuneate or slightly rounded, margin flat, apex obtuse or retuse, rarely acute, thickly to thinly leathery, glabrous or sparsely pubescent above, glabrous or glabrescent below, midvein impressed above, prominent below. Inflorescences solitary flowers or cymose, 4–8 mm, 1(–3)-flowered. Pedicels 2–3 mm, sparsely pubescent. Flowers 8–10 mm across. Hypanthium campanulate, pubescent outside at least initially. Sepals ovate-triangular, apex obtuse, pubescent. Petals spreading, white, suborbicular, 2–4 mm long, apex obtuse, glabrous. Stamens 15–20, shorter than petals; anthers pink to brownish or violet-black. Styles 2. Fruit scarlet-red or crimson, globose or depressed globose, 5–6(–10) mm in diameter; nutlets (1 or)2(or 3).
Fig. 40b.

Distribution: Nepal, W Himalaya, E Himalaya, Tibetan Plateau, Assam-Burma and E Asia.

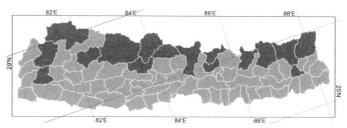

Altitudinal range: 900–5400 m.

Ecology: Open places, rocks, slopes, high mountain areas, thickets, roadsides.

Flowering: May–June(–August). **Fruiting:** August–October(–November).

This widespread and common Nepalese species is often grown in horticulture in temperate regions.
   Ripe fruits are eaten fresh and are used medicinally. Leaves are used as incense. The plant is planted for fences and is also used as firewood.

303

7. *Cotoneaster integrifolius* (Roxb.) G.Klotz, Wiss. Z. Martin-Luther-Univ. Halle-Wittenberg, Math.-Naturwiss. Reihe 12(10): 779 (1963).
Crataegus integrifolia Roxb., Fl. Ind., ed. 1832, 2: 509 (1832); *Cotoneaster buxifolius* var. *marginatus* Lindl. ex Loudon; *C. congestus* Baker; *C. lanatus* Jacques; *C. linearifolius* (G.Klotz) G.Klotz; *C. marginatus* (Lindl. ex Loudon) Schltdl.; *C. meuselii* G.Klotz; *C. microphyllus* forma *lanatus* Dippel; *C. microphyllus* forma *linearifolius* G.Klotz; *C. microphyllus* var. *buxifolius* Dippel; *C. microphyllus* var. *thymifolius* (Baker) Koehne; *C. poluninii* G.Klotz; *C. prostratus* Baker; *C. prostratus* var. *lanatus* (Dippel) Rehder; *C. rotundifolius* var. *lanatus* (Dippel) C.K.Schneid.; *C. thymifolius* Baker.

Shrubs, evergreen or semi-evergreen, small, (0.2–)0.5–1.5(–5) m tall, divaricate or squarrose, with regular branching; branches rigid, erect to decumbent. Branchlets spiralled, purple-black or green, terete, initially strigose. Stipules lanceolate, strigose, caducous. Petioles 1–5 mm, strigose. Leaves spiralled, shiny and pale to dark green above, lanceolate to oblong or ovate, 0.4–1.7(–2.9) cm × (1.5–)3–8(–17) mm, base cuneate, margin ± revolute, apex acute or acuminate, sometimes obtuse, leathery or papery, subglabrous above, tomentose, pubescent or glabrescent below, midvein impressed above, prominent below. Inflorescences solitary flowers or cymose, 5–10 mm, 1–10-flowered or more. Pedicel 3–7 mm. Hypanthium campanulate, strigose. Sepals ovate-triangular, acute or obtuse, sometimes mucronulate, pubescent or glabrous. Petals spreading, white or pinkish white, suborbicular, ca. 5 mm wide, glabrous, apex obtuse. Stamens 20, erect; anthers reddish purple or violet-black. Styles 2(or 3). Fruit red or crimson, depressed-globose or globose, 7–9 mm; nutlets (1 or)2, rarely 3.

Distribution: Nepal, W Himalaya, E Himalaya, Tibetan Plateau, Assam-Burma and E Asia.

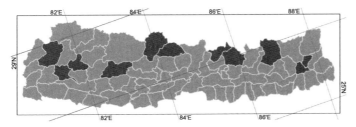

Altitudinal range: 1100–3500(–3900) m.

Ecology: High mountain areas.

Flowering: May–July. **Fruiting:** October–November.

The treatment of this taxon follows Dickoré & Kasperek (Willdenowia 40: 28. 2010), who noted that this species has been frequently confused with other species of *Cotoneaster*, particularly *C. microphyllus* Wall. ex Lindl. It is certainly under-recorded in Nepal.

8. *Cotoneaster rotundifolius* Wall. ex Lindl., Edward's Bot. Reg. 15: pl. 1229 (1829).
Cotoneaster microphyllus var. *rotundifolius* (Wall. ex Lindl.) Wenz.; *C. microphyllus* var. *uva-ursi* Lindl.; *C. uva-ursi* (Lindl.) G.Don.

Shrubs, evergreen, to 4 m tall, with ± erect branches. Branchlets spreading, distichous or spiralled, greyish brown to blackish brown, terete, initially appressed villous, glabrous when old. Stipules lanceolate, puberulous, persistent or caducous. Petioles 1–3 mm, pilose. Leaves shiny, dark green above, suborbicular to broadly ovate or broadly obovate, 0.8–2(–2.5) cm × 6–10(–16) mm, base broadly cuneate to rounded, margin revolute, apex obtuse or emarginate, sometimes acute and mucronate, subleathery, glabrous or rarely puberulous above, pilose below, midvein impressed above, prominent below. Inflorescences solitary flowers or cymose, 4–7 mm, 1–3-flowered. Bracts linear-lanceolate, caducous. Pedicels 2–4 mm, pubescent. Flowers ca. 1 cm across. Hypanthium campanulate, pilose outside. Sepals triangular, apex acute, glabrous or pilose. Petals spreading, white or tinged reddish, broadly ovate or obovate, 4–5 mm and nearly as broad, apex obtuse or emarginate. Stamens 20, about as long as or somewhat shorter than petals; anthers white. Styles 2 or 3, nearly equalling or shorter than stamens. Fruit red to bright scarlet, obovoid to turbinate, 7–9 mm in diameter; nutlets 2 or 3(or 4).

Distribution: Nepal, W Himalaya, E Himalaya, Tibetan Plateau and E Asia.

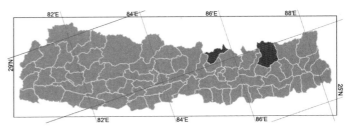

Altitudinal range: 1200–4000(–4300) m.

Ecology: Grassy slopes, rocks, mountain summits.

Flowering: May–June(–July). **Fruiting:** August–September(–December).

9. *Cotoneaster nitidus* Jacques, J. Soc. Imp. Centr. Hort. 5: 516 (1859).
Cotoneaster cavei G.Klotz; *C. cordifolius* G.Klotz; *C. distichus* Lange; *C. distichus* var. *parvifolius* T.T.Yu; *C. encavei* J.Fryer & B.Hylmö; *C. milkedandaensis* J.Fryer & B.Hylmö; *C. nitidus* subsp. *cavei* (G.Klotz) H.Ohashi; *C. nitidus* subsp. *taylorii* (T.T.Yu) H.Ohashi; *C. nitidus* var. *parviflorus* (T.T.Yu) T.T.Yu; *C. rupestris* D.Charlton; *C. taylori* T.T.Yu.

Shrubs, deciduous or semi-evergreen, erect, to 2.5 m tall. Branchlets ± distichous, greyish brown when old, terete, initially densely yellow strigose, gradually glabrescent. Stipules lanceolate, pilose, persistent. Petioles 1–3 mm, villous. Leaves spiralled or distichous, dark green above, broadly ovate

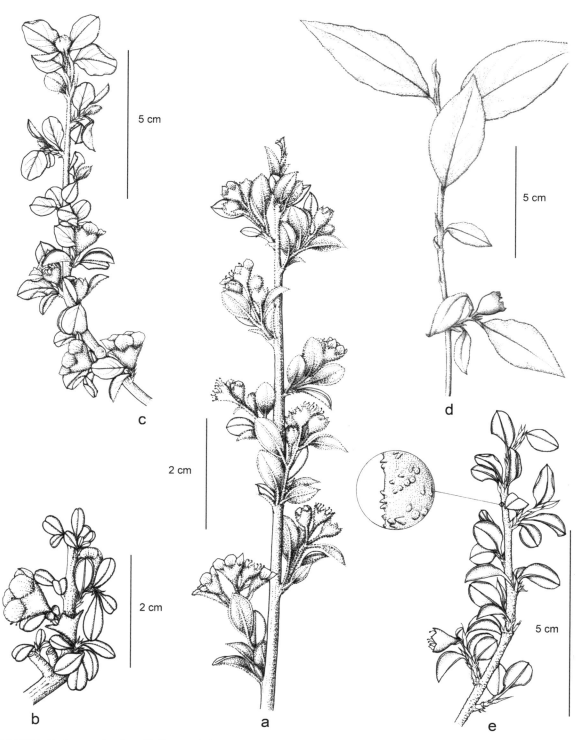

FIG. 40. Rᴏsᴀᴄᴇᴀᴇ. **Cotoneaster buxifolius**: a, inflorescence and leaves. **Cotoneaster microphyllus**: b, inflorescence and leaves. **Cotoneaster adpressus**: c, inflorescence and leaves. **Cotoneaster acuminatus**: d, infructescence and leaves. **Cotoneaster verruculosus**: e, inflorescence and leaves.

or broadly obovate, (0.4–)0.8–1.5(–2.5) × (0.5–)0.7–1.3(–1.8) cm, base rounded or broadly cuneate, apex acute and mucronate, rarely obtuse, subleathery or thickly papery, both surfaces appressed villous, more densely so above, midvein impressed above, prominent below. Inflorescences cymose, 5–10 mm, 2–6-flowered; rachis strigose or subglabrous. Pedicels 1–2(–5) mm. Flowers 5–7 mm across. Hypanthium campanulate, glabrous outside. Sepals broadly triangular, apex acute to obtuse, glabrous or villous. Petals erect, white or stained reddish, ovate or suborbicular, 3–4 mm wide, apex obtuse, glabrous. Stamens (10 or)20, shorter than petals; anthers white. Styles (2 or)3(or 4). Fruit red or scarlet, broadly obovoid, 7–8 mm in diameter, shortly puberulous; nutlets (2 or)3(or 4).

Distribution: Nepal, E Himalaya, Tibetan Plateau, Assam-Burma and E Asia.

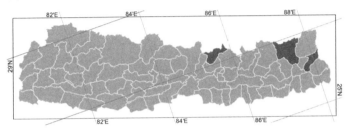

Altitudinal range: (1600–)2500–2600(–4000) m.

Ecology: Thickets, forests, grassy slopes, mountain regions and valleys.

Flowering: June–July. **Fruiting:** September–October.

Several authors (e.g. Hooker, Fl. Brit. Ind. 2: 386. 1878) have partially misapplied *Cotoneaster rotundifolia* Wall. ex Lindl. to this species.

10. *Cotoneaster symondsii* T.Moore, Proc. Roy. Hort. Soc. London, n.s. 1: 298 (1861).
Cotoneaster simonsii Baker.

Shrubs, deciduous or semi-evergreen, erect, 1–4(–5) m tall. Branchlets divaricate, spiralled, erect to spreading, terete, initially strigose-tomentose, late glabrescent. Stipules caducous. Petioles 1–3 mm, strigose-pilose. Leaves spiralled, deep green above, pale green below, broadly elliptic or rhombic-elliptic to suborbicular, rarely obovate-elliptic, 0.9–2.5(–3.5) × 0.5–1.7(–2) cm, base obtuse or broadly cuneate, margin revolute, apex acute, rarely obtuse or shortly acuminate, mucronate, papery to subleathery, sparsely strigose-pilose or subglabrous above, strigose-pilose below, midvein impressed above, prominent below. Inflorescence cymose, 5–15 mm, 2–6-flowered; rachis strigose. Hypanthium campanulate, strigose-villous. Sepals triangular, apex acuminate, densely villous at margin. Petals erect, red, glabrous. Stamens 15–20, somewhat shorter than petals, anthers white. Fruit bright scarlet, obovoid-ellipsoid or turbinate, 8–9.5 mm, ca. 6 mm in diameter, nutlets 3 or 4.

Distribution: Nepal, E Himalaya and Assam-Burma.

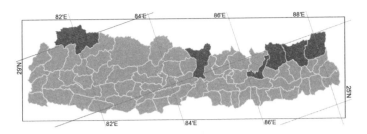

Altitudinal range: 3000–3800 m.

Ecology: In shrubby areas, river valleys.

Flowering: May–July. **Fruiting:** October.

11. *Cotoneaster adpressus* Bois, Frutic. Vilmor.: 116 (1904).
Cotoneaster distichus var. *duthieanus* C.K.Schneid.; *C. duthieanus* (C.K.Schneid.) G.Klotz; *C. horizontalis* var. *adpressus* (Bois) C.K.Schneid.; *C. nitidus* var. *duthieanus* (C.K.Schneid.) T.T.Yu.

Shrubs, deciduous, prostrate, to 30(–100) cm high, irregularly branched; branches divaricate. Branchlets distichous, reddish brown or greyish brown to greyish black, terete, slender, initially strigose, glabrous when old. Stipules subulate, caducous. Petioles 1–2(–4) mm, glabrous or pilose. Leaves spiralled, dull green above, broadly ovate or obovate, rarely elliptic, 5–25 × 4–18(–20) mm, base cuneate or obtuse, margin flat or revolute, apex obtuse to acute and shortly mucronate, margin undulate, thinly papery, sparsely pubescent or glabrous below, glabrous above, midvein impressed above, prominent below. Inflorescences solitary flowers or cymose, 5–10 mm, 1- or 2(–4)-flowered. Flowers subsessile, 7–8 mm across. Hypanthium campanulate, sparsely pubescent outside. Sepals ovate-triangular, shorter than petals, apex acute, glabrous or pilose. Petals erect, pink or red, obovate, 4–5 mm and nearly as broad, apex emarginate or obtuse, glabrous. Stamens ca. 10–15, shorter than petals; anthers white or pink to red. Styles 2. Fruit bright red, subglobose or oblong, 7–9 mm in diameter; nutlets (1 or)2, rarely 3.
Fig. 40c.

Distribution: Nepal, E Himalaya, Tibetan Plateau, Assam-Burma and E Asia.

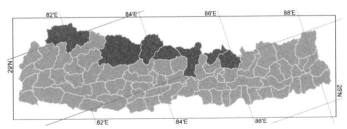

Altitudinal range: (1900–)3000–4500 m.

Ecology: Mixed forests on slopes, rocky places.

Flowering: May–June. **Fruiting:** August–September.

12. *Cotoneaster acuminatus* Lindl., Trans. Linn. Soc. London 13: 101 (1822).

Cotoneaster bakeri G.Klotz; *C. bisramianus* G.Klotz; *C. kongboensis* G.Klotz; *C. mucronatus* Franch.; *C. nepalensis* André later homonym, non K.Koch; *C. paradoxus* G.Klotz; *C. sanguineus* T.T.Yu; *C. staintonii* G.Klotz; *C. stracheyi* G.Klotz; *Crataegus acuminata* (Lindl.) Desf. ex Steud.; *Mespilus acuminatus* (Lindl.) Lodd.

ढल्के फूल Dhalke phul (Nepali).

Shrubs, deciduous, erect, 2–3(–4) m tall. Branchlets spiralled, greyish brown to brownish, or greyish black when old, terete, initially yellowish strigose, glabrous when old. Stipules lanceolate or linear-lanceolate, villous, partly caducous. Petioles 3–7 mm, villous or pilose. Leaves spiralled or distichous, shiny, pale to bright green above, lanceolate to ovate or elliptic, (1–)2–6.5(–8) × (0.7–)2–3 cm, base broadly cuneate or rounded, rarely obtuse, margin undulate, apex acuminate or acute, rarely obtuse or rounded, mucronulate, papery, both surfaces villous or pilose, more densely so below, glabrescent, veins impressed or flat above, prominent below. Inflorescences cymose, 1–2 cm, 1(–3)-flowered, sometimes more; rachis and pedicels yellowish villous. Bracts lanceolate or linear. Pedicels to 3–5 mm. Flowers (5–)6–8(–10) mm across. Hypanthium campanulate, appressed villous or subglabrous outside. Sepals triangular, apex acute or obtuse, glabrous or pubescent. Petals erect to spreading, white or pink, rarely red, ovate to obovate, (2.5–)3–4(–5) mm and nearly as broad, apex obtuse or erose, glabrous. Stamens (10 or)20, shorter than petals; anthers white, sometimes purple. Styles 2. Fruit red, bright red or scarlet, ellipsoid or subglobose, 8–10 × 7–8 mm; nutlets (1 or)2(or 3).
Fig. 40d.

Distribution: Nepal, W Himalaya, E Himalaya, Tibetan Plateau, Assam-Burma and E Asia.

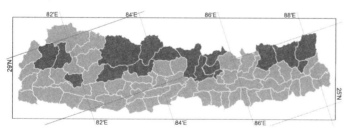

Altitudinal range: (1500–)2500–3700(–4200) m.

Ecology: Mixed forests, thickets, fields, open sunny places, slopes, gullies.

Flowering: (April–)May–July(–August). **Fruiting:** (August–) September–October.

Cotoneaster sanguineus is treated here as synonymous with this taxon, differing only in having red petals and small leaves (at the low end of the range of leaf size but comparable, e.g., *Delavay 3738*).
  Wood is used to make walking sticks.

13. *Cotoneaster verruculosus* Diels, Notes Roy. Bot. Gard. Edinburgh 5: 272 (1912).
Cotoneaster distichus var. *verruculosus* (Diels) T.T.Yu; *C. sandakphuensis* G.Klotz.

Shrubs, semi-evergreen or evergreen, prostrate to erect, (0.2–)0.6–1.2(–2) m tall; branches erect or ascending. Branchlets spiralled or distichous, initially dark reddish brown to greyish brown, dark grey when old, terete, initially densely yellow strigose, glabrate, conspicuously verruculose when old. Stipules lanceolate, pubescent, persistent. Petioles 2–5 mm, pubescent. Leaves spiralled, dark green above, suborbicular or broadly ovate to broadly obovate, 0.7–1.4 cm × (4–)6–12 mm, base broadly cuneate or rounded, margin revolute, apex emarginate or partly mucronulate, both surfaces appressed pilose, glabrescent, subleathery, midvein impressed above, prominent below. Inflorescences usually 1-flowered. Flowers 7–8 mm across, nearly sessile. Hypanthium campanulate, glabrous outside. Sepals ovate-triangular, apex ± obtuse, glabrescent. Petals erect, pink, suborbicular, 4–4.5 mm wide, glabrous. Stamens 20, shorter than petals; anthers white. Styles 2. Fruit red, globose, 7–8 mm in diameter; nutlets 2(or 3).
Fig. 40e.

Distribution: Nepal, E Himalaya, Tibetan Plateau, Assam-Burma and E Asia.

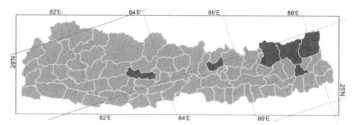

Altitudinal range: (2800–)3000–3800 m.

Ecology: Mixed forests on dry slopes, grasslands, fields, roadsides.

Flowering: May–June. **Fruiting:** September–October.

Further research is needed on the identity of *Cotoneaster sandakphuensis* and the distributional range of *C. verruculosus*.

6. *Pyracantha* M.Roem., Fam. Nat. Syn. Monogr. 3: 104, 219 (1847).

Colin A. Pendry

Thorny evergreen shrubs or small trees. Stipules minute, free, caducous. Leaves simple, alternate, fascicled on short shoots in the axils of the spines, shortly petiolate, margin crenulate. Inflorescences compound corymbs. Flowers bisexual. Bracts solitary; bracteoles 2. Hypanthium short, cupulate. Sepals 5, spreading. Petals 5, spreading, white, base shortly clawed. Stamens 20, anthers yellow. Ovary semi-inferior, carpels 5, lower half adnate to hypanthium, apically free, each with 2 ovules. Styles 5, free. Pome red or orange at maturity, globose, with persistent erect sepals at apex. Pyrenes (nutlets) 5, 1-seeded.

Worldwide three species from SE Europe and Asia. One species in Nepal.

1. *Pyracantha crenulata* (D.Don) M.Roem., Fam. Nat. Syn. Monogr. 3: 220 (1847).
Mespilus crenulata D.Don, Prodr. Fl. Nepal.: 238 (1825); *Crataegus crenulata* (D.Don) Roxb.

घङ्गारु Ghangaru (Nepali).

Plants to 3(–5) m. Young twigs and leafy short shoots tomentose, older twigs and branches dark red-brown, glabrous. Thorns lateral, to 2 cm, or terminating short side branches. Petioles 3–5 mm, glabrous. Leaves oblong or obovate, rarely elliptic, 1.5–6 × 0.7–2.5 cm, base broadly cuneate or slightly rounded, apex obtuse, margin crenulate or sparsely so, glabrous. Corymb to 3 cm. Peduncle glabrescent. Bracts oblong-elliptic, ca. 2 mm, sparsely hairy. Pedicels 3–10 mm, glabrous. Flowers 6–9 mm across. Hypanthium 1–2 × 2–3 mm, glabrous. Sepals spreading, ca. 1 mm, rounded. Petals broadly ovate, 3–4 × 2.5–3 mm, apex rounded, sometimes notched, sparsely hairy at base within. Stamens 2–3 mm. Carpels 0.5–1 mm, densely long white pubescent apically. Styles nearly as long as stamens. Pome 3–8 mm in diameter.
Fig. 41a.

Distribution: Nepal, W Himalaya, E Himalaya, Tibetan Plateau, Assam-Burma and E Asia.

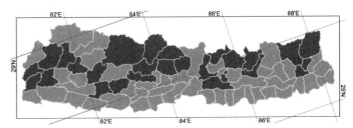

Altitudinal range: 800–2800 m.

Ecology: Roadsides, forest margins and abandoned cultivated areas.

Flowering: April–August. **Fruiting:** July–September.

Ripe fruits are eaten fresh. Powdered dry fruits are taken in case of bloody dysentery. Branches are used as walking sticks.

7. *Photinia* Lindl., Bot. Reg. 6: pl. 491 (1820).

Stranvaesia Lindl.

Mark F. Watson

Evergreen shrubs or trees. Twigs unarmed. Stipules subulate, minute, caducous. Leaves alternate, simple, unlobed, leathery, midvein pronounced below. Inflorescence a terminal corymbose panicle. Flowers bisexual. Bracts and bracteoles linear, caducous. Hypanthium cup-shaped. Sepals 5, small, persistent in fruit. Petals 5, white or cream, contorted or imbricate in bud, obovate or ovate, base clawed, apex rounded entire or irregularly toothed, glabrous. Stamens ca. 20. Ovary semi-inferior to almost inferior, 2–5-celled, ovules 2 per locule. Styles 2–5, fused at least at the base, stigmas truncate. Fruit a pome, globose, somewhat fleshy, 2–5-celled, calyx rim set back from apex. Seeds (1 or)2, testa leathery.

Worldwide about 40 species, from the Himalaya to Japan and in SE Asia south to Sumatra. Two species in Nepal.

The Nepalese species are sometimes confused with *Cotoneaster* which have similar leaves and fruit, but *Cotoneaster* fruit have seeds in woody nutlets. Following overwhelming evidence from taxonomic research in recent decades, *Stranvaesia nussia* is now treated within *Photinia* (see also commentary under *Stranvaesia* in Fl. China 8: 119. 2003).

Key to Species

1a   Leaf margin entire. Hypanthium, sepals and inflorescence branches glabrous or almost so. Petals up to 3 mm. Styles 2..**1. _P. integrifolia_**

b   Leaf margin obscurely dentate or crenate, at least towards apex. Hypanthium, sepals and inflorescence branches densely pilose. Petals longer than 4.5 mm. Styles (4–)5 ....**2. _P. nussia_**

---

1. _Photinia integrifolia_ Lindl., Trans. Linn. Soc. London, Bot. 13: 103, pl. 10 (1821).
Eriobotrya integrifolia (Lindl.) Kurz; _Photinia eugenifolia_ Lindl.; _P. integrifolia_ var. _notoniana_ (Wight & Arn.) J.E.Vidal; _P. notoniana_ Wight & Arn.; _Pyrus integerrima_ D.Don.

लहरे बाकलपाते Lahare bakalpate (Nepali).

Shrub or tree, 2–9(–15) m, often epiphytic, sometimes climbing. Branchlets brown or blackish, becoming grey and lenticellate with age. Petioles (0.5–)1–2 cm, glabrous. Leaves elliptic or narrowly obovate, 6–12(–15) × (2–)3–5 cm, base cuneate, apex abruptly and shortly acuminate, tip 0.5–1.5 cm and bent downwards, margin entire, upper surface very sparsely hairy, lower surface densely minutely papillose, essentially glabrous. Inflorescence diffuse, 7–12(–18) × 8–15(–25) cm, branches glabrous or sparsely pubescent, lenticels absent. Very early flowering material (in tight bud) with small bracts and bracteoles ca. 1.5 mm, apex with white tufts of hairs, very early caducous. Pedicels 3–5 mm. Flowers 4–6 mm across. Hypanthium 1–1.5 mm, glabrous. Sepals ca. 1 mm, broadly triangular, tips pubescent otherwise glabrous. Petals white or cream, orbicular, ca. 1.5–2 × 1.5–2 mm, apex rounded. Stamens ca. 1.5 mm, almost equal. Ovary semi-inferior, densely pubescent. Styles 2, ca. 0.75 mm, fused at base, glabrous. Fruit dark reddish brown becoming purplish black, subglobose, 4–5 × 4–4.5 mm, glabrous or with some hairs remaining at apex within the calyx remains, stamens caducous, styles persistent.

Distribution: Nepal, W Himalaya, E Himalaya, Assam-Burma, E Asia and SE Asia.

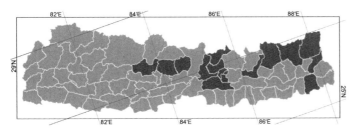

Altitudinal range: 1300–3000 m.

Ecology: Free standing or epiphytic on trees or boulders in warm temperate mixed evergreen forests with _Castanopsis_, _Quercus_, _Rhododendron_ and _Acer_, also in disturbed open places.

Flowering: April–May. **Fruiting:** August–November.

The abruptly acuminate downward pointing leaf tips and entire margins are characteristic of this species. _Photinia arguta_ Lindl. is recorded from the Tarai in Darjeeling district and may occur in far eastern border regions of Nepal. It is readily distinguished by the serrate margins to the leaves (but also compare with _Eriobotrya_) and branches which are strongly warted with lenticels.

2. _Photinia nussia_ (Buch.-Ham. ex D.Don) Kalkman, Blumea 21(2): 429 (1973).
Pyrus nussia Buch.-Ham. ex D.Don, Prodr. Fl. Nepal.: 237 (1825); _Cotoneaster affinis_ Lindl. ex Wall. nom. nud.; _Crataegus glauca_ Wall. nom. nud.; _C. glauca_ Wall. ex G.Don; _Stranvaesia glaucescens_ Lindl.; _S. nussia_ (Buch.-Ham. ex D.Don) Decne.

जूरे मयल Jure mayal (Nepali).

Shrub or medium-sized tree, 2–10(–15) m. Branches purplish brown when old, densely pilose when young, glabrescent. Petioles ca. 1 cm, densely pilose. Leaves lanceolate to narrowly obovate, 6–10 × 2–4 cm, base cuneate, apex acute, margin obscurely dentate or crenate, sometimes very shallowly so, but usually with at least some teeth towards the apex, upper surface glabrous except pilose along impressed midvein, lower surface pilose with long hairs at least along the midvein. Corymbs flat-topped, compact, 6–9 × 8–13 cm, inflorescence branches densely pilose, bracts and bracteoles 2–3 mm, pubescent. Pedicels 5–10 mm, densely pilose, at least at apex. Flowers ca. 10 mm across. Hypanthium 3–4 mm, densely pilose. Sepals broadly triangular, 1–1.5 × 1.5–1.75 mm, densely pilose. Petals white, broadly obovate, 4.5–5 × 3–3.5 mm, apex irregularly shallowly lobed. Stamens unequal, 2–3 mm. Ovary semi-inferior, densely pilose. Styles (4–)5, ca. 2.7 mm, fused for half their length, glabrous. Fruit orange-red, globose ca. 8 × 8 mm, hairy when young, later glabrous, styles and filaments persistent. Fig. 41b.

Distribution: Nepal, W Himalaya, Tibetan Plateau, Assam-Burma, E Asia and SE Asia.

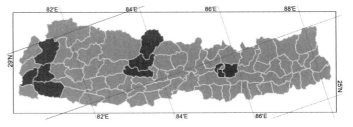

Altitudinal range: 700–2200 m.

Ecology: Degraded areas and open subtropical to warm temperate mixed hill forests with _Schima_, _Shorea_, _Engelhardia_, Lauraceae and _Quercus_.

Rosaceae

Flowering: April–May. **Fruiting:** August–January.

Sometimes confused with the morphologically similar

Cotoneaster frigidus Wall. ex Lindl. which differs in having 2 styles and narrower, less leathery and less shiny leaves.

8. *Eriobotrya* Lindl., Trans. Linn. Soc. London 13: 96, 102 (1821).

Colin A. Pendry & Cathy King

Small to medium-sized evergreen trees. Twigs unarmed. Stipules persistent or caducous. Leaves simple, alternate, coriaceous, petiolate, more or less serrate, secondary veins prominent, glabrous or tomentose beneath. Inflorescence a tomentose panicle or cymose thyrse. Flowers numerous, sweet-scented, bisexual. Bracts solitary; bracteoles 2. Hypanthium cup-shaped or funnel-shaped, tomentose or villous. Sepals 5, triangular. Petals 5, white or yellow, orbicular, clawed. Stamens 15–20. Ovary inferior, 2–5 locular, with 2 ovules per locule. Styles 2–5, connate at base, glabrous or villous. Fruit an elongated fleshy pome with persistent calyx remains. Seeds large, 1 or 2.

Worldwide about 20 species from the Himalaya to China and Malesia. Three species native to Nepal with another species cultivated.

The Japanese native *Eriobotrya japonica* (Thunb.) Lindl. (syn. *Mespilus japonica* Thunb.) is cultivated in villages of Nepal and widely in temperate areas of the northern hemisphere. It is included in the key to species, but not otherwise treated.

Key to Species

1a	Twigs glabrous, 2–4 mm thick	**1. *E. dubia***
b	Twigs tomentose, 5–9 mm thick	2
2a	Leaves densely tomentose below. Stipules persistent, subulate. Cultivated	***E. japonica***
b	Leaves glabrous or glabrescent below. Stipules caducous. Wild	3
3a	Petioles 16–50 mm. Leaves usually only slightly toothed towards apex	**2. *E. elliptica***
b	Petioles 5–9 mm or leaves subsessile. Leaves serrate with hooked teeth in at least apical three quarters of margin	**3. *E. hookeriana***

1. *Eriobotrya dubia* (Lindl.) Decne., Nouv. Arch. Mus. Hist. Nat. 10: 145 (1874).
Photinia dubia Lindl., Trans. Linn. Soc. London 13: 104, pl. 10 (1821); *Crataegus shicola* Buch.-Ham. ex D.Don nom. inval.; *C. shicola* Buch.-Ham. ex Lindl. nom. inval.; *Mespilus tinctoria* D.Don nom. illegit.

जूरे काफल Jure kaphal (Nepali).

Trees to 10 m. Twigs 2–3 mm thick, glabrous. Stipules oblong, 2–3 mm, caducous. Petioles 5–10 mm, glabrous. Leaves elliptic to narrowly obovate, 6–11 × 2–4 cm, base cuneate to attenuate, apex acute to acuminate, margins serrate, the serrations denser towards the apex, glabrous, secondary veins 9–14 pairs. Inflorescence an open panicle to 7 cm, branches tomentose. Bracts 2–2.5 mm, bracteoles 1–2 mm. Hypanthium shallowly cup-shaped, ca. 1 mm, tomentose. Sepals narrowly triangular, 1 mm, tomentose outside, glabrous within. Petals white. Stamens 15–20, ca. 2 mm. Styles 2, ca. 1 mm, villous at base. Fruit ellipsoid, 10–12 mm. Fig. 41c.

Distribution: Nepal, W Himalaya and E Himalaya.

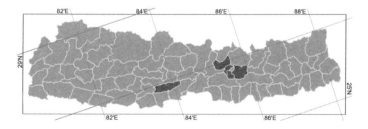

Altitudinal range: 1300–2800 m.

Ecology: Moist evergreen forest.

Flowering: September–November. **Fruiting:** February–April.

The small leaves, slender, glabrous twigs and rather small inflorescence of *Eriobotrya dubia* readily distinguish it from the other species in Nepal.
Ripe fruits are eaten fresh.

2. *Eriobotrya elliptica* Lindl., Trans. Linn. Soc. London 13: 102 (1821).
Mespilus cuila Buch.-Ham. ex D.Don nom. inval.; *M. cuila* Buch.-Ham. ex Lindl. nom. inval.

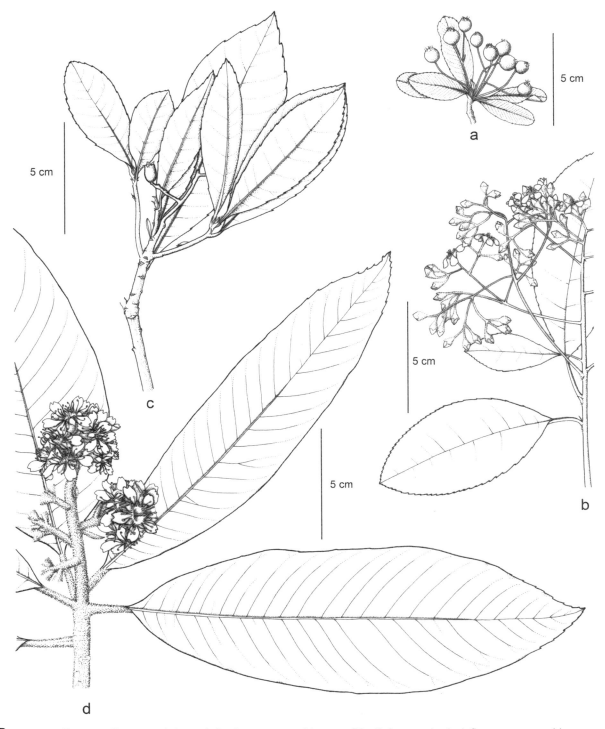

FIG. 41. ROSACEAE. **Pyracantha crenulata**: a, infructescence and leaves. **Photinia nussia**: b, inflorescence and leaves. **Eriobotrya dubia**: c, infructescence and leaves. **Eriobotrya elliptica**: d, inflorescence and leaves.

माया Maya (Nepali).

Trees to 15 m. Twigs 6–8 mm thick, densely tomentose when young. Stipules triangular, 2–3 mm, caducous. Petioles 16–40 mm, tomentose, glabrescent. Leaves elliptic to oblong or slightly ovate or obovate, 12–22(–30) × 4–9(–13) cm, base rounded to cuneate, apex acute to acuminate, margins entire and obscurely repand, somewhat serrate towards the apex, rarely sparsely serrate towards base, glabrous, secondary veins 12–24 pairs. Inflorescence a cymose thyrse to 9 cm, densely rusty villous. Bracts 6–8 mm, bracteoles ca. 4 mm. Flowers subsessile or on pedicels to 3 mm. Hypanthium funnel-shaped, 5 mm, rusty villous. Sepals triangular, 4 mm, villous outside, glabrous within. Petals white or pale yellow, 8 mm, glabrous, slightly villous at base within. Stamens ca. 20, 5–6 mm. Styles 5, 4–5 mm, villous. Fruit obovoid-globose, 8–12 mm.
Fig. 41d.

Distribution: Nepal and Tibetan Plateau.

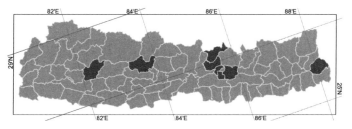

Altitudinal range: 500–2800 m.

Ecology: Moist evergreen forest.

Flowering: April. **Fruiting:** June.

The large leaves of *Eriobotrya elliptica* with their long petioles are unlike any of the other Nepalese species. The cultivated *E. japonica* (Thunb.) Lindl. has similarly sized leaves, but they are subsessile or only shortly petiolate and densely tomentose below.

3. *Eriobotrya hookeriana* Decne., Nouv. Arch. Mus. Hist. Nat. 10: 146 (1874).

Trees to 10 m. Twigs 4–6 mm thick, tomentose. Stipules elliptic to oblong, 3–7 mm. Petioles 5–9 mm, tomentose, or leaves subsessile. Leaves elliptic to oblong or narrowly obovate, 12–25 × 4–8 cm, base cuneate to rounded, apex acuminate, margins serrate, sometimes entire towards base, young leaves densely tomentose, soon glabrescent except sometimes tomentose on veins above and below, secondary veins (12–)16–34 pairs. Inflorescence an open panicle to 12 cm, densely tomentose. Bracts 4 mm, bracteoles 2 mm. Flowers subsessile. Hypanthium campanulate, 2 mm, tomentose. Sepals triangular, 1.5–2 mm, tomentose on margins only. Petals white, 2–3 mm, glabrous. Stamens ca. 20, 2 mm. Styles 2, 1 mm, glabrous, villous at base. Fruit ellipsoid to almost globose, to 14 mm.

Distribution: Nepal and E Himalaya.

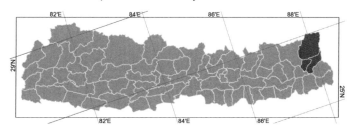

Altitudinal range: 1400–2400 m.

Ecology: Mixed evergreen broad-leaved forest.

Flowering: October–November. **Fruiting:** March–June.

The combination of shortly petiolate leaves with dense serrations along almost the whole margin is found only in *Eriobotrya hookeriana*.

9. *Sorbus* L., Sp. Pl. 1: 477 (1753).

Aria (Pers.) Host.

Mark F. Watson & Vidya K. Manandhar

Trees or shrubs, all Nepalese species deciduous. Twigs unarmed. Stipules present, often caducous. Leaves alternate, simple or imparipinnate, margins mostly serrate, sometimes entire, glabrous or variously hairy, especially below. Inflorescences terminal and lateral compound corymbs, glabrous, pubescent or tomentose. Flowers bisexual. Bracts solitary; bracteoles 2, early caducous. Hypanthium campanulate. Sepals 5. Petals 5, free, usually clawed at base. Stamens 15–25, equal or somewhat unequal. Ovary semi-inferior or inferior. Styles 2–5, free or connate, glabrous or hairy. Fruit a pome, ovoid or globose, with or without persistent sepals, locules 2–5. Seeds brown, 1 or 2 per locule.

Worldwide about 100 species in temperate regions of N hemisphere. 13 species in Nepal.

Himalayan *Sorbus* is taxonomically problematic with apomictic reproduction and polyploidy contributing to the creation of locally distinct variants and the blurring of species boundaries across the region. The simple and pinnate-leaved species form natural groups, and are sometimes recognized as separate genera (*Aria* and *Sorbus*). This division is gaining support in recent molecular studies, but as the classification for the genus as a whole has not been completed a broad generic concept is adopted here.

Identification of simple-leaved species is usually straightforward, as long as material is gathered from mature plants and not juvenile growth. Pinnate-leaved species are more complex as three types of shoots are produced: short reproductive shoots, short vegetative shoots and long vegetative shoots, and these have somewhat different leaf and stipule morphologies. The short shoots are most useful for identification, ideally combining short vegetative shoots (best for leaf and stipule characters) with short reproductive shoots (good for flower and fruit characters, but sometimes with atypical stipules). Long vegetative shoots are found on young plants and as suckers on old plants, and these should be avoided as the stipules and leaflets are often very atypical.

Key to Species

1a   Leaves simple, sometimes shallowly lobed 2
   b   Leaves pinnate 7

2a   Leaves glabrous or brown pubescent on the veins below, never white tomentose. Sepals deciduous in fruit 3
   b   Leaves white or cream tomentose below. Sepals persistent in fruit 4

3a   Leaves (7–)10–12 cm, lower surface glabrous, lateral veins 12–15 pairs, strongly raised. Petioles 1–1.5 cm
       ....**1. S. rhamnoides**
   b   Leaves 4–8 cm, lower surface sparsely rusty brown pubescent when young, lateral veins 7–10 pairs, not raised.
       Petioles 0.5–1 cm..**2. S. thomsonii**

4a   Leaves un-lobed, sometimes doubly serrate 5
   b   Leaves more or less lobed or lobulate 6

5a   Leaf ovate to rounded, width to length ratio 0.58–0.68(–0.82), veins 11–16 pairs. Petioles short, 0.5–1.5 cm, 4–9% of
       lamina length. Petals sparsely hairy within. Fruit 0.9–1.5 cm across, lenticels dense, prominent ....**3. S. sharmae**
   b   Leaf lanceolate-ovate, width to length ratio 0.33–0.57, veins 6–11 pairs. Petioles long, 1–2.5 cm, 9–12% of lamina
       length. Petals woolly within. Fruit 1.5–2.0 cm across, lenticels sparse, not prominent ....**4. S. vestita**

6a   Leaf midrib and lateral veins with distinct brown tomentum below, lobes shallow, acute. Styles white woolly only
       at base. Fruit 1.0–1.3 cm across, with lenticels ....**5. S. hedlundii**
   b   Leaf midrib and lateral veins without brown hairs below, lobes deep, rounded. Styles white woolly all over. Fruit
       1.3–2.0 cm, without lenticels ....**6. S. lanata**

7a   Leaflet pairs 3–8(–9) 8
   b   Leaflet pairs 8–14(–18) 10

8a   Leaflet pairs 3–5. Leaflets more than 5 cm..**7. S. insignis**
   b   Leaflet pairs 4–8(–9). Leaflets less than 3 cm 9

9a   Leaflet pairs 4–6. Leaflets toothed in apical half, shiny above, lower surface not prominently papillose. Stipules on
       short shoots ovate-lanceolate, acuminate or caudate with a toothed margin....................................**8. S. kurzii**
   b   Leaflets pairs (6–)7–8(–9). Leaflets entire or almost so, matt above, white hairy above and below when young,
       older leaves glabrescent, lower surface densely papillose. Stipules on short shoots subulate ....**9. S. wallichii**

10a   Leaflet pairs 12–14(–18), very rarely only 8 or 9. Leaflets (1.0–)1.5–2(–2.4) × (0.3–)0.5–0.8(–1) cm, rarely more
       than 2.4 cm. Stipules small, 3–4 mm long, subulate-lanceolate, dark brown, caducous ....**10. S. microphylla**
   b   Leaflet pairs 8–12(–13). Leaflets 2.5–6 × 0.7–1.5 cm. Stipules at least 5 mm long, lanceolate, brown and caducous
       or broad, green and persistent. 11

11a   Stipules narrowly lanceolate or subulate, brown and caducous. Largest leaflets to 3.5 cm. Terminal branchlets
       slender, less than 4 mm thick. Styles hairy at base..**11. S. himalaica**
   b   Stipules broadly lanceolate to suborbicular, green and persistent. Largest leaflets usually more than 3.5 cm.
       Terminal branchlets stout, more than 4 mm thick. Styles glabrous or hairy at base. 12

12a   Stipules broadly ovate, stalked and clasping, 5–15 mm broad. Leaflets entire or only remotely toothed. Styles
       glabrous..**12. S. arachnoidea**
   b   Stipules ovate or lanceolate, 2–3 mm broad. Leaflets serrate in apical half or quarter. Styles hairy at base
       ....**13. S. foliolosa**

1. *Sorbus rhamnoides* (Decne.) Rehder, in Sarg., Pl. Wilson. 2(2): 278 (1915).
Micromeles rhamnoides Decne., Nouv. Arch. Mus. Hist. Nat. 10: 59 (1874); *Aria rhamnoides* (Decne.) H.Ohashi & H.Iketani; *Pyrus rhamnoides* (Decne.) Hook.f.; *Sorbus sikkimensis* var. *oblongifolia* Wenz.

Shrub or tree, 3–5(–10) m, sometimes epiphytic. Branchlets dark brown, 2.5–5 mm thick, glabrous, young shoots red-brown. Buds globose 2–2.5 mm, apex obtuse, scales brown, glabrous. Stipules linear, ca. 4 × 0.5 mm, caducous. Petioles 1–1.5 cm, glabrous. Leaves simple, oblong-elliptic, (7–)10–12 × 2.5–4 cm, base broadly cuneate, apex acuminate, margin finely serrate, lateral veins 12–15 pairs, almost parallel, strongly raised, upper surface glabrous or glabrescent, lower surface glabrous, white tomentose along veins when young. Inflorescence ca. 7 cm diameter, rachis and pedicels sparsely white tomentose with conspicuous lenticels, glabrous in fruit. Bracts red-brown, linear, 6–8 mm, glabrous. Flowers 5–6 mm diameter. Pedicels 4–5 mm. Sepals oblong, 2–3 mm, apex acute, outer surface sparsely white tomentose towards apex. Petals white, oblong-ovate, 4–5 × 2–3 mm, base shortly clawed, apex obtuse or rounded, glabrous. Stamens 20, cream, 5–5.5 mm. Styles 2, 4–4.5 mm, glabrous, connate at base. Fruit pale yellow, globose or slightly ovoid, solid, 3.5–7 mm diameter, lenticels absent, sepals deciduous in fruit leaving a ring-shaped scar.

Distribution: Nepal, E Himalaya, Assam-Burma and E Asia.

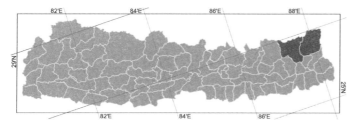

Altitudinal range: 2500–3000(–3500) m.

Ecology: In mixed forests.

Flowering: May–June. **Fruiting:** September–October.

This species is similar to *Sorbus thomsonii* (King ex Hook.f.) Rehder, but it is easily distinguished by longer petioles, strongly raised veins on lower surface of leaves and fruits not spotted with lenticels. Records from the far eastern Himalayan region require further taxonomic study.

2. *Sorbus thomsonii* (King ex Hook.f.) Rehder, in Sarg., Pl. Wilson. 2(2): 277 (1915).
Pyrus thomsoni King ex Hook.f., Fl. Brit. Ind. 2[5]: 379 (1878); *Aria thomsonii* (King ex Hook.f.) H.Ohashi & H.Iketani.

Tree or large shrub, ca. 10 m. Branchlets grey brown, 2.5–5 mm thick, glabrous. Buds ovoid, ca. 5 × 2.5 mm, apex acute, bud scales pale brown, glabrous. Petioles 5–7 mm, glabrous, white tomentose when young. Stipules brown, lanceolate, 1–3 × 0.5–1 mm, sometimes minute, glabrous,

caducous. Leaves simple, ovate-elliptic or oblong-elliptic, 4–8 × 1.8–2.8 cm, base cuneate, apex acute or short acuminate, margin minutely serrate, apically minutely toothed, lateral veins 7–10 pairs, upper surface glabrous, lower surface sparsely rusty brown hairy when young. Inflorescence 2.5–4 cm diameter, rachis and pedicels white pubescent. Pedicels 3–7 mm. Bracts brown, linear, 3–4 mm, glabrescent. Flowers 8–10 mm diameter. Sepals triangular-ovate ca. 2 mm, apex obtuse, outer surface pubescent, inner glabrous. Petals creamy white, obovate-elliptic, 4–5.5 × 3–4 mm, base clawed, apex obtuse, sparsely hairy at base within. Stamens 20, unequal, longer 3.5–4.5 mm, shorter 1.5–2 mm, filaments pale brown, anthers dark brown. Styles 3 or 4, 4–4.5 mm, glabrous, connate at base. Fruit light pink, globose, solid, 10–12 mm diameter, with few lenticels, sepals deciduous.

Distribution: Nepal, E Himalaya, Tibetan Plateau, Assam-Burma and E Asia.

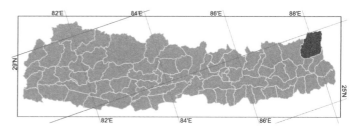

Altitudinal range: 1700–2000 m.

Ecology: In warm temperate habitats.

Flowering: April–May. **Fruiting:** August–September.

3. *Sorbus sharmae* M.F.Watson, V.Manandhar & Rushforth, Int. Dendrol. Soc. Year Book 2009: 79 (2010).

Tree, 10–15 m. Branchlets grey brown, (4–)5–8 mm thick, young shoots white tomentose, later glabrescent. Buds ovoid with acute apex, 5–8 mm, scales red-brown, glabrous, apex pubescent. Stipules yellow brown, linear-lanceolate, ca. 9 × 0.8 mm, white hairy within, caducous. Petioles 0.5–1.5 cm, white tomentose, later glabrescent. Leaves simple, broadly elliptic, 8–22 × 3.5–11 cm, base cuneate, apex acute or short acuminate, margin double serrate, teeth rounded with a forward beaked tip, lateral veins 11–16 pairs, almost parallel, upper surface dark green, glabrous, lower surface white tomentose. Inflorescence 4–7 cm diameter, white tomentose. Pedicels 2–3.5 mm, white hairy. Flowers 1–1.2 cm diameter, white tomentose. Sepals triangular, 2–3.5 mm, apex acute or acuminate with red tip. Petals white or creamy yellow with brown veins, spatulate, 5–7 × 3–5 mm, base shortly clawed, apex obtuse, glabrous, adaxially sparsely hairy towards apex. Stamens 15–20, ca. 3.5 mm, filaments white, anthers crimson. Styles 2–5, 4–4.5 mm, connate and densely hairy at base. Fruit red-brown, ovoid, solid, 0.9–1.5(–2) cm diameter, lenticels dense, sepals persistent.

Distribution: Endemic to Nepal.

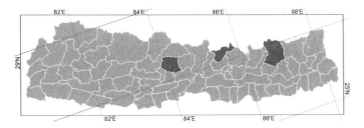

Altitudinal range: 2500–3200 m.

Ecology: Mixed and open forest with *Pinus* and *Salix*, sometimes isolated trees on degraded hillsides.

Flowering: May–August. **Fruiting:** June–September.

Recent research has revealed that Nepalese plants formerly treated as *Sorbus thibetica* (Cardot) Hand.-Mazz. (syn. *Pyrus thibetica* Cardot) are quite different from the Chinese type specimens in some leaf characters but especially in the abundant lenticels on the fruit (*S. thibetica* has no lenticels). Nepalese material has now been recognized as a new endemic species *Sorbus sharmae*. This species is similar to *S. vestita* (Wall. ex G.Don) Lodd. and in Nepal it is often confused with it, but it differs in leaf characters (see key) as well as the densely lenticellate fruits. The species are also distinguished by their distributions, with *S. sharmae* occuring predominantly eastwards from Central Nepal (Langtang), and *S. vestita* replacing it to the west.

4. Sorbus vestita (Wall. ex G.Don) Lodd., Cat. Pl., ed. 16: 66 (1836).
Pyrus vestita Wall. ex G.Don, Gen. Hist. 2: 647 (1832); *Aria vestita* (Wall. ex G.Don) M.Roem.; *Crataegus cuspidata* Spach nom. illegit.; *Pyrus crenata* Lindl. later homonym, non D.Don; *P. vestita* Wall. nom. nud.; *Sorbus crenata* K.Koch nom. superfl.; *S. crenata* S.Schauer; *S. cuspidata* (Spach) Hedl. nom. illegit.

Tree, 6–10(–13) m. Branchlets glabrous, brown, 5–7 mm thick, young shoots white tomentose, lenticels ellipsoid. Buds ovoid, 0.7–1.4 cm, scales brown, glabrous. Stipules brown, lanceolate, 0.8–1 cm, entire, tomentose, caducous. Petioles 1–2.5 cm, white pubescent. Leaves simple, broadly elliptic or lanceolate-ovate, 12–20 × 7–11 cm, base rounded or broadly cuneate, apex acute or obtuse, margin regularly serrate or crenate, lateral veins 6–11 pairs, almost parallel, upper surface dark green, glabrous, pubescent when young, lower surface dense white tomentose. Inflorescence 4–7 cm diameter, young rachis and pedicels white tomentose, later glabrescent, lenticels small, obscured by indumentum. Bracts reddish, linear, 10–12 mm, ciliate, caducous. Pedicels 5–7 mm. Flowers 5–6 mm diameter. Sepals triangular, 2.5–3 × 1.5–2.5 mm, apex acute, white woolly. Petals white or cream, oblong-obovate, 5.5–7.5 × 4–5 mm, base shortly clawed, apex rounded, glabrous on outside, variably hairy within from densely white woolly to almost glabrous. Stamens ca. 20, 5–6 mm, filaments brown, anthers pale purple. Styles 3 or 4, ca. 4 mm, connate, white woolly at base. Fruit yellow-green flushed red to dark purple-red, globose, solid, 1.5–2 cm diameter, lenticels sparse, sepals persistent.

Distribution: Nepal and W Himalaya.

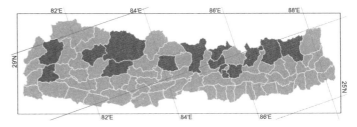

Altitudinal range: (1350–)2000–3700 m.

Ecology: Mixed forests with *Pinus*, *Taxus*, *Quercus*, *Acer*, *Rhododendron* and *Corylus*, and open rocky or grassy slopes.

Flowering: (April–)May–June. **Fruiting:** (July–) August–September.

Lindley (Bot. Reg. 20: t. 1655. 1835) misapplied the name *Pyrus crenata* D.Don to this species.
   Sorbus vestita has been reported further east into E Himalaya and China, but recent research (Rushforth, Int. Dendr. Soc. Yearb. 2009: 74. 2010) suggests that these records refer to other taxa and true *S. vestita* is restricted to central Nepal to NW India. See also note under *Sorbus sharmae* M.F.Watson, V.Manandhar & Rushforth.

5. Sorbus hedlundii C.K.Schneid., Ill. Handb. Laubholzk. 1: 685 (1906).
Pyrus hedlundii (C.K.Schneid.) Lacaita.

नाझील Najhil (Nepali).

Tree, 10–15 m. Branchlets dark brown, 5–8(–8) mm thick, glabrous, youngest shoots white tomentose. Buds ovoid, 5–6 mm diameter, scales brown, glabrous, inner surface white tomentose. Stipules brown, linear, 1.5–2 cm, white tomentose, caducous. Petioles 5–18 mm, white tomentose. Leaves simple, broadly elliptic, 18–30 × 4–20 cm, base cuneate, apex acute or rounded, margin serrate, sometimes shallowly lobulate, lateral veins 12–15 pairs, upper surface dark green, glabrous or glabrescent, lower surface white tomentose with midrib and veins prominently brown tomentose. Inflorescence 5–8 cm diameter, rachis and pedicels brown hairy, lenticels present, small. Bracts brown, linear, 1.8–2.5 cm, white ciliate, caducous. Flowers 5–7 mm diameter. Sepals 5–7 mm, linear-triangular, reflexed, apex acuminate, brown hairy. Petals white, obovate, 6–7 × 3–4 mm, base clawed, apex rounded often with irregular toothing, glabrous. Stamens 20, 3–4 mm, filaments pale brown, anthers pale yellow. Styles 3–5, 3–4 mm, connate, white woolly at base, later glabrescent. Fruits yellowish, globose, solid, 1–2 cm diameter, lenticels present, sepals persistent.
Fig. 42f.

Distribution: Nepal and E Himalaya.

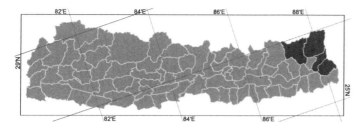

Altitudinal range: (2500–)2700–3050(–3400) m.

Ecology: Mixed forests with *Tsuga*, *Betula*, *Quercus*, *Rhododendron*, forest remnant in degraded areas.

Flowering: May–June. **Fruiting:** (September–) October–November.

In leaf this species with tomentose leaves and lobulate margins can been confused with *Sorbus lanata* (D.Don) Schauer, however, it is readily distinguished by the brown hairs along the veins on the lower leaf surface of mature growth.
    Ripe fruits are eaten fresh.

6. Sorbus lanata (D.Don) Schauer, Übers. Arbeiten Veränd. Schles. Gs. Vaterl. Kult., 1847: 292 (1848).
Pyrus lanata D.Don, Prodr. Fl. Nepal.: 237 (1825); *Aria lanata* (D.Don) Decne.; *Pyrus kumaonensis* Wall. nom. nud.

Tree, 5–15 m. Branchlets shiny brown, 4–7(–9) mm thick, young shoots densely white woolly, at least at first. Buds ovoid with obtuse or acute apex, scales brown, glabrous except apex hairy. Stipules pale brown, linear-lanceolate, 5–7 × 1 mm, caducous, white tomentose. Petioles 1.5–2 cm, white tomentose. Leaves simple, broadly oblong, 6–13 × 3–9 cm, base cuneate, apex acute, margin lobulate, lobules serrulate, lateral veins 9–15, almost parallel, upper surface glabrescent, lower surface white woolly. Inflorescences ca. 6 cm diameter, white woolly. Bracts linear, pale brown, 6–7 mm, ciliate, caducous. Pedicels 4.5–5.5 mm, densely white woolly. Flowers 12–15 mm diameter. Sepals brown with pale tip, ovate, 4–5 mm, apex acute, white woolly. Petals creamy white, obovate, 3.5–4.5 mm, base shortly clawed, apex obtuse, glabrous. Stamens 20, ca. 2 mm, filaments cream to pale brown, anthers pale yellow or brownish. Styles 3, ca. 2 mm, connate at base, white woolly all over. Fruit red or dark brown, globose, solid, 1.3–2(–3) cm diameter, sepals persistent. Fig. 43f–g.

Distribution: Nepal and W Himalaya.

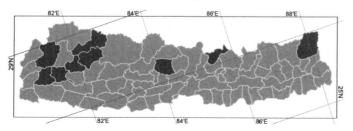

Altitudinal range: 2200–3400 m.

Ecology: In mixed forests with *Abies*, *Picea*, *Taxus*, *Tsuga*, *Aesculus*, *Acer*, *Betula* and *Quercus*.

Flowering: (February–)April–May. **Fruiting:** (May–) June–August(–September).

Morphologically this species can be difficult to distinguish from other white-tomentose simple-leaved *Sorbus*, although the pronounced lobulate leaf margin is usually characteristic. In fruiting specimens it is easily separated by the lack of lenticels on the fruit, and in flowering stage by the connate styles completely covered by woolly white hairs. In Nepal the geographic locality may be helpful to distinguish from other similar species (except *Sorbus vestita* (Wall. ex G.Don) Lodd.) as it is a predominantly W Himalayan species.
    Ripe fruits are eaten fresh. Leafy branches are lopped for fodder.

7. Sorbus insignis (Hook.f.) Hedl., Kongl. Vetensk. Acad. Handl. 35: 32 (1901).
Pyrus insignis Hook.f., Fl. Brit. Ind. 2[5]: 377 (1878).

Tree, 4–10 m. Branchlets greyish brown, stout, to 9 mm thick, glabrous, rusty brown tomentose when young. Buds ovoid, 10–15 × 5–10 mm, apex acute, scales red-brown, margin and apex brown ciliate. Stipules green, suborbicular, 1–2 cm, persistent, sometimes lobed, margin entire or serrate with pointed teeth, glabrous. Petioles (2–)3–5 cm, glabrous. Leaves pinnate, 20–30 cm, rachis grooved above, sparsely tomentose becoming glabrescent. Leaflets 3–5 pairs, sessile, oblong, 5.5–10.5 × 1.5–2.5 cm, base obliquely rounded, apex acute or obtuse, margin finely serrate but recurved and serrations obscured, upper surface dark green, glossy, lower surface brown glaucous, sparsely white hairy when young soon becoming glabrescent apart from tufts of brown hairs at base. Inflorescence 10–15 cm diameter, rachis and pedicels stiff, sparsely pubescent, lenticels present. Bracts brown, linear, 6–8 mm, sparsely hairy within. Flowers 5–6 mm diameter. Calyx lobes triangular, ca. 1.5 mm, glabrous or slightly puberulous. Petals creamy white, ovate or suborbicular, ca. 4 × 3 mm, base clawed, apex rounded with mucronate tip, glabrous. Stamens ca. 20, 2–3 mm, colour not recorded. Styles 2–3, ca. 2 mm, glabrous, lower half connate. Fruit somewhat fleshy, white or crimson when mature, globose or ovoid, 5–8 mm diameter, glabrous, lenticels present, calyx lobes persistent.

Distribution: Nepal, E Himalaya, Tibetan Plateau and Assam-Burma.

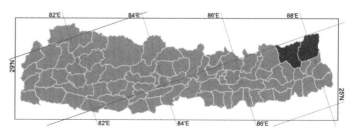

Altitudinal range: (2800–)2900–3200 m.

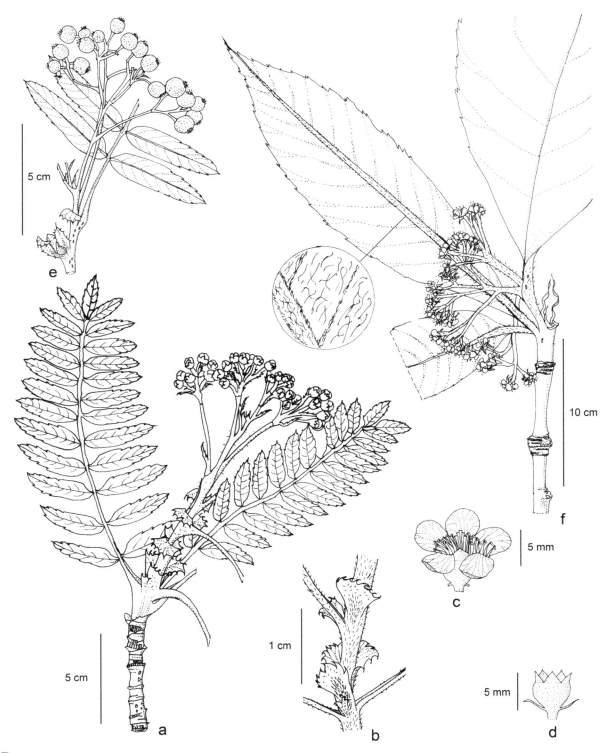

FIG. 42. Rosaceae. **Sorbus arachnoidea**: a, inflorescence and leaves; b, stipules; c, flower; d, hypanthium; e, infructescence and leaves. **Sorbus hedlundii**: f, inflorescence and leaves.

Ecology: In mixed broad leaved forests with *Acer* and *Rhododendron*.

Flowering: May–June. **Fruiting:** August–October.

8. Sorbus kurzii (Watt ex Prain) C.K.Schneid., Bull. Herb. Boissier, Sér 2 6: 315 (1906).
Pyrus kurzii Watt ex Prain, J. Asiat. Soc. Bengal 73: 203 (1904).

Tree, 4–5 m, sometimes epiphytic. Branchlets greyish brown, slender, 3–5 mm thick, glabrous or almost so. Buds ovoid, 0.5–0.8 × 0.2–0.3 mm, apex acute, scales red-brown, glabrous. Stipules green, ovate-lanceolate, ca. 10 × 5 mm, persistent, apex acuminate or caudate with entire or toothed margin, glabrous. Petiole 1.2–2.8 cm, glabrous. Leaves pinnate, 7–11 cm, rachis and petioles slightly grooved above, glabrous. Leaflets 4–6 pairs, subsessile (petiolules 1–1.5 mm), obovate, 1.8–3 × 0.8–1.8 cm, base obliquely round, apex rounded or obtuse, margin entire in lower portion, apical portion finely toothed, glabrescent, upper surface dark green, shiny, lower surface glaucous. Inflorescence 2–3 cm diameter, rachis and pedicels brown, glabrous in fruit, lenticels present. Calyx lobes triangular, 2–3 mm, glabrous to minutely pubescent. Petals white, obovate, 2.5–3 × 1.8–2.2 mm, base clawed, apex rounded, glabrous. Stamens ca. 20, ca. 2.5 mm, colour not recorded. Styles 3–5, 1.5–2 mm, free or slightly connate at base, glabrous. Fruit fleshy, red-pink, later fading almost to white, globose, 5–6 mm diameter, lenticels absent, calyx lobes persistent.

Distribution: Nepal, E Himalaya, Tibetan Plateau and Assam-Burma.

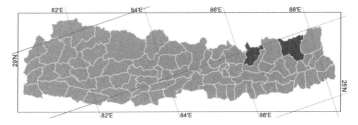

Altitudinal range: 2800–3700 m.

Ecology: In mixed broad leaved forests with *Abies* and *Rhododendron*.

Flowering: May–June. **Fruiting:** July–September.

The persistent green, ovate stipules are characteristics of this species which help to distinguish it from other species that can have low leaflet numbers, such as *Sorbus wallichii* (Hook.f.) T.T.Yu. Literature records of lower altitudes in Nepal (down to 2300 m) have not yet been verified by herbarium collections.

9. Sorbus wallichii (Hook.f.) T.T.Yu, in T.T.Yu, Fl. Reipubl. Popularis Sin. 36: 329 (1974).
Pyrus wallichii Hook.f., Fl. Brit. Ind. 2[5]: 376 (1878).

Shrub or tree, 4–7 m, often epiphytic at lower altitudes. Branchlets greyish brown, 2–4(–6) mm thick, densely white woolly when young, glabrescent. Buds ovoid, ca. 5 × 3 mm, apex acute, scales brown, glabrous, apically sparsely ciliate. Stipules brownish green, linear-lanceolate or subulate, 3–5 mm, persistent, entire, glabrous. Leaves pinnate, 9–13 cm, rachis grooved above, glabrescent. Petioles 1–2.5(–3.5) cm, densely white hairy at first, glabrescent. Leaflets (6–)7–8(–9) pairs, sessile, oblong-ovate, 2–4 × 0.8–1.2 cm, base obliquely round, apex acute or rounded, mucronate, margin entire or obscurely serrate with few minute teeth towards apex, upper surface dark green, matt, not shiny, glabrous, lower surface pale green, densely white woolly when young, later glabrous. Inflorescence 4–6 cm diameter, rachis and pedicels brown tomentose, becoming glabrous and densely lenticellate with age. Pedicels 4–5 mm. Calyx lobes brown, triangular, 1–1.5 mm, apex dark brown, acute, glabrescent. Petals white, broadly obovate, 1.5–2 × 1.3–1.8 mm, base shortly clawed, apex rounded, glabrous. Stamens ca. 20, 1.5–2 mm, filaments cream, anthers pink. Styles 3, 2.5–3 mm, free, glabrous. Fruits, green ripening to red when mature, soft not fleshy, globose or ovoid, 2–2.5 mm diameter, bearing an apical ring of persistent dark purple-black calyx lobes, lenticels absent.

Distribution: Nepal, E Himalaya, Assam-Burma and E Asia.

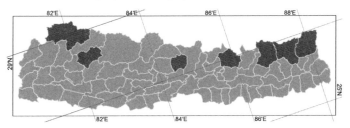

Altitudinal range: 1900–4100 m.

Ecology: In mixed forests with *Pinus*, *Abies*, *Castanopsis*, *Quercus*, *Betula*, *Rhododendron* and *Lyonia*.

Flowering: April–May(–June). **Fruiting:** (July–) September–October(–November).

Schneider (Ill. Handb. Laubholzk. 1: 680. 1906) misapplied the name *Sorbus foliolosa* (Wall.) Spach to this species.
   Sorbus wallichii a very distinctive species with interesting habit and morphological characteristics. At low altitudes, it is often epiphytic while at higher altitudes it can grow to a small tree. When in flower, young shoots and leaves of this species are densely covered with a white woolly indumentum and plants become glabrescent with age. *Sorbus* specimens are usually collected in full flower or with mature fruit, and conspicuous changes in characters with age, such as the loss of the woolly indumentum, can confuse accurate identification. Flowering material of *S. wallichii* is easily recognized by the indumentum which is quite unlike any other Nepalese pinnate-leaved species. Fruiting material is characterized by the densely lenticellate infrutescence branches and dark purple-black hypanthium and sepals.

10. *Sorbus microphylla* Wenz., Linnaea 38: 76 (1873).
Pyrus microphylla (Wenz.) Wall. ex Hook.f.; *P. microphylla*
Wall. nom. nud.; *Sorbus khumbuensis* McAll.

बझार Bajhar (Nepali).

Shrub or tree, 2–5(–7) m. Branchlets greyish brown, reddish
brown, 2–4(–6) mm thick, puberulous when young. Buds
narrowly ovoid, ca. 6 × 1.5 mm, apex acute, scales brown,
glabrous. Stipules dark brown, narrowly lanceolate, ca. 4 ×
2 mm, caducous, apex acuminate. Petiole 1–2 cm, sparsely
brown hairy. Leaves pinnate 9–14 cm, rachis grooved above,
glabrous, brown puberulous when young. Leaflets 12–14(–18)
pairs, very rarely only 8 or 9 pairs, sessile, ovoid-elliptic to
oblong, (1.0–)1.5–2(–2.4) × (0.3–)0.5–0.8(–1) cm, rarely more
than 2.4 cm, base rounded, apex obtuse or acute, margin
serrate with acute teeth, both surfaces glabrous or brown
hairy along veins when young. Inflorescence 2–4 cm diameter,
glabrous, lenticels present on rachis and pedicels. Bracts
linear-lanceolate, 2–3 mm, caducous. Pedicels 6–8 mm.
Flowers 7–10 mm diameter, glabrous. Sepals triangular,
1.5–2 mm, apex acute. Petals white flushed pink to deep pink
with a white margin, suborbicular, 3–4 mm, base clawed, apex
rounded, sometimes irregularly toothed, glabrous. Stamens
20, 2.5–3.5 mm, filaments white, anthers pink or purple.
Styles 5, ca. 3.5 mm, free or slightly connate, pubescent at
base. Fruits white, flushed pink or crimson, globose or ovoid,
fleshy, 8–10 mm diameter, glabrous, lenticels absent, sepals
persistent.
Frontispiece g.

Distribution: Nepal, W Himalaya, E Himalaya, Tibetan
Plateau, Assam-Burma and E Asia.

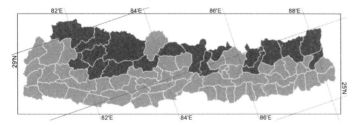

Altitudinal range: 2000–4200 m.

Ecology: In several mixed forest types (*Pinus*, *Abies*, *Picea*,
Tsuga, *Quercus*, *Populus*, *Rhododendron* and *Betula*),
open woods and tall shrubs, along river banks and streams,
moraines, stony and grassy slopes.

Flowering: (May–)June–July(–August). **Fruiting:** (July–)
August–October.

Long (Fl. Bhutan 1: 597. 1987) considered *Sorbus rufopilosa*
Rushforth as a distinct species but with doubtful status. This E
Himalayan and Chinese species is characterized by densely
rusty red pubescent inflorescence branches, smaller (less than
10 mm) and more numerous leaflets (14–17 pairs) with fewer
teeth and smaller mature fruit (less than 10 mm). In NE Nepal
specimens of *S. microphylla* with numerous small leaflets are
reminiscent of *S. rufopilosa*, and are sometimes determined
as such. However, we consider that they form a part of
variation within *S. microphylla* agg. as herbarium specimens

cannot be confidently determined as *S. rufopilosa*. Similarly,
S. khumbuensis with many pairs (12–19) of small leaflets is
not morphologically discontinuous and is not recognized here.
Hence, all Nepalese and Bhutanese plants of this group are
considered as a part of an aggregate *S. microphylla*.
Leafy branches are lopped for fodder.

11. *Sorbus himalaica* Gabrieljan, Bot. Zhurn. [Moscow &
Leningrad] 56: 658, pl. 1 – 2 (1971).

Shrub or tree, 2–6(–8) m. Branchlets greyish brown, ca. 4 mm
thick, glabrous or glabrescent. Buds ovoid, ca. 8 × 3 mm,
apex obtuse, scales brown, rusty brown pubescent. Stipules
brown, linear, 5–9 mm, densely brown hairy, caducous. Petiole
1.2–2.5(–3.5) mm, densely brown hairy, becoming glabrescent.
Leaves 8–15 cm, pinnate, rachis grooved above, winged, rusty
puberulous. Leaflets 8–12(–13) pairs, sessile, oblong-elliptic,
2.5–3 (–3.5) × 0.8–1.2 cm, base obliquely rounded, apex
obtuse or acute, margin serrate to below middle, upper surface
dark green, glabrescent, lower surface pale green, rusty brown
hairy, denser along the veins. Inflorescence 4–10 cm diameter,
rachis and pedicels rusty brown hairy, lenticels present.
Pedicels 2–5 mm. Bracts linear, ca. 4 mm, brown hairy.
Flowers 6–10 mm in diameter. Sepals 1.5–2 mm, triangular,
red-brown, hairy on outside, glabrous within, apex obtuse or
acute. Petals red or pink, broadly ovate, 4–5 × 2–2.5 mm,
base clawed, apex rounded, glabrous. Stamens 20, 2.5–3 mm,
filaments pink, anthers brown. Styles 1.5–2 mm, free, white
hairy at base. Fruits red or pink, globose, fleshy, 5–9mm
diameter, with a few lenticels, sepals persistent.
Fig. 43a–e, Frontispiece a–f.

Distribution: Nepal and E Himalaya.

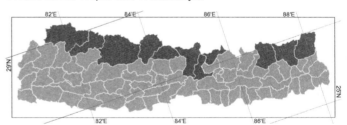

Altitudinal range: 2500–4100 m.

Ecology: Mixed forest with *Abies*, *Betula* and *Rhododendron*.

Flowering: May–June. **Fruiting:** August–September(–October).

Sorbus foliolosa (Wall.) Spach and *S. himalaica* are high
altitude plants with similar morphological characteristics and
are difficult to distinguish. Stipule characters can usually be
helpful to separate these species: in *S. himalaica*, stipules
are long subulate or toothed, pubescent with brown hairs,
never green and leafy, while in *S. foliolosa* they are ovate or
lanceolate and reddish brown hairy. These species often grow
with *S. microphylla* Wenz.

12. *Sorbus arachnoidea* Koehne, Repert. Spec. Nov. Regni Veg. 10: 514 (1912).

Shrub or tree, 4–8(–10) m. Branchlets greyish brown, ca. 5 mm thick, densely rusty brown pubescent when young, glabrescent. Buds ovoid, ca. 9 × 3 mm, apex acute, scales brown, glabrous or rusty puberulous apically. Stipules green, broadly ovate or suborbicular, 6–7 × 5.5–10 mm, persistent, entire or remotely toothed, apex mucronate, glabrous. Petioles 1.4–2.3 mm. Leaves pinnate, 12–18 cm, rachis winged, glabrescent. Leaflets 7–12 pairs, oblong-elliptic, sessile, 2.3–5.0 × 0.5–1.2 cm, base obliquely cuneate, apex mucronate, margin finely serrate towards apex, upper surface dark green, glabrous, lower surface pale green, brown pubescent along veins. Inflorescence 4.5–7.5 cm diameter, rachis and pedicels brown tomentose, lenticels present, inconspicuous. Pedicels 0.5–1 mm. Bracts linear, 8–10 mm, sparsely brown hairy. Flowers ca. 3 mm diameter, glabrous. Sepals triangular, 1–1.2 mm, apex obtuse to acute, glabrous. Petals white or pink, oblong-ovate, 2–3 × 1.8–2.2 mm, base clawed, apex rounded, glabrous. Stamens 20, 1.5–2 mm, filaments pale pink, anthers pale brown. Styles 4–5, 1.25–2 mm, free, glabrous. Fruits initially crimson, becoming pink to white flushed pink, globose, fleshy, 9–10 mm diameter, slightly brown pubescent, lenticels absent, sepals persistent. Fig. 42a–e.

Distribution: Nepal and E Himalaya.

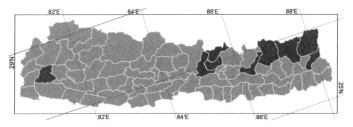

Altitudinal range: 2800–4200 m.

Ecology: In mixed forests with *Abies*, *Quercus*, *Rhododendron* and *Juniperus*, often at margins or in open areas and shrubland.

Flowering: May–June(–July). **Fruiting:** (June–) July–August(–September).

Sorbus arachnoidea has very distinctive stout twigs, dark green leaflets and large, broad, green leaf-like stipules which often clasp the stem. However, this species is often overlooked and usually confused with *S. foliolosa* (Wall.) Spach. but it can be readily distinguished by stipule characters. Although this is predominantly an E Himalayan species, one specimen (*Shrestha 4219*, KATH) has been collected from W Nepal.

13. *Sorbus foliolosa* (Wall.) Spach, Hist. Nat. Veg. 7: 96 (1839).

Pyrus foliolosa Wall., Pl. Asiat. Rar. 2[8]: 81, pl. 189 (1831); *Photinia foliolosa* (Wall.) Koehne; *Pyrus ursina* Wall. nom. nud.; *Sorbus foliolosa* var. *ursina* Wenz.; *S. ursina* (Wall. ex G.Don) S.Schauer; *S. ursina* (Wall. ex G.Don) Decne. nom. superfl.; *S. ursina* var. *wenzigiana* C.K.Schneid.; *S. wenzigiana* (C.K.Schneid.) Koehne.

Shrub or small tree, 2–5(–9) m. Branchlets greyish brown, fairly stout, ca. 5.5 mm thick, glabrescent, densely rusty pubescent when young. Buds ovoid, ca. 9 × 2.5 mm, apex obtuse, scales brown, glabrous or rusty puberulous apically. Stipules greenish-brown, ovate or lanceolate, 4–8 × 2–3 mm, margin entire or serrate with a few apical teeth, reddish brown hairy, persistent. Petioles 1.4–2.3 cm, densely pubescent becoming glabrescent. Leaves 7–14 cm, pinnate, rachis grooved above, slightly rusty puberulous, winged. Leaflets 6–9 pairs, oblong-elliptic, sessile, (2.5–)3–4.5 × 0.8–1.4 cm, base obliquely rounded, apex obtuse, margin sharp toothed apically, upper surface dark green, glabrescent, lower surface pale green, papillose, rusty brown hairy along the veins. Inflorescence 4.5–20 cm diameter, rachis and pedicels rusty brown hairy, lenticels present. Pedicels 3–5 mm. Bracts linear, ca. 5 mm, brown hairy. Flowers 6–9 mm diameter. Sepals triangular, 3–5 mm, apex obtuse or acute, glabrous. Petals white or flushed pink (especially on the reverse), broadly ovate, 2–3 × 1.5–2 mm, base clawed, apex rounded, glabrous. Stamens 20, ca. 2 mm, filaments white, anthers pink. Styles 5, 1.5–2 mm, free, white hairy at base. Fruits red, yellowish-red or pale red, globose, 4–6 mm diameter, glabrous, lenticels very sparse, sepals persistent.

Distribution: Nepal, W Himalaya, E Himalaya, Tibetan Plateau, Assam-Burma and E Asia.

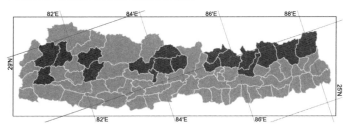

Altitudinal range: 2200–4300 m.

Ecology: Mixed forests with *Abies*, *Juniperus*, *Quercus*, *Betula*, *Rhododendron*, *Prunus* and *Salix*, often near and above the tree line, usually near streams and on north-facing slopes.

Flowering: (April–)May–June(–July). **Fruiting:** (July–) August–September(–October).

Sorbus foliolosa (Wall.) Spach has been misapplied in the Indian and Nepalese literature to this species.

It is not possible to follow Lu & Spongberg (Fl. China 8: 154. 2003) in distinguishing 'wenzigiana' at any rank in Nepal, and we agree with Long (Fl. Bhutan 1: 598. 1987) both in recognizing one broad species and in treating *S. ursina* as a synonym of *S. foliolosa*.

A paste of the leaves is applied to treat boils.

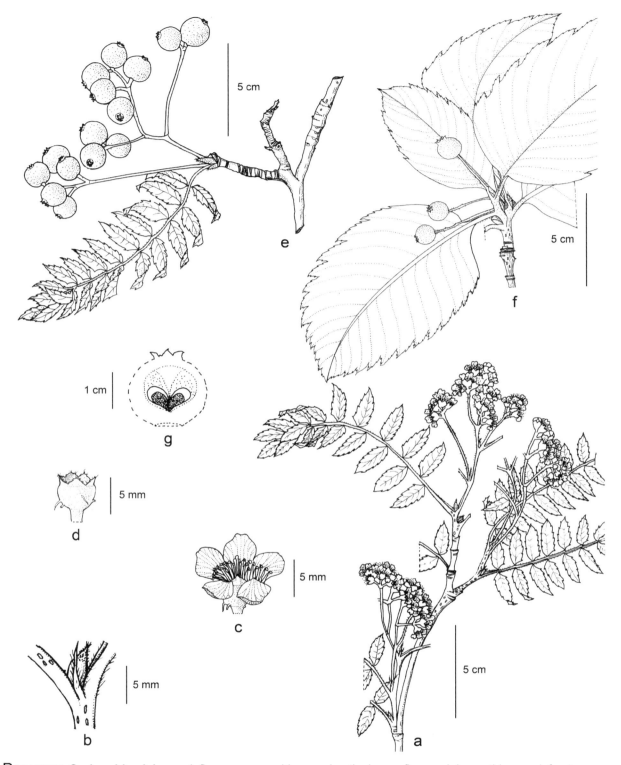

FIG. 43. ROSACEAE. **Sorbus himalaica**: a, inflorescence and leaves; b, stipules; c, flower; d, hypanthium; e, infructescence and leaf. **Sorbus lanata**: f, fruiting branch; g, longitudinal section of fruit.

10. *Docynia* Decne., Nouv. Arch. Mus. Hist. Nat. 10: 125, 131 (1874).

Colin A. Pendry

Trees, semi-evergreen or deciduous. Twigs unarmed. Stipules caducous. Leaves simple, alternate, stipulate, petiolate, margin toothed at least towards apex, sometimes slightly lobed. Inflorescences terminal, flowers solitary or 2–5-fascicled, emerging with the leaves. Flowers bisexual. Bracts small, caducous. Hypanthium funnel-shaped to campanulate, densely pale tomentose outside. Sepals 5, lanceolate. Petals 5, white, base shortly clawed. Stamens ca. 30, in 2 whorls. Ovary inferior, 5-loculed, with 3–10 ovules per locule. Styles 5, connate at base, villous. Fruit a subglobose or ellipsoid pome with persistent erect or incurved sepals. Seeds 10–20.

Worldwide 15 species from the Himalaya to China and SE Asia. One species in Nepal.

1. *Docynia indica* (Colebr.) Decne., Nouv. Arch. Mus. Hist. Nat. 10: pl. 14 (1874).
Pyrus indica Colebr., in Wall., Pl. Asiat. Rar. 2[8]: 56 (1831); *Docynia docynioides* (C.K.Schneid.) Rehder; *D. griffithiana* Decne.; *D. hookeriana* Decne.; *D. rufifolia* (H.Lév.) Rehder.

पासी Paasi (Nepali).

Trees to 2–3 m tall. Branchlets purplish brown or blackish brown when old, terete, initially densely tomentose, glabrescent. Stipules lanceolate, 5–6 mm, apex acuminate. Petiole 1–2.5 cm, pubescent, glabrescent. Leaf blade elliptic or oblong-lanceolate, 6–14 × 2.5–5 cm, base broadly cuneate or rounded, sometimes oblique, apex acute or acuminate, margin serrate or shallowly crenate, rarely only toothed towards apex, glabrous and lustrous above, sparsely pubescent or subglabrous below, secondary veins 6–9 pairs. Bracts lanceolate. Flowers ca. 2.5 cm across. Pedicel short or nearly absent, pubescent. Hypanthium 6–10 mm long. Sepals narrowly triangular, 5–11 mm, apex acute or acuminate, margin entire, tomentose on both surfaces. Petals oblong or oblong-obovate, 15–25 × 8–12 mm. Stamens 10–12 mm. Styles about as long as stamens, connate and pubescent at base. Pome yellow, 2–3 cm in diameter, slightly pubescent when young.
Fig. 44a–d.

Distribution: Nepal, E Himalaya, Assam-Burma, E Asia and SE Asia.

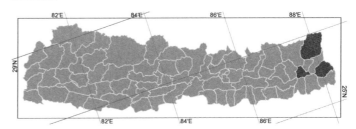

Altitudinal range: 700–2000 m.

Ecology: Open forests, stream-sides, thickets.

Flowering: March–May. **Fruiting:** August–December.

Propagated by seed and cuttings. Closely related to *Cydonia* which differs in having free styles. *Cydonia oblonga* Mill. is occasionally cultivated in Nepal for its edible fruit (quince).
   Ripe fruits are edible.

11. *Pyrus* L., Sp. Pl. 1: 479 (1753).

Mark F. Watson

Deciduous trees, usually unarmed, but juvenile vegetative shoots sometimes with short, spine-tipped lateral branches. Stipules subulate, caducous. Leaves alternate, simple, unlobed, clustered on short side shoots. Inflorescence a corymb, appearing before or with young leaves. Flowers bisexual. Bracts and bracteoles linear, tomentose, caducuous. Hypanthium urn-shaped, constricted at the ovary apex then expanded above. Sepals 5. Petals 5, white, obovate, base clawed, apex rounded. Stamens 25–30, slightly shorter than petals, filaments white, anthers deep pink at first, becoming purple-black. Ovary inferior, 3–5 locular, ovules 2 per locule. Styles 3–5, about as long as stamens, free, usually pubescent at base. Fruit a pome, ellipsoid or obovoid, fleshy, juicy, granular with stone cells, 3–5-celled, endocarp (core) cartilaginous, sepals deciduous or somewhat persistent. Seeds almost black, 1(–2) per locule.

Worldwide about 15–25 species in Europe and Asia (especially SW Asia). One species native in Nepal.

Pyrus crenata Buch.-Ham. ex D.Don (Prodr. Fl. Nepal.: 237. 1825) is only known from the type collection made by Buchanan-Hamilton at Swayambhunath (Kathmandu Valley), flowering in July 1802. Unfortunately the specimen that Don would have used has not yet been found in the BM herbarium, nor is there a duplicate in the Smith herbarium (LINN-SM). Don's description of leaves ovate-acute, crenate, long petioles, glabrous above, young branches snowy tomentose, corymbs lanuginose and sepals ovate-acute, is insufficient to confirm identification and so it is currently impossible to apply this name. July is very late for a *Pyrus* or *Malus* to be in flower, and it is more likely that this name refers to a species of *Photinia*, *Cotoneaster* or *Eriobotrya*.

1. *Pyrus pashia* Buch.-Ham. ex D.Don, Prodr. Fl. Nepal.: 236 (1825).
Pyrus kumaoni Decne. ex Hook.f.; *P. nepalensis* Hort. ex Decne. nom. inval.; *P. nepalensis* Lodd. ex Loudon nom. nud.; *P. nepalensis* Buch.-Ham. ex Hook.f. nom. inval.; *P. pashia* var. *kumaoni* (Decne. ex Hook.f.) Stapf; *P. variolosa* Wall. nom. nud.; *P. variolosa* Wall. ex G.Don; *Sorbus variolosa* (Wall. ex G.Don) S.Schauer.

मयल Mayal (Nepali).

Tree, 5–10(–15) m. Branchlets densely white hairy when young, soon glabrescent, bark purplish brown becoming almost black. Stipules 4–10 mm, base fused to petiole. Petioles 1.8–4(–5.5) cm, initially white pilose, soon glabrescent. Leaves ovate, 4.5–9(–12) × 2.5–6 cm, base rounded or slightly cordate, rarely cuneate, apex acuminate, margin finely obtusely serrate, sparsely white tomentose at first, glabrescent, upper surface usually glossy. Corymbs umbellate, 3–8-flowered, inflorescence branches and pedicels densely tomentose at first, soon glabrescent. Pedicels 1–2(–3) cm, elongating to 3.5 cm in fruit. Flowers 2–5 cm diameter. Hypanthium 2.5–3 mm. Sepals shortly triangular, 2.5–4 × 2–2.5 mm, margin with glands often obscured by hairs, brownish or white tomentose within and on margins, variably pubescent or glabrescent on outside. Petals 8–17 × 6–11 mm. Ovary pubescent. Pomes grey-brown, densely pale spotted, ellipsoid or globose, 1.8–2.5(–3) × 1.7–2.5(–3) cm, essentially glabrous, sepals deciduous.
Fig. 44e–f.

Distribution: Nepal, W Himalaya, E Himalaya, Tibetan Plateau, Assam-Burma, E Asia and SE Asia.

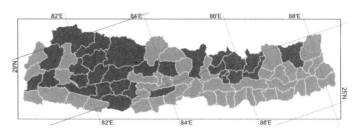

Altitudinal range: 700–3100 m.

Ecology: Hillsides and near rivers in mixed temperate evergreen and deciduous forests (with *Castanopsis*, *Schima*, *Quercus*, Lauraceae), and open shrubland. Isolated trees often found in degraded habitats, around cultivation and near habitation.

Flowering: March–April. **Fruiting:** (June–)July–October(–February).

A tree with remarkably wide altitudinal range. Although flowering is mostly restricted to March and April, it occasionally flowers out of season in October or November, the edible fruit retained on tree until the spring flowering. The cultivated pear, *Pyrus communis* L., has broadly ovate leaves (up to 6 cm broad) with abruptly acute apex, crenately serrate or subentire margins, longer petioles (up to 6 cm), and larger fruit (4–6 cm diameter) borne on short (2.5–3 cm), much scarred shoots.
   Pyrus pashia is often used as a rootstock for grafting cultivated pear varieties. Juice of ripe fruits is put in the eyes of cattle to treat conjunctivitis. Wood is used for making walking sticks.

12. *Malus* Mill., Gard. Dict. Abr., ed. 4, 2: [835] (1754).

Mark F. Watson

Deciduous trees or shrubs, spines absent. Stipules linear-ovate, caducous. Leaves alternate, simple, unlobed, clustered on short side shoots. Corymbs terminal on short lateral shoots, appearing with the young leaves. Flowers bisexual. Bracts and bracteoles caducous. Hypanthium ellipsoid narrowed at top of ovary, widening to cup-shaped above. Sepals 5, about equalling the hypanthium. Petals 5, white or pink, elliptic to broadly so or obovate, base clawed, apex rounded. Stamens 15–20(–50), of unequal length, filaments white, anthers yellow or orange-yellow. Ovary inferior, 4–6-locular, ovules 2 per cell. Styles 3–5, fused at least at base. Fruit a pome, fleshy, juicy, pithy but not granular (stone cells absent), 3–6-celled, endocarp (core) cartilaginous, sepals persistent or deciduous. Seeds dark brown, 1 or 2 per cell.

Worldwide about 40–50 species in temperate zones of N America, Europe and Asia. Three species in Nepal.

Key to Species

1a   Pedicels short and thick, 1–2 cm. Fruit thickly fleshy, more than 4 × 2.5 cm ....**1. *M. pumila***
   b   Pedicels long and slender, 3–4 cm. Fruit thinly fleshy less than 1.5 × 1 cm 2

2a Underside of leaves, petioles, pedicels and calyx lobes densely white pubescent, at least at first. Corymbs 4–20-flowered. Pomes ca. 1.5 × 1 cm, calyx persistent .... **2. *M. sikkimensis***

b Undersides of leaves at most finely pubescent along veins and other parts not densely white pubescent. Corymbs 3–7-flowered. Pomes 7–8 × 4–5 mm, calyx deciduous .... **3. *M. baccata***

1. *Malus pumila* Mill., Gard. Dict., ed. 8: Malus n. 3 (1768). *Malus communis* Poir.; *M. domestica* Borkh.; *M. domestica* subsp. *pumila* (Mill.) Likhonos; *M. sylvestris* Mill.; *Pyrus malus* L.; *P. malus* var. *sylvestris* L.

स्याउ Syau (Nepali).

Shrub or tree, 4–10(–15)m. Branchlets densely tomentose when young, soon glabrescent, purplish brown with age. Stipules green, 3–5 mm, margin eglandular, densely puberulous. Petioles (1.5–)2–3 cm, stout, puberulous. Leaves elliptic or broadly so, 3.5–8 × 2–5 cm, base rounded to broadly cuneate, apex acuminate, margin sharply serrate, pubescent when young, especially on lower surface, glabrescent. Corymbs 4–7-flowered, 4–6 cm across. Pedicels (0.5–)1–1.5(–2.5) cm, short and stout. Flowers 3–4 cm diameter. Hypanthium 4–5(–8) mm, glabrous or almost so. Sepals erect, triangular-ovate or narrowly so, about equalling tube, apex acuminate, tomentose on both surfaces. Petals white or tinged pink, deep pink on reverse, elliptic to broadly so, 1–2 × 0.7–1 cm, glabrous. Stamens 20, about half as long as petals. Ovary 5-locular, styles 5, slightly longer than stamens, fused at base, pubescent at base. Pome green, yellow or red at maturity, ovoid-ellipsoid or subglobose, 4–8 × 2.5–7 cm, thickly fleshy, floral parts somewhat persistent.

Distribution: Cosmopolitan.

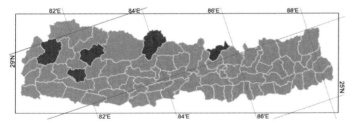

Altitudinal range: 2000–3300 m.

Ecology: Open shrubland or degraded areas, often near villages or former areas of habitation, probably originally planted.

Flowering: April–May. **Fruiting:** August–October.

The domestic apple, native to SW Asia and Europe, widely cultivated in temperate regions of SW Asia and Europe, China, NW Himalaya. Under recorded in Nepal.
Fruit edible.

2. *Malus sikkimensis* (Wenz.) Koehne, Gatt. Pomac.: 27 (1890).
Pyrus pashia var. *sikkimensis* Wenz., Linnaea 38: 49 (1873); *Malus baccata* subsp. *sikkimensis* (Wenz.) Likhonos; *Pyrus sikkimensis* (Wenz.) Hook.f.

Tree 6–9 m, branchlets grey tomentose when young, soon glabrescent, red-brown becoming blackish with age. Stipules green, 6–8 mm, margin glandular, apex acuminate. Petioles 2–3 cm, slender, densely white pubescent at first. Leaves elliptic, ovate or ovate-lanceolate, 5–8(–13) × 2–4(–6.5) cm, base rounded to broadly cuneate, apex acuminate, margin sharply serrate, lower surface white or greyish pubescent, especially on veins. Pedicels (1.5–)3.5–4.5(–5), long and slender, densely white pubescent at first, glabrescent in fruit. Corymbs (4–)7–10(–15)-flowered, 5–9 cm across. Flowers 2.5–3 cm diameter. Hypanthium 3–4 mm, outside densely white pubescent at first, glabrescent in fruit. Sepals recurved, narrowly triangular-ovate, 4–7 × 1.8–2.2 mm, apex acute, densely white pubescent at first. Petals white, pink on reverse, suborbicular, 12–15 mm, pubescent at base. Stamens 25–30, about half as long as petals. Ovary 5–6-locular, styles 5(–6), ca. 7 mm, slightly longer than stamens, fused to half way, fused section exerted above hypanthium, glabrous. Pomes in clusters of 3–7, dark red with white spots, ellipsoid to obovoid, 1.5–2 × 1–1.5 cm, thinly fleshy, sepals persistent.

Distribution: Nepal, E Himalaya, Tibetan Plateau, Assam-Burma and E Asia.

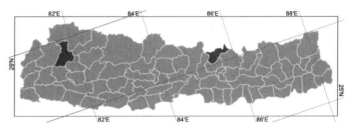

Altitudinal range: 2400–3000 m.

Ecology: Open mixed forest, often in moist valleys.

Flowering: April–May. **Fruiting:** August.

Rarely collected in Nepal, more common in the E Himalaya, especially Bhutan. Flowering material is most easily distinguished from *Malus baccata* (L.) Borkh. by the dense pubesence on the pedicels and hypanthium, and in fruit by the persistent sepals.

3. *Malus baccata* (L.) Borkh., Theor. Prakt. Handb. Forstbot. 2: 1280 (1803).
Pyrus baccata L., Mant. Pl.: 75 (1767); *Malus baccata* subsp. *himalaica* (Maxim.) Likhonos; *M. baccata* subsp. *mandshurica* (Maxim.) Likhonos; *M. baccata* subsp. *sachalinensis* (Juz.) Likhonos; *M. baccata* var. *himalaica* (Maxim.) C.K.Schneid.; *M. baccata* var. *mandshurica* (Maxim.) C.K.Schneid.; *M. baccata* var. *sibirica* (Maxim.) C.K.Schneid.; *M. mandshurica* (Maxim.) Kom. ex Juz.; *M. pallasiana* Juz.; *M. rockii* Rehder;

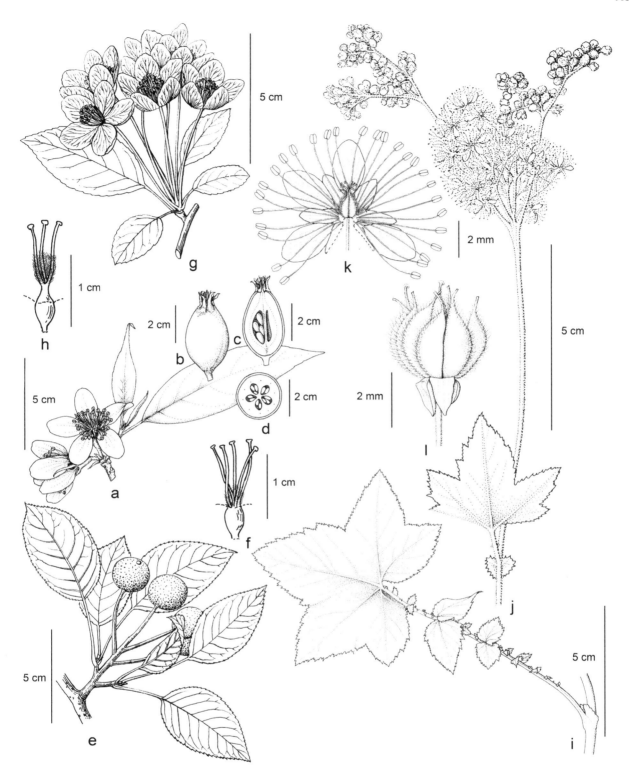

FIG. 44. ROSACEAE. **Docynia indica**: a, inflorescence and leaves; b, fruit; c, longitudinal section of fruit; d, transverse section of fruit. **Pyrus pashia**: e, infructescence and leaf; f, ovary. **Malus baccata**: g, inflorescence and leaves; h, ovary. **Filipendula vestita**: i, lower leaf; j, inflorescence; k, opened flower; l, fruit.

M. sachalinensis Juz.; *M. sibierica* Borkh.; *Pyrus baccata* var. *himalaica* Maxim.; *P. baccata* var. *mandshurica* Maxim.; *P. baccata* var. *sibirica* Maxim.

जंगली स्याउ Jungali syau (Nepali).

Shrub or tree, 3–6(–10) m, branches often arching and pendulous, bark red-brown darkening with age, glabrous. Stipules red-tipped, 3–5 mm, margin glandular, sparsely pubescent. Petioles 0.6–4(–5) cm, slender, moderately to sparsely hairy at first, glabrescent. Leaves narrowly elliptic, sometimes broadly so, (1.6–)3–8.5(–9) × (1–)1.5–3.5(–5) cm, base cuneate or rounded, apex acute to acuminate, margin sharply finely serrate, sparsely hairy along midrib at first (especially above), glabrescent. Corymbs umbellate, 3–7-flowered, 5–7 cm across. Pedicels (1.5–)3–4(–5) cm, long and slender, glabrous or almost so. Flowers 1.7–3 cm in diameter. Hypanthium 2–3 mm, outside glabrous or almost so. Sepals erect, lanceolate, 5–6(–7) × 1.5–2.5 mm, apex acute, densely white hairy within, outside glabrous or almost so. Petals white or tinged pink (especially in bud), obovate, (10–)12–13.5(–15) × 6–9(–10) mm, glabrous. Stamens 15–20, half as long as petals. Ovary 4–5-locular, styles 4–5, ca. 10 mm, conspicuously longer than stamens, fused at base, densely white pubescent at base. Pomes in clusters of (2–)3–6, red or yellowish, globose at maturity (ovoid when young), 7–11 × (4–)7–11 mm, thinly fleshy, floral parts deciduous at maturity, floral scar ca. 2 mm diameter. Fig. 44g–h.

Distribution: Nepal, W Himalaya, E Himalaya, Tibetan Plateau, Assam-Burma, E Asia and N Asia.

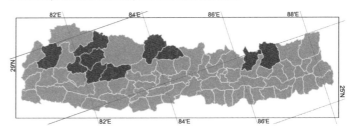

Altitudinal range: 1800–4100 m.

Ecology: Occasional tree in mixed evergreen or deciduous forests, often solitary trees found in degraded situations or in cultivated areas around habitation where it has possibly been planted for its sweet fleshy fruit.

Flowering: April–June. **Fruiting:** (July–)August–October.

A beautiful ornamental tree widely grown in temperate gardens for its showy flowers and colourful fruit. Commonly used as a rootstock for grafting the cultivated apple (*Malus pumila* Mill.). A paste of the fruits is applied to the forehead to relieve headaches.

13. *Filipendula* Mill., Gard. Dict. Abr., ed. 4, 1: 512 (1754).

Hiroshi Ikeda

Perennial herbs. Rhizomes stout. Stipules leaf-like, adnate basally. Radical leaves pinnate, terminal leaflets large, palmately lobed, toothed; lateral leaflets small or sometimes obsolete. Inflorescence a cyme. Flowers bisexual or rarely unisexual. Bracts and bracteoles absent. Hypanthium short, flat. Epicalyx absent. Sepals 4 or 5, reflexed. Petals 4 or 5, white or pinkish. Stamens numerous, deciduous after anthesis; filaments filiform. Ovary superior, carpels free, 5–15, stipitate or subsessile; ovules 1 or 2 in each carpel. Style short, terminal. Achenes glabrous or ciliate, laterally flattened, stipitate or sessile.

Worldwide about 15 species in temperate to cool regions of the N hemisphere. One species in Nepal.

1. *Filipendula vestita* (Wall. ex G.Don) Maxim., Trudy Imp. S.-Peterburgsk. Bot. Sada 6: 248 (1879).
Spiraea vestita Wall. ex G.Don, Gen. Hist. 2: 521 (1832); *S. camtschatica* var. *himalensis* Lindl.

Perennial herbs. Stems erect, 70–140 cm, slightly angled, sparsely or densely pubescent. Stipules semicordate, margin serrate. Leaves lyrately pinnate with 3–5 pairs of lateral leaflets and intercalary segments, 10–30 × 5–15 cm, glabrescent above, densely greyish white tomentose below, rusty brown pubescent on veins; terminal leaflet ovate to nearly orbicular, 3–12 × 4–15 cm, palmately 3–5-lobed, lobes narrowly ovate, margin doubly serrate or incised-toothed, apex acute to acuminate, base subcordate; lateral leaflets often obscure in upper leaves; petioles and rachis sparsely or densely pubescent. Inflorescence to 10 cm. Peduncles and pedicels densely tomentose. Flowers 5–7 mm across. Hypanthium densely tomentose outside. Sepals ovate, 1.9–2.1 × 1–1.3 mm, apex obtuse or semiacute, margin entire or sparsely serrate near apex, sparsely pubescent outside, glabrous inside. Petals 5, white, elliptic to obovate, 3–4 × 2–2.5 mm, margin entire or slightly undulate, apex rounded, base cuneate to widely cuneate, short-clawed. Filaments 3–4 mm; anthers ellipsoid, 0.6–0.8 mm. Carpels oblanceolate, subsessile, 1–1.3 × 0.5–0.6 mm, strigose-ciliate along outer side; style slightly curved near base, 0.6–0.8 mm; stigma inflated, papillate. Achenes narrowly oblong, sessile, 2–3 × 1–1.2 mm, strigose-ciliate along inner and outer sides. Fig. 44i–l.

Distribution: Nepal, W Himalaya, E Asia and SW Asia.

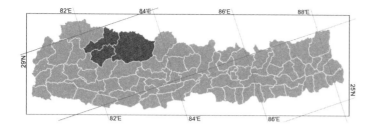

Altitudinal range: 2200–3200 m.

Ecology: Open moist meadows and river banks or under forest.

Flowering: July–August. **Fruiting:** August–September.

Wallich (Numer. List: 21 n. 704. 1829) misapplied the name *Spiraea camtschatica* Pall. (as '*kamtschatica*') to this species.

14. *Rubus* L., Sp. Pl. 1: 492 (1753).

David E. Boufford, Mohan Siwakoti & Colin A. Pendry

Shrubs, or subshrubs, deciduous, rarely evergreen or semi-evergreen, or perennial creeping dwarf herbs. Stems erect, climbing, arching or prostrate, glabrous or hairy, usually with prickles or bristles, sometimes with glandular hairs, rarely unarmed. Stipules either ± adnate to petiole basally, persistent and undivided or occasionally lobed, or free, near base of petiole or at junction of stem or petiole, persistent or caducous, mostly dissected, occasionally entire. Leaves alternate, petiolate, simple, palmately or pinnately compound, divided or undivided, toothed, glabrous or pubescent, sometimes with glandular hairs, bristles or glands. Inflorescences cymose panicles, racemes or corymbs, or flowers several in clusters or solitary. Flowers bisexual, rarely unisexual and plants dioecious. Bracts solitary, at base of pedicel; bracteoles absent. Hypanthium short, broad. Epicalyx absent. Sepals (4 or)5(–8), sometimes unequal, erect, spreading or reflexed, persistent. Petals usually 5, rarely more, occasionally absent, white, pink or red, glabrous or hairy, margin entire, rarely premorse. Stamens numerous, sometimes few, inserted at mouth of hypanthium; filaments filiform; anthers 2-locular. Ovary superior, carpels 4–100 or more, inserted on convex torus, each carpel becoming a drupelet or drupaceous achene; locule 1; ovules 2, pendulous, only 1 developing. Style filiform, subterminal, glabrous or hairy; stigma simple, capitate. Drupelets or drupaceous achenes aggregated on hemispherical, conical or cylindrical torus, forming an aggregate fruit, separating from torus and aggregate hollow, or adnate to torus and falling with torus attached at maturity and aggregate solid.

Worldwide about 700 species, particularly abundant in the temperate N hemisphere with a few species extending into the S hemisphere. 32 species in Nepal.

The order of species is based mainly on their sequence in Lu & Boufford (Fl. China 9. 2003) and does not necessarily reflect a phylogenetic sequence. Specimens of *Rubus* from Nepal are relatively few, and some species appear to be rare, making complete information difficult to obtain for some species. *Rubus hibiscifolius* Focke was described from Nepal based on material in Copenhagen, but as for Hara (Enum. Fl. Pl. Nepal 2: 145. 1979) we were unable to trace any material corresponding to it and so it is not included in this account.

Key to Species

1a	Herbs or herb-like, prostrate or creeping, rooting at nodes	2
b	Shrubs, plants erect, arching, climbing or scandent, not rooting at nodes	6
2a	Leaves simple or both simple and 3-foliolate	3
b	Leaves compound, 3- or 5-foliolate	4
3a	Leaves all simple	**18.** *R. calycinus*
b	Leaves both simple and 3-foliolate	**32.** *R. x seminepalensis*
4a	At least some leaves pedately 5-foliolate	**29.** *R. franchetianus*
b	All leaves 3-foliolate	5
5a	Stems and petioles pilose, without bristles or prickles, sometimes with intermixed small glandular hairs	**30.** *R. fockeanus*
b	Stems and petioles with bristles, pilosulose and hispidulous	**31.** *R. nepalensis*
6a	Leaves both simple and 3-foliolate	**32.** *R. x seminepalensis*
b	Leaves simple or leaves pinnately, palmately or pedately compound, plants without both simple and compound leaves	7

7a	Leaves simple	8
b	Leaves pinnately, palmately or pedately compound	17
8a	Leaves unlobed, elliptic to elliptic-lanceolate, oblong to narrowly ovate, ovate-oblong or lanceolate	9
b	Leaves shallowly to conspicuously 3–11 lobed, narrowly ovate to suborbicular	11
9a	Leaves subcoriaceous to coriaceous, abruptly acuminate. Petioles 0.5–1 cm long	**22. R. hamiltonii**
b	Leaves membranaceous, long acuminate to caudate. Petioles 0.2–2 cm long	10
10a	Petiole 1–2 cm long. Leaves ovate-oblong or lanceolate. Fruit red	**23. R. acuminatus**
b	Petiole 0.2–1 cm long. Leaves elliptic to elliptic-lanceolate to oblong-lanceolate. Fruit black	**24. R. griffithii**
11a	Inflorescences terminal panicles, 8–40 × 5–35 cm. Leaves mostly longer than wide, some leaves unlobed but at least a few on each stem with very shallow lobes	12
b	Inflorescences axillary clusters or terminal racemes or slender panicles on short axillary shoots, 5–9 × 3–6 cm. Leaves ovate to suborbicular, some often as wide as long, obviously but sometimes shallowly 3–11 lobed	13
12a	Petioles and branchlets without glandular hairs. Petals 6–8 mm	**19. R. paniculatus**
b	Petioles and branchlets with at least a few glandular hairs. Petals 4–5 mm	**20. R. glandulifer**
13a	Stems and petioles densely stipitate glandular and with sparse needle-like prickles	**25. R. treutleri**
b	Stems and leaves variously pubescent, but without stipitate glands	14
14a	Stems, petioles and abaxial midrib of leaves unarmed or with few weak straight or recurved prickles	**26. R. kumaonensis**
b	Stems, petioles and abaxial midrib often with conspicuous straight, recurved or reflexed hard prickles	15
15a	Petioles 1.5–2.5 mm. Sepals 4 mm	**21. R. efferatus**
b	Petioles 2.5–11 mm. Sepals 8–15 mm	16
16a	Pedicels 2–15 mm long. Sepals 11–15 mm long, margins lacerate, apex acute or sometimes lacerate	**27. R. calycinoides**
b	Pedicels 2–7 mm long. Sepals 8–12 mm long, margins entire, apex acute to acuminate	**28. R. rugosus**
17a	Leaves palmately, digitately or pedately compound. Petiolules of terminal and lateral leaflets ± equal in length or all leaflets sessile. Leaflets 3 or 5	18
b	Leaves pinnately compound. Petiolule of terminal leaflet longer than petiolules of lateral leaflets. Leaflets 3–11	21
18a	Leaflets with up to 14 pairs of secondary veins	19
b	Leaflets with 20 or more pairs of secondary veins	20
19a	Calyx lobes 8–18 mm long, glabrous except for velutinous band along margin. Petals pink	**14. R. thomsonii**
b	Calyx lobes 3–5 mm long, with short straight spines. Petals white	**15. R. pentagonus**
20a	Branchlets, petioles, inflorescence branches and pedicels softly tomentose with soft appressed hairs, without long patent hairs. Leaves mostly 5-foliolate	**16. R. lineatus**
b	Branchlets, petioles, inflorescence branches and pedicels hispidulous with long patent gland tipped hairs. Leaves mostly 3-foliolate	**17. R. splendidissimus**
21a	Leaflets 3	22
b	Leaflets 5–11, some leaves with 3 leaflets	27
22a	Calyx lobes 3–5 mm long, with short straight spines, but otherwise glabrous. Petals white	**15. R. pentagonus**
b	Calyx lobes 4–20 mm long, variously pubescent or glabrous, with or without spines or prickles, if glabrous, then without spines. Petals white or pink	23
23a	Branchlets, petioles and inflorescence branches hispid and with dense to sparse sharply unguiculate spines. Leaflets widely elliptic to obovate	**3. R. ellipticus**
b	Branchlets petioles and inflorescence branches glabrous, pilose, softly stipitate glandular or with appressed hairs, with sparse straight or slightly unguiculate spines or prickles. Leaflets narrowly to widely ovate, rarely narrowly elliptic	24

24a   Lower surface of leaves glabrous or slightly softly hairy, or pubescent only along midvein, green ...... **11. *R. macilentus***
  b   Lower surface of leaves white velutinous or densely grey or yellowish grey tomentose...................................25

25a   Calyx and sepals glabrous. Lower surface of leaves densely grey or yellowish grey tomentose. Base of terminal leaflet broadly cuneate to rounded. Petals white, suborbicular, longer than sepals ....**5. *R. biflorus***
  b   Calyx and sepals pubescent. Lower surface of leaves white pubescent. Base of terminal leaflet cuneate, rounded, truncate to shallowly cordate. Petals white or pink, shorter than sepals 26

26a   Base of terminal leaflet cuneate, rounded or truncate; petals white, ca. 6 mm long.............................**6. *R. alexeterius***
  b   Base of terminal leaflet rounded to shallowly cordate; petals pink 4 × 3-4 mm ....**7. *R. hoffmeisterianus***

27a   Leaflets 3 or 5 28
  b   Leaflets (3–)7–11...29

28a   Inflorescences terminal on short axillary branches, 2–7 flowered or flowers solitary ....**2. *R. foliolosus***
  b   Inflorescences terminal and axillary, several to more than 20-flowered ....**4. *R. mesogaeus***

29a   Lower surface of leaves white or grey tomentose. Leaflets 5–11 ....**1. *R. niveus***
  b   Lower surface of leaves green or dark green, pubescent, glabrous or slightly pubescent along midvein. Leaflets 3–11 30

30a   Calyx and base of calyx lobes with dense straight spines...**10. *R. pungens***
  b   Calyx variously hairy, but without spines...31

31a   Inflorescences axillary, 1 or 2 flowered. Flowers 3–5 cm across...32
  b   Inflorescences terminal, corymbose, 3-several flowered. Flowers up to 2 cm across.....................................33

32a   Leaves without yellow glands abaxially softly hairy and with small prickles along veins, adaxially glabrous or with sparse appressed hairs ....**9. *R. amabilis***
  b   Leaves with yellow glands on both surfaces, abaxially pilose to subglabrescent and with sparse minute prickles along midvein, adaxially pilose to subglabrescent ....**13. *R. rosifolius***

33a   Glandular hairs absent. Aggregate fruit globose..**8. *R. inopertus***
  b   Glandular hairs present on branchlets, petioles, inflorescence rachis, pedicels and calyx. Aggregate fruit oblong or oblong-conic...**12. *R. sumatranus***

1. *Rubus niveus* Thunb., Rubo: 9 (1813).
Rubus bonatii H.Lév.; *R. bouderi* H.Lév.; *R. distans* D.Don; *R. foliolosus* var. *incanus* (K.Sasaki ex Y.C.Liu & T.Y.Yang) S.S.Ying; *R. incanus* K.Sasaki ex Y.C.Liu & T.Y.Yang; *R. lasiocarpus* Sm.; *R. lasiocarpus* var. *ectenothyrsus* Cardot; *R. lasiocarpus* var. *membranaceus* Hook.f.; *R. lasiocarpus* var. *micranthus* (D.Don) Hook.f.; *R. lasiocarpus* var. *pauciflorus* (Wall. ex Lindl.) Hook.f.; *R. longistylus* H.Lév.; *R. mairei* H.Lév.; *R. micranthus* D.Don; *R. mysorensis* F.Heyne ex Roth; *R. niveus* var. *micranthus* (D.Don) H.Hara; *R. niveus* var. *pauciflorus* (Wall. ex Lindl.) Focke; *R. pauciflorus* Wall. ex Lindl.; *R. pinnatus* D.Don; *R. pyi* H.Lév.; *R. rosiflorus* Roxb.; *R. rosiflorus* Roxb. nom. nud.; *R. tongchouanensis* H.Lév.

कालो ऐंसेलु Kalo ainselu (Nepali).

Scandent shrubs, 1–2.5 m. Branchlets purple or green, tomentose when young, soon glabrous, with whitish bloom, with sparse prickles. Stipules linear-lanceolate, 5–8 mm, persistent, softly hairy. Leaves (5–)7–9(–11)-foliolate, softly hairy along veins or glabrate above, white or grey tomentose below. Petiole 1.5–4 cm, petiolule of terminal leaflet 0.5–1.5 cm, lateral leaflets subsessile, petiolule and rachis tomentose, with sparse minute curved prickles. Blade of leaflets elliptic, ovate-elliptic or rhombic-elliptic, terminal leaflet ovate to elliptic, slightly longer than lateral leaflets, 2.5–6(–8) × 1–3(–4) cm, base cuneate to rounded, apex acute, rarely obtuse, terminal leaflet sometimes acuminate, margin coarsely serrate or doubly serrate, sometimes terminal leaflet 3-lobed. Inflorescences terminal or axillary, terminal corymbs, rarely short thyrses, many-flowered, 4–6 cm, axillary corymbs 1–few-flowered, rachis and pedicels tomentose. Bracts lanceolate or linear, ca. 5 mm, pubescent. Pedicels 0.5–1 cm. Flowers to 1 cm across. Outer surface of calyx densely tomentose, with intermixed soft hairs. Sepals erect, triangular-ovate or triangular-lanceolate, 5–8 × 2–3 mm, apex acute or abruptly pointed, rarely shortly acuminate. Petals red, suborbicular, 3–5 mm in diameter, sparsely hairy, shorter than sepals, base shortly clawed. Stamens nearly as long as petals; filaments broadened basally. Pistils ca. 55–70, nearly as long as stamens. Ovary grey tomentose. Styles purplish red, base densely grey tomentose. Aggregate fruit dark red when immature, black at maturity, semiglobose, 0.8–1.2 cm in diameter, enclosed by calyx, densely grey tomentose, style persistent. Pyrenes shallowly rugose.

Distribution: Nepal, E Himalaya, Tibetan Plateau, Assam-Burma, E Asia, SE Asia and SW Asia.

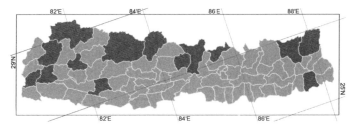

Altitudinal range: 1000–3100 m.

Ecology: Thickets on slopes, sparse forests, montane valleys, streamsides, flood plains.

Flowering: May–July. **Fruiting:** August–September.

Ripe fruits are eaten fresh. Leaf juice is used to treat fever.

2. Rubus foliolosus D.Don, Prodr. Fl. Nepal.: 256 (1825).
Rubus concolor Wall. nom. inval.; *R. gracilis* Roxb. later homonym, non J.Presl & C.Presl; *R. gracilis* Roxb. nom. nud.; *R. gracilis* var. *chiliacanthus* Hand.-Mazz.; *R. gracilis* var. *pluvialis* Hand.-Mazz.; *R. hypargyrus* Edgew.; *R. hypargyrus* var. *concolor* (Hook.f.) H.Hara; *R. hypargyrus* Edgew. var. *hypargyrus*; *R. hypargyrus* var. *niveus* (Wall. ex G.Don) H.Hara; *R. microphyllus* D.Don later homonym, non L.f.; *R. niveus* Wall. ex G.Don later homonym, non Thunb.; *R. niveus* Wall. nom. nud.; *R. niveus* var. *concolor* Hook.f.; *R. niveus* var. *hypargyrus* (Edgew.) Hook.f.; *R. niveus* var. *pedunculosus* (D.Don) Hook.f.; *R. pedunculosus* D.Don; *R. pedunculosus* var. *concolor* (Hook.f.) Kitam.; *R. pedunculosus* var. *hypargyrus* (Edgew.) Kitam.; *R. roylei* Klotzsch.

कालो ऐंसेलु Kalo ainselu (Nepali).

Straggling shrubs, 1–2 m. Branchlets brownish to reddish brown, initially pubescent, gradually glabrescent, with few prickles or nearly unarmed. Stipules linear or linear-lanceolate, 5–8 mm, persistent, softly hairy. Leaves 3-foliolate, rarely 5-foliolate, pubescent above, densely persistently white tomentose below, glabrescent. Petiole 3–5 cm, rachis pubescent, with sparse, minute prickles. Blade of leaflets ovate or broadly ovate, rarely ovate-lanceolate, 2.5–6 × 2–5 cm, base rounded to shallowly cordate, apex acute to acuminate, margin irregularly incised and roughly sharply doubly serrate. Inflorescences terminal on short axillary branches, corymbose, 2–5 cm, 2–7-flowered or flowers solitary. Pedicels 1.5–3.5 cm, softly hairy. Flowers 1.5–2 cm across. Calyx to ca. 1.5 cm, outer surface pubescent. Hypanthium pelviform. Sepals erect after anthesis, narrowly ovate to ovate-lanceolate, 1–1.2 cm × 3–6 mm, margin tomentose, apex long acuminate or caudate. Petals pink or red, ovate, 7–10 × 4–6 mm, shorter than sepals, base shortly clawed, apex slightly incised or entire. Stamens many. Pistils somewhat shorter than or nearly as long as

stamens. Ovary sericeous, softly hairy. Aggregate fruit initially yellowish orange, red or black at maturity, ovoid-globose.

Distribution: Nepal, W Himalaya, E Himalaya and Tibetan Plateau.

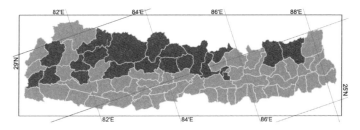

Altitudinal range: 1600–3200 m.

Ecology: Forested slopes, thickets, degraded woodland.

Flowering: March–November. **Fruiting:** April–November.

Smith (in Rees, Cylc. 30: Rubus n. 21. 1815) and Wallich (Numer. List.: 22 n. 736. 1829) misapplied the name *Rubus parvifolius* L. to this species.
   Rubus foliolosus is treated as a synonym of *R. niveus* Thunb. by Lu & Boufford (Fl. China 9: 205. 2003).
   Ripe fruits are eaten fresh. Immature fruits are chewed to relieve headache.

3. Rubus ellipticus Sm., in Rees, Cycl. 30(2): Rubus no.16 (1815).
Rubus ellipticus forma *obcordatus* Franch.; *R. ellipticus* subsp. *fasciculatus* (Duthie) Focke; *R. ellipticus* var. *fasciculatus* (Duthie) Masam.; *R. ellipticus* var. *obcordatus* (Franch.) Focke; *R. erythrolasius* Focke; *R. fasciculatus* Duthie; *R. flavus* Buch.-Ham. ex D.Don; *R. gowreephul* Roxb.; *R. gowryphul* Roxb. ex Wall. nom. nud.; *R. gowryphul* Roxb. nom. nud.; *R. obcordatus* (Franch.) Thuan; *R. pinfaensis* H.Lév. & Vaniot; *R. rotundifolius* Wall. nom. inval.; *R. wallichianus* Wight & Arn.

ऐंसेलु Ainselu (Nepali).

Scandent shrubs, 1–3 m. Branchlets purplish brown or brownish, hispid, with sparse curved prickles and dense purplish brown bristles or glandular hairs. Stipules linear, 7–11 mm, caducous, pubescent, with intermixed glandular hairs. Leaves 3-foliolate, pubescent along midvein above, densely tomentose below, with purplish red bristles along prominent veins. Petiole 2–6 cm, petiolule of terminal leaflet 2–3 cm, lateral leaflets subsessile, petiolule and rachis purplish red hispid, pubescent, with minute prickles. Blade of leaflets elliptic or obovate, 4–8(–12) × 3–6(–9) cm, terminal leaflet much larger than lateral leaflets, base rounded, apex acute abruptly pointed, shallowly cordate or subtruncate, margin unevenly minute sharply serrate. Inflorescences terminal, dense glomerate racemes, (1.5–)2–4 cm, flowers several–10 or more, or flowers several in clusters in leaf axils, rarely flowers solitary; rachis and pedicels hispid. Bracts linear, 5–9 mm, pubescent. Pedicels 4–6 mm. Flowers 1–1.5 cm across. Outer surface of calyx pubescent, intermixed yellowish tomentose, sparsely hispid. Sepals erect, ovate, 4–5(–6) ×

2–3(–4) mm, densely yellowish grey tomentose without, apex acute and abruptly pointed. Petals 6–8 × 3–5 mm, white or pink, spatulate, longer than sepals, margin premorse, densely pubescent, base clawed. Stamens shorter than petals; filaments broadened and flattened basally. Ovary pubescent. Styles glabrous, slightly longer than stamens. Aggregate fruit golden yellow, subglobose, ca. 1 cm in diameter, glabrous or drupelets pubescent at apex, styles persistent. Pyrenes triangular-ovoid, densely rugulose.

Distribution: Nepal, W Himalaya, E Himalaya, Tibetan Plateau, Assam-Burma, S Asia, E Asia and SE Asia.

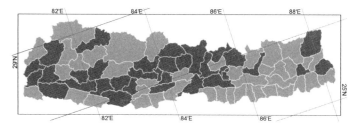

Altitudinal range: 300–2600 m.

Ecology: Dry slopes, montane valleys, sparse forests, thickets.

Flowering: March–April. **Fruiting:** April–May.

Rubus ellipticus may be separated into var. *ellipticus* and var. *obcordatus*. The former has elliptic leaflets with acute apices and hispid pedicels and calyces, while the latter has obovate leaflets with shallowly cordate or subtruncate apices and pedicels and calyces with few bristles.

A common plant of ruderal habitats with sweet tasting fruit that is sometimes preserved as a jam. The plant is astringent and tonic, and a concentrated decoction of the ripe fruit is taken for dysentery. Root paste is applied to wounds.

4. *Rubus mesogaeus* Focke, Bot. Jahrb. Syst. 29: 399 (1900).
Rubus eous Focke; *R. euleucus* Focke; *R. idaeus* var. *exsuccus* Franch. & Sav.; *R. illudens* H.Lév.; *R. kinashii* H.Lév. & Vaniot; *R. kinashii* forma *macrophyllus* Cardot; *R. kinashii* forma *microphyllus* Cardot; *R. mesogaeus* forma *floribusroseis* Focke; *R. mesogaeus* var. *incisus* Cardot; *R. niveus* var. *microcarpus* Hook.f.; *R. occidentalis* var. *exsuccus* (Franch. & Sav.) Makino; *R. occidentalis* var. *japonicus* Miyabe; *R. rarissimus* Hayata.

Shrubs, scandent, 1–4 m. Branchlets reddish brown or purplish brown, softly hairy, with sparse needle-like prickles or nearly unarmed. Old branches greyish brown, with sparse prickles broadened basally. Stipules linear, to 1.2 cm, pubescent, margin entire. Leaves often 3-foliolate, rarely 5-foliolate, appressed pubescent or glabrescent above, densely grey tomentose below. Petiole 3–7 cm, petiolule of terminal leaflet 1.5–4 cm, lateral leaflets shortly petiolulate or subsessile, petiolules and rachis pubescent, with sparse minute curved prickles. Blade of leaflets variable in shape, terminal leaflet broadly rhombic-ovate or elliptic-ovate,

base rounded to subcordate, apex acuminate, margin often pinnate-lobed, lateral leaflets obliquely elliptic or ovate, 4–9(–11) × 3–7(–9) cm, base cuneate to rounded, apex acute, margin unevenly coarsely serrate, often lobed. Inflorescences terminal or axillary, corymbose, (2–)3–4.5 cm, shorter than petiole, several- to more than 20-flowered; rachis and pedicels pubescent, with sparse needle-like prickles. Bracts linear, 7–10 mm; pubescent. Pedicels 6–12 mm. Flowers ca. 1 cm or more across. Outer surface of calyx densely pubescent. Sepals often reflexed after anthesis, lanceolate, (4–)5–8 × 3–4 mm, inner sepals with tomentose margin, apex acute to shortly acuminate. Petals white or pink, obovate, suborbicular or elliptic, 5–6 × 4–5 mm, premorse, base slightly pubescent and shortly clawed. Stamens about as long as petals. Ovary pilose. Styles glabrous. Aggregate fruit purplish black, compressed globose, 6–8 mm in diameter, glabrous. Pyrenes triangularly ovoid-globose, rugose.

Distribution: Nepal, E Himalaya, Tibetan Plateau, E Asia and N Asia.

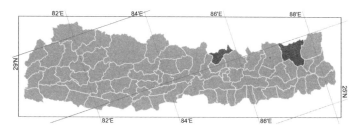

Altitudinal range: 900–3600 m.

Ecology: Slopes, forests in montane valleys, river banks.

Flowering: April–May. **Fruiting:** July–August.

Hooker (Fl. Brit. Ind. 2: 335. 1878) partially misapplied the name *Rubus niveus* Wall ex. G.Don to this species.

The plants in Nepal are of the typical variety, *R. mesogaeus* vars *glabrescens* T.T.Yu & L.T.Yu and var. *oxycomus* Focke are recorded from China.

5. *Rubus biflorus* Buch.-Ham. ex Sm., in Rees, Cycl. 30(2): Rubus no.9 (1815).
Rubus biflorus forma *parceglanduligera* Focke; *R. biflorus* var. *adenophorus* Franch.; *R. biflorus* var. *quinqueflorus* Focke.

सानु गुलाफ Sanu gulaph (Nepali).

Shrubs, scandent, 1–3 m tall. Branchlets purplish brown to brownish, glabrous, with sparse robust curved prickles and glaucous bloom, flowering branches are angled. Old branches terete. Stipules narrowly lanceolate, 6–8 mm, persistent, pubescent, with few stipitate glands. Leaves 3-foliolate, rarely 5-foliolate, appressed pubescent above, densely grey or yellowish grey tomentose, with sparse minute prickles along midvein below. Petiole 2–4(–5) cm; petiolule of terminal leaflet 1–2.5 cm, lateral leaflets subsessile, glabrous, rarely pilose, sparsely stipitate glandular. Blade of terminal leaflet broadly ovate or suborbicular, larger, 4–5 × 3–5 cm; blade of

lateral leaflets ovate or elliptic, 2.5–5 × 1.5–4(–5) cm, base broadly cuneate to rounded, apex acute or acuminate, margin unevenly coarsely serrate or doubly serrate, often deeply 3-lobed on terminal leaflet. Inflorescences terminal, rarely axillary, corymbs, 4–6 cm, often 4–8-flowered, or flowers 2– several in clusters in leaf axils; rachis and pedicels glabrous or pubescent, with needle-like prickles. Bracts linear or narrowly lanceolate, 4–7 mm, glabrous, rarely pilose. Pedicels (1–)2–3 cm. Flowers 1.5–2 cm across. Outer surface of calyx glabrous. Sepals erect, spreading at anthesis, broadly ovate or orbicular-ovate, 6–8 × 5–7 mm, apex acute, apiculate. Petals white, suborbicular, 7–8 mm in diameter, longer than sepals. Stamens shorter than petals; filaments linear, broader at base. Pistils somewhat shorter than stamens. Base of style and apex of ovary densely grey tomentose. Aggregate fruit enclosed in calyx, yellow, globose, 1–1.5(–2) cm in diameter, glabrous; drupelets apically with persistently tomentose styles. Pyrenes reniform, densely rugulose.
Fig. 45a–c.

Distribution: Nepal, W Himalaya, E Himalaya, Tibetan Plateau, Assam-Burma and E Asia.

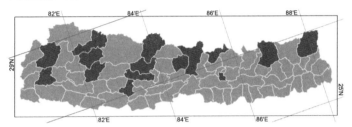

Altitudinal range: 1500–3500 m.

Ecology: Valleys, riversides, mixed forests and forest margins.

Flowering: May–June. **Fruiting:** July–August.

Two varieties may be recognized in *Rubus biflorus*: var. *biflorus* has glabrous petioles, pedicels and calyces, usually without stalked glands, while var. *adenophorus* has tomentose petioles, pedicels and calyces, sparsely set with stalked glands.
  Ripe fruits are eaten fresh.

6. *Rubus alexeterius* Focke, Notes Roy. Bot. Gard. Edinburgh 5: 75 (1911).
Rubus acaenocalyx H.Hara; *R. alexeterius* var. *acaenocalyx* (H.Hara) T.T.Yu & L.T.Lu.

Scandent shrubs, 1–2 m tall. Old branches reddish brown, glabrous, glaucous and 6–8 mm curved prickles. Flower-bearing branchlets short, densely villous, with minute curved prickles. Stipules linear or subulate, 4–6 mm, villous, persistent. Leaves 3-foliolate, rarely 5-foliolate, appressed villous above, densely grey tomentose below. Petiole 2.5–3.5 cm, petiolule of terminal leaflet 0.5–1 cm, lateral leaflets subsessile, petiolules and rachis densely villous, with sparse minute curved prickles. Terminal leaflet rhombic, rarely ovate, larger than others, 4–5 × 2.5–3.5 cm, base cuneate, rounded or truncate, apex acute, incised lobed; lateral

leaflets ovate or elliptic, 3–4(–5) × 1.5–3 cm, base cuneate to rounded, apex acute, margin irregularly sharply serrate or incised doubly serrate above middle, sometimes 3-lobed or incised-lobed on terminal leaflet. Inflorescences clusters of 3 or 4 flowers at apex of short lateral branchlets, or flowers solitary in leaf axils. Pedicels 1–2(–3) cm, villous, with slender prickles. Flowers 1.5–2 cm across. Calyx 1.5–2 cm long, villous without, with needle-like prickles; tube pelviform. Sepals erect, spreading, ovate-lanceolate or lanceolate, unequal in length, longer sepals to 1.5 cm, shorter sepals 6–10 mm, occasionally with sparse glandular hairs, apex caudate and enlarged, sometimes divided. Petals white, suborbicular, ca. 6 mm, shorter than sepals, base shortly clawed. Stamens shorter than petals; filaments linear or basally somewhat broadened. Pistils many, shorter than stamens. Ovary glabrous or tomentose only at apex. Styles densely white tomentose basally. Aggregate fruit yellow, globose, 1.2–1.5(–2) cm in diameter, glabrous, enclosed in calyx; drupelets apically with persistently tomentose styles. Pyrenes reniform, shallowly rugulose.
Fig. 45d–f.

Distribution: Nepal, E Himalaya, Tibetan Plateau and E Asia.

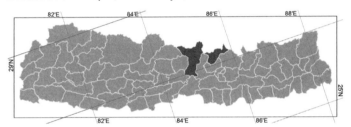

Altitudinal range: 2000–3700 m.

Ecology: Montane valleys, streamsides, degraded slopes, forest clearings and forest margins.

Flowering: April–May. **Fruiting:** June–July.

Two varieties can be recognized in *Rubus alexeterius*: var. *alexeterius* has petioles, pedicels and calyces, without glandular hairs, while var. *acaenocalyx* has petioles, pedicels and calyces with prominent glandular hairs.

7. *Rubus hoffmeisterianus* Kunth ex Bouché, Index Seminum Hort. Bot. Berol. 1847: 14.
Rubus niveus var. *aitchisoni* Hook.f.

Shrubs, erect or reclining, 1–1.5 m. Branchlets green, pilose and stipitate glandular. Stipules linear, ca. 1.5 mm, sericeous and stipitate glandular. Leaves pinnately trifoliolate, with sparse, erect straight hairs above, below white velutinous, midrib with long straight hairs, with or without few stiff, straight or hooked prickles. Petiole 2–5 cm; petiolule of terminal leaflet 1–5 cm; petiolule of lateral leaflets 1–3 mm, white sericeous and stipitate glandular. Blade of terminal leaflet ovate to widely ovate, shallowly few-lobed or not, 3–8 × 2.5–7 cm, base rounded to shallowly cordate; blade of lateral leaflets ovate, only slightly oblique, 1.5–5 × 1–4 cm, base cuneate to widely cuneate, apex acute, margin doubly dentate. Inflorescences

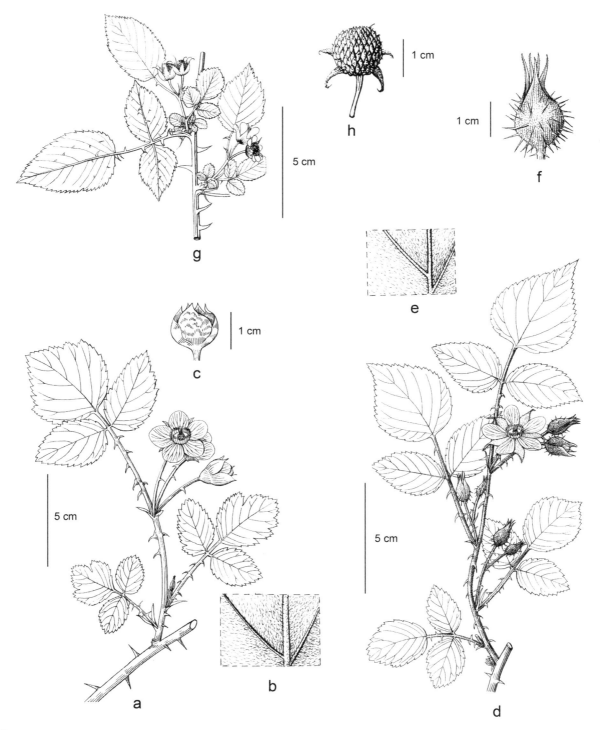

FIG. 45. Rosaceae. **Rubus biflorus**: a, inflorescence and leaves; b, lower surface of leaf; c, fruit with persistent calyx. **Rubus alexeterius**: d, inflorescence and leaf; e, lower surface of leaf; f, fruit with persistent calyx. **Rubus macilentus**: g, inflorescence with fruits and leaves; h, fruit with persistent calyx.

axillary and terminal, racemes or panicles, 2–7 cm, 2–15-flowered, rachis and pedicels pilose and stipitate glandular and with stiff hooked prickles. Bracts linear, white tomentose and stipitate glandular. Pedicels 2–8 cm. Flowers ca. 1.2 cm across. Outer surface of calyx white tomentose and pilose. Sepals spreading at anthesis, ovate, 1–1.5 cm, apex acuminate. Petals pink, widely obovate to rounded, clawed, ca. 4 × 3–4 mm, shorter than sepals. Stamens about equalling petals; filaments linear, tapering from base to apex. Pistils longer than stamens. Base of style and apex of ovary white sericeous. Aggregate fruit enclosed by calyx, orangish yellow, ellipsoid, 1–2 cm long, style of drupelets persistent, white pilose. Pyrenes ovoid, reticulate.

Distribution: Nepal, W Himalaya and SW Asia.

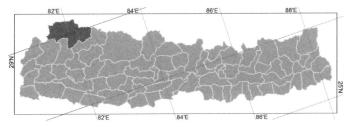

Altitudinal range: 1600–2300 m.

Ecology: Shaded slopes, clearings.

Flowering: May–June. **Fruiting:** August–September.

8. *Rubus inopertus* (Focke) Focke, Biblioth. Bot. 72: 182 (1911).
Rubus niveus subsp. *inopertus* Focke, Bot. Jahrb. Syst. 29(3-4): 400 (1900); *R. lasiocarpus* var. *rosifolius* Hook.f.; *R. niveus* var. *rhodophyllos* Focke; *R. niveus* var. *rosifolius* (Hook.f.) H.Hara.

Shrubs scandent, climbing to 2 m. Branchlets brownish to purplish brown, slender, angled, puberulous or glabrescent, armed with sparse, backward-curved prickles ca. 5 mm. Stipules linear-lanceolate, entire, 2–4(–5) mm, glabrous or margin puberulous. Leaves imparipinnate, (5–)7–9(–11)-foliolate, pubescent along veins above, pilose below. Petiole (1.5–)3.5–6.5 cm, purplish brown, petiolule of terminal leaflet 0.6–3 cm, lateral leaflets subsessile, petiolules and rachis moderately pubescent, armed with sparse, curved prickles ca. 2 mm. Leaflet blade ovate to narrowly ovate, (1.4–)3.5–7(–9) × (0.7–)1–3(–4) cm, base rounded or subtruncate, apex acute to acuminate, margin coarsely sharply double serrate. Terminal leaflets larger and more acuminate than laterals, sometimes imperfectly separated from last pair of lateral leaflets or appearing 3-lobed. Inflorescences terminal or axillary, several-flowered corymbs or clusters, 2–3 cm in diameter, sometimes flowers solitary in axils, rachis and pedicels densely puberulous. Bracts linear-lanceolate, 1.5–3 mm, densely puberulous. Pedicels 4–7 mm. Flowers 0.8–1.2 cm across. Outer surface of calyx glabrous. Sepals erect to spreading at anthesis, reflexed in fruit, narrowly ovate or triangular-ovate, 4–7 × 2–4 mm, inner sepals apex

acute to acuminate, margin white tomentose. Petals pink to dark pink, broadly obovate, 3–4 × 4–5 mm, base shortly clawed, margin and apex irregularly toothed, glabrous. Stamens numerous, 2–4 mm, filaments linear, sometimes broadened at base. Pistils 4–5 mm. Ovary and base of style densely white woolly pubescent. Aggregate fruit purplish black, globose, 8–9 mm in diameter, thinly pubescent. Pyrenes rugulose.

Distribution: Nepal, E Asia and SE Asia.

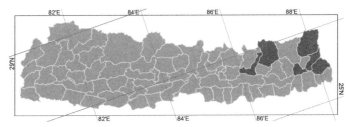

Altitudinal range: 2400–2600 m.

Ecology: River banks.

Flowering: June. **Fruiting:** July.

A poorly understood and under-collected species in Nepal, only known from a few specimens. The Nepalese material differs slightly from the description given by Lu & Boufford (Fl. China 9: 218. 2003) as the stems and inflorescence branches are moderately to densely puberulent. The Nepalese plants match the type specimen, *Hooker s.n.* (K) from Sikkim. Description of the fruit has been taken from Chinese material. Amongst the synonyms the spelling of the epithet as '*rosaefolius*' is often found in older literature

9. *Rubus amabilis* Focke, Bot. Jahrb. Syst. 36(Beibl.82): 53 (1905).

Shrubs, 1–3 m. Branches purplish brown or dark brown, glabrous, with sparse prickles. Flower-bearing branchlets short, softly hairy, with minute prickles. Stipules linear-lanceolate, 5–8 mm, softly hairy. Leaves 7–11-foliolate, glabrous or with sparse appressed hairs above, softly hairy and with small prickles along veins below. Petiole 1–3 cm; petiolule of terminal leaflet absent or to ca. 1 cm, lateral leaflets subsessile, petiolule and rachis pubescent when young, gradually glabrescent, glabrous or subglabrescent, with small sparse prickles. Blade of leaflets ovate or ovate-lanceolate, 1–5.5 × 0.8–2.5 cm, usually upper ones larger than lower ones, base rounded, sometimes subcuneate on terminal leaflet, apex acute, often acuminate on terminal leaflet, margin incised to doubly serrate, occasionally terminal leaflet 2- or 3-lobed or fused basally with upper lateral leaflets. Flowers solitary, terminal on lateral branchlets, pendent. Pedicels 2.5–6 cm, softly hairy, with sparse minute prickles, sometimes with intermixed sparse stipitate glands. Flowers 3–4 cm across. Calyx green, tinged red, outer surface pubescent, unarmed, rarely with sparse short needle-like prickles or stalked glands; tube pelviform. Sepals spreading, broadly ovate, 1–1.5 × 0.5–0.8 cm, apex acuminate or abruptly pointed. Petals white, suborbicular, 1–1.7 cm in diameter, longer than or nearly as long as sepals, base shortly clawed and pubescent. Stamens slightly shorter than petals; filaments whitish, linear, broadened basally. Pistils shorter than stamens. Ovary pubescent. Styles greenish, glabrous. Aggregate fruit red, oblong, rarely ellipsoid, 1.5–2.5 × 1–1.2 cm, sparsely pubescent when young, glabrescent. Pyrenes reniform, somewhat reticulate.

Distribution: Nepal, Tibetan Plateau and E Asia.

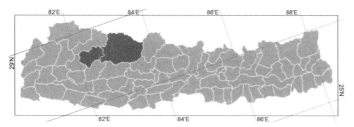

Altitudinal range: 1000–3800 m.

Ecology: Ravines, montane valleys, slopes, forests, thickets, forest margins, roadsides.

Flowering: April–May. **Fruiting:** July–August.

The fruits are edible.

10. Rubus pungens Cambess., in Jacquem., Voy. Inde 4(1): 48 (1841).
Rubus pungens var. *discolor* Prochanov; *R. pungens* var. *fargesii* Cardot.

Shrubs, to 3 m. Branchlets brownish to reddish brown, terete, pubescent, gradually glabrescent, glabrous when young, usually with dense needle-like prickles. Stipules linear, 4–7 mm, pubescent. Leaves (3–)5–7(–9)-foliolate, pubescent above, pubescent below especially along midvein and veins. Petiole (2–)3–6 cm; petiolule of terminal leaflet 0.5–1 cm, lateral leaflets subsessile, petiolule and rachis pubescent or subglabrous, with sparse minute prickles and with glandular hairs. Blade of leaflets ovate, triangular-ovate or ovate-lanceolate, (1–)2–5(–6) × 1–3 cm, base rounded to subcordate, apex acute to shortly acuminate, but usually acuminate on terminal leaflet, margin sharply or incised doubly serrate, terminate leaflet often pinnately lobed. Inflorescences terminal or axillary, with flowers solitary or rarely corymbose and 2–4-flowered; rachis and pedicels pubescent, with minute prickles. Bracts linear, 4–6 mm, pubescent. Pedicels 2–3.5 cm. Flowers 1–2 cm across. Outer surface of calyx pubescent, with dense needle-like prickles; tube semiglobose. Sepals erect, rarely reflexed, linear to lanceolate or triangular-lanceolate, 0.8–1.2(–2) × 0.2–0.4 cm, apex long acuminate. Petals white, oblong, obovate or suborbicular, 7–9 × 5–6 mm, much shorter than sepals, glabrous, base clawed. Stamens unequal in length, longer ones longer than petals, shorter ones nearly as long as or slightly shorter than petals; filaments somewhat broadened and flattened basally. Pistils numerous. Ovary softly hairy or subglabrous. Styles glabrous or sparsely softly hairy basally. Aggregate fruit red, subglobose, 1–1.5 cm in

diameter, pubescent or subglabrous. Pyrenes ovoid-globose, conspicuously rugose.

Distribution: Nepal, W Himalaya, E Himalaya, Tibetan Plateau, Assam-Burma and E Asia.

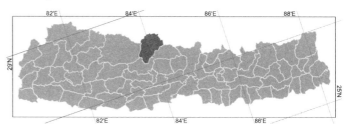

Altitudinal range: 2200–3300 m.

Ecology: Forested slopes, forest margins, riversides.

Flowering: April–May. **Fruiting:** July–August.

Represented by var. *pungens* in Nepal with several additional varieties in China, Japan and Korea.

11. Rubus macilentus Cambess., in Jacquem., Voy. Inde 4(1): 49 (1841).
Rubus minensis Pax & K.Hoffm.; *R. trichopetalus* Hand.-Mazz.; *R. uncatus* Wall. nom. nud.

जोगी ऐंसेलु Jogi ainselu (Nepali).

Shrubs, 1–2 m tall. Branchlets brown or reddish brown, terete, villous, with unequal long straight, reflexed or recurved prickles. Stipules linear or linear-lanceolate, 4–6 mm, softly hairy. Leaves 3-foliolate, rarely simple, glabrous above or slightly softly hairy, glabrous below or slightly softly hairy, with sparse minute prickles only along veins. Petiole 0.8–1(–1.5) cm; petiolule of terminal leaflet 0.5–1 cm, lateral leaflets subsessile, pubescent, with minute prickles. Blade of leaflets lanceolate, ovate-lanceolate or ovate, terminal leaflet much longer than lateral leaflets, 3–5 × 1–2.5 cm, lateral leaflets 1–2 × 0.7–1.4 cm, base rounded or broadly cuneate, margin sharply serrate, apex acute, rarely obtuse, often shortly acuminate on terminal leaflet. Inflorescences terminal on short lateral branchlets, 1–3-flowered. Bracts lanceolate or linear-lanceolate, somewhat smaller than stipules, softly hairy. Pedicels 0.6–1 cm, villous, sometimes with sparse minute prickles. Flowers ca. 1 cm across. Outer surface of calyx villous. Sepals erect, rarely spreading, lanceolate or triangular-lanceolate, 6–8 × 2–3(–4) mm, apex short caudate. Petals white, broadly ovate to oblong, slightly longer than or nearly as long as sepals, both surfaces softly hairy, base clawed. Stamens shorter than sepals; filaments broadened. Pistils many, slightly shorter than stamens. Base of style and upper part of ovary sparsely villous. Aggregate fruit orange or red, subglobose, glabrous or somewhat softly hairy, enclosed in calyx. Pyrenes globose, deeply reticulate.
Fig. 45g–h.

Distribution: Nepal, W Himalaya, E Himalaya, Tibetan Plateau and E Asia.

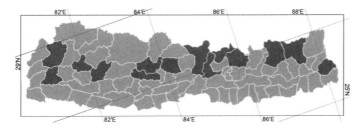

Altitudinal range: 900–3300 m.

Ecology: Slopes, roadsides, near watercourses, forest margins.

Flowering: April–May. **Fruiting:** July–August.

The plants in Nepal are the typical variety: var. *angulatus* Franch. has prominently angled branchlets and is found in Yunnan.

Ripe fruits are eaten fresh.

12. *Rubus sumatranus* Miq., Fl. Ned. Ind., Eerste Bijv. [2]: 307 (1860).
Rubus asper Wall. ex D.Don later homonym, non J.Presl & C.Presl; *R. asper* var. *grandifoliolatus* (H.Lév.) Focke; *R. asper* var. *myriadenus* (H.Lév. & Vaniot) Focke; *R. asper* var. *pekanius* Focke; *R. dolichocephalus* Hayata; *R. indotibetanus* Koidz.; *R. myriadenus* H.Lév. & Vaniot; *R. myriadenus* var. *grandifoliolatus* H.Lév.; *R. rosifolius* subsp. *sumatranus* (Miq.) Focke; *R. rosifolius* var. *asper* (Wall. ex D.Don) Kuntze; *R. somai* Hayata; *R. sorbifolius* Maxim.; *R. takasagoensis* Koidz.

Shrubs, erect or scandent, to 2 m. Branchlets brownish to dark reddish brown, terete, long softly hairy, usually with scattered setose purplish red glandular hairs and curved prickles; glandular hairs and prickles unequal in length, glandular hairs to 4–5 mm, prickles to 8 mm. Stipules lanceolate or linear-lanceolate, 6–8 mm, softly hairy, with intermixed glandular hairs. Leaves to 15 cm, 5–7-foliolate, rarely 3-foliolate, villous above, especially along midvein, villous below with gland-tipped hairs and small prickles along midvein. Petiole 3–5 cm; petiolule of terminal leaflet to 1 cm; lateral leaflets shortly petiolulate, petiolule and rachis softly hairy, with intermixed glandular hairs and minute curved prickles. Blade of leaflets ovate-lanceolate to lanceolate, 3–8 × 1.5–3 cm, base rounded, apex acuminate, margin irregularly sharply serrate to doubly serrate, sometimes 3-lobed on terminal leaflet. Inflorescences terminal, corymbose, 4–7 cm, 3- to several-flowered, rarely flower solitary; rachis and pedicels villous, with intermixed long glandular hairs and minute prickles. Bracts lanceolate or linear, 5–7 mm, softly hairy, with intermixed glandular hairs. Pedicels 2–3 cm. Flowers 1–2 cm across. Outer surface of calyx with soft hairs, intermixed with long, unequal, gland-tipped hairs. Sepals reflexed in fruit, lanceolate or oblong-lanceolate, 7–12 × 2–4 mm; apex long caudate. Petals white, narrowly obovate or spatulate, slightly shorter than sepals, base clawed. Stamens shorter than petals. Pistils many. Style and ovary glabrous. Torus raised, oblong, base shortly stalked. Aggregate fruit orange-red, oblong to oblong-conic, 1.2–1.8 × 0.7–1.1 cm, glabrous. Pyrenes reticulate.

Distribution: Nepal, E Himalaya, Tibetan Plateau, Assam-Burma, E Asia and SE Asia.

Altitudinal range: 700–2500 m.

Ecology: Forests, forest margins, thickets, bamboo forests, grasslands.

Flowering: April–June. **Fruiting:** July–August.

Hooker (Fl. Brit. Ind. 2: 34. 1878) partially misapplied the name *Rubus rosifolius* Sm. to this species.
Rubus sumatranus was reported from Nepal by Lu & Boufford (Fl. China 9: 225. 2003). Although we have not seen specimens to substantiate that report, *R. sumatranus* has been collected in Sikkim and Darjeeling is very likely also to be in Nepal and so is in included here as a full entry without distribution map. Information for the description is taken from Fl. China and E Himalayan specimens.

13. *Rubus rosifolius* Sm., Pl. Icon. Ined.(3): pl. 60 (1791).
Rubus coronarius (Sims) Sweet; *R. rosifolius* forma *coronarius* (Sims) Focke; *R. rosifolius* forma *coronarius* (Sims) Kuntze; *R. rosifolius* var. *coronarius* Sims.

रातो ऍसेलु Rato ainselu (Nepali).

Shrubs, erect or climbing, 2–3 m. Branchlets greyish brown or dark reddish brown, terete, softly hairy or subglabrous, with straight to curved prickles and yellowish glands. Stipules linear or lanceolate to ovate-lanceolate, 8–12 × 1.5–3.5 mm, sparsely softly hairy. Leaves usually 7–11-foliolate, pilose to subglabrescent above, with yellow glands, pilose to subglabrescent below and with sparse minute prickles along midvein, with yellow glands. Petiole 2–3 cm; petiolule of terminal leaflet 0.8–1.5 cm, lateral leaflets subsessile, petiolule and rachis softly hairy, with sparse, minute prickles, sometimes subglabrous, with yellowish glands. Blade of leaflets ovate or ovate-elliptic to lanceolate, 4–7(–10) × 1.5–5 cm, base rounded, apex acuminate, margin sharply incised doubly serrate or coarsely doubly serrate. Inflorescences terminal or in leaf axils, 1- or 2-flowered. Bracts linear or lanceolate, 5–9 mm, puberulous. Pedicels (1–)2–3.5 cm, ± softly hairy, with sparse, minute prickles, sometimes glandular. Flowers 3–5 cm across. Outer surface of calyx softly hairy and glandular. Sepals erect before anthesis, reflexed after anthesis, triangular-lanceolate or ovate-lanceolate, 0.8–1.2(–1.4) × 0.4–0.6 cm, apex long caudate. Petals white, oblong, narrowly obovate or suborbicular, 0.8–1.5 × 0.8–1.2 cm, short hairy outside, base clawed, apex obtuse. Stamens shorter than petals; filaments broad. Pistils to 2 mm, shorter than stamens. Ovary glabrous, sometimes glandular. Styles glabrous. Torus shortly stalked. Aggregate fruit red, ovoid-globose or narrowly obovoid to oblong, 1–1.5 × 0.8–1.2 cm, glabrous, with few glands. Pyrenes deeply foveolate. Fig. 46a–c.

Distribution: Nepal, E Himalaya, Assam-Burma, E Asia, Africa and Australasia.

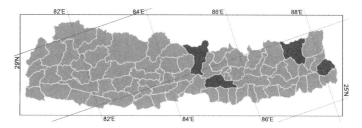

Altitudinal range: 600–2500 m.

Ecology: Mixed forests, forest margins, grassy slopes, roadsides, landslides. Often cultivated.

Flowering: March–May. **Fruiting:** June–July.

Originally published using the spelling 'rosaefolius', and sometimes listed as this in the older literature. The double-flowered form, with fragrant flowers, originally described from cultivated plants in England, and now occasionally cultivated for ornamental use throughout SE Asia, is named *Rosa rosifolius* forma *coronarius* (Fl. Bhutan 1: 562. 1987). *Rosa rosifolius* may not be truly wild in Nepal.

14. *Rubus thomsonii* Focke, Abh. Naturwiss. Vereine Bremen 4: 198 (1874).

ऐंसेलु Ainselu (Nepali).

Shrubs, prostrate, climbing or arching, height unknown. Branchlets green, glabrous or strigillose, often in lines, with recurved or straight hard prickles. Stipules linear to linear-lanceolate, entire or with a few long lacerations, ca. 5 mm, sparsely pubescent. Leaves pinnately 3-foliolate, sparsely strigillose above, nearly glabrous below with a few hooked prickles along midvein. Petiole 1–4 cm; petiolule of terminal leaflet 2–4 mm; petiolule of lateral leaflets ca. 1 mm, strigillose, with a few stiff hooked prickles. Blade of terminal leaflet rhomboid to elliptic, 6–12 × 3–5 cm, base cuneate, apex acute to long acuminate; blade of lateral leaflets obliquely ovate, 2.5–5 × 1–2.5 cm, base cuneate, apex acute to long acuminate, margin irregularly and coarsely doubly dentate to doubly serrate. Inflorescences axillary, panicles, racemes or flowers solitary, 2–5 cm, (1–)3–20-flowered; rachis and pedicels pilosulose, with few stiff recurved prickles. Bracts linear, entire or with 1–few lacerations, glabrous or with few hairs. Pedicel 0.7–2.5 cm. Flowers 1.2–2.4 cm across. Outer surface of calyx glabrous except for velutinous band along margin. Sepals spreading at anthesis, ovate to broadly ovate, 3–5 mm, apex acuminate. Petals pink, broadly elliptic to widely ovate, ca. 3 × 2 mm, shorter than sepals. Stamens equalling or slightly longer than petals; filaments linear, uniform in width, except narrower below anther. Pistils longer than stamens. Base of style and ovary pilose. Aggregate fruit not enclosed by calyx, red, globose or depressed globose, 0.5–0.8 cm in diameter, pilose; style of drupelets persistent, pilose. Pyrenes oblong to lunate, thickened, surface smooth or with a few raised lines.

Distribution: Nepal and E Himalaya.

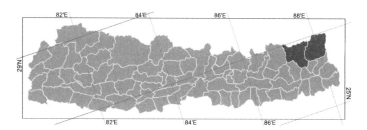

Altitudinal range: 2600–3700 m.

Ecology: Forests, thickets, clearings.

Flowering: August–October. **Fruiting:** October–December.

Ripe fruits are edible.

15. *Rubus pentagonus* Wall. ex Focke, Biblioth. Bot. 72: 145 (1911).
Rubus alpestris Blume.

लेख ऐंसेलु Lekh ainselu (Nepali).

Shrubs, scrambling, 1.5–3 m. Branchlets climbing, brownish to dark brown, slightly pubescent when young, later glabrous, with minute or needle-like prickles and stalked glands. Stipules linear-lanceolate, 7–10 mm, persistent, puberulous, margin with shortly stalked glands, margin entire or deeply 2-laciniate. Leaves palmately 3- or 5-foliolate, pilose along veins above, sparsely pubescent below. Petiole 2–4 cm, sparsely pubescent, with stalked glands and minute prickles, rarely without stalked glands, leaflets sessile. Blade of leaflets rhombic-lanceolate, 3–8(–11) × 1.5–4 cm, base cuneate, apex acuminate to caudate, margin coarsely incised to doubly serrate. Inflorescences terminal, corymbose, 2–3 cm long, 2- or 3-flowered or with flowers solitary in leaf axils. Bracts linear-lanceolate, 6–9 mm, often with stalked glands, margin entire or 2- or 3-laciniate. Pedicel 1.5–2.5 cm, glabrous, with sparse gland-tipped hairs and small prickles. Flowers 1.5–2 cm across. Calyx glabrous, outer surface glandular pubescent, with minute prickles. Sepals erect, spreading, lanceolate or ovate-lanceolate, 8–15 × 5–8 mm, margin entire or 3-laciniate, inner sepals with tomentose margin, apex acuminate or caudate. Petals white, elliptic or oblong, much shorter than sepals, base shortly clawed. Stamens with broad filaments. Pistils 10–15, slightly shorter than stamens. Ovary and style glabrous. Aggregate fruit red or orange-red, subglobose, to 2 cm in diameter, glabrous, enclosed in calyx. Pyrenes reniform, to 4 mm, rugose.

Distribution: Nepal, W Himalaya, E Himalaya, Tibetan Plateau, Assam-Burma, E Asia and SE Asia.

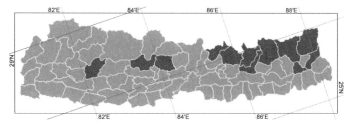

Altitudinal range: 2000–3100 m.

Ecology: Evergreen forests, mixed forests, thickets.

Flowering: April–May. **Fruiting:** June–July.

Hooker (Fl. Brit. Ind. 2: 332. 1878) misapplied the name *Rubus alpestris* Blume to this species.

Nepalese plants are the typical variety, several additional varieties occur in China.

Ripe fruits are eaten fresh.

16. *Rubus lineatus* Reinw., in Blume, Bijdr. Fl. Ned. Ind. [17]: 1108 (1827).
Rubus pulcherrimus Hook.

घ्याम्पे ऐंसेलु Ghyampe ainselu (Nepali).

Shrubs, 1–2 m, much-branched. Branchlets brownish to greyish brown, terete, with sparse minute prickles, with dense appressed silvery-grey or yellowish grey silky hairs, glabrescent. Stipules caducous, free, lanceolate or ovate-oblong, 1.2–2 cm, sometimes to 2–3 cm on sterile branchlets, often broad, membranaceous, abaxially densely sericeous, not divided. Leaves palmately compound, 3–5-foliolate, glabrous above or long hairy along midvein, densely silvery-grey or yellowish grey appressed-sericeous below. Petiole 2–5 cm, sericeous, lateral leaflets sessile or subsessile. Blade of leaflets oblong or lanceolate to oblanceolate, 8–12 × 1.5–3.5 cm, base cuneate, apex acuminate to caudate, margin sharply serrate to doubly serrate, pinnately veined with (20–)30–50 pairs of parallel lateral veins terminating at margin, midvein and lateral veins impressed above, prominent below. Inflorescences terminal and in axils of upper leaves, cymose panicles, ca. 15–20-flowered, sometimes flowers in clusters in leaf axils; rachis and pedicels sericeous or glabrescent. Bracts lanceolate or ovate-oblong, smaller than stipules, sericeous. Pedicels 1–2 cm. Flowers ca. 1.5 cm across. Outer surface of calyx densely silvery-grey or yellowish grey, sericeous or glabrescent. Sepals ovate-triangular to ovate-lanceolate, 6–12 × 3–7 mm, apex acuminate to caudate, margin entire. Petals white or greenish white, elliptic or obovate, much smaller than sepals, glabrous, base not distinctly clawed. Stamens somewhat shorter or as long as petals. Pistils ca. 80–100 or more, shorter than stamens. Lower part of style and apical part of ovary long hairy. Aggregate fruit orange to red at maturity, semiglobose or globose-ovoid, 7–10 mm in diameter, sericeous when young, glabrescent. Pyrenes distinctly rugose.

Distribution: Nepal, E Himalaya, Tibetan Plateau, Assam-Burma, E Asia and SE Asia.

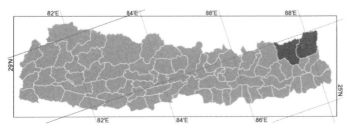

Altitudinal range: 1400–3100 m.

Ecology: Slopes, valleys, forests, margins of forests, fallow fields.

Flowering: July–August. **Fruiting:** September–October.

The plants in Nepal are all the typical variety. The other two varieties are restricted to Yunnan.

17. *Rubus splendidissimus* H.Hara, J. Jap. Bot. 40: 327 (1965).
Rubus andersonii Hook.f. later homonym, non Lefèvre.

चाँदे ऐंसेलु Chande ainselu (Nepali).

Shrubs, erect or scrambling, to 3 m tall. Branchlets green or purplish, pubescence glandular hispid and velutinous-sericeous. Stipules lanceolate, 0.7–2 cm, outer surface sericeous and glandular hispid, inner surface glabrous. Leaves pinnately 3-foliolate, with scattered erect hairs above, white or silvery sericeous below, unarmed along midvein. Petiole 5–10 cm; petiolule of terminal leaflet 0.1–0.7 cm; petiolule of lateral leaflets absent to 2 mm, glandular hispid and sericeous. Blade of terminal leaflet elliptic to widely elliptic to obovate, 5–18 × 4–11 cm, base cuneate with straight or slightly concave sides, apex acuminate, margin doubly serrate; blade of lateral leaflets obliquely elliptic to obliquely widely elliptic or obliquely oblong, 5.5–10 × 2–6 cm, midrib with appressed straight hairs, base rounded, apex acuminate, margin doubly serrate, many prominent parallel veins. Inflorescences terminal and in upper axils, panicles or racemes, 3–15 cm, 2–many-flowered; rachis and pedicels glandular hispid and sericeous-velutinous, unarmed. Bracts persistent, elliptic or lanceolate-elliptic, 5–15 mm, outer surface glandular hispid and velutinous, glabrous within. Pedicels 0.6–2.5 cm. Flowers 2–3 cm across. Outer surface of calyx glandular hispid and velutinous. Sepals spreading at anthesis, triangular to narrowly triangular or triangular-lanceolate, 1.5–2 cm, apex acuminate to caudate. Petals white, widely elliptic to obovate, ca. 0.7–1cm in diameter, shorter than sepals. Stamens shorter than petals; filaments linear, broadened toward base. Pistils longer than stamens. Base of style and apex of ovary glabrous. Aggregate fruit exposed, orange, orangish red or red, globose, 1–1.3 cm in diameter, glabrous; style of drupelets persistent, glabrous. Pyrenes half round or lunate, thickened, surface shallowly reticulate.

Distribution: Nepal and E Himalaya.

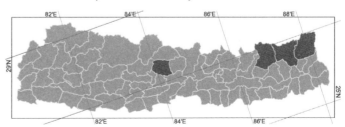

Altitudinal range: 2200–3100 m.

Ecology: Thickets, forests.

Flowering: June–October. **Fruiting:** September–November.

Ripe fruits are edible.

18. *Rubus calycinus* Wall. ex D.Don, Prodr. Fl. Nepal.: 235 (1825).
Dalibarda calycinus (Wall. ex D.Don) Ser.; *Rubus lobatus* Wall. nom. nud.

Herbs, creeping, 15–20 cm, main stems creeping, to 2–3 m long, rooting at nodes, with erect, sparsely branched or unbranched lateral branches. Stems with sparse needle-like prickles or nearly unarmed, with sparse hairs. Stipules ovate, rarely obovate, 8–13 × 6–11 mm, persistent, margin shallowly coarsely serrate, rarely entire. Leaves simple, pilose above when young, gradually glabrescent, pilose below when young, gradually glabrescent, hairy only along veins in age, with needle-like prickles along veins. Petiole 5–10 cm, villous, with needle-like prickles. Leaf blade orbicular-ovate or suborbicular, 2.5–6 cm in diameter, base deeply cordate, apex obtuse or rounded, margin undulate or shallowly 3–5-lobed, irregularly coarsely serrate. Inflorescences usually terminal, 1- or 2-flowered. Bracts ovate, 6–10 × 5–9 mm, coarsely serrate, rarely entire. Pedicel 3–5 cm, usually villous, with needle-like prickles. Flowers to 3 cm across. Outer surface of calyx softly hairy, with straight subulate prickles; tube broadly pelviform, 3.5–5 mm in diameter. Sepals leaf-like, narrowly ovate to elliptic, 0.8–1.4 × 0.6–1.1 cm, outer sepals broader, pinnately lobed or incised-serrate, inner sepals narrower, apex or margin coarsely incised-serrate, sometimes entire. Petals ca. 8 × 5 mm, white, obovate to elliptic, equalling or slightly shorter than sepals, puberulous outside, base clawed. Stamens shorter than petals; filaments to 6 mm; anthers ca. 1 mm. Pistils ca. 30–50(–70), slightly shorter than stamens. Ovary glabrous. Styles to 5 mm long, glabrous. Aggregate fruit red to dark red, globose, 0.9–1.4 cm in diameter, consisting of few drupelets, enclosed in persistent calyx. Pyrenes rugulose. Fig. 46d–e.

Distribution: Nepal, W Himalaya, E Himalaya, Assam-Burma, E Asia and SE Asia.

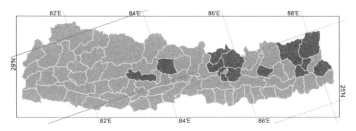

Altitudinal range: 1200–3000 m.

Ecology: Slopes, forests, forest margins.

Flowering: May–June. **Fruiting:** July–August.

Ripe fruits are eaten fresh.

19. *Rubus paniculatus* Sm., in Rees, Cycl. 30(2): Rubus no.41 (1815).
Rubus cordifolius D.Don later homonym, non Noronha; *R. paniculatus* forma *tiliaceus* (Sm.) H.Hara; *R. tiliaceus* Sm.

भालु ऐंसेलु Bhalu ainselu (Nepali).

Shrubs, scandent, to 3 m. Branchlets brownish or reddish brown, terete, yellowish grey tomentose-villous, gradually glabrescent, with sparse minute prickles. Stipules oblong or ovate-lanceolate, to 8–11 mm, villous, margin laciniate lobed above middle, lobes linear. Leaves simple, villous above, more densely so along veins, densely yellowish grey to grey tomentose below, villous along veins or glabrescent. Petiole 2–4 cm, yellowish grey or grey tomentose-villous, usually unarmed. Leaf blade ovate to narrowly ovate, 9–15 × 6–10 cm, base cordate, apex acuminate, margin undulate or inconspicuously lobed, irregularly coarsely serrate to doubly serrate, lateral veins 5–7 pairs. Inflorescences terminal or axillary, terminal inflorescences cymose panicles, laxly spreading, 10–24 cm, axillary inflorescences smaller, racemose with few flowers; rachis and pedicels yellowish grey or grey tomentose-villous. Bracts elliptic or oblong to lanceolate, 7–9 mm, villous, apex lobed or not divided. Pedicel to 1.5 cm. Flowers to 1.8 cm across. Outer surface of calyx tomentose and villous. Sepals ovate to lanceolate, 5–7 × 2–4 mm, apex acute to caudate-acuminate, outer sepals lobed, inner sepals entire. Petals white to yellowish white, oblong, 6–8 mm in diameter. Stamens shorter than petals; filaments linear, glabrous. Pistils longer than stamens, glabrous. Aggregate fruit dark red to blackish purple, globose, glabrous, ca. 1 cm in diameter, enclosed by calyx. Pyrenes distinctly rugose, style present.

Distribution: Nepal, W Himalaya, E Himalaya, Tibetan Plateau, Assam-Burma and E Asia.

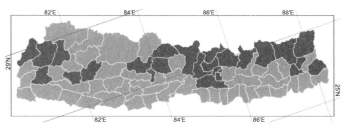

Altitudinal range: 1500–3200 m.

Ecology: Mixed forests on slopes, ravines, stream-sides.

Flowering: June–September. **Fruiting:** August–October.

Ripe fruits are edible. A paste of the bark is applied to scabies and other rashes. A paste of the leaves is applied to sprains.

20. *Rubus glandulifer* N.P.Balakr., J. Bombay Nat. Hist. Soc. 67: 58 (1970).
Rubus lanatus Wall. nom. nud.; *R. lanatus* Wall. ex Hook.f. later homonym, non Focke.

Shrubs, scandent, to 8 m. Branchlets yellowish grey, densely tomentose, with few to many glandular hairs and sparse prickles to 1.5 mm, broadened basally or nearly unarmed. Stipules linear, to 5 mm, densely pubescent, margin 3–5-toothed. Leaves simple, glabrous or pubescent on veins above, densely grey tomentose below. Petiole 1.5–4.5 cm, densely tomentose, with or without glandular hairs, rarely with small prickles. Leaf blade ovate to broadly ovate, 5–10 × 4–8 cm, base cordate, apex acute to shortly acuminate, margin obscurely to shallowly 3–5-lobed, irregularly finely serrate. Inflorescences terminal, paniculate, 8–15 cm, 20–40-flowered, rachis and pedicels softly pale tomentose, with or without glandular hairs. Bracts linear, 5–10 mm, 2–7-fid, pubescent. Pedicels 5–12 mm. Flowers ca. 2 cm or more across. Outer surface of calyx densely tomentose. Sepals spreading to reflexed after anthesis, lanceolate, 8–12 × 3–5 mm, apex acute to shortly acuminate. Petals white, obovate, or elliptic, 4–5 × 2–2.5 mm, base shortly clawed. Stamens about as long as petals. Styles glabrous. Ovary pilose. Fruit purplish black, globose, ca. 1 cm in diameter, glabrous. Pyrenes ovoid, foveolate.

Distribution: Nepal and W Himalaya.

Altitudinal range: 900–2500 m.

Ecology: Forests, thickets.

Flowering: May–June. **Fruiting:** June.

The presence of this W Himalayan species in C Nepal requires confirmation. *Rubus glandulifer* is currently known only from *Wallich 746* (BM), the type of *R. lanatus*, which lacks accurate locality data and so cannot be mapped. It was probably collected during Wallich's stay in the Kathmandu Valley, but could have been collected along the route to Kathmandu.

21. *Rubus efferatus* Craib, Fl. Siam. 1(4): 570 (1931).
Rubus ferox Wall. nom. nud.; *R. ferox* Wall. ex Kurz later homonym, non Vest ex Tratt.; *R. kurzii* N.P.Balakr. nom. superfl.; *R. moluccanus* var. *ferox* Kuntze; *R. sterilis* Kuntze.

Shrubs, scandent. Branchlets brownish grey, densely villous, with numerous prickles ca. 1.5 mm, broadened basally. Stipules linear, 4–5 mm, densely pubescent, margin 5–7-toothed. Leaves simple, sparsely pubescent above, glabrous below, except rather densely villous on veins above and below, occasional small prickles on midrib. Petiole 1.5–2.5 cm, densely villous, with several small prickles. Leaf blade ovate to broadly ovate, 2.5–9 × 1.5–6 cm, base cordate, apex acute to shortly acuminate, margin obscurely to shallowly 3–7-lobed, irregularly finely serrate. Inflorescences terminal and axillary, racemose, 4–7 cm, 5–7-flowered, rachis and pedicels densely villous. Bracts linear, ca. 4 mm, long-toothed, pubescent. Pedicels 6–8 mm. Flowers ca. 1 cm across. Outer surface of calyx densely tomentose. Sepals erect, broadly triangular, ca. 4 × 4 mm, apex acute, fimbriate-dentate. Petals white, elliptic, ca. 4 × 3 mm, base shortly clawed. Stamens about as long as petals. Styles glabrous. Ovary pilose. Fruit not seen.

Distribution: E Himalaya, Assam-Burma and SE Asia.

Altitudinal range: ca. 1200 m.

Ecology: Thickets.

Flowering: ?June.

Rubus efferatus is currently known in Nepal only from *Wallich 724* (BM), which lacks accurate locality data and so cannot be mapped. It is assumed to have been collected during Wallich's stay in the Kathmandu Valley, but it could also be from the journey to or from Kathmandu.

22. *Rubus hamiltonii* Hook.f., Fl. Brit. Ind. 2[5]: 328 (1878).

Shrubs, to 8 m. Branchlets green, velutinous, with few slightly reflexed short prickles. Stipules not seen. Leaves simple, subglabrous above or with ascending or erect hairs along main veins, below with sparse long straight hairs, unarmed along midvein. Petiole 0.5–1 cm. Leaf blade oblong to narrowly ovate, 6–9 × 2.5–4 cm, base rounded, apex shortly acuminate, margin finely denticulate to finely serrulate. Inflorescences terminal, panicles, 2–15 cm, 5–50 or more-flowered; rachis and pedicels velutinous or softly pilosulose, unarmed. Bracts ovate, entire or apically laciniate, 2–9 mm long, velutinous-pilose. Pedicels 5–8 cm. Flowers 0.9–1.2 cm across. Outer surface of calyx velutinous. Sepals spreading at anthesis, ovate, 4–5 mm, acute to shortly acuminate, spreading at anthesis. Petals white, obovate to oblanceolate, shorter than sepals. Stamens longer than petals; filaments linear, not broadened at base. Pistils shorter than stamens. Base of style and apex of ovary glabrous. Fruit not enclosed by calyx, globose, 0.8–1cm in diameter, glabrous; style of drupelets persistent or deciduous, glabrous. Pyrenes lunate, thickened or obliquely obovoid, with few coarse reticulations.

Distribution: Nepal, E Himalaya and Assam-Burma.

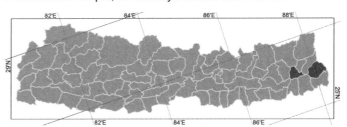

Altitudinal range: 300–700 m.

Ecology: Forest margins, clearings.

Flowering: August–November. **Fruiting:** October–December.

23. *Rubus acuminatus* Sm., in Rees, Cycl. 30(2): Rubus no. 43 (1815).
Rubus betulinus D.Don.

रातो ऐंसेलु Rato ainselu (Nepali).

Shrubs, scandent, to 8 m. Branchlets brown or reddish brown, slightly angled, thinly pubescent, glabrescent, with sparse

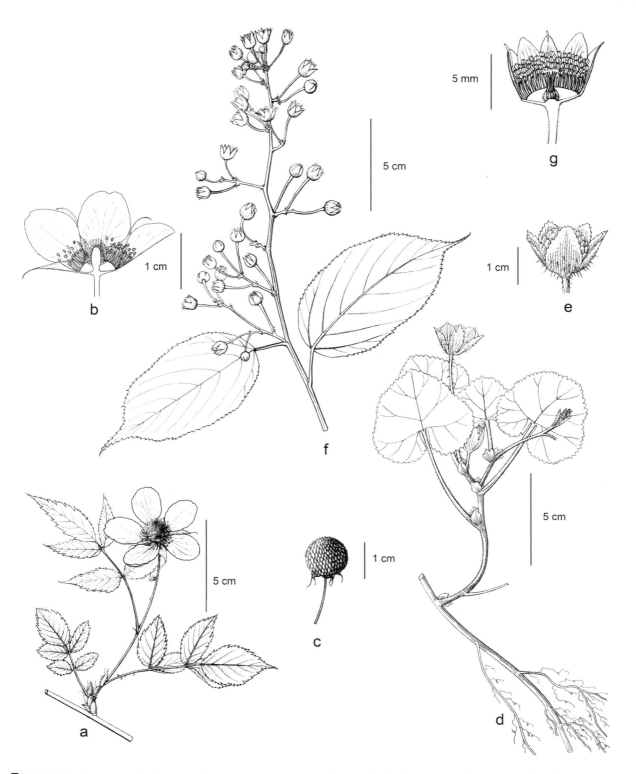

FIG. 46. ROSACEAE. **Rubus rosifolius**: a, inflorescence and leaves; b, longitudinal section of flower; c, fruit with persistent calyx. **Rubus calycinus**: d, flowering plant; e, fruit with persistent calyx. **Rubus acuminatus**: f, inflorescence and leaves, g, longitudinal section of flower.

minute prickles. Stipules usually caducous, free, linear or lanceolate, 4–6 mm, margin entire or toothed. Leaves simple, glabrous above, glabrous below or sparsely pubescent only along veins. Petiole 1–2 cm, slightly puberulous, with sparse minute prickles. Leaf blade ovate-oblong to lanceolate, 7–12 × 3–5 cm, base rounded, apex caudate, margin sharply serrulate, lateral veins 6–8 pairs, slightly raised below. Inflorescences terminal or axillary, terminal inflorescences narrow cymose panicles (6–)10–16 cm long, axillary inflorescences racemose, sometimes subcorymbose, less than 8 cm long; rachis and pedicels usually glabrous, occasionally sparsely pubescent. Bracts linear or lanceolate, 3–5 mm, margin entire or dentate, pubescent. Pedicels 1–2 cm. Flowers ca. 1 cm across. Outer surface of calyx glabrous; tube shallowly cupular. Sepals erect in fruit, ovate, 4–6 × 2–4 mm, inner sepals tomentose marginally, margin undivided, apex shortly acuminate. Petals white, broadly elliptic or oblong, 4–5 × 2–3 mm, equalling or shorter than sepals, glabrous, base shortly clawed. Stamens shorter than petals; filaments slightly broadened basally. Pistils ca. 10–20, shorter than stamens. Ovary glabrous. Aggregate fruit red, glabrous, with a few large drupelets enclosed in persistent calyx. Pyrenes ovoid, rugose. Fig. 46f–g.

Distribution: Nepal, E Himalaya, Assam-Burma, E Asia and SE Asia.

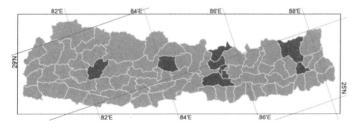

Altitudinal range: 1000–2500 m.

Ecology: Bamboo thickets, streamsides, roadsides.

Flowering: July–September. **Fruiting:** October–November.

Ripe fruits are eaten fresh. Juice of the bark is give to ease indigestion.

24. *Rubus griffithii* Hook.f., Fl. Brit. Ind. 2[5]: 327 (1878).

भुइँ ऐसेलु Bhuin ainselu (Nepali).

Shrubs, erect to ascending, ca. 3 m. Branchlets green or reddish brown, finely pubescent with more or less appressed straight hairs. Stipules subulate, ca. 10 mm, persistent, softly hairy. Leaves simple, glabrous above, below with a few short hairs between veins, veins with dense ascending straight hairs, midvein unarmed. Petiole 2–10 mm. Leaf blade elliptic to elliptic-lanceolate to oblong-lanceolate, 5–14 × 3–6 cm, base truncate to shallowly cordate, apex acuminate, margin singly or doubly serrulate-serrate. Inflorescences terminal or axillary, panicles or racemes, 5–12 cm, (1–)4–many-flowered; rachis and pedicels velutinous or softly pubescent with short curved hairs, unarmed. Bracts linear, serrate, 5 mm, softly

pubescent. Pedicels 8–19 mm. Flowers 1.5–3 cm across. Outer surface of calyx minutely velutinous. Sepals spreading at anthesis, narrowly lanceolate to lanceolate or narrowly ovate, 1.2–1.5 cm, apex long acuminate to caudate. Petals white, spatulate, ca. 3 mm, much shorter than sepals. Stamens much longer than petals; filaments linear, gradually widened towards base. Pistils longer than stamens. Base of style and apex of ovary with few sparse straight hairs. Aggregate fruit exserted, black, globose, 8–12 mm in diameter, apically with a few hairs. Styles persistent, with a few long hairs. Pyrenes thickly lunate to thickly semicircular, with conspicuous rounded horizontal ridges.

Distribution: Nepal and E Himalaya.

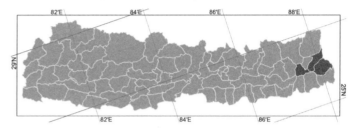

Altitudinal range: 1400–2400 m.

Ecology: Woodland margins, clearings, thickets.

Flowering: August–December. **Fruiting:** October-January.

Hara & Ohashi (Fl. E. Himalaya: 129. 1966) misapplied the name *Rubus hexagynus* Roxb. to this species.

Rubus griffithii is a near endemic to Nepal, also found in adjacent areas of Sikkim. *R. hexagynus* Roxb. has a more easterly distribution and is not found further west than Assam.

Ripe fruits are eaten fresh.

25. *Rubus treutleri* Hook.f., Fl. Brit. Ind. 2[5]: 331 (1878). *Rubus arcuatus* Kuntze; *R. reticulatus* Wall. nom. nud.; *R. rosulans* Kuntze; *R. tonglooensis* Kuntze.

ठूलो ऐसेलु Thulo ainselu (Nepali).

Scandent shrubs, 0.5–1 m. Branches greyish brown, brown or blackish brown, villous, purplish red stipitate glandular, with sparse needle-like prickles. Stipules free, 1–1.5 cm, palmatipartite nearly to base; lobes linear to linear-lanceolate, villous, stipitate glandular. Leaves simple, sparsely villous above but more densely so along veins, with sparse stipitate glands along veins, below densely tomentose-villous when young, glabrate in age, sparsely stipitate glandular along veins. Petiole 2.5–5.5 cm, with dense purplish red stipitate glands, long hairs and sparse needle-like prickles. Leaf blade suborbicular, 6–12 cm in diameter, base deeply cordate, margin 3–5-lobed, terminal lobe slightly larger than lateral lobes, apex acute, lateral lobes obtuse, rarely acute, margin irregularly coarsely serrate. Inflorescences terminal, racemose, 3–4 cm or slightly longer, several- to more than 10-flowered, or flowers few in clusters in leaf axils; rachis, pedicels and outer surface of calyx densely villous, stipitate glandular. Bracts 1–1.3 cm, palmatiparted; lobes linear, villous, stipitate glandular. Pedicels 5–10 mm. Flowers 1.5–2(–2.5) cm across. Calyx tube cup-shaped, ca. 5 mm. Sepals narrowly ovate or narrowly ovate-lanceolate, 8–15 × 5–7 mm, apex acuminate, outer sepals leaf-like, margin laciniate or pinnately divided into lanceolate lobes, inner sepals entire. Petals pink, suborbicular, 8–11 mm in diameter, barely clawed. Stamens shorter than petals; filaments broad, flat. Pistils nearly as long as stamens. Ovary and style glabrous. Aggregate fruit red, globose, 8–10 mm in diameter, enclosed in calyx. Pyrenes densely rugulose.

Distribution: Nepal, E Himalaya, Tibetan Plateau and E Asia.

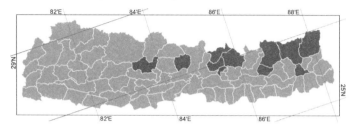

Altitudinal range: 2300–3700 m.

Ecology: Forests, forest margins.

Flowering: June–July. **Fruiting:** August–September(–October).

Ripe fruits are edible.

26. Rubus kumaonensis N.P.Balakr., J. Bombay Nat. Hist. Soc. 67: 58 (1970).
Rubus reticulatus Wall. ex Hook.f. later homonym, non A.Kern.

Shrubs, scandent, to 1 cm tall. Branchlets brown, grey to yellowish grey tomentose, softly hairy, with sparse minute curved prickles, rarely unarmed. Stipules subflabellate, pectinately lobed; lobes divided again, with lanceolate lobules, tomentose, softly hairy. Leaves simple, pilose above, densely so along veins, densely yellowish grey tomentose below, with long hairs along veins. Petiole 4–9 cm, grey to yellowish grey tomentose, softly hairy, with sparse minute prickles. Leaf blade broadly ovate to suborbicular, 12–20 × 10–18 cm, base cordate, margin distinctly 5-lobed, lobes apically acute or ± obtuse, unevenly densely serrate, palmately 5-veined, lateral veins 5–7 pairs, prominent below. Inflorescences terminal and axillary, terminal inflorescences lax narrow, racemose or cymose-paniculate, ca. 6 cm, axillary inflorescences racemose or a few flowers in clusters in leaf axils; rachis and pedicels densely grey or yellowish brown tomentose and villous. Bracts subflabellate, margin entire or apically divided, thin tomentose, softly hairy. Pedicels 1.5–3(–4) cm. Flowers 1–1.5 cm across. Calyx densely tomentose and villous. Sepals broadly ovate, 4.5–5 mm and about as broad, margin entire or slightly lobed, apex acuminate. Petals white or yellowish white, obovate to suborbicular, ca. 5 × 5 mm. Stamens glabrous; filaments linear. Pistils numerous. Style and ovary glabrous. Aggregate fruit red, globose, 1–1.5 cm in diameter, enclosed by calyx.

Distribution: Nepal, W Himalaya, E Himalaya and Tibetan Plateau.

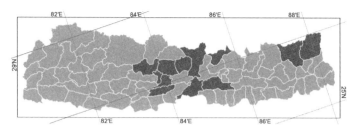

Altitudinal range: 1100–2800 m.

Ecology: Broadleaved evergreen forests in montane valleys, ravines, thickets on slopes.

Flowering: July–August. **Fruiting:** September–October.

Ripe fruits are edible. Juice of the fruits is given in case of stomach disorders.

27. Rubus calycinoides Kuntze, Meth. Sp.-Beschr. Rubus: 67, 78, 83 (1879).
Rubus bhotanensis Kuntze; *R. darschilingensis* Kuntze; *R. diffisus* Focke; *R. himalaicus* Kuntze.

Shrubs, erect or arching,1–2 m or more. Branchlets green, velutinous or tomentose. Stipules lanceolate to ovate, lacerate, 1–1.5 × 0.2–0.5 cm, caducous, sparsely velutinous or subglabrous. Leaves simple, velutinous or subglabrous above, without prickles, velutinous below, midrib with or without minute straight or hooked prickles. Petiole 3–11 cm, velutinous or minutely tomentose. Leaf blade ovate to widely ovate or suborbicular in outline, 5–20 × 5.5–22 cm, base shallowly to deeply cordate, apex acute to shortly acuminate, shallowly 3- or 5-lobed, lobes acute or rounded, margins minutely singly or doubly serrulate to dentate-serrate, finely and more or less conspicuously reticulate above and below. Inflorescences axillary and terminal, racemose or paniculate, 2–15 cm, 2–20-flowered, flowers clustered in axils and along main axis of inflorescence; rachis and pedicels velutinous or slightly tomentose, with sparse hooked or straight prickles. Bracts broadly ovate to rounded, 5–15 mm, sparsely velutinous. Pedicels 2–15 mm, variable in length within a single inflorescence. Flowers 2–2.5 cm across. Outer surface of calyx velutinous to tomentose. Sepals erect or spreading at anthesis, lanceolate to widely ovate, 11–15 mm, margins sometimes with few lacerations, apex acute, sometimes with three narrowly lanceolate lacerations. Petals white, obovate to rounded, clawed, ca. 7 mm in diameter, shorter than or equalling sepals. Stamens ± equalling petals; filaments linear, gradually and slightly broadened toward base. Pistils longer than stamens. Base of style and apex of ovary glabrous. Aggregate fruit enclosed by calyx or slightly exserted, red, globose to ovoid, 1.2–1.5 cm in diameter, glabrous; style of drupelets persistent, glabrous. Pyrenes thickly lunate, horizontally ribbed.

Distribution: Nepal, E Himalaya and Assam-Burma.

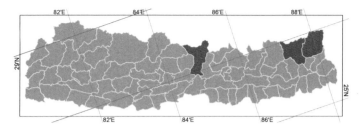

Altitudinal range: 1200–2200 m.

Ecology: Clearings, thickets, woodlands, woodland margins.

Flowering: June–October. **Fruiting:** October–December.

Plants frequently offered in the nursery trade under the name *Rubus calycinoides* Hayata ex Koidzumi (a later homonym of *R. calycinoides* Kuntze) are *R. rolfei* Vidal.

28. *Rubus rugosus* Sm., in Rees, Cycl. 30(2): Rubus no.34 (1815).
Rubus hamiltonianus Ser.; *R. rugosus* Buch.-Ham. ex D.Don nom. superfl.

गोरु ऐंसेलु Goru ainselu (Nepali).

Shrubs, climbing or scandent, to 5 m. Branchlets green, velutinous. Stipules ca. 1.5 cm, ovate, entire or serrate, with long straight hairs. Leaves simple, with sparse straight hairs above, below rugose or bullate-like, shortly pilose and arachnoid, midvein with or without sharp, recurved prickles. Petiole 2.5–10 cm, strigose or strigillose, with sharp recurved spines. Leaves ovate to suborbicular, 6–12 × 5–12 cm, base cordate to shallowly cordate, apex acute to rounded, 3 or 5(or 7)-angled or lobed, margin irregularly denticulate. Inflorescences terminal and axillary clusters or racemes with flowers clustered at nodes, 3–20 cm long, 2–30-flowered; rachis and pedicels densely pilose, with a few straight or slightly curved prickles. Bracts ovate to broadly elliptic, ca. 5 mm, pilose. Pedicels 4–7 mm. Flowers 0.7–1 cm across. Outer surface of calyx tomentose and pilose. Sepals erect at anthesis, ovate, 0.8–1.2 mm, apex acute to acuminate. Petals white, obovate, ca. 8 mm in diameter, shorter than to equalling sepals. Stamens shorter than petals; filaments linear, broadest at base and tapering gradually to apex. Pistils longer than stamens. Base of style and apex of ovary glabrous. Aggregate fruit half enclosed by calyx, red, black at maturity, globose, 1–1.4 cm in diameter, glabrous; style of drupelets persistent, glabrous. Pyrenes lunate, thickened, surface raised reticulate.

Distribution: Nepal, E Himalaya and Assam-Burma.

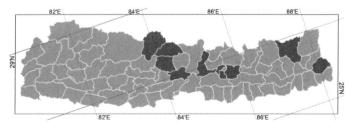

Altitudinal range: 600–2700 m.

Ecology: Thickets, clearings, disturbed forests.

Flowering: May–September. **Fruiting:** July–September.

Hooker (Fl. Brit. Ind. 2: 330. 1878) partly misapplied the name *Rubus moluccanus* L. to this species.
Ripe fruits are eaten fresh.

29. *Rubus franchetianus* H.Lév., Bull. Acad. Int. Geogr. Bot. 20: 71 (1909).
Rubus arcticus var. *fragarioides* (Bertol.) Focke nom. illegit.; *R. fragarioides* Bertol. later homonym, non (Michx.) Dum. Cours.

Herbs, perennial, prostrate or creeping. Stems greenish brown or brownish, woody, softly hairy. Stipules free, ovate or elliptic, 7–10 × 5–7 mm, apex acute to obtuse. Leaves 3–5-foliolate, usually glabrous above and below. Petiole 3–9 cm, slightly softly hairy; leaflets shortly petiolulate or subsessile. Blade of leaflets obovate to suborbicular, 2–5 × 1.5–3.5 cm, base cuneate, apex acute or obtuse, margin usually lobed, incised or coarsely sharply serrate or doubly serrate. Inflorescences usually terminal, with 1 or 2 flowers, rachis and pedicels softly hairy. Flowers 1–2 cm across. Pedicels 1–3(–5) mm. Outer surface of calyx glabrous or softly hairy; tube shallowly cup-shaped. Sepals ovate-lanceolate, 6–10 × 2–4 mm, margin entire, apex long acuminate to caudate. Petals white, obovate, 0.8–1.2 × 0.5–0.6 cm, apex acute, Stamens unequal in length, shorter than petals; filaments dilated in lower part. Pistils 4–6, shorter than stamens. Ovary usually glabrous. Aggregate fruit 7–9 mm in diameter, with several drupelets, persistent calyx lobes erect; drupelets to 4 mm. Pyrenes rugose.

Distribution: Nepal, E Himalaya, Tibetan Plateau, Assam-Burma and E Asia.

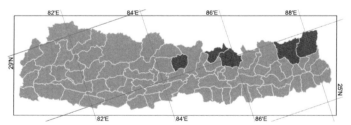

Altitudinal range: 3000–4200 m.

Ecology: High mountains, grasslands on slopes, forests, forest margins.

Flowering: May–July. **Fruiting:** July–September.

Nepalese material refers to the typical variety, other varieties occur in China.

30. *Rubus fockeanus* Kurz, J. Asiat. Soc. Bengal 44(2): 206 (1875).
Rubus allophyllus Hemsl.; *R. loropetalus* Franch.; *R. nutans* var. *fockeanus* (Kurz) Kuntze; *R. radicans* Focke.

Herbs, perennial, prostrate or creeping, rooting at nodes. Branchlets pilose, without bristles or prickles, sometimes with intermixed small glandular hairs. Stipules free, elliptic, 5–7 × 3–5 mm, persistent, membranaceous, margin entire, rarely toothed, apex obtuse or shortly pointed. Leaves 3-foliolate, pilose above, pilose below along veins. Petiole 2–5 cm, pubescent, terminal leaflet shortly petiolulate, lateral leaflets subsessile. Blade of leaflets suborbicular to broadly obovate, 1–4 × 1–5 cm, base broadly cuneate to rounded, apex obtuse, margin unevenly coarsely obtusely serrate; base of lateral leaflets oblique. Flowers solitary or in pairs, terminal. Bracts elliptic, 3–5 × 2–2.5 mm, smaller than stipules, membranaceous, puberulous, margin entire or toothed. Pedicels 2–5 mm, pubescent, sometimes bristly. Outer surface of calyx pubescent, or intermixed with sparse reddish brown bristles. Sepals ovate-lanceolate to narrowly lanceolate, 7–10 × (2–)3–4 mm, undivided, rarely shallowly laciniate, apex long acuminate. Petals white, obovate-oblong to linear-oblong, 7–11 × 3–5 mm, base clawed. Stamens shorter than petals; filaments enlarged towards base. Pistils 4–20, shorter than stamens. Ovary glabrous or slightly puberulous. Styles glabrous or basally slightly pubescent. Aggregate fruit red, globose, glabrous, with few semiglobose drupelets. Pyrenes ellipsoid, rugose.

Distribution: Nepal, E Himalaya, Tibetan Plateau, Assam-Burma and E Asia.

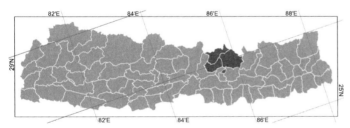

Altitudinal range: 2000–4000 m.

Ecology: Grassy slopes, forests.

Flowering: May–June. **Fruiting:** July–August.

Ripe fruits are eaten fresh.

31. *Rubus nepalensis* (Hook.f.) Kuntze, Meth. Sp.-Beschr. Rubus: 125 (1879).
Rubus nutans var. *nepalensis* Hook.f., Fl. Brit. Ind. 2[5]: 334 (1878); *R. barbatus* Edgew. later homonym, non Fritsch; *R. nutans* Wall. ex G.Don later homonym, non Vest; *R. nutans* Wall. nom. nud.; *R. nutantiflorus* H.Hara.

न्ह्यालाङ्ग Nhyalang (Sherpa).

Herbs, creeping or prostrate. Branchlets green or purplish, pilosulose and hispidulous. Stipules lanceolate to ovate,

5–15 mm, persistent, softly hispid and sometimes also tomentose, entire or apically lacerate. Leaves 3-foliolate, strigose or strigillose above, strigillose along midvein below. Petiole 1.5–7.5 cm; petiolule of terminal leaflet 0.3–0.7 cm; petiolule of lateral leaflets 0.2–0.4 cm, softly pilosulose and hispidulous. Blade of terminal leaflet slightly larger than lateral leaflets, 1–4.5 × 1–4 cm, rhombic to obovate, base cuneate to broadly cuneate, sides straight or slightly concave to rachis, apex rounded, sometimes shallowly lobed below middle, margins finely doubly denticulate to doubly serrulate. Blade of lateral leaflets oblique, 1–4.5 × 0.8–4 cm, base obliquely cuneate, apex rounded, margin finely doubly denticulate to doubly serrulate. Inflorescences terminal on short braches but appearing axillary, racemose or flowers solitary, 3–12 cm long, up to 6-flowered; rachis and pedicels tomentose and sparsely hispidulous. Bracts elliptic to narrowly ovate, 5–9 mm, margins with a few soft hairs. Pedicels 4–9 cm. Flowers 1.5–3.8 cm across. Outer surface of calyx hispidulous and minutely tomentose. Sepals spreading at anthesis, lanceolate to ovate or broadly ovate, 1–1.4 cm, apex acuminate to caudate. Petals white, obovate to widely obovate or rounded, 0.8–1.3 mm, equalling or longer than sepals, glabrous. Stamens shorter than petals; filaments slightly flattened, broader at base. Pistils shorter than stamens. Base of style and apex of ovary with few erect straight hairs. Aggregate fruit exposed, red or orangish red, globose, ca. 1.5 cm in diameter, glabrous; style of drupelets persistent, glabrous. Pyrenes lunate, smooth or with few faint raised lines.

Distribution: Nepal, W Himalaya and E Himalaya.

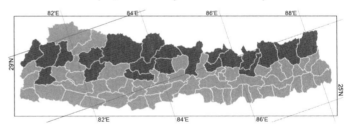

Altitudinal range: 1800–3500 m.

Ecology: Thickets, among boulders on open slopes, clearings, mixed and deciduous forests.

Flowering: July–October. **Fruiting:** October–December.

Ripe fruits are edible.

32. *Rubus* x *seminepalensis* Naruh., J. Jap. Bot. 65(6): 187 (1990).

Shrubs, prostrate. Branchlets densely hispid and sparsely villous. Stipules digitate-pectinate, 8–12 mm, villous. Petioles 5–7 cm, hispid and villous. Leaves simple or 3-foliolate, sparsely villous above, white hirsute below along veins, unarmed along midvein, margins velutinous; petiolule of lateral leaflets 1–2 mm, villous. Blade of terminal leaflet obovate to elliptic, 6–7 × 3–3.5 cm, base cuneate, apex acute, blade of lateral leaflets obliquely deltoid-ovate, base cuneate, apex acute to acuminate, margin serrulate. Blade of simple leaves deltoid-orbicular, trilobed or triparted, base cordate, apex acute. Inflorescences axillary and terminal, racemose, 1–4-flowered. Pedicels 1.5–3.5 cm, hispid and villous. Sepals lanceolate, ca. 10 × 5 mm, apex aristate to 3-parted, hispid and villose, margins and between veins tomentose. Petals purplish red, ovate, ca. 10 mm, shorter than sepals. Stamens ca. 5 mm, shorter than petals; filaments linear, dilated. Pistils ca. 4.5 mm, longer than stamens. Fruit unknown.

Distribution: Endemic to Nepal.

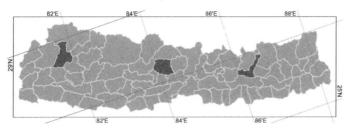

Altitudinal range: 2300–3100 m.

Ecology: Forests.

Flowering: July–August.

Rubus x *seminepalensis* is an apparently sterile hybrid between the compound-leaved *R. nepalensis* (Hook.f.) Kuntze and the simple-leaved *R. treutleri* Hook.f. Naruhashi (J. Jap. Bot. 65: 186-191. 1990), from whom this description was largely taken, noted differences and similarities between *R.* x *seminepalensis* and the two parents.

15. *Geum* L., Sp. Pl. 1: 500 (1753).

Hiroshi Ikeda

Perennial herbs with stout rootstock. Radical leaves pinnate, petiolate, stipulate; stipules adnate to petiole; leaflet margin toothed or cleft; cauline leaves few, reduced in size towards top. Flowers terminal, 1–3 in cymes. Hypanthium cupulate to turbinate. Sepals 5, valvate. Episepals 5, alternate with and much smaller than sepals. Petals 5, yellow or white. Stamens many, inserted at mouth of hypanthium. Carpels many, densely hirsute or glabrescent, apically tapering into persistent style; style terminal; ovule semi-basal. Fruit an achene.

Worldwide about 70 species, widespread in temperate zones in N and S hemispheres. Three species in Nepal.

Key to Species

1a	Style jointed, deciduous above joint after flowering	**3. *G. roylei***
b	Style not jointed, persistent	2
2a	Terminal leaflet of radical leaves ovate to broadly elliptic, similar to lateral ones in size. Petals yellow, rarely crimson	**1. *G. elatum***
b	Terminal leaflet of radical leaves broadly ovate to suborbicular, much larger than lateral ones. Petals white to pale pink	**2. *G. sikkimense***

1. *Geum elatum* Wall. ex G.Don, Gen. Hist. 2: 256 (1832). *Acomastylis elata* (Wall. ex G.Don) F.Bolle; *A. elata* var. *leiocarpa* (W.E.Evans) F.Bolle; *Geum adnatum* Wall. nom. nud.; *G. elatum* Wall. nom. nud.; *G. elatum* var. *humile* (Royle) Hook.f.; *G. elatum* var. *leiocarpum* W.E.Evans; *Potentilla adnata* Wall. ex Lehm.; *Sieversia elata* (Wall. ex G.Don) Royle; *S. elata* var. *humile* Royle.

बेलोचन Belochan (Nepali).

Rosulate herb, 10–40 cm. Rootstock stout, terete. Flowering stems from axils of radical leaves, pubescent, sometimes mixed with multicellular glandular hairs. Radical leaves broadly linear in outline, 8–20(–30) × 1.5–5(–7) cm, with 9–13 pairs of leaflets, interrupted with alternating smaller leaflets, pubescent or pilose, rarely glabrescent on both surfaces except on

veins; petiole 1–4 cm, pilose or glabrescent; terminal leaflets broadly obovate to semiorbicular, 7–30 × 5–30 mm, base broadly cuneate, apex rounded; cauline leaves reduced in size toward top, upper ones bract-like, oblanceolate, deeply parted. Inflorescence terminal, in cyme, 1–3(–5)-flowered. Flowers 2.5–3.5(–4) cm across; pedicel pubescent, sometimes with multicellular glandular hairs. Episepals green or deep purple outside, green inside, lanceolate to narrowly ovate, much smaller than sepals, 2–3 × 1–1.2 mm, apex acute, sparsely or densely hirsute outside, glabrescent inside. Sepals green or deep purple outside, green inside, ovate-triangular, 6–8 × 3–5 mm, apex acute or obtuse, sparsely or densely hirsute outside, glabrescent inside. Petals yellow or bright red, obovate to broadly obovate, 1.2–1.5 × 0.8–1.2 cm, glabrous, apex emarginate. Filaments 3.5–5 mm; anthers dark green,

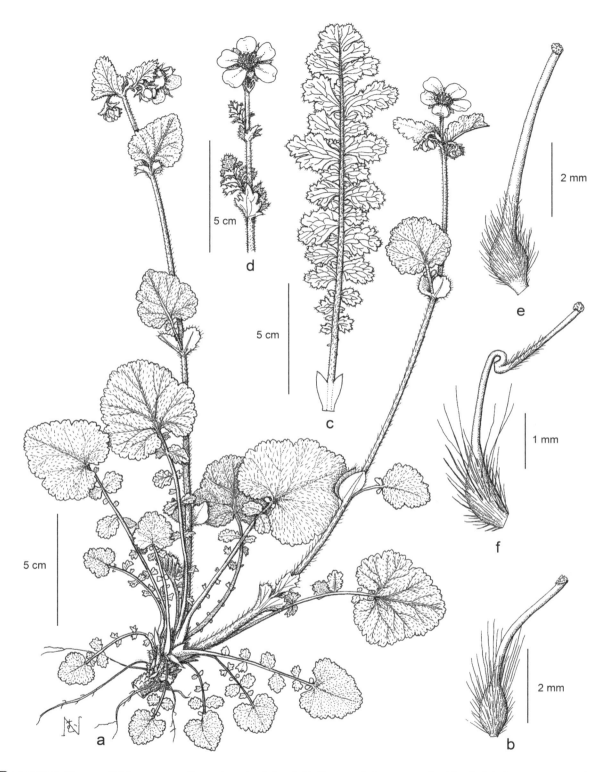

FIG. 47. Rosaceae. **Geum sikkimense**: a, flowering plant; b, pistil. **Geum elatum**: c, radical leaf; d, inflorescence and cauline leaves; e, pistil. **Geum roylei**: f, pistil.

ellipsoid, ca. 1.2 × ca. 0.9 mm. Ovary glabrous to densely hirsute, 1.5–2 × ca. 0.8 mm; style not twisted, glabrous or sparsely pilose near base; stigma minute. Achenes ovoid, 3–3.3 × 1.5–1.8 mm excluding persistent style. Fig. 47c–e.

Distribution: Nepal, W Himalaya, E Himalaya, Tibetan Plateau and E Asia.

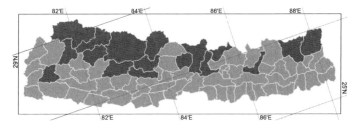

Geum elatum var. *humile* (Royle) Hook.f. is a dwarf form of *G. elatum*, but the variation of morphological characters between the two overlap and they are indistinguishable.

Leaves are pounded and applied to wounds. It is also useful in diarrhoea, for sore throats, and leucorrhoea.

1a Petals yellow .... forma *elatum*
 b Petals crimson.................................. forma *rubrum*

Geum elatum Wall. ex G.Don forma *elatum*

Petals yellow.

Distribution: Nepal, W Himalaya, E Himalaya, Tibetan Plateau and SW Asia.

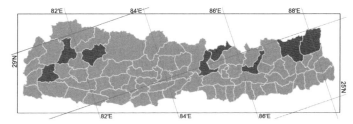

Altitudinal range: 2900–4500 m.

Ecology: Alpine meadows.

Flowering: June–August. **Fruiting:** August–October.

Geum elatum forma *rubrum* Ludlow, Bull. Brit. Mus. (Nat. Hist.), Bot. 5: 276, pl. 30B (1976).

Petals crimson.

Distribution: Endemic to Nepal.

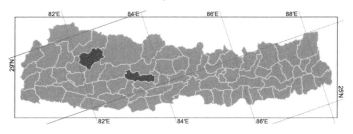

Altitudinal range: 3000–4400 m.

Ecology: Open grassy slopes.

Flowering: May–July.

2. Geum sikkimense Prain, J. Asiat. Soc. Bengal, Pt. 2, Nat. Hist. 73: 200, pl. 7 (1904).
Acomastylis sikkimensis (Prain) F.Bolle; *Geum versipatella* C.Marquand.

Rosulate herbs, 5–40 cm. Rootstock stout, terete. Flowering stems from axils of radical leaves, pubescent with mixed with long and short hairs. Radical leaves lyrately pinnate, lanceolate in outline, 4–25 × 1.5–7.5 cm, with 5–10 opposite or alternate pairs of smaller leaflets, sparsely or densely pilose on both surfaces; petiole 1–3 cm, pubescent, mixed with long and short hairs; terminal leaflets lyrate, broadly obovate to semiorbicular, 1.5–7 × 1.2–7 cm, base shallowly to deeply cordate, apex rounded, margin shallowly incised-serrate, serrate tips obtuse; cauline leaves reduced in size towards top, simple or with a few smaller leaflets, oblong to ovate. Inflorescence terminal, in cyme, 1–3-flowered. Flowers nodding, 1.5–2.5 cm across; pedicel pubescent. Episepals broadly lineate to lanceolate, much smaller than sepals, 2–3 × 0.5–1.2 mm, apex acute or obtuse, sparsely hirsute outside, glabrescent inside. Sepals ovate-triangular, 4–8(–11) × 3–5(–6) mm, apex acute or obtuse, sparsely hirsute outside, glabrescent inside. Petals white to pale pink, obovate, 5–9 × 4–7 mm, sparsely hairy in lower half outside, glabrous inside, apex rounded or shallowly emarginate. Filaments 3–4 mm, pilose; anthers ellipsoid, ca. 0.7 × ca. 0.5 mm. Ovary densely hirsute, 1.5–2 × ca. 1 mm; style not twisted, 3.5–4 mm, glabrous, slightly hooked near stigma; stigma minute. Fig. 47a–b.

Distribution: Nepal and E Himalaya.

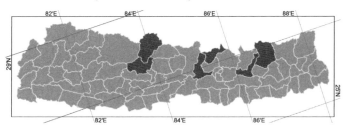

Altitudinal range: 3000–4400 m.

Ecology: Alpine meadows.

Flowering: June–August.

3. Geum roylei Wall. ex Bolle, Repert. Spec. Nov. Regni Veg. Beih. 72: 66 (1933).
Geum roylei Wall. nom. nud.

Rosulate herb, 30–50 cm. Stems pubescent with spreading hispid hairs. Radical leaves oblanceolate to narrowly obovate in outline, 10–20 × 5–10 cm; lateral leaflets 3–5 pairs with

alternate smaller secondary leaflets, uppermost pair much larger than others; terminal leaflet, rhombic-obovate to broadly ovate, 3–6 × 2.5–6 cm, apex acute or obtuse, base broadly cuneate, margin irregularly toothed, both surfaces sparsely hispid; petiole 3–6 cm, erect-hispid. Cauline leaves short-petiolate, 3(–5)-foliolate. Inflorescence terminal or from axils of upper cauline leaves, in cymes, 2–5-flowered. Flowers 1.5–2 cm across. Episepals much smaller than sepals, linear to narrowly lanceolate, 2–3 × ca. 0.5 mm, margin entire or sometimes deeply divided into two lobes, sparsely hispid outside, glabrescent inside, apex obtuse. Sepals ovate-triangular, 6–9 × 3–5 mm, apex acuminate to semi-caudate, sparsely hispid outside, sparsely puberulent inside. Petals yellow, oblong to narrowly ovate, 7–9 × 4–5 mm. Filaments 1.5–2 mm; anthers flattened globose, ca. 1 mm across. Ovary obliquely oblong, 0.8–1 × ca. 0.5 mm, densely hispid; style jointed, glabrous basally, strigose above joint, deciduous above joint after flowering. Achenes 2–2.5 mm × ca. 1 mm, hispid.
Fig. 47f.

Distribution: Nepal, W Himalaya and SW Asia.

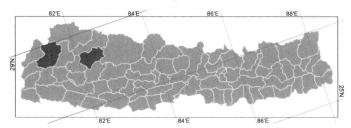

Altitudinal range: 1900–3500 m.

Ecology: Forest margins, thickets.

Flowering: June–July. **Fruiting:** July–August.

Hooker (Fl. Brit. Ind. 2: 342. 1878) and Nepalese authors (Bull. Dept. Med. Pl. Nepal 7: 78. 1976) misapplied *Geum urbanum* L. to this species. *Geum urbanum* is a species from Europe and the Middle East.

16. *Potentilla* L., Sp. Pl. 1: 495 (1753).

Duchesnea Sm.

Hiroshi Ikeda

Perennial, rarely annual herbs or subshrubs. Stems erect, ascending or prostrate, sometimes stoloniferous. Stipules adnate to petioles, upper part (auricle) free or more or less connate, membranous or leaf-like. Leaves radical or cauline, pinnate, trifoliolate or palmately compound. Inflorescences cymose or flowers solitary. Flowers usually bisexual, rarely unisexual. Bracts at base of pedicel, leafy or membranous; bracteoles absent. Hypanthium cupulate. Episepals 5, alternating with sepals. Sepals 5, valvate. Petals 5, yellow, rarely orange or bright red or white with crimson base. Stamens usually 20 in 3 whorls, rarely fewer or more, anthers 4-locular. Ovary superior, carpels many, crowded on a dome-shaped receptacle. Styles subterminal, lateral or basal. Achenes on a dry or swollen, somewhat spongy receptacle, sepals and episepals persistent.

Worldwide about 500 species in temperate, arctic, and alpine zones, mostly in the N hemisphere, several in the S hemisphere. 31 species in Nepal.

In the following treatments three species groups are tentatively used for aggregates of morphologically similar species: *Potentilla griffithii* agg., *P. multifida* agg., and *P. saundersiana* agg. Within these species groups species delimitation is extremely difficult because of their morphological plasticity (plant size, leaf shape, leaflet number, and indumentum), because of the ease of hybridization amongst species and consequent apomictic reproduction of their progeny. Further systematic studies are needed to clarify the taxonomic status of entities within these species groups.

Several putative natural hybrids among *Potentilla* species have been described from Nepal. *Potentilla* x *microcontigua* H.Ikeda & H.Ohba, J. Jap. Bot. 68: 35 (1993) (=*P. contigua* Soják x *P. microphylla* D.Don); *P.* x *micropeduncularis* H.Ikeda & H.Ohba, J. Jap. Bot. 71: 253 (1996) (=*P. microphylla* D.Don x *P. peduncularis* D.Don); *P.* x *polyjosephiana* H.Ikeda & H.Ohba, Bot. J. Linn. Soc. 112: 183 (1993) (=*P. josephiana* H.Ikeda & H.Ohba x *P. polyphylla* Wall. ex Lehm. var. *polyphylla*). The following three putative hybrids were also noted by Ikeda and Ohba (Bot. J. Linn. Soc. 112: 184. 1993): *P. festiva* Soják x *P. lineata* Trev.; *P. festiva* Soják x *P. polyphylla* Wall. ex Lehm; *P. lineata* Trev. x *P. polyphylla* Wall. ex Lehm.

Key to Species

1a	Shrubs or subshrubs with woody stems	2
b	Perennial or annual herbs. Stems not woody	5
2a	Shrubs with stems 20–100 cm tall	**1. *P. fruticosa***
b	Plant rhizomatous or cushion-forming	3

3a   Plant cushion-forming. Leaves 5–7-foliolate ....**4. *P. biflora***

b   Plant rhizomatous. Leaves trifoliolate 4

4a   Leaflets 3-dentate..**2. *P. cuneata***

b   Leaflets shallowly to deeply 3–5 divided..**3. *P. eriocarpa***

5a   Petals orange, bright red or white with crimson base 6

b   Petals yellow 8

6a   Petals white with crimson base. Leaves pinnate, leaflets deeply divided into linear segments ....**28. *P. coriandrifolia***

b   Petals orange or crimson. Leaves trifoliolate or palmate...7

7a   Leaves trifoliolate ....**23. *P. argyrophylla***

b   Leaves palmate...**27. *P. nepalensis***

8a   Leaves trifoliolate or palmate 9

b   Leaves pinnate...13

9a   Plant stoloniferous. Flowers on stolons...**31. *P. indica***

b   Plant not stoloniferous. Flowers (inflorescences) from axils of radical leaves...10

10a   Stems erect 11

b   Stems creeping or ascending..12

11a   Stems usually more than 20 cm. Leaves trifoliolate ....**23. *P. argyrophylla***

b   Stems usually less than 20 cm. Leaves usually palmate, rarely trifoliolate...................................**24. *P. saundersiana***

12a   Leaves trifoliolate. Flowers terminal, solitary. In alpine zone..**22. *P. monanthes***

b   Leaves palmate. Flowers in cymes. In temperate zone..**30. *P. sundaica***

13a   Stipules of radical leaves adnate to lateral side of petiole...14

b   Stipules of radical leaves adnate to ventral side of petiole..18

14a   Rhizomes long-creeping, branched...**5. *P. bifurca***

b   Rhizomes short, not or little-branched 15

15a   Stems much-branched at base, decumbent or repent. Flowers small, 6–8 mm across...............................**29. *P. supina***

b   Stems erect or ascending. Flowers larger, more than 1 cm across...16

16a   Leaflets deeply divided, segments linear to narrowly lanceolate...**26. *P. multifida***

b   Leaflets oblong to elliptic with serrate margin...17

17a   Stems ascending, usually less than 20 cm. Leaflets less than 1 cm...**24. *P. saundersiana***

b   Stems erect, usually more than 30 cm. Leaflets more than 1 cm ....**25. *P. griffithii***

18a   Plant stoloniferous..**21. *P. anserina***

b   Plant not stoloniferous..19

19a   Cauline leaves more than 4 20

b   Cauline leaves 1–3...23

20a   Auricles of stipules of radical leaves 2, connate from base to middle ....**6. *P. festiva***

b   Auricles of stipules of radical leaves 2, free from each other 21

21a   Base of uppermost pair of leaflets decurrent ....**7. *P. josephiana***

b   Base of uppermost pair of leaflets cuneate..22

22a   Peduncles and hypanthium with gland-tipped multicellular hairs..**8. *P. lineata***

b   Peduncles and hypanthium without gland-tipped multicellular hairs..**9. *P. polyphylla***

1. *Potentilla fruticosa* L., Sp. Pl. 1: 495 (1753).
Dasiphora fruticosa (L.) Rydb.

Small shrubs, erect or sometimes prostrate. Stipules membranous, lower half adnate to petiole, glabrous inside, sparsely lanate outside, surrounding stem. Petioles sparsely lanate or glabrescent. Leaves imparipinnate with 2–3 pairs of lateral leaflets, rarely trifoliolate; base of uppermost leaflet pair decurrent; terminal leaflet linear-oblong to elliptic, base cuneate, apex obtuse to acute, margin entire, often revolute, pubescent on both sides. Flowers 2–4 cm across, terminal, solitary. Pedicels white-hairy. Episepals leafy, oblong to elliptic, apex acute, sometimes two-lobed, often revolute, sparsely or densely pubescent on both sides. Sepals oblong-ovate, apex acute, glabrescent inside, sparsely or densely pubescent outside. Petals yellow, obovate to nearly orbicular, apex rounded. Stamens 20–35, anthers compressed cordate-oblong. Carpels many, on dome-shaped receptacle, styles subbasal, tapering toward base; ovary with dense long hairs.

Distribution: Asia, Europe and N America.

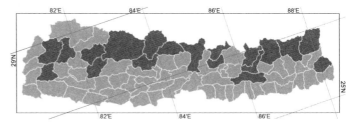

Var. *fruticosa* is widespread in the N hemisphere and is distinguished from the other varieties by its leaflets which are sparsely pilose or subglabrous below, with inconspicuous venation and flat margins.

1a   Stems prostrate, 5–20 cm...................... var. ***pumila***

   b   Stems erect, 30–120 cm.. 2

2a   Leaves sparsely or densely lanate above, glabrescent or lanate on veins beneath .... var. ***arbuscula***

   b   Leaves densely white hairy above, white tomentose except midvein beneath.................... var. ***ochreata***

Potentilla fruticosa var. *arbuscula* (D.Don) Maxim., Mélanges Biol. Bull. Phys.-Math. Acad. Imp. Sci. Saint-Pétersbourg 9: 158 (1873).
Potentilla arbuscula D.Don, Prodr. Fl. Nepal.: 256 (1825); *P. fruticosa* var. *rigida* (Wall. ex Lehm.) Wolf; *P. nepalensis* D.Don later homonym, non Hook.; *P. rigida* Wall. ex Lehm.; *P. rigida* Wall. nom. nud.

Stems erect, 50–120 cm. Petioles 0.5–1.5 cm. Leaves usually with two pairs of lateral leaflets, rarely trifoliolate, 1–3.5 × 0.8–2.5 cm, sparsely or densely lanate above, glabrescent or lanate on veins beneath, terminal leaflet 0.5–1.7 × 0.2–0.7 cm. Flowers 2–4 cm across. Episepals 5–10 × 2–4 mm. Sepals 4–8 × 3–4 mm. Petals 10–15(–20) × 9–12(–17) mm. Anthers 1–1.2 × ca. 1 mm. Carpels ca. 0.5 × 0.3 mm, styles 1.2–1.5 mm.
Fig. 49a–b, front cover.

Distribution: Nepal, W Himalaya, E Himalaya and Tibetan Plateau.

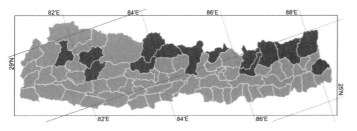

Altitudinal range: 2700–6000 m.

Ecology: Open meadows, thickets, forest edges, rocky slopes.

Flowering: July–September.

Potentilla fruticosa var. *ochreata* Lindl. ex Lehm., Revis. Potentill.: 17 (1856).
Potentilla ochreata Lindl. ex Wall. nom. nud.

Stems erect, 30–100 cm. Petioles 0.2–0.8 cm. Leaves with 2–3 pairs of lateral leaflets, lowermost pair of lateral leaflet often deeply divided into two lobes, 0.5–1.5 × 0.3–1.2 cm, densely white hairy above, white tomentose except midvein beneath, terminal leaflet 2–8 × 0.5–2.5 mm. Flowers 2–2.5 cm across. Episepals 4–6 × 1.5–2.5 mm. Sepals 4–6 × 3–4 mm. Petals 8–12 × 6–8 mm. Anthers ca. 1.5 × 1 mm. Carpels ca. 0.6 × 0.3 mm, styles ca. 1.5 mm.

Distribution: Nepal and Tibetan Plateau.

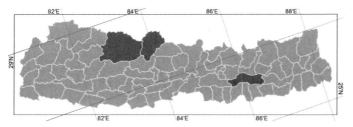

Altitudinal range: 3300–4600 m.

Ecology: Open meadows, rocky slopes, sandy river banks.

Flowering: July–September.

Potentilla fruticosa var. *pumila* Hook.f., Fl. Brit. Ind. 2[5]: 348 (1878).
Potentilla arbuscula var. *pumila* (Hook.f.) Hand.-Mazz.

Stems prostrate, 5–20 cm. Petioles 0.3–0.8 cm. Leaves with two pairs of lateral leaflets, 0.8–1.5 × 0.8–1.3 cm, sparsely or densely lanate above, glabrescent or lanate on midvein beneath, terminal leaflet 0.3–0.8 × 0.2–0.3 cm. Flowers 2–3 cm across. Episepals 5–6 × 2–4 mm. Sepals 5–7 × 3–4 mm. Petals 9–15 × 8–13 mm. Anthers 1–1.2 × ca. 1 mm. Carpels ca. 0.5 × 0.3 mm, styles ca. 1.5 mm.

Distribution: Nepal, W Himalaya, E Himalaya and Tibetan Plateau.

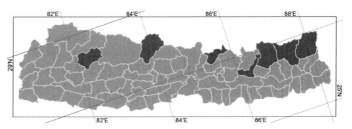

Altitudinal range: 3800–6000 m.

Ecology: Open meadows, thickets, forest edges, rocky slopes, sandy river banks.

Flowering: July–September.

2. *Potentilla cuneata* Wall. ex Lehm., Nov. Stirp. Pug. 3: 34 (1831).
Potentilla ambigua Cambess.; *P. cuneata* Wall. nom. nud.; *P. cuneifolia* Bertol.; *P. dolichopogon* H.Lév.

Prostrate mat-forming subshrubs to 10cm. Stems much branched, slender, woody. Stipules of radical leaves membranous, brown, glabrous inside, spreading pilose outside. Leaves trifoliolate, 2–3(–4) × 0.8–2 cm; leaflets subsessile or shortly petiolulate, obovate to narrowly obovate, 5–12 × 3–7 mm, base cuneate, apex acute 3-dentate, margin entire, densely or sparsely pilose on both surfaces. Flowering stems erect or ascending, 2–10 cm, with appressed or ascending hairs. Stipules of cauline leaves leaf-like, ovate to lanceolate, glabrescent inside, sparsely pilose outside. Flowers 1.6–2.5 cm across, solitary or in 2–3-flowered cymes. Episepals oblong to elliptic, 3–5 × 2–2.5 mm, entire or notched at apex, apex obtuse, sparsely pilose on both sides. Sepals triangular-ovate, 4–7 × 2.5–3.5 mm, apex acute or obtuse, glabrous except short brown hairy near apex and margin inside, sparsely pilose outside. Petals yellow, broadly obovate, 6–10 × 7–11 mm, apex rounded or slightly emarginate. Stamens ca. 20, anthers compressed ellipsoid to ovoid, 0.6–0.8 × 0.8–1 mm. Carpels obliquely ellipsoid, 0.6–0.8 ×

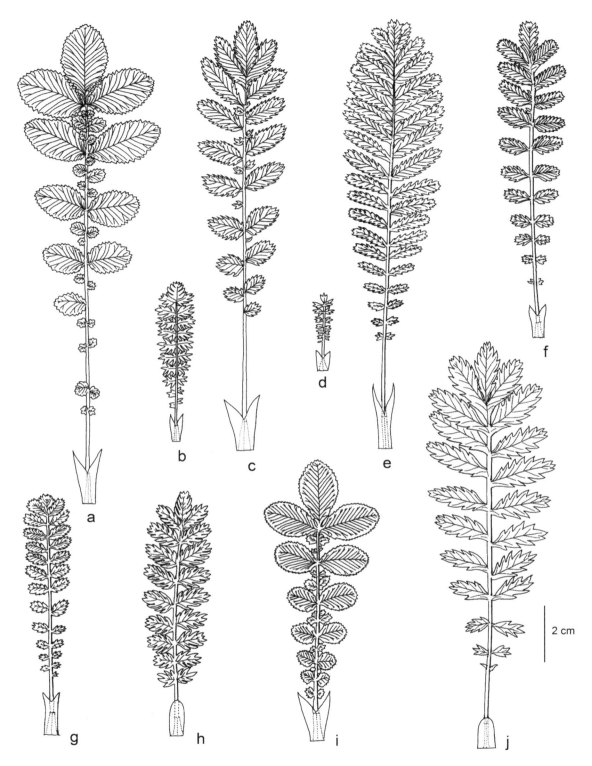

FIG. 48. ROSACEAE. Radical leaves of *Potentilla* species. a, **Potentilla polyphylla** var. **polyphylla**; b, **Potentilla aristata**; c, **Potentilla josephiana**; d, **Potentilla microphylla**; e, **Potentilla contigua**; f, **Potentilla leuconota**; g, **Potentilla tristis**; h, **Potentilla commutata** var. **commutata**; i, **Potentilla lineata**; j, **Potentilla peduncularis** var. **peduncularis**.

0.4–0.5 mm, with 3–4 mm hairs, styles subbasal, filiform, 3.5–4.5 mm, stigmas slightly inflated, papillate. Achenes ellipsoid-ovoid, ca. 1 × 0.8 mm, long hairy.

Distribution: Nepal, W Himalaya, E Himalaya, Tibetan Plateau and E Asia.

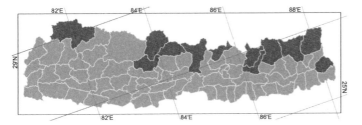

Altitudinal range: 1900–4900 m.

Ecology: Open meadows, sandy river banks, forest edges, rock crevices.

Flowering: May–September. **Fruiting:** July–October.

3. Potentilla eriocarpa Wall. ex Lehm., Nov. Stirp. Pug. 3: 35 (1831).
Potentilla eriocarpa Wall. nom. nud.

Prostrate mat-forming subshrubs to 12 cm. Stems slender or robust, woody, densely covered with remains of old stipules. Stipules of radical leaves membranous, brown, abaxially white-villous. Radical leaves dark green above, green below, palmately ternate, including petiole 3–7 cm; leaflets short-petiolulate or subsessile, obovate-elliptic or rhombic-elliptic, base cuneate, apically serrately or dentately (3–)5–7(–9)-divided, teeth ovate or elliptic-ovate, sometimes narrowly lanceolate, apex acute or acuminate, lower margin entire, sparsely pilose or glabrescent above, sparsely white-villous along veins below, other parts glabrescent later. Flowering stems erect or ascending, 4–12 cm, sparsely white-villous or glabrescent. Cauline leaves absent or only with bracteole leaf, or occasionally 3-foliolate. Stipules of cauline leaves herbaceous, ovate-elliptic, entire or slightly serrate, apex acuminate. Flowers 1.8–2.8 cm across, in 1–3(–4)-flowered cymes. Pedicels 10–30 mm, white-villous or glabrescent. Episepals oblong-elliptic or elliptic-lanceolate, entire or apically bifid, adaxially sparsely pilose or subglabrous. Sepals triangular-ovate, apically acuminate. Petals yellow, broadly obovate to nearly rounded, apically emarginate or rounded. Stamens ca. 20, filaments filiform, anthers globose, bright red or sometimes pale orange before dehiscent. Carpels oblong-ellipsoid, dense long white hairy; styles subterminal, filiform, stigmas slightly dilated.

Distribution: Nepal, W Himalaya, E Himalaya, Tibetan Plateau and E Asia.

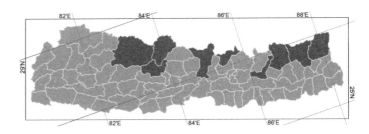

1a   Stems slender, 2–5(–8) mm in diameter. Leaves herbaceous. Terminal leaflet dentately 3–7-divided. Petals bright yellow with orange base, apparently emarginate .... var. **eriocarpa**

 b   Stems robust, 5–10 mm in diameter. Leaves somewhat coriaceous. Terminal leaflet serrately 5–9-divided.Petals creamy yellow with orange base, apically rounded or slightly emarginate .... var. **major**

Potentilla eriocarpa Wall. ex Lehm. var. **eriocarpa**

Stems slender, 2–5(–8) mm in diameter. Leaves herbaceous, terminal leaflet dentately 3–7-divided. Episepals 3.3–6 × 1.5–3 mm. Sepals 5–7.5 × 3.5–4 mm. Petals bright yellow with orange base, 8–12 mm long, 7–13 mm wide, apically apparently emarginate. Longer stamens 4.5–7 mm, anthers 1–1.2 mm across. Carpels 0.7–0.9 mm × ca. 0.4 mm, styles 2.3–3.3 mm.
Fig. 49c–d.

Distribution: Nepal, W Himalaya, E Himalaya, Tibetan Plateau and E Asia.

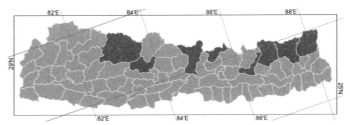

Altitudinal range: 3500–5100 m.

Ecology: Rocky crevices.

Flowering: July–September.

Potentilla eriocarpa var. **major** Kitam., Acta Phytotax. Geobot. 15(5): 130 (1954).

Stems robust, 5–10 mm in diameter. Leaves somewhat coriaceous, terminal leaflet serrately 5–9-divided. Episepals 6.8–10 mm × 2.5–4 mm. Sepals 9.5–10.5 × 5–7 mm. Petals creamy yellow with orange base, 8–10 × 7–11 mm, apically rounded or slightly emarginate. Longer stamens 6.5–8.5 mm, anthers 1.2–1.8 mm across. Carpels 0.8–1.1 × 0.4–0.5 mm, styles 4.2–5.2 mm long.

Distribution: Endemic to Nepal.

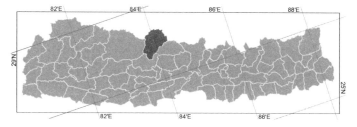

Altitudinal range: 3400–3900 m.

Ecology: Rocky crevices.

Flowering: July–September.

4. *Potentilla biflora* Willd. ex Schltdl., Ges. Naturf. Freunde Berlin Mag. Neuesten Entdeck. Gesammten Naturk. 7: 197 (1816).
Potentilla fruticosa var. *inglisii* (Royle) Hook.f.; *P. inglisii* Royle; *P. inglisii* var. *lahulensis* Wolf.

Perennial cushion-forming herbs. Rhizomes stout, much branched near ground, covered with remains of old leaves. Petioles 1–2 cm, lower part adnate with stipules. Radical leaves crowded, pinnately or subpalmately 5–7-foliolate, 1.5–2.5 × 0.4–1 cm; leaflets broadly linear, 0.6–1.2 × 0.1–0.2 cm, base narrowly cuneate, apex acute to obtuse, margin revolute, densely strigose on both sides. Flowering stems from radical leaves, 1–2 cm, densely villous with white hairs. Flowers 1.2–1.8 cm across, solitary or rarely in 2–3-flowered cymes. Episepals oblong-elliptic, 4–6 × 2–3 mm, entire, apex acute or obtuse, sparsely strigose on both sides. Sepals ovate to widely ovate, 5–7 × 4–6 mm, entire, apex acute or obtuse, glabrous except shortly lanate near apex inside, sparsely strigose outside and margin. Petals yellow, broadly elliptic to obovate, 5–7 × 5–8 mm, apex rounded or slightly emarginate. Stamens 20, anthers ellipsoid, ca. 0.8 × 0.6 mm. Carpels oblong-ellipsoid, ca. 1 × 0.6 mm, glabrous but concealed by long hairs from receptacle, styles subterminal, filiform, 2–2.5 mm, stigmas slightly inflated.

Distribution: Nepal, W Himalaya, Tibetan Plateau, N Asia and N America.

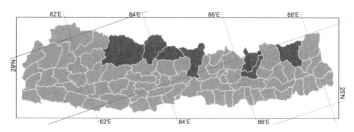

Altitudinal range: 4000–5000 m.

Ecology: Alpine meadows, rocky crevices, among gravel.

Flowering: June–October. **Fruiting:** June–October.

5. *Potentilla bifurca* L., Sp. Pl. 1: 497 (1753).

Perennial dioecious herbs. Rhizomes woody, much branching, slender and long-creeping. Stipules of lower leaves membranous, brown, glabrous inside, pilose outside; stipules of upper leaves herbaceous, ovate to lanceolate, glabrescent inside, sparsely pilose outside. Leaves pinnate, with 3–8 pairs of lateral leaflets, 1–5 × 0.6–1.5 cm; leaflets oblong to elliptic, sessile or uppermost 2 or 3 pairs decurrent, 3–12 × 2–5 mm, base cuneate, apex acute, margin entire or dentately divided into two or three lobes at apex, glabrescent above, sparsely appressed pilose below and on margin. Flowers 0.7–1.0 cm across, terminal, solitary or several in cymes. Episepals oblong-elliptic, 2–3 × 1–1.5 mm, entire or notched at apex, apex acute or obtuse, glabrescent inside, sparsely pilose outside. Sepals triangular-ovate, 3–4 × 2–2.5 mm, apex acute or obtuse, glabrous except shortly pubescent near apex inside, sparsely pilose outside. Petals yellow, broadly elliptic to obovate, 4–5 × 3–4 mm, apex rounded or slightly emarginate. Stamens ca. 20, anthers compressed ellipsoid, 0.5–0.8 × 0.5–0.7 mm. Carpels obliquely ellipsoid, 0.4–0.5 × ca. 0.3 mm, glabrous, styles lateral to subbasal, rod-shaped, tapered toward base, ca. 1.5 mm long, stigmas inflated, papillate. Achenes ellipsoid-ovoid, ca. 1.5 × 1.2 mm, smooth.

Distribution: Asia and Europe.

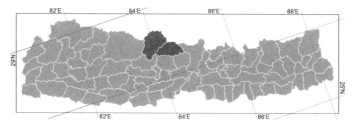

Altitudinal range: 3100–5200 m.

Ecology: Sandy river banks, open meadows, among gravel.

Flowering: May–September. **Fruiting:** May–September.

6. *Potentilla festiva* Soják, Candollea 43: 166 (1988).

Perennial rosulate herbs to 30 cm. Rhizomes stout, terete. Stipules of radical leaves with membranous auricles, connate in basal part. Petioles 1–2 cm. Radical leaves oblanceolate, 4–15 × 2–4 cm, pinnate with 5–15 pairs of lateral leaflets, with alternating smaller leaflets; base of uppermost leaflet pair cuneate; terminal leaflet subsessile, oblong to narrowly obovate, 1–3 × 0.5–1.2 cm, sharply serrate with 18–38 teeth, lanate above, white sericeous below. Flowering stems from radical leaves, 5–30 cm tall, with patent or ascending white hairs. Stipules of cauline leaves with leafy auricles, serrate with 5–15 teeth. Cauline leaves similar to radical leaves, reducing in size upwards, lower leaves with 5–10 pairs of leaflets, upper leaves with 2–3 pairs of leaflets, smaller leaflets absent. Flowers 0.7–1.2 cm across, in 5–9-flowered cymes. Pedicels 0.5–2 cm, with patent or ascending hairs. Episepals oblong to obovate, 3–5 × 1.5–3 mm, entire or with 3 teeth, apex acute or obtuse, sparsely strigose above, sericeous below. Sepals elliptic to ovate, 4–6 × 2.5–4.5 mm, entire, apex acute or obtuse, glabrous except lanate near tips above, strigose below. Petals yellow, oblong to elliptic, 5–7 mm × 4.0–5.5 mm, apex rounded. Stamens 20, anthers globose to ellipsoid, 0.6–1.0 × 0.4–0.9 mm. Carpels ellipsoid, 0.6–0.8 × 0.4–0.6 mm, glabrous, styles lateral, slender, 0.9–1.3 mm, stigmas slightly inflated.

Distribution: Nepal, W Himalaya, E Himalaya, Tibetan Plateau, Assam-Burma and E Asia.

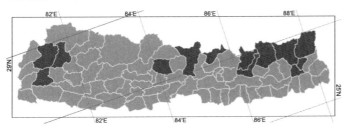

Altitudinal range: 2200–3800 m.

Ecology: Open meadows, forest margins, roadsides.

Flowering: June–October. **Fruiting:** June–October.

Ohashi (Enum. Fl. Pl. Nepal 2: 140) partially misapplied the name *Potentilla fulgens* Wall. ex Hook. to this species.

7. *Potentilla josephiana* H.Ikeda & H.Ohba, Bot. J. Linn. Soc. 112(2): 168 (1993).
Potentilla fulgens var. *intermedia* Hook.f.; *P. lineata* var. *intermedia* (Hook.f.) R.D.Dixit & Panigrahi.

Perennial rosulate herbs to 35 cm. Rhizomes stout, terete. Stipules of radical leaves with free, membranous auricles. Petioles 2–5 cm. Radical leaves oblanceolate, 4–20 cm × 2–5 cm, pinnate with 5–15 pairs of lateral leaflets alternating with smaller leaflets; base of uppermost leaflet pair decurrent; leaflets appressed short hairy above, pubescent, especially villous on main and lateral veins below; terminal leaflet sessile, broadly lanceolate to oblong, 1–3 × 0.5–1.0 cm, sharply serrate with 20–30 teeth. Flowering stems from radical leaves, 5–35 cm, with lanate hairs. Stipules of cauline leaves with leaf-like auricles, serrate with 7–12 teeth. Cauline leaves similar to radical leaves, reducing in size upwards, lower leaves with 5–12 pairs of lateral leaflets, upper leaves with 1 or 2 pairs, smaller leaflets absent. Flowers 1–1.5 cm across. Pedicels 2–4 cm, with appressed or ascending white hairs. Episepals oblong to elliptic, 3–6 × 2–4 mm, entire or with 3–6 teeth, apex acute to acuminate, villous on both sides. Sepals ovate to broadly ovate, 3–6 × 3–5 mm, entire, apex acute to obtuse, glabrous except lanate near tips above, villous below and margin. Petals yellow, obovate to broad obovate, 5–10 × 5–8 mm, apex rounded. Stamens 20, anthers globose, 0.6–1.0 × 0.6–1.0 mm. Carpels ellipsoid to ovoid, 0.5–1.0 × 0.4–0.8 mm, glabrous, styles lateral, 0.8–1.4 mm, stigmas slightly inflated.
Fig. 48c.

Distribution: Nepal, W Himalaya and E Himalaya.

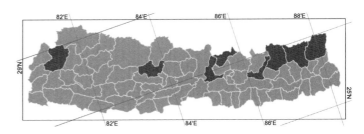

Altitudinal range: 3200–4100 m.

Ecology: Alpine meadows, forest margins.

Flowering: June–August. **Fruiting:** June–August.

8. *Potentilla lineata* Trevir., Index Sem. Hort. Bot. Vratisl.: unpaginated (1822).
Potentilla fulgens Wall. ex Hook.; *P. fulgens* var. *acutiserrata* (T.T.Yu & C.L.Li) T.T.Yu & C.L.Li; *P. lineata* subsp. *exortiva* Soják; *P. martinii* H.Lév.; *P. siemersiana* Lehm.; *P. siemersiana* var. *acutiserrata* T.T.Yu & C.L.Li; *P. splendens* Wall. ex D.Don later homonym, non Ramond.

बज्रदन्ती Bajradanti (Nepali).

Perennial rosulate herbs to 50 cm. Rhizomes stout, terete. Stipules of radical leaves with free, membranous auricles. Petioles 2–5 cm. Radical leaves oblanceolate, 4–30 × 2–8 cm, pinnate with 5–20 pairs of lateral leaflets with alternating smaller leaflets; base of uppermost leaflet pair cuneate; leaflets sparsely or densely lanate above, white sericeous below; terminal leaflet subsessile, oblong to broadly obovate, 1.5–4 × 0.8–1.5 cm, sharply serrate with 20–40 teeth. Flowering stems from radical leaves, (5–)10–50 cm, white hairy throughout, mixed with glandular hairs upper part. Stipules of cauline leaves with leaf-like auricles, serrate with 10–20 teeth Cauline leaves similar to radical leaves, reducing in size upwards, lower leaves with 5–13 pairs of leaflets, upper leaves with 1–2 pairs, smaller leaflets absent. Flowers 1–1.5 cm across. Pedicels 2–4 cm, white hairy with glandular hairs. Episepals oblong to elliptic, 5–7 × 3–6 mm, entire with acute tips, lanate above, white sericeous below. Sepals elliptic to ovate, 5–9 × 4–6 mm, entire or with 2 or 3 teeth, apex acute to obtuse, lanate near tips but otherwise glabrous above, white sericeous beneath. Petals obovate to broadly obovate with round or retuse apex, 8–15 × 7–13 mm, margin often recurved. Long stamens 3–5 mm, anthers ovoid, 1–1.3 × 0.6–1.2 mm. Carpels narrowly ovoid, 1.2–2 × 0.8–1.4 mm, glabrous, styles lateral, 1.4–2 mm, stigmas not inflated.
Fig. 48i.

Distribution: Nepal, W Himalaya, E Himalaya, Tibetan Plateau, Assam-Burma, E Asia and SE Asia.

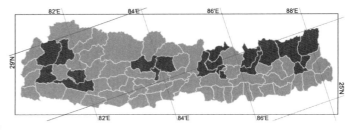

Altitudinal range: 1700–3700 m.

Ecology: Open meadows, forest margins, roadsides.

Flowering: June–October. **Fruiting:** June–October.

Roots are used to relieve throat infection. Root paste is stomachic and is used to treat peptic ulcers, coughs and colds.

9. ***Potentilla polyphylla*** Wall. ex Lehm., Nov. Stirp. Pug. 3: 13 (1831).
Potentilla mooniana Wight; *P. polyphylla* Wall. nom. nud.; *P. polyphylla* var. *barbata* Lehm.; *P. sordida* Klotzsch.

Perennial rosulate herbs to 40 cm. Rhizomes stout, terete. Stipules of radical leaves with membranous, free auricles. Petioles 2–5 cm. Radical leaves oblanceolate; lateral leaflets 5–20 pairs with alternating smaller leaflets; base of uppermost leaflet pair cuneate; terminal leaflet petiolulate or subsessile, margin sharply serrate. Flowering stems from radical leaves, 5–40 cm, with patent or ascending white hairs. Stipules of cauline leaves with leaf-like auricles, serrate with 10–20 teeth. Cauline leaves similar to radical leaves, reducing in size upwards, lower leaves with 5–13 pairs of leaflets, upper leaves with 1–2 pairs, smaller leaflets absent. Flowers 1–3 cm across. Episepals oblong to obovate, usually serrate with 3–5 teeth, rarely entire, apex acute, glabrescent or sparsely strigose inside, strigose outside. Sepals elliptic to ovate, entire, apex acute to obtuse, lanate inside, strigose outside and margin. Petals yellow, obovate to broadly obovate, apex rounded or shallowly emarginate. Stamens 20, anthers globose. Carpels ellipsoid to ovoid, glabrous, styles nearly basal, fusiform or rod-shaped, stigmas slightly inflated.

Distribution: Nepal, W Himalaya, E Himalaya, Tibetan Plateau, Assam-Burma, S Asia, E Asia and SE Asia.

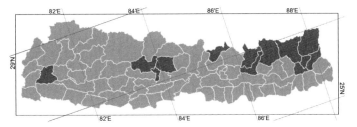

1a Radical leaves 16–30 cm. Terminal leaflets oblong to oblong-oblanceolate, densely villous on both sides.. var. ***interrupta***
 b Radical leaves 4–20(–25) cm. Terminal leaflets elliptic to obovate, sparsely strigose on both sides.. 2

2a Petals 9–11 × 9–12 mm. Episepals 4.5–7.5 mm .... var. ***himalaica***
 b Petals 5–9 × 4.5–9 mm. Episepals 2.5–5 mm .... var. ***polyphylla***

Potentilla polyphylla var. ***himalaica*** H.Ikeda & H.Ohba, Bot. J. Linn. Soc. 112: 179 (1993).

Radical leaves 8–20(–25) × 2–5 cm, leaflets sparsely strigose on both sides, terminal leaflet oblong to obovate, 1–3 × 1–2 cm, serrate with 20–25 teeth. Flowers 1.5–3 cm across. Episepals 4–7 × 3–6 mm. Sepals 4–6 × 3–4.5 mm. Petals 9–11 × 9–12 mm. Anthers 0.9–1.2 × 1–1.2 mm. Carpels 0.7–0.8 × 0.4–0.6 mm, styles 1–1.4 mm, rod-shaped.

Distribution: Nepal and E Himalaya.

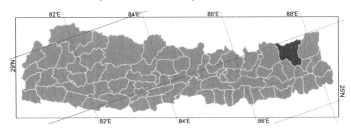

Altitudinal range: 3400–4000 m.

Ecology: Open alpine meadows.

Flowering: July–September. **Fruiting:** July–September.

Potentilla polyphylla var. ***interrupta*** (T.T.Yu & C.L.Li) H.Ikeda & H.Ohba, Bot. J. Linn. Soc. 112(2): 179 (1993).
Potentilla interrupta T.T.Yu & C.L.Li, Acta Phytotax. Sin. 18: 8 (1980).

Radical leaves 16–30 × 2–5 cm, leaflets densely villous on both sides, terminal leaflet oblong to oblong-oblanceolate, 3–6 × 1.5–2.5 cm, serrate with 25–50 teeth. Flowers 1.5–2 cm across. Episepals 3–5 × 2–3 mm. Sepals 4–5.5 × 3–4 mm. Petals 8–9.5 × 6–8.5 mm. Anthers 0.6–1 × 0.5–0.8 mm. Carpels 0.7–0.9 × 0.5–0.6 mm, styles 1–1.3 mm, rod-shaped.

Distribution: Nepal, E Himalaya, Tibetan Plateau and E Asia.

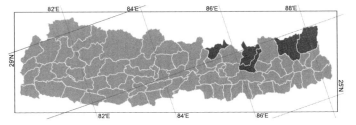

Altitudinal range: 3500–4200 m.

Ecology: Open alpine meadows.

Flowering: July–September. **Fruiting:** July–September.

Potentilla polyphylla Wall. ex Lehm. var. ***polyphylla***

Radical leaves 4–20(–25) × 2–6 cm, leaflets sparsely strigose on both sides, terminal leaflet elliptic to obovate, 1.5–4 × 0.8–1.5 cm, serrate with 14–24 teeth. Flowers 1–1.5 cm across. Episepals 2.5–5 × 1.5–3.5 mm. Sepals 2.5–5.5 × 2–4 mm. Petals 5–9 × 4.5–9 mm. Anthers 1–1.3

× 0.6–1.2 mm. Carpels 0.6–0.8 × 0.4–0.6 mm, styles 0.9–1.2 mm, fusiform or rod-shaped.
Fig. 48a.

Distribution: Nepal, W Himalaya, E Himalaya, Assam-Burma, S Asia and SE Asia.

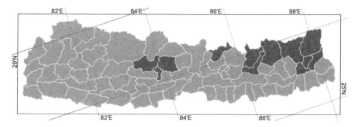

Altitudinal range: 1400–4000 m.

Ecology: Open meadows, forest margins, roadsides.

Flowering: July–September. **Fruiting:** July–September.

10. *Potentilla microphylla* D.Don, Prodr. Fl. Nepal.: 231 (1825).

Potentilla microphylla var. *depressa* Wall. ex Lehm.; *P. microphylla* var. *glabriuscula* Wall. ex Lehm.; *P. microphylla* var. *latifolia* Wall. ex Lehm.; *P. microphylla* var. *latiloba* Hook.f.

ताछेर Tachher (Tibetan).

Perennial cushion-forming herbs to 10 cm. Stipules of radical leaves with membranous, free auricles. Peticles 0.5–4 cm. Radical leaves oblanceolate, 0.5–6 × 0.4–1 cm, with 2–10 pairs of lateral leaflets, smaller leaflets absent; base of uppermost pair of leaflets cuneate; leaflets glabrous or sparsely strigose above, strigose or hirsute below; terminal leaflet sessile or subsessile, oblong to narrowly obovate, 2–4 × 1.5–2.5 mm, deeply serrate with 3–9 teeth. Flowering stems slender, much-branched near ground, 1–10 cm. Stipules of cauline leaves with entire or rarely 3-lobed, membranous auricles. Cauline leaves simple, entire or tri-lobed. Pedicel 2–12 mm, densely villous. Flowers 1–1.5 cm across, solitary or in 2- or 3-flowered cymes. Episepals lanceolate to oblong, 1.5–3.5 × 0.5–1 mm, entire or deeply divided into two lobes, apex acute or obtuse, sparsely strigose on both sides. Sepals elliptic to ovate, 3.5–4.5 × 1.5–3 mm, entire, apex acute or obtuse, glabrescent above, sparsely strigose below and margin. Petals yellow, oblong to elliptic, 3–8 × 2.5–6 mm, apex rounded or slightly emarginate. Stamens 20, anthers globose to ellipsoid, 0.4–0.7 × 0.4–0.7 mm. Carpels ellipsoid, 0.5–0.8 × 0.4–0.5 mm, glabrous, styles lateral, 0.5–1.3 mm, stigmas slightly inflated.
Fig. 48d.

Distribution: Nepal, W Himalaya, E Himalaya, Tibetan Plateau and Assam-Burma.

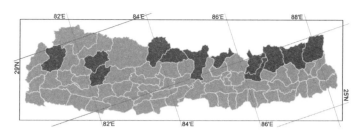

Altitudinal range: 3400–5200 m.

Ecology: Open meadows, gravelly slopes, moist silty ground.

Flowering: July–September. **Fruiting:** July–September.

11. *Potentilla makaluensis* H.Ikeda & H.Ohba, J. Jap. Bot. 67(3): 149 (1992).

Perennial rhizomatous herbs to 5 cm. Rhizomes slender, 1.5–3.5 cm, with sparse degenerated leaves. Stipules of radical leaves with membranous auricles connate from base to top, apex rounded. Petioles 4–10 mm. Radical leaves oblanceolate, 2–4 cm with 4–5 pairs of lateral leaflets, without smaller leaflet; base of uppermost pair of leaflets cuneate; leaflets glabrescent above, strigose below; terminal leaflet subsessile, oblong to obovate, 3–7 × 3–4.5 mm, deeply serrate with 5–9 teeth. Flowering stems from radical leaves, 1.5–5 cm tall. Stipules of cauline leaves with leaf-like auricles. Cauline leaves simple, oblanceolate, apically 1–3 lobulate. Flowers 1–1.5 cm across, solitary. Episepals oblong to elliptic, 2–3 × 1–1.5 mm, entire or shallowly incised, apex acute, glabrescent above, strigose below and margin. Sepals elliptic to ovate, 3–4 × 2–3 mm, entire, apex acute to obtuse, glabrous except puberulent near apex above, strigose below and margin. Petals yellow, obovate to broadly obovate, 5.5–7 × 5–7 mm, apex rounded or shallowly emarginate. Stamens 20, anthers orbicular, 0.7–1 × 0.6–0.9 mm. Carpels ellipsoid to ovoid, 0.7–0.9 × 0.6–1 mm, glabrous, styles lateral, 1–1.2 mm, stigmas slightly or apparently inflated.

Distribution: Endemic to Nepal.

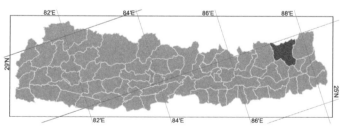

Altitudinal range: ca. 4000 m.

Ecology: Wet, silty ground.

Flowering: July–August. **Fruiting:** July–August.

Known from several collections around the Shipton La, Sankhuwasabha District.

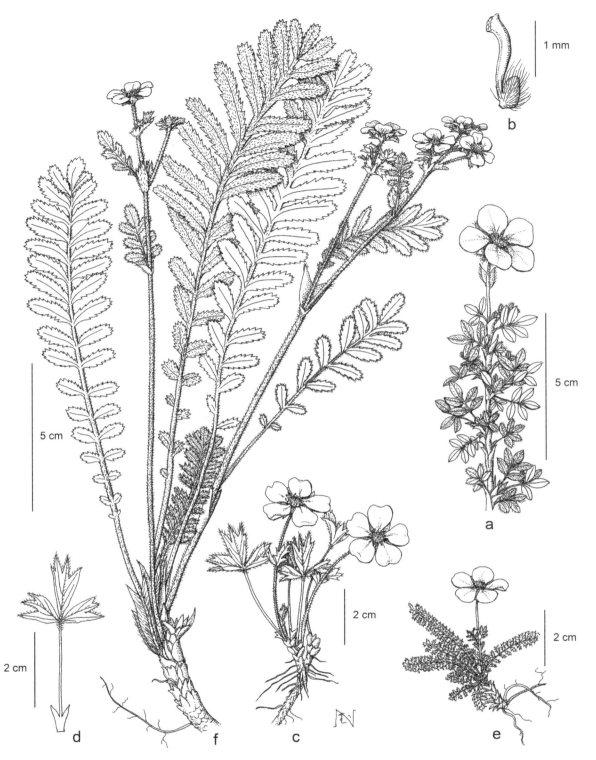

1 mm

5 cm

5 cm

2 cm

2 cm

2 cm

FIG. 49. ROSACEAE. **Potentilla fruticosa** var. **arbuscula**: a, flowering shoot; b, pistil. **Potentilla eriocarpa** var. **eriocarpa**: c, flowering plant; d, radical leaf. **Potentilla aristata**: e, flowering plant. **Potentilla contigua**: f, flowering plant.

12. *Potentilla glabriuscula* (T.T.Yu & C.L.Li) Soják, Candollea 43: 453 (1988).
Sibbaldia glabriuscula T.T.Yu & C.L.Li, Acta Phytotax. Sin. 19: 516 (1981).

Perennial rosulate herbs to 2.5 cm. Rhizomes stout, short, few-branched. Stipules of radical leaves with membranous auricles, connate from base to top, apex rounded. Petioles 3–12 mm. Radical leaves oblanceolate, 1.8–5.0 × 0.5–1.0 cm, with 3–7 pairs of lateral leaflets, without smaller leaflets; base of uppermost pair of leaflets cuneate; sparsely strigose on both sides of leaflets and rachis; terminal leaflet subsessile, oblong to narrowly obovate, 3.5–4.2 × 2–4 mm, serrate with 3–7 teeth. Flowering stems from radical leaves, 0.5–2.5 cm, with appressed or ascending hairs. Stipules of cauline leaves with entire auricles. Cauline leaves simple, entire or 3-lobed. Flowers 0.4–1.2 cm across, solitary. Episepals lanceolate to oblong, 1.8–3.0 × 0.4–1.2 mm, entire, apex acute or obtuse, glabrescent above, sparsely strigose below. Sepals elliptic to ovate, 2.0–3.0 × 1.0–1.8 mm, entire, apex acute or obtuse, glabrous except puberulent near apex above, sparsely strigose below and margin. Petals yellow, oblong to elliptic, 3.0–5.0 × 2.0–4.0 mm, apex rounded. Stamens 5–10, anthers globose to ellipsoid, 0.5–0.7 × 0.4–0.8 mm. Carpels ellipsoid, 0.6–0.8 × 0.4–0.5 mm, glabrous, styles lateral, 0.6–0.8 mm, stigmas slightly inflated.

Distribution: Nepal, E Himalaya, Tibetan Plateau, Assam-Burma and E Asia.

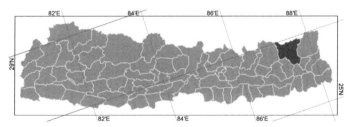

Altitudinal range: 3500–5000 m.

Ecology: Alpine meadows, moist silty ground.

Flowering: June–October. **Fruiting:** June–October.

13. *Potentilla aristata* Soják, Candollea 43: 159, pl. 1 fig. 7-9 (1988).
Potentilla microphylla var. *achilleifolia* Hook.f.

Perennial rosulate herbs to 4 cm. Rhizomes short, with several slender roots. Stipules of radical leaves with membranous, free auricles. Petioles 5–8 mm. Radical leaves oblanceolate, 1.5–5(–9) × 0.5–1.2 cm, with 13–16 pairs of lateral leaflets, without alternating smaller leaflets; base of uppermost pair of leaflets cuneate; leaflets glabrescent above, strigose beneath and margins. Terminal leaflet subsessile, oblong to narrowly obovate, 2–3 × 1.5–2.5 mm, deeply serrate with 8–12 teeth. Flowering stems from radical leaves, 1–4 cm, with appressed white hairs. Stipules of cauline leaves with entire, leaf-like auricles. Cauline leaves simple to with 4 or 5 pairs of leaflets. Flower 1–1.5 cm across, solitary. Episepals

lanceolate to oblong, 2–3 × 1.5–2.5 mm, entire or deeply divided into two lobes, apex acute or obtuse, sparsely strigose on both sides. Sepals elliptic to ovate, 2.5–4 × 1.5–2.5 mm, entire, apex acute or obtuse, glabrescent above, strigose below. Petals yellow, oblong to elliptic, 6–7 × 4.5–6 mm, apex rounded. Stamens 20, anthers globose to ellipsoid, 0.5–0.7 × 0.5–0.7 mm. Carpels glabrous except usually sparsely strigose near top, ellipsoid, 0.7–1.2 × 0.5–0.8 mm, styles lateral, 1.3–1.8 mm, stigmas slightly inflated.
Figs 48b, 49e.

Distribution: Nepal, E Himalaya and Tibetan Plateau.

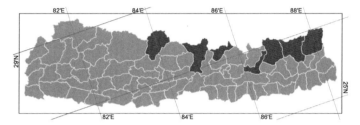

Altitudinal range: 3600–4700 m.

Ecology: Open meadows.

Flowering: June–August. **Fruiting:** June–August.

14. *Potentilla turfosoides* H.Ikeda & H.Ohba, in H.Ohba, Himal. Pl. 3: 62 (1999).
Potentilla turfosa var. *caudiculata* Soják.

Perennial rosulate herbs to 20 cm. Rhizomes stout, terete. Stipules of radical leaves with free auricles. Petioles 1–2 cm. Radical leaves oblanceolate, 3.5–11 × 1.2–2 cm, with 5–8 pairs of lateral leaflets, with smaller leaflets between uppermost and next uppermost pair of leaflets; base of uppermost pair of leaflets cuneate; leaflets glabrescent or sparsely strigose above, glabrescent except villous along veins below; terminal leaflet subsessile, oblong to narrowly obovate, 6–10 × 4–7 mm, serrate with 12–18 teeth. Flowering stems from radical leaves, 3–20 cm, with appressed or ascending hairs. Stipules of lower cauline leaves with entire auricles connate in lower half, those of upper leaves entire or serrate with 2–4 teeth. Cauline leaves similar to radical leaves, reducing in size upwards, lower leaves with 2–4 pairs of leaflets, upper leaves simple or trifoliolate. Flowers 1–1.5 cm across, solitary or in 2(rarely 3)-flowered cymes. Episepals oblong to obovate, 2–3 × 0.8–1.2 mm, entire or deeply divided into two lobes, apex acute or obtuse, glabrescent or sparsely strigose on both sides. Sepals elliptic to ovate, 3.0–3.5 × 2–3 mm, entire, apex acute or obtuse, glabrous except puberulent near tips above, sparsely strigose below. Petals yellow, oblong to elliptic, 5–6.5 × 4.5–6 mm, apex rounded or shallowly emarginate. Stamens 20, anthers globose to ellipsoid, 0.5–0.8 × 0.6–0.7 mm. Ovary glabrous, ellipsoid, 0.7–0.9 × 0.5–0.6 mm, styles lateral, 0.7–0.9 mm, stigmas slightly inflated.

Distribution: Endemic to Nepal.

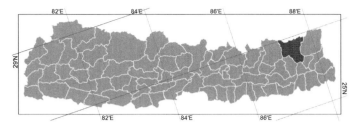

Altitudinal range: 3500–4100 m.

Ecology: Open meadows, gravelly slopes.

Flowering: July–August. **Fruiting:** July–August.

Known from several collections around the Shipton La, Sankhuwasabha District.

15. *Potentilla peduncularis* D.Don, Prodr. Fl. Nepal.: 230 (1825).
Potentilla peduncularis var. *subcontigua* Soják; *P. velutina* Wall. nom. nud.

मूला झार Mula jhaar (Nepali).

Perennial rhizomatous herbs. Rhizomes stout, terete, long-creeping underground. Stipules of radical leaves with auricles connate from base to top, apex rounded. Petioles 0.5–5 cm. Radical leaves oblanceolate, imparipinnate with 10–18 pairs of lateral leaflets, usually without alternating smaller leaflets; base of uppermost pair of leaflets decurrent; terminal leaflet sessile, oblong to narrowly obovate, margin serrate with 10–15 teeth. Flowering stems from radical leaves, 10–25 cm, with dense appressed white hairs. Stipules of cauline leaves with leaf-like auricles, entire or serrate with 3–5 teeth. Cauline leaves simple or with 1 or 2 pairs of leaflets. Flowers 2–3.5 cm across, in 2–4(–7)-flowered cymes. Episepals oblong to obovate, entire or 3-dentate, apex acute or obtuse. Sepals elliptic to ovate, entire, apex acute or obtuse. Petals yellow, oblong to widely elliptic, apex rounded or slightly emarginate. Stamens 20–25, anthers globose to ellipsoid. Carpels glabrous, ellipsoid to narrowly obovoid, styles lateral, rod-shaped, stigmas slightly inflated.

Distribution: Nepal, E Himalaya, Tibetan Plateau and E Asia.

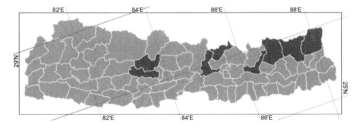

A paste of roots is given to treat profuse menstruation. It is taken with milk to treat diarrhoea.

1a Leaflets sparsely lanate beneath .... var. ***ganeshii***
 b Leaflets sericeous beneath .... var. ***peduncularis***

Potentilla peduncularis var. ***ganeshii*** H.Ikeda & H.Ohba, J. Jap. Bot. 71: 252, pl. 1 fig. i (1996).

Radical leaves 4–20 × 1.5–3 cm, lateral leaflets 10–20 pairs, terminal leaflets 0.8–1.5 × 0.4–0.8 cm, with 5–13 teeth, sparsely lanate on both sides. Flowers 2–2.5 cm across. Episepals 4–6 × 2.5–3.5 mm, sparsely strigose on both sides. Sepals 5–6.5 × 3.5–4.5 mm, glabrescent inside, sparsely strigose outside. Petals 12–14 × 12–13 mm. Anthers 1–1.5 × 1–1.6 mm. Carpels ca. 1 × 0.7 mm, styles 1.5–2 mm.

Distribution: Endemic to Nepal.

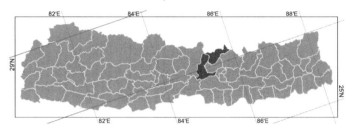

Altitudinal range: 3400–4300 m.

Ecology: Open meadows.

Flowering: July–August. **Fruiting:** July–August.

Potentilla peduncularis D.Don var. ***peduncularis***

Radical leaves 10–25 × 2–6 cm, lateral leaflets 10–15 pairs, terminal leaflets 1.5–3 × 0.5–1 cm, with 10–15 teeth, sparsely or densely lanate above, sericeous beneath. Flowers 2–3.5 cm across. Episepals 6–7 × 2.5–3.5 mm, sparsely strigose inside, densely strigose outside. Sepals 6–7 × 4–5 mm, glabrescent except shortly lanate near tips inside, sparsely to densely strigose outside. Petals 12.5–14.5 × 12–13.5 mm. Anthers 1.2–1.7 × 1–1.5 mm. Carpels 1.2–1.5 × 0.8–1 mm, styles 1.8–2.2 mm.
Fig. 48j.

Distribution: Nepal, E Himalaya and Tibetan Plateau.

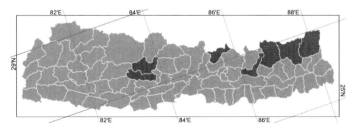

Altitudinal range: 3000–4700 m.

Ecology: Open meadows, thickets, gravelly slopes.

Flowering: June–September. **Fruiting:** June–September.

16. *Potentilla contigua* Soják, Candollea 43: 160, pl. 3 fig. 1 (1988).
Potentilla peduncularis var. *clarkei* Hook.f.

Perennial rhizomatous herbs to 30 cm. Rhizomes stout, terete, creeping underground. Stipules of radical leaves with free auricles. Radical leaves oblanceolate, 10–20 × 1.5–4.5 cm, with 13–16 pairs of lateral leaflets, usually without alternating smaller leaflets; base of uppermost pair of leaflets decurrent; leaflets densely or sparsely villous above, densely villous below; terminal leaflet sessile, oblong to narrowly obovate, 1–2 × 0.5–0.7 cm, serrate with 11–15 teeth. Flowering stems from radical leaves, 6–30 cm, with patent or ascending pale brown hairs. Stipules of cauline leaves with leaf-like auricles, serrate with 3–6 teeth. Cauline leaves similar to radical ones, with 3–5 pairs of leaflets. Flowers 1.5–2.5 cm across, in (1 or)2–5-flowered cymes. Pedicels 1.5–4.0 cm. Episepals oblong to obovate, 4–5 × 2–3 mm, entire or with 3(or 5) teeth, apex acute or obtuse, villous on both sides. Sepals elliptic to ovate, 4.5–6.0 × 3.5–4.5 mm, entire, apex acute or obtuse, glabrous except puberulent near tips above, villous below. Petals yellow, oblong to elliptic, 9.5–10.5 × 9.5–11.0 mm, apex rounded or slightly emarginate. Stamens 20–30, anthers globose to ellipsoid, 1.1–1.3 × 0.9–1.1 mm. Carpels glabrous, ellipsoid, 1.0–1.3 × 0.8–1.0 mm, styles lateral, 1.6–2.0 mm, stigmas slightly inflated. Achenes obliquely ellipsoid, 2–2.5 × 1.3–1.5 mm, rugose.
Figs 48e, 49f.

Distribution: Nepal, E Himalaya and Tibetan Plateau.

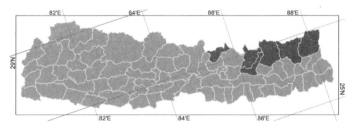

Altitudinal range: 3500–4700 m.

Ecology: Open meadows, thickets.

Flowering: June–September. **Fruiting:** June–September.

Ohashi (Enum. Fl. Pl. Nepal 2: 140. 1979) partially misapplied the name *Potentilla peduncularis* D.Don to this species.

17. *Potentilla cardotiana* Hand.-Mazz., Acta Horti Gothob. 13: 322 (1939).

Potentilla cardotiana var. **nepalensis** H.Ikeda & H.Ohba, in H.Ohba, Himal. Pl. 3: 72 (1999).

Perennial rosulate herbs to 20 cm. Rhizomes stout, terete, short. Stipules of radical leaves with free auricles. Petioles 2–4 cm. Radical leaves oblanceolate, 7–16 cm × 2–3 cm,

with 10–22 pairs of lateral leaflets, usually with alternating smaller leaflets; base of uppermost pair of leaflets decurrent; leaflets strigose with appressed or ascending hairs above, densely villous below; terminal leaflet sessile, lanceolate or narrowly oblong, 1–2 × 0.4–0.7 cm, serrate with 15–38 teeth. Flowering stems from radical leaves, 15–20 cm tall, with appressed white hairs. Stipules of cauline leaves with leaf-like auricles, serrate with 3–6 teeth. Cauline leaves with 1 or 2 pairs of lateral leaflets. Flowers 1–1.5 cm across, in 3–6-flowered cymes. Pedicels 1.5–2 cm. Episepals oblong to obovate, 3–5 × 1.5–3 mm, entire or with 3 teeth, apex acute or obtuse, glabrescent above, densely villous below. Sepals elliptic to ovate, 4–6 × 2.5–4.5 mm, entire, apex acute or obtuse, glabrous except puberulent near tips above, densely villous below. Petals yellow, oblong to elliptic, apex rounded, 5–7 × 4–5.5 mm. Stamens 20, anthers globose to ellipsoid, 0.4–1 × 0.4–0.9 mm. Carpels glabrous, ellipsoid, 0.5–0.8 × 0.4–0.6 mm, styles lateral, 0.5–1.3 mm, stigmas slightly inflated.

Distribution: Endemic to Nepal.

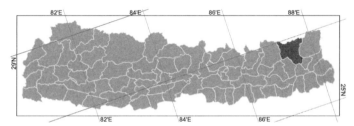

Altitudinal range: ca. 3800 m.

Ecology: Open meadows, thickets.

Flowering: July–August. **Fruiting:** July–August.

Potentilla cardotina var. *nepalensis* is restricted to the upper Arun Valley, E Nepal whilst var. *cardotiana* is distributed in E Asia (Yunnan) and Assam-Burma. The varieties differ in their stature and indumentum, with var. *nepalensis* smaller with silvery hairs on the undersides of the leaves, while var. *cardotiana* has golden hairs.

18. *Potentilla tristis* Soják, Preslia 63: 333 (1991).
Potentilla peduncularis var. *obscura* Hook.f.; *P. tristis* forma *ciliata* H.Ikeda & H.Ohba.

Perennial rosulate herbs to 12 cm. Rhizomes stout, terete, few-branched. Stipules of radical leaves with auricles connate from base to middle. Petioles 0.5–1 cm. Radical leaves oblanceolate, 2–7 × 1–1.5 cm, with 8–13 pairs of lateral leaflets, usually with alternating smaller leaflets; base of uppermost pair of leaflets decurrent; leaflets villous above, villous with patent hairs below, especially on veins; terminal leaflet sessile, oblong to obovate, 4–5 × 4–5 mm, serrate with 6–11 teeth. Flowering stems from radical leaves 3–12 cm tall, with ascending hairs. Stipules of cauline leaves with leaf-like auricles, serrate with 2–4 teeth. Cauline leaves with 1 or 2 pairs of leaflets. Flowers 0.8–1.2 cm across, solitary or in 2–3-flowered cymes. Episepals oblong to obovate, 2–3.5 ×

1.2–2.3 mm, entire or with 3 teeth, apex acute or obtuse, sparsely villous above, villous below. Sepals elliptic to ovate, 2–3.5 × 1.5–3 mm, entire, apex acute or obtuse, glabrous except puberulent near tips above, villous below. Petals yellow, oblong to elliptic, 4–5.2 × 4–4.5 mm, apex rounded or emarginate. Stamens 20, anthers globose to ellipsoid, 0.5–0.7 × 0.4–0.7 mm. Carpels glabrous, ellipsoid, 0.6–0.8 × 0.4–0.6 mm, styles lateral, 0.9–1.3 mm, stigmas slightly inflated. Achenes compressed globose, ca. 2 mm across, slightly rugose.
Fig. 48g.

Distribution: Nepal and W Himalaya.

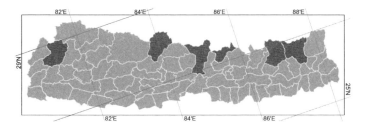

Altitudinal range: 3300–5000 m.

Ecology: Open meadows, gravelly slopes.

Flowering: July–September. **Fruiting:** July–September.

19. *Potentilla commutata* Lehm., Nov. Stirp. Pug. 3: 16 (1831).
Potentilla microphylla var. *commutata* (Lehm.) Hook.f.

Perennial rosulate herbs to 10 cm. Rhizomes stout, terete, few-branched. Stipules of radical leaves with membranous auricles, connate from base to top, apex rounded. Radical leaves oblanceolate, imparipinnate, without smaller leaflets; base of uppermost pair of leaflets decurrent; leaflets villous above, densely sericeous below; terminal leaflet sessile or subsessile, oblong to narrowly obovate. Flowering stems from radical leaves, 2–10 cm, with dense pale brown hairs. Stipules of cauline leaves with leaf-like auricles, entire or with 2 or 3 teeth Cauline leaves simple or with one pair of leaflets, entire or 3-lobed. Flowers 0.5–1.2 cm across, solitary or occasionally in 2-flowered cymes. Episepals lanceolate to oblong, entire or with 3 teeth, apex acute or obtuse, sparsely strigose above, densely villous below. Sepals elliptic to ovate, entire, apex acute or obtuse, glabrous except puberulent near apex above, densely villous below. Petals yellow, oblong to elliptic, apex rounded. Stamens 10–20, anthers globose to ellipsoid. Carpels glabrous, ellipsoid, styles subterminal, stigmas slightly inflated. Achenes obliquely ellipsoid, ca. 1.5 × 1 mm, smooth or slightly rugose.

Distribution: Nepal, W Himalaya, E Himalaya and Tibetan Plateau.

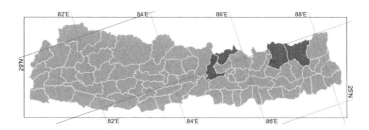

1a Radical leaves 2–4 cm. Lateral leaflets 5–9 pairs. Stamens 10–14 .... var. ***commutata***
 b Radical leaflets 3–8 cm. Lateral leaflets 10–15 pairs. Stamens around 20 .... var. ***polyandra***

Potentilla commutata Lehm. var. ***commutata***

Radical leaves with 3–5 mm petioles, 2–4 × 0.6–0.8 cm, lateral leaflets 5–9 pairs; terminal leaflet 3.5–5 × 2–3 mm, with 7–12 teeth. Peduncles 0.5–1.5 cm. Pedicels 3–7 mm. Flowers 0.5–0.8 cm across. Hypanthium 4–6 mm across. Episepals 1.5–2.5 × 1.5–2.5 mm, with 3 teeth. Sepals 1.5–2.2 × 1.5–2 mm. Petals 3.5–4.5 × 3–4.2 mm. Stamens 10–14, anthers 0.5–0.7 × 0.5–0.7 mm. Ovaries 0.6–0.8 × 0.4–0.5 mm, styles 0.4–0.5 mm.
Fig. 48h.

Distribution: Nepal, W Himalaya and E Himalaya.

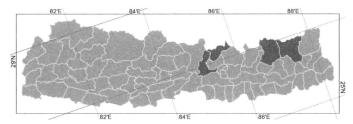

Altitudinal range: 3500–4500 m.

Ecology: Open meadows, gravelly slopes.

Flowering: July–August. **Fruiting:** July–August.

Potentilla commutata var. ***polyandra*** Soják, Bot. Jahrb. Syst. 116: 38 (1994).
Potentilla mieheorum Soják.

Radical leaves with 0.5–1 cm petioles, 3–8 × 1–1.5 cm, lateral leaflets 10–15 pairs, terminal leaflet 5–8 × 3–4 mm, with 7–11 teeth. Peduncles 2–5 cm. Pedicels 1–2.5 cm. Flowers 0.8–1.2 cm across. Hypanthium 6–8 mm across. Episepals 1.5–2 × 0.7–1 mm. Sepals 2–2.5 × 1.5–2 mm. Petals 3–4 × 2.6–3.3 mm. Stamens ca. 20, anthers 0.4–0.5 × 0.4–0.5 mm. Ovaries 0.5–0.6 × 0.4–0.5 mm, styles 0.5–0.7 mm.

Distribution: Nepal, W Himalaya, E Himalaya and Tibetan Plateau.

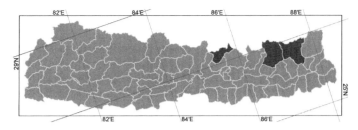

Altitudinal range: 4000–4500 m.

Ecology: Open meadows, gravelly slopes.

Flowering: July–August. **Fruiting:** July–August.

20. *Potentilla leuconota* D.Don, Prodr. Fl. Nepal.: 230 (1825).

सक्कली झारे जरो Sakkali jhare jaro (Nepali).

Perennial rosulate herbs to 20 cm. Rhizomes stout, terete. Stipules of radical leaves with auricles connate from base to middle. Petioles 1–4 cm. Radical leaves oblanceolate, 2.5–20 × 1–4 cm, with 6–18 pairs of lateral leaflets, usually with alternating smaller leaflets; base of uppermost pair of leaflets decurrent; leaflets sparsely sericeous with appressed hairs above, densely sericeous below; terminal leaflet sessile, oblong to narrowly obovate, 0.5–2 × 0.3–1 cm, serrate with 10–25 teeth. Flowering stems from radical leaves, (2–)5–20 cm with dense appressed white hairs. Stipules of cauline leaves with leaf-like auricles, serrate with 5–10 teeth. Cauline leaves similar to radical leaves, with 2–10 pairs of leaflets. Flowers 0.5–0.8 cm across, in 3–12-flowered umbel-like cymes. Episepals lanceolate to oblong, 1.5–2 × 0.7–1 mm, entire, apex acute or obtuse, sparsely strigose above, white sericeous beneath. Sepals elliptic to ovate, 2–2.5 × 1.5–2 mm, entire, glabrous above except puberulent near apex, white sericeous beneath. Petals yellow, oblong to elliptic, 3–4 × 2.6–3.3 mm, apex rounded. Stamens 20, anthers globose to ellipsoid, 0.4–0.5 mm across. Carpels glabrous, ellipsoid, 0.5–0.6 × 0.4–0.5 mm, styles lateral, 0.5–0.7 mm, stigmas slightly inflated. Achenes ellipsoid, ca. 1 × 0.7 mm, smooth. Fig. 48f.

Distribution: Nepal, W Himalaya, E Himalaya, Tibetan Plateau, Assam-Burma and E Asia.

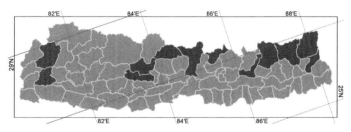

Altitudinal range: 2700–4500 m.

Ecology: Alpine meadows, forest margins, thickets.

Flowering: June–September. **Fruiting:** June–September.

An infusion of the roots is given to ease indigestion.

21. *Potentilla anserina* L., Sp. Pl. 1: 495 (1753).

येचुरुक Yechuruk (Tibetan).

Perennial stoloniferous herb. Stolons from axils of radical leaves, glabrous or sparsely hairy. Roots from nodes often with fusiform tubers. Stipules of radical leaves with membranous auricles, connate from base to top, apex rounded. Petioles 1–2 cm. Radical leaves 3–12 cm with 4–10 pairs of lateral leaflets, sometimes with alternating smaller leaflets; sparsely pilose or glabrescent; leaflets oblong to narrowly obovate, reduced in size towards leaf base, petiolulate or subsessile, base cuneate, apex obtuse or rounded, margin serrate or deeply incised, glabrescent or sparsely hairy above, silvery-white sericeous below; terminal leaflet 0.5–1.2 × 0.4–1 cm, with 13–19 teeth. Flowers 1.5–2.5 cm across, solitary from axils of radical leaves or on stolons. Pedicels 2–5 cm, glabrous. Episepals oblong to obovate, 3–8 × 1–4 mm, entire or with 2–4 teeth, apex acute or obtuse, sparsely hairy above, silvery-white sericeous below. Sepals elliptic to ovate, 3–8 × 2–5 mm, entire, apex acute or obtuse, glabrous except puberulent near apex above, silvery-white sericeous below. Petals yellow, oblong to obovate, 5–12 × 3–7 mm, apex rounded or slightly emarginate. Stamens 20–30, anthers compressed ellipsoid, 0.8–1.2 × 0.4–0.8 mm. Carpels glabrous, ellipsoid, 0.8–1.2 × 0.4–0.8 mm, styles lateral, 2–4 mm, stigmas slightly inflated.

Distribution: Asia, Europe, N America, S America and Australasia.

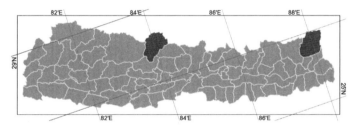

Altitudinal range: 3100–4600 m.

Ecology: Open meadows, gravelly roadsides, sandy slopes or river banks.

Flowering: June–September. **Fruiting:** June–September.

The plant is astringent and tonic, root nodules are edible and dried flowers and leaves are taken to relieve nausea, vomiting and heart-burn.

22. *Potentilla monanthes* Wall. ex Lehm., Nov. Stirp. Pug. 3: 33 (1831).
Potentilla cryptantha Klotzsch; *P. monanthes* Wall. nom. nud.; *P. monanthes* var. *alata* Soják.

साप्लङ्ग Saplang (Nepali).

Perennial prostrate herb. Rhizomes not developed. Stipules of radical leaves membranous. Radical leaves trifoliolate; leaflets subsessile or short-petiolulate, oblong-obovate to obovate, base cuneate, margin bluntly serrate, apex obtuse, pilose on both sides, often with sessile glands; petiole sparsely or densely pilose. Stipules of cauline leaves leaf-like, entire or 2- or 3-incised at apex. Flowering stems many, tufted, prostrate or ascending, sparsely or densely pilose, often with short glandular hairs. Cauline leaves similar to radical leaves, trifoliolate. Flowers in cyme, congested terminally. Pedicels densely villous, with short glandular hairs. Episepals lanceolate or elliptic-lanceolate, apex obtuse or acute, pilose with subsessile glands on both sides. Sepals triangular-ovate, apex acute, sparsely pilose with subsessile glands outside, glabrous except lanate near apex inside. Petals yellow, obovate, apex emarginate. Stamens 15–20, anthers globose. Carpels many, on globose receptacle glabrous, styles subterminal, base thickened, stigmas slightly dilated.

Distribution: Nepal, W Himalaya and E Himalaya.

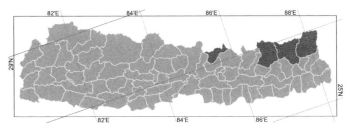

Murata (Fl. E. Himalaya: 124. 1966) misapplied the name *Potentilla monanthes* var. *sibthorpioides* Hook.f. to the typical variety of this species.

1a Radical leaves 2–6(–8) cm. Terminal leaflet 0.7–1.5(–2.5) cm. Flowers 8–12 mm across. Stamens ca. 20 .... var. ***monanthes***
   b Radical leaflets 0.8–2(–4.5) cm. Terminal leaflets 0.3–0.6 cm. Flowers 6–7 mm across. Stamens ca. 15 .... var. ***sibthorpioides***

Potentilla monanthes Wall. ex Lehm. var. ***monanthes***

Radical leaves with 1.5–3.5(–5.5) cm petioles, 2–6(–8) × 1.2–3(–3.5) cm, terminal leaflet 0.7–1.5(–2.5) × 0.5–0.8(–1.2) cm. Flowers 0.8–1.2 cm across. Episepals 3–3.5 × 1.8–2.3 mm. Sepals 3–3.5 × 2.5–3 mm. Petals 5–6 × 4.5–6 mm. Stamens ca. 20, anthers 0.6–0.7 mm across. Ovaries 0.5–0.6 × 0.4–0.5 mm, styles ca. 1 mm.

Distribution: Nepal, W Himalaya and E Himalaya.

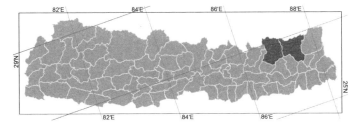

Altitudinal range: 3000–4800 m.

Ecology: Open meadows.

Flowering: July–August. **Fruiting:** July–August.

Potentilla monanthes var. ***sibthorpioides*** Hook.f., Fl. Brit. Ind. 2[5]: 358 (1878).

Radical leaves with 0.3–1(–4) cm petioles, 0.8–2(–4.5) × 0.6–1.2 cm, terminal leaflet 3–6 × 2.5–6 mm. Flowers 0.6–0.7 cm across. Episepals 1.8–2.2 × 1–1.4 mm. Sepals 2–2.5 × ca. 1.5 mm. Petals 3–4 × 2.5–3.5 mm. Stamens ca. 15, anthers ca. 0.4 mm across. Ovaries ca. 0.5 × 0.4 mm, styles 0.6–0.7 mm.

Distribution: Nepal and E Himalaya.

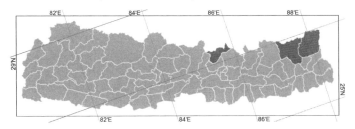

Altitudinal range: 4100–4700 m.

Ecology: Open meadows.

Flowering: June–July. **Fruiting:** June–July.

23. *Potentilla argyrophylla* Wall. ex Lehm., Nov. Stirp. Pug. 3: 36 (1831).
Potentilla argyrophylla Wall. nom. nud.; *P. argyrophylla* var. *leucochroa* Hook.f.; *P. atrosanguinea* var. *cataclines* (Lehm.) Wolf; *P. cataclines* Lehm.; *P. insignis* Royle ex Lindl.; *P. jacquemontiana* Cambess.; *P. leucochroa* Lindl. ex Wall. nom. nud.; *P. nivea* var. *himalaica* Kitam.; *P. venusta* Soják.

आटे Aate (Nepali).

Perennial rosulate herbs to 40 cm. Rootstock stout, short, simple or sometimes branched. Radical leaves trifoliolate; petiole pubescent with appressed white hairs; terminal leaflet oblong to narrowly obovate, base cuneate, apex obtuse, margin acutely or obtusely serrate, densely or sparsely appressed hairy above, densely sericeous below. Flowering stems from axils of radical leaves, pubescent with appressed or ascending hairs. Cauline leaves reducing in size upwards, lower leaves trifoliolate, upper leaves simple. Flowers 1.2–2.2 cm across. Pedicels pubescent. Episepals lanceolate to narrowly ovate, apex acute, densely or sparsely hirsute on both sides. Sepals ovate-triangular, apex acute, glabrescent inside, densely or sparsely hirsute outside. Petals yellow, orange or crimson, obovate to broadly obovate, apex emarginate. Stamens 20, anthers compressed globose to ellipsoid. Carpels ellipsoid, glabrous, styles subterminal, slightly swollen at base, stigmas slightly inflated.

Distribution: Nepal, W Himalaya, E Himalaya, Tibetan Plateau and SW Asia.

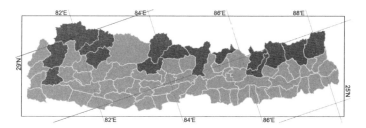

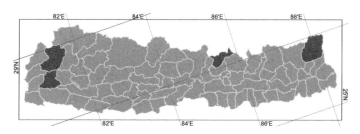

A paste or juice of the root is applied to treat toothache.

1a Petals yellow .... var. ***argyrophylla***
 b Petals crimson or orange .... var. ***atrosanguinea***

Potentilla argyrophylla Wall. ex Lehm. var. ***argyrophylla***

Radical leaves with (0.5–)1–15 cm petioles, 1–20 × 0.8–7 cm, terminal leaflet 0.5–4.5 × 0.3–2.5 cm. Flowering stems 3–40 cm. Flowers 1.2–2 cm across. Episepals 3–5(–7) × 1–3(–4) mm. Sepals 4–6(–8) × 2–3.5 mm. Petals yellow, (4–)5–8 × 4–8 mm. Anthers ca. 1 mm across. Carpels 0.6–0.7 × ca. 0.4 mm, styles ca. 2 mm.

Distribution: Nepal, W Himalaya, E Himalaya, Tibetan Plateau and SW Asia.

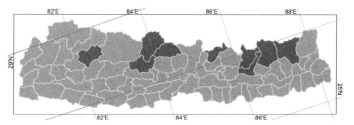

Altitudinal range: 3600–4800 m.

Ecology: Open meadows, sandy river banks.

Flowering: May–September. **Fruiting:** May–September.

Potentilla argyrophylla var. ***atrosanguinea*** (Lodd.) Hook.f., Fl. Brit. Ind. 2[5]: 357 (1878).
Potentilla atrosanguinea Lodd., Bot. Cab. 8(9): pl. 786 (1823); *Potentilla cautleyana* Royle.

Radical leaves with 2–8 cm petioles, 5–12 × 3–5.5 cm, terminal leaflet 2–4 × 1.2–2.5 cm. Flowering stems 15–30 cm. Flowers 1.5–2.2 cm across. Episepals 4–6 × 2–3 mm. Sepals 4.5–7 × 3–5 mm. Petals crimson or orange, 6–8 × 6–9 mm. Anthers ca. 1 mm across. Carpels ca. 0.7 × 0.5 mm, styles ca. 2 mm.

Distribution: Nepal, W Himalaya and Tibetan Plateau.

Altitudinal range: 2900–4600 m.

Ecology: Open meadows, sandy river banks.

Flowering: June–September. **Fruiting:** June–September.

24. *Potentilla saundersiana* Royle, Ill. Bot. Himal. Mts. [6]: 207, pl. 41 fig. I (1835).
Potentilla caespitosa Lehm.; *P. caliginosa* Soják; *P. forrestii* W.W.Sm.; *P. forrestii* var. *caespitosa* (Wolf) Soják; *P. forrestii* var. *segmentata* Soják; *P. illudens* Soják; *P. jacquemontii* (Franch.) Soják; *P. multifida* var. *saundersiana* (Royle) Hook.f.; *P. potaninii* Th.Wolf; *P. saundersiana* var. *caespitosa* Wolf; *P. saundersiana* var. *segmentata* (Soják) Soják; *P. thibetica* Cardot; *P. williamsii* Soják.

Perennial rosulate herbs to 20 cm. Rootstock stout, short, with several slender roots. Petioles 0.5–3.5(–6) cm, pubescent with patent or appressed hairs. Radical leaves palmately 5-foliolate, sometimes mixed with trifoliolate leaves, 1–6(–8) × 0.7–2.5 cm; terminal leaflet oblong to narrowly obovate, 0.5–2 × 0.3–1 cm, base cuneate, apex obtuse, margin serrate or bluntly incised, densely or rarely sparsely lanate above, densely white sericeous below; cauline leaves reduced in size towards top, usually trifoliolate. Flowering stems from axils of radical leaves, 2–10(–20) cm, densely covered with patent or ascending hairs. Flowers 1–1.6 cm across. Pedicels pubescent. Episepals lanceolate to oblong, 3–5 × 1.5–3 mm, apex acute, densely or sparsely hirsute on both sides. Sepals ovate-triangular, 3–4.5(–6) × 1.5–2.5 mm, apex acute, glabrescent inside except shortly lanate near apex, densely hirsute outside. Petals yellow with orange base, obovate to broadly obovate, 5–7 × 5–7 mm, apex emarginate. Stamens 20, anthers compressed ellipsoid, ca. 0.8 × 0.7 mm. Carpels ellipsoid, ca. 0.5 × 0.3 mm, glabrous, styles subterminal, ca. 1 mm.

Distribution: Nepal, E Himalaya, Tibetan Plateau and E Asia.

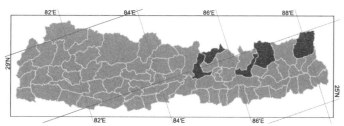

Altitudinal range: 3100–4900 m.

Ecology: Open meadows, thickets, gravelly slopes, sandy riversides.

366

Flowering: May–August. **Fruiting:** May–August.

Kitamura (Fauna Fl. Pl. Nepal Himalaya: 149. 1955) misapplied *Potentilla argentea* L. to this species.

25. *Potentilla griffithii* Hook.f., Fl. Brit. Ind. 2[5]: 351 (1878). *Potentilla griffithii* var. *metallica* Soják; *P. leschenaultiana* Ser.; *P. sikkimensis* Th.Wolf; *P. spodiochlora* Soják.

Perennial rosulate herbs to 50 cm. Rootstock stout, short, with several slender roots. Petioles 3.5–8 cm, pubescent with patent or appressed hairs. Radical leaves imparipinnate with 2–5 pairs of lateral leaflets, 5–15(–20) × 2–4.5 cm; terminal leaflet oblong to narrowly oblong, 1.5–3.5 × 0.8–2 cm, base cuneate, apex obtuse, margin acutely serrate, densely or rarely sparsely lanate above, densely white sericeous below. Flowering stems from axils of radical leaves, 10–40(–50) cm tall, densely covered with patent or ascending hairs. Cauline leaves reduced in size towards top, lower leaves 5-foliolate, upper leaves 3-foliolate. Flowers 1.2–1.8 cm across. Pedicels pubescent. Episepals lanceolate to elliptic, 3–7 × 2–4 mm, margin entire, often revolute, apex acute, densely or sparsely hirsute on both sides. Sepals ovate-triangular, 5–6 × 2.5–3.5 mm, apex acute to acuminate, glabrescent inside except shortly lanate near apex, densely hirsute outside. Petals yellow with orange base, obovate to broadly obovate, 6–8 × 5–8 mm, apex emarginate. Stamens 20, anthers compressed ellipsoid, ca. 0.8 × 0.5 mm. Carpels ellipsoid, ca. 0.3 × 0.2 mm, glabrous, styles subterminal, ca. 1.5 mm, slightly swollen at base.

Distribution: Nepal, E Himalaya, Tibetan Plateau and E Asia.

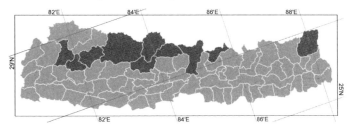

Altitudinal range: 1600–4000 m.

Ecology: Open meadows, forest margins, thickets.

Flowering: July–September. **Fruiting:** July–September.

26. *Potentilla multifida* L., Sp. Pl. 1: 496 (1753). *Potentilla exigua* Soják; *P. ornithopoda* Tausch; *P. plurijuga* Hand.-Mazz.; *P. plurijuga* var. *lhasana* Soják.

Perennial rosulate herbs to 30 cm. Rootstock stout, terete. Petioles 0.5–3(–6) cm, pubescent with patent or appressed hairs. Radical leaves imparipinnate with 2–5 pairs of lateral leaflets, 2–10 × 1–2.5 cm; leaflets divided into linear lobes; terminal leaflet oblanceolate to oblong, 1–3 × 0.6–0.8 cm, base narrowly cuneate, apex obtuse, margin deeply incised into linear lobes, sparsely lanate above, densely white

sericeous below. Flowering stems from axils of radical leaves, (5–)10–30 cm, densely covered with patent or appressed hairs. Cauline leaves reducing in size upwards. Flowers 0.8–1.2 cm across; pedicel pubescent. Episepals lanceolate to oblong, 2–4 × 1–1.5 mm, apex acute, densely or sparsely hirsute on both sides. Sepals ovate-triangular, 3.5–5 × 2.5–3.5 mm, apex acute, glabrescent except shortly lanate near apex inside, densely hirsute outside. Petals yellow with orange base, obovate, 5–6 × 5–6 mm, apex emarginate. Stamens 20, anthers compressed ellipsoid to ovoid, ca. 0.8 × 0.6 mm. Carpels ellipsoid, ca. 0.6 × 0.4 mm, glabrous, styles subterminal, ca. 1.5 mm.

Distribution: Asia and Europe.

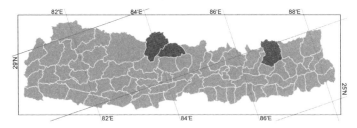

Altitudinal range: 2900–4500 m.

Ecology: Open meadows, forest margins, roadsides.

Flowering: June–August. **Fruiting:** June–August.

27. *Potentilla nepalensis* Hook., Exot. Fl. 2(7): pl. 88 (1824). *Potentilla formosa* D.Don nom. illegit.; *P. formosa* Sweet; *P. gulielmi-waldermarii* Klotzsch.

Perennial rosulate herbs to 80 cm. Rootstock stout, short, with several slender roots. Petioles (0.5–)1–4 cm, pubescent with patent hairs. Radical leaves palmately 5-foliolate, 3–8 × 1–5 cm; terminal leaflet oblong to oblong-obovate, 0.5–3.5 × 0.4–1.5 cm, base cuneate, apex obtuse, margin acutely or obtusely serrate, pilose on both surfaces, sometimes glabrescent above, densely appressed villous below. Flowering stems from axils of radical leaves, (5–)20–60(–80) cm, pubescent with patent or ascending hairs. Cauline leaves reducing in size upwards, lower leaves 5-foliolate, upper leaves 3-foliolate. Flowers 1.2–1.6 cm across. Pedicels pubescent. Episepals lanceolate to narrowly ovate, 3–7 × 1–3 mm, apex acute, sparsely hirsute on both sides. Sepals ovate-triangular, 3.5–6(–8) × 1.5–3(–4) mm, apex acute to acuminate, glabrescent inside, sparsely hirsute outside. Petals crimson, obovate to broadly obovate, 5–7 × 4–6 mm, apex shallowly emarginate. Stamens 20, anthers compressed ellipsoid, ca. 1.2 × 0.9 mm. Carpels ellipsoid, ca. 0.6 × 0.3 mm, glabrous, styles subterminal, ca. 1.5 mm, slightly swollen at base.

Distribution: Nepal and W Himalaya.

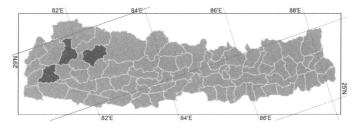

Altitudinal range: 2000–2600 m.

Ecology: Open meadows, forest margins, roadsides.

Flowering: May–August. **Fruiting:** May–August.

28. *Potentilla coriandrifolia* D.Don, Prodr. Fl. Nepal.: 232 (1825).
Potentilla meifolia Wall. ex Lehm.; *P. meifolia* Wall. nom. nud.

Perennial rosulate herbs to 12 cm. Rootstock robust, terete, somewhat thickened, blackish, simple or occasionally branched. Stipules of radical leaves membranous, brown. Radical leaves elongated lanceolate, 3–10 cm, pinnate with 3–8 pairs of lateral leaflets; leaflets opposite, subsessile, palmately divided into linear lobes, apex acuminate, sparsely appressed pilose or glabrescent inside, appressed villous or glabrescent outside. Flowering stems from axils of radical leaves, erect or ascending, 3–12 cm, sparsely pilose or glabrescent. Stipules of cauline leaves leaf-like, glabrescent inside, appressed villous outside Cauline leaves 1 or 2, similar but much smaller than radical leaves, dissected into linear lobes. Flowers 1.0–1.7 cm across, solitary or in 2–3(–5)-flowered cymes. Pedicels 1.2–3 cm, appressed or ascending pubescent. Episepals lanceolate, 2–4 × 1–1.5 mm, apex acute or acuminate, glabrescent inside, sparsely pilose outside. Sepals triangular-ovate, 3–5 × 2–3 mm, apex acute or acuminate, glabrescent inside, lanate near apex, sparsely pilose outside. Petals white with crimson base, obovate to broadly obovate, 4–7 × 3–7 mm, apex emarginate. Stamens 20, anthers globose, ca. 0.5 mm across. Carpels ellipsoid to obovoid, glabrous, smooth, ca. 1 × 0.6–0.8 mm, styles subterminal, 1.2–1.5 mm. Achenes obovoid, smooth, ca. 1.5 × 1 mm.

Distribution: Nepal, W Himalaya, E Himalaya and Tibetan Plateau.

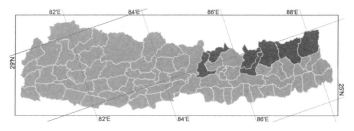

Altitudinal range: 3700–5600 m.

Ecology: Open meadows, silty slopes.

Flowering: June–September. **Fruiting:** June–September.

29. *Potentilla supina* L., Sp. Pl. 1: 497 (1753).
Comarum flavum Buch.-Ham. ex Roxb.; *C. flavum* Buch.-Ham. ex Roxb. nom. nud.; *Potentilla heynii* Roth; *P. paradoxa* Nutt.; *P. supina* var. *paradoxa* (Nutt.) Th.Wolf.

बज़रदन्ती Bajradanti (Nepali).

Annual or biennial herbs to 30 cm. Stipules of radical leaves membranous, brown, glabrous inside, pilose outside. Radical leaves pinnate with 2–4 pairs of lateral leaflets, 2–8 cm long; leaflets alternate or opposite, subsessile, oblong to obovate, 5–12 × 2–8 mm, base cuneate except those uppermost pair decurrent, margin incised-serrate, apex obtuse, pilose or glabrescent upper, sparsely to densely pilose lower. Flowering stems erect or ascending, (6–)10–30 cm, pubescent. Stipules of cauline leaves leaf-like, obliquely ovate to lanceolate, sparsely pilose or glabrescent inside, sparsely to densely pilose outside. Cauline leaves similar to radical leaves, but with fewer leaflets. Flowers 0.5–0.8 cm across, solitary terminal. Pedicel 3–10 mm long, pubescent. Episepals oblong-elliptic, 2–3 × 0.8–1.2 mm, entire or divided into two lobes, apex acute or obtuse. Sepals triangular-ovate, 2.5–4 × 1.5–2.5 mm, apex acute or obtuse, glabrescent inside, sparsely to densely pilose outside. Petals yellow, oblong to obovate, 1.5–4 × 0.8–3 mm, apex rounded or slightly emarginate, usually shorter than sepals. Stamens 15–20, filaments filiform, anthers compressed globose, ca. 0.2 × 0.4 mm. Carpels many, crowded on dome-shaped receptacle, obliquely ellipsoid, ca. 0.3 × 0.2 mm, glabrous, styles subterminal, ca. 0.5 mm long, stigmas slightly dilated. Achenes oblong to narrowly ovoid, ca. 0.4 × 0.3 mm, smooth except ridged margin.

Distribution: Asia and Europe.

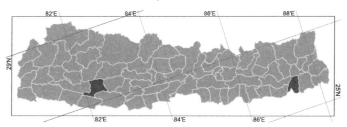

Altitudinal range: 100–2000 m.

Ecology: Field margins, river banks, moist meadows, gravelly slopes.

Flowering: March–July. **Fruiting:** March–July.

Root paste is applied to boils. Pieces of root are chewed to relieve toothache.

30. *Potentilla sundaica* (Blume) Kuntze, Revis. Gen. Pl. 1: 219 (1891).
Fragaria sundaica Blume, Bijdr. Fl. Ned. Ind. [17]: 1106 (1827); *Potentilla anemonefolia* Lehm.; *P. bodinieri* H.Lév.; *P. kleiniana* Wight; *P. wallichiana* Delile ex Lehm. later homonym, non Ser.

Perennial herbs to 50 cm, ascending or prostrate, often

rooting at nodes. Stipules of radical leaves membranous, pale brown, glabrous inside, sparsely pilose or glabrescent outside. Radical leaves subpedately 5-foliolate, 3–12 cm long; leaflets subsessile or short-petiolulate, oblanceolate to oblong-obovate, 5–30 × 4–17 mm, base cuneate, margin serrate, apex obtuse, sparsely pilose upper, appressed villous lower, especially along veins. Flowering stems 10–50 cm, pilose or spreading villous. Stipules of cauline leaves leaf-like, ovate or ovate-lanceolate, glabrous inside, sparsely villous outside. Cauline leaves similar to radical leaves; lower cauline leaves 5-foliolate, upper leaves 3-foliolate. Flowers 0.8–0.9 cm across, crowded terminally in many-flowered cymes. Pedicels 5–12 mm, appressed or ascending villous. Episepals lanceolate or oblong-lanceolate, 2.5–4 × 1–1.3 mm, apex acute or acuminate, glabrescent inside, pilose outside. Sepals triangular-ovate, 3–4 × 1.5–2.2 mm, apex acute or acuminate, glabrescent inside, pilose outside. Petals yellow, oblong-oblanceolate, 3.5–5 × 2.5–3 mm, apex slightly emarginate. Stamens ca. 20. Carpels many, crowded on dome-shaped receptacle obliquely ellipsoid, smooth, 0.3–0.4 × 0.2–0.3 mm, styles subterminal, 0.7–0.8 mm long, base thickened, stigmas slightly dilated. Achenes ovoid, 0.6–0.7 × 0.4–0.5 mm, rugose.

Distribution: Nepal, W Himalaya, E Himalaya, Tibetan Plateau, Assam-Burma, S Asia, E Asia and SE Asia.

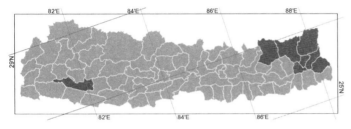

Altitudinal range: 1000–2400 m.

Ecology: Field margins, river banks, moist meadows, forest margins.

Flowering: March–September(–December). **Fruiting:** March–September(–December).

Roots and stems are considered toxic and they are pounded and applied to abscesses and the bites of snakes and centipedes.

31. *Potentilla indica* (Andrews) Th.Wolf, in Asch. & Graebn., Syn. Mitteleur. Fl. 6,1(Lief 6): 661 (1904).
Fragaria indica Andrews, Bot. Repos. 7: pl. 479 (1807); *Duchesnea fragiformis* Sm.; *D. indica* (Andrews) Focke; *Potentilla denticulosa* Ser.; *P. wallichiana* Ser.

भुइँ काफल Bhuin kaphal (Nepali).

Perennial stoloniferous herbs. Rhizomes not developed. Stolons from radical leaves, usually rooting at nodes after flowering. Stipules of radical leaves membranous, free. Radical leaves with appressed or ascending hairy petioles, trifoliolate or rarely 5-foliolate, sparsely hairy upper, sparsely or densely hairy lower, especially densely hairy on veins; terminal leaflet rhombic-oblong to elliptic, apex acute, margin usually single, sometimes doubly serrate, (0.7–)1–4 × (0.5–)0.7–2.5 cm. Flowers 1.2–2 cm across, on stolons. Episepals obovate to broadly obovate, 5–8 × 4–9 mm, apex 3–5-lobed, densely hairy outside, sparsely hairy inside. Sepals triangular-ovate, 5–9 × 3–4 mm, apex acute to acuminate, densely hairy outside, glabrous lower half, lanate upper half inside. Petals yellow, narrowly obovate, 6–9 × 4–6 mm. Carpels ellipsoid, glabrous, styles subterminal, ca. 1 mm. Stamens 20, anthers compressed ellipsoid, 0.6–0.8 mm. Carpels on globose receptacle, ellipsoid, glabrous, 0.3–0.4 mm, styles subterminal, ca. 1 mm. Fruiting receptacle spongy, red, globose, glabrous, 7–13 mm across. Achenes crowded on receptacle, red, 1–1.3 × 0.5–0.8 mm, nearly smooth.

Distribution: Nepal, W Himalaya, E Himalaya, E Asia, SE Asia and SW Asia.

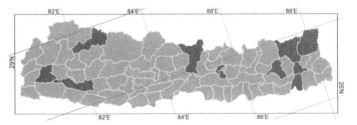

Altitudinal range: 900–2500 m.

Ecology: Field margins, river banks, moist meadows, forest margins, thickets.

Flowering: April–June. **Fruiting:** May–October.

Fruits are taken to treat blisters on the tongue.

17. *Sibbaldia* L., Sp. Pl. 1: 284 (1753).

Hiroshi Ikeda

Perennial herbs, stems often woody at base. Stipules adnate to petioles, upper part (auricle) free or more or less connate, membranous or leaf-like. Leaves pinnate or palmately 3–5-foliolate; leaflets serrate or dentate at apex. Inflorescences cymose or flowers solitary, terminal or from axils of radical leaves. Flowers bisexual or rarely unisexual. Bracts at base of pedicel, leafy or membranous; bracteoles absent. Hypanthium cupulate or saucer-shaped. Episepals (4 or)5. Sepals (4 or)5, alternate, persistent. Petals (4 or)5, yellow, dark red or white. Stamens (4 or)5(–10); anthers 2-locular. Ovary superior, carpels free, 4–20; ovule usually ascending. Style subbasal, lateral or subterminal. Achenes 4–20, inserted on somewhat swollen receptacle.

Rosaceae

Worlwide about 20 species in arctic or alpine regions of N hemisphere. Eight species in Nepal.

Key to Species

1a	Radical leaves pinnate	2
b	Radical leaves trifoliolate or palmately 5-foliolate	4
2a	Rhizomes slender, long-creeping under ground	**1. *S. adpressa***
b	Rhizomes short. Flowering stems decumbent, above ground	3
3a	Petals red. Auricles of stipules of radical leaves connate from base to top	**3. *S. emodi***
b	Petals yellow or pale orange. Auricles of stipules of radical leaves free	**5. *S. micropetala***
4a	Radical leaves palmately 5-foliolate	**7. *S. purpurea***
b	Radical leaves trifoliolate	5
5a	Flowers usually tetramerous, rarely pentamerous	**8. *S. tetrandra***
b	Flowers usually pentamerous, rarely tetramerous	6
6a	Petals red	**4. *S. sikkimensis***
b	Petals white or yellow	7
7a	Rhizomes short, aerial stems woody, tufted on ground. Petals yellow	**2. *S. cuneata***
b	Rhizomes slender, long-creeping under ground. Petals white	8
8a	Petals shorter or as long as sepals	**1. *S. adpressa***
b	Petals longer than sepals	**6. *S. perpusilloides***

1. *Sibbaldia adpressa* Bunge, Fl. Altaic. 1: 428 (1829).
Sibbaldia minutissima Kitam.

Perennial herbs. Rhizomes woody, much-branched, slender and long-creeping. Stipules adnate to peticles in lower half, auricles triangular to linear, entire, apex acute to acuminate, glabrous inside, sparsely strigose outside and on margin. Petiole 1.0–22 mm long, with appressed or ascending hairs. Radical leaves trifoliolate or imparipinnate with 5 leaflets, oblanceolate or obovate, 0.5–5.5 × 0.3–2.2 cm; leaflets almost glabrous on upper surface, strigose on lower surface and margin. Lateral leaflets lanceolate to narrowly elliptic, 2.2–13 × 0.9–3.2 mm, apex acute, entire; base of uppermost leaflet pair decurrent. Terminal leaflet narrowly obovate to oblanceolate, 2.3–13 × 1.1–6.2 mm, base cuneate, apex tridentate. Cauline leaves 1 or 2, similar to radical leaves. Flowers hermaphrodite, solitary or 2–5 in cymes, 3.5–7.0 mm in diameter. Peduncles and pedicels with appressed hairs. Hypanthia shallowly cupulate, strigose outside, sparsely strigose inside. Episepals 5 (rarely 4), ovate to broadly lanceolate, 0.8–1.2 × 0.7–1.0 mm, apex obtuse to acute, strigose with minute glandular hairs outside, glabrous or sparsely hairy inside. Sepals triangular to broadly ovate, 1.5–2.2 × 1.3–1.7 mm, apex obtuse, strigose with minute glandular hairs outside, short villous on upper portion inside. Petals 5 (rarely 4), white to creamy white, spathulate to narrowly obovate, 1.7–2.5 × 1.0–1.3 mm. Stamens 10 (rarely 8); filaments glabrous, 0.3–0.5 mm; anthers semi-orbicular, 0.3–0.4 mm across. Carpels 10–16; ovaries ovoid, 0.5–0.6 × 0.4–0.5 mm, glabrous; styles lateral, slightly fusiform, 1.0–1.2 mm; stigmas slightly inflated, papillate.

Distribution: Nepal, Tibetan Plateau, E Asia and N Asia.

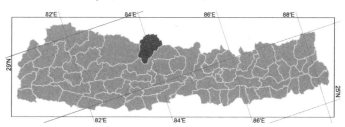

Altitudinal range: 3900–4400 m.

Ecology: Open meadows, roadside dumps.

Flowering: July–August.

Kitamura described *Sibbaldia minutissima* from a single specimen collected by Nakao at Sangda, Mustang District (KYO). Ikeda and Ohba collected specimens near the type locality and concluded that *S. minutissima* was a small form of *S. adpressa*.

2. *Sibbaldia cuneata* Hornem. ex Kuntze, Linnaea 20: 59 (1847).
Potentilla sibbaldi Hook. f.; *Sibbaldia parviflora* Edgew. later homonym, non Willd.

Prostrate or tufted shrublets. Rhizomes woody, much-branched, upper part covered with old stipules and petioles. Flowering stems 5–14 cm long, erect or ascending, with appressed or ascending hairs. Stipules of radical leaves adnate to petioles in lower half, auricles membranous, brown to light brown, apex acuminate, stipules of cauline leaves herbaceous, apex acute to acuminate. Petioles 1.0–6.0(–7.0) cm, with appressed or ascending hairs. Radical leaves trifoliolate, 1.5–8(–10) cm; sparsely or densely pilose on both surfaces. Terminal leaflet short-petiolulate or subsessile, broadly obovate to obovate, 5–15 × 4–13 mm wide, base cuneate, apex dentate with 3–5 teeth. Cauline leaves 1 or 2, similar to radical leaves. Inflorescence a terminal, compact cyme. Flowers 5–7 mm across. Episepals 5, lanceolate to narrowly elliptic, shorter than sepals, 1.5–3 × 0.5–1.0 mm, apex acute to acuminate, pilose with appressed or ascending hairs outside. Sepals 5, ovate or oblong, 2.0–4.0 × 1.5–2.0 mm, apex acute. Petals 5, yellow, narrowly obovate, 0.8–1.0 × 0.5–0.7 mm, apex rounded or shallowly retuse, as long as or a little longer than sepals. Stamens 5. Style lateral. Achenes glabrous.

Distribution: Nepal, W Himalaya, E Himalaya, Tibetan Plateau, E Asia, N Asia and SW Asia.

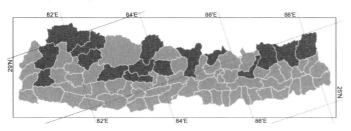

Altitudinal range: 3000–4900 m.

Ecology: Alpine meadows, rocky crevices.

Flowering: June–August. **Fruiting:** August–October.

3. *Sibbaldia emodi* H.Ikeda & H.Ohba, J. Jap. Bot. 71: 188 (1996).

Rosulate herbs with prostrate flowering stems. Stipules of radical leaves connate with membranous auricles, apex rounded or 2–3-divided Radical leaves interrupted-imparipinnate with 7–17-foliolate, 3–6 × 1.2–2.4 cm, leaflets sparsely hairy on upper surface, sericeous on lower surface. Terminal leaflet 0.8–1.4 × 0.5–0.8 cm, serrate with 5–9 teeth. Cauline leaves 3–5-foliolate, sparsely hairy on upper surface, sericeous on lower surface. Flowering stems axillary from radical leaves. Flowers solitary, terminal; pedicels 2–8 mm long, with appressed white hairs. Episepals lanceolate to ovate, 1.5–3.0 × 1.0–1.3 mm, densely sericeous outside, sparsely sericeous with strigose hairs inside. Sepals ovate to triangular, 1.8–3.1 × 1.2–1.9 mm, apex obtuse, sericeous outside with long and minute hairs, glabrous inside except densely pilose in upper half. Petals deep red, narrowly elliptic to lanceolate, 1.5–2.0 × 1.0–1.5 mm. Stamens 5; filaments 0.4–0.6 mm, pale red; anthers globose, 0.3–0.5 mm across, dark yellow before dehiscence. Ovaries ovoid, 0.5–0.8 mm long, 0.3–0.5 mm wide, with short multicellular stalked

glands, pale green; styles subbasal, 0.8–1.1 mm long, pale green; stigmas inflated and papillate; ovule single. Achenes pale to dark brown, 1.8–2.2 mm long, rugose with swollen appendages of spongy tissue, 1-seeded. Seeds smooth.

Distribution: Endemic to Nepal.

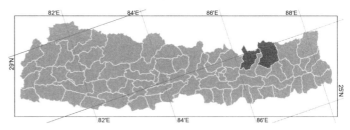

Altitudinal range: 4000–4100 m.

Ecology: Open meadows.

Flowering: August. **Fruiting:** August.

Sibbaldia emodi is similar to *S. micropetala* (D.Don) Hand.-Mazz. in gross appearance, but differs in having deep red petals compared with the usually yellow petals of *S. micropetala*. *Sibbaldia emodi* also differs from *S. micropetala* in the stipules of its radical leaves which have connate upper parts while those of *S. micropetala* are free.

4. *Sibbaldia sikkimensis* (Prain) Chatterjee, Notes Roy. Bot. Gard. Edinburgh 19: 327 (1938).
Potentilla sikkimensis Prain, J. Asiat. Soc. Bengal, Pt. 2, Nat. Hist. 73(5): 201 (1904); *Sibbaldia melinotricha* Hand.-Mazz.

Rosulate perennial herbs. Roots robust, terete, upper part covered with old petioles and stipules. Flowering stems erect or ascending, 5–20 cm tall, with patent or ascending hairs. Stipules of radical leaves membranous, light to dark brown, pilose outside, auricles lanceolate to triangular, apex acute to acuminate. Petioles 1–8 cm long, with patent or ascending hairs. Radical leaves trifoliolate, 3–10 × 1.3–4 cm; leaflets sparsely pilose on both surfaces. Terminal leaflet short-petiolulate or subsessile, obovate to broadly elliptic, 7–23 × 5–17 mm, base broadly cuneate or subrounded, coarsely dentate-serrate with 5–9 teeth in upper half, acute or obtuse at tips. Cauline leaves 1–2, similar to radical leaves but smaller in size. Stipules of cauline leaves herbaceous, auricles ovate to triangular, pilose outside. Inflorescence terminal, umbel-like. Flowers 5-8 mm across. Pedicel 5–12 mm long, with patent or ascending hairs. Episepals lanceolate to narrowly ovate, usually shorter than sepals, 1.5–2.5 × 0.8–1.2 mm, pilose. Sepals ovate to triangular, 2–4 × 1–1.5 mm wide, apex acute.Petals 5 (rarely 6), dark red, obovate, 4–6 × 3–4 mm, apex rounded, truncate or shallowly retuse, nearly as long as sepals. Stamens 5 (rarely 6); filaments ca. 0.5 mm; style subterminal, 0.4–0.6 mm. Ovaries ovoid, 0.6–0.7 × 0.4–0.5 mm. Achenes ovoid, light brown, 1.2–1.5 mm long, 0.7–1.0 mm wide, glabrous.

Distribution: Nepal, E Himalaya, Assam-Burma and E Asia.

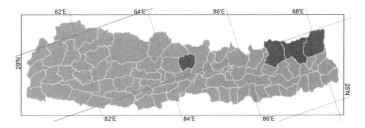

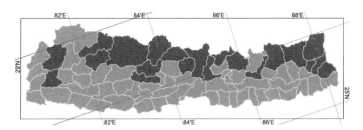

Altitudinal range: 3300–4800 m.

Ecology: Open alpine meadows or at edge of shrubberies.

Flowering: June–July. **Fruiting:** July–September.

Chinese plants have been treated by some authors as a distinct species *Sibbaldia melinotricha*, but they are not distinguishable from *S. sikkimensis*.

5. Sibbaldia micropetala (D.Don) Hand.-Mazz., Vegetationsbilder 22(8): 6 (1932).
Potentilla micropetala D.Don, Prodr. Fl. Nepal.: 231 (1825); *P. albifolia* Wall. ex Hook.f.; *P. albifolia* Wall. nom. nud.; *Sibbaldia potentilloides* Cambess.

Rosulate perennial herbs. Roots robust, terete. Flowering stems repent or ascending, 5–30 cm long, white tomentose or glabrescent. Stipules of radical leaves membranous, pale brown, adnate to the lower side of petioles in lower half, auricles free, apex acute to acuminate. Radical leaves imparipinnate, 5–11-foliolate and sometimes with alternate smaller leaflets, 2–20 × 1–6 cm wide, sparsely pilose on upper surface and margin, white tomentose on lower surface. Terminal leaflet sessile or subsessile, elliptic to obovate, 0.5–2.5 × 0.5–2.5 cm wide, margin coarsely dentate-serrate with 10–20 teeth, teeth tips acute. Stipules of cauline herbaceous, auricles obliquely ovate to lanceolate, margin entire or with several teeth. Cauline leaves similar to radical leaves but reduced in number and size, leaflets glabrescent or sparsely pilose on upper surface and white tomentose on lower surface. Flowers solitary, terminal, 4–7 mm across. Episepals narrowly lanceolate to narrowly elliptic, usually shorter than sepals, sparsely or densely white tomentose outside. Sepals narrowly to broadly ovate, 2.5–4 × 1.5–2 mm wide, apex acute to acuminate. Petals yellow or pale orange, oblong to lanceolate, 1.4–1.5 × 0.8–1.0 mm, apex obtuse or shallowly retuse, usually shorter than sepals. Stamens 5, alternate with sepals; filaments, ca. 0.5 mm long. Ovaries ovoid, 0.5–0.6 × 0.3–0.4 mm, style lateral to subbasal. Achene ovoid, with swollen appendage on outer side, 1.3–1.5 mm long, 1.1–1.2 mm wide, brown with shallow grooves except smooth appendage.
Fig. 50a–b.

Distribution: Nepal, W Himalaya, E Himalaya, Tibetan Plateau and E Asia.

Altitudinal range: 2300–4800 m.

Ecology: Open meadows.

Flowering: June–August. **Fruiting:** July–September.

6. Sibbaldia perpusilloides (W.W.Sm.) Hand.-Mazz., Symb. Sin. 7(Leif 3): 520 (1933).
Potentilla perpusilloides W.W.Sm., Rec. Bot. Surv. India 4: 188 (1911).

Perennial herbs. Rhizomes slender, slightly woody, branching with fine roots. Flowering stems short, 3–15 mm tall, with sparse appressed hairs. Stipules membranous, pale to dark brown, subglabrous except ciliate margin; glabrous or sparsely hairy. Leaves radical, trifoliolate, 5–15 × 4–10 mm; leaflets subglabrous on both surfaces except sparsely pilose on margin; terminal leaflet short-petiolulate or subsessile, obovate to broadly obovate, 2–4 × 1.5–3 mm, 3–5-dentate near apex, teeth with acute tips. Flowers hermaphrodite, solitary, terminal, 6–10 mm across. Episepals lanceolate, oblong or oblanceolate, a little shorter than sepals, 1.5–2 × 0.6–1 mm, entire or shallowly to deeply divided, apex acute to obtuse, sparsely hairy or subglabrous outside, glabrous inside, ciliate on margin. Sepals ovate, 2–2.5 × 0.8–1.2 mm, apex acute or obtuse. Petals 5, white, obovate to broadly obovate, apex rounded or shallowly retuse, longer than sepals. Stamens 10 or fewer; filaments 0.5–0.6 mm. Ovaries ovoid, smooth style lateral. Achenes ovoid, 1.2–1.3 × 0.8–0.9 mm, glabrous.

Distribution: Nepal, E Himalaya, Tibetan Plateau and E Asia.

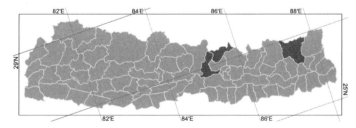

Altitudinal range: 4000–4200 m.

Ecology: Moist rocky crevices, often with mosses.

Flowering: June–August. **Fruiting:** July–September.

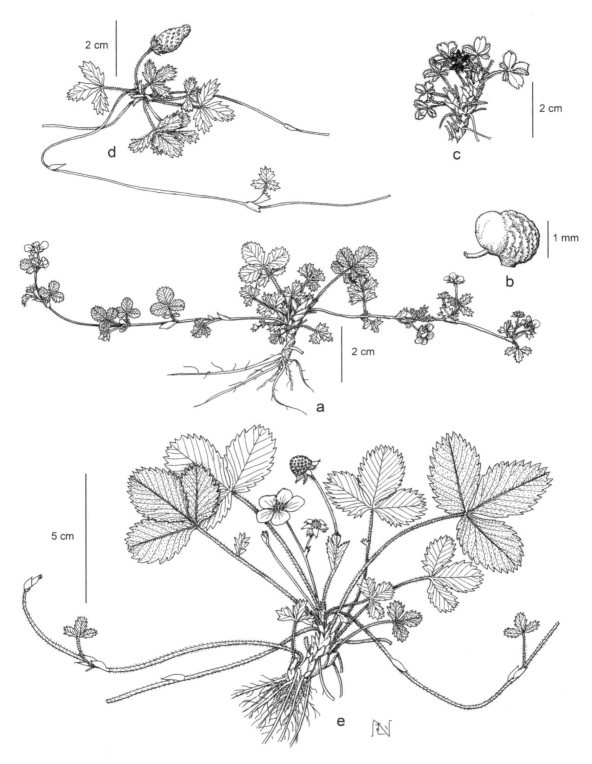

FIG. 50. ROSACEAE. **Sibbaldia micropetala**: a, flowering plant; b, achene. **Sibbaldia purpurea**: c, flowering plant. **Fragaria daltoniana**: d, fruiting plant. **Fragaria nubicola**: e, flowering plant.

7. *Sibbaldia purpurea* Royle, Ill. Bot. Himal. Mts. [4]: pl. 40 fig. 3 (1834).
Potentilla purpurea (Royle) Hook.f.; *Sibbaldia macropetala* Murav.

Dioecious, prostrate, tufted shrublets. Rhizomes woody, much branched, mat-forming, upper part covered with old stipules and petioles. Flowering stems short, erect or ascending, 0.5–2 cm tall, with appressed hairs. Stipules membranous, pale purple-brown when young, turned to dark brown, sparsely pilose outside; petioles with appressed hairs. Radical leaves palmately 5-foliolate, 0.8–2 × 0.5–1.5 cm; leaflets sparsely appressed villous on upper surface, densely villous on lower surface.Terminal leaflet sessile or subsessile, obovate or obovate-oblong, 3–10 × 2–6 mm wide, base cuneate, apex obtuse, entire except for 2 or 3 teeth at apex. Flowers solitary, unisexual, axillary from radical leaves, pedicels shorter or as long as radical leaves. Flowers 4–6 mm across. Episepals lanceolate to narrowly oblong, shorter than sepals, 1–1.5 × 0.5–0.7 mm, apex acute to obtuse, sparsely pilose outside. Sepals ovate to triangular-ovate, 0.8–1 × 0.6–0.7 mm, apex acute to obtuse. Petals 5, purple-red, oblanceolate to narrowly obovate, 1.5–2 × 1–1.2 mm, apex rounded or shallowly retuse; nectary disk developed, purple-red; male flowers with 5 stamens alternating with petals; filaments ca. 0.7 mm. Female flowers with 10–15 carpels style lateral to subterminal. Achenes ovoid, glabrous, purple-brown.
Fig. 50c.

Distribution: Nepal, W Himalaya, E Himalaya and Tibetan Plateau.

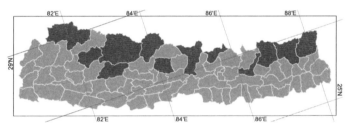

Altitudinal range: 3600–5700 m.

Ecology: Open meadows, rocky crevices.

Flowering: June–August. **Fruiting:** July–September.

Sibbaldia macropetala Muraj. (from Sikkim, Bhutan, SW China) is sometimes treated as a variety of *S. purpurea*, but it differs from *S. purpurea* by having flowers usually in corymbs which exceed the radical leaves, and hermaphrodite flowers.

8. *Sibbaldia tetrandra* Bunge, Verz. Altai Pfl.: 17 (1836).
Dryadanthe bungeana Ledeb.; *D. tetrandra* (Bunge) Juz.; *Potentilla tetrandra* (Bunge) Bunge ex Hook.f.

Low, tufted, perennial herbs. Usually dioecious. Rhizomes robust, terete, much branched, mat-forming, upper part covered with old leaves. Flowering stems short, 0.5–1 cm tall. Stipules membranous, pale to dark brown, adnate to petioles, pilose outside. Petioles dilated in lower half, upper half with pilose hairs. Leaves trifoliolate, 0.5–1.5 × 0.4–1 cm; leaflets pilose on both surfaces. Terminal leaflet sessile to subsessile, oblong to obovate, 4–8 × 3–4 mm wide, base cuneate, apex retuse, 3-dentate. Flowers 1 or 2, 4–8 mm across. Episepals 4, lanceolate to narrowly ovate, nearly as long as or shorter than sepals, 1.5–2.5 × 0.5–1 mm, apex acute to obtuse. Sepals 4, ovate to triangularly ovate, 1.5–2 × 1–1.2 mm wide, apex obtuse to subrounded. Petals 4, creamy white to pale yellow, oblong to narrowly obovate, 1.5–2 × 0.7–1 mm, alternate with sepals, nearly as long as or a little longer than sepals. Male flowers with 4 stamens, opposite to sepals; filaments 0.8–1 mm; nectary disk developed. Female flowers with 4 or 5 carpels; ovaries ovoid, ca. 1 mm × 0.5 mm; style lateral to subterminal, 1–1.2 mm; stigma slightly inflated. Achenes glabrous.

Distribution: Nepal, W Himalaya, E Himalaya, Tibetan Plateau, N Asia and C Asia.

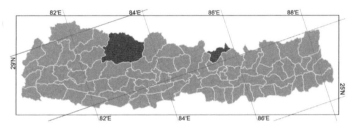

Altitudinal range: 4500–6000 m.

Ecology: Rocky slopes or rock crevices, meadows.

Flowering: May–July. **Fruiting:** July–September.

18. *Fragaria* L., Sp. Pl. 1: 494 (1753).

Hiroshi Ikeda

Perennial herbs with slender prostrate stolons. Stipules partly adnate to base of petioles. Leaves rosulate, alternate, 3(or 5)-foliolate, toothed, petiolate. Flowers solitary, rarely 2- or 3-flowered, axillary from rosulate leaves. Flowers bisexual. Bracts at base of pedicel; bracteoles present or absent. Hypanthium shallowly cupulate. Episepals 5. Sepals 5. Petals 5, usually white, rarely pinkish. Stamens 20–30. Ovary superior, carpels many, free, inserted on a semi-globose receptacle. Style lateral. Achenes small, inserted on a cavity of enlarged fleshy receptacle.

Worldwide about 20 species in the temperate and warmer regions of the N hemisphere and in S America. Three species in Nepal.

Key to Species

1a Mature fruiting receptacle conical. Terminal leaflet oblanceolate to obovate, with 3–6 teeth on each side, shiny deep green on upper surface ....**1. F. daltoniana**

b Mature fruiting receptacle subglobose. Terminal leaflet narrowly to broadly obovate, with 4–13 teeth on each side, not shiny on upper surface 2

2a Petioles and flowering stems with brownish patent hairs. Petals 4–6 mm ....**2. F. nilgerrensis**

b Petioles and flowering stems with whitish appressed hairs. Petals 7–10 mm...**3. F. nubicola**

1. _Fragaria daltoniana_ J.Gay, Ann. Sci. Nat., Bot., Sér. 4 8: 204 (1857).
Fragaria rubiginosa Lacaita; _F. sikkimensis_ Kurz; _F. vesca_ var. _collina_ Hook.f.

कर्णफूल झार Karnaphul jhar (Nepali).

Stolons subglabrous or sparsely appressed hairy. Radical leaves trifoliolate, (1–)1.5–7(–10) × (0.7–)1–3(–4) cm; petiole slender, 0.5–4(–9) cm, whitish appressed hairy; leaflets shortly petiolulate, oblanceolate to obovate, (0.5–)1–2(–2.5) × 0.5–1.2(–1.5) cm, base of terminal leaflet cuneate, base of lateral leaflets obliquely cuneate, apices obtuse or acute, margin incised serrate, 3–6 teeth on each side, shiny deep green above, sparsely appressed hairy when young, glabrous when old, appressed hairy below, especially along midvein and lateral veins. Flower solitary; scape 1–5 cm, with white appressed hairs. Episepals oblong, 3.5–5 × 1.5–2 mm, glabrescent or sparsely white appressed hairy above, white appressed hairy below, apex 2–4-toothed, acute to acuminate. Sepals ovate, 3–5 × 2–2.5 mm, glabrescent above, white appressed hairy below. Petals white, broadly obovate to semiorbicular, 4–7 × 3–6.5 mm. Stamens 15–20; filaments 1–1.5 mm; anthers flattened ellipsoid, ca. 1 × ca. 0.7 mm. Ovaries ellipsoid, ca. 0.5 × ca. 0.3 mm; style ca. 0.8 mm. Fruiting receptacle globose when young, becoming conical when mature, reddish, pink or whitish, 0.7–2.5(–3) × 0.5–1(–1.5) cm; persistent sepals spreading. Achenes smooth, glabrous, ca. 1 × ca. 0.6 mm. Fig. 50d.

Distribution: Nepal, W Himalaya, E Himalaya, Tibetan Plateau and Assam-Burma.

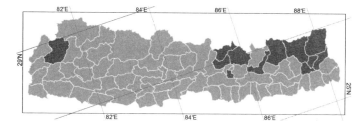

Altitudinal range: 2000–4000 m.

Ecology: Forest margins, moist banks, open meadows.

Flowering: June–July. **Fruiting:** July–August.

Fruits are edible. Juice of the roots is taken to treat fever.

2. _Fragaria nilgerrensis_ Schltdl. ex J.Gay, Ann. Sci. Nat., Bot., Sér. 4 8: 206 (1857).

Stolons brownish patent hairy. Radical leaves trifoliolate, (1–)1.5–10(–15) × (1.2–)1.5–4(–6) cm; petiole slender, 0.8–8(–12) cm, brownish patent hairy; leaflets shortly petiolulate, oblanceolate to obovate, (0.8–)1–3(–4) × (0.5–)1–2(–3) cm, base of terminal leaflet cuneate, base of lateral leaflets obliquely cuneate, apices obtuse or acute, margin incised serrate, 4–10 teeth on each side, sparsely or densely brownish patent hairy above, adaxially brownish patent hairy below, especially along midvein and lateral veins, . Flower 1 or 2(or 3); scape 1–5(–10) cm, with brownish patent hairs. Episepals lanceolate to oblong, 4–5 × 1.5–2 mm, sparsely brownish ascending hairy above, brownish patent hairy below, entire, apex acute. Sepals oblong-ovate, 4–5 × 2–2.5 mm, glabrescent above, brownish patent hairy below. Petals white, broadly obovate to semiorbicular, 4–6 × 3–6 mm. Stamens around 20; filaments 0.7–2 mm; anthers flattened, narrowly ovoid, 0.8–1 × ca. 0.6 mm. Ovaries ellipsoid, ca. 0.6 × ca. 0.4 mm; style fusiform, ca. 1 mm. Fruiting receptacle subglobose, whitish to pink, 0.8–1.5 cm in diameter; persistent sepals appressed to aggregate fruit. Achenes smooth, glabrous, ca. 1 × ca. 0.8 mm.

Distribution: Nepal, E Himalaya, Assam-Burma and E Asia.

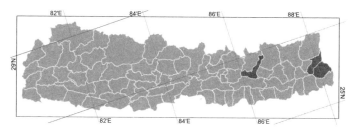

Altitudinal range: 2800–4200 m.

Ecology: Forest margins and moist meadows.

Flowering: May–June. **Fruiting:** June–July.

3. _Fragaria nubicola_ (Hook.f.) Lacaita, J. Linn. Soc., Bot. 43: 467 (1916).
Fragaria vesca var. _nubicola_ Hook.f., Fl. Brit. Ind. 2[5]: 344 (1878); _F. nubicola_ Lindl. ex Wall. nom. nud.

Stolons subglabrous or sparsely appressed hairy. Radical leaves trifoliolate, sometimes bearing 2 additional minor leaflets; terminal leaflet 1.5–10(–15) × 1.5–4(–5) cm; petiole slender, 1–8(–12) cm, whitish appressed hairy; leaflets shortly petiolulate or subsessile, narrowly to broadly obovate, 1–4(–5) × (0.7–)1–2(–3) cm, base of terminal leaflet cuneate, base of lateral leaflets obliquely cuneate, apices obtuse or acute, margin incised serrate, 5–13 teeth on each side, light green above, sparsely white appressed hairy, white appressed hairy below, especially along midvein and lateral veins. Flowers 1 or 2; scape 2–8(–13) cm, with white appressed hairs. Episepals oblong to spathulate, 3–5(–7) × 1–2 mm, glabrescent above, sparsely white appressed hairy below, apex 2–4-toothed, sometimes deeply divided into two lobes, acute to acuminate. Sepals ovate, 3–5(–7) × 1.5–3(–4) mm, glabrescent above, white appressed hairy below. Petals white, broadly obovate to semiorbicular, 7–10 × 5–10 mm. Stamens 20–25; filaments 2–3 mm; anthers flattened ovoid, ca. 1.2 × ca. 1 mm. Ovaries oblong, ca. 0.8 × ca. 0.3 mm; style ca. 1.2 mm. Fruiting receptacle subglobose, bright red, 0.8–1.2 cm in diameter; persistent sepals appressed to aggregate fruit. Achenes smooth, glabrous, ca. 1 × ca. 0.8 mm. Fig. 50e.

Distribution: Nepal, W Himalaya, E Himalaya and Tibetan Plateau.

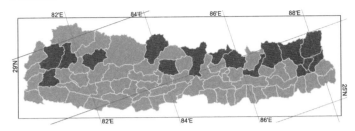

Altitudinal range: 1600–4000 m.

Ecology: Forest margins, moist banks, open meadows.

Flowering: April–July(–October). **Fruiting:** May–July.

Rao (Indian Forester 93: 47. 1967) misapplied the name *Fragaria vesca* L. to this species.

The juice of the plant is given to give relief from profuse menstruation. Root juice is taken to treat fever. Unripe fruits are chewed to treat blemishes on tongue.

19. *Rosa* L., Sp. Pl. 1: 491 (1753).

Hideaki Ohba & Colin A. Pendry

Shrubs, erect and diffuse or climbing, mostly prickly, bristly, or rarely unarmed. Stipules persistent and adnate to petiole or free except base and caducous, rarely absent. Leaves alternate, odd pinnate, rarely with a leaf-like bract with the appearing of a simple leaf. Flowers solitary or in corymbs, umbels or fascicles, usually bracteate, usually 5-, rarely 4-merous. Flowers bisexual. Bracts solitary, at base of pedicel; bracteoles absent or rarely present. Hypanthium ovoid to globose or obovoid. Epicalyx absent. Sepals 4 or 5, entire or with variously pinnately lobed margins, inner surface usually densely hairy. Petals 4 or 5 or numerous (double flowers), imbricate, inserted at mouth of hypanthium. Stamens numerous, in several whorls. Ovary superior, carpels free, numerous, rarely few, inserted at margin or base of hypanthium, ovules pendulous. Styles terminal or lateral, free or connate and column-like. Fruit a hip, formed from fleshy hypanthium with numerous (rarely few) achenes. Seed pendulous.

Worldwide 100 to 150 species from N temperate regions. Five species native to Nepal, with another eight species reported to be cultivated.

Rosa taxonomy is complicated because of hybridization, and many species may actually be ancient hybrids of complex and disputed taxonomy (Mabberley, Plant Book: 745-747. 2008). There is a huge commercial rose industry which produces numerous new hybrids every year, combining different cultivars to produce the desired combination of traits. Eight species have been reported to be cultivated in Nepal and are included in the key but not treated further. *Rosa chinensis* Jacq. is the basis for the 'China roses'. *Rosa* x *odorata* (Andrews) Sweet, the 'tea rose', is a hybrid between *R. chinensis* and *R. gigantea* Collett ex Crép. *Rosa moschata* Herrm. is the 'musk rose', whose exact origins are uncertain, but thought to be Himalayan or SW Asian; it is cultivated primarily for its scent and is the source of rose water. *Rosa gallica* L. is native to S Europe and W Asia and has long been cultivated there. *Rosa* x *damascena* Mill., the 'damask rose', is the natural hybrid between *R. moschata* and *R. gallica*. The three climbing species cultivated in Nepal are *R. banksiae* Aiton, *R. laevigata* Michx. and *R. multiflora* Thunb. var. *carnea* Thory, and all originate from China.

Key to Species

1a	Stipules predominantly free from petiole, caducous	2
b	Stipules predominantly adnate to petiole, usually persistent	4
2a	Branchlets villous. Leaflets 9–11. Flowers with leaf-like bracts	**5. *R. clinophylla***
b	Branchlets glabrous. Leaflets 3–5. Flowers without bracts	3

3a Flowers numerous, in corymbs, ca. 2 cm across. Pedicels with minute spine-like hairs, hypanthium
   glabrous.. *R. banksiae*
 b Flowers solitary, 6–8 cm across. Pedicel and hypanthium with many needle-like spines .... *R. laevigata*

4a Styles connate, column-like, conspicuously exserted from hypanthium, as long as stamens.......................................5
 b Styles free or slightly exserted from hypanthium, shorter than stamens 7

5a Stipules entire..**4. *R. brunonii***
 b Stipules dentate or pectinate..6

6a Styles villous... *R. moschata*
 b Styles glabrous.. *R. multiflora* var. *carnea*

7a Petals 4 or 5. Sepals erect or spreading, persistent in fruit. Wild..8
 b Petals numerous. Sepals reflexed, caducous. Cultivated or naturalized...10

8a Flowers 4-merous, without bract, white, always solitary ....**1. *R. sericea***
 b Flowers 5-merous, bracteate, pink or red, solitary or up to 5 in a fascicle 9

9a Petals reddish or pink, broadly obovate, base cuneate. Leaflets 5–9, 0.6–2 cm, upper surface glabrous or
   sparsely puberulous along veins, apex mostly rounded-obtuse. Pedicel 1–1.5 cm ....**2. *R. webbiana***
 b Petals deep red, obtriangular-obovate, base broadly cuneate. Leaflets usually 9–11, 2.2–6 cm, upper surface
   villous, apex mostly acute. Pedicel 1.5–3.5 cm ....**3. *R. macrophylla***

10a Hypanthium glandular. Sepal margin variously lobed or toothed..11
 b Hypanthium glabrous. Sepal margin usually entire...12

11a Leaflets simply serrate. Flowers solitary or in groups of 2–4... *R. gallica*
 b Leaflets doubly serrate. Flowers in clusters of up to 12 .... *R. x damascena*

12a Flowers pink to crimson. Leaflets downy on midvein beneath.. *R. chinensis*
 b Flowers white, yellow or pale pink. Leaflets completely glabrous.. *R. x odorata*

---

1. *Rosa sericea* Lindl., Ros. Monogr.: 105, pl. 12 (1820).
Rosa sericea var. *hookeri* Regel; *R. sericea* var. *omeiensis*
(Rolfe) Rowley; *R. sericea* var. *pteracantha* A.R.Bean; *R.
tetrasepala* Royle nom. nud.; *R. wallichii* Tratt.

दारिमपाते Darimpate (Nepali).

Erect shrubs to 1–2 m. Twigs robust, smooth to villous or
glandular-hairy with bristles throughout. Prickles usually in
pairs below leaves and scattered or rarely absent, to 1.2 cm
long, robust, abruptly flaring to broad base, or rarely wing-like.
Stipules mostly adnate to petiole, free parts auriculate, hairy
or glabrous, margin glandular. Leaves including petiole
3.5–8 cm, rachis and petiole hairy. Leaflets (5–)7–11, ovate or
obovate to obovate-oblong, 0.8–2 × 0.5–0.8 cm, base broadly
cuneate, apex rounded-obtuse or sometimes acute, margin
serrate in upper part, entire below, glabrous to sparsely villous
above, sericeous-villous or glabrous below, sometimes with
glandular hairs. Flowers solitary, axillary, 2.5–5 cm across.
Bracts absent. Pedicel 1–2 cm, glabrous. Hypanthium obovoid
or globose, nearly glabrous. Sepals 4, ovate-lanceolate,
8–15 mm, apex acuminate or acute, margin entire, outer
surface glabrous to sparsely hairy, sometimes rather densely
glandular, inner surface appressed hairy. Petals 4, white or
often with creamy base, rarely pale yellow, broadly obovate,
base broadly cuneate, apex emarginate. Styles free, shorter
than stamens, slightly exserted, villous. Hip red or purple-
brown, shiny, obovoid or globose, 8–15 mm across, glabrous,
with persistent, erect sepals.
Fig. 51a–c.

Distribution: Nepal, W Himalaya, E Himalaya, Tibetan
Plateau, Assam-Burma and E Asia.

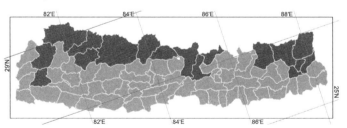

Altitudinal range: 2100–4600 m.

Ecology: Open woods, forest margins, scrub, dry sunny
places, sometimes gregarious.

Flowering: May–August. **Fruiting:** July–September.

A distinctive 4-petalled rose, common throughout Nepal. The
shape of the hip is variable, with globose hips common in W
and C Nepal, and obovate hips in E Nepal. Four varieties are
recognized in Nepal, but Grierson (Fl. Bhutan 1: 586. 1987)

noted that their characters overlap; var. *sericea* with slender, green fruiting pedicels; var. *omeiensis* with thickened, fleshy, reddish fruiting pedicels; var. *pteracantha* with broad prickles up to 2 cm along the base and var. *hookeri* with small prickles, bristles and glands on its stems.

Ripe fruits are edible. A paste of flowers is applied to treat headaches and also given for liver complaints.

2. Rosa webbiana Wall. ex Royle, Ill. Bot. Himal. Mts. [4]: pl. 42 fig. 2 (1834).
Rosa webbiana Wall. nom. nud.

Shrubs 1–2 m. Twigs slender, purple-brown. Prickles in pairs below leaves, and scattered, straight, to 1 cm, stout, gradually tapering with broad base. Stipules mostly adnate to petiole, free parts ovate, apex acute, margin glandular-hairy. Leaves including petiole 3–4 cm, rachis and petiole glabrous. Leaflets 5–9, suborbicular, obovate, or broadly elliptic, 0.6–2 × 0.4–1.2 cm, base rounded or cuneate, apex rounded-obtuse, rarely acute, margin simply serrate in upper part, entire below, glabrous above, glabrous or sparsely hairy below, especially along veins. Flowers solitary, rarely 2 or 3 in fascicles, 3.5–5 cm across. Bracts ovate, margin glandular serrate, upper surface with conspicuous midvein and lateral veins. Pedicel 1–1.5 cm, glabrous or glandular-hairy. Hypanthium subglobose or ovoid, glabrous or glandular-hairy. Sepals 5, triangular-lanceolate, 10–15 mm, apex elongate, margin entire, outer surface glandular-hairy. Petals 5, reddish or pink, broadly obovate, base cuneate, apex emarginate. Styles free, shorter than stamens, hairy. Hip nodding, bright red, subglobose or ovoid, 1.5–2 cm across, glabrous, with persistent, spreading sepals.
Fig. 51d–f.

Distribution: Nepal, W Himalaya, Tibetan Plateau and N Asia.

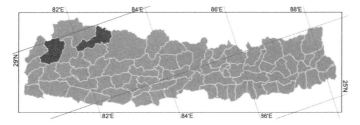

Altitudinal range: 2200–2900 m.

Ecology: Scrub, forest margins, grassy places, valley slopes.

Flowering: June–August. **Fruiting:** July–September.

3. Rosa macrophylla Lindl., Ros. Monogr.: 35, pl. 6 (1820).
Rosa hookeriana Wall. nom. nud.; *R. macrophylla* var. *hookeriana* Hook.f.

जंगली गुलाब Jungali gulab (Nepali).

Shrubs 1.5–3 m. Twigs robust, purple-brown. Prickles in pairs below leaves or absent, straight. Stipules mostly adnate

to petiole, free parts triangular to narrowly ovate, usually glabrous, apex shortly acuminate, margin glandular-serrate. Leaves including petiole 7–15 cm, rachis and petiole villous, often with sparse glandular hairs, sometimes with small prickles. Leaflets (7 or)9–11(–13), narrowly oblong or elliptic-ovate, 2.2–5(–7) × 0.9–2.5(–3) cm, base rounded, rarely broadly cuneate, apex acute, rarely rounded-obtuse, margin nearly entire or doubly serrate, glabrous above, glabrous or sometimes sparsely glandular below, with prominent villous veins. Flowers 3.5–7 cm across, solitary, or 2 or 3 in fascicles. Pedicel 1.5–3.5 cm, usually with dense stipitate glandular hairs. Bracts 1 or 2, ovate, 1.5–2.5 cm, apex acuminate, margin glandular-hairy, upper surface puberulous along midvein or glabrous, with prominent veins. Hypanthium ovoid or narrowly ovoid, with dense stipitate glandular hairs. Sepals 5, narrowly triangular to narrowly ovate, 2–3.5(–5) cm, apex long caudate, often expanded into elliptic apical appendages to 1.2 cm, margin entire, outer surface stipitate glandular hairy, inner surface densely hairy. Petals 5, deep red, obtriangular-obovate or obovate, base broadly cuneate, apex rounded or emarginate. Styles free, much shorter than stamens, hairy. Hip purple-red, shiny, oblong-ovoid or narrowly ovoid, 1.5–3 × ca. 1.5 cm, often with stipitate glandular hairs, apex shortly necked, with persistent, erect sepals.
Fig. 51g–h.

Distribution: Nepal, W Himalaya, E Himalaya and Tibetan Plateau.

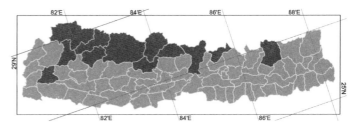

Altitudinal range: 2100–4400 m.

Ecology: Scrub, forest margins, slopes.

Flowering: June–July. **Fruiting:** August–November.

Paste of the fruits is regarded as a tonic, and good for eyesight, and is also useful in fever, diarrhoea, and bile problems.

4. Rosa brunonii Lindl., Ros. Monogr.: 120, pl. 14 (1820).
Rosa clavigera H.Lév.; *R. moschata* var. *nepalensis* Lindl.; *R. pubescens* Roxb.

भैंसीकाँडा Bhainsikanda (Nepali).

Climbing or scandent shrubs, 4–6 m. Twigs, red- or purple-brown, terete, glabrate. Prickles scattered, curved, to 5 mm long, flat, gradually tapering from broad base. Stipules mostly adnate to petiole, free parts lanceolate, apex acuminate, margin glandular, both surfaces hairy. Leaves including petiole 6–9 cm, rachis and petiole densely hairy, with scattered small prickles. Leaflets 5 or 7, oblong or oblong-lanceolate,

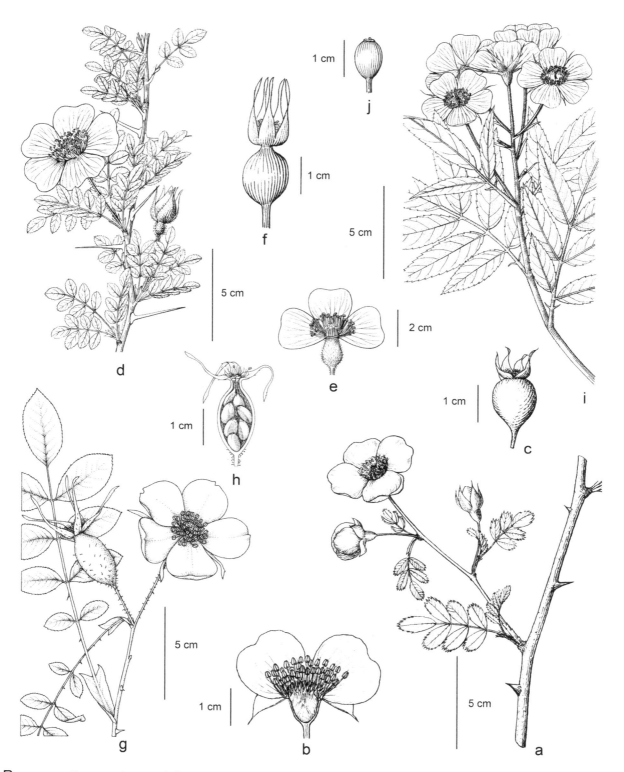

FIG. 51. ROSACEAE. **Rosa sericea**: a, inflorescence and leaves; b, longitudinal section of flower; c, fruit with persistent calyx. **Rosa webbiana**: d, inflorescence and leaves; e, flower with sepals and two petals removed; f, fruit with persistent calyx. **Rosa macrophylla**: g, inflorescence with fruit and leaves; h, longitudinal section of fruit with persistent calyx. **Rosa brunonii**: i, inflorescence and leaves; j, fruit.

3–8 × 1–2.6 cm, broadly cuneate to rounded truncate, apex acuminate or acute, margin serrate, glabrous above, glabrous or sparsely hairy below, sometimes with glandular hairs along veins. Flowers 3–5 cm across, numerous in compound corymbs. Pedicel 2.8–3.5 cm, hairy with sparse glandular hairs. Bracts sometimes a unifoliolate leaf. Bracteoles tiny or absent. Hypanthium obovoid, outer surface hairy. Sepals 5, deciduous, lanceolate, 1.5–2 cm, apex acuminate, often with 1 or 2 pairs of lobes, both surfaces hairy. Petals 5, white, fragrant, broadly obovate, base broadly cuneate, apex rounded or emarginate. Stamens numerous, filaments pale yellow, anther yellow or orange. Styles connate into column, exserted, slightly longer than stamens, hairy. Hip purple-brown or dark red, shiny, ovoid 0.7–1.2 cm across, glabrous. Fig. 51i–j.

Distribution: Nepal, W Himalaya, E Himalaya, Tibetan Plateau, Assam-Burma and E Asia.

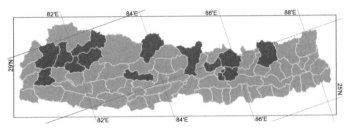

Altitudinal range: 1300–3000 m.

Ecology: Margins of thickets, scrub, newly opened slopes.

Flowering: April–June. **Fruiting:** July–November.

Hooker (Fl. Brit. Ind. 2: 367. 1878) and later authors in India have misapplied the name *Rosa moschata* Mill. to this species.

Rosa multiflora Thunb. var. *carnea* Thory (in Redouté, Roses 2: 70, t. 1821) is a climbing rose cultivated as an ornamental. It is easily distinguished from *R. brunonii* by its pinkish, double-petalled flowers, fimbriate stipules and glabrous styles.

5. *Rosa clinophylla* Thory, Roses 1: 43 (1817).
Rosa involucrata Roxb. ex Lindl.; *R. involucrata* Roxb. nom. nud.; *R. lyellii* Lindl.; *R. palustris* Buch.-Ham. ex Lindl.

Erect shrubs, to 6 m. Stems and twigs silky-villous. Prickles in pairs below stipules. Stipules shortly adnate to petiole, free parts linear-lanceolate, margin fimbriate, apex acuminate. Leaves rachis and petiole densely pubescent, with glandular-villous hairs and minute prickles. Leaflets 9–11, sessile or nearly so, oblong-elliptic or elliptic, 1.5–3.5 × 0.5–1.5 cm, base rounded-cuneate, apex acute to obtuse, margin serrate, glabrous and shiny above, villous below throughout or on venation. Flowers usually solitary on leafy branchlets, 3–5 cm across. Bract a unifoliolate leaf. Pedicel 2.8–3.5 cm, villous. Hypanthium globose, densely villous. Sepals 5, long triangular, 13–17 mm, apex acuminate, outer surface villous, sometimes with minute straw-coloured prickles, inner surface villous. Petals 5, white, base often with yellow shade, broadly obovate, base broadly cuneate, apex emarginate. Styles connate into column, exserted, as long as stamens, glabrous. Hip globose, ca. 1 cm across, villous.

Distribution: Nepal, W Himalaya, Assam-Burma, S Asia and SE Asia.

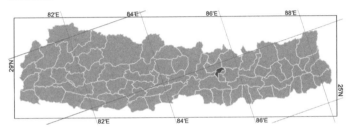

Altitudinal range: 100–1000 m.

Ecology: Often growing on river banks or in marshy conditions.

Flowering: June. **Fruiting:** July–November.

Hooker (Fl. Brit. Ind. 2: 365. 1878) described *Rosa clinophylla* (as *R. involucrata*) as the "common rose of the Bengal plains and foot of the Himalaya and the only really tropical species of India". It is probably much more common in the Tarai region of Nepal than suggested by the distribution map.

20. *Agrimonia* L., Sp. Pl. 1: 448 (1753).

Hiroshi Ikeda

Perennial herbs. Rootstock robust. Stipules adnate to petioles in lower part, free in upper part. Leaves imparipinnate with small accessory leaflets, petiolate; leaflets serrate at margin. Inflorescence a many-flowered terminal raceme. Flowers bisexual, shortly pedicellate. Bracts solitary, at base of pedicel; bracteoles 2, at apex of pedicel. Hypanthium turbinate, constricted at throat, with many hooked prickles. Epicalyx absent. Sepals 5. Petals 5. Stamens 5–15, inserted in throat of hypanthium; filaments free, filiform; anthers 2-locular. Ovary superior, carpels 2, enclosed in hypanthium; ovule 1, pendulous. Style subterminal, filiform; stigma dilated. Achene dry, enclosed in hardened hypanthium.

Worldwide about 10 species in temperate zones of the N hemisphere. One species in Nepal.

1. Agrimonia pilosa Ledeb., Index Seminum Dorpat. Suppl.: 1 (1823).

Agrimonia pilosa var. **nepalensis** (D.Don) Nakai, Bot. Mag. [Tokyo] 47: 247 (1933).
Agrimonia nepalensis D.Don, Prodr. Fl. Nepal.: 229 (1825); *Agrimonia eupatoria* var. *nepalensis* (D.Don) Kuntze; *A. lanata* Wall. ex Wallr.; *A. lanata* Wall. nom. nud.

बोक्रोमरन Bokromaran (Tamang).

Stems erect, 30–100 cm tall, sparsely pilose and pubescent, and densely hirsute in lower part. Radical leaves remaining and often withered at anthesis, 7–20 × 3–8 cm, interrupted pinnate with 2–4 pairs of leaflets. Petioles 1–5 cm, hirsute; free part of stipules membranous, obliquely lanceolate to narrowly ovate; leaflets subopposite, deep green above, powdery green below, subsessile, oblong to narrowly obovate, 2–5 × 1–3 cm, base cuneate, apex acute or obtuse, margin serrate. Cauline leaves similar to radical leaves; stipules of cauline leaves leafy, falcate, margin dentato-serrate. Inflorescence 10–30 cm; bracts deeply divided into three lobes. Hypanthium sparsely hairy outside. Sepals narrowly ovate, 1.2–1.5 × 0.6–0.8 mm, margin entire, glabrous above, sparsely hirsute and glandular below. Petals yellow, oblong to narrowly oblong, 2.5–3 × 1.3–2 mm. Filaments 1–1.2 mm; anthers globose, ca. 0.6 mm in diameter. Pistils 2; ovary oblong, 0.5–0.7 × 0.3–0.4 mm;

style 1, 0.8–1 mm. Hypanthium hardened in fruit, grooved, 4–5 × 3–4 mm, patent hairy, with persistent sepals and many hooked prickles, containing 1 or 2 achenes. Fig. 52a–c.

Distribution: Nepal, W Himalaya, E Himalaya, Tibetan Plateau, Assam-Burma and E Asia.

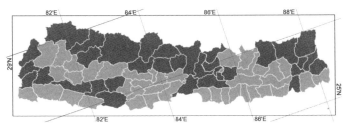

Altitudinal range: 1000–3900 m.

Ecology: Open meadows and forest edges.

Flowering: July–September. **Fruiting:** August–October.

A decoction of plant is given for abdominal pain, diarrhoea, dysentery, sore throats, coughs, colds and tuberculosis. Ash of the plant is applied to wounds. Roots are used to treat bowel complaints and as an anti-venom.

21. Sanguisorba L., Sp. Pl. 1: 116 (1753).

Hiroshi Ikeda

Perennial herb. Rootstock robust. Stipules adnate to petioles in lower part, free in upper part. Leaves imparipinnate, petiolate; leaflets serrate. Inflorescence terminal or lateral from axils of upper cauline leaves, densely capitate. Flowers bisexual. Bracts solitary, at base of pedicel; bracteoles absent. Hypanthium with constricted throat. Epicalyx absent. Sepals 4, petaloid, dark brown to tinged green. Petals absent. Stamens 2, inserted in throat of hypanthium; filaments free; anthers 2-locular. Ovary superior, carpel 1, included in hypanthium; ovule pendulous. Style terminal, filiform; stigmas 2, penicillate. Aggregate fruit capitate, achenes solitary, dry, each included in hardened, winged hypanthium.

Worldwide about 30 species in Asia, Europe and N America. One species in Nepal.

1. Sanguisorba diandra (Hook.f.) Nordborg, Opera Bot. 11(2): 60 (1966).
Poterium diandrum Hook.f., Fl. Brit. Ind. 2[5]: 362 (1878); *Sanguisorba diandra* Wall. nom. nud.; *S. dissita* T.T.Yu & C.L.Li.

Stems 25–90 cm, glabrescent or sparsely glandular pubescent. Radical leaves remaining or often withered at anthesis, 10–25 × 3–7 cm, with 4–8 pairs of leaflets. Petiole 0.5–2 cm, glabrescent or sparsely glandular pubescent. Leaflets alternate or subopposite, light to deep green above, powdery green below, petiolulate, narrowly ovate to elliptic, 0.6–3 × 0.5–2.5 cm, base cordate to truncate, apex obtuse or rounded, margin dentato-serrate. Cauline leaves similar to radical leaves; stipules of cauline leaves leafy, falcate, margin dentato-serrate. Inflorescence capitate, terminal or lateral

from axils of upper cauline leaves, 5–8 mm in diameter, with 3–5 bracts at base; bracts lanceolate, membranous, margin ciliate. Flowers bisexual, subsessile. Hypanthium narrowly ellipsoid to cylindrical, 4-ridged, 2 × 1 mm, glabrescent or with sparsely short glandular hairs. Sepals 4, dark brown to tinged green, oblong to elliptic, 1.5–2.0 × 0.8–2.3 mm, margin entire, glabrous above, with short tufted hairs near tips below. Stamens 2, exserted from hypanthium; filaments filiform, 1.6–2.0 mm; anthers globose, ca. 1 mm in diam. Ovary oblong, 1.0–1.2 × 0.7–0.8 mm; style ca. 1 mm. Aggregate fruit 1–1.5 cm in diameter, hypanthium 5–7 × 3–5 mm, glabrous, with persistent sepals, containing 1 seed; wings 1.2–2 mm wide.
Fig. 52d–g.

Distribution: Nepal, W Himalaya, E Himalaya and Tibetan Plateau.

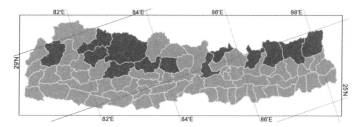

Altitudinal range: 2400–4400 m.

Ecology: Meadows, forest margins and thickets.

Flowering: June–August. **Fruiting:** August–September.

22. *Alchemilla* L., Sp. Pl. 1: 123 (1753).

Hiroshi Ikeda & Mark F. Watson

Deciduous perennial rosulate herbs. Flowering stems from radical leaves. Stipules present, toothed or lobed, adnate to petioles. Leaves alternate, simple, orbicular, margin palmately lobed, palmately veined, stem leaves often less lobed than basal leaves. Inflorescence cymose, a much branched corymb, bracts and bracteoles absent. Flowers numerous, very small, inconspicuous, pale green, bisexual, shortly pedicellate. Bracts solitary, leafy; bracteoles absent. Hypanthium urceolate, throat slightly constricted, persistent. Episepals 4, alternating with sepals. Sepals 4, valvate in bud. Petals absent. Nectariferous disk lining hypanthium with thickened margin. Stamens 4, filaments free, short. Ovary superior, carpel 1, sessile or almost so, ovule 1, basal. Style basal, filiform, glabrous, stigma capitate. Fruit an achene.

Worldwide 13–300 species (or more: see below), alpine and cool temperate regions of Asia, Europe, Africa and N America, and mountains in tropical areas of S America and Africa, most diverse in mountainous regions of Europe and the Caucasus. One species in Nepal.

A taxonomically complex genus with identification often problematic because of blurred morphological boundaries between species. Ancient hybridization events coupled with obligate apomictic reproduction and high polyploid levels contribute to this. Local morphologically uniform populations are maintained through apomixy and have resulted in some authors recognizing 1000 or more microspecies. Summer-season, well-grown material is preferred for identification as late-season growth can be atypical.

Alchemilla subcrenata Buser *sensu* Hara (Enum. Fl. Pl. Nepal 2: 133. 1979) was reported from Nepal from a single collection, *Pande 75* (BM). This collection is almost certainly associated with the Botanical Garden at Godavari and not considered native nor naturalized, so is excluded from this account. Hara cited this collection as *A. subcrenata* based on an identification by S.M. Walters, however, the BM specimen only bears a single determination slip by Walters, as *A. trollii* Rothm. dated June 1980. *Alchemilla trollii*, part of the *A. vulgaris* L. complex, has previously been considered restricted to Kashmir and Pakistan (Dhar & Kachroo, Alp. Fl. Kashmir Him. 1983).

1. *Alchemilla vulgaris* L., Sp. Pl. 1: 123 (1753).

Medium-sized green herbs 10–25(–30) cm. Rhizome 5–9 mm thick. Stems slender, erect or ascending; stem, petioles and most of plant moderately pilose with spreading straight rather coarse hairs some slightly downward-pointing. Lower leaf stipule pale brown, narrowly ovate, 6–10 mm, apex acute or lacerate, papery. Petiole (5–)10–20 cm. Blade subreniform to orbicular, 2–4(–6) × 2.5–5(–7) cm, base caudate, apex rounded, margin 5–7(–9)-lobed, lobes deep (to one third length of leaf) and broad with coarse teeth each side, both surfaces usually moderately pilose with long coarse hairs, densely so at the margin, or glabrescent. Cauline leaves 3 or 4, similar to basal leaves but reducing in size upwards, less deeply lobed, more shallowly caudate, with shorter petioles (0.5–3 cm) and conspicuous, green, leaf-like stipules. Inflorescences few, narrow, little-branched, ascending, ca. 7 × 3 cm, glabrescent to pilose with spreading hairs. Pedicel 1–2.5 mm, glabrous. Flowers 2.5–3.5 mm diameter, in few-flowered loose heads or solitary. Hypanthium 1–1.2 mm, glabrous or pilose with coarse hairs. Episepals linear-lanceolate, ca. 1.2 × 0.4 mm, apex acute, outer surfaces sparsely pilose with coarse hairs towards apex. Sepals ovate, ca. 1.2 × 1.0 mm, apex acute, outer surfaces sparsely pilose with coarse hairs on outside towards apex. Stamens 0.5–1 mm. Achenes pale brown, ovoid, ca. 1.5 × 1 mm.
Fig. 52h–i.

Distribution: Nepal, W Himalaya, SW Asia and Europe.

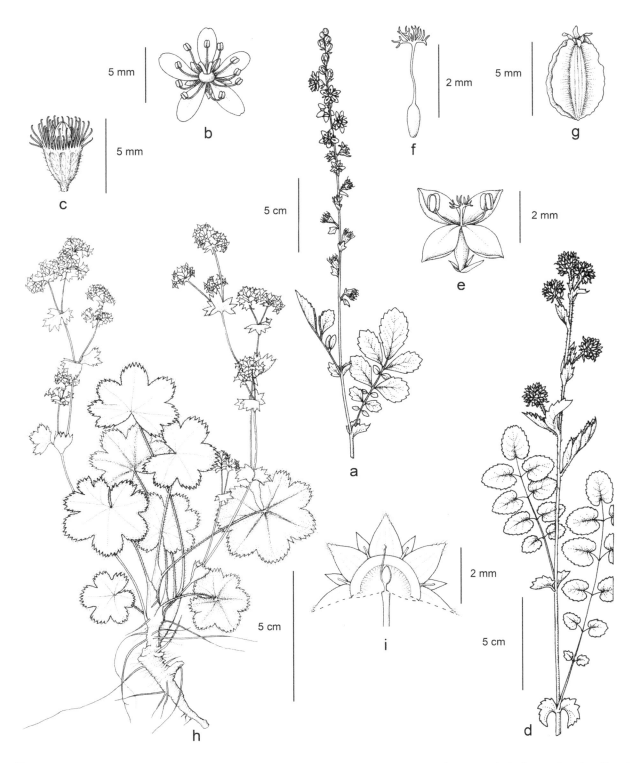

FIG. 52. ROSACEAE. **Agrimonia pilosa** var. **nepalensis**: a, inflorescence and leaves; b, flower; c, fruit. **Sanguisorba diandra**: d, inflorescence and leaves; e, flower; f, pistil; g, fruit. **Alchemilla vulgaris**: h, flowering plant; i, opened flower.

383

Rosaceae

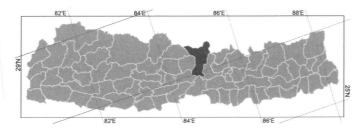

Altitudinal range: 4000–4300 m.

Ecology: Amongst tall herbs and between rocks, humus rich soils in alpine gullies.

Flowering: June–August. **Fruiting:** August.

This *Alchemilla* species was found in Manaslu Himalaya in 2008 (*Ikeda et al. 20815161, 20811203, 20811218* in E, KATH, and TI). Unlike a previous record which is almost certainly of cultivated origin, this is the first record of an *Alchemilla* indigenous to Nepal. However, the identity of this plant is currently uncertain, and awaits revision along with other *Alchemilla* species in the Pan-Himalayan region. Several morphologically similar species in the *A. vulgaris* aggregate species complex have been described in NW India, of which the Manaslu material most closely matches *A. ypsilotoma* Rothm.

23. *Prinsepia* Royle, Ill. Bot. Himal. Mts. [3]: pl. 38 fig. 1. a-g (1834).

Colin A. Pendry

Deciduous thorny shrubs. Thorns sometimes with a few leaves. Perulate winter buds absent; axillary buds 1 or 2. Stipules minute, soon caducous. Leaves petiolate, alternate, simple, margins serrulate. Inflorescences below thorns, racemose, solitary or fascicled. Bracts solitary, minute, persistent; bracteoles 2. Flowers bisexual, pedicellate. Hypanthium cup-shaped with an annular disk at mouth. Sepals 5, unequal, 3 large and 2 small, persistent in fruit. Petals 5, base shortly clawed. Stamens ca. 30, in 2–3 whorls, inserted on hypanthium rim. Ovary superior, glabrous, 1-locular, ovules 2. Style lateral; stigma capitate. Fruit a drupe; mesocarp fleshy.

Worldwide about five species in China and the Himalaya. One species in Nepal.

1. *Prinsepia utilis* Royle, Ill. Bot. Himal. Mts. [6]: 206 (1835).

धटेलो Dhatelo (Nepali).

Shrubs to 4 m. Young twigs greenish brown, tomentose, soon glabrescent, older twigs greyish green, robust, glabrous. Thorns 2–5 cm. Stipules 0.3 × 1 mm. Petioles 5–11 mm, glabrous. Leaves narrowly ovate to elliptic, 3–6 × 1–2.5 cm, base broadly cuneate, sometimes unequal, margin serrulate, apex acute to acuminate, veins 3–5 pairs, glabrous above and below except sometimes tomentose on midrib and veins. Racemes 2–5(–7) cm. Peduncle brown pubescent, soon glabrescent. Bracts and bracteoles lanceolate, brown pubescent, glabrescent. Pedicels 6–8 mm. Flowers ca. 1 cm across. Hypanthium cup-shaped, glabrous. Sepals semi-orbicular to broadly ovate, small sepals ca. 2 mm, large sepals ca. 4 mm. Petals white, reflexed, broadly obovate or suborbicular, 5–6 mm, base shortly clawed, margin apically erose. Stamens 3–4 mm. Ovary ca. 1 mm, glabrous. Style 2–3 mm. Drupe purplish, oblong to obovoid-oblong, 10–15 mm.
Fig. 53a–b.

Distribution: Nepal, W Himalaya, E Himalaya, Tibetan Plateau and E Asia.

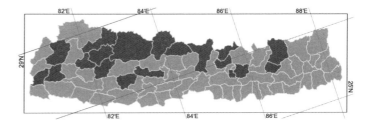

Altitudinal range: 1100–3400 m.

Ecology: Along trail-sides and in disturbed open areas.

Flowering: (October–)March–April. **Fruiting:** May–June.

Seed oil is applied externally in cases of rheumatism, muscular pain, coughs and colds. Heated oil cake is applied as a poultice to the abdomen to relieve pain. Seed oil is used for cooking and lighting. A deep purple pigment is obtained from fruits and is used for colouring windows and walls. The spiny branches make a formidable stock-proof fence.

24. *Prunus* L., Sp. Pl. 1: 473 (1753).

Hideaki Ohba, Colin A. Pendry & Sangeeta Rajbhandary

Trees or shrubs, deciduous or evergreen. Twigs unarmed or rarely spiny. Perulate winter buds present or absent, leaving a ring of scars around twig; axillary buds 1–3. Stipules membranous, often early caducous, margin entire or toothed. Leaves simple, alternate, petiolate, often with glands present at the apex of the petiole or base of leaf, margins variously crenate, rarely entire or undulate, the teeth often gland-tipped. Flowers solitary or in few-flowered fascicles or umbels or many-flowered racemes. Inflorescences terminal or on short axillary branches, with or without an involucre of persistent floral bud scales, bracteate if more than 1-flowered; bracteoles absent. Flowers bisexual, opening before or after the leaves, subsessile to pedicellate. Hypanthium campanulate to cup-shaped or urceolate. Sepals and petals 5, imbricate. Stamens 10–100, in 1 or 2 whorls. Ovary superior, 1-locular; ovules 2, collateral, pendulous. Style terminal, elongated. Fruit a drupe, glabrous, often glaucous, with or without a longitudinal groove; mesocarp fleshy; endocarp bony, ovoid or globose, sometimes laterally compressed, smooth or variously grooved, pitted or rugose. Seed solitary.

Worldwide about 200 species. 15 species in Nepal, some of which are cultivated.

The subgenera *Amygdalus* (peach), *Armeniaca* (apricot), *Cerasus* (cherry), *Laurocerasus* and *Padus* are sometimes treated as distinct genera (e.g. Fl. China), but recent molecular studies (Bortiri *et al.*, Syst. Bot. 26: 797–807. 2001) do not support their separation and they are not recognised here.

Key to Species

1a Flowers in elongate racemes with more than 10 flowers................2
b Flowers solitary or in short racemes. Corymbs or umbels with fewer than 10 flowers................6

2a Leaves evergreen, coriaceous (subgenus *Laurocerasus*)................3
b Leaves deciduous, herbaceous (subgenus *Padus*)................4

3a Leaf margin entire or rarely with a few teeth. Racemes solitary or 2–4 in a fascicle, 5–40 cm long. Stamens 3–4 mm long. Ovary hairy................**14. P. undulata**
b Leaf margin serrate throughout. Racemes solitary, 4–5 cm long. Stamens 2.5–3 mm long. Ovary glabrous ....**15. P. jajarkotensis**

4a Racemes without basal leaves. Sepals persistent in fruit................**13. P. venosa**
b Racemes with basal leaves. Sepals caducous in fruit................5

5a Peduncles and pedicels not thickened and without lenticels in fruit. Nectaries at petiole apex, conspicuous. Leaf margin sparsely serrulate. Peduncles velutinous at flowering. Drupes ca. 0.8 cm across................**11. P. cornuta**
b Peduncles and pedicels thickened with conspicuous lenticels in fruit. Nectaries near base of blade, inconspicuous. Leaf margin coarsely serrate or sometimes undulate. Peduncles glabrous or subglabrous at flowering. Drupes 1–3 cm across................**12. P. napaulensis**

6a Ovary and fruit hairy. Pedicels less than 1 cm................7
b Ovary and fruit glabrous, usually shiny. Pedicels more than 2 cm................10

7a Leaves broadly ovate to orbicular-ovate. Axillary bud 1 (subgenus *Armeniaca*)................**5. P. armeniaca**
b Leaves narrowly ovate to elliptic or oblong. Axillary buds 3 (subgenus *Amygdalus*)................8

8a Endocarp smooth, with shallow furrows, without pits. Leaf margin crenate................**2. P. mira**
b Endocarp deeply furrowed and pitted. Leaf margin serrate................9

9a Upper surfaces of leaves glabrous. Outside of calyx glabrous. Mesocarp thin and rather dry. Endocarp usually not laterally compressed, apex obtuse................**1. P. davidiana**
b Upper surfaces of leaves sparsely hairy in vein axils, rarely glabrous. Outside of calyx hairy. Mesocarp thick and succulent. Endocarp laterally compressed, apex acuminate................**3. P. persica**

10a   Fruit more than 4× longer than pedicel or nearly sessile. Leaves with 5–7 pairs of secondary veins (subgenus *Prunus*) ....**4. *P. cerasifera***

   b   Fruits less than one third the length of pedicel. Leaves with 9–15 pairs of secondary veins (subgenus *Cerasus*) 11

11a   Peduncles hairy...**7. *P. himalaica***

   b   Peduncles glabrous..12

12a   Petiole with 2–4 glands towards apex. Petals usually emarginate. Leaf margin acutely biserrulate or biserrate, also serrate (the apex of teeth acuminate ending in a minute capitate gland to 0.2 mm). Flowers 1–4 in umbel, opening before or at the same time as leaves..**6. *P. cerasoides***

   b   Petiole eglandular, glands sometimes present at base of leaf. Petals entire. Leaf margin sharply serrulate (the apex of teeth acuminate ending in a conspicuous conical or capitate gland to 0.4 mm). Flowers usually 1 or 2 in umbel, opening at same time as the leaves 13

13a   Petals glabrous. Leaves ovate or narrowly ovate, lower surface with white or pale brown, straight hairs, margins sharply doubly serrate. Calyx-tube urceolate...**9. *P. topkegolensis***

   b   Petals sparsely pubescent. Leaves elliptic or narrowly obovate to obovate-elliptic, lower surface with even indumentum of brown, crisped hairs or villous only on veins. Calyx tube tubular to campanulate 14

14a   Hypanthium 11–15 mm long. Leaves 2.8–5 cm wide, villous along veins below ....**8. *P. rufa***

   b   Hypanthium 7–10 mm long. Leaves 0.8–2 cm wide, lower surface with dense indumentum of brown, crisped hairs..**10. *P. taplejunica***

1. *Prunus davidiana* (Carrière) N.E.Br., Suppl. Johnson's Gard. Dict.: 991 (1882).
Persica davidiana Carrière, Rev. Hort. 1872: 74, pl. 10 (1872).

झुसेआरु Jhuse aaru (Nepali).

Deciduous trees to 10 m. Twigs brown, glabrous. Stipules subulate, 3–5 mm, toothed, caducous. Axillary winter buds (2 or)3, lateral ones flower buds, central one a leaf bud, terminal winter buds present. Petioles 1–2 cm, glabrous, usually with glands. Leaves ovate-lanceolate, 5–13 × 1.5–4 cm, base cuneate to rounded, apex acuminate, margin acutely or obtusely serrate, glabrous on both surfaces, secondary veins ca. 15 pairs, fine with smaller intermediate veins. Flowers solitary, opening before leaves, 2–3 cm across, subsessile. Bud scales 2–4 mm caducous. Pedicels 1–2 mm, 3–4 mm in fruit, glabrous. Hypanthium campanulate, glabrous outside. Sepals ovate to ovate-oblong, apex obtuse, glabrous outside. Petals pink, obovate to suborbicular, 1–1.5 × 0.8–1.2 cm, apex obtuse or rarely emarginate. Stamens nearly as long as petals. Ovary hairy. Style longer or about as long as stamens. Drupe yellowish, globose to ellipsoid or oblong, 2.5–3.5 cm across, base truncate, apex obtuse, densely hairy; mesocarp thin, not splitting when ripe; endocarp not compressed on both sides, surface with numerous furrows and pits, separating from mesocarp.

Distribution: Nepal, Tibetan Plateau and E Asia.

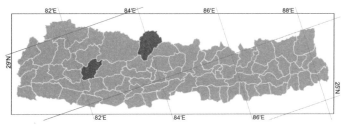

Altitudinal range: 2100–2700 m.

Ecology: Cultivated.

Flowering: March–April. **Fruiting:** July–August.

Serafimov considered *Polunin, Sykes & Williams 3893* (BM) to be a hybrid between *P. davidiana* and *P. amygdalus* Batsch (*Prunus dulcis* (Mill.) Rchb., the almond).
     Fruits are useful in lung complaints, eye troubles, and for wounds. Seed oil is used as a hair tonic.

2. *Prunus mira* Koehne, Pl. Wilson. 1(2): 272 (1912).
Amygdalus mira Ricker; *Persica mira* (Koehne) Kovalev & Kostina.

Prunus mira subsp. ***nepalensis*** (Seraf.) H.Hara, Enum. Fl. Pl. Nepal 2: 142 (1979).
Persica mira subsp. *nepalensis* Seraf., Dokl. Bolg. Akad. Nauk 27(6): 835, pl. 1 fig. 1 (1974).

Deciduous trees to 10 m. Twigs first green, then greyish brown, glabrous. Stipules unknown. Axillary winter buds (2 or)3, lateral ones flower buds, central one a leaf bud; terminal winter buds present. Petioles 0.8–1.5 cm, glabrous, often with flattened nectaries. Leaves narrowly ovate, 5–11 × 1.5–4 cm, base broadly cuneate to subrounded, apex acuminate, margin shallowly crenate, with gland-tipped teeth, upper surface glabrous, lower surface hairy along midvein, secondary veins 12–15 pairs, fine with smaller intermediate veins. Flowers solitary, opening before leaves, 2.2–3 cm across. Pedicel 1–3 mm, 4–5 mm in fruit, glabrous, surrounded at base by persistent bud scales ca. 2 mm. Hypanthium campanulate, purplish green, glabrous. Sepals ovate to narrowly ovate, apex obtuse. Petals pink, broadly obovate, 1–1.5 cm, apex emarginate. Stamens much shorter than petals. Ovary hairy. Style longer or about as long as stamens. Drupe subglobose, ca. 3 cm across, densely hairy, base subtruncate and slightly asymmetric, apex acute; mesocarp fleshy, not splitting when ripe; endocarp compressed ovoid-globose, ca. 2 cm, slightly flattened on both sides, surface smooth with few longitudinal shallow furrows only on dorsal and ventral sides.

Distribution: Nepal, Tibetan Plateau and E Asia.

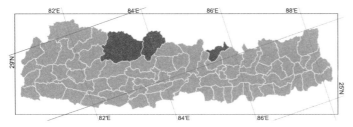

Altitudinal range: 2700–4000 m.

Ecology: Mountain slopes in thickets.

Flowering: March–April. **Fruiting:** August–September.

According to Lu & Bartholemew (Fl. China 9: 395. 2003) *Prunus mira* is cultivated for its fruits and seeds and it is often uncertain whether collections are wild, cultivated or naturalised.

3. *Prunus persica* (L.) Batsch, Beytr. Entw. Gewächsreich: 30 (1801).
Amygdalus persica L., Sp. Pl. 1: 472 (1753); *Persica vulgaris* Mill.; *Prunus persica* var. *compressa* (Loudon) Bean; *P. persica* var. *platycarpa* (Decne.) L.H.Bailey; *P. vulgaris* Mill.; *P. vulgaris* var. *compressa* Loudon.

आरु Aaru (Nepali).

Deciduous trees to 3–8 m. Twigs reddish green, glabrous, lustrous. Stipules narrowly triangular, 5–8 mm, toothed, caducous. Axillary winter buds (2 or)3, lateral ones flower buds, central one a leaf bud; terminal winter buds present. Petioles 1–2 cm, with or without 1 to several glands. Leaves blade oblong- or elliptic-lanceolate to obovate-oblanceolate, 7–15 × 2–4 cm, base broadly cuneate, apex acuminate, margin finely to coarsely serrate, upper glabrous, lower surface with or without a few hairs in vein axils, secondary veins 10–14 pairs, fine with smaller intermediate veins. Flowers solitary, opening before leaves, 2–3.5 cm across, subsessile, pedicel to 2–4 mm in fruit, glabrous. Flower bud scales 2–5 mm, caducous. Hypanthium shortly campanulate, 3–5 mm, glabrous, gland-spotted. Sepals ovate to oblong, about as long as hypanthium, apex obtuse, densely tomentose outside. Petals pink or white, oblong-elliptic to broadly obovate, 1–1.7 × 0.9–1.2 cm. Stamens 20–30; anthers purplish red. Ovary hairy. Style nearly as long as stamens. Drupe ovoid to broadly ellipsoid or compressed globose, (3–)5–7(–12) cm across, apex acuminate, usually densely hairy, with ventral suture; mesocarp succulent,

fragrant; endocarp ellipsoid to suborbicular, compressed on both sides, surface with longitudinal and transverse furrows and pits.

Distribution: Nepal, E Himalaya and E Asia.

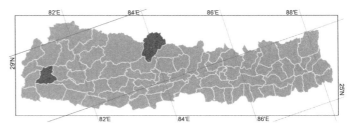

Altitudinal range: 1100–3600 m.

Ecology: Cultivated and possibly naturalized.

Flowering: March–April. **Fruiting:** August–September.

The peach. Cultivated and not known in the wild, though thought to originate in N China. It is possibly a cultigen derived from *Prunus davidiana* (Mabberley, Plant Book: 704. 2008). Under represented by herbarium collections.
    Ripe fruits are edible. Seed oil is used for cooking and lighting. Juice of the leaves is applied to wounds. Flowers are diuretic and purgative.

4. *Prunus cerasifera* Ehrh., Beitr. Naturk. 4: 17 (1789).
Prunus cerasifera subsp. *myrobalana* (L.) C.K. Schneid.; *P. domestica* L.; *P. domestica* var. *myrobalana* L.

आरुबखरा Aarubakhara (Nepali).

Deciduous shrubs or trees to 8 m. Twigs dark red, glabrous, sometimes spiny. Stipules narrowly elliptic, 3–6 mm, glandular-toothed. Axillary winter buds solitary; perulate terminal winter buds present. Petioles 0.6–1.2 cm, without glands. Leaves elliptic to ovate or obovate or rarely elliptic-lanceolate, 3–6 × 2–5 cm, base cuneate to subrounded, apex acute, margin crenate or sometimes doubly crenate, upper surface dark green, glabrous, lower surface hairy on midvein, secondary veins 5–7 pairs. Flowers solitary, opening before leaves, rarely 2 in a fascicle, 2–2.5 cm across. Pedicel 1–2.2 cm, glabrous or sparsely hairy, at base with persistent involucre of bud scales ca. 1 mm. Hypanthium campanulate, glabrous. Sepals narrowly to broadly ovate, apex obtuse, glabrous, margin shallowly serrate. Petals white, oblong to spatulate, base cuneate, apex obtuse, margin undulate. Stamens 25–30. Ovary villous. Style slightly shorter than stamens, stigma disc-shaped. Drupe subglobose to ellipsoid, 2–3 cm across, slightly glaucous; endocarp ellipsoid to ovoid, smooth or scabrous, sometimes pitted.

Distribution: Nepal, W Himalaya, E Himalaya, E Asia, C Asia, SW Asia and Europe.

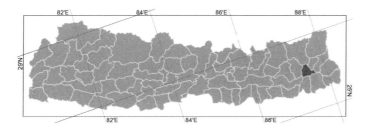

Altitudinal range: ca. 1800 m.

Ecology: Sometimes cultivated.

Flowering: April. **Fruiting:** August.

The plum. Cultivated for its edible fruits.
   Ripe fruits are eaten fresh and have laxative properties. Foliage is cut for fodder.

5. *Prunus armeniaca* L., Sp. Pl. 1: 474 (1753).
Armeniaca vulgaris Lam.; *Prunus ansu* (Maxim.) Kom.; *P. armeniaca* var. *ansu* Maxim.

खुरपानी Khurpani (Nepali).

Deciduous trees to 5–8(–12) m. Twigs purplish brown, glabrous. Stipules minute, rounded, to 0.5 mm, persistent. Axillary winter buds solitary; perulate terminal winter buds present. Petioles 2–3.5 cm, glabrous or hairy, with 1–6 glands. Leaves broadly ovate to orbicular-ovate, 5–9 × 4–8 cm, base cuneate to rounded or subcordate, with several glands, apex acute to shortly acuminate, margin crenate, upper surface hairy with white hairs, lower surface hairy in vein axils, secondary veins 5–8 pairs. Flowers opening before leaves, 2–4.5 cm across. Pedicel 1–3 mm, hairy, at base with persistent involucre of bud scales to 2 mm. Hypanthium cylindrical, pubescent towards base. Sepals purplish green, ovate to ovate-oblong, 3–5 mm, apex usually acute, reflexed after flowering. Petals white or pink, orbicular to obovate, 0.8–1.2 cm, base unguiculate, apex rounded. Stamens 20–100, slightly shorter than petals; filaments white. Ovary hairy. Style slightly longer to nearly as long as stamens, basally hairy. Drupe globose to ovoid or rarely obovoid, 1.5–2.5 cm across, base symmetric or rarely asymmetric, apex obtuse to rounded, hairy, usually glaucous; mesocarp succulent, not splitting when ripe; endocarp globose or ellipsoid, compressed laterally, with keel-like ribs on ventral side, surface scabrous or smooth.

Distribution: Cosmopolitan.

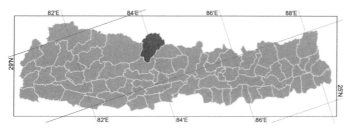

Altitudinal range: 2900–3500 m.

Ecology: Cultivated and sometimes naturalizing.

Flowering: March–April. **Fruiting:** June–July.

The apricot. Widely cultivated for its edible fruits and seed oil, it is probably originally from C Asia.
   The ripe fruits are eaten fresh or dried and have laxative properties. Seeds are edible and anthelmintic, and are used to treat liver diseases, piles, earache and deafness. Seed oil is used for cooking.

6. *Prunus cerasoides* D.Don, Prodr. Fl. Nepal.: 239 (1825). *Cerasus cerasoides* (D.Don) S.Ya.Sokolov; *C. cerasoides* var. *rubea* (Ingram) T.T.Yu & C.L.Li; *C. phoshia* Buch.-Ham. ex D.Don; *C. puddum* Roxb. ex Ser.; *Maddenia pedicellata* Hook.f.; *Prunus carmesina* H.Hara; *P. cerasoides* var. *majestica* (Koehne) Ingram; *P. cerasoides* var. *rubea* Ingram; *P. majestica* Koehne; *P. puddum* (Roxb. ex Ser.) Brandis.

पैंयू Painyu (Nepali).

Deciduous trees to 3–10 m. Twigs green, hairy, glabrescent. Stipules linear, branched basally, 5–15 mm, fimbriate, glandular-toothed. Axillary winter buds solitary, perulate terminal winter buds present. Petioles 1.2–2 cm, with 2–4 glands towards apex. Leaves ovate to oblong or obovate, 6–12 × 3–5 cm, base rounded, apex acuminate to long acuminate, margin acutely biserrulate to biserrate or serrate, the teeth with a 0.1–0.2 mm capitate apical gland, upper surface dark green, lower surface pale green, glabrous or villous along veins, secondary veins 10–15 pairs. Flowers 1–4 in an umbel, opening at the same time or before leaves. Bud scales ca. 2 mm, caducous. Peduncles 1–1.5 cm, glabrous. Bracts suborbicular, 1–1.2 cm, apically divided, margin glandular serrate, withered after anthesis. Pedicels 1–2.3 cm, to 3 cm and apically thickened in fruit. Hypanthium campanulate to widely campanulate, glabrous. Sepals usually reddish, triangular, 0.4–5.5 mm, apex acute to obtuse, erect. Petals white or pink, ovate to obovate, apex usually emarginate. Stamens 32–34, shorter than petals. Ovary glabrous. Style as long as stamens, glabrous; stigma disciform. Drupe purplish black, ovoid, 1.2–1.6 × 0.8–1.2 cm, apex obtuse; endocarp ovoid, laterally deeply furrowed and pitted.
Fig. 53c–e.

Distribution: Nepal, W Himalaya, E Himalaya, Tibetan Plateau, Assam-Burma, E Asia and SE Asia.

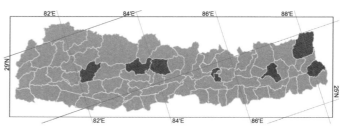

Altitudinal range: 1300–2700 m.

Ecology: Evergreen *Quercus* forest, thickets on slopes.

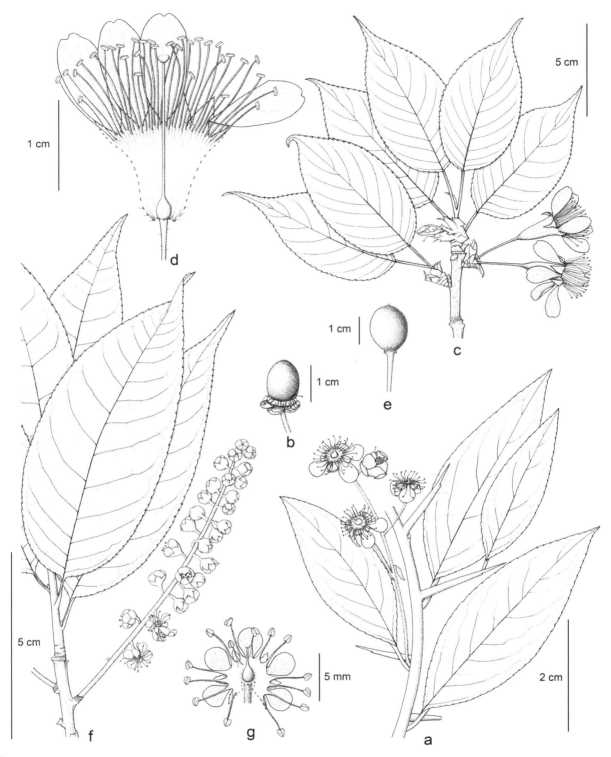

FIG. 53. Rosaceae. **Prinsepia utilis**: a, inflorescence and leaves; b, fruit. **Prunus cerasoides**: c, inflorescence and leaves; d, opened flower; e, fruit. **Prunus venosa**: f, inflorescence and leaves; g, opened flower.

Flowering: October–January. **Fruiting:** January–March.

Hara (Enum. Fl. Pl. Nepal 2: 141. 1979) described *Prunus carmesina* for specimens flowering before emergence of the leaves, with crimson hypanthium and deep pink, erect petals. Although this species was recognized by Grierson (Fl. Bhutan 1: 541. 1987) the distinction was not maintained by Li & Boufford (Fl. China 9: 418. 2003).

Ripe fruits are eaten fresh. Juice of the bark is applied to treat back pain. Leafy branches are lopped for fodder. Branches are used for making walking sticks.

7. *Prunus himalaica* Kitam., Acta Phytotax. Geobot. 15: 131 (1954).

Deciduous trees to ca. 2 m. Branches purplish, with red-brownish hairs. Stipules not known. Perulate terminal winter buds present. Petioles 0.9–1 cm, with brownish hairs. Leaves elliptic, 6–8 × 4–5 cm, base rounded, apex caudate-acuminate, margin acutely double-serrate, upper surface green, sparsely minutely hirsute, lower surface pale, with dense brown hairs in vein axils, secondary veins 9–13 pairs. Flowers 1 or 2 in an umbel. Peduncle 1 cm, hairy. Bracts oblong, 1.5–2.7 cm, apex acute, margin serrulate. Pedicels 3.5–4.5 cm, sparsely hairy. Hypanthium urceolate, glabrous. Sepals ovate, 4 mm, apex acute, margin with glandular serrations. Petals pale pink. Stamens ca. 45. Ovary and fruits unknown.

Distribution: Endemic to Nepal.

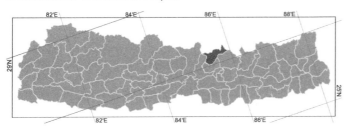

Altitudinal range: ca. 3900 m.

Ecology: Unknown.

Flowering: Probably autumn. **Fruiting:** Probably winter.

Known only from the incomplete type specimen collected by Nakao, 3 July 1953, at Shin Gompa, Langtang (KYO). Kitamura reported that it was close to *Prunus rufa* Hook.f. with its densely pubescent twigs and 'urceolate' hypanthium, but differing from it in its doubly serrate leaves which are rounded at the base.

8. *Prunus rufa* Hook.f., Fl. Brit. Ind. 2[5]: 314 (1878). *Cerasus rufa* Wall. nom. nud.; *C. rufa* T.T.Yu & C.L.Li; *Prunus imanishii* Kitam.; *P. rufa* Steud. nom. inval.; *P. rufa* var. *trichantha* (Koehne) H.Hara; *P. trichantha* Koehne; var. *rufa* (Koehne) T.T.Yu & C.L.Li.

Deciduous trees to 2–10 m. Twigs purplish brown, rufous

tomentose initially, glabrate. Axillary winter buds solitary, perulate terminal winter buds present. Stipules linear, 4–13 mm, prominently glandular-toothed. Petioles 0.5–1 cm, densely hairy, without glands. Leaves narrowly obovate to obovate-elliptic, 3–12 × 2.5–5 cm, base cuneate, apex acuminate to caudate-acuminate, margin sharply serrulate, the teeth with prominent 0.2–0.4 mm conical or capitate glands, upper surface pilose but soon glabrescent, lower surface villous along veins, secondary veins 9–11 pairs. Flowers 1 or 2(rarely more) in umbel, opening at same time as leaves. Peduncles very short. Bracts oblong-spathulate, 1–1.2 cm, margin glandular. Pedicels 1–2.2 cm, glabrous. Flowers 1–1.3 cm across. Hypanthium tubular to campanulate, 1–1.5 × ca. 0.4 cm, hairy outside. Sepals straight or spreading, triangular-ovate, 2–4 mm, margin sparsely glandular. Petals white or pink, ovate to obovate, 5–6 × 4–6 mm, entire or apically emarginate, outside pilose. Stamens ca. 45. Ovary basally hairy. Style as long as stamens, basally hairy, stigma obtuse, not lobed. Drupe ellipsoid to obovoid, 1.2–1.8 cm, apex obtuse, endocarp ovoid, laterally deeply furrowed and pitted.

Distribution: Nepal, E Himalaya and Tibetan Plateau.

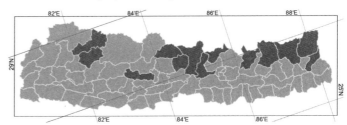

Ecology: Mixed forests with *Rhododendron*, *Betula utilis* and *Abies spectabilis*.

Flowering: May–June. **Fruiting:** July–August.

A collection from Kangchenjunga, *KEKE 657* (E), keys out to *P. rufa*, but has relatively broader, more coarsely biserrate leaves and may be *Prunus latidentata* Koehne or an allied species.

9. *Prunus topkegolensis* H.Ohba & S.Akiyama, Bull. Natl. Sci. Mus., Tokyo, B. 36: 134 (2010).

Deciduous shrub to 2 m. Twigs slender, spreading; glabrous, initially greenish, later pale red-purplish. Axillary winter buds solitary, perulate terminal winter buds present. Stipules linear, ca. 5 mm, prominently glandular-toothed. Petioles 6–11 mm. Leaves ovate to narrowly ovate, 4.2–6.6 × 2.1–3 cm at flowering, base rounded to cuneate, apex acuminate, margin sharply doubly serrate, mostly with gland at apex of tooth, lower surface with sparse to moderate white or pale brown straight hairs near base. Flowers solitary in axil, opening at same time as leaves. Pedicels ca. 1.3 cm. Hypanthium urceolate, ca. 7 mm. Sepal triangular-ovate, ca. 2.5 mm, with sparse pale brown crisped hairs, margin entire or serrate. Petals glabrous, white with faint pink markings, erect at flowering, suborbicular, ca. 5.5 × 3 mm. Style exserted, glabrous apically. Fruits unknown.

Distribution: Endemic to Nepal.

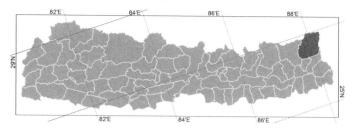

Altitudinal range: ca. 3700 m.

Ecology: In thickets along steep gorge.

Flowering: June. **Fruiting:** probably August.

Prunus topkegolensis differs from *P. rufa* Hook.f. in having deeply double serrate margins, glabrous leaves except pilose nerves on both sides, solitary relatively small flowers with urceolate calyx tube. It is close to the Japanese species *P. incisa* Franch. & Sav.

10. *Prunus taplejunica* H.Ohba & S.Akiyama, Bull. Natl. Sci. Mus., Tokyo, B. 36: 135 (2010).

Deciduous shrub, ca. 2 m. Twigs glabrous, slender, spreading, initially greenish, later pale red-purplish. Axillary winter bud solitary. Petioles glabrous, 4–14 mm. Leaves elliptic, 2.6–5.1 × 0.8–2 cm, base rounded to cuneate, with a pair of glands, apex acute to acuminate, margin serrate, lower surface with brown crisped dense hairs near base. Flowers usually solitary or 2 in axillary inflorescence. Peduncles 1–2 mm, pedicels 0.5–2.8 cm. Hypanthium 7–11 mm. Sepals triangular-ovate, 2–4 mm, with brown crisped dense hairs, margin serrate. Petals white, suborbicular, 5–6 × 3–4 mm, with very sparse brown hairs. Style exserted. Fruits unknown.

Distribution: Endemic to Nepal.

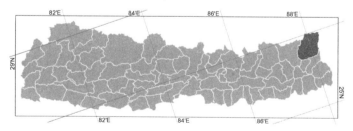

Altitudinal range: 2400–3600 m.

Ecology: Open thickets on hills.

Flowering: May–June. **Fruiting:** Unknown.

Prunus taplejunica was collected from close to the locality of *P. topkegolensis* H.Ohba & S.Akiyama. The elliptic leaves, gradually narrowing toward both ends and with conspicuously double serrate margins, distinguish *P. taplejunica* from *P. topkegolensis*.

11. *Prunus cornuta* (Wall. ex Royle) Steud., Nomencl. Bot., ed. 2, 2(3): 403 (1841).
Cerasus cornuta Wall. ex Royle, Ill. Bot. Himal. Mts. [6]: 205, pl. 38 (1835); *C. cornuta* Wall. nom. nud.; *Padus cornuta* (Wall. ex Royle) Carrière; *P. cornuta* var. *glabra* Fritsch ex C.K. Schneid.; *P. cornuta* var. *villosa* G.Singh; *Prunus cornuta* forma *villosa* (H.Hara) H.Hara; *P. cornuta* var. *villosa* H.Hara.

लेख आरु Lekh aaru (Nepali).

Deciduous trees to 3–15 m. Twigs purplish brown, glabrous or sometimes velutinous. Stipules linear, 8–15 mm, margin glandular serrate. Axillary winter buds solitary; perulate terminal winters bud present. Petioles 1–2.3 cm, usually glabrous, apex with 2 glands. Leaves narrowly elliptic to oblong, rarely oblong-ovate or obovate, 6–11 × 3–5 cm, base subcordate to broadly cuneate, apex shortly acuminate to shortly caudate, margin sparsely serrulate, upper surface dark green, glabrous, lower surface pale green, glabrous or with tufts of hair at vein axils, secondary veins 10–14 pairs. Flowers opening after the leaves. Racemes solitary, 8–16 cm, many-flowered, basally with 1–3 leaves, rachis velutinous. Bracts narrowly ovate, 6–8 mm, glandular-toothed, very early caducous. Flowers 6–8 mm across. Pedicels 5–7 mm, velutinous. Hypanthium campanulate, velutinous at base outside. Sepals triangular-ovate, apex obtuse, margin glandular serrulate, soon caducous, glabrous outside. Petals white, obovate, base cuneate and shortly clawed, margin apically erose. Stamens 20–25. Ovary glabrous. Style slightly shorter than stamens, stigma disc-shaped. Drupe blackish brown, ovoid-globose, ca. 8 mm in diameter, glabrous.

Distribution: Nepal, W Himalaya, E Himalaya, Tibetan Plateau and SW Asia.

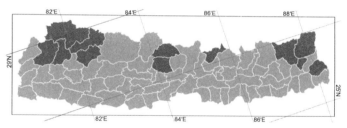

Altitudinal range: 2100–3500 m.

Ecology: Secondary forest, thickets.

Flowering: April–May. **Fruiting:** May–October.

Brandis (Forest Fl. N.W. India: 194. 1874) and Hooker (Fl. Brit. Ind. 2: 315. 1878) misapplied the name *Prunus padus* L. to this species.

The ripe fruits are eaten and used for brewing local liquor. The foliage is cut for fodder.

12. *Prunus napaulensis* (Ser.) Steud., Nomencl. Bot., ed. 2, 2(3): 403 (1841).
Cerasus napaulensis Ser., Prodr. 2: 540 (1825); *Cerasus glaucifolia* Wall. nom. nud.; *Padus napaulensis* (Ser.) C.K. Schneid.

जंगली आरु Jungali aaru (Nepali).

Deciduous trees to 27 m. Twigs reddish brown, glabrous. Axillary winter buds solitary; perulate terminal winter buds present. Stipules linear, to 15 mm, margin glandular serrate. Petioles 0.8–1.5 cm, glabrous, without glands. Leaves narrowly elliptic to ovate-elliptic, 6–14 × 2–6 cm, base cuneate, apex acute to shortly acuminate, margin coarsely serrate or sometimes undulate, upper surface dark green, glabrous, lower surface pale green, glabrous or very rarely sparsely hairy when young, secondary veins 9–15 pairs. Flowers opening after the leaves. Racemes solitary, 7–14 cm, basally with 2 or 3 leaves, rachis glabrous or subglabrous, later thickened, with pale lenticels. Bracts linear-elliptic, ca. 2 mm, toothed, very early caducous. Flowers ca. 1 cm across. Pedicel 4–6 mm, conspicuously thickened in fruit, velutinous, with pale lenticels in fruit. Hypanthium cup-shaped, outside usually hairy. Sepals triangular-ovate, apex acuminate, soon caducous, outside hairy, margin serrulate. Petals white, obovate-oblong, base cuneate and shortly clawed, apically erose. Stamens 22–27. Ovary glabrous. Style longer than stamens; stigma disc-shaped. Drupe dark purple to black, ovoid, 1–2 cm across, glabrous.

Distribution: Nepal, W Himalaya, E Himalaya, Tibetan Plateau, Assam-Burma and E Asia.

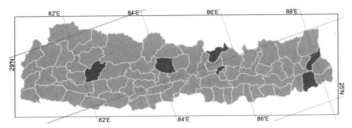

Altitudinal range: 1600–3000 m.

Ecology: In forest on hill slopes.

Flowering: April–June. **Fruiting:** July–October.

The narrowly ovate-elliptic leaves are very characteristic of *Prunus napaulensis*, as are the thickened, lenticellate pedicels of the fruit. Varient spellings of the name in the literature include 'nepaulensis' and 'nepalensis'.

Ripe fruits are eaten fresh. Wood is used as timber and firewood. Young leaves and shoots are poisonous to cattle.

13. *Prunus venosa* Koehne, Pl. Wilson. 1(2): 60 (1912).
Prunus buergeriana var. *nudiuscula* Koehne; *P. undulata* forma *venosa* (Koehne) Koehne.

Deciduous trees to 6–12(–25) m. Twigs purplish brown, usually glabrous. Axillary winter buds solitary, perulate terminal winter buds present. Stipules linear, ca. 5 mm, margin glandular-serrate, very early caducous. Petioles 1–1.5 cm, eglandular, usually glabrous. Leaves elliptic, oblong- or rarely obovate-elliptic, 4–10 × 2.5–5 cm, base rounded to widely cuneate, apex caudate-acuminate to shortly acuminate, margin crenulate, upper surface dark green, glabrous, lower

surface pale green, glabrous, secondary veins 7–10 pairs. Flowers opening after the leaves. Racemes solitary, 6–10 cm, 20–30-flowered, without basal leaves, rachis more or less hairy. Bracts not seen, very early caducous. Flowers 5–7 mm across. Pedicel ca. 2 mm, to 3 mm in fruit, hairy, glabrescent. Hypanthium campanulate, outside glabrous. Sepals triangular-ovate, apex acute, margin irregularly serrulate, persistent in fruit, subglabrous outside. Petals white, broadly ovate, base cuneate to shortly clawed, apically erose. Stamens 10–14. Ovary glabrous. Style ca. half as long as stamens, stigma discoid to semi-rounded. Drupe blackish brown, subglobose to ovoid, ca. 5 mm across, glabrous.
Fig. 53f–g.

Distribution: Nepal, W Himalaya, E Himalaya, Tibetan Plateau, Assam-Burma and E Asia.

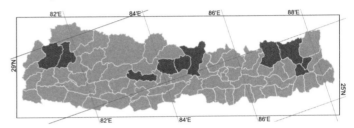

Altitudinal range: 1600–3200 m.

Ecology: Forest or thickets with *Pinus wallichiana*, *Rhododendron arboreum*, *Alnus* or *Corylus*.

Flowering: April–May. **Fruiting:** May–October.

Hooker (Fl. Brit. Ind. 2: 316. 1878) and Koehne (Bot. Jahrb. Syst. 52: 285. 1915) misapplied the name *Prunus undulata* Buch.-Ham. ex D.Don to this species.

The fruits with the remains of the sepals and stamens persisting at the base are very distinctive. The young foliage is poisonous to goats.

14. *Prunus undulata* Buch.-Ham. ex D.Don, Prodr. Fl. Nepal.: 239 (1825).
Cerasus acuminata Wall.; *C. acuminata* Wall. nom. nud.; *C. capricida* G.Don nom. superfl.; *C. capricida* Wall. nom. nud.; *C. undulata* (Buch.-Ham. ex D.Don) Ser.; *C. wallichii* (Steud.) M.Roem.; *Laurocerasus acuminata* (Wall.) M.Roem.; *L. undulata* (Buch.-Ham. ex D.Don) M.Roem.; *Prunus acuminata* (Wall.) D.Dietr.; *P. wallichii* Steud.

खोसीनी Khoshini (Nepali).

Evergreen shrubs or trees to 5–16 m. Twigs greyish brown to purplish brown, glabrous. Axillary winter buds solitary; perulate terminal winter buds absent. Stipules early caducous, not seen. Petioles 0.5–1 cm, glabrous, eglandular. Leaves elliptic to oblong-lanceolate, 6–16.5 × 3–5 cm, base broadly cuneate to subrounded, apex acuminate, margin entire or rarely with a few teeth, papery to thinly leathery, glabrous above and below; usually with a pair of flat glands near base and sometimes additional small glands above; secondary veins 6–9 pairs. Flowers opening after the leaves. Racemes solitary or 2–4 in a fascicle, 5–10 cm; rachis glabrous. Bracts 1–2 mm, soon caducous, basal ones sometimes with a tridentate apex. Pedicels 2–5 mm, glabrous. Flowers 3–5 mm across. Hypanthium broadly campanulate, glabrous outside. Sepals ovate-triangular, apex obtuse, glabrous outside. Petals creamy white, elliptic to obovate, 2–4 mm. Stamens 10–30, 3–4 mm. Ovary hairy. Style shorter than stamens. Drupe purplish black, ovoid-globose to ellipsoid, 1–1.6 × 0.7–1.1 cm, apex acute to obtuse, glabrous; endocarp thin, smooth.

Distribution: E Himalaya, Assam-Burma, E Asia and SE Asia.

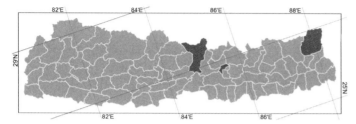

Altitudinal range: 1500–2600 m.

Ecology: Slopes or stream-sides in evergreen broad-leaved forests.

Flowering: August–October. **Fruiting:** December–March.

15. *Prunus jajarkotensis* H.Hara, J. Jap. Bot. 52(12): 355, fig. 2 (1977).

Evergreen trees to 6–7.5m tall. Twigs greenish brown, glabrous. Axillary winter buds solitary, perulate terminal winter buds absent. Stipules early caducous, not seen. Petioles 1–1.5 cm, glabrous, without glands. Leaves oblong to narrowly oblong-ovate, 5–18 × 2.5–4 cm, base broadly cuneate, apex acuminate, margin crenulate, coriaceous, both surfaces glabrous, secondary veins 6–8 pairs. Flowers opening after the leaves. Racemes solitary, 4–5 cm, rachis glabrous. Bracts small, soon caducous. Pedicels 5–10 mm, glabrous. Flowers ca. 5 mm across. Hypanthium broadly campanulate, glabrous. Sepals long-triangular, 2–2.5 mm, outside glabrous, reflexed in flowering. Petals cream, oblong, 2–2.5 mm long, margin hairy. Stamens 25–35, 2.5–3 mm; anthers pale yellow. Ovary green, glabrous. Style 2–2.5 mm. Drupe unknown.

Distribution: Endemic to Nepal.

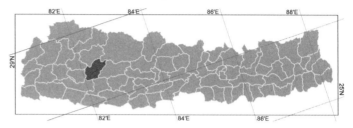

Altitudinal range: ca. 1000 m.

Ecology: Unknown.

Flowering: Autumn (October). **Fruiting:** Probably December–March.

Known only from the type collection, *Polunin, Sykes & Williams 5784* (BM) from Jajarkot.

5. *Pygeum* Gaertn., Fruct. Sem. Pl. 1: 218 (1788).

Colin A. Pendry

Evergreen trees. Twigs unarmed. Perulate winter buds absent; axillary buds solitary. Stipules free, soon caducous. Leaves simple, alternate, petiolate, usually with a pair of flat glands near base below, margin entire. Inflorescences unbranched axillary racemes. Bracts small, soon caducous; bracteoles absent. Flowers bisexual, pedicellate. Hypanthium funnel-shaped. Perianth segments 10, small, sepals and petals rather similar in shape and texture. Stamens 25–30, in 1 series inserted on rim of hypanthium. Ovary superior, 1-locular, glabrous; ovules 2. Style terminal, stigma capitate. Fruit a drupe, dry, usually transversely oblong to ellipsoid, obscurely didymous, at the base with the remains of the circumsessile hypanthium persisting as a ring of tissue. Seed solitary.

Worldwide about 40 species in tropical Africa, S and SE Asia, NE Australia, New Guinea and the Pacific Islands. One species in Nepal.

Although molecular evidence indicates that *Pygeum* should be included within *Prunus* (Jun Wen *et al.*, J. Syst. Evol. 46: 322–332. 2008), it has been considered distinct in recent Nepalese floristic publications and Lu & Bartholomew (Fl. China 9: 430–432. 2003), and so this approach is followed here.

1. *Pygeum zeylanicum* Gaertn., Fruct. Sem. Pl. 1: 218, pl. 46 (1788).
Polydontia ceylanica Wight; *Prunus ceylanica* (Wight) Miq.; *Pygeum acuminatum* Colebr.; *P. glaberrimum* Hook.f.

Trees to 25 m. Twigs glabrous, lenticellate. Stipules ca. 1 mm.

Petioles 1–1.5 cm, grooved above. Leaves elliptic to ovate, 6–15 × 2.5–6 cm, base cuneate to rounded, apex acute to acuminate, glabrous, secondary veins 7–8 pairs, midrib and secondary venation impressed above. Racemes to 8 cm, tomentose, glabrescent. Bracts narrowly triangular, to 1 mm. Flowers white. Pedicels ca. 2mm. Hypanthium to 1.5 mm, glabrous within, sericeous outside. Sepals triangular, ca. 1 mm, glabrous within, sericeous outside. Petals elliptic, ca. 1 mm, glabrous within, sericeous outside. Stamens to 5 mm. Ovary globose, ca. 1 mm, glabrous. Style 5 mm. Fruit ca. 1.5 × 2–2.5 cm, glabrous, on 8–10 mm pedicel. Fig. 54a.

Distribution: Nepal and S Asia.

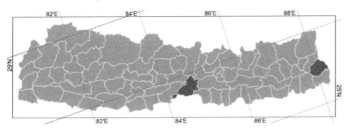

Altitudinal range: 300–500 m.

Ecology: Subtropical forest.

Flowering: August–November. **Fruiting:** December–April.

26. *Maddenia* Hook.f. & Thomson, Hooker's J. Bot. Kew Gard. Misc. 6: 381, pl. 12 (1854).

Colin A. Pendry

Trees or shrubs, deciduous. Twigs unarmed. Perulate winter buds present, leaving a ring of scars around twig; axillary buds solitary. Stipules large, persistent. Leaves alternate, simple with toothed margins, the teeth simple or glandular. Inflorescences terminal, racemose, many-flowered. Bracts large, soon caducous; bracteoles absent. Flowers reddish green, bisexual or functionally female with anthers absent or non-functional. Hypanthium campanulate. Perianth segments 10–12, sepals and petals indistinguishable. Stamens or staminodes 20–30, irregularly inserted on rim of hypanthium. Ovary superior, with 1 or occasionally 2 carpels, each with 2 pendulous ovules. Style equalling stamens, stigma capitate. Fruit a drupe with a thin, fleshy mesocarp. Seeds 1 or 2.

Worldwide about seven species in the Himalaya and China. One species in Nepal.

Although molecular evidence indicates that *Maddenia* should be included within *Prunus* (Jun Wen *et al.*, J. Syst. Evol. 46: 322–332. 2008), it has been considered distinct in recent Nepalese floristic publications and Gu & Bartholomew (Fl. China 9: 432–434. 2003), and so this approach is followed here.

1. *Maddenia himalaica* Hook.f. & Thomson, Hooker's J. Bot. Kew Gard. Misc. 6: 381, pl. 12 (1854).
Maddenia himalaica var. *glabrifolia* H.Hara.

Trees or shrubs to 4 m. Current year's twigs more or less densely brown villous, older twigs purple-brown, smooth. Stipules strap-shaped to very narrowly ovate, to 20 × 6 mm, sparsely sericeous, with or without glandular hairs on margins. Petioles 2–3 mm, villous. Leaves ovate to oblong, 5–12 × 2–5 cm, base rounded to cordate, apex acuminate, margins serrulate to biserrate, the teeth gland-tipped towards the base, veins 12–19 pairs, glabrous above, villous below, especially on the veins. Inflorescences to 5 cm, peduncle brown villous. Bracts ca. 10 mm, sericeous, margins with glandular hairs towards base. Pedicels 2–5 mm. Hypanthium 2–3 mm, villous outside, glabrous within. Perianth segments 2–3 mm, villous outside. Stamens and staminodes cream, 5–6 mm. Carpels 2–3 mm, glabrous or sparsely villous at apex. Styles 6–8 mm. Drupes dark red, ovoid, ca. 1 cm. Fig. 54b–d.

Distribution: Nepal, E Himalaya, Tibetan Plateau and Assam-Burma.

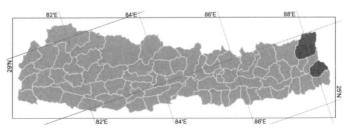

Altitudinal range: 2500–3400 m.

Ecology: Coniferous forests with *Rhododendron*.

Flowering: April–May. **Fruiting:** July.

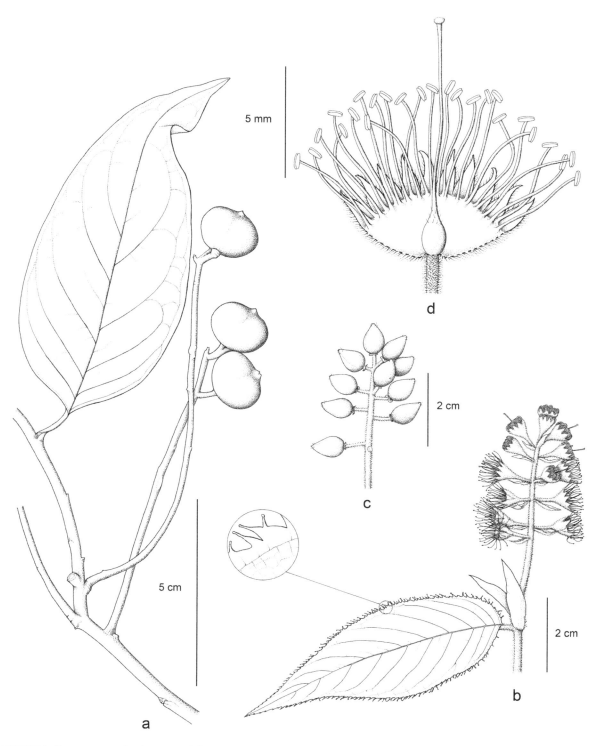

FIG. 54. ROSACEAE. **Pygeum zeylanicum**: a, infructescence and leaf. **Maddenia himalaica**: b, inflorescence and leaf; c, infructescence; d, opened flower.

Illustration Accreditation

The editors are pleased to credit the artwork from the following artists and sources used by Bhaskar Adhikari when composing the plate illustrations used in this volume. 'FOB' refers to *Flora of Bhutan* (Grierson, Long & Noltie, 1983–2002. Royal Botanic Garden Edinburgh); 'FOCI' refers to *Flora of China Illustrations* (Wu, Raven & Hong, 1998–ongoing. Science Press (Beijing) & Missouri Botanical Garden Press); and 'FRPS' refers to *Flora Reipublicae Popularis Sinicae* (1959–2004. Science Press (Beijing)). The copyright holders of these three publications, Science Press (Beijing), Missouri Botanical Garden Press, and Royal Botanic Garden Edinburgh, are thanked for permission to reproduce these illustrations, and for their generosity in making the images available in digital format.

Fig. 1
a–c	FOB 1(2): fig. 20. Mary Bates
d, e	Louise Olley
f	FOCI 7: 91. FRPS 30(1): pl. 40. Deng Yingfeng
g, h	FOCI 7: 87. FRPS 30(1): pl. 39. Deng Yingfeng
i, j	FOCI 7: 90. FRPS 30(1): pl. 41. Deng Yingfeng
k, l	FOCI 7: 87. FRPS 30(1): pl. 39. Deng Yingfeng
m	FOCI 7: 71. FRPS 30(1): pl. 35. Deng Yingfeng. Redrawn Louise Olley

Fig. 2
a–f	FOCI 7: 37. FRPS 30(1): pl. 70. Deng Yingfeng
g–i	FOCI 7: 41. FRPS 30(1): pl. 75. Deng Yingfeng
j–l	FOCI 7: 49. FRPS 30(1): pl. 77. Deng Yingfeng. Modified Bhaskar Adhikari
m–o	FOB 1(2): fig. 20. Mary Bates

Fig. 3 Claire Banks

Fig. 4 Louise Olley

Fig. 5 Louise Olley

Fig. 6
a–c	FOCI 7: 294. FRPS 32: pl. 15. Zhang Baofu
d, e	FOCI 7: 302. FRPS 32: pl. 17. Zhang Baofu
f	Bhaskar Adhikari
g	FOCI 7: 302. FRPS 32: pl. 17. Zhang Baofu

Fig. 7
a, b	FOCI 7: 307. FRPS 32: pl. 99. Wu Xilin
c	FOCI 7: 395. FRPS 32: pl. 71. Wu Xilin, redrawn Wang Ling, modified Zhang Libing
d–g	Bhaskar Adhikari

Fig. 8
a	FOCI 7: 392. FRPS 32: pl. 71. Wu Xilin, redrawn Wang Ling
b, c	Bhaskar Adhikari
d	FOCI 7: 388. FRPS 32: pl. 58 Li Xichou, redrawn Wang Ling
e, f	Bhaskar Adhikari
g–i	FOCI 7: 412. FRPS 32: pl. 53. Wu Xilin
j, k	Bhaskar Adhikari

Fig. 9
a	FOCI 7: 437. FRPS 32: pl. 46. Li Xichou
b, c	Bhaskar Adhikari
d–f	FOCI 7: 296. FRPS 32: pl. 18. Zhang Baofu

Fig. 10
a, b	FOCI 7: 290. FRPS 32: pl. 12. Yang Jiankun
c, d	Louise Olley
e–i	FOB 1(2): fig. 32. Mary Bates

Fig. 11
a	Louise Olley
b, c	FOCI 7: 279. FRPS 32: pl. 3. Yang Jiankun

Fig. 12
a–e	Louise Olley
f, g	FOCI 7: 283. FRPS 32: pl. 7. Li Xichou

Fig. 13
a–c	Claire Banks
d–h	FOCI 7: 485. FRPS 32: pl. 134. Zhang Hanwen

Fig. 14
a–f	FOCI 7: 486. FRPS 32: pl. 135. Zhang Hanwen
g	FOB 1(2): fig. 33. Mary Bates

Fig. 15 Jane Nyberg

Fig. 16
a–f	Jane Nyberg
g	FOCI 8: 81. FRPS 33: pl. 78. Chen Rongdao
h–o	Jane Nyberg
p	FOCI 8: 85. FRPS 33: pl. 81. Zhang Taili

Fig. 17
a	Jane Nyberg
b, c	FOCI 8: 76. FRPS 33: pl. 79. Shi Weiqing
d, e	FOCI 8: 77. FRPS 33: pl. 76. Wei Lisheng
f–i	FOCI 8: 32. FRPS 33: pl. 39. Chen Rongdao
j–l	Jane Nyberg

Fig. 18
a–c	FOCI 8: 3. FRPS 33: pl. 4. Zhang Chunfang
d–f	FOCI 8: 81. FRPS 33: pl. 78. Chen Rongdao
g–i	FOCI 8: 76. FRPS 33: pl. 79. Shi Weiqing

Fig. 19
a	FOCI 8: 70. FRPS 33: pl 66. Wei Lisheng
b, c	FOCI 8: 44. FRPS 33: pl 44. Shi Weiqing
d–g	FOCI 8: 62. FRPS 33: pl 57. Chen Rongdao
h–i	Jane Nyberg
j	FOCI 8: 89. FRPS 33: pl 86. Shi Weiqing
k	Jane Nyberg
l–o	FOCI 8: 96. FRPS 33: pl 92. Shi Weiqing

Fig. 20
a–c	FOCI 8: 127. FRPS 33: pl 123. Wei Guangzhou, redrawn Li Aili
d, e	FOCI 8: 85. FRPS 33: pl 81. Zhang Taili
f, g	FOCI 8: 120. FRPS 33: pl 19. Wu Zhanghua
h, i	Jane Nyberg
j	FOCI 8: 129. FRPS 33: pl 124. Zhang Chunfang

Fig. 21
a, b	FOCI 8: 66. FRPS 33: pl 63. Chen Rongdao
c–d	Jane Nyberg
e, f	FOCI 8: 18. FRPS 33: pl 18. Wu Zhanghua
g, h	Jane Nyberg

Fig. 22
a–e   FOCI 8: 137. FRPS 34(1): pl 3. Yu Hanping
f, g   FOCI 8: 144. FRPS 34(1): pl. 10. Yu Hanping. Modified Bhaskar Adhikari
h, i   FOCI 8: 141. FRPS 34(1): pl. 6. Yu Hanping
j–l   FOB 1(3): fig. 36. Mary Bates
m–q   FOB 1(3): fig. 34. Mary Bates

Fig. 23
a–e   FOCI 8: 146. FRPS 34(1): pl. 12. Qian Cunyuan
f–i   Claire Banks
j–n   FOCI 8: 146. FRPS 34(1): pl. 12. Qian Cunyuan
o–q   FOCI 8: 153. FRPS 34(1): pl. 29. Qian Cunyuan
r–w   Claire Banks

Fig. 24
a–f   FOCI 8: 173. FRPS 34(1): pl. 38. Qian Cunyuan
g–i   FOCI 8: 151. FRPS 34(1): pl. 17. Cai Shuqin
j–o   FOCI 8: 169. FRPS 34(1): pl. 34. Cai Shuqin
p–s   FOCI 8: 170. FRPS 34(1): pl. 35. Zhong Shiqi. Modified Bhaskar Adhikari
t–v   FOCI 8: 174. FRPS 34(1): pl. 40. Qian Cunyuan

Fig. 25
a–d   FOCI 8: 162. FRPS 34(1): pl. 27. Qi Shizhang. Modified Bhaskar Adhikari
e–g   FOCI 8: 162. FRPS 34(1): pl. 27. Qi Shizhang
h–k   FOCI 8: 165. FRPS 34(1): pl. 31. Cai Shuqin
l–o   FOCI 8: 155. FRPS 34(1): pl. 20. Qi Shizhang
p–s   FOCI 8: 159. FRPS 34(1): pl. 23. Qi Shizhang & Qian Cunyuan
t–v   FOCI 8: 164. FRPS 34(1): pl. 30. Cai Shuqin
w–z   FOCI 8: 160. FRPS 34(1): pl. 24. Qi Shizhang. Modified Bhaskar Adhikari

Fig. 26
a–c   FOB 1(3): fig. 35. Mary Bates
d–g   Claire Banks
h–j   FOB 1(3): fig. 35. Mary Bates

Fig. 27
a–c   FOB 1(3): fig. 35. Mary Bates
d–g   FOCI 8: 185. FRPS 34(2): pl. 8. Pan Jintang & Wang Ying
h–k   FOCI 8: 190. FRPS 34(2): pl. 17. Pan Jintang & Liu Jinjun

Fig. 28
a–d   FOCI 8: 194. FRPS 34(2): pl. 32. Pan Jintang & Liu Jinjun
e–i   FOCI 8: 212. FRPS 34(2): pl. 19. Pan Jintang & Liu Jinjun
j–n   FOCI 8: 205. FRPS 34(2): pl. 53. Pan Jintang & Liu Jinjun
o–r   FOCI 8: 205. FRPS 34(2): pl. 53. Pan Jintang & Liu Jinjun. Modified Bhaskar Adhikari
s–w   FOCI 8: 209. FRPS 34(2): pl. 21. Pan Jintang & Liu Jinjun

Fig. 29
a–f   FOCI 8: 209. FRPS 34(2): pl. 21. Pan Jintang & Liu Jinjun
g–j   FOCI 8: 215. FRPS 34(2): pl. 40. Pan Jintang & Liu Jinjun
k–n   FOCI 8: 216. FRPS 34(2): pl. 39. Pan Jintang & Liu Jinjun
o–s   FOCI 8: 209. FRPS 34(2): pl. 21. Pan Jintang & Liu Jinjun

Fig. 30
a–d   FOCI 8: 216. FRPS 34(2): pl. 39. Pan Jintang & Liu Jinjun
e–h   FOCI 8: 214. FRPS 34(2): pl. 38. Pan Jintang & Liu Jinjun
i–m   FOCI 8: 231. FRPS 34(2): pl. 52. Pan Jintang & Liu Jinjun
n–r   FOCI 8: 235. FRPS 34(2): pl. 58. Pan Jintang & Liu Jinjun

Fig. 31
a–d   FOCI 8: 230. FRPS 34(2): pl. 59. Pan Jintang & Liu Jinjun
e–i   FOCI 8: 229. FRPS 34(2): pl. 51. Pan Jintang & Liu Jinjun

j–n   FOCI 8: 235. FRPS 34(2): pl. 58. Pan Jintang & Liu Jinjun
o–x   FOCI 8: 230. FRPS 34(2): pl. 59. Pan Jintang & Liu Jinjun

Fig. 32
a–e   FOCI 8: 236. FRPS 34(2): pl. 55. Pan Jintang & Liu Jinjun
f–i   FOCI 8: 222. FRPS 34(2): pl. 46. Pan Jintang & Liu Jinjun
j–n   FOCI 8: 226. FRPS 34(2): pl. 43. Pan Jintang & Liu Jinjun
o–r   FOCI 8: 205. FRPS 34(2): pl. 53. Pan Jintang & Liu Jinjun

Fig. 33
a–c   FOCI 8: 232. FRPS 34(2): pl. 16. Pan Jintang & Liu Jinjun
d–h   FOCI 8: 235. FRPS 34(2): pl. 58. Pan Jintang & Liu Jinjun
i–m   FOCI 8: 234. FRPS 34(2): pl. 56. Pan Jintang & Liu Jinjun
n–r   FOCI 8: 233. FRPS 34(2): pl. 57. Pan Jintang & Liu Jinjun
s–v   FOCI 8: 236. FRPS 34(2): pl. 55. Pan Jintang & Liu Jinjun

Fig. 34   Bhaskar Adhikari

Fig. 35   Claire Banks

Fig. 36   Claire Banks

Fig. 37
a–c   Cliodhna Ní Bhroin
d–f   FOB 3(1): fig. 38. Mary Bates
g–p   Cliodhna Ní Bhroin

Fig. 38   Claire Banks

Fig. 39   Claire Banks

Fig. 40   Claire Banks

Fig. 41   Claire Banks

Fig. 42   Neera Joshi Pradhan

Fig. 43   Neera Joshi Pradhan

Fig. 44
a–d   FOB 3(1): fig. 38. Mary Bates
e, f   FOCI 9: 76. FRPS 36: pl. 50. Wu Zhanghua & Liu Jingmian
g, h   FOCI 9: 77. FRPS 36: pl. 51. Zhao Baoheng, redrawn Cai Shuqin
i–l   Claire Banks

Fig. 45
a–f   FOCI 9: 88. FRPS 37: pl. 7. Wang Jinfeng
g, h   FOCI 9: 90. FRPS 37: pl. 9. Anonymous

Fig. 46
a–c   FOCI 9: 91. FRPS 37: pl. 10. Liu Chunrong
d, e   FOCI 9: 110. FRPS 37: pl. 29. Wang Jinfeng
f, g   FOCI 9: 104. FRPS 37: pl. 16. Liu Chunrong

Fig. 47   Mutsuko Nakajima

Fig. 48   Mutsuko Nakajima

Fig. 49   Mutsuko Nakajima

Fig. 50   Mutsuko Nakajima

Fig. 51
a–c   FOCI 9: 140. FRPS 37: pl. 59. Zhang Taili
d–f   FOCI 9: 147. FRPS 37: pl. 66. Wang Jinfeng
g, h   FOB 1(3): fig. 37. Mary Bates
i, j   FOCI 9: 149. FRPS 37: pl. 68. Wang Jinfeng

Fig. 52
a–g   FOB 1(3): fig. 37. Mary Bates
h, i   Claire Banks

Fig. 53   Claire Banks

Fig. 54   Claire Banks

Authors of Accounts

Bhaskar **Adhikari**
Royal Botanic Garden Edinburgh
20a Inverleith Row, Edinburgh
EH3 5LR, Scotland, UK

Shinobu **Akiyama**
National Museum of Nature and Science
4-1-1 Amakubo, Tsukuba
Ibaraki 305-0005, Japan

Ihsan A. **Al-Shehbaz**
Missouri Botanical Garden
PO Box 299, St. Louis, MO 63166-0299,
USA

J. Crinan M. **Alexander** (retired)
Royal Botanic Garden Edinburgh
20a Inverleith Row, Edinburgh
EH3 5LR, Scotland, UK

Yumiko **Baba**
Australian Tropical Herbarium
James Cook University
Smithfield, QLD 4878, Australia

Stephen **Blackmore**
Royal Botanic Garden Edinburgh
20a Inverleith Row, Edinburgh
EH3 5LR, Scotland, UK

David E. **Boufford**
Harvard University Herbaria
22 Divinity Avenue, Cambridge
MA 02138, USA

Anthony R. **Brach**
Harvard University Herbaria
22 Divinity Avenue, Cambridge
MA 02138, USA

Sajan **Dahal** (retired)
National Herbarium and Plant
Laboratories
Department of Plant Resources
GPO Box 3708, Kathmandu, Nepal

Paul A. **Egan**
Botany Department, Trinity College
Dublin
Dublin 2, Ireland

Jyoti P. **Gajurel**
C/o Central Department of Botany
Tribhuvan University, Kirtipur
Kathmandu, Nepal

Anjana **Giri**
Nepal Academy of Science & Technology,
GPO Box 3323, Khumaltar, Lalitpur,
Nepal

Richard J. **Gornall**
Department of Biology
University of Leicester, Leicester
LE1 7RH, UK

Hiroshi **Ikeda**
Department of Botany
University Museum, University of Tokyo
Hongo 7-3-1, Tokyo, 113-0033 Japan

Nirmala **Joshi**
National Herbarium and Plant
Laboratories
Department of Plant Resources
GPO Box 3708, Kathmandu, Nepal

Cathy **King**
Royal Botanic Garden Edinburgh
20a Inverleith Row, Edinburgh
EH3 5LR, Scotland, UK

Puran P. **Kurmi** (retired)
National Herbarium and Plant
Laboratories
Department of Plant Resources
GPO Box 3708, Kathmandu, Nepal

Magnus **Lidén**
Uppsala Universitet, Botaniska
Trädgården
Villavägen 8, 752 36 Uppsala, Sweden

Kamal **Maden**
C/o Central Department of Botany
Tribhuvan University, Kirtipur
Kathmandu, Nepal

Vidya K. **Manandhar** (retired)
National Herbarium and Plant
Laboratories
Department of Plant Resources
GPO Box 3708, Kathmandu, Nepal

Cliodhna D. **Ní Bhroin**
Royal Botanic Garden Edinburgh
20a Inverleith Row, Edinburgh
EH3 5LR, Scotland, UK

Hideaki **Ohba** (retired)
Department of Botany
University Museum, University of Tokyo
Hongo 7-3-1, Tokyo, 113-0033, Japan

Colin A. **Pendry**
Royal Botanic Garden Edinburgh
20a Inverleith Row, Edinburgh
EH3 5LR, Scotland, UK

Ram C. **Poudel**
Kunming Institute of Botany
Heilongtan, Kunming 650204
Yunnan, People's Republic of China

Martin R. **Pullan**
Royal Botanic Garden Edinburgh
20a Inverleith Row, Edinburgh
EH3 5LR, Scotland, UK

Keshab R. **Rajbhandari** (retired)
National Herbarium and Plant
Laboratories
Department of Plant Resources
GPO Box 3708, Kathmandu, Nepal

Sangeeta **Rajbhandary**
Central Department of Botany
Tribhuvan University, Kirtipur
Kathmandu, Nepal

Bandana **Shakya**
International Centre for Integrated
Mountain Development (ICIMOD)
GPO Box 3226, Khumaltar, Lalitpur,
Nepal

Krishna K. **Shrestha**
Central Department of Botany
Tribhuvan University, Kirtipur
Kathmandu, Nepal

Sangita **Shrestha**
Nepal Academy of Science & Technology
GPO Box 3323, Khumaltar, Lalitpur,
Nepal

Sirjana **Shrestha**
Central Department of Botany
Tribhuvan University, Kirtipur
Kathmandu, Nepal

Mohan **Siwakoti**
Central Department of Botany
Tribhuvan University, Kirtipur
Kathmandu, Nepal

Gordon C. **Tucker**
Deptartment of Biological Sciences
Eastern Illinois University
Charlestonm IL 61920, USA

Mark F. **Watson**
Royal Botanic Garden Edinburgh
20a Inverleith Row, Edinburgh
EH3 5LR, Scotland, UK

Index to Vernacular Names (Devanagari)

Index to Vernacular Names (Transliteration)

Index to Scientific Names

Entries in **bold** type refer to main citations in the text, *italic* type refers to synonyms, and roman type refers to illustrations, secondary mentions and misapplied names.

R

flavus Buch.-Ham. ex D.Don	*330*	
fockeanus Kurz	**345**	
foliolosus D.Don	**330**	
var. *incanus* (K.Sasaki ex Y.C.Liu & T.Y.Yang) S.S.Ying	*329*	
fragarioides Bertol.	*344*	
franchetianus H.Lév.	**344**	
glandulifer N.P.Balakr.	**339**	
gowreephul Roxb.	*330*	
gowryphul Roxb.	*330*	
gowryphul Roxb. ex Wall.	*330*	
gracilis Roxb.	*330*	
var. *chiliacanthus* Hand.-Mazz.	*330*	
var. *pluvialis* Hand.-Mazz.	*330*	
griffithii Hook.f.	**342**	
hamiltonianus Ser.	*344*	
hamiltonii Hook.f.	**340**	
hexagynus Roxb.	342	
hibiscifolius Focke	327	
himalaicus Kuntze	*343*	
hoffmeisterianus Kunth ex Bouché	**332**	
hypargyrus Edgew.	*330*	
var. *concolor* (Hook.f.) H.Hara	*330*	
var. *hypargyrus*	*330*	
var. *niveus* (Wall. ex G.Don) H.Hara	*330*	
idaeus var. *exsuccus* Franch. & Sav.	*331*	
illudens H.Lév.	*331*	
incanus K.Sasaki ex Y.C.Liu & T.Y.Yang	*329*	
indotibetanus Koidz.	*336*	
inopertus (Focke) Focke	**334**	
kinashii H.Lév. & Vaniot	*331*	
forma *macrophyllus* Cardot	*331*	
forma *microphyllus* Cardot	*331*	
kumaonensis N.P.Balakr.	**343**	
kurzii N.P.Balakr.	*340*	
lanatus Wall.	*339*	
lanatus Wall. ex Hook.f.	*339*	
lasiocarpus Sm.	*329*	
var. *ectenothyrsus* Cardot	*329*	
var. *membranaceus* Hook.f.	*329*	
var. *micranthus* (D.Don) Hook.f.	*329*	
var. *pauciflorus* (Wall. ex Lindl.) Hook.f.	*329*	
var. *rosifolius* Hook.f.	*334*	
lineatus Reinw.	**338**	
lobatus Wall.	*339*	
longistylus H.Lév.	*329*	
loropetalus Franch.	*345*	
macilentus Cambess.	333, **335**	
mairei H.Lév.	*329*	
mesogaeus Focke	**331**	
forma *floribus-roseis* Focke	*331*	
var. glabrescens T.T.Yu & L.T.Yu	331	
var. *incisus* Cardot	*331*	
var. *oxycomus* Focke	*331*	
micranthus D.Don	*329*	
microphyllus D.Don	*330*	
minensis Pax & K.Hoffm.	335	
moluccanus L.	344	
var. *ferox* Kuntze	*340*	
myriadenus H.Lév. & Vaniot	*336*	
var. *grandifoliolatus* H.Lév.	*336*	
mysorensis F.Heyne ex Roth	*329*	
nepalensis (Hook.f.) Kuntze	**345**, 346	
niveus Thunb.	**329**, 330	
subsp. *inopertus* Focke	*334*	
var. *aitchisoni* Hook.f.	*332*	

var. *concolor* Hook.f.	*330*	
var. *hypargyrus* (Edgew.) Hook.f.	*330*	
var. *micranthus* (D.Don) H.Hara	*329*	
var. *microcarpus* Hook.f.	*331*	
var. *pauciflorus* (Wall. ex Lindl.) Focke	*329*	
var. *pedunculosus* (D.Don) Hook.f.	*330*	
var. *rhodophyllos* Focke	*334*	
var. *rosifolius* (Hook.f.) H.Hara	*334*	
niveus Wall.	*330*	
niveus Wall. ex G.Don	*330*, 331	
nutans Wall.	*345*	
nutans Wall. ex G.Don	*345*	
var. *fockeanus* (Kurz) Kuntze	*345*	
var. *nepalensis* Hook.f.	*345*	
nutantiflorus H.Hara	*345*	
obcordatus (Franch.) Thuan	*330*	
occidentalis		
var. *exsuccus* (Franch. & Sav.) Makino	*331*	
var. *japonicus* Miyabe	*331*	
paniculatus Sm.	**339**	
forma *tiliaceus* (Sm.) H.Hara	*339*	
parviflorus L.	330	
pauciflorus Wall. ex Lindl.	*329*	
pedunculosus D.Don	*330*	
var. *concolor* (Hook.f.) Kitam.	*330*	
var. *hypargyrus* (Edgew.) Kitam.	*330*	
pentagonus Wall. ex Focke	**337**	
pinfaensis H.Lév. & Vaniot	*330*	
pinnatus D.Don	*329*	
pulcherrimus Hook.	*338*	
pungens Cambess.	**335**	
var. *discolor* Prochanov	*335*	
var. *fargesii* Cardot	*335*	
pyi H.Lév.	*329*	
radicans Focke	*345*	
rarissimus Hayata	*331*	
reticulatus Wall.	*342*	
reticulatus Wall. ex Hook.f.	*343*	
rolfei Vidal	344	
rosiflorus Roxb.	*329*	
rosifolius Sm.	336, **336**, 341	
forma *coronarius* (Sims) Focke	*336*	
forma *coronarius* (Sims) Kuntze	*336*	
subsp. *sumatranus* (Miq.) Focke	*336*	
var. *asper* (Wall. ex D.Don) Kuntze	*336*	
var. *coronarius* Sims	*336*	
rosulans Kuntze	*342*	
rotundifolius Wall.	*330*	
roylei Klotzsch	*330*	
rugosus Buch.-Ham. ex D.Don	*344*	
rugosus Sm.	**344**	
somai Hayata	*336*	
sorbifolius Maxim.	*336*	
splendidissimus H.Hara	**338**	
sterilis Kuntze	*340*	
sumatranus Miq.	**336**	
takasagoensis Koidz.	*336*	
thomsonii Focke	**337**	
tiliaceus Sm.	*339*	
tongchouanensis H.Lév.	*329*	
tonglooensis Kuntze	*342*	
treutleri Hook.f.	**342**, 346	
trichopetalus Hand.-Mazz.	*335*	
uncatus Wall.	*335*	
wallichianus Wight & Arn.	*330*	
x seminepalensis Naruh.	xvii, **345**	

T

Index to Gymnosperm and Flowering Plant Families in Volumes of *Flora of Nepal*

Pteridophytes are all in Volume 1